MOLECULAR BASES
OF NEURAL DEVELOPMENT

THE NEUROSCIENCES INSTITUTE
of the Neurosciences Research Program

Gerald M. Edelman, *Director*
W. Einar Gall, *Research Director*
W. Maxwell Cowan, *Chairman,*
Scientific Advisory Committee

The Neurosciences Institute was founded in 1981 by the Neurosciences Research Program to promote the study of scientific problems within the broad range of disciplines related to the neurosciences. It provides visiting scientists with facilities for planning and review of experimental and theoretical research with emphasis on understanding the biological basis for higher brain function.

The Institute has initiated an active publishing program. This volume, the second in a continuing series of edited volumes, provides current views and experimental results on the complex pattern of temporal and spatial interactions that give rise to the adult nervous system, with emphasis on the description of cellular interactions at the molecular level.

Support for the Neurosciences Research Foundation, Inc., which makes the Institute's programs possible, has come in part from generous gifts by The Vincent Astor Foundation, Lily Auchincloss, Francois de Menil, Sibyl & William T. Golden Foundation, Lita Annenberg Hazen, Lita Annenberg Hazen Charitable Trust, The IFF Foundation, Inc., Johnson & Johnson, Harvey L. Karp, John D. & Catherine T. MacArthur Foundation, Rockefeller Brothers Fund, van Ameringen Foundation, The G. Unger Vetlesen Foundation, and The Vollmer Foundation.

The Neurosciences Institute Publications Series

Neurophysiological Approaches to Higher Brain Functions
Edward V. Evarts, Yoshikazu Shinoda, and Steven P. Wise

Protein Phosphorylation in the Nervous System
Eric J. Nestler and Paul Greengard

Dynamic Aspects of Neocortical Function
Gerald M. Edelman, W. Einar Gall, and W. Maxwell Cowan, Editors

Molecular Bases of Neural Development
Gerald M. Edelman, W. Einar Gall, and W. Maxwell Cowan, Editors

MOLECULAR BASES
OF NEURAL DEVELOPMENT

Edited by

GERALD M. EDELMAN

W. EINAR GALL

W. MAXWELL COWAN

A Neurosciences Institute Publication
JOHN WILEY & SONS
New York • Chichester • Brisbane • Toronto • Singapore

Copyright © 1985 by Neurosciences Research Foundation, Inc.

Published by John Wiley & Sons, Inc.

All rights reserved, provided that parts of this work which may have been written as part of the official duties of an employee of the United States Government may not be copyrighted under the 1976 United States Copyright Act. Requests for further information should be addressed to the Neurosciences Research Foundation, Inc., 1230 York Avenue, New York, New York 10021.

Published simultaneously in Canada.

Library of Congress Cataloging in Publication Data:

Main entry under title:

Molecular bases of neural development.

"A Neurosciences Institute publication."
Includes index.
1. Developmental neurology. 2. Molecular biology.
I. Edelman, Gerald M.   II. Gall, W. Einar.   III. Cowan, W. Maxwell.   IV. Neurosciences Institute (New York, N.Y.)
[DNLM: 1. Neurons—cytology.   WL 102.5 M718]
QP363.5.M65   1985   596'.0188   84-29928
ISBN 0-471-81561-6

Printed in the United States of America
10 9 8 7 6 5 4 3 2 1

Preface

This is the second in a continuing series of reports of the annual symposia of The Neurosciences Institute of the Neurosciences Research Program. The choice of subject matter was made by the Scientific Advisory Committee of The Neurosciences Institute (NSI) consisting of: Floyd E. Bloom, Research Institute of Scripps Clinic; W. Maxwell Cowan (Chairman), The Salk Institute; Gerald M. Edelman, The Rockefeller University; John J. Hopfield, California Institute of Technology; Eric R. Kandel, Columbia University; Alvin M. Liberman, The Haskins Laboratories; and Vernon B. Mountcastle, The Johns Hopkins University. They chose to follow the first meeting on dynamic aspects of neocortical function with one on neural development. It seemed appropriate to pick this important field of neuroscience at a time when it is beginning to be transformed by new molecular approaches and the hypotheses they engender.

The volume is organized into six sections, ranging from molecular and functional aspects of early development through aspects of neural migration, glial guidance, and synapse formation, to the culminating challenge of the development of neural maps. It ends with a consideration of genes and behavior.

It would be too much to expect that complete coverage could be achieved across so wide a range of subdisciplines. Nonetheless, the volume does highlight key issues, brings some of the style and flavor of new molecular approaches into conjunction with those of classical approaches, and indicates the range of diverse systems that may be studied to illuminate central problems. In a few cases, it is fair to say that whole avenues have been opened up by molecular approaches; in others, such as the retinotectal map, one observes a convergence of views from careful studies in several laboratories pointing to a partial solution of a classical problem.

One final word of disclaimer may be in order: This is not a book on general principles of developmental biology. It is largely restricted to regulative development and to the nervous system. There is, however, much here to provoke thought in fields concerned with embryogenesis, histogenesis, and disorders of neural development.

We are grateful to Susan Hassler, Editor of the NSI, and the members of the Scientific Advisory Committee for their central role in helping to assemble this book. Again, we recognize the hospitality of The Salk Institute, and thank the contributors whose stay during the symposium meeting was graced by many pleasant and insightful exchanges.

G. M. E.
W. E. G.
W. M. C.

Contents

Preface		v
Introduction		1
W. Maxwell Cowan		

SECTION 1	EARLY DEVELOPMENT: CELL SURFACE MOLECULES AND SUBSTRATES	9
Chapter 1	Factors Controlling the Early Development of the Nervous System	11
	Anne E. Warner	
Chapter 2	Molecular Regulation of Neural Morphogenesis	35
	Gerald M. Edelman	
Chapter 3	Identifying the Components of Muscle Fiber Basal Lamina that Aggregate Acetylcholine Receptors at Regenerating Neuromuscular Junctions	61
	Justin R. Fallon, Ralph M. Nitkin, Bruce G. Wallace, and U. Jackson McMahan	
Chapter 4	The Control of Development of Neuronal Excitability	67
	Nicholas C. Spitzer	

SECTION 2	GLIAL DEVELOPMENT AND NEURONAL GUIDANCE	89
Chapter 5	Macroglial Cell Lineages	91
	Sergey Fedoroff	
Chapter 6	Glial Growth Factor and the Neuronal Control of Cell Division in Amphibian Limb Regeneration	119
	Chris R. Kintner, Greg Erwin Lemke, and Jeremy P. Brockes	

Chapter 7	Mechanisms of Neuronal Migration in Developing Cerebellar Cortex	139
	Pasko Rakic	

SECTION 3	THE NEURAL CREST AND DEVELOPMENT OF THE PERIPHERY	161
Chapter 8	*In Vivo* and *In Vitro* Analysis of the Differentiation of the Peripheral Nervous System in the Avian Embryo	163
	Nicole M. Le Douarin	
Chapter 9	Gangliogenesis in the Avian Embryo: Migration and Adhesion Properties of Neural Crest Cells	181
	Jean-Paul Thiery, Gordon C. Tucker, and Hirohiko Aoyama	
Chapter 10	Neuronal Determination in the Enteric Nervous System	213
	Michael D. Gershon and Taube P. Rothman	
Chapter 11	Development of Motor Innervation in Vertebrate Limbs	243
	Margaret Hollyday	

SECTION 4	FORMATION OF NEURITES AND SYNAPSES	265
Chapter 12	Axonal Growth and Guidance	269
	Paul C. Letourneau	
Chapter 13	Cell Recognition During Neuronal Development in Grasshopper and *Drosophila*	295
	Corey S. Goodman, Michael J. Bastiani, Jonathan A. Raper, and John B. Thomas	
Chapter 14	Developmental Interactions between Neurons in Insects	317
	John S. Edwards and Mark R. Meyer	
Chapter 15	Changes of State during Neuronal Development: Regulation of Axon Elongation	341
	Mark B. Willard, Karina Meiri, and Marcie Glicksman	

Chapter 16	Cholinergic Development and Identification of Synaptic Components for Chick Ciliary Ganglion Neurons in Cell Culture	363
	Darwin K. Berg, Michele H. Jacob, Joseph F. Margiotta, Rae Nishi, Jes Stollberg, Martin A. Smith, and Jon M. Lindstrom	
SECTION 5	MAP FORMATION IN THE RETINOTECTAL SYSTEM	385
Chapter 17	The Development of the Retinotectal Projection: An Overview	389
	W. Maxwell Cowan and R. Kevin Hunt	
Chapter 18	The Continuous Formation of the Retinotectal Map in Goldfish, with Special Attention to the Role of the Axonal Pathway	429
	Stephen S. Easter, Jr.	
Chapter 19	Factors Involved in Retinotopic Map Formation: Complementary Roles for Membrane Recognition and Activity-Dependent Synaptic Stabilization	453
	John T. Schmidt	
Chapter 20	Cell Interactions Involved in Neuronal Patterning: An Experimental and Theoretical Approach	481
	Scott E. Fraser	
SECTION 6	GENE EXPRESSION, PRIMARY PROCESSES, AND BEHAVIOR	509
Chapter 21	Gene Expression in *Aplysia* Peptidergic Neurons	513
	Richard H. Scheller	
Chapter 22	Hormonal Approaches to the Study of Cell Death in a Developing Nervous System	531
	James W. Truman	
Chapter 23	Immunohistochemical and Genetic Studies of Serotonin and Neuropeptides in *Drosophila*	547
	Kalpana White and Ana M. Valles	

Chapter 24 Neural and Developmental Implications of the Genetic
           and Molecular Analysis of Behavior   565
           Jeffrey C. Hall

CONTRIBUTORS AND PARTICIPANTS   589

INDEX   593

Introduction

W. MAXWELL COWAN

Developmental neuroscience has now entered what promises to be one of the most exciting epochs in its long and distinguished history. For while much remains to be done at the descriptive level (as some of the chapters in this volume clearly attest), it seems probable that many of the problems that for so long have dominated the field can now be approached in a more direct way, and that solid data, which permit unequivocal interpretation, will replace the often doubtful observations that have clouded rather than clarified the issues, to say nothing of the many speculative proposals that have arisen from these uncertain observations. It is the bane of a young science to be trammeled by experiments that are difficult (if not impossible) to interpret, and by hypotheses that for the most part are untestable. To say this is not to cast aspersion on the long history of developmental studies, but rather to assert that the forward thrust of the field must be in the direction of isolating and characterizing the critical molecules and genes involved—as indeed it must be in most areas of biology. But before looking toward the future of the field, it is perhaps worth looking back on its history, if only to ensure that we perceive it in context, and to remind ourselves that many of the concepts that inform our current thinking antedate the discovery of the double helix, the cloning of the first gene, and even the isolation of the first enzyme.

Unfortunately, the history of developmental neurobiology (as it is now called) does not lend itself readily to easy division into major epochs. Its roots lie largely concealed within the underbrush of both developmental biology and neuroscience. And indeed, for much of its history its principal contributors were drawn from one or the other of these two areas; it is only relatively recently that the two fields have come together and that a new cadre of workers who style themselves as developmental neurobiologists (or are so called by others) has come into being.

Given its dual origin it is not surprising that some are inclined to view the field as being simply an extension of the great era of developmental biology associated with the names of Weissman, Boveri, Driesch, and Roux, or that others, perhaps viewing it more narrowly, see it as having its origins in the descriptive studies of the great nineteenth century neurologists Waldeyer, His, Schaper, and preeminently, Langley and Ramón y Cajal. Both views, of course, are correct. Most of the central issues in neural development are essentially the same as those in the development of all other organs and tissues: Where and when does the tissue first appear? How do such complex structures arise from such seemingly simple antecedents? To what extent is the development of the organ or tissue due to the progressive acquisition of new properties and to what extent is it due to the progressive restriction of potentialities? Such concepts as "embryonic fields," the establishment of polarities, regulative versus mosaic development, inductive interactions, and so forth—many of which are still central to our understanding of the development of the nervous system—were first formulated and often bitterly fought over during the classical era of embryology that spanned the turn of the century. But at the same time, their relevance to the first appearance of neural tissue, to the processes of cell proliferation and migration in the nervous system, to the regional determination of different parts of the brain and spinal cord, and perhaps most importantly, to the question of how different regions of the nervous system form highly selective patterns of connections with each other, was, for the most part, the overriding interest of those who were drawn to the field from a neurological background. With a few notable exceptions, these specific issues were of only passing interest to embryologists, but were rightly seen by neurologists to occupy the central stage.

Among the abiding contributions from this era are the detailed descriptions of the gross morphology of the brain at successive stages in its development, the identification of the major classes of cells in the nervous system (both the many varieties of neurons and certain of the glial cells), the correct interpretation of the mode of cell proliferation in the neural tube and its derivatives, and most important of all, the establishment of the neuron doctrine. It is perhaps a commentary on the rate of progress in this field that it was almost 70 years after the formulation of the cell theory by Schleiden and Schwann before the majority of neurobiologists came to accept the notion that the nervous system was not some form of syncytium but, like other tissues, was composed of discrete cells.

Although this period was largely concerned with what are essentially anatomical issues, and the figures most commonly mentioned are the great descriptive morphologists, it would be inappropriate not to call attention to the seminal experimental studies of the English physiologist J. N. Langley. It is to Langley that we are principally indebted for the first demonstration that regenerating axons (in the autonomic nervous system) can not only grow back to their target field, but also selectively reinnervate the subclass of neurons with which they were previously in contact. And it was Langley who, in considering this aspect of neuronal specificity, first put forward the notion that neurons might be distinguished from each other chemically, and that these chemical distinctions might provide the basis for selective neuronal innervation and reinnervation (Langley, 1897).

It is difficult to point with certainty to the close of this classical era. But we would not be far off if we were to relate it to the growing acceptance of the work of Santiago Ramón y Cajal and to the emergence of the two figures who were to dominate the first third of this century, Hans Spemann and Ross Harrison, with Ramón y Cajal providing the link between the classical and the new era. For although much of Ramón y Cajal's work was in the descriptive tradition of classical morphology, like Langley's, it reached forward into the modern period.

As he tells us in his fascinating autobiography, it was while trying to unravel the bewildering complexity of the adult brain that it suddenly occurred to him that the key to understanding the nervous system might better come from studying its development:

> *Since the full-grown forest turns out to be impenetrable and indefinable, why not revert to the study of the young wood, in the nursery stages as we might say . . . . [If] the brain and other adult organs . . . are too complex to permit scrutinizing their structural plan . . . why not apply [the available methods] systematically to lower vertebrates and to the early stages of ontogenetic development, in which the nervous system should present a simple and, so to speak, diagrammatic organization.*
>
> Ramón y Cajal (1937)

It is unnecessary to detail just how successful he was in this endeavor; it will suffice to mention only some of the major contributions that flowed from his skilled and thoughtful application of the Golgi method to embryonic neural tissue. These include: the first identification of the growth cone; the discovery of radial glial processes and their possible role in neuronal migration; the first clear account of the growth of dendrites; the recognition that many more neuronal processes are formed than survive to maturity; the finding that positional and targeting errors occur during neural development, and the prediction that most such errors would be eliminated at later stages in development. And of course, in addition to all these general discoveries, there were his careful descriptive accounts of the development of most regions of the brain and the peripheral nervous system in a broad spectrum of vertebrates. So large was his canvas and so sure were his brush strokes that it is difficult, even today, to approach any region of the nervous system without being aware of his genius and his towering contributions.

Spemann and Harrison, on the other hand, were in the great tradition of *Entwicklungsmechanik* that will always be associated with the name of Wilhelm Roux. Of the two, Spemann's contributions to neurobiology were less numerous but certainly no less important than those of Harrison. Spemann's discoveries of the mechanism of primary induction and of the organizer (admirably summarized in his 1938 Silliman Lectures) are rightly considered to be among the greatest discoveries in developmental biology of this century. Unfortunately, their significance was somewhat diminished by the failure of a generation of biochemists who optimistically believed that they could readily isolate and characterize the inducing agent(s). It is now evident that their efforts were both premature and misguided; to a large extent they were abandoned in frustration when Johannes Holtfreter reported that killed tissues, and even a variety of nonbiological substances, could under some conditions serve as effective inducers.

Harrison always considered himself a general developmental biologist; certainly, many of his most important discoveries dealt with structures other than the nervous system—limb buds, ears, and so forth. However, his contributions to developmental neurobiology have had a lasting influence on the field. They fall broadly into three categories. First were the definitive discoveries that laid to rest a number of troublesome issues. Among these were: the first unequivocal demonstration (using living tissues—and incidentally involving the invention of the method of tissue culture) that neural processes are extensions of nerve cells and are not formed by other cellular elements as some had argued; the direct demonstration of the role of growth cones in process elongation; the discovery that sensory ganglion cells, peripheral glia (including Schwann cells), and pigment cells are all derived from the neural crest; and the discovery that motor axons can grow out from the spinal cord and innervate muscle fibers in the complete absence of Schwann cells. Second was the creation of the conceptual framework for so much of the work that has followed. For example, his work on the development of polarity in the embryonic ear and limb has laid the foundation for almost all later thinking on how tissues acquire and express positional information. And third, through the training of a succession of brilliant students—Detwiler, Stone, Piatt, Nicholas, and Twitty to name only the most prominent—he established the first, and for many years the most influential, school of developmental neurobiology.

This "golden era," if we may so call it, came to an end some time before World War II, and from that time until the late 1970s developmental studies came to be concerned largely with two (not unrelated) issues: (1) the relationship of center and periphery in neural development; and (2) structuralist versus functionalist views of the development of neuronal circuits.

While it is perhaps invidious to single out individual contributors from an ever growing body of workers, it would not be unfair to identify Samuel Detwiler, Viktor Hamburger, and Rita Levi-Montalcini as the foremost contributors to the first issue. Although others had earlier shown a reduction in size (or even the degeneration) of certain neural structures following the early extirpation of peripheral body parts, it was Detwiler's seminal study of the effects of limb ablations on the development of the sensory ganglia in *Ambystoma* that paved the way for this singularly productive approach to the study of neural development. Its appeal lay in the fact that peripheral manipulations were relatively easy to perform (compared to, say, comparable manipulations in the brain or spinal cord), and the analysis of the resulting changes in the nervous system seemed to be quite straightforward. In any event, the analysis of the changes that follow peripheral ablations or the addition of supernumerary tissues turned out to be rather difficult, and it was not until 1949 that the underlying mechanisms began to be clarified. The key finding that when a peripheral target is ablated the corresponding neural center is either absent or much reduced in size was initially interpreted in terms of a diminished recruitment of undifferentiated cells into the relevant neural center. It was only after Levi-Montalcini's early studies on the motor columns of the chick spinal cord, and her later work with Hamburger on the development of the spinal ganglia, that it became generally accepted that such peripheral manipulations lead to a regression of previously differentiated neurons (see Hamburger, 1976). Incidentally, Hamburger and Levi-Montalcini's 1949 study also served to draw attention to the importance of cell death during normal neural development, a subject that later work has established as a major regulative mechanism in determining the definitive sizes of most neuronal populations. Their observations also led, almost serendipitously, to what we can now recognize as one of the first and most significant contributions to molecular neurobiology—the discovery of the celebrated nerve growth factor (NGF).

The way in which the complex circuitry of the nervous system comes to be established was rightly perceived from the outset to be the central problem of developmental neuroscience, and from the outset two opposing views were put forward. According to the first, the initial wiring of the nervous system is largely, if not completely, random, and it is from this random system of connections that use and function select, enhance, or in some way stabilize those patterns that are most appropriate for the behavior of the organism. We can see a foreshadowing of this notion in the "reticularist" view of the nervous system advocated by Golgi and others at the turn of the century, but it is with the name of Paul Weiss that the functionalist position is most closely associated. It is not necessary to go over the various lines of evidence that were adduced to support this position, or the various intellectual contortions that had to be made to maintain it in the face of evidence to the contrary (see Gaze, 1970). Suffice it to say that it was one of Weiss's own students, Roger Sperry, who in a series of brilliantly conceived experiments gradually overturned the functionalist position between 1938 and 1960, and at the same time laid a new foundation upon which most of our current ideas about the way in which connections are formed have been built. Sperry's strategy was to put the functionalist position to direct test by altering or malpositioning various body parts or tissues, and asking whether under these conditions the animal could learn from experience how to correct for the experimentally induced aberrations. Again and again he found that when flexor nerves were forced to innervate extensor muscles, when various nerves were cross-united, or when the entire eye was surgically rotated, the resulting maladaptive behavior persisted throughout the animal's life. Neither learning nor experience could compensate for the anatomical misarrangement (see Hunt and Cowan, 1984).

It was from his work on the regeneration of the optic nerve that Sperry was first led to postulate that, during development, nerve cells acquire distinctive chemical labels that define their positions within the fields in which they lie, and that it is the presence of these labels on their outgrowing axons that enables neurons to identify and make lasting connections with their respective targets. The formulation of what is now generally referred to as the "chemoaffinity hypothesis" stands as one of the great intellectual contributions of developmental neurobiology. In part this is because the theory is a rather general one—and in its most general form relevant to the fieldlike differentiation of all cellular populations and the establishment of most cell-to-cell contacts. And in part, its strength lies in the absence of particularities. This has meant that the hypothesis has been able to withstand innumerable assaults, and more than thirty years after its formulation it remains as the only hypothesis for the establishment of nerve connections that has not been refuted. However, as is discussed elsewhere in this volume, the generality of the hypothesis is also perceived as its greatest weakness: The nature and number of the postulated chemical labels is unknown; how they are distributed among the cells of any given population is not clear; how they arise and come to be distributed upon the surfaces of cells, or on the surfaces and tips of growing axons, is not evident.

What is not widely acknowledged is that the chemoaffinity hypothesis (as broadly conceived) owes much to the earlier work of Holtfreter on tissue affinities among embryonic cells. Indeed, Holtfreter's demonstration that dissociated embryonic cells can selectively reaggregate—ectodermal cells adhering to ectodermal cells, mesodermal to mesodermal—in tissue-specific ways, may rightly be viewed as the starting point of all later work on selective cell adhesion. Most later work has also drawn heavily on the theoretical framework that Holtfreter and his colleague Townes developed to account for their cell reaggregation experiments. While considering various alternative explanations (many of which were to be clarified and championed by later workers such as Steinberg, Moscona, and Trinkaus), they concluded that the most plausible interpretation for their findings was that all cells carry on their surfaces specific molecules, that these molecules are distinctive for each type of tissue, and that it is some form of molecular interaction at their surfaces that causes cells of like-kind to selectively adhere to each other. Wisely, Holtfreter seems to have recognized that the time was not ripe to attempt to isolate the relevant molecules. But arguably, his experiments can be judged to be the first significant molecular contributions to developmental biology—and certainly the experimental paradigm he established has resulted in selective neuronal aggregation being the first of the primary processes in neural development to yield to successful molecular analysis (see Edelman's chapter in this volume).

Chronologically, of course, pride of place must be given to the discovery of nerve growth factor (NGF) by Hamburger and Levi-Montalcini in the early 1950s as the first substantive molecular contribution to developmental neurobiology. The events that led to this discovery have been recounted in several places (see Levi-Montalcini, 1966; Hamburger, 1980) and need not be repeated here, except to point out that NGF is still the only neuronal trophic factor to have been isolated and fully characterized, and that it will always serve as the paradigm for the study of the many other trophic and tropic factors in the nervous system that remain to be discovered.

It is interesting to ask why it has taken almost thirty years since the discovery of NGF for other comparable molecular contributions to be made in developmental neurobiology. The answer, of course, is that the discovery of NGF was largely fortuitous, and serendipity played a large part in its isolation and characterization. It is unlikely that fortune will smile again upon the field in quite so generous a fashion. As the many attempts to isolate the factor(s) involved in neural induction in the 1920s and 1930s showed all too poignantly, real advances in biology can only be made when the field as a whole has achieved a certain maturity. By maturity, in this context, I mean when the subject has progressed to the point that the critical issues have become clear, when the essential questions can be asked, and when the necessary methods and analytical tools are available. As we have seen in the case of developmental neurobiology, the majority (but by no means all) of the key issues have been known for a long time; indeed, some go back to before the turn of the century, and many of the central questions were correctly posed by the great figures of that era. But it is only in the past decade or so that the requisite methods have been perfected to make a frontal assault on the problems possible. It is also not without significance that nearly all of these methods have been developed outside of neurobiology. For, by and large, the complexity of the nervous system does not lend itself readily to the perfection of generalizable methods. The high technologies of neurobiology—like single unit and intracellular recording, voltage and patch clamping—have for the most part been developed and have remained outside the mainstream of biology. But recent developments in other areas, notably in immunology, in biochemistry and molecular biology, and especially in molecular genetics, promise to change all this, and the first fruits of what promises to be a very rich harvest from the applications of these methods have already appeared.

Some of these first fruits are evident in this volume and point the way to what we may realistically expect in the coming years. As several chapters indicate, immunological methods have been imaginatively used to isolate hitherto unrecognized membrane components involved in selective cell aggregation, in axon fasciculation, and possibly in a number of inductive events; polyclonal and monoclonal antibodies have been widely used to identify cells of different lineages and/or different functional types, and to establish the time of appearance (and in some cases, the disappearance) of the relevant antigens; the identification of proteins involved in axonal growth and axon–substrate interactions has drawn heavily on conventional biochemical methods, as have attempts to isolate new growth factors and the factors involved in the maturation and maintenance of the neuromuscular junction; and for the first time recombinant DNA technology is being used to clone the genes for some of the peptides involved in the development and function of the invertebrate nervous system. These are impressive accomplishments—as much for what they promise as for what they have already disclosed.

SELECTED REFERENCES

Gaze, R. M. (1970) *The Formation of Nerve Connections*, Academic, New York.

Hamburger, V. (1976) The developmental history of the motor neuron. *Neurosci. Res. Progr. Bull.* **15**:1–37.

Hamburger, V. (1980) Trophic interactions in neurogenesis: A personal historical account. *Ann. Rev. Neurosci.* **3**:269–278.

Hamburger, V. and R. Levi-Montalcini (1949) Proliferation, differentiation, and degeneration in the spinal ganglia of the chick embryo under normal and experimental conditions. *J. Exp. Zool.* **111**:457–501.

Hunt, R. K. and W. M. Cowan (1984) The chemoaffinity hypothesis: An appreciation of Roger Sperry's contributions to developmental biology. In: *Brain Circuits and Functions of the Mind*, C. Trevarthen, ed., Cambridge Univ. Press, Cambridge (in press).

Langley, J. N. (1897) On the regeneration of pre-ganglionic and post-ganglionic visceral nerve fiber. *J. Physiol. (Lond.)* **22**:215–230.

Levi-Montalcini, R. (1966) The nerve growth factor. *Harvey Lectures* **60**:217–259.

Ramón y Cajal, S. (1937) *Recollections of My Life*, trans. by E. Horne Craigie, MIT Press, Cambridge, Mass.

Spemann, H. (1938) *Embryonic Development and Induction*, Yale Univ. Press, New Haven.

Section 1

Early Development:
Cell Surface Molecules and Substrates

Development to yield form is characterized by a series of epigenetic sequences — interactions of cellular events that must follow each other in a historical order so that other interactions become possible. Regardless of order, these events can be grouped into primary processes: cell division, cell movement, cell adhesion, cell differentiation, and cell death. Understanding the control of the onset and the relative predominance of each of these processes at different epochs is the main challenge to modern molecular embryology. In regulative development, this challenge is sharply posed: Cells of different history are brought together by morphogenetic movements to result in embryonic induction. What controls these movements and the gene expressions that accompany induction?

In this section, several aspects of these questions are addressed. Warner attacks an issue of primary or neural induction that results from interactions between cells of the dorsal ectoderm and those of the underlying mesoderm. Her data suggest that, early in this process, activation of additional sodium pumps in the neuroectoderm with reduction of intracellular sodium concentration is a key requirement for this fundamental process. This work, done on amphibians, is likely to have general extensions to other species, and it serves as a paradigm for studies on early signals initiating induction.

One of the key elements in determining neural morphogenesis is the regulation of cellular movement, the formation and stabilization of epithelial tissue sheets, and the discrimination of the various germ layer derivatives that arise early as collectives of cells. By the very nature of regulative development, this control of movement is important in inductive events, and the adhesive interactions that govern movement must themselves be regulated in turn. In his chapter, Edelman describes the discovery, binding mechanisms, and distribution of cell adhesion molecules (CAMs). These cell surface glycoproteins of different structure and binding specificities mediate cell adhesion in all vertebrate species examined, and they undergo changes in prevalence, distribution, and chemical binding properties in the course of development. Such changes are orderly, are particularly prominent at times of primary and secondary induction, and can be perturbed to alter tissue ordering and pattern. Thus, the primary process of cell–cell adhesion is now beginning to be analyzed in terms of its central molecules. There is reason to hope that an attack on the means by which genes for these molecules are expressed might greatly illuminate the coordination of primary processes and the problem of embryonic induction.

Several other specific issues bear upon an understanding of the development of neural structure. Of these, one of the most important relates to synapse structure and function. In their chapter, McMahan and his colleagues describe an approach to understanding the specific aggregation of postsynaptic acetylcholine receptors in muscle cells opposite the axon terminal. By analyzing components of the basal lamina of the *Torpedo* electric organ, an insoluble fraction was isolated that causes aggregation of receptors on cultured chick muscle cells. The fraction was rich in extracellular matrix components and structures resembling basal lamina. Progress is being made in extracting the relevant active protein and in relating it to the structure of the basal lamina. When successfully completed, this research should make it possible to compare the initial interactions of nerve and muscle (mediated by N-CAM, as pointed out in the previous chapter) with those involved in regeneration.

Development is characterized by the appearance of neural functions and properties and these feed back on further development of the phenotype in a temporally ordered way. In his contribution, Spitzer has studied this orderly sequence in cultured neurons. He has shown that RNA and protein syntheses are both necessary for the appearance of mature sodium-dependent action potentials. On the other hand, development of functional synaptic contacts does not require expression of calcium-dependent action potentials. Continued exploration of this approach should make it possible to determine the causal biochemical sequences related to functional expression of neural phenotypic characters.

As all of these studies illustrate, molecular approaches to key aspects of regulation, induction, and phenotypic expression are both feasible and productive. They tie in well with the studies of cell interactions, cell lineages, and cell migrations that are the subjects of the next section.

Chapter 1

Factors Controlling the Early Development of the Nervous System

ANNE E. WARNER

ABSTRACT

The development of the nervous system is initiated by an interaction between dorsal ectoderm cells and underlying mesoderm cells. This process, termed neural induction, is very poorly understood. Likely methods of transfer of the inducing signal are discussed and the possibility raised that there are two inductive signals involved in initiating subsequent development of the nervous system. The inductive signal has three consequences: morphogenetic movements that lead to formation of the neural tube begin; the future anatomical organization of the nervous system is laid out on the neural plate; and the physiological properties of neural plate cells change. Evidence is presented that suggests that the physiological changes, which are manifested as the activation of additional sodium pumps within the neuroectoderm, are essential for subsequent expression of the neuronal phenotype. The activation of additional sodium pumps reduces the intracellular sodium concentration in neuroectoderm cells; this decrease in intracellular sodium is closely involved in activating subsequent events in the morphological and physiological differentiation of neurons.

The primary event in the development of the vertebrate nervous system is an interaction between mesoderm cells brought into position by gastrulation movements and the overlying ectoderm cells. This interaction, termed primary induction, takes place at the time of establishment of the three primordial germ layers. It is supposed that the mesoderm cells release a signal, or signals, which initiate a program leading to the development of nervous tissue in ectoderm cells. This is initially manifested by the morphogenetic movements of neurulation (see Figure 1, which illustrates neurulation in the amphibian embryo). First the neural groove appears in the midline, followed by the lifting of the neural folds. The neural folds mark the site of eventual fusion of the forming neural tube in the midline. The area enclosed within the neural folds is the neural plate, which will give rise to both neuronal and many glial elements of the central nervous system. At the top of the neural folds lies the neural crest, which will give rise to the sensory ganglia, the sympathetic ganglia, the enteric nervous system, the Schwann cells, and a variety of other derivatives. Once the neural tube has closed, neural crest cells migrate to their final destination before differentiating. The development of this part of the nervous system is considered further by Le

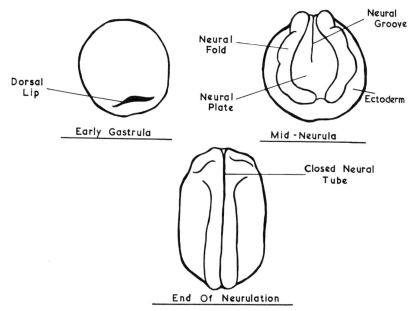

Figure 1. *Diagram illustrating formation of the nervous system in the amphibian embryo.* The dorsal lip appears in the marginal zone of the early gastrula and marks the site of invagination of mesoderm and endoderm cells to form the three germ layers. The neural plate forms the neurons and supporting cells of the central nervous system. Neural crest cells lie at the top of the neural folds. They can be seen between the neural tube and ectoderm layers on the dorsal side after the neural tube closes.

Douarin and by Thiery (this volume). The first morphologically and physiologically differentiated neurons appear in the neural tube very shortly after it closes. These are Rohon-Beard neurons (primitive sensory cells) and primary motoneurons. Spitzer (this volume) deals with the early events in the physiological differentiation of these cells.

In both amphibians and birds the experimental evidence implicating the mesoderm as the generator of the inducing signal is good. In the amphibian embryo, for example, if gastrulation fails to occur so that mesoderm cells do not reach their normal position beneath the dorsal ectoderm, the nervous system does not develop. Experiments using the grafting techniques of classical embryology have revealed a number of features of this interaction in both species. Although the bulk of the evidence is derived from experiments on amphibians, it is likely that many characteristics of the process of neural induction will prove to be common to all vertebrates. The most important of these features, which must be understood if we are to explain the mechanism of neural induction, are listed below.

1. All ectoderm cells, not just those on the dorsal surface which normally give rise to the nervous system, are able to respond to the inducing signal provided by the mesoderm.
2. A very small number of mesoderm cells is sufficient for transmission of the signal.
3. Ectoderm cells, once induced, are able to induce other ectoderm cells.

4. The time over which ectoderm cells can respond to the signal given by the mesoderm is limited, and thus ectoderm cells are said to be "competent" to respond for only a limited period of time during development.

These findings pose a number of very clear questions about primary induction that will be considered in turn.

WHAT IS THE NATURE OF THE INDUCING SIGNAL?

At the present time it is fair to say that we have no idea as to the identity of the inducing signal. The observation that many substances can act as artificial inducers of nervous tissue, particularly in the amphibian embryo, has made this search difficult. It has led to the suggestion that neural induction may be a permissive, rather than an instructive, event, but there is no good evidence to separate these possibilities. Attempts to extract inducing molecules from early embryos and to relate them to the natural signal have been, by and large, unsuccessful. These difficulties with the identification of inducing molecules have further led Jacobson (1982) to question the whole concept of neural induction and to propose that the nervous system arises as a consequence of processes occurring before gastrulation. Jacobson's criticisms were mainly directed against experiments carried out by Spemann early this century (see Spemann, 1938). One of the most important experiments indicating that the mesoderm was responsible for initiating the development of the nervous system came from Spemann's grafts of an additional dorsal lip (the site where the invagination of mesoderm and endoderm cells is initiated; see Figure 1) into a host embryo. Spemann's results showed that in these circumstances a second nervous system, derived from host tissue, was formed. Jacobson, who used the intracellular marker horseradish peroxidase to identify graft and host tissue, claimed that the second nervous system was formed from the graft. However, subsequent experiments (Gimlich and Cooke, 1983; Smith and Slack, 1983) using the same technique as Jacobson's have confirmed Spemann's observations. The concept of neural induction by mesoderm cells therefore seems to be well founded.

HOW IS THE SIGNAL TRANSMITTED?

The method of transfer of the signal would seem more amenable to experimental attack, but again, at the time of writing no clear consensus has emerged. There are two consequences of neural induction, which often are not separated. One is the initiation of morphogenetic movements that lead to the formation of the neural groove and neural folds and culminate in the closure of the neural tube. The second is the gradual conversion of epithelial cells into neuroepithelial cells, which then differentiate into neurons and glial cells. If these two events are initiated independently, there may be two inducing substances, rather than one.

Inducing substances could be released into the intercellular space by mesoderm cells and then diffuse to their target, the overlying ectoderm. Or they could be transferred directly from mesoderm to ectoderm cells through some specialized pathway, such as the gap junction. This structure is known to link the interiors of adjacent cells and to provide a pathway for the transfer of ions and other small molecules directly from one cell to the next without recourse to the extracellular space (see Wolpert, 1978, for a review).

If morphogenetic movements are initiated by a signal different from that which initiates differentiation within neural plate cells, it ought to be possible to separate these two consequences of neural induction. There is some evidence to suggest that this may be the case. Jacobson (1962) showed that if the neural plate is surgically removed from axolotl embryos at early neural plate stages, the morphogenetic movements of the lateral ectoderm continue and the folds still come together in the midline. This shows that the morphogenetic signal has effects well beyond the borders of the neural plate. Morphogenetic movements continue unimpaired when the program for neuronal differentiation within the neural tube is artificially interrupted (see below). A separation of morphogenetic movements and neuronal differentiation has also been observed in recent experiments (L. Breckenridge, unpublished observations) in which the consequences of treating early amphibian embryos with lithium ions were reexamined. Lithium ions are classically regarded as a "vegetalizing" agent because treatment of cleavage-stage amphibian embryos generates tadpoles with pronounced bellies and shortened neural axes; frequently, their neural tubes fail to form. Such results have been interpreted as indicating that neural induction has failed because the neural tube is absent. If the dorsal ectoderm that would have formed the nervous system is dissected from such "vegetalized" embryos, dissociated, and placed into tissue culture, it is found that a substantial number of neurons differentiates from embryos treated with lithium ions. Quantitative analysis (for method, see Messenger and Warner, 1979; Breckenridge and Warner, 1982) reveals that the number of neurons differentiating from lithium-treated embryos is rarely less, and often greater, than from untreated embryos, sometimes by as much as 100%. These results suggest two possible conclusions: (1) Neural induction, assessed in terms of primary differentiation in tissue culture, can proceed in the absence of the morphogenetic movements of neurulation; (2) the classical view of lithium as a vegetalizing agent may not be entirely correct.

In experiments on neural induction in the newt (Toivonen and Wartiovaara, 1975; Toivonen et al., 1975), the method of transfer of the inducing signal was assessed by interposing nucleopore filters of varying pore sizes between ectoderm and mesoderm layers. To assess how far neural induction could proceed, ectoderm cells were stripped from mesoderm cells after various contact times and the subsequent differentiation of the dorsal ectoderm assessed. When the filters were interposed over the time of primary induction (gastrula–early-neurula stages), the ectoderm differentiated into neuroid structures. Electron-microscope examination of ectoderm–mesoderm sandwiches showed processes from ectoderm cells entering, but not traversing, the filters. The authors concluded that neuroid development of the ectoderm was brought about by the release of a diffusible substance into the extracellular space and did not require direct cell-to-cell contact. This would seem to be an experiment clearly indicating that neural induction is brought about by an extracellular factor. However, neural differentiation in the ectoderm was always very primitive and no evidence for morphologically differentiated neurons was obtained. To examine whether longer contact between ectoderm and mesoderm, under similar conditions, could complete neural induction, a further series of experiments was carried out. In this series, the filter was interposed when the neural groove was first formed, classically supposed to be at the end of neural induction, and the ectoderm and mesoderm cells were left in contact for a period covering the later stages of neurulation. This set of experiments again showed that direct contact between ectoderm and mesoderm cells did not take place through the filter. However, they also demonstrated that neural differentiation proceeded no further than that found in ectoderm cells induced through the filter by mesoderm cells during the early gastrula–early-neurula stages. Clearly, neural induction could not be completed in the absence of direct contact between ectoderm and mesoderm cells.

All these experiments are consistent with the view that there are two signals involved in the process of neural induction. The first is released by mesoderm cells into the extracellular space, has a widespread influence on ectoderm cells, and is responsible for initiating the morphogenetic movements involved in neurulation. For neuronal differentiation, a second signal is necessary; this could be transmitted directly from mesoderm to ectoderm, perhaps through gap junctions. Electrical coupling (one sensitive indicator of the presence of a pathway allowing small ions to move directly from one cell to another) has been noted between ectoderm cells and underlying notochord cells at midneural fold stages in both chick (Sheridan, 1968) and amphibian (Warner, 1973) embryos. Although the body of evidence for there being two inducing signals rather than one is becoming suggestive, clear evidence for or against this hypothesis has yet to be obtained.

WHAT DEFINES THE BORDER OF THE NEURAL PLATE?

Given that ectoderm cells, once induced, can further induce other ectoderm cells, it is pertinent to ask how the inducing signal is restricted to those cells within an area delineated by the neural folds. One possibility could be that gap junctions, known to be widespread within presumptive ectoderm and neural cells at pregastrula stages, are lost at the edge of the neural fold once neural induction is over. This possibility was examined by Warner (1973). The results showed clearly that ionic current flow between neural plate cells and lateral ectoderm cells remained until the neural tube closed. An experiment showing current spread from ectoderm to neural plate at the late neural fold stage is illustrated in Figure 2. This means that the interaction across the edges of the neural folds is still possible through the gap junctional pathway until well beyond the time usually supposed to mark the end of neural induction. The experiments of Toivonen et al., described above, seem to indicate that essential interactions between ectoderm and mesoderm continue beyond the early neurula stage. Thus, it may be that the interaction in the dorsal midline is complete relatively soon after gastrulation, but that it takes time for the signal to spread throughout the neural plate to its lateral borders. If that is the case, interactions within the neuroectoderm may continue well into the neural fold stages. Electrical coupling between lateral ectoderm and neuroectoderm is lost when the neural tube closes (Warner, 1973), and this may be sufficient to limit the further spread of the signal

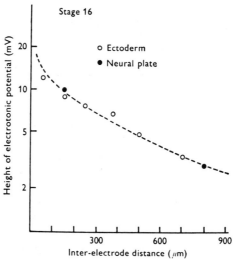

Figure 2. *Electrical coupling between ectoderm and neuroectoderm is still present at the late neural fold stage.* Ordinate: Voltage deflection produced by injection of a hyperpolarizing current pulse into an ectoderm cell. Abscissa: Distance between current injection and voltage recording electrodes. Open symbols: Voltage recording electrode in ectoderm cells. Solid symbols: Voltage recording electrode in neural plate cells. Note that current spreads with equal ease into ectoderm or neural plate. (From Warner, 1973.)

into the lateral ectoderm. Such a possibility has a clear consequence that should be testable: If the neural tube is prevented from closing, and thus from breaking gap junctions at the neural tube–ectoderm border, the amount of tissue devoted to the nervous system should increase.

WHEN ARE THE FIRST CONSEQUENCES OF NEURAL INDUCTION MANIFESTED IN ECTODERM CELLS?

There are two visually obvious consequences of the process of neural induction: the appearance of the neural groove and the conversion of cells in the dorsal ectoderm from flat epithelial cells to columnar neuroepithelial cells. The relation of these morphological changes to alterations in cell physiology that may arise consequent to neural induction is not known. The role of these changes is likely, at least in part, to be related to the mechanical process of generation of the neural tube. The shape-changes within the ectoderm cells destined to become the neural tube will reduce the surface area within the neural plate and thus contribute to the overall shape-change observed within the whole embryo. However, these shape-changes alone are insufficient to bring about the profound morphogenetic alterations observed during neurulation. It has been suggested that contractile material located at the dorsal side of the neuroepithelial cells takes an active part in pulling neuroepithelial cells together (see Shroeder, 1970). Attachment to the mesodermal structure, the notochord, which forms immediately below the neural plate, has also been suggested to be important (Burnside and Jacobson, 1968).

Since morphologically and physiologically differentiated neurons are found in the neural tube almost immediately after the tube closes, these alterations in shape of both neuroepithelial cells and the neural plate must be accompanied by other changes within the cells. Such changes might be a direct consequence of events leading to differentiation within the neural plate, rather than the accompanying morphogenetic events. Recent experiments have shown that alterations in cell physiology can be recognized within neural plate cells from the midneural fold stages onward (Warner, 1973; Blackshaw and Warner, 1976a; Messenger and Warner, 1979; Breckenridge and Warner, 1982). These experiments, which also give support to the view that morphogenesis and differentiation may be under separate control, are outlined below.

In the course of making measurements of electrical coupling between the neuroectoderm and lateral ectoderm, Warner (1973) noted that the resting membrane potential of cells residing in the neural plate and ventral ectoderm gradually diverged as neurulation proceeded. When the time course of these alterations was examined more closely (Blackshaw and Warner, 1976a), it was found that the membrane potential of neural plate cells began to increase at about the midneural fold stage. Measurements in two axolotl embryos illustrating this point are shown in Figure 3A. Figure 3B shows, for comparison, membrane potential measurements made in ventral ectoderm cells over the same developmental stages. Several points emerge from these results. At the beginning of neurulation, membrane potentials in all areas of the ectoderm are closely similar, in the region of -20 to -30 mV. As neurulation proceeds, neural plate cells gradually acquire a more negative resting potential, approaching -65 mV by the end of the midneural fold stage. Such changes do not occur in the ventral ectoderm; they are characteristic of neural plate cells. Furthermore, neural plate cells acquire a more negative resting potential despite being in electrical communication with the surrounding lateral ectoderm. This makes it unlikely that the increase in intracellular negativity is generated by a change in the passive permeability properties of neuroectoderm cells, because a potential difference within an electrically coupled system would then be transient rather than maintained. It is more probable, therefore, that the rise in resting potential is brought about by some active transport mechanism. One obvious candidate for such a role is a change in the amount of sodium pumping carried out by neural plate cells over these stages of development.

It is well known that the sodium pump can make an electrogenic contribution to the resting membrane potential (see Thomas, 1972, for a review) because three sodium ions are usually exchanged for two potassium ions, so that as the sodium pump cycles, positive charge is removed from the interior of the cell. Early amphibian embryonic cells possess some sodium pumping capacity at very early cleavage stages (Slack and Warner, 1973). This means that there are three possible ways in which an increase in the electrogenic contribution of the sodium pump could come about: an increase in membrane resistance, so that sodium pump current generates a larger voltage to be added to the membrane potential; an increase in the number of sodium pumps in the membrane of neural plate cells; a change in the stoichiometry of the pump, so that fewer potassium ions are taken up in exchange for sodium. Before considering these alternatives, it is necessary to establish whether sodium pumping is responsible for the increase

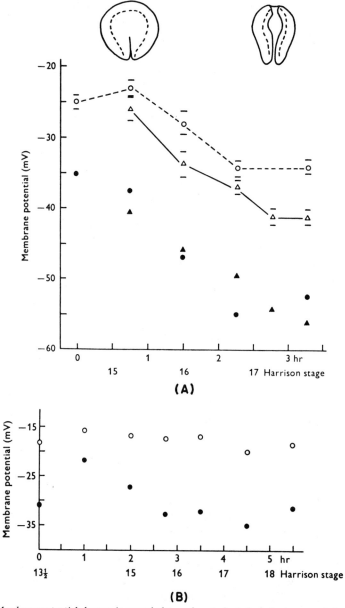

Figure 3. *Membrane potential changes in neural plate and ventral ectoderm during neurulation in the axolotl embryo.* A: Measurements in the neural plate. Circles and triangles give results for two different embryos. Note increase in both mean and maximum potentials as neurulation proceeds. B: Measurements in the ventral ectoderm. Note that membrane potentials remain in the region of −30 mV throughout neurulation. Ordinates: Membrane potential (mV). Abscissae: Developmental stage or time after beginning measurements. The diagrams above the graph in Figure 3A indicate the superficial appearance of the embryo over the period of all the experiments. Open symbols: Mean of 20 determinations at each time interval during neurulation; bars give standard error of the mean (SEM). Filled symbols: Maximum membrane potential recorded in each group. Because the cells are small and undergoing vigorous morphogenetic movements, some electrode damage is inevitable. Maximum potential recorded is therefore closest to the true membrane potential. (From Blackshaw and Warner, 1976a.)

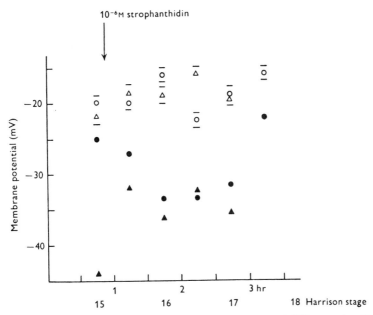

Figure 4. *Treatment of neurulating axolotl embryos with the sodium pump inhibitor strophanthidin abolishes the increase in resting potential of neural plate cells.* Axes as in Figure 2. (From Blackshaw and Warner, 1976a.)

in resting membrane potential in the neural plate. This can be done by applying one of the cardiotonic steroids, which are known to be highly specific inhibitors of the sodium pump (see Glynn, 1957). Figure 4 shows the consequence of treating neurulating embryos with one such steroid, strophanthidin, at a concentration known to bring about substantial inhibition of the sodium pump. In order to ensure that the drug penetrated into the intercellular spaces, a small hole was made in the belly ectoderm at the beginning of each experiment. It can be seen from Figure 4 that under such circumstances the increase in resting potential normally observed in neural plate cells no longer takes place. This finding suggests that the sodium pump is closely involved in the generation of the rise in resting potential in the neural plate. Measurements of the ease with which current spreads within the neural plate over these stages (Warner, 1973; Blackshaw and Warner, 1976a) indicate that there is no substantial change in surface membrane resistance, and studies on the properties of the sodium pump in HeLa cells (Baker and Willis, 1972) suggest that alterations in stoichiometry of the sodium pump are unusual. It is therefore most probable that there is an increase in the number of sodium pumps in the membranes of neural plate cells at these stages of development. Since Blackshaw and Warner (1976a) observed that an increase in resting membrane potential only occurred in embryos which subsequently completed neurulation, it seems likely that this alteration in membrane properties of neural plate cells is in some way directly related to the process of neurulation. It then becomes important to know whether the increase in resting membrane potential, or the activation of the sodium pump that it reflects, is in any way essential for the normal development of the nervous system. Embryos treated with strophanthidin complete neurulation, suggesting that completion of morphogenesis does not require the change in resting membrane potential.

The consequences for further neural development were examined by Messenger and Warner (1979). *Xenopus* embryos were exposed to strophanthidin while they were neurulating. As the neural tube closed, the hole in the belly well was reopened and the drug washed away. The embryos were then left in dilute salt solution (10% Ringer's solution) for two days. By this time, control tadpoles are able to swim and have well-developed brains and eyes. Comparison of control and treated tadpoles showed clearly that inhibiting the sodium pump during neurulation had catastrophic consequences for the developing brain and eyes. Control tadpoles at this stage show the normal, orderly arrangement of ependymal, mantle, and marginal layers in the brain, with the white matter already substantial. The eyes contain a lens, outer pigment layer, and developing retinal ganglion layers. In tadpoles treated with strophanthidin, a very different picture emerged. The cells were extremely disordered in both brain and eyes, with no sign of developing layers of cells. There was little or no white matter, indicating that the axon outgrowth had not taken place. Other structures in the tadpoles looked relatively normal. These experiments alone suggest that the presence of a functional sodium pump during neurulation is in some way essential for the subsequent correct development of the nervous system. However, quantitative assessment of the effects of drug treatment is extremely difficult in the whole embryo because of inherent variability in the sensitivity of embryos both within and between batches. In order to overcome this problem, Messenger and Warner (1979) developed a method that allows the degree of neuronal differentiation to be measured directly.

The technique, which is described in detail in Messenger and Warner (1979), relies on the ability of amphibian neurons to complete primary differentiation in tissue culture. By using a number of embryos to prepare each culture, variability from embryo to embryo can be smoothed out. Cultures are prepared from *Xenopus* embryos taken just as the neural tube closes. The neural tube, along with the underlying notochord and somites, is dissected out, treated for 2–3 min with calcium-free medium containing 1 mM EGTA, and then disaggregated into single cells by mechanical trituration. Material from three embryos is dispensed into each culture dish and the cultures left overnight at 24°C in culture medium consisting of Ringer's solution together with 10% fetal calf serum. Amphibian embryonic cells contain yolk and will differentiate in the absence of serum, but the addition of fetal calf serum to the culture medium improves the reproducibility from batch to batch. This is important because it allows comparison of experiments carried out on different batches of embryos.

Figure 5 is a photograph of such a culture and shows a typical microscope field used for quantitative assessment of the proportion of nerve that differentiates under different conditions. The arrows indicate cells that we identify as neurons. The criteria used are that the cell should have a phase-bright cell body, neurites, and growth cones. How reliable is this identification? Penetration with an intracellular microelectrode shows that cells with this morphology generate action potentials (see also Spitzer, this volume). If in contact with striated muscle cells, they form functional neuromuscular junctions and can drive end-plate activity and muscle contraction. They stain specifically with tetanus toxin (Vulliamy and

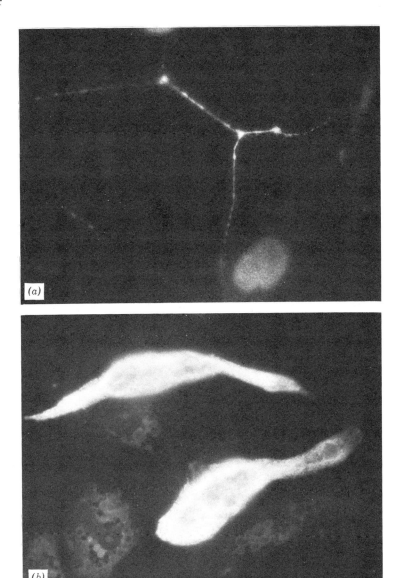

Figure 5. *a: Neuronal processes stained with antineurofilament antibody.* Note staining of nuclei in nonneural cells. *b*: Muscle cell stained with a monoclonal antibody that specifically recognizes muscle. Other cells in the field are not stained.

Messenger, 1981) and with an antibody directed against neurofilament protein (Figure 5A). The neurons also stain for the enzyme cholinesterase, as would be expected because many of them will be primary motoneurons. Other neurons contain monoamines and can be recognized by using the glyoxylic acid technique. Only two other cell types in these cultures can be recognized unequivocally. The first, striated muscle cells, form bipolar, mononucleate cells with a prominent nucleolus, striate after about 24 hours in culture, and contract if innervated or if stimulated electrically. When several muscle cells differentiate in close proximity to one another they align and are electrically coupled, as found *in vivo* (Blackshaw and Warner, 1976b). The muscle cells can be stained specifically with a monoclonal antibody raised by J. Brockes and C. Kintner. An example is shown in Figure 5B. After 36–48 hours in culture, the second recognizable cell type, melanocytes, begin to pigment. The remaining cells are largely mesenchymal cells and fibroblasts.

For quantitative assessment of neuronal differentiation, control and strophanthidin-treated embryos are taken just as the neural folds close. Neural tube, notochord, and somites are dissected out, dissected portions from three embryos are put into each Petri dish in culture medium, dispersed into single cells, and left for about 18 hours. Cultures from control and strophanthidin-treated embryos both undergo differentiation in the same solution (i.e., in the absence of the cardiotonic steroid), so that any effect on the number of neurons that differentiate must be the consequence of strophanthidin treatment *in vivo* during neurulation.

Eighteen to 24 hours after plating, a monolayer of differentiated cells has formed. Twenty to 30 microscope fields are randomly selected from the cultures, and in each field the number of nerve cells, the number of muscle cells, and the total number of cells are counted. The fields contain between 100 and 500 cells. For each field, the number of nerve cells is expressed as a percentage of the total number of cells. A similar calculation is made for the number of muscle cells to provide a control for the effect of drug treatment on another excitable cell. Figure 6 plots frequency histograms determined in this way for the number of neurons differentiating from control embryos and embryos treated with strophanthidin during neurulation. It is clear that neurons occur much less frequently in cultures prepared from embryos treated with strophanthidin. Thus, in control cultures, neurons made up at least 4% and up to 18% of the differentiated cells in each field, with the median of the frequency distribution lying at 9.6%. In the cultures from treated embryos, there were fields that contained no neurons, the maximum lay at 8%, and the median at 3.6%. Comparison of the two frequency histograms using the Mann–Whitney U test showed them to be significantly different ($p < .001$). The total number of cells that differentiate from the two sets of embryos was the same, and the percentage of muscle cells that differentiate was also unaffected by strophanthidin treatment. These findings suggest that only the neuronal population is influenced when the sodium pump is inhibited during neurulation, in good agreement with conclusions drawn from examination of whole embryos (see above).

Figure 7 shows a dose–response curve for the degree of inhibition of neuronal differentiation, expressed as the ratio of the median of the treated population to the median of the control population. A ratio of 1.0 means there has been no effect on the number of neurons; a ratio of 0.0 means that all morphologically recognizable neurons have been abolished from the cultures. The ratio of the medians is plotted against the concentration of strophanthidin that was present during neurulation. This figure illustrates two important points. First, the effective concentration range lies between 10^{-7} and 10^{-5} M strophanthidin, closely similar to the range over which strophanthidin inhibits the sodium pump in assays using the flux of sodium or potassium driven by the sodium pump in red cells (Glynn, 1957). Second, maximally effective concentrations of strophanthidin never abolish all neurons from the cultures. The neurons remaining (approximately 30%) are probably derived from neural crest neuronal precursors, because they

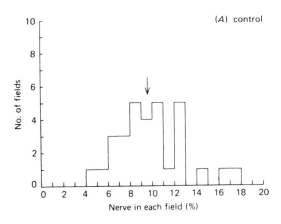

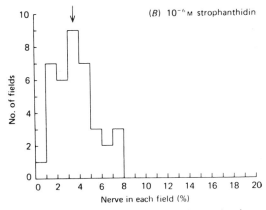

Figure 6. *Frequency histograms for the number of neurons differentiating from control embryos (A) and embryos treated with strophanthidin during neurulation (B).* Median of control population: 9.6%. Median of population from treated embryos: 3.6%. The two distributions are significantly different from each other (Mann–Whitney U test: $p < .001$). (From Messenger and Warner, 1979.)

can be largely abolished by the addition of an antibody to nerve growth factor (see Messenger and Warner, 1979). This implies that only neurons derived from the neural plate population are affected when the sodium pump is inhibited during neurulation. Neural crest neurons may go through a similar period of sensitivity once they have migrated to their final destination within the embryo. Since this does not take place until some time after the neural tube closes, such neurons would be expected to escape the consequences of sodium pump inhibition during neurulation alone.

Further evidence for the view that these effects on the ability of neurons to differentiate are the consequence of inhibiting the sodium pump comes from a number of findings. Lowering the potassium concentration in the intercellular spaces during neurulation to less than 1 mM also reduces the number of neurons that subsequently differentiate (Breckenridge and Warner, 1982). Conversely, raising extracellular potassium in the presence of strophanthidin protects neurons from the inhibitory effect (Messenger and Warner, 1979). Both these results are to be expected from the known sensitivity of sodium pump activation to extra-

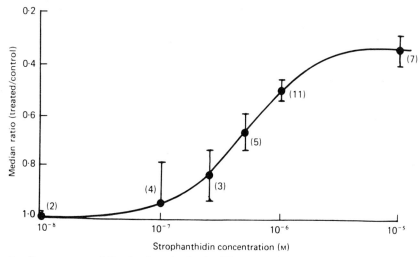

Figure 7. *Dose–response relation for the reduction in differentiated neurons after treatment with different concentrations of strophanthidin during neurulation.* Ordinate: Ratio of the median of the treated population to the median of the control population. 1.0 means no effect. Abscissa: Concentration of strophanthidin present during neurulation. (From Messenger and Warner, 1979.)

cellular potassium and the competition between cardiotonic steroids and extracellular potassium. Analogues of cardiotonic steroids that are poorly effective at inhibiting the sodium pump also do not affect neuronal differentiation.

It seems therefore that inhibition of the sodium pump while the embryo is neurulating has a profound effect on the ability of neurons to express their developmental fate, whether differentiation takes place *in vivo* or *in vitro*. How does inhibition of the sodium pump achieve this effect? There are three obvious consequences of sodium pump inhibition during neurulation. The membrane potential changes no longer take place. The gradual abolition of the gradient for sodium ions between intracellular and extracellular phases normally maintained by the activity of the sodium pump will not only influence the ion gradients for sodium and potassium, but also will lead to accumulation of calcium ions within the cell (Blaustein, 1974). An increase in the intracellular free-calcium concentration is known to reduce the permeability of gap junctions (Rose and Loewenstein, 1976). The whole of the ectoderm, both neural and nonneural, is linked by gap junctions during the time that inhibiting the sodium pump achieves its effect on neuronal differentiation, so it could be that this treatment uncouples gap junctions. This is known to happen during cardiotonic steroid treatment in heart muscle (Weingart, 1977). To understand the mechanism by which inhibiting the sodium pump achieves its effects, it is necessary to eliminate these possibilities in turn.

The possibility that strophanthidin prevents subsequent neuronal differentiation because of a reduction in the permeability of gap junctions can be examined by measuring whether neural plate cells remain electrically coupled to each other during drug treatment. It turns out that even after four hours of exposure to strophanthidin, electrical coupling remains within the neural plate (Messenger and Warner, 1979). However, electrical coupling is an extremely sensitive measure and only determines the ability of small ions to flow directly from cell to cell; the transfer of larger molecules may nevertheless be impeded. An alternative strategy is to see whether the effects of strophanthidin treatment can be modulated by altering the extracellular calcium ion concentration. Weingart (1977) has shown that in heart muscle the effect of cardiotonic steroids on cell-to-cell transfer of small ions is potentiated when calcium is increased from its normal extracellular level of 2 mM to 6 or 10 mM. When embryos are exposed to strophanthidin along with 10 mM calcium while they are neurulating and then assessed for neuronal differentiation either *in vivo* or *in vitro*, it turns out that the high calcium concentration greatly reduces the efficacy of strophanthidin treatment (see Table 1; Breckenridge and Warner, 1982). This makes it extremely unlikely that a reduction in the permeability of gap junctions is responsible for the reduction of neuronal differentiation following sodium pump inhibition.

Raising extracellular calcium while the sodium pump is inhibited prevents the cardiotonic steroid from exerting its effect, and this finding allows us to test whether the membrane potential during neurulation is important for subsequent neuronal differentiation. If the membrane potential is determined in embryos neurulating in the presence of strophanthidin along with 10 mM extracellular calcium, it turns out that the increase in resting membrane potential of neural plate cells, characteristic of control embryos, does not take place (Breckenridge and Warner, 1982). This means that the high extracellular calcium is able to oppose the effect of inhibiting the sodium pump without restoration of the increase in membrane potential. We can conclude that the absolute level of the membrane potential in neural plate cells plays no role in determining whether neuronal differentiation follows neurulation.

We are therefore left with the supposition that the reduction in neural differentiation achieved by inhibition of the sodium pump during the neural plate stages is related to alterations in ion content which ensue when the pump is inhibited. Can we identify which ion is likely to be responsible? An important clue can be obtained from the finding that raising extracellular calcium has a protective effect. When external calcium is increased, particularly if the sodium

Table 1. The Ability of Divalent Cations to Protect Differentiating Nerve Cells from the Inhibitory Effect of Strophanthidin[a]

n	Strophanthidin/ Control Median Ratio	Ca^{2+}	Sr^{2+}	Mg^{2+}	Mn^{2+}	$Ca^{2+} + Mn^{2+}$
8	0.53 ± 0.03	0.84 ± 0.02	—	—	—	—
5	0.55 ± 0.02	—	0.72 ± 0.03	—	—	—
4	0.58 ± 0.03	—	—	0.55 ± 0.06	—	—
4	0.65 ± 0.06	—	—	—	0.61 ± 0.05	—
6	0.61 ± 0.04	0.88 ± 0.04	—	—	—	0.60 ± 0.06

[a] Embryos were treated with 5×10^{-6} M strophanthidin alone, or together with 10 mM of the appropriate divalent cation, between the early and late neural fold stages. Cultures were prepared as the neural tube closed. Each column gives the median ratio (treated population/control) mean ± SEM. When the effect of calcium was compared to that of calcium plus manganese, each cation was present at 6 mM. n is the number of independent cultures.

pump is inhibited, intracellular sodium falls (Deitmer and Ellis, 1978). A number of other experimental manipulations will also lead to a fall in intracellular sodium even if the sodium pump is inhibited. By testing whether any manipulation that will reduce intracellular sodium influences the outcome of inhibiting the sodium pump, it should be possible to see whether the internal concentration of sodium plays a role in ensuring that neurons differentiate after neural induction. The details of such manipulations are given in Deitmer and Ellis (1978). They allow a number of clear predictions that should be fulfilled if the intracellular concentration of sodium ions is an important factor:

1. Strontium ions, as well as calcium ions, should be capable of protecting neuronal differentiation from the consequences of inhibiting the sodium pump. Table 1 shows the outcome of adding 10 mM strontium to the bathing medium in the presence of strophanthidin. Strontium ions are clearly capable of protecting neuronal differentiation, but to a lesser degree than calcium (Breckenridge and Warner, 1982). This would be expected on the basis of their ability to lower internal sodium (see Deitmer and Ellis, 1978).
2. Magnesium ions should not be able to oppose the consequences of sodium pump inhibition, as they do not bring about a fall in intracellular sodium. Table 1 shows that magnesium ions do not oppose the consequences of treatment with strophanthidin.
3. Manganese ions also should be unable to prevent the reduction in subsequent neuronal differentiation when the sodium pump is inhibited. This prediction proved to be correct (Table 1; Breckenridge and Warner, 1982).
4. If manganese ions are present along with an increased concentration of calcium, the protection afforded by raising extracellular calcium should be abolished. Again this prediction is upheld (Table 1; Breckenridge and Warner, 1982).
5. Replacing extracellular sodium with another monovalent cation should rescue neurons from the consequences of inhibiting the sodium pump. Under these conditions, intracellular sodium cannot rise when the pump is inhibited because sodium ions are not available to enter passively down their concentration gradient. In 10 experiments to examine the ability of neurons to differentiate in culture after treatment with strophanthidin during neurulation, in the presence and absence of sodium ions, the ratio of the distributions of cultures from both control and treated embryos (5×10^{-6} M strophanthidin/control) was 0.57 ± 0.04 (mean $\pm$ SEM) in the presence of extracellular sodium and 0.89 ± 0.06 (mean $\pm$ SEM) in the absence of extracellular sodium. Clearly, removing sodium ions from the bathing medium is an extremely effective way of preventing the consequences of sodium pump inhibition.

Thus all these predictions to test whether internal sodium is important are fulfilled. Furthermore, the manipulations had similar consequences whether the outcome was assessed *in vitro* or *in vivo* (Breckenridge and Warner, 1982). Tadpoles treated with high calcium along with strophanthidin grew up with normal brains and eyes, as did those in which sodium was lowered while the embryos were treated with the cardiotonic steroid. Conversely, when manganese ions were present along with calcium, strophanthidin was able to exert its usual effect.

In the light of these findings, it seems important to examine whether alterations in intracellular sodium in neural plate cells accompany normal neurulation. Figure 8 illustrates how the intracellular concentration of sodium ions changes during neurulation in two batches of axolotl embryos. Intracellular sodium was measured with sodium-sensitive intracellular microelectrodes (for method, see Breckenridge and Warner, 1982). Intracellular sodium in neural plate cells is about 30 mM shortly after neural induction. As the embryos progress through neurulation, intracellular sodium falls, with approximately the same time course as the increase in membrane potential (cf. Figure 2). By the late neural fold stage, internal sodium has fallen by a factor of three to around 10 mM. At this stage of development, sodium in ventral ectoderm cells is still in the region of 25 mM. Thus substantial changes in internal sodium occur as the additional sodium pump capacity of neural plate cells is activated during the midneural fold stages. Measurements of intracellular sodium made in the presence of strophanthidin confirm that internal sodium rises when the sodium pump is inhibited, with intracellular sodium approaching the level found at the beginning of neurulation, that is, close to 30 mM, after two hours' exposure to the pump inhibitor (Breckenridge and Warner, 1982). The experiments indicating that high external calcium can protect neural plate cells from the consequences of inhibiting the sodium pump suggested that internal sodium fell under these conditions despite inhibition of the pump. Figure 9 shows measurements in four embryos that confirm that internal sodium does fall under these conditions, providing additional strength to the hypothesis that intracellular sodium plays a key role during the neural plate stages.

Can we eliminate the possibility that calcium ions exert the controlling role, rather than sodium? This is an important consideration, as the cellular mechanisms controlling internal calcium and sodium are closely interlinked. Comparison of the effects of inhibiting the sodium pump alone with the effects of inhibiting the pump in the presence of high extracellular calcium allows us to draw the inference that sodium levels are more important than those of calcium. In both cases, intracellular calcium will rise (see Bers and Ellis, 1982), but internal sodium rises with pump inhibition alone and falls when calcium is present along with the pump inhibitor. Because neurons derived from the neural plate differentiate perfectly well as long as external calcium is high, despite inhibition of the sodium pump, it is most likely that sodium, rather than calcium, exerts control.

Before finally concluding that intracellular sodium levels are most concerned, it is pertinent to consider whether the reciprocal alterations in intracellular potassium, which must occur both when the sodium pump is inhibited and when it is activated, might also be important. A reduction in internal sodium from 30 to 10 mM, as happens during normal development of the amphibian nervous system, will be accompanied by an equivalent increase in intracellular potassium. Thus, while internal sodium falls by a factor of three, internal potassium will rise by about 20%. There is evidence that the level of intracellular potassium is important for protein synthesis (Ledbetter and Lubin, 1977), but a change of

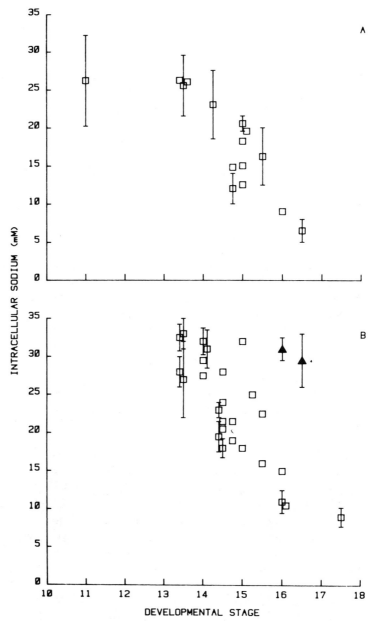

Figure 8. *Intracellular sodium in neural plate cells during neurulation.* A and B show measurements in two separate batches of axolotl embryos. Where error bars are shown, the mean ± SEM is given. Otherwise, points give average of three measurements in each embryo. Open symbols: Measurements in neural plate. Solid symbols: Measurements in ventral ectoderm. (From Breckenridge and Warner, 1982.)

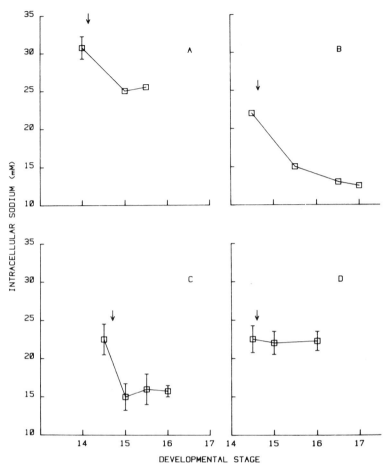

Figure 9. *High extracellular calcium allows intracellular sodium to fall in neural plate cells despite inhibition of the sodium pump.* Measurements in four embryos are illustrated. In all cases, the ordinate gives the intracellular sodium activity (mM) and the abscissa gives the time after commencing the measurements. Where error bars are shown, points give mean ± SEM; otherwise, points give average of three measurements. First measurement taken in Ringer's solution. At the arrow, Ringer's solution containing 10 mM calcium and 5×10^{-6} M strophanthidin was flushed into the bath. Note that internal sodium either falls or is held at a low level, despite inhibition of the sodium pump. (From Breckenridge and Warner, 1982.)

20% is unlikely to be great enough to have a marked effect. Alterations in internal potassium when intracellular sodium is reduced by raising external calcium are also likely to be small, suggesting that the sodium concentration is the dominant factor. The fall in internal sodium relative to the internal concentration of potassium, which may be the relevant parameter, is even more substantial, close to 30-fold, and it is entirely possible that a change of this magnitude in the combined ion levels could be sufficient to play an important part in the control of events following neural induction. At present we know too little about changes in ion levels as intracellular activators to speculate how these changes in ion concentrations might exert their effect. We also do not know whether such changes act as an essential cofactor, allowing the expression of other changes in cell metabolism that occur as a consequence of neural induction, or as a trigger, initiating subsequent steps in the differentiation of neurons.

Study of the detailed timing of the alterations in sodium pumping activity in neural plate cells might help to define some of these questions more closely. Information on the time when the additional pumping sites are first inserted into the membrane of neural plate cells can be derived from experiments by Blackshaw and Warner (1976a), who examined the potassium sensitivity of the membrane potential at different times during neurulation. It is well known that the degree of activation of the sodium pump is powerfully influenced by the concentration of potassium in the extracellular medium. Blackshaw and Warner found that substantial increases in resting membrane potential, up to 30 mV, could be induced in neural plate cells at early neural fold stages by raising external potassium from its resting level of 2.5 to 20 mM before the natural increase in membrane potential took place. The increase in resting potential produced by potassium initially appeared shortly after the neural groove was first visible on the dorsal surface of the embryo. Once the natural increase in membrane potential was under way, it no longer occurred. The potassium-induced hyperpolarization was blocked by adding either ouabain or strophanthidin, both specific inhibitors of the sodium pump. Cells in the ventral ectoderm did not show any rise in membrane potential in response to an increase in extracellular potassium. The concentration of potassium which causes pump activation at early neural fold stages is much higher than normally considered to produce maximal activation of the sodium pump (5 mM; Glynn, 1956). This suggests that the additional sodium pumps are first inserted into neural plate cells with a very low affinity for extracellular potassium and that they become effective at the midneural fold stage as a consequence of an increase in affinity for potassium ions. There have been no studies of the properties of sodium pumps immediately after insertion into the cell membrane, so it is not possible to say whether a change in affinity for potassium with time is a normal feature of newly inserted pumps, or whether this behavior of sodium pumps inserted into neural plate cells is part of a developmental sequence only. Nevertheless, these findings indicate that the signal for insertion (and possibly synthesis) of extra sodium pumps into the membrane of neural plate cells is a very early consequence of neural induction.

The period when neural plate cells are sensitive to sodium pump inhibition turns out to be very restricted. Messenger and Warner (1979) examined the period of sensitivity of nervous system development more precisely and found that inhibiting the sodium pump at the midneural fold stage only, that is, over the period when the membrane potential is increasing, was sufficient to achieve a maximal effect. In the *Xenopus* embryo, this corresponds to a two-hour window. If intracellular sodium in neural plate cells is not reduced during this time period, subsequent differentiation of neurons from the neural plate fails. This suggests that the fall in intracellular sodium is linked to some irrevocable event which commits cells in the neural plate to follow a developmental pathway leading to the differentiation of neurons. This conclusion inevitably leads to consideration of the path followed by cells which have been diverted from the neuronal lineage. Careful examination by light and electron microscopy of the cells found within the neural tube of intact embryos after sodium pump inhibition shows them to be apparently healthy, and the number of cells within the brain is not significantly reduced below that found in controls (see Messenger and Warner, 1979). Because it is known that glial cells, as well as neurons, arise from the neural plate, one interesting possibility is that cells have been shifted into a glial, rather than a neuronal, lineage. The use of markers for glial cells, such as antibodies recognizing glial fibrillar acidic protein, together with antibodies to neurofilament protein, might help to resolve this question.

WHEN IS THE ANATOMICAL ORGANIZATION OF THE NERVOUS SYSTEM LAID DOWN?

The experiments described in the previous section suggest that alterations in cell physiology related to the commitment of cells in the neural plate to a neuronal lineage are complete by the end of the midneural fold stage of development. The anatomical arrangement of the nervous system is equally important in ensuring that the correct wiring diagram is set up during development, as the position of neurons within the nervous system largely determines the possible synaptic connections that any functional neuron may make. Jacobson (1964) addressed this question, using grafting techniques, in the developing axolotl embryo. He rotated small portions of the neural plate of known developmental fate through 180° at various times during neurulation and then examined the organization of the structures formed by the graft in adult animals. When the graft was made at the early neural fold stage, the subsequent organization of the nervous system was always normal, indicating that the graft had developed according to its new position. Grafts made at or after the late neural fold stage developed in a reversed position. Grafts at the midneural fold stages sometimes developed normally and sometimes in reverse orientation. Jacobson concluded that the anatomical organization of the nervous system was laid out on the neural plate during the midneural fold stages. Because it takes between one and two hours for such grafts to reestablish electrical coupling with their neighbors in the neural plate (see Warner, 1981), Jacobson's estimates of timing may not be entirely accurate. Nevertheless, the correlation between estimates of commitment to functional differentiation as neurons and of commitment to form a particular part of the nervous system is remarkably good. Both put the important time at the midneural fold stages of development.

It is unlikely that the changes in cell physiology described here play an integral part in the overall anatomical patterning of the nervous system, because inhibition of the sodium pump does not interfere with the morphological arrangement of brain and eye structures; it only prevents differentiation of neurons within the brain and eye. It may turn out that the electrical coupling between cells in the neural plate, which was almost certainly unaffected by the manipulations which interfere with differentiation, is important for ensuring the overall pattern of structures within the nervous system. Until a method of interfering with gap junction permeability that does not affect other aspects of cell metabolism is available, it will not be possible to resolve this issue.

CONCLUSIONS

In this chapter I have tried to bring together current information on the very early events associated with the differentiation of the nervous system. Many questions remain to be resolved, but our understanding of this important period of development is becoming sufficiently clear to allow the issues which need to be understood to be spelled out more precisely. A major question is still the mechanism of neural induction. I have suggested that it may be profitable to examine this process in terms of two signaling events, one which initiates morphogenetic changes leading to the formation of the neural tube, the other responsible for setting in hand the changes in cell physiology which lead to physiological and morphological differentiation of neurons. Although neither the identity nor the method of transfer of such signals is known, some progress has been made in unraveling subsequent events within the neural plate. Thus it is now clear that an early consequence of neural induction, which is essential if neuronal differentiation is to take place, is the insertion and subsequent activation of additional sodium pumps within the membranes of neural plate cells. These sodium pumps are essential because they bring about a substantial reduction in intracellular sodium; the evidence currently available suggests that this fall in intracellular sodium is the critical event. There still exists an important gap in our understanding of how these alterations in sodium concentration are related to the subsequent morphological and physiological differentiation of neurons. It is also not clear why an event which is experienced by the whole of the neural plate, destined to give rise to both neuronal and glial elements, should apparently specifically influence the fate of the neuronal elements of the nervous system. The anatomical patterning of the nervous system takes place over the same developmental stages as the alterations in cell physiology that precede neuronal differentiation. It seems that both the anatomical organization of the nervous system and the commitment to form neurons from neural plate cells are determined by events within the neural plate at the midneural fold stages of development. The means by which anatomical patterning of the nervous system is set up within the neural plate is not known.

A more general question raised by these findings relates to whether similar changes in cell physiology prior to differentiation occur in other developing neurons, such as those derived from the neural crest. We can also ask whether alterations in the properties of the cell membrane presage differentiative events in other cell types, such as muscle.

Most of the experimental evidence relating to the early development of the nervous system discussed in this chapter has been derived from experiments on amphibian embryos. Experiments relating to neural induction itself suggest that the process is very similar among the vertebrate species, and it is to be hoped that this will also prove to be true for other events during the early development of the nervous system.

ACKNOWLEDGMENTS

I am indebted to Sarah Guthrie and Lorna Breckenridge for their comments on the manuscript. Lorna Breckenridge also allowed me to quote unpublished results.

Jeremy Brockes and Chris Kintner kindly gave us samples of their monoclonal antibody that recognizes muscle cells, and Brian Anderton gave us samples of antineurofilament antibody. Much of the work described in this chapter was made possible by grants from the Medical Research Council.

REFERENCES

Baker, P. F., and J. Willis (1972) Binding of the cardiac glycoside ouabain to intact cells. *J. Physiol. (Lond.)* **224**:441–462.

Bers, D. M., and D. Ellis (1982) Intracellular calcium and sodium activity in sheep heart Purkinje fibres: Effect of changes in external sodium and intracellular pH. *Pflugers Arch.* **393**:171–178.

Blackshaw, S. E., and A. E. Warner (1976a) Alterations in resting membrane properties during neural plate stages of development of the nervous system. *J. Physiol. (Lond.)* **225**:231–247.

Blackshaw, S. E., and A. E. Warner (1976b) Low resistance junctions between mesoderm cells during development of the trunk muscles. *J. Physiol. (Lond.)* **255**:209–230.

Blaustein, M. P. (1974) The inter-relationship between sodium and calcium fluxes across cell membranes. *Rev. Physiol. Biochem. Exp. Pharmacol.* **70**:33–82.

Breckenridge, L. J., and A. E. Warner (1982) Intracellular sodium and the differentiation of amphibian embryonic neurones. *J. Physiol. (Lond.)* **332**:393–413.

Burnside, M., and A. G. Jacobson (1968) Analysis of morphogenetic movements in the neural plate of the newt *Tarichia torosa*. *Dev. Biol.* **18**:537–552.

Deitmer, J. W., and D. Ellis (1978) Changes in the intracellular sodium activity of sheep heart Purkinje fibres produced by calcium and other divalent cations. *J. Physiol. (Lond.)* **277**:437–453.

Gimlich, R., and J. E. Cooke (1983) Cell lineage and the induction of second nervous systems in amphibian development. *Nature* **306**:471–473.

Glynn, I. M. (1956) Sodium and potassium movements in human red cells. *J. Physiol. (Lond.)* **134**:278–310.

Glynn, I. M. (1957) The action of cardiac glycosides on sodium and potassium movements in human red cells. *J. Physiol. (Lond.)* **136**:148–173.

Jacobson, C.-O. (1962) Cell migration in the neural plate and the process of neurulation in the axolotl larva. *Zool. Bidr. Upps.* **35**:433–449.

Jacobson, C.-O. (1964) Motor nuclei, cranial nerve roots and fibre pattern in the medulla oblongata after reversal experiments on the neural plate of axolotl larvae. *Zool. Bidr. Upps.* **36**:73–160.

Jacobson, M. (1982) Origins of the nervous system in amphibia. In *Neuronal Development*, N. C. Spitzer, ed., pp. 45–99, Plenum, New York.

Karfunkel, P. (1971) The role of microtubules and microfilaments in neurulation of *Xenopus*. *Dev. Biol.* **25**:30–56.

Ledbetter, M. S., and M. Lubin (1977) Control of protein synthesis in human fibroblasts by intracellular potassium. *Exp. Cell Res.* **105**:223–236.

Messenger, E. A., and A. E. Warner (1979) The function of the sodium pump during differentiation of amphibian embryonic neurones. *J. Physiol. (Lond.)* **292**:85–105.

Rose, B., and W. R. Loewenstein (1976) Permeability of a cell junction and the local cytoplasmic free ionized calcium concentration: A study with aequorin. *J. Membr. Biol.* **28**:87–119.

Sheridan, J. D. (1968) Electrophysiological evidence for low resistance electrical connections between cells of the chick embryo. *J. Cell Biol.* **37**:650–659.

Shroeder, T. E. (1970) Neurulation in *Xenopus laevis*. *J. Embryol. Exp. Morphol.* **26**:543–570.

Slack, C., and A. E. Warner (1973) Intracellular and intercellular potentials in early embryos of *Xenopus laevis*. *J. Physiol. (Lond.)* **232**:313–330.

Smith, J., and J. M. W. Slack (1983) Dorsalization and neural induction: Properties of the organizer in *Xenopus laevis*. *J. Embryol. Exp. Morphol.* **78**:299–317.

Spemann, H. (1938) *Embryonic Induction and Development*. Reprinted in 1967. Hafner, New York.

Thomas, R. C. (1972) Electrogenic sodium pump in nerve and muscle cells. *Physiol. Rev.* **52**:563–594.

Toivonen, S., and J. Wartiovaara (1975) Mechanisms of cell interaction during primary embryonic induction studied in transfilter experiments. *Differentiation* **5**:61–66.

Toivonen, S., D. Tarin, L. Saxen, P. J. Tarin, and J. Wartiovaara (1975) Transfilter studies on neural induction in the newt. *Differentiation* **4**:1–7.

Vulliamy, T., and E. A. Messenger (1981) Tetanus toxin: A marker of amphibian neuronal differentiation in vitro. *Neurosci. Lett.* **22**:87–90.

Warner, A. E. (1973) The electrical properties of the ectoderm in the amphibian embryo during induction and early development of the nervous system. *J. Physiol. (Lond.)* **235**:267–286.

Warner, A. E. (1981) The early development of the nervous system. In *Development of the Nervous System*, D. R. Garrod and J. Feldman, eds., pp. 109–127, Cambridge Univ. Press, Cambridge.

Weingart, R. (1977) The actions of ouabain on intercellular coupling and conduction velocity in mammalian ventricular muscle. *J. Physiol. (Lond.)* **264**:341–366.

Wolpert, L. (1978) Gap junctions: Channels for communication in development. In *Intercellular Junctions and Synapses*, J. Feldman, N. B. Gilula, and J. D. Pitts, eds., pp. 23–56, Chapman & Hall, London.

Chapter 2

Molecular Regulation of Neural Morphogenesis

GERALD M. EDELMAN

ABSTRACT

The isolation and biochemical study of cell adhesion molecules (CAMs) has allowed an analysis of their spatial and temporal distribution and mechanism of action during various epochs of neural development. The neural cell adhesion molecule (N-CAM), which binds homophilically in calcium-independent fashion, and the liver cell adhesion molecule (L-CAM), the binding of which is calcium-dependent, are both distributed over the entire chick blastoderm. At neural induction, the area of the neural plate loses L-CAM and the bordering nonneural ectoderm shows a sharp decrease in N-CAM. During cell migrations, such as those of neural crest cells, N-CAM is lost from the cell surface and reappears as cells aggregate to form ganglia. At 3½ days of chick development, a new CAM that mediates neuron–glia interactions appears on neurons but is not seen in glia.

Composite fate maps relating CAM distribution to classical chick fate map boundaries show that N-CAM will appear in central regions of presumptive neuroectoderm, notochord, somites, and parts of lateral-plate mesoderm. L-CAM appears in surrounding regions of nonneural ectoderm and endoderm, as well as in that portion of lateral-plate mesoderm corresponding to kidney precursors. The appearance of CAMs at the cell surface in regions of secondary inductions is highly dynamic, showing increases, decreases, and disappearances at specific developmental times. The data on early neural histogenesis suggest that at least two CAMs of different specificities (N-CAM and L-CAM) are required to discriminate the epithelia involved in neural induction. At later times, the primary N-CAM and the secondary neuron–glia CAM (Ng-CAM) change in relative amounts during cell migration and tract formation.

As fiber tracts are consolidated in perinatal neural development, N-CAM undergoes a distinct change: the microheterogeneous embryonic (E) form of the molecule that contains a high amount of polysialic acid in a domain adjacent to the amino-terminal binding domain is replaced by two less heterogeneous adult (A) forms of the molecule that contain one-third as much sialic acid. Although sialic acid is not required for binding, it has been shown that this E-to-A conversion results in a large increase in the binding rates of lipid vesicles containing N-CAM. An even larger increase occurs in these rates upon increasing the surface density of N-CAM. Moreover, E-to-A conversion occurs on different schedules in different brain regions and is delayed and incomplete in staggerer *mutant mice that show connectional defects in the cerebellum.*

The accumulated data suggest that the expression of CAM genes on a defined temporal schedule and the occurrence of local cell surface modulation altering the prevalence, polarity, or chemistry of CAMs are major modes of regulation of the key neural morphogenetic events resulting from cell migration, tissue sheet folding, process extension, and milieu-dependent differentiation.

The problem of how fiber tracts and maps are established in the central nervous system involves a variety of mechanisms that must be described in terms of both microscopic anatomy and cellular physiology. Nevertheless, as Sperry (1963) correctly perceived, a clear-cut solution to the problem of neural patterning must ultimately include a description of each of the key morphogenetic mechanisms at the molecular level. The pivotal issue concerns both the nature and the number of molecular markers responsible for cell–cell interactions. The description of the action of these markers must, in addition, be related to the very earliest emergence of neural tissue. A major issue is to determine the number of different molecular species that are required to establish the connectivity of neural circuits and to distinguish whether the connectivity is prespecified by these different molecules (Edelman, 1984a). Prespecification would imply the existence of a relatively large number of evolutionarily selected gene products acting singly or in combination at the cell surface to determine neural patterning.

To reduce this central problem in neural morphogenesis to operational terms, it may be framed in two related questions: (1) How many different specific molecules are necessary during very early development to distinguish neural cells as a collection (e.g., the neural plate) from other kinds of cells? (2) How many different kinds of molecules are necessary later in development to link differentiating neurons into complex neural networks of a particular pattern? Cell adhesion molecules or CAMs (Edelman, 1983) are among the main candidates for the molecular species involved in these fundamental events of neural morphogenesis. The purpose of this chapter is to describe some approaches to the questions posed above by summarizing progress over the last decade in isolating these molecules, characterizing their functions, and relating their expression to morphogenetic events (Edelman, 1984b). The major impressions gained from surveying the evidence on the mediation of cell interactions by CAMs are of the subtlety of molecular-adhesion mechanisms at the cell surface, of a great dynamism in CAM expression, and of the lack of evidence supporting prespecification of large numbers of different molecules as a major feature of neural patterning.

CELL SURFACE MODULATION AS A DEVELOPMENTAL MECHANISM

A survey of the various cellular processes leading to neural patterns (Cowan, 1978) also conveys an impression of extraordinary plasticity, selectivity, and dynamism during development. Given these impressions, it is reasonable to consider that cell patterning might be mediated by a dynamic mechanism that is parsimonious with respect to the number of cell surface gene products required to establish cell patterns (Edelman, 1983; Edelman, 1984c). In this hypothesis, it is supposed that a relatively small number of different specific cell surface molecules responsible for cell adhesion could generate intricate morphogenetic patterns, provided that these molecules underwent certain changes in expression or in chemical properties during development. These changes, collectively termed local cell surface modulation (Edelman, 1976), consist of modifications in the number, distribution, or chemical structure of each particular cell adhesion molecule (Figure 1A). By changing the binding behavior of CAMs at the cell surface, the various modulation events would directly or indirectly alter the dynamics and

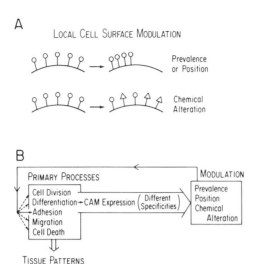

Figure 1. *Modulation of CAMs in neural morphogenesis.* A: Schematic representation of local cell surface modulation. Various elements represent a specific glycoprotein (for example, N-CAM) on the cell surface. The upper sequence shows modulation by alteration of both the prevalence of a particular molecule and its distribution on the cell surface. The lower sequence shows modulations by chemical modification resulting in the appearance of new or related forms (triangles) of the molecule with altered activities. Local modulation is distinct from global modulation, which refers to alterations in the whole membrane that affect a variety of different receptors independent of their specificity (see Edelman, 1976). B: Regulatory effects of CAM expression and modulation. Although the major effect is on the adhesion process, CAM expression and modulation may alter the sequence and extent of each of the other primary processes, either indirectly or (as suggested by the dotted arrows) directly. These effects may lead, in turn, to different tissue patterns.

interactions of other primary processes of development such as cell movement, differentiation, division, and death. Alterations of these processes would in turn result in a change in form (Figure 1B). The effects of modulation upon selective cell–cell patterning would thus depend upon three variables: (1) the number of CAMs of different specificities that are actually present; (2) the particular binding mechanisms of these different CAMs; (3) the change of both of these variables with time or cellular position. An adequate modulation theory must show that the action of all of these variables could yield sufficient selectivity to give rise to a particular neural structure. In order to understand how this might occur, a short review of the assays developed for CAMs and of CAM structures and binding mechanisms is required.

CELL ADHESION ASSAYS: THE DETECTION AND PERTURBATION OF ADHESION

The search for CAMs was guided by several logistic assumptions. It was reasonable to assume, for example, that such molecules were proteins on the cell surface and, because they would be scarce, that cells from a large number of embryos would be required to isolate the molecules for chemical characterization. Indeed, over the last decade, more than half a million chick embryos were used for this

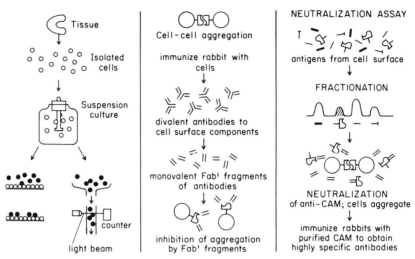

Figure 2. *Immunologically based adhesion assays.* Chick retinal cells are dissociated by light trypsin digestion and allowed to resynthesize their surface molecules in suspension culture (left panel). Adhesion is assayed by counting labeled cells bound to a layer of fixed cells or by shaking cells together and measuring the disappearance of single cells in an automatic counter. Antibodies to cell surface components are assessed for the ability to block cell–cell adhesion following conversion to Fab' fragments (center panel). Cell-surface antigens are fractionated, and fractions that neutralize inhibition by these antibody fragments (right panel) are used to reimmunize rabbits and thus obtain anti-CAMs of higher specificity.

purpose in our laboratory. The basic strategy was to devise two kinds of assay: an immunological detection assay based on the recognition of CAMs on single cells as they collided and interacted over short times (Brackenbury et al., 1977), and a series of perturbation assays (Buskirk et al., 1980; Rutishauser et al., 1983) in which antibodies known to be directed against particular CAMs could be shown to disrupt tissue patterns by binding to cells in developing structures.

The detection assays will be considered first (Figure 2). Cells in a tissue were dissociated by digestion of their surface protein molecules with trypsin. After a recovery period during which these cells were allowed to resynthesize their surface proteins, they were stirred together to allow collision and adhesion over short times in arrangements that depended on the particular assay. In one detection assay, for example, a portion of the cells was fixed in a layer and another labeled portion of cells was thrown onto this layer; the cells that adhered to the layer could then be counted. In another assay (more rapid but without facilities for watching the interaction of individual cells), the disappearance of single cells into aggregates was determined in an automatic cell counter (Brackenbury et al., 1977).

How may one detect the molecules responsible for the actual ligation of cell surfaces? Taking a cue from earlier work by Gerisch (Gerisch and Malchow, 1976), who used antibodies to perturb adhesion in slime molds, we decided to search for antibodies to CAMs that would specifically block adhesion. Rabbits were immunized with chick brains and retinas and the antibodies were scanned for their ability to block adhesion after they were cleaved into univalent Fab' fragments. (Uncleaved antibodies are divalent and, instead of blocking adhesion, would actually bind two cells together by their CAMs.) After specific adhesion-blocking antibodies were found, a dilemma had to be faced. The antibodies in the rabbit sera were mixed with those directed against other cell surface molecules and could not be used as specific probes to identify CAMs or isolate them. In order to reduce the heterogeneity of the antibody population, a neutralization assay was therefore devised (Brackenbury et al., 1977). In this assay, fractions of surface protein antigens were tested for their ability to compete with cells for binding to the anti-CAM antibody. If such fractions contained CAMs, they were then used to reimmunize rabbits. By iteration of this procedure, antigenic fractions of high specificity were obtained that elicited highly specific anti-CAM antibodies in the rabbits. Of course, once a means had been found for identifying such antigenic CAM fractions, modern techniques of monoclonal antibody pro-

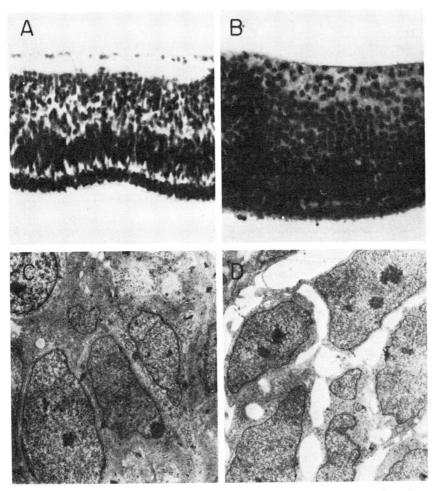

Figure 3. *An example of a perturbation assay.* Chick neural retina in organ culture for three days after removal from the embryo on day six. A: Culture in presence of Fab' fragments from normal rabbit IgG; the structure shows normal layering ($\times 650$). B: Culture in presence of anti-N-CAM Fab' showing evident disruption ($\times 650$). C: Cells from A indicating close normal apposition of membranes ($\times 5800$). D: Cells from B showing large areas of extracellular space and few areas of cell–cell contact ($\times 5800$).

duction could be used to make very specific antibodies to CAMs in different animal species (Hoffman et al., 1982).

With these various antibodies in hand, we could check for the distribution of CAMs in tissues and attempt specifically to disrupt the development of tissue patterns (Figure 3). Using anti-N-CAM antibodies labeled by a fluorescent marker, it was found that this molecule was present on all neurons in the central and peripheral nervous systems. Furthermore, anti-CAM antibodies could disrupt the orderly development of neural tissues, such as those of the retina (Buskirk et al., 1980), as they grew in tissue culture. By similar approaches employing detection and perturbation assays (Bertolotti et al., 1980; Gallin et al., 1983; Grumet and Edelman, 1984) several other adhesion molecules (L-CAM from liver and neuron–glia CAM or Ng-CAM, a molecule on neurons that binds them to glia) were identified; as we shall see later, knowledge of these other molecules (even those not specific to the neuronal and glial systems) was essential in analyzing the role of adhesion in early neural histogenesis.

While tissue perturbation assays were useful in convincing us that we had truly identified a particular CAM, it became clear that deep insights into the relation of adhesion molecules to neural development could not rely solely on this phenomenological approach; such insights could come only from combining a variety of more fundamental approaches. These included an analysis of the chemical structure, specificity, and binding mechanism of each particular CAM in very early development and a study of CAM function and genetic control in later stages of development as well as in adult life.

STRUCTURE AND SPECIFICITY OF DIFFERENT CAMs

By means of classical chromatographic fractionation techniques and affinity chromatography, both N-CAM and L-CAM were purified to the point at which a chemical analysis of their structures could be carried out (Thiery et al., 1977; Hoffman et al., 1982; Gallin et al., 1983). Both molecules are glycoproteins, but N-CAM is unusual in that it contains extraordinarily large amounts of sialic acid, a complex negatively charged sugar that is present in an unusual polymerized form (Rothbard et al., 1982). Although sialic acid is found attached in much smaller amounts on other cell surface proteins, it has not previously been seen as polysialic acid on proteins in the vertebrate species except in N-CAM (Figure 4).

Once the molecules had been purified, various cleavages with proteolytic enzymes could be used to construct topographic or linear maps of N-CAM (Cunningham et al., 1983) and L-CAM (Gallin et al., 1983; Figure 4). Examination of these maps indicated that the molecules were structurally different. The polypeptide chain of N-CAM (which, exclusive of sugar, had a maximal molecular weight of 160 kD) consisted of three domains linked by stretches of polypeptide chain susceptible to attack by proteolytic enzymes. The amino-terminal domain projected away from the cell and was found to contain the binding region. A middle domain contained the great bulk of the sialic acid, and the carboxy-terminal domain was associated with the cell membrane. While not conclusive, the chemical evidence suggested that a portion of this membrane-associated

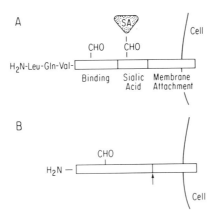

Figure 4. *Linear structures of N-CAM and L-CAM.* A: Three structural and functional regions of N-CAM deduced from studies of the intact molecule and a series of fragments. The amino-terminal region includes a specific binding domain and carbohydrate (CHO), but little, if any, sialic acid; the neighboring region is very rich in sialic acid (SA), present mainly as polysialic acid; the carboxy-terminal region is associated with the plasma membrane. B: Linear structure of L-CAM obtained by comparing the intact molecule [molecular weight (M_r) = 124 kD] released by detergent extraction with a fragment (M_r = 81 kD) released by trypsin (see arrow). In both N-CAM and L-CAM the carbohydrate is attached at several sites.

domain was actually inserted into the lipid bilayer of the membrane. This was inferred from the requirements for extracting N-CAM from cell membranes, as well as from the fact that the intact chain could become readily associated with artificial lipid vesicles. Such vesicles are extremely useful in following the binding behavior of the molecule. For example, the vesicles could be attached to cells; if the cells only were treated first with anti-N-CAM Fab' and washed, it was found that the vesicles, which contained only N-CAM and lipid, would not bind (Cunningham et al., 1983). This suggested that the binding mechanism is homophilic: N-CAM on the membrane of cell 1 binds to N-CAM on the membrane of cell 2. The idea that there may be functional and structural domains in N-CAM has recently received support from preliminary experiments in which N-CAM molecules were visualized by electron microscopy; the pictures suggested regions of folded polypeptide separated by bends that might correspond to interdomain polypeptide chains (Edelman et al., 1983b).

It is useful to describe briefly two other CAMs, the detailed structures of which are less well worked out but which are known to be different from each other and from that of N-CAM. The first is L-CAM, a molecule with a molecular weight of 124 kD (Gallin et al., 1983), first isolated in lower molecular-weight forms from embryonic liver (Bertolotti et al., 1980). L-CAM is a normal glycoprotein and, unlike N-CAM, will not mediate cell adhesion unless calcium ions are present. Indeed, unless the L-CAM molecule binds calcium, it is susceptible to rapid proteolysis or cleavage by enzymes. It is not definitely known whether L-CAM binding is homophilic. Consistent with these differences in binding functions, detailed structural and cleavage studies of L-CAM indicate that the molecule differs considerably from N-CAM. Although L-CAM was isolated originally from liver, it was found to play a fundamental role in early developing tissues in conjunction with N-CAM. We shall consider this early role at length in another section.

A third adhesion molecule, isolated most recently from neural tissue, is Ng-CAM, which mediates the binding of neurons to glial cells (Grumet and Edelman, 1984). Ng-CAM has a molecular weight of 135 kD (with minor components of molecular weights 200 and 80 kD). It is found on neurons, but not on glia, and is therefore probably bound to a glial adhesion molecule that is structurally different from Ng-CAM; that is, the binding mechanism is likely to be heterophilic. An important fact is that N-CAM, L-CAM, and Ng-CAM do not appear to be cross-specific; that is, in linking cells, no one binds effectively to any of the others. Despite the fact that their binding specificities differ, however, it has been shown that N-CAM and Ng-CAM share an antigenic determinant (Grumet et al., 1984). Several monoclonal antibodies have been found that react with both molecules, which therefore probably have a similar stretch of amino acid sequence or have similar carbohydrate structures. The alternative that the antibodies may simply recognize a similar peptide fold or very short sequence has not, however, been excluded.

LOCAL CELL SURFACE MODULATION AND CAM BINDING

One key question dominated the studies of the relation of N-CAM structure to function: Does the unusual carbohydrate—particularly the sialic acid—play a direct role in cell–cell binding? The answer turned out to be no: N-CAM molecules from which all of this sugar had been specifically removed by the enzyme neuraminidase still would bind specifically to cells. Nonetheless, two observations (Rothbard et al., 1982) showed an unexpected but very important function for this sugar. The first was that N-CAM from embryonic brains had 30 g sialic acid/100 g polypeptide and migrated on electrophoretic gels as a diffuse band with a broad molecular weight distribution. In contrast, N-CAM from adults had only 10 g sialic acid/100 g polypeptide and migrated as two or three sharp bands (Figure 5). Thus, at some time during development, the embryonic (E) form of the molecule must be converted to or exchanged for the adult (A) forms. The second observation was that the A-forms appeared to bind more effectively than did the E-form. We will discuss this in some detail, for it turns out that it bears strongly on mechanisms that might govern the selectivity of cell adhesion in the developing nervous system.

Changes either in prevalence of CAMs at the cell surface or in their individual binding strength through E-to-A conversion would be expected to lead to different interactions among the cells that were subject to other primary processes, such as migration. These altered interactions might in turn lead to the formation of different cellular patterns. To demonstrate that such modulatory changes could actually affect binding, a direct measurement had to be carried out. This was done by using artificial lipid vesicles into which N-CAM was inserted; the vesicles were then tested for their relative rates of aggregation (Hoffman and Edelman, 1983). Both the E-form and A-forms could be inserted separately into different vesicle fractions (Figure 6). The prediction was that the order of rates would be E–E < E–A < A–A, based on the idea (Edelman, 1983) that the sialic acid in the

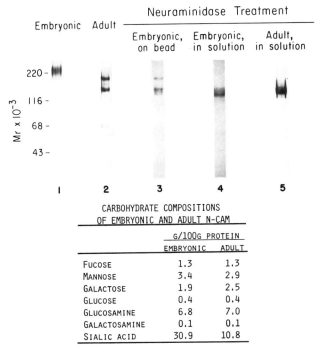

Figure 5. *Electrophoretic patterns and carbohydrate compositions of embryonic (E) and adult (A) forms of N-CAM.* On polyacrylamide gel electrophoresis in SDS, the E-form shows a diffuse microheterogeneous smear varying from 200 to 250 kD (lane 1), whereas the A-form shows two sharp bands at 180 kD and 140 kD (lane 2). After treatment with neuraminidase to remove sialic acid completely, the two forms migrate similarly (cf. lanes 4 and 5). As shown in the table, the E-form has three times as much sialic acid as the A-form.

middle domain (Figure 3) was either altering the shape of the binding domain or directly repelling the opposing N-CAM molecule from another cell. Thus, vesicles containing the A-form, which had lesser amounts of charged sugar, should aggregate more rapidly. As shown in Figure 6, this prediction was clearly confirmed.

Even more striking was the effect of increasing the amount of N-CAM of a given form in the membrane. A twofold increase led to a greater than 30-fold increase in binding rates (Hoffman and Edelman, 1983). Thus, both E-to-A conversion and surface-prevalence changes *in vivo* would be expected to lead to large changes in rates of binding, a necessary condition for any kinetically constrained model of pattern formation. It is particularly important to note in addition that both of these modulatory changes are essentially continuous and therefore that they define a very large number of possible binding states for N-CAM.

EARLY NEUROGENESIS AND THE CAM FATE MAP

With this information on structure and binding, we are in a position to consider the role of CAMs in early developmental events, such as neural induction. Antibodies against the known CAMs may be used to detect whether these

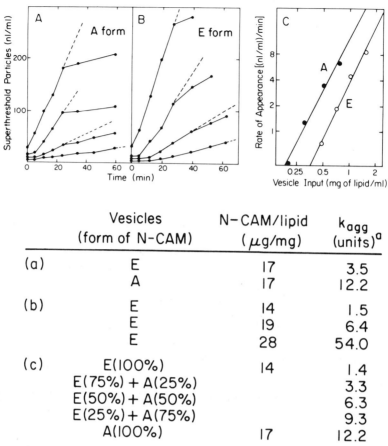

Figure 6. *Aggregation of reconstituted vesicles containing E- or A-forms of N-CAM.* Vesicles were reconstituted from purified N-CAM and lipid, and their aggregation was analyzed in a particle counter. The concentration of superthreshold particles is plotted as a function of time for four concentrations of vesicles. Apparent initial rates of aggregation are calculated from the initial slopes (broken lines) of the aggregation curves. A: A-form vesicles (16.3 µg of N-CAM per mg of lipid) aggregated at 0.76, 0.50, 0.30, and 0.20 mg of lipid per ml (curves proceeding from left to right). B: E-form vesicles (16.9 µg of N-CAM per mg of lipid) aggregated at 1.55, 1.03, 0.71, and 0.48 mg of lipid per ml. C: Log-log plot of the rate of appearance of superthreshold A-form (closed circles) and E-form (open circles) particles versus input concentration. Below: Table comparing effects on aggregation rates of E- and A-forms (a); of alterations in surface amounts of E-form (b); and of effects of mixtures of vesicles containing either E- or A-forms (c). k_{agg}: One unit is the rate of aggregation of a sample (measured in nl of superthreshold product per ml per min) divided by the square of the concentration of vesicles in the sample (see Hoffman and Edelman, 1983).

molecules are present in very early embryos (Thiery et al., 1982; Edelman et al., 1983a,b). So far, two of the CAMs, N-CAM and L-CAM, have been found early in the chick; the earliest stage at which they have been detectable in this species is just prior to the formation of the germ layers. At this stage, the epiblast and hypoblast stain with fluorescent antibodies to both CAMs. Later on, as the primitive streak develops, the middle layer cells that are migrating do not stain. At this point, a remarkable transition occurs: L-CAM appears on Hensen's node and remains with it for its existence; cells that will become the neural plate lose evidence of L-CAM and stain very strongly with antibodies to N-CAM (Figure 7). Reciprocally, at the border of the ectoderm and neuroectoderm, the ectodermal cells stain strongly with antibodies to L-CAM, as do the endodermal cells. This emerging pattern of CAM segregation and border formation accompanies the key event of neural induction.

In order to see whether the topology defined by these early events is conserved in later patterns, we employed composite fate maps (Edelman et al., 1983a) constructed from classical fate maps of the chick embryo and from data on identification of tissues that stained with specific anti-CAM antibodies. Fate maps summarize what will become of cells in each embryonic region in terms of shape and the structures to which they will give rise for a defined period of time or epoch. A fate map is only a summary of events that occur in such a time period, and a series of such maps is usually necessary to cover several epochs, particularly if one wishes to follow the whole line of development to maturity. Fate maps are similar from individual to individual but differ in precision, depending both upon the time chosen and the degree to which individual cells in a species give rise to only one exact line of descendants. It is important to realize that a future four-dimensional distribution (time + three spatial dimensions) of cells and CAM markers is being mapped back onto a two-dimensional surface consisting of the sheet of blastoderm-progenitor cells. This results in a topological map rather than a topographic one (i.e., exact details of structure are sacrificed to reveal connectedness and neighboring relationships). It is these relationships that are particularly important in understanding embryonic induction events.

By tracing the descendants of blastodermal cells in the chick fate map, as revised by Vakaet (see Figure 8A), for their expression of different kinds of CAMs, a composite CAM fate map was constructed (Figure 8B). This map has a number of intriguing features. The first striking feature to be noted in interpreting the map is that the calcium-independent N-CAM regions that will give rise to

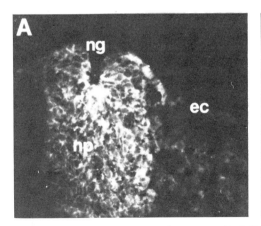
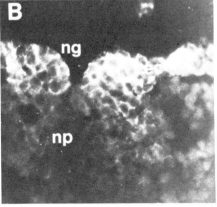

Figure 7. *Change in distribution of N-CAM and L-CAM at formation of the neural plate and groove (neural induction). A: Cross section of the neural plate (np) as the neural groove (ng) forms. N-CAM is present in large amounts in the chordamesoderm and neural ectoderm and in small amounts in ectoderm (ec) and adjoining mesoderm. B: L-CAM staining disappears from the neural ectoderm in the neural groove and becomes restricted to the nonneural ectoderm (ec). Just before these events, N-CAM and L-CAM were present in all regions of the blastoderm that give rise to these structures.*

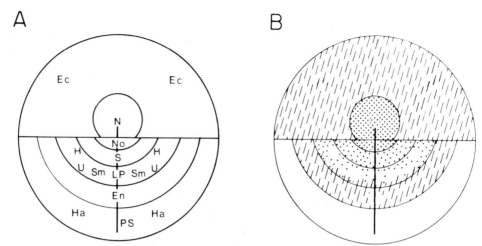

Figure 8. *A: A fate map of the blastodisc* (L. Vakaet, 1984). Areas of cells that will give rise to differentiated tissues are indicated by letters; see below for designations. *B:* Map of cells that will express CAMs. The distribution of N-CAM (stippled) or L-CAM (slashed) in tissues at 5–14 days (stages 26–40) as determined by immunofluorescence staining is mapped back onto the blastodisc fate map. Cells that will give rise to the urinary tract (U) express both L-CAM and N-CAM. Smooth muscle (Sm) and hemangioblastic (Ha) tissues express neither N-CAM nor L-CAM; areas giving rise to these tissues are blank on this map. The vertical bar represents the primitive streak (PS). Ec, intraembryonic and extraembryonic ectoderm; En, endoderm; H, heart; LP, lateral plate (splanchno-somatopleural mesoderm); N, nervous system; No, prechordal and chordamesoderm; S, somite.

the neural plate, to the notochord, and to certain parts of the lateral plate mesoderm are completely surrounded by a contiguous, simply connected ring of regions that will express the calcium-dependent L-CAM—regions that together comprise the rest of the ectoderm and the endoderm. Second, there is a cephalocaudal coarse gradient of N-CAM antibody-staining that is most intense in the region of the neural plate. In underlying regions such as the notochord, the staining is less intense and the pattern is dynamic—at first, there is no staining for N-CAM, then the notochord (a mesodermal structure) stains intensely, and in later stages the stain disappears. A similar dynamic pattern of N-CAM appearance is seen in the somites just as they segregate into segments. Moreover, the early sequence—L-CAM staining plus N-CAM staining followed by loss of L-CAM staining—is seen in placodes that will form ganglia. Both CAMs are also seen at the apical ectodermal ridge of the limb bud, a key structure necessary for formation of the appendages. A third detailed feature of the map is seen in the kidney elements that come from the mesoderm: Both L-CAM and N-CAM appear and disappear in sequences corresponding to the reciprocal embryonic induction of the so-called mesonephric mesenchyme by the tubular structure called the Wolffian duct. L-CAM first appears on the Wolffian duct, then N-CAM appears on mesonephric tubules as they organize; this in turn is replaced by L-CAM as collecting tubules extend. A fourth feature of the map is that there are regions of the lateral plate mesoderm that will give rise to smooth muscle and regions containing the hemangioblastic precursors that do not stain for either CAM. This raises the possibility that at least one other primary CAM mediates the adhesive interactions of these areas in early development; it is too early to say how many others will be found.

Two large generalizations emerged from these observations on the early embryo: (1) CAMs undergo dynamic changes in distribution, sequence, and amount at regions of primary and secondary induction; (2) wherever epithelia are converted to mesenchyme, CAMs appear to be lost from the cell surface. A striking example of this is the prevalence modulation of N-CAM in neural crest cells. When crest cells first appear, they stain for N-CAM (Thiery et al., 1982). As soon as they begin their migration, however, they lose this staining, and fibronectin (a substrate adhesion molecule or SAM) appears in their path (Thiery, 1983). When they reach their destination, and just before they form ganglia, N-CAM reappears on their surface. This classical example of the conversion of a sheet of cells to a mesenchyme consisting of loosely attached mobile cells is clearly correlated with the occurrence of strong CAM modulation.

The evidence reviewed so far suggests that at least two primary CAMs of different specificity (N-CAM and L-CAM) are required to form the distinct epithelial sheets defining the topology of primary induction and neural plate formation. What is required is to show directly that perturbation of these CAMs will alter and block morphogenetic development. Although it is not possible to say exactly how many CAMs will be discovered, it appears likely that the number required for early embryogenesis will not be large. A related question is whether additional or secondary set CAMs will be necessary in later epochs of development in order to account for refined tissue interactions such as those seen in the later stages of brain development. The answer for the brain appears to be that while N-CAM continues to be used in these later stages, at least one new specificity (Ng-CAM) is required for heterotypic interactions of neurons with glia and possibly for neuronal migration. Moreover, in even later periods it was observed that not only changes in the prevalence of Ng-CAM and N-CAM but also E-to-A conversion are called into play, particularly at just those times at which various neuronal fiber tracts are being mapped. We turn to a detailed consideration of these facts, which indicate that the primary CAMs are used in later development.

BRAIN ORGANOGENESIS, NEURAL CONNECTIONS, AND NEURON–GLIA INTERACTIONS

If the modulation hypothesis (Figure 1) is correct, one should see evidence during tract formation not only of continuing changes in the surface prevalence of N-CAM, such as those seen in the early fate map, but also variations in the extent of E-to-A conversion at different times in structurally different parts of the brain. This is a reasonable expectation because E-to-A conversion is a later event and, like changes in prevalence, it changes CAM binding rates (Figure 6).

An analysis of E-to-A conversion in different gross brain regions (Figure 9) did, in fact, show strong differences in pattern (Chuong and Edelman, 1984). This suggests that the rate of conversion, its time of initiation, or its degree in different cell types may differ in these histologically different regions. Conversion (see Figure 5) is an epigenetic event that results either from enzymatic cleavage of the sialic acid from surface N-CAM or from turnover of E-forms followed by

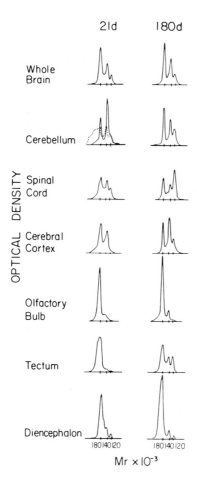

Figure 9. *Differential expression of N-CAM forms by different brain regions of 21- and 180-day-old mice.* Each panel shows a densitometric scan of autoradiographs of immunoblots with anti-N-CAM of brain extracts fractionated by SDS-polyacrylamide gel electrophoresis. Ordinate, optical density; abscissa, apparent molecular weight. The total area under each tracing has been normalized to the same value. The three A-forms (M_r = 180,000, 140,000 and 120,000) can be seen. Different brain regions are presented from top to bottom in order of decreasing rates of E-to-A conversion. The dotted line in cerebellum at 21 days shows the profile from homozygous *staggerer* mice and reveals a delay in E-to-A conversion as compared to the normal profile.

replacement by A-forms of N-CAM that have lesser amounts of sialic acid. In the latter case, an intracellular enzyme or sialyl transferase responsible for linking sialic acid to N-CAM would be implicated. Whatever the mechanism turns out to be, the significant observation is that during organogenesis, and particularly in the perinatal period, grossly different amounts of E- and A-forms are present in different regions. As indicated by the studies of binding kinetics (Hoffman and Edelman, 1983), this would be expected to change the binding efficiencies of various cells in these regions in different ways during histogenesis, and thus alter neural structure during cell differentiation and movement.

While E-to-A conversion is epigenetic and depends upon enzymatic activity, the likelihood nevertheless existed that it might be altered in one or more of the connectional defects seen in mutant animals. The granuloprival mutants (Sidman, 1974; Caviness and Rakic, 1978) showing cerebellar disorders provided an opportunity to test this hypothesis. These mice all have major defects in cerebellar development; the major symptom is ataxia, which begins to appear in the perinatal period. Of the three extensively investigated mutants, *staggerer* has a connection defect in neurons, while *reeler* and *weaver* have disorders that also involve the Bergmann glial cells, which play a key role in the migration of nerve fibers

(Caviness and Rakic, 1978). *Staggerer* is expressed only in homozygous animals; the cerebella of these animals show faults in the formation of synapses between parallel fibers and Purkinje cells, which do not appear to have normal tertiary dendritic branches. After failure to make these synapses, the granule cells die in great quantities; the consequence of these anomalies is an ataxic animal with a small and disordered cerebellum destined to die without further care at about one month after birth.

We surmised that a defect in N-CAM modulation was most likely to be observed in *staggerer*. This was found to be the case (see Figure 9, cerebellum at 21 days): E-to-A conversion was greatly delayed in the cerebella of the homozygotes, but the N-CAM polypeptide appeared to be normal. On the other hand, *reeler* and *weaver* mutants had normal schedules of E-to-A conversion (Edelman and Chuong, 1982). While these findings do not provide an explanation for the cause of disease in *staggerer*, they suggest that one consequence of the genetic defect is a failure in the activity or in the expression of enzymes responsible for E-to-A conversion. It may well be that this leads in turn to a failure to terminate certain key cellular binding events, with a consequent failure in coordination of the various aspects of neural process formation, migration, and synapse formation.

As *reeler* and *weaver* show, however, not all connectional defects (particularly those involving glia) need arise from failures of neuron–neuron adhesion. Indeed, it would be expected that, in order to bind to glia, neurons have molecules with specificities different from that of N-CAM. As mentioned earlier, recent experiments in our laboratory have revealed the existence on chick neurons of Ng-CAM, the components of which have molecular weights of 80, 135, and 200 kD; unlike N-CAM, Ng-CAM does not alter radically after treatment with neuraminidase (Grumet and Edelman, 1984; Grumet et al., 1984). Ng-CAM mediates the binding of neurons to glial cells (probably astrocytes, although the evidence is not yet complete). It also may be associated with N-CAM at the cell surface (Grumet et al., 1984), raising the possibility that it also undergoes modulation. This idea is reinforced by observations (Thiery et al., 1984b) that its distribution is favored on processes of neurons in the central nervous system (i.e., that it undergoes polar modulation).

Although there is no evidence so far that Ng-CAM on one cell binds effectively to N-CAM on another, Ng-CAM does share some structural features with N-CAM; these features can be recognized by monoclonal antibodies that specifically bind to both molecules (Grumet et al., 1984). Ng-CAM and N-CAM, therefore, may either have come from the same evolutionary precursor or share some carbohydrate structures in common. It is important to note that Ng-CAM appears relatively late in development of the chick, at about 3½ days, just before the definitive appearance of glia in the central nervous system. Most significantly, unlike N-CAM and L-CAM, Ng-CAM would not be represented in an early CAM fate map (i.e., one constructed prior to four days of development); if placed in a later map, it would be found only in the neural region. It will be instructive to examine Ng-CAM in *reeler* and *weaver* mutants of the mouse, for it is possible that it or a complementary CAM on glia may be involved in one or both of these disorders. This hypothesized glial molecule (Gn-CAM) is likely to be involved in a heterophilic interaction, for Ng-CAM itself is not present on glia. Should this turn out to be the case, it would not be surprising: if the heterotypic neuron–

glia interaction were not mediated by a heterophilic mechanism, complete confusion might occur during development in distinguishing between neuron–neuron and neuron–glia interactions.

Despite the power and the variety of modulation mechanisms (prevalence, chemical, and polarity modulation), we now know that there are certain histogenetic circumstances, such as glial interactions and neurulation, that are likely to require CAMs of different specificity. We conclude that it is the combination of both specificity and modulation that is important in morphogenetic change. The need during morphogenesis for CAMs of different specificities is clearly context-dependent, that is, it reflects potentially conflicting interactions among various cell types as cytodifferentiation proceeds *pari passu* with cell adhesion and motion. Within a given tissue, the expression and function of different CAMs depend upon timing, modulation, concurrent presence of different cell types, and interactions with the other primary processes of development.

Table 1. Neural Patterning via Modulation Compared to Strict Addressing by Individual Markers

	Local Cell Surface Modulation (Edelman, 1976, 1983)	Chemoaffinity and Strict Neural Addressing (Sperry, 1963)
Number of molecules	Tens	Up to millions
	Each CAM is specific to the binding partner; no more than dozens of different specificities	Highly refined: Sufficient to distinguish each marker pair at a cellular location
	Homophilic: N-CAM to N-CAM	
	Heterophilic: Ng-CAM	
Affinity changes	Very great	Not explicitly considered but presumably reflected to some degree in different specificities
Same molecules in early and late development?	Yes, with a few new additions	Possibly, but with many new additions
Main mode of pattern generation?	Selective constraints on interactions of primary processes	Specific recognition of the appropriate complementary cell marker
Evolutionary basis of altered neural form	Altered regulatory gene pattern for CAMs and primary processes	Change in number or in type of markers by mutation and gene expression

We are now in a position to consider the second question posed at the beginning of this chapter. How many different molecular specificities are likely to be required in forming fiber tracts and maps in the central nervous system? Until an exact causal analysis is made of the role of CAMs in neural mapping *in vivo*, interpretation must be guarded. Nonetheless, a number of observations suggest that whatever new molecules are found as markers, they will not have the function of local cellular addresses (Sperry, 1963) directly responsible for pattern (see Table 1). These observations can be listed as follows: (1) It is necessary and sufficient to block only N-CAM binding to disrupt neural patterns in a variety of tissues. Neural CAMs change binding efficacy by modulation; it does not seem likely that they will represent a very large family (i.e., greater than 100) of different molecular specificities. Indeed, the same N-CAM that plays a key role in early neurogenesis is ubiquitous on differentiated neurons and plays a key role in later histogenesis. It should be pointed out that this does not imply that other CAMs do not play a role in neural patterning. (2) Neural patterning depends critically upon glial interactions as well as upon neural interactions; just before the appearance of glia, Ng-CAM shows modulation in its expression, rather than a host of different specificities. (3) A number of primary processes of development contribute to neural patterning in a dynamic fashion that is responsive in part to CAM expression. The most striking example is the modulation of N-CAM on the surface of neural crest cells in which both cell movement and loss of CAM expression are temporally correlated (Thiery, 1985; Thiery et al., 1984a). The basis of pattern in tissues formed by these cells appears to be cell surface modulation, with the changes in selectivity of molecules of a relatively small number of fixed specificities acting in concert with differentiation and movement. It should be emphasized again that the graded nature of surface and modulation events at the molecular level can lead to a very large number of cell binding states; no loss of discriminability would ensue even if strict neural addressing by local markers is absent (Table 1).

CAMs IN THE ADULT

Adult life may be considered an epoch of development in which most morphogenetic processes are considerably regulated and damped by the time of maturity. In considering the histological role of CAMs, it is important to determine their continuing function in morphologically established tissues, such as those of the adult. In certain animals capable of regeneration (such as *Xenopus*), portions of morphogenetic processes may be reactivated to form tissues although their sequences will not be identical to those of original development. Questions concerning both the distribution and the function of CAMs in adult life arise naturally from these considerations.

The detailed comparison of the distribution of N-CAM and L-CAM in three epochs (Edelman et al., 1983a; Table 2), including adult life, is revealing. The CAMs appear in adult tissues in distributions that follow from those seen in embryonic life. But in most cases, they constitute a smaller proportion of the tissues, appear only in certain limited tissue locations (Thiery et al., 1977; Damsky et al., 1982; Thiery et al., 1982; Ogou et al., 1983), and in some cases (such as

Table 2. Distribution of L-CAM and N-CAM in Three Epochs

0–3-day Embryo	5–13-day Embryo	Adult
L-CAM		
Ectoderm		
Upper layer	Epidermis	Skin: Stratum
Epiblast	Extraembryonic ectoderm	germinativum
Presumptive epidermis		
Placodes		
Mesoderm		
Wolffian duct	Wolffian duct	Epithelium of:
	Ureter	Kidney
	Most meso- and	Oviduct
	metanephric epithelium	
Endoderm		
Endophyll	Epithelium of:	Epithelium of:
Hypoblast	Esophagus	Tongue
Gut primordium and	Proventriculus	Esophagus
buddings	Gizzard	Proventriculus
	Intestine	Gizzard
	Liver	Intestine
	Pancreas	Liver
	Lung	Pancreas
	Thymus	Lung
	Bursa	Thymus
	Thyroid	Thyroid
	Parathyroid	Parathyroid
	Extraembryonic	Bursa
	endoderm	
N-CAM		
Ectoderm		
Upper layer	Nervous system	Nervous system
Epiblast		
Neural plate		
Placodes		
Mesoderm		
Notochord	Striated muscle	[a]
Somites	Adrenal cortex	Cardiac muscle
Dermomyotome	Gonad cortex	Testis
Somato- and	Some mesonephric and	
splanchnopleural	metanephric epithelia	
mesoderm	Somato- and	
Heart	splanchnopleural	
Mesonephric	elements	
primordium	Heart	

[a] It is not yet known whether adult striated muscle contains N-CAM.

the pancreas) are found distributed on cells in a polar fashion (Edelman et al., 1983b; Gallin et al., 1983). The sparser CAM distribution is perhaps not surprising, for adult tissues contain increased amounts of derivatives of connective tissue and show evidence for cellular interaction with substrate adhesion molecules (SAMs; molecules that include collagen, laminin, fibronectin, glycosoaminoglycans, etc.); cells of adult tissues also have formed specialized junctions of various types mediated by cell junctional molecules (CJMs). The central nervous system is an exception: fiber tracts do not have CJMs, and SAMs are present in lower amounts than in nonneural tissue. It is not surprising that CAMs therefore predominate.

While the major determinants of tissue specificity derive from the selectivity of CAM modulation, the different SAMs and CJMs also play important parts in morphogenesis. The function of SAMs is related to intermodulation with CAMs and cell migration as seen, for example, in neural crest cells (Thiery et al., 1982; Edelman et al., 1983b; Thiery, 1983), tissue partition, and the development of hard tissues (Hay, 1981). The function of CJMs is cell connection and communication as seen in gap junctions or in the sealing of the surfaces of epithelial sheets (Gilula, 1978; Lane, 1984). Based on the evidence accumulated so far, CAMs, SAMs, and CJMs seem to be completely separate families of molecules with disparate but conjugate functions.

The main point is that no view of the molecular basis of morphogenesis would be complete without considering the potential interactions of molecules in all of these groups. Such interactions already are revealed clearly in the case of neural crest cells, as mentioned above. Gap junctions can be found in a variety of nervous systems (Lane, 1984), as well as in the earliest embryos. Indeed, it is an open question whether coordination of gap junction formation by CJMs and the expression of a molecule called uvomorulin, isolated by François Jacob and his colleagues (Hyafil et al., 1980) are not related in the earliest embryos. Uvomorulin, which appears to be identical with the 81-kD fragment of L-CAM (see Figure 4), is involved in the compaction of the morula in the mouse, and therefore this is a major early function of L-CAM. Furthermore, SAMs have been implicated in certain embryonic induction events and it is reasonable to surmise that their role in these events may be related to their support of cell migration (Hay, 1981). It remains to be seen how the temporal expression of all three functional families of cell-binding molecules is regulated in particular tissues during embryogenesis. One intriguing possibility is that such SAMs as fibronectin and laminin may induce global cell surface modulation (Edelman, 1976), which in turn may alter CAM prevalence, itself a local modulation event.

Although the precise functions of CAMs in adult life remain unknown, they are likely to be connected with regeneration, repair, and the maintenance of adhesion. CAMs may also be necessary to keep certain cell populations together for metabolic and regulatory purposes. This is almost certainly the case in the central nervous system, where, as stated previously, there is little room in fiber tracts for SAMs and no evidence whatsoever in such tracts for CJMs.

In view of the widespread distribution of CAMs and their fundamental adhesive roles, there is a high probability that they will be implicated in the pathogenesis of a variety of general and tissue-specific diseases. Almost certainly, they must have a role in the metastasis of particular kinds of cancer cells. Recent studies in our laboratory (Greenberg et al., 1984) have shown, for example, that transformation of embryonal neural cells by such tumor viruses as the Rous sarcoma virus leads to a loss of N-CAM mediated adhesivity. This is a clear-cut demonstration that tumor transformation can be related to loss of a defined CAM and concomitantly of cell adhesion. It suggests one likely basis for the invasiveness of tumor cells into their parent tissues, as well as for their metastasis or spread to other tissues; the relationship of this modulation to alterations in fibronectin and substrates remains to be defined.

SEQUENCES OF CAM EXPRESSION IN PATTERN FORMATION AND MORPHOGENESIS

We can now look back at the accumulated data on the CAMs and relate them to some of the issues raised at the beginning of this chapter. The majority of the evidence supports the idea that local cell surface modulation is a major contributor to morphogenesis. We now know that CAMs of different structure and specificity exist and have different requirements for binding. The best studied of these, N-CAM, has a homophilic binding mechanism and, like Ng-CAM, does not require calcium ions for binding. In contrast, L-CAM, which has a completely different specificity, is calcium-dependent but has not yet been shown to be homophilic. These three CAMs are the only ones isolated to date as purified molecules, and the question of the total number of different CAMs remains open. Undoubtedly, more CAMs and various molecular interactions among them will be discovered. Nonetheless, a very large repertoire (i.e., greater than 100) of CAMs of different specificity would not be required even for complex tissue formation, particularly inasmuch as different modulation mechanisms greatly alter the selectivity of CAMs during development. According to theories that pose large numbers of markers for cell recognition in patterning, specificity is a property of each marker and therefore of the marked cells; selectivity for different patterns is a necessary consequence of this property, as we have discussed previously (see Table 1). But prior specification is not necessary: A dynamic pattern-forming system with graded modulation can be highly selective despite the fact that it does not possess a large range of different molecular specificities.

Both the expression of genes for different CAMs in various sequences during embryonic development and the mechanisms for transport and epigenetic modulation of CAMs at cell surfaces appear to be fundamental in morphogenesis. The known sequences can be summarized in a temporal diagram (Figure 10) which shows N-CAM and L-CAM (primary set) appearing first together, followed by divergence of their spatial distribution at neural induction, and subsequent increases, decreases, or disappearances of the CAMs that are characteristic of each tissue. Placodes echo the differential CAM expression; both CAMs are initially present, but subsequently, in placodes destined for neural structures, L-CAM disappears and N-CAM increases. Ng-CAM genes (secondary set, Figure 10) are expressed later, leading to the appearance of this molecule on neurons. The fact that this second neuronal CAM, which appears to be involved in interactions of neurons with glia, is not expressed during early embryogenesis but only at 3½ days in the chick embryo (just before the appearance of glial

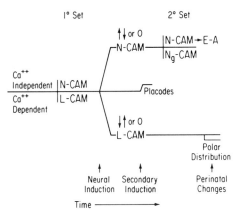

Figure 10. *Schematic diagram showing the temporal sequence of expression of CAMs.* After an initial differentiation event, N-CAM and L-CAM diverge in cellular distribution and are then modulated in prevalence (↑ ↓) within various regions of inductions, or actually disappear (0) when mesenchyme appears or cell migration occurs. Note that placodes, which have both CAMs, echo the events seen for neural induction. Just before appearance of glia, a secondary set CAM (Ng-CAM) emerges; unlike the other two CAMs, this CAM would not be found in the map shown in Figure 8 before 3½ days. In the perinatal period, a series of epigenetic modulations occurs: E-to-A conversion for N-CAM and polar redistribution for L-CAM. The diagrammed events are based mainly on work on the chick. The regulator hypothesis states that activation of a set of regulatory genes controlling a small number of CAM structural genes alters cell–cell adhesion and therefore morphogenetic cell movements. This controls embryonic induction and activation of other regulatory genes, particularly those concerned with cytodifferentiation. Regulation of CAM genes is thus considered to be prior to and relatively independent of cytodifferentiation schedules until later histogenesis.

cells in the central nervous system), suggests that temporal control of CAM gene expression is also critical in histogenesis. A reasonable interpretation is that CAM genes for secondary-set molecules, such as those for Ng-CAM, would be called into play sequentially to provide for the more complex cellular interactions that are necessary during later differentiation of different organs. Even later, in the perinatal period, E-to-A conversion occurs in N-CAM while polar redistributions of L-CAM take place on cells of certain tissues, such as those of the pancreas. Additional epigenetic alterations arising from E-to-A conversion of N-CAM and initiated by enzymes that are themselves under the control of additional regulatory genes would provide further components of selectivity to this dynamic picture of CAM gene expression during organogenesis.

On the basis of this sequence and the evidence derived from the fate map, an hypothesis on the general role of the CAMs in morphogenesis has been put forward (Edelman, 1984c). While this is not the place to consider this so-called regulator hypothesis at length, a brief survey of its main arguments may provide a frame for some concluding remarks on neural morphogenesis. This is particularly appropriate because the evidence suggests that, while mechanisms vary at different stages of embryogenesis in different tissues, at all stages there is both a continuity in the types of molecules expressed and an emphasis on highly dynamic changes during their expression.

The key assumptions and conclusions of the CAM regulator hypothesis may be summarized as follows:

1. CAMs play a central role in morphogenesis by acting through adhesion as steersmen or regulators for other primary processes, particularly morphogenetic movements. CAM expression must play a direct role in the control of motion, which is a result of the play among cellular motility, tension in tissue sheets, and adhesion. CAMs exercise their role as regulators by means of local cell surface modulation.
2. Genes for CAMs are expressed in schedules that are prior to and relatively independent of those for particular networks of cytodifferentiation in different organs. This appears to be reflected in the fact that a CAM in the composite fate map is expressed early and remains in structures that span classical map boundaries not only within a germ layer (e.g., somites, heart, kidney), but also across germ layers (all three for L-CAM; ectoderm and mesoderm for N-CAM).
3. The control of CAM structural genes by regulatory genes is responsible for the body plan as seen in fate maps. In the chick, this plan is reflected in a topological order: a simply connected central region of N-CAM surrounded by a contiguous, simply connected ring of cells expressing L-CAM.
4. Morphogenetic movements result from the inherent motility of cells and from CAM expression as it is coordinated with the expression of SAMs and such substrates as fibronectin. These movements, which are regulated by CAM modulation, are responsible for bringing cells of different history together to result in various embryonic inductions, including neural induction.
5. Natural selection acts to eliminate inappropriate movements by selecting against organisms that express CAM genes in sequences leading to failure of induction. On the other hand, any variant combination of movements and covariant timing of CAM gene expression (resulting from variation in regulatory genes) that leads to appropriate inductive sequences will, in general, be evolutionarily selected. This allows for great variation in the details of fate maps from species to species but at the same time would tend to conserve the basic body plan.
6. Small changes in CAM regulatory genes that do not abrogate this principle of selection could nevertheless lead to large changes in form in relatively short evolutionary times.

It has been pointed out (Edelman, 1984c) that this model is parsimonious—the total number of genes involved need not be large. Because of the wide dynamic range of cell surface modulation effects (Edelman, 1983; Hoffman and Edelman, 1983) and their temporal relationships (Thiery et al., 1984a), the developmental and evolutionary effects of the variant expression of a rather small number of genes related to cell adhesion could be momentous. The various predictions and the contrary results that would falsify the hypothesis have been reviewed elsewhere (Edelman, 1984a,c), but in the present context one prediction is particularly germane: if the topology of CAM expression in neural induction is not the same in various animals with similar body plans but different morphogenetic movements, then the hypothesis must be abandoned. For example, the frog should show a composite CAM fate map topology similar to that of the chicken. The "symmetry breaking" pattern seen in Figures 7 and 8 and in the sequences of Figure 10 should be generally maintained, at least at early developmental stages.

Whatever its merits or inadequacies, the regulator hypothesis focuses attention upon a key and determinant issue of both early and late neurogenesis: What regulates the expression of CAM genes themselves and how are the other primary processes controlled? It remains an open question whether the action of CAM regulatory genes is triggered by morphogens or is triggered by feedback from CAM interactions, spatial asymmetries, or interaction by global surface modulation (Edelman, 1976) induced by binding of SAMs, as suggested earlier in this chapter. Furthermore, it is not known whether CAM gene expression and the ensuing cell–cell adhesion events that irreversibly alter cell movements actually feed back directly to affect other differentiation or division events (see Figure 1). It seems likely that, at the least, differential CAM expression will have indirect effects which change the proportional contribution of each primary process to morphogenesis.

These issues of molecular regulation remain unresolved, but their importance is underlined by results reviewed here, such as the defined CAM expression sequence in neural induction and the delay in E-to-A conversion in *staggerer* mice. In this latter case, it is not yet possible to decide whether the failure in conversion is a consequence of failure to form functioning synapses because of a structural defect unrelated to CAM or because of primary genetic alterations affecting the enzymes responsible for conversion. If conversion failure is shown to be a consequence of failure in synaptic function, one might reasonably assume that there is a regulatory loop linking neural function and CAM modulation.

The clarification of these issues will depend on cloning of genes for N-CAM and other CAMs (Murray et al., 1984), as well as on further *in vitro* experiments on CAM expression and modulation at the cell surface. Accumulation of evidence on the sequence of expression of SAMs and on the interdependence of CAM and SAM families will be equally important. New CAMs and additional modulatory mechanisms will undoubtedly be discovered, requiring extension and modification of this preliminary view of the relationship of adhesion to neural morphogenesis. Enough is known already to say that the CAMs of different specificity that appear differentially during development by gene expression and surface modulation can act as kinetic constraints upon other primary processes of development and thus lead to alterations of early and late neural form. As this survey has indicated, no single static factor such as CAM specificity can account for histogenesis. Nevertheless, while CAMs cannot be sufficient for the establishment of form, the evidence suggests that they are absolutely necessary and that they play pivotal roles along with SAMs and CJMs in constraining other morphogenetic processes that lead to the formation of neural patterns.

REFERENCES

Bertolotti, R., U. Rutishauser, and G. M. Edelman (1980) A cell surface molecule involved in aggregation of embryonic liver cells. *Proc. Natl. Acad. Sci. USA* **77**:4831–4835.

Brackenbury, R., J.-P. Thiery, U. Rutishauser, and G. M. Edelman (1977) Adhesion among neural cells of the chick embryo. I. An immunological assay for molecules involved in cell–cell binding. *J. Biol. Chem.* **252**:6835–6840.

Buskirk, D. R., J.-P. Thiery, U. Rutishauser, and G. M. Edelman (1980) Antibodies to a neural cell adhesion molecule disrupt histogenesis in cultured chick retinae. *Nature* **285**:488–489.

Caviness, V. S., Jr., and P. Rakic (1978) Mechanisms of cortical development. A view from mutations in mice. *Annu. Rev. Neurosci.* **1**:297–326.

Chuong, C.-M., and G. M. Edelman (1984) Alterations in neural cell adhesion molecules during development of different regions of the nervous system. *J. Neurosci.* **4**:2354–2368.

Cowan, W. M. (1978) Aspects of neural development. *Int. Rev. Physiol. Neurophysiol. III.* **17**:149–191.

Cunningham, B. A., S. Hoffman, U. Rutishauser, J. J. Hemperly, and G. M. Edelman (1983) Molecular topography of N-CAM: Surface orientation and the location of sialic acid-rich and binding regions. *Proc. Natl. Acad. Sci. USA* **80**:3116–3120.

Damsky, C. H., J. Richa, K. Knudsen, D. Solter, and C. A. Buck (1982) Identification of a cell–cell adhesion glycoprotein from mammary tumor epithelium. *J. Cell Biol. (Abstr.)* **95**:22.

Edelman, G. M. (1976) Surface modulation in cell recognition and cell growth. *Science* **192**:218–226.

Edelman, G. M. (1983) Cell adhesion molecules. *Science* **219**:450–457.

Edelman, G. M. (1984a) Cell modulation and marker multiplicity in neural patterning. *Trends Neurosci.* **7**:78–84.

Edelman, G. M. (1984b) Modulation of cell adhesion during induction, histogenesis, and perinatal development of the nervous system. *Annu. Rev. Neurosci.* **7**:339–377.

Edelman, G. M. (1984c) Cell adhesion and morphogenesis: The regulator hypothesis. *Proc. Natl. Acad. Sci. USA* **81**:1460–1464.

Edelman, G. M., and C.-M. Chuong (1982) Embryonic to adult conversion of neural cell adhesion molecules in normal and *staggerer* mice. *Proc. Natl. Acad. Sci. USA* **79**:7036–7040.

Edelman, G. M., W. J. Gallin, A. Delouvée, B. A. Cunningham, and J.-P. Thiery (1983a) Early epochal maps of two different cell adhesion molecules. *Proc. Natl. Acad. Sci. USA* **80**:4384–4388.

Edelman, G. M., S. Hoffman, C.-M. Chuong, J.-P. Thiery, R. Brackenbury, W. J. Gallin, M. Grumet, M. E. Greenberg, J. J. Hemperly, C. Cohen, and B. A. Cunningham (1983b) Structure and modulation of neural cell adhesion molecules in early and late embryogenesis. *Cold Spring Harbor Symp. Quant. Biol.* **48**:515–526.

Gallin, W. J., G. M. Edelman, and B. A. Cunningham (1983) Characterization of L-CAM, a major cell adhesion molecule from embryonic liver cells. *Proc. Natl. Acad. Sci. USA* **80**:1038–1042.

Gerisch, G., and D. Malchow (1976) Cyclic AMP receptors and the control of cell aggregation in *Dictyostelium*. *Adv. Cyclic Nucleotide Res.* **7**:49–68.

Gilula, N. B. (1978) Structure of intercellular junctions. In *Intercellular Junctions and Synapses*, Vol. 2, J. Feldman, N. B. Gilula, and J. D. Pitts, eds., pp. 3–22, Chapman & Hall, London.

Greenberg, M. E., R. Brackenbury, and G. M. Edelman (1984) Alteration of N-CAM expression following neuronal cell transformation by Rous sarcoma virus. *Proc. Natl. Acad. Sci. USA* **81**:969–973.

Grumet, M., and G. M. Edelman (1984) Heterotypic binding between neuronal membrane vesicles and glial cells is mediated by a specific neuron–glial cell adhesion molecule. *J. Cell Biol.* **98**:1746–1756.

Grumet, M., S. Hoffman, and G. M. Edelman (1984) Two antigenically related neuronal CAM's of different specificities mediate neuron–neuron and neuron–glia adhesion. *Proc. Natl. Acad. Sci. USA* **81**:267–271.

Hay, E. D. (1981) Collagen and embryonic development. In *Cell Biology of Extracellular Matrix*, E. D. Hay, ed., pp. 379–409, Plenum, New York.

Hoffman, S., and G. M. Edelman (1983) Kinetics of homophilic binding by E and A forms of the neural cell adhesion molecule. *Proc. Natl. Acad. Sci. USA* **80**:5762–5766.

Hoffman, S., B. C. Sorkin, P. C. White, R. Brackenbury, R. Mailhammer, U. Rutishauser, B. A. Cunningham, and G. M. Edelman (1982) Chemical characterization of a neural cell adhesion molecule purified from embryonic brain membranes. *J. Biol. Chem.* **257**:7720–7729.

Hyafil, F., D. Morello, C. Babinet, and F. Jacob (1980) A cell surface glycoprotein involved in the compaction of embryonal carcinoma cells and cleavage stage embryos. *Cell* **21**:927–934.

Lane, N.J. (1984) A comparison of the construction of intercellular junctions in the CNS of vertebrates and invertebrates. *Trends Neurosci.* **7**:95–99.

Murray, B. A., J. J. Hemperly, W. J. Gallin, J. S. MacGregor, G. M. Edelman, and B. A. Cunningham (1984) Isolation of cDNA clones for the chick neural cell adhesion molecule (N-CAM). *Proc. Natl. Acad. Sci. USA* **81**:5584–5588.

Ogou, S. I., C. Yoshida-Noro, and M. Takeichi (1983) Calcium-dependent cell–cell adhesion molecules common to hepatocytes and teratocarcinoma stem cells. *J. Cell Biol.* **97**:944–948.

Rothbard, J. B., R. Brackenbury, B. A. Cunningham, and G. M. Edelman (1982) Differences in the carbohydrate structures of neural cell adhesion molecules from adult and embryonic chicken brains. *J. Biol. Chem.* **257**:11064–11069.

Rutishauser, U., M. Grumet, and G. M. Edelman (1983) N-CAM mediates initial interactions between spinal cord neurons and muscle cells in culture. *J. Cell Biol.* **97**:145–152.

Sidman, R. L. (1974) Contact interaction among developing mammalian brain cells. In *Cell Surface in Development*, A. Moscona, ed., pp. 221–253, Wiley, New York.

Sperry, R. W. (1963) Chemoaffinity in the orderly growth of nerve fiber patterns and connections. *Proc. Natl. Acad. Sci. USA* **50**:703–710.

Thiery, J.-P. (1985) Roles of fibronectin in embryogenesis. In *Fibronectin*, D. Mosher, ed., Academic, New York (in press).

Thiery, J.-P., R. Brackenbury, U. Rutishauser, and G. M. Edelman (1977) Adhesion among neural cells of the chick embryo. II. Purification and characterization of a cell adhesion molecule from neural retina. *J. Biol. Chem.* **252**:6841–6845.

Thiery, J.-P., J.-L. Duband, U. Rutishauser, and G. M. Edelman (1982) Cell adhesion molecules in early chick embryogenesis. *Proc. Natl. Acad. Sci. USA* **79**:6737–6741.

Thiery, J.-P., A. Delouvée, W. J. Gallin, B. A. Cunningham, and G. M. Edelman (1984a) Ontogenetic expression of cell adhesion molecules: L-CAM is found in epithelia derived from the three primary germ layers. *Dev. Biol.* **102**:61–78.

Thiery, J.-P., A. Delouvée, M. Grumet, and G. M. Edelman (1984b) Initial appearance and regional distribution of the neuron-glia cell adhesion molecule (Ng-CAM) in the chick embryo. *J. Cell Biol.* (in press).

Vakaet, L. (1984) Early development of birds. In *Chimeras in Developmental Biology*, N. M. Le Douarin and A. McLaren, eds., pp. 71–87, Academic, New York.

Chapter 3

Identifying the Components of Muscle Fiber Basal Lamina That Aggregate Acetylcholine Receptors at Regenerating Neuromuscular Junctions

JUSTIN R. FALLON
RALPH M. NITKIN
BRUCE G. WALLACE
U. JACKSON McMAHAN

ABSTRACT

During the formation of the postsynaptic membrane in developing and regenerating muscles, acetylcholine receptors in the myofiber membrane aggregate just opposite the axon terminal. It is clear that receptor aggregation depends on communication between the axon terminal and the myofiber, but the molecular basis of this interaction is a mystery. We are studying components of the neuromuscular junction and the Torpedo electric organ with a view toward identifying and characterizing the molecules that cause receptor aggregation at the regenerating neuromuscular synapse.

The high density of acetylcholine receptors (AChRs) in the muscle fiber plasma membrane at neuromuscular junctions is crucial for optimal neuromuscular transmission. Detailed *in vitro* studies on the development of this postsynaptic specialization have demonstrated that AChR aggregation is caused by the presence of the axon terminal and occurs, at least in part, by movement to the synaptic site of receptors that previously had a wider distribution in the myofiber plasma membrane (Anderson and Cohen, 1977; Frank and Fischbach, 1979). The influence of the nerve terminal on receptor migration does not appear to depend on neuromuscular transmission; axon-induced AChR aggregation occurs even in the presence of cholinergic antagonists (Anderson and Cohen, 1977). Soluble extracts of nervous tissue can cause the aggregation of AChRs on cultured myofibers, and the active components have an apparent molecular weight greater than that of acetylcholine (see Godfrey et al., 1984, for a review). These findings raise the possibility that the axon terminal releases molecules other than acetylcholine that cause AChR aggregation.

We have been examining the factors that direct AChR aggregation at synaptic sites on regenerating muscle fibers *in vivo*. If muscles are damaged in ways that spare the basal lamina sheaths of the muscle fibers, new myofibers develop within the sheaths and neuromuscular junctions form on them at the original synaptic sites. At the regenerated neuromuscular junctions, as at the original ones, the muscle fiber is characterized by a high concentration of AChRs. Several experiments conducted in this laboratory lead to the conclusion that the synaptic portion of the muscle fiber's basal lamina contains molecules that direct AChR aggregation on regenerating myofibers (Burden et al., 1979; McMahan and Slater, 1984). For example, freezing the junctional region of muscles in adult frogs results in the disintegration of all cellular components of the neuromuscular junction— the muscle fiber and the axon terminal with its Schwann cell cap—but the basal lamina sheath of the muscle fiber, including the portion that occupied the synaptic cleft, survives. Reinnervation can then be prevented. Under these conditions, new muscle fibers develop within the old sheaths, and the AChRs in their plasma membrane aggregate preferentially at the spots where the membrane is apposed to the original synaptic sites on the sheaths, despite the absence of cells on the presynaptic side of the basal lamina (McMahan and Slater, 1984).

The finding that components of the synaptic basal lamina cause AChR aggregation on regenerating myofibers raises several questions. From which of the cells at the neuromuscular junction are the basal lamina AChR-aggregating molecules derived? What is the nature of their association with other basal lamina components? How are these extracellular molecules regulated? How do they cause AChR aggregation? Are they similar to the molecules that cause AChR aggregation during neuromuscular development in the embryo? One approach to answering these questions is to identify and characterize the active molecules in the basal lamina and make specific markers for them. We have undertaken a series of studies with this aim in mind; a summary of our progress to date follows (for a more detailed account, see Nitkin et al., 1983).

AChR-AGGREGATING MOLECULES

Basal laminae and other extracellular matrix constituents are insoluble in isotonic saline and detergent solutions, so one might expect that insoluble detergent-extracted fractions from tissues receiving cholinergic input would provide an enriched source of the basal lamina AChR-aggregating molecules. Indeed, we found that such an insoluble fraction from *Torpedo* electric organ, a tissue akin to muscle but with a much higher concentration of cholinergic synapses, causes AChR aggregation on cultured chick muscle cells. Our initial experiments (Godfrey et al., 1984) revealed that this insoluble fraction, which was rich in extracellular matrix constituents and contained structures resembling basal lamina, caused a three- to 20-fold increase in the number of AChR clusters. The number and size of the myotubes were unaffected by the treatment. The increase was first seen 2–4 hours after the insoluble fraction was added and was maximal by 24 hours. The AChR-aggregating effect was dose-dependent and was due, at least in part, to lateral migration of AChRs present in the muscle-cell plasma membrane at the time the fraction was applied. The activity in this fraction was destroyed by heat or by trypsin, indicating that the active molecule, or molecules, is proteinaceous. The active component could be extracted from the insoluble fraction with high ionic strength or pH 5.5 buffers. These extracts increased the number of AChR clusters on myotubes without affecting the number or degradation rate of surface AChRs. An antiserum developed against the solubilized material bound to the active component and blocked its effect on AChR distribution.

Insoluble fractions of *Torpedo* liver and muscle did not cause AChR aggregation on cultured myotubes. However, a low level of activity was detected in pH 5.5 extracts from the muscle fraction. The active component, or components, in the muscle extract could be immunoprecipitated by the antiserum against the material extracted from the electric organ insoluble fraction. Moreover, this antiserum bound to the extracellular matrix in frog muscles, including the junctional and extrajunctional regions of the myofiber basal lamina sheath (Figure 1; Godfrey et al., 1984). These findings demonstrate that the insoluble fraction of the *Torpedo* electric organ is rich in AChR-aggregating molecules and that biochemically and functionally related components are also present in *Torpedo* muscle. They also show that components of the electric organ insoluble fraction are antigenically related to molecules in the frog myofiber basal lamina.

We further purified and characterized the electric organ AChR-aggregating molecules (Nitkin et al., 1983). The pH 5.5 extract was fractionated by gel filtration; the majority of the activity was recovered in fractions corresponding to proteins of 50–100 kD. When the active fractions were pooled and applied to an ion-exchange column (DEAE) at pH 6.0, about 60–70% of the activity passed through the column; this material had a specific activity more than 1000-fold higher than did the initial electric organ homogenate. Only nanogram amounts of protein from the most purified fraction were required to cause detectable AChR aggregation on cultured myotubes. If the AChR-aggregating molecule has a molecular weight of 50–100 kD, as indicated by gel filtration, it would be acting at a concentration of 10^{-10} M or less. This is in the range of effectiveness of many pure growth factors and hormones. When we injected our most purified fraction into rabbits, an antiserum that blocked AChR aggregation and immunoprecipitated the active molecules was generated. Further, immunofluorescence studies on frozen sections showed that this antiserum bound specifically to components of the neuromuscular junction in frog muscles (Figure 2). Thus, antigens that are related to specific components of the neuromuscular junction copurify with the AChR-aggregating activity.

Analysis by SDS-polyacrylamide gel electrophoresis of our most purified material revealed several bands. In preliminary studies, briefly described by Nitkin et al. (1983), we electroeluted individual polypeptide bands from SDS gels and injected them into rabbits. Serum from one rabbit injected with material from the 80-kD region of a gel blocked the electric organ AChR-aggregating activity when added to cultures along with the AChR-aggregating molecules. However, this antiserum was of very low titer and did not immunoprecipitate the active molecules. Further, we have been unable to generate blocking antisera in other rabbits injected with 80-kD material. We are continuing our immunological and biochemical studies to establish which polypeptide in our most purified electric organ fraction mediates the AChR-clustering activity, and whether that polypeptide is antigenically related to molecules in the synaptic basal lamina at the neuromuscular junction.

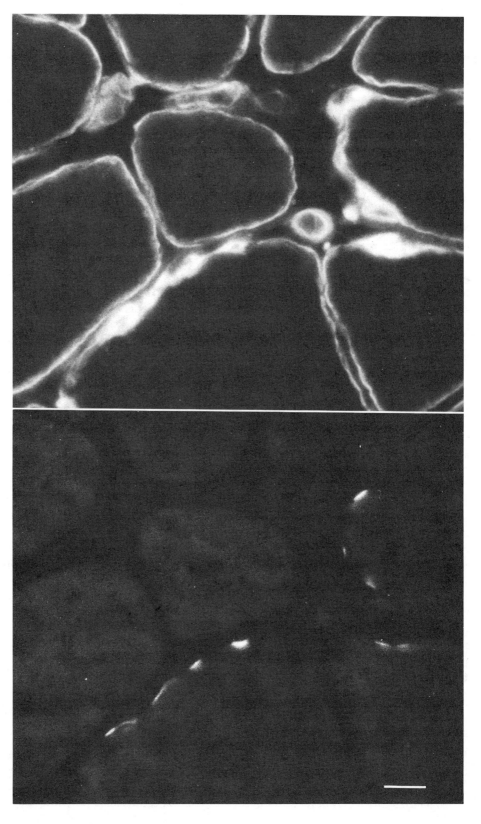

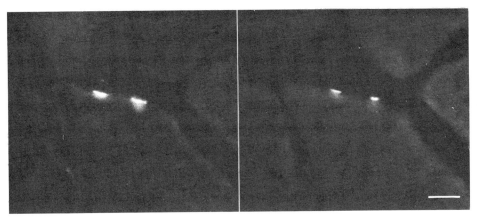

Figure 2. *An antiserum against a greater than 1000-fold purified preparation of* Torpedo *AChR-aggregating molecules binds specifically to the neuromuscular junction.* A frozen section of frog muscle was double-stained, as described in Figure 1, except the first antiserum was raised against the ion-exchange–purified fraction of *Torpedo* electric organ. The same field is shown (left) under fluorescein optics (antibody labeling) and (right) under rhodamine optics (α-bungarotoxin labeling). Calibration bar = 10 μm.

We noted above that soluble extracts that cause AChR aggregation on cultured myotubes have been obtained by others from a variety of neuron-rich tissues. These include fetal rat brain (Salpeter et al., 1982), embryonic pig brain (Olek et al., 1983), embryonic chick brain and spinal cord (Jessell et al., 1979), cultured nerve cells (Schaffner and Daniels, 1982), and *Torpedo* electric lobe and organ (Buc-Caron et al., 1983). Purified laminin, a component of both junctional and extrajunctional myofiber basal lamina (Sanes, 1982) also has detectable AChR-aggregating activity (Vogel et al., 1983). It remains to be learned whether any or all of the active molecules examined cause AChR aggregation *in vivo* and whether they share structural, as well as functional, similarities. The importance of such detailed characterization is emphasized by the finding that nonphysiological agents, such as latex beads (Peng and Cheng, 1982), can cause AChR aggregation on cultured myotubes.

REFERENCES

Anderson, M. J., and M. W. Cohen (1977) Nerve-induced and spontaneous redistribution of acetylcholine receptors on cultured muscle cells. *J. Physiol. (Lond.)* **268**:757–773.

Buc-Caron, M., P. Nystrom, and G. D. Fischbach (1983) Induction of acetylcholine receptor synthesis and aggregation: Partial purification of low-molecular-weight activity. *Dev. Biol.* **95**:378–386.

Figure 1. *An antiserum against a high salt extract of electric organ insoluble material labels frog muscle basal lamina and neuromuscular synapses.* A frozen section of frog muscle was incubated with rabbit antibody raised against a 2 M MgCl$_2$ extract of *Torpedo* electric organ insoluble fraction, and then with a second layer containing fluorescein-coupled goat anti-rabbit antibody and rhodamine-conjugated α-bungarotoxin (to label AChRs at synaptic sites). The same field is viewed (top) with fluorescein optics and (bottom) with rhodamine optics to visualize the antibody staining and α-bungarotoxin staining, respectively. Calibration bar = 10 μm.

Burden, S. J., P. B. Sargent, and U. J. McMahan (1979) Acetylcholine receptors in regenerating muscle accumulate at original synaptic sites in the absence of nerve. *J. Cell Biol.* **83**:412–425.

Frank, E., and G. D. Fischbach (1979) Early events in neuromuscular junction formation *in vitro*. *J. Cell Biol.* **83**:143–158.

Godfrey, E. W., R. M. Nitkin, B. G. Wallace, L. L. Rubin, and U. J. McMahan (1984) Components of *Torpedo* electric organ and muscle that cause aggregation of acetylcholine receptors on cultured muscle cells. *J. Cell Biol.* (in press).

Jessell, T. M., R. E. Siegel, and G. D. Fischbach (1979) Induction of acetylcholine receptors on cultured skeletal muscle by a factor extracted from brain and spinal cord. *Proc. Natl. Acad. Sci. USA* **76**:5397–5401.

McMahan, U. J., and C. R. Slater (1984) The influence of basal lamina on the accumulation of acetylcholine receptors at synaptic sites in regenerating muscles. *J. Cell Biol.* **98**:1453–1473.

Nitkin, R. M., B. G. Wallace, M. E. Spira, E. W. Godfrey, and U. J. McMahan (1983) Molecular components of the synaptic basal lamina that direct differentiation of regenerating neuromuscular junctions. *Cold Spring Harbor Symp. Quant. Biol.* **48**:653–665.

Olek, A. J., P. A. Pudimat, and M. P. Daniels (1983) Direct observation of the rapid aggregation of acetylcholine receptors on identified cultured myotubes after exposure to embryonic brain extract. *Cell* **34**:255–264.

Peng, H. B., and P.-C. Cheng (1982) Formation of postsynaptic specializations induced by latex beads in cultured muscle cells. *J. Neurosci.* **2**:1760–1774.

Salpeter, M. M., S. Spanton, K. Holley, and T. R. Podleski (1982) Brain extract causes acetylcholine receptor redistribution which mimics some early events at developing neuromuscular junctions. *J. Cell Biol.* **93**:417–425.

Sanes, J. R. (1982) Laminin, fibronectin, and collagen in synaptic and extrasynaptic portions of muscle fiber basement membrane. *J. Cell Biol.* **93**:442–451.

Schaffner, A. E., and M. P. Daniels (1982) Conditioned medium from cultures of embryonic neurons contains a high molecular weight factor which induces acetylcholine receptor aggregation on cultured myotubes. *J. Neurosci.* **2**:623–632.

Vogel, Z., C. N. Christian, M. Vigny, H. C. Bauer, P. Sonderegger, and M. P. Daniels (1983) Laminin induces acetylcholine receptor aggregation on cultured myotubes and enhances the receptor aggregation activity of a neuron factor. *J. Neurosci.* **3**:1058–1068.

Chapter 4

The Control of Development of Neuronal Excitability

NICHOLAS C. SPITZER

ABSTRACT

Study of the early differentiation of spinal neurons has led to the formulation of a developmental timetable charting the time course of the appearance of neuronal characters. Development in vivo is paralleled by development in culture with respect to aspects of neurite outgrowth, the shift in the ionic dependence of the action potential, and the onset of sensitivity to neurotransmitters. The mechanisms involved in the control of this ordered sequence of phenotypes in cultured neurons are being pursued. The roles of phenotypic expression, translation, and transcription have been examined. The expression of the calcium-dependent action potential is not required for several aspects of development, including the formation of functional synaptic contacts. In contrast, RNA and protein synthesis are necessary for the appearance of the mature sodium-dependent action potential.

Recent years have seen the publication of a number of developmental timetables for different populations of neurons in a variety of species. These timetables are the synoptic compilations of the results obtained by extensive examination of particular sets of neurons. They include information about the time of first appearance of phenotypic characters and any subsequent changes that occur during maturation. This information is presented in the context of the other events occurring in the differentiation of the embryo. The timetables that have been assembled so far vary in the number of differentiated characters that are listed. Nonetheless, they define the process of cellular development with the level of detail necessary for analysis of the underlying molecular events. One such timetable is shown in Figure 1.

In our studies of the membrane properties of Rohon-Beard neurons of the amphibian spinal cord, we have examined the development of the action potential and the development of the neurons' sensitivity to neurotransmitters. The cells are initially insensitive to depolarization of their membrane—they do not make impulses. But at about the time of closure of the neural tube, action potentials are produced that depend largely on an influx of calcium ions and are long in duration, lasting hundreds of milliseconds. A little later, a prominent sodium component of this action potential appears, and during the next few days the calcium component disappears. The cells are also initially insensitive to neu-

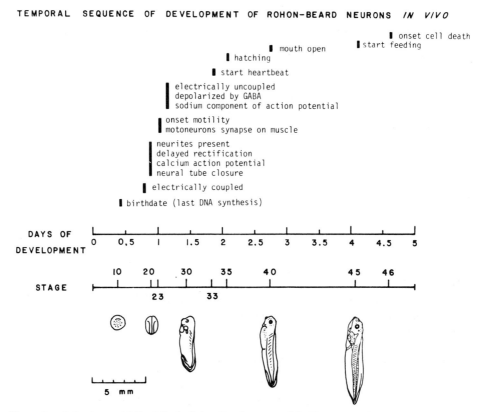

Figure 1. *A developmental timetable for Rohon-Beard neurons of the* Xenopus *embryo*. The time of onset or first appearance of a phenotype in these neurons is indicated by a vertical bar. The morphology of the embryo at different stages is shown below. (From Spitzer and Lamborghini, 1981.)

rotransmitters, but can be depolarized by γ-aminobutyric acid (GABA) at the time that the cells acquire the sodium component of their action potential. We hope to extend this timetable and look at the development of additional membrane and cytoplasmic properties.

The sequence information presented in developmental timetables focuses attention on several questions that logically seem to be the next ones to address. The first of these concerns the mechanisms by which newly expressed phenotypes appear. Given the evidence for variable gene activity during development in other eukaryotic cells, one may ask if the appearance of a new function in these developmentally expressed neuronal sequences invariably involves transcription of new messenger RNA, or whether control is exerted at the levels of translation of new proteins or of posttranslational modification of existing molecules. This class of questions can be approached by the use of specific metabolic inhibitors to determine the level of control. When specific probes are available, one may use them to define more precisely and reliably the timing and location of the relevant macromolecular syntheses. For example, antibodies directed to specific proteins and DNA complementary to specific messenger RNAs will be particularly useful.

It is appealing to imagine that studying membrane properties may introduce a simplification in the study of early neuronal development, in that a very small number of membrane proteins may be needed to assemble an ion channel. This may be contrasted with neurite extension, where actin, tubulin, and a host of important molecules (e.g., fodrin and growth-associated proteins) are known to be involved.

A second question concerns the control of sequencing in the developmental timetable, that is, how the expression of phenotypes in a stereotyped and invariant order (at least within cells of a single class) is achieved. It is as if development were governed by a clock that ensures the correct timing for the expression of the phenotypes by the mechanisms alluded to above. Particular interest is attached to understanding what governs the clockwork underlying a developmental timetable. One may envisage two principal classes of clocks. The clock could be an independent entity that is then coupled to the expression of different phenotypes. The simple possibility that genes are transcribed serially as a consequence of their position along the genome, as in some viruses and prokaryotes, would fall in this class. This possibility can be readily excluded by information about the chromosomal locations of the relevant genes, and it inherently seems unlikely. The other class of clock is dependent on the very processes it implements and could in fact consist of these processes. According to this view, some aspect of the transcription of one gene (or its translation, posttranslational modification, or expression of the resulting phenotype) would provide the necessary signal for initiating the events leading to the expression of the next gene. This possibility is attractive because it avoids the necessity of invoking the existence of an independent clock; furthermore, it can be evaluated by a variety of experimental tests. Selective perturbations involving interruptions of synthesis of a particular RNA or protein, or some form of molecular modification, might be expected to disrupt the expression of the developmental timetable. Alternatively, blockade of the functional expression of a phenotype could prevent the appearance of other phenotypes that normally occur later.

It seems likely that these questions, which are of broad general interest, will be tackled in a number of laboratories. Different preparations will favor somewhat different approaches. Such questions provide motivation for the studies that my colleagues and I are pursuing, and although we do not yet have complete answers, some observations are of interest for the possible mechanisms they exclude.

NEURONAL DEVELOPMENT *IN VITRO* CAN PARALLEL DEVELOPMENT *IN VIVO*

The neurons of the *Xenopus* spinal cord constitute an attractive system in which to tackle these questions. Rohon-Beard neurons in particular are amenable to examination *in situ*. Furthermore, it has been possible to study their development, in conjunction with that of other neurons, in dissociated cell culture, allowing us to perform various experimental manipulations that are not possible *in vivo*.

Rohon-Beard cells are located on the dorsal aspect of the spinal cord (Figure 2) where they are readily exposed by a simple dissection. Their large size (25 μm) has allowed them to be impaled with intracellular microelectrodes and their

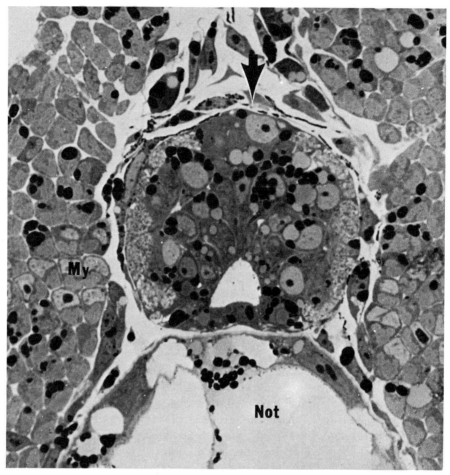

Figure 2. *The size and position of a Rohon-Beard cell (large arrow) in the spinal cord of an early larval stage (day 2) Xenopus. Light micrograph of a transverse section through the midtrunk region. The spinal cord is flanked laterally by myotomal muscles (My) and ventrally by the notochord (Not). (From Lamborghini et al., 1979.)*

membrane properties examined. These cells are primary sensory neurons and fill the role of dorsal root ganglion neurons in the early embryo. They all die at about the time when the dorsal root ganglion cells appear (Lamborghini, 1981); this is an interesting problem in itself, but one which we have not pursued. One wonders here, as elsewhere, what the cause of cell death in embryogenesis may be. The phenomenon is accentuated because *all* of the cells disappear. There are about 150 Rohon-Beard neurons in each spinal cord and they have a very early birthdate, along with some of the primary motoneurons. These two populations are among the first to undergo their final DNA synthesis in the embryo (and the first in the spinal cord). This occurs during the midgastrula stage of development, prior to the appearance of the neural plate and the neural tube (Lamborghini, 1980). These cells have been studied electrophysiologically from stages just prior to closure of the neural tube up to the time of their death.

Early in our studies of Rohon-Beard neurons it seemed clear that it would be advantageous to grow them in culture. We have done this, and the procedure we have followed is simple. Cultures are prepared from the neural plate stage of the embryo (Nieuwkoop and Faber, 1956, stage 15) by dissociating the presumptive spinal cord tissue in a saline solution that lacks divalent cations; no proteolytic enzymes are needed. The cells are grown in a defined medium that consists of four salts and a buffer. The cells are not fed, and we rely on the natural intracellular inclusions of yolk protein and lipid granules to serve as a nutrient supply while the cells differentiate in culture. This procedure affords us the opportunity to control the environment in a detailed way. However, it has the drawback of not allowing the cells to survive for extended periods of time; most cultures do not last longer than three or four days. At present, this is not a serious problem because the features of differentiation we have been studying are all expressed during the first three days. The cells are plated at low density (one neural plate per dish), and neurons differentiate as shown in Figure 3. Phase-contrast optics reveal a neuron's cell body, filled with yolk, rendering it highly refractile. The neurite is seen extending away from it, with fine lateral filopodia and a terminal growth cone. We impale these cells with microelectrodes and examine their membrane properties.

It has been possible to show that the differentiation of these neurons in culture parallels in three ways the differentiation of Rohon-Beard neurons *in vivo*. The time of onset of primary neurite outgrowth from the cell body, the change in the ionic dependence of the action potential, and the appearance and development of neurotransmitter sensitivity are the same for the neurons in these two rather different environments.

Before addressing the evidence, some comment about the identity of the cultured neurons is necessary. It is not possible to identify Rohon-Beard cells in culture by morphological criteria. Furthermore, the cultures are heterogeneous

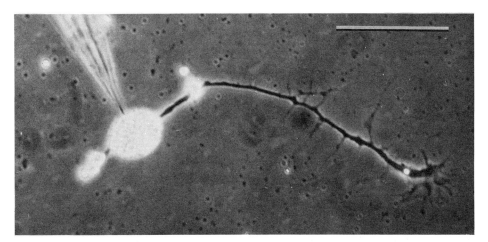

Figure 3. *A neuron from a dissociated* Xenopus *neural plate, grown in standard culture medium for 16 hours and viewed with phase-contrast optics.* Note phase-dark process with filopodia and growth cone; phase-bright yolk granules obscure the nucleus in the cell body, which is indicated by the microelectrode. Calibration bar = 50 μm. (From Spitzer, 1979.)

in two respects. First, they contain not only neurons, but muscle cells, pigment cells, fibroblasts, and cells from the notochord. The density of neurons is low, probably in part because the cultures are prepared when the number of neurons that have undergone their final mitosis is still small. Second, there is heterogeneity with respect to the neuronal composition of the cultures; two lines of evidence indicate that the neurons are principally Rohon-Beard or primary motoneurons. The first is that as many as 80% of the neurons have the early birthdate, which means that they must be one or the other of these two cell classes (or one of the much smaller class of extramedullary neurons that probably resemble the Rohon-Beard cells). The second evidence as to the identity of these neurons comes from an examination of their neurotransmitter sensitivity. We now know, as described below, a substantial amount about the sensitivity of Rohon-Beard neurons. The neurotransmitter sensitivity of primary motoneurons is also well known from the work of others. Most of the neurons we impale and examine have sensitivities to neurotransmitters that are compatible with their belonging to one of these two classes. We therefore believe that the cultures contain not only Rohon-Beard neurons, but one other major neuronal population as well. This is unfortunate in the sense that we do not have a pure population of cultured Rohon-Beard neurons; on the other hand, it is useful in that observations about the development of this broader population of neurons have a more general significance.

The initial outgrowth of neurites from Rohon-Beard neurons in the spinal cord has recently been described in a scanning electron-microscope analysis (Taylor and Roberts, 1983). This occurs at stage 22, which is the equivalent of six hours in culture. At this time we observe the initial outgrowth of neurites from cells that can be identified as Rohon-Beard-like cells (Spitzer and Lamborghini, 1976; Bixby and Spitzer, 1984a). This convergence of evidence indicates that neurite outgrowth occurs on the same time schedule in culture as it does *in vivo*.

The development of the action potential in these cultured neurons has been investigated (Spitzer and Lamborghini, 1976; Willard, 1980; O'Dowd, 1981, 1983a,b; Blair, 1983). The same change in the ionic dependence of the impulse is seen as that described for Rohon-Beard neurons *in vivo* (Baccaglini and Spitzer, 1977). One can distinguish three stages in the development of excitability. Very early—shortly after they have become excitable—impulses of long duration are elicited, and ion substitution experiments indicate that most of the inward current is carried by calcium (Figure 4). Later, the inward current of the impulse depends on an influx of both sodium and calcium. Finally, after 2–3 days in culture, the impulse is largely dependent on an influx of sodium ions. The maturation of the impulse occurs in the same way, and over the same time course, in neurons differentiating in culture and *in vivo*. This shift in the ionic dependence of the action potential during development is emerging as a general feature of excitable cells. A survey of neurons and muscle cells that have been examined in this regard is shown in Table 1. Often the shift involves a change from calcium- to sodium-ion dependence for the inward current, but this is not universal. A counterexample is provided by tunicate skeletal muscle, in which the shift is in the opposite direction, and the sodium-dependent component of the impulse is lost during differentiation.

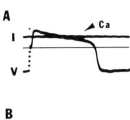
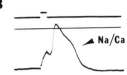
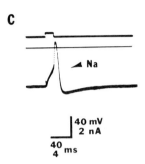

Figure 4. *The shift in ionic dependence of the action potential during the course of development of* Xenopus *spinal neurons.* Records are typical of Rohon-Beard cells differentiating *in vivo* and a more diverse population of neurons differentiating in dissociated cell culture. A–C: The inward current of the impulse changes from primarily calcium to primarily sodium during a period of several days. (Modified from Spitzer, 1981.)

Before turning to the perturbation experiments we have performed recently, the study of the normal development of cultured neurons can be concluded by reviewing what is currently known about the development of their neurotransmitter sensitivity. Initially, the cells are insensitive to bath application of a variety of different putative neurotransmitters, but they become sensitive to GABA at 8–10 hours *in vitro* (Bixby and Spitzer, 1984a). Some produce depolarizing responses, and a fraction of these are also depolarized by glycine. These cells thus behave like Rohon-Beard neurons *in vivo*, with respect to both the time of onset of sensitivity and the agents to which they are sensitive; GABA causes an increase in sodium and potassium conductance (Bixby and Spitzer, 1982). The other neurons produce hyperpolarizing responses to GABA and to glycine and are depolarized by glutamate, as are motor neurons *in vivo* (Nicoll et al., 1976). In contrast to the developmental change in ionic dependence of the impulse, no change in ionic dependence for the response to GABA is elicited from Rohon-Beard cells *in vivo* or from Rohon-Beard-like neurons in culture. The response to iontophoretically applied GABA can be used to determine the reversal potential for the neurotransmitter, which is a measure of the underlying ionic conductances. This also does not change during development (Figure 5). The constancy of the ionic dependence of responses to neurotransmitters during development also emerges as a general feature of the development of excitable cells. A survey of cells that have been analyzed in this respect is presented in Table 2. An apparent exception is presented by cardiac muscle cells, but it may be that the populations of cells sampled at different ages were not the same.

Table 1. Developmental Changes in the Ionic Dependence of Action Potentials

Cells	Animal	Change	Investigators
Rohon-Beard neurons *in vivo*	*Xenopus*	Ca→Na	Spitzer and Baccaglini (1976); Baccaglini and Spitzer (1977)
Dorsal root ganglion neurons *in vivo*	*Xenopus*	Ca→Na	Baccaglini (1978)
Olfactory neurons *in vivo*	Bullfrog	Ca→Na	Strichartz et al. (1980)
Dorsal unpaired median (DUM) neurons *in vivo*	Grasshopper	Ca or Na→Ca/Na	Goodman and Spitzer (1979, 1981)
Spinal neurons *in vitro*	*Xenopus*	Ca→Na	Spitzer and Lamborghini (1976); Willard (1980)
Dorsal root ganglion neurons *in vitro*	Mouse	Ca/Na→Na	Matsuda et al. (1978)
Neuroblastoma *in vitro*	Mouse	Ca→Na	Miyake (1978)
Cerebral cortical neurons *in vitro*	Chick	Ca/Na→Na	Mori-Okamoto et al. (1983)
Dorsal root ganglion neurons *in vitro*	Adult guinea pig	Ca→Na	Fukuda and Kameyama (1979)

Giant axons *in vivo*	Adult cockroach	Ca→Na	Meiri et al. (1981)
Striated trunk muscle *in vivo*	Tunicate	Ca/Na→Ca	Takahashi et al. (1971) Miyazaki et al. (1972, 1974)
Striated leg muscle *in vivo*	Chick	Ca→Na	Kano (1975)
Striated leg muscle *in vitro*	Chick	Ca→Na	Kano et al. (1972) Kano and Shimada (1973) Kano and Yamamoto (1977)
Clonal striated muscle cell line (L6) *in vitro*	Rat	Na→Na/Ca	Kidokoro (1973, 1975)
Cardiac muscle *in vivo*	Chick	Ca/Na→Na	Pappano (1972a) Sperelakis and Shigenobu (1972)
Cardiac muscle *in vitro*	Chick	Ca/Na→Na	McDonald and Sachs (1975) DeHaan et al. (1976)
Cardiac muscle *in vivo*	Rat	Ca/Na→Na	Bernard et al. (1969) Bernard and Gargouïl (1970) Bernard (1976)

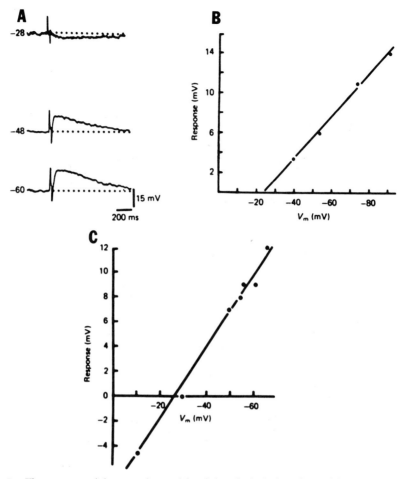

Figure 5. *The constancy of the reversal potential and thus the ionic dependence of the response to a neurotransmitter, GABA, from the time of onset of sensitivity in* Xenopus *spinal neurons.* Records are typical of Rohon-Beard cells differentiating *in vivo* and a population of Rohon-Beard-like cells differentiating in dissociated cell culture. A: Responses of a neuron to iontophoresis of GABA at different membrane potentials. B: The responses from a mature neuron plotted as a function of the membrane potential; the extrapolated reversal potential is −23 mV. C: The responses of a newly sensitive neuron are plotted as a function of the membrane potential; the reversal potential is −26 mV. The mean reversal potential for both newly sensitive and mature cells is about −30 mV. (From Bixby and Spitzer, 1982.)

The observation that neuronal development proceeds in dissociated cell culture as it does *in vivo*, at least in some respects, is important for the studies being undertaken. It gives experimental results some credibility for relevance to differentiation in the animal. Furthermore, it suggests that there is substantial autonomy in early cellular differentiation and that it may proceed independent of extracellular cues. For example, a role for cell-contact–mediated interactions in development of single cell membrane properties is excluded by these findings. A role for soluble factors is rendered unlikely by the low density of the cultures and the use of a defined medium, but cannot be ruled out fully. The result is to focus attention on intracellular programming in an effort to understand the basis for the expression of the developmental timetable.

Table 2. Developmental Constancy[a] of the Ionic Dependence of Responses to Neurotransmitters

Cells	Animal	Neurotransmitter	Investigators
Dorsal unpaired median (DUM) neurons *in vivo*	Grasshopper	ACh, GABA	Goodman and Spitzer (1979, 1980)
Rohon-Beard neurons *in vivo*	*Xenopus*	GABA	Bixby and Spitzer (1982)
Spinal neurons *in vitro*	*Xenopus*	GABA	Bixby and Spitzer (1984a)
Striated trunk muscle *in vivo*	*Xenopus*	ACh	Cohen and Kullberg (1974); Blackshaw and Warner (1976)
Striated trunk muscle *in vivo*	Tunicate	ACh	Ohmori and Sasaki (1977)
Striated leg muscle *in vitro*	Rat	ACh	Fambrough and Rash (1971); Ritchie and Fambrough (1975)
Striated pectoral, leg muscle *in vitro*	Chick	ACh	Fischbach (1972); Harris et al. (1973)
Clonal striated muscle cell line (L6) *in vitro*	Rat	ACh	Steinbach (1975)
EXCEPTION:			
Cardiac muscle *in vivo*	Chick	ACh (Na/K→K dependence)	Pappano (1972b)

[a] Constant reversal potential.

EARLY DIFFERENTIATION IN THE ABSENCE OF CALCIUM-DEPENDENT IMPULSES

This culture system has been used for perturbation experiments that we believe are illuminating in trying to understand the factors that control the developmental timetable. Control could be exerted in many ways, as indicated above, and one is at the level of expression of the phenotype itself. This hypothesis has been tested by asking what the role of the long-duration, calcium-dependent action potential might be during development (Bixby and Spitzer, 1984b). This impulse allows a substantial influx of calcium into a cell. Calcium is a ubiquitous intracellular regulator, so we wondered what the consequences might be if the cells' ability to make such action potentials were blocked during development. A search for a compound that would specifically block voltage-dependent calcium channels in these neurons was unsuccessful; D-600, nifedipine, and other frequently effective compounds had no blocking action. We resorted to a different strategy by simply removing calcium from the culture medium, replacing it with magnesium, and adding 1 mM EGTA to chelate any residual calcium. In many experiments, tetrodotoxin (TTX) was added to block the sodium component of the impulse when it appeared. We then examined the differentiation of the neurons. Examination of neurons developing in normal medium revealed that replacement of calcium with magnesium and the addition of TTX fully blocked the generation of action potentials. How did growth in a calcium-free medium affect differentiation?

In at least three respects, differentiation is not retarded by the removal of calcium. First, neurite outgrowth begins on the same time course as in normal medium (which, as we have said, is the same as that seen *in vivo*), and the extension of neurites on the tissue-culture plastic is not impeded by the absence of extracellular calcium. In fact, neurite outgrowth is accelerated in this medium, and neurites achieved a mean length 2.8 times that of neurites of cells grown in normal medium containing calcium. There is no obvious, consistent difference in neurite or growth-cone morphology. This is not a new finding; others have noted the ability of neurons to extend processes when calcium influx is blocked (Letourneau and Wessells, 1974; D. Bray, unpublished results). These findings do not exclude a role for intracellular stores of calcium. Second, the removal of calcium from the medium has no detectable effect on the normal developmental shift in ionic dependence of the impulse, which was examined during perfusion of cultures with a standard saline. The duration of the impulse, which has been shown to be a useful measure of its ionic dependence, declines in the same way as that described for neurons grown in normal culture medium (Figure 6). Third, the onset of neurotransmitter sensitivity, assessed by examining the ability of neurons to respond to bath applications of GABA, is unaffected by growth in calcium-free medium (Table 3). Cells initially become sensitive at the same time and at the same rate as neurons grown in normal culture medium. However, later maturation of sensitivity is retarded: some cells do not become sensitive; of those that give responses, some are less sensitive than are neurons grown in normal medium; finally, some lose their initial sensitivity to GABA.

These results are useful, in part because they exclude some developmental mechanisms. They indicate that calcium- and sodium-ion influx through voltage-

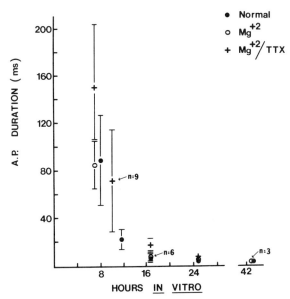

Figure 6. *Differentiation in calcium-free medium does not alter the normal shift in ionic dependence of the neuronal action potential, although voltage-dependent calcium influx is blocked. The duration of the action potential, which is a measure of its ionic dependence, was determined in calcium-containing saline and is plotted as a function of age in culture. The duration of the impulse decreases as the ionic dependence shifts from calcium to sodium, and this decrease is unaffected by the absence of calcium and the presence of tetrodotoxin. Each point is the mean ± SEM for 10 or more cells unless otherwise indicated. (From Bixby and Spitzer, 1984b.)*

dependent channels is not necessary for some aspects of early differentiation of these amphibian spinal neurons. The expression of calcium- and sodium-dependent action potentials is not required for the development of some phenotypes; the possibility that these impulses affect the differentiation of later phenotypes is unresolved. In particular, interest is focused on the mechanism by which the sensitivity to GABA is reduced. In contrast, substantial evidence exists from other systems that later aspects of development are dependent on electrical activity (Thompson et al., 1979; Bergey et al., 1981; Oppenheim, 1981).

These findings also render less likely one simple hypothesis for the clock mechanism underlying the ordered expression of the developmental timetable that relates the timing of phenotypic expression to the exhaustion of metabolic reservoirs. Growth in calcium-free medium leads to more rapid extension of

Table 3. Development of GABA Sensitivity in Amphibian Spinal Neurons in Dissociated Cell Culture[a]

Culture Medium	< 8 hr	8–10 hr	19–25 hr	> 25 hr
Control	0%	38%	96%	100%
Mg^{2+} or Mg^{2+}/TTX	0%	35%	69%	44%

[a] Percentage of cells responding to bath application of 100 μM GABA.

neurites, as indicated above. In addition, there is an accelerated depletion of intracellular yolk platelets, perhaps as a result of rapid neurite outgrowth, and early death of the cultured neurons. In spite of this change in metabolic processing, the timing of initial neurite outgrowth, the development of the action potential, and the onset of neurotransmitter sensitivity are unaffected.

FORMATION OF SYNAPTIC CONTACTS IN THE ABSENCE OF IMPULSE-EVOKED TRANSMITTER RELEASE

Another potential role for impulses at early stages of development could be to promote formation of functional synaptic contacts. Although the cues that allow synaptogenesis are presently unknown, one possibility is that some of the well-studied features of transmission at mature synapses are necessary for the initial formation of connections. It now seems unlikely that postsynaptic acetylcholine (ACh) receptors play a role in formation of neuromuscular junctions because connections can be made in the presence of reversible blockers (Cohen, 1972). Attention has recently been focused on a role for impulse-evoked release of neurotransmitter and associated molecules, or the exposure of vesicular antigens, after the demonstration that ACh or ACh-like material can be released from the growth cones of neurons in culture (Hume et al., 1983; Young and Poo, 1983). We have taken advantage of the ability to grow embryonic amphibian nerve and muscle cells in dissociated cell culture in the absence of calcium—a condition under which impulse-evoked release of transmitter is blocked (del Castillo and Katz, 1954)—and have asked if synaptic contacts could be formed in this environment (Henderson et al., 1984).

Morphological observations of the frequency of nerve–muscle contacts in both calcium-free and normal media suggested that the formation of functional contacts might be impaired. Although neurons extended longer neurites, the probability of contacting myocytes was reduced in the absence of calcium to 70% of that of controls. Furthermore, the frequency with which neurites terminated on contacted myocytes was reduced to 33%.

Intracellular recordings from neurons and contacted myocytes grown in calcium-free medium were made in standard saline containing calcium. The frequency with which an impulse evoked in a neuron elicited an excitatory postsynaptic potential from a myocyte was 32% of that seen in normal medium. This reduced frequency is unlikely to have been the result of damage to the neuron upon impalement, as the frequency of observation of functional contacts, assessed by single electrode recordings of spontaneous postsynaptic potentials in contacted myocytes, was roughly the same (27% of that in normal medium). Evoked potentials recorded from myocytes grown in calcium-free medium were smaller in amplitude and slower in rise time than those from cultures in normal medium (Figure 7), and the frequency of spontaneous postsynaptic potentials was reduced by nearly an order of magnitude. These potentials were blocked by curare, consistent with their production by the release of ACh. Significantly, the impulse-evoked postsynaptic potentials recorded in standard saline containing calcium were reversibly abolished by perfusion with saline in which magnesium replaced calcium. This observation helped to validate the initial assumption—that removal

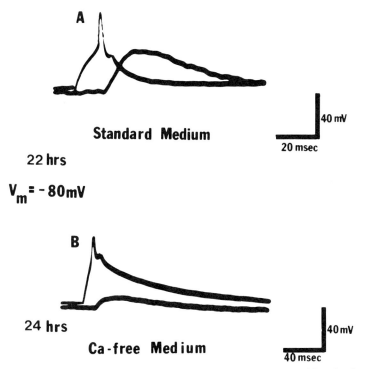

Figure 7. *Differentiation in calcium-free medium does not prevent the formation of functional synaptic contacts between nerve and muscle cells, although impulse-evoked release of acetylcholine is blocked. When examined in calcium-containing saline, action potentials in neurons could elicit postsynaptic potentials from myocytes contacted by their processes. A: Cells cultured in normal medium. B: Cells cultured in calcium-free medium.* (From Henderson et al., 1984.)

of calcium blocks the impulse-evoked release of neurotransmitter, even when neurons have been grown in the absence of calcium. In contrast, this magnesium-containing saline did not always block spontaneous ACh release, which was detected in a small number of myocytes at a very low frequency.

These findings suggested that the calcium-free medium might be causing a decrease in the presynaptic release of neurotransmitter or a decrease in the postsynaptic sensitivity to ACh. The distribution of ACh receptors was examined by fluorescent α-bungarotoxin binding and by iontophoresis. The toxin binding assays revealed an absence of hot spots on the myocytes grown in calcium-free medium, in contrast to their presence on 88% of myocytes grown in normal medium. Further, these experiments demonstrated that, during the time the cells were examined physiologically in calcium-containing saline, neither reaggregation nor new synthesis of ACh receptors could account for the results. Measurements of ACh sensitivity confirmed these findings, revealing a uniform level of receptors on myocytes grown in the absence of calcium and regions of high sensitivity on those grown in normal medium. Thus, at least part of the impairment of synaptic transmission observed between neurons and myocytes grown in calcium-free medium is due to postsynaptic effects. The reduced frequency of spontaneous postsynaptic potentials indicates that there are presynaptic changes as well.

Extensive studies have been conducted on the formation of neuromuscular contacts in culture in normal calcium-containing medium (Anderson and Cohen, 1977; Anderson et al., 1977, 1979; Kidokoro et al., 1980; Gruener and Kidokoro, 1982; Kidokoro and Yeh, 1982). In comparison, the contacts formed in calcium-free medium achieve only a limited level of differentiation. However, the cells are not simply retarded in their development, since myocytes develop electrical excitability and contractile apparatus and synthesize reasonable levels of ACh receptors, whereas neurons develop normally in several respects (see above). The later consolidation and maturation of synaptic contacts into differentiated synaptic connections may be arrested by growth in a calcium-free medium.

These results demonstrate that neurons can develop the ability to synthesize, store, and release neurotransmitter in the absence of voltage-dependent calcium influx. Further, motoneurons appear to make their normal transmitter (ACh) under these circumstances. Although the time of onset of neurotransmitter synthesis is unknown, the onset of electrical excitability occurs well after cells have been plated in culture; they were grown in calcium-free medium, so they never had the opportunity to generate impulses. In contrast, an increase in calcium influx can affect the developmental choice of transmitter synthesized by neurons (Walicke and Patterson, 1981). The possibility of such control by calcium in this system remains to be examined.

It seems clear that the absence of calcium in the culture medium blocks impulse-evoked release of ACh but does not block the formation of neuromuscular synaptic contacts: Neither electrical activity nor calcium influx associated with long-duration impulses is necessary. A role for spontaneous or extravesicular release, however, is not ruled out. Many aspects of the mechanisms by which the frequency of formation and efficacy of synaptic contacts are reduced remain unclear, but the results demonstrate that the formation can proceed in the absence of the evoked release of transmitter.

ROLES OF RNA AND PROTEIN SYNTHESIS IN THE DEVELOPMENT OF ELECTRICAL EXCITABILITY

A useful beginning to the analysis of the roles of specific macromolecular syntheses in the development of excitable membranes can be achieved by the application of specific inhibitors, which allow definition of the class of synthesis that is required. Their application for varying times during development reveals the period(s) during which the expression of a phenotype is sensitive to inhibition. These experiments pinpoint the times of interest, to which more focused analyses can be addressed, and have been used to investigate neurite outgrowth and neurotransmitter synthesis by cultured ganglionic neurons of *Drosophila* and the mouse (Donady et al., 1975; Seecof, 1977; Bloom and Black, 1979), as well as neurite extension by pheochromocytoma (PC12) cells (Burstein and Greene, 1978).

Our recent efforts have been concerned with the development of the sodium-dependent impulse in cultured amphibian spinal neurons. This is one of the first phenotypes whose development in culture was described, and it is one of the simplest electrical properties to examine. The strategy has been to determine the dose of inhibitor effective in producing a substantial reduction of RNA or protein synthesis, assessed by the level of incorporation of radiolabeled precursors. The effect of chronic inhibition on the development of the impulse was then examined; inhibitors were added at various times after plating. The effects of actinomycin D, which inhibits RNA synthesis, and of puromycin and cycloheximide, which inhibit protein synthesis, have been described (Blair, 1983; O'Dowd, 1983a).

When actinomycin D is added to the cultures several hours after they are prepared, the development of the action potential is blocked, and the normal shift from a long-duration, principally calcium-dependent impulse to a brief, largely sodium-dependent impulse does not take place. Neurite extension proceeds normally, implying that the necessary RNA synthesis has been completed. (This is not the case if the inhibitor is added at the time of plating, when it blocks the morphological differentiation of the cells.) The neurons have been examined after as much as three days' exposure to actinomycin D, and the impulse retains its immature form (Figure 8).

Three lines of evidence suggest that this is a relatively specific effect, rather than the consequence of a general developmental arrest. First, the resting potentials and input resistances of these cells, reflecting housekeeping functions, are not affected. Second, other aspects of neuronal development are not perturbed. Neurite outgrowth is unaffected by the treatment, as noted above. Furthermore, the development of delayed rectification, reflecting the presence of voltage-dependent outward potassium currents, occurs normally in the presence of the inhibitor. Third, if the inhibitor is applied about 10 hours after the latest time at which other experiments show it to be completely effective, the sodium-dependent action potential develops normally, presumably as a result of the completion of the necessary RNA synthesis. Initiation of inhibition at intermediate times leads to partial expression of the phenotype.

These results indicate the existence of a sensitive period—a time during which the development of the phenotype is sensitive to RNA synthesis inhibition. One possibility is that the messenger RNA coding for the voltage-dependent sodium channel is blocked, but other interpretations have not been excluded (e.g., modification of existing channels). Voltage-clamp analysis of the inward currents underlying the impulse reveals that an increase in the sodium current occurs during normal development and is blocked by the application of actinomycin D (O'Dowd, 1983b, and personal communication). This technique has not revealed a developmental decrease in the calcium current during the time examined. The normal decrease in the calcium component of the action potential may be due to the increased activation of outward potassium currents that repolarize the cell membrane and shut off the calcium conductance.

Application of protein synthesis inhibitors produced a similar effect on the development of the sodium-dependent impulse (Blair, 1983). Cycloheximide proved more useful than puromycin, allowing better neuronal survival during substantial inhibition. The appearance of the sodium-dependent impulse was blocked, and long-duration, calcium-dependent action potentials persisted for as long as the cells could be maintained (about 24 hours). The effect of inhibition was specific, by the same criteria as defined above. Again, the existence of a sensitive period was demonstrated, after which the application of inhibitor had

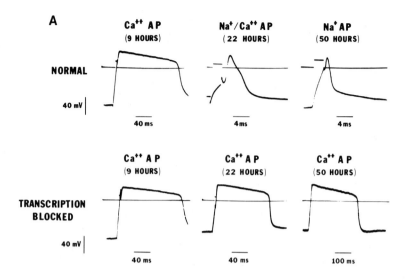
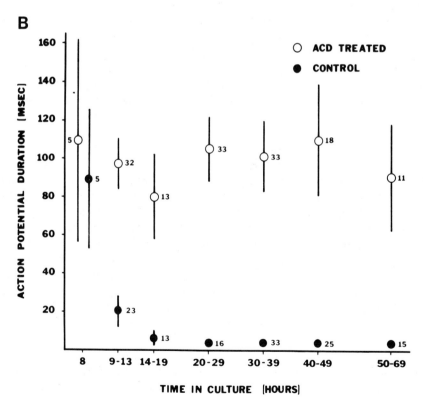

Figure 8. *Differentiation of the mature sodium-dependent action potential in cultured* Xenopus *spinal neurons is prevented by blocking RNA synthesis with actinomycin D (ACD).* A: The action potential is illustrated at three different ages in control (upper) and treated (lower) cultures. B: The duration of the impulse declines with age in control cultures (filled circles) but not in treated cultures (open circles). Each point is the mean ± SEM for the number of cells indicated. (From O'Dowd, 1981, 1983a.)

no detectable effect on the sodium-dependent action potential. Comparison of the results of these two studies suggests that the translation of the relevant proteins is completed roughly three hours after transcription of the necessary RNAs.

CONCLUSIONS

The studies reviewed here expand our knowledge of the mechanisms by which some early phases of neuronal differentiation occur. RNA and protein synthesis are required for the development of the sodium-dependent action potential, and the timing of these requirements has been defined. Attention must now be focused on the identity of molecules synthesized during this time. It will also be of interest to discover if there is a role for posttranslational modification (e.g., glycosylation) in the differentiation of the phenotype. Similar studies of the mechanisms underlying the expression of neurotransmitter receptors will begin to define the basis of differentiation of this form of neuronal excitability.

The mechanisms underlying the control of sequencing in the developmental timetable remain obscure. In a limited sense, control by phenotypic expression has been rendered less likely, because blockade of calcium- and sodium-dependent action potentials has little effect on several aspects of early neuronal differentiation. Neurite outgrowth, development of the action potential mechanism, the onset of neurotransmitter sensitivity, and formation of functional synaptic contacts are not prevented by this treatment. However, the maturation of neuronal sensitivity to neurotransmitters is reduced, and the basis of this effect merits examination. The effect on other, later-arising phenotypes remains to be assessed. Some insight into the flexibility of the developmental program may be obtained by determining if sensitive periods for RNA and protein synthesis are, in fact, critical periods. If a reversible inhibitor is removed at the end of a sensitive period and synthesis is resumed, will a phenotype (e.g., the sodium-dependent action potential) be expressed, after some delay, or will it be permanently absent? Finally, when they ultimately become available, more specific probes that functionally inactivate particular messenger RNAs or the resulting proteins may allow experiments that will determine the molecular level at which control of the developmental timetable is expressed.

ACKNOWLEDGMENTS

I thank Leslie Henderson and Janet Lamborghini for their critical comments on the manuscript. My research is currently supported by the NIH (NS15918).

REFERENCES

Anderson, M. J., and M. W. Cohen (1977) Nerve-induced and spontaneous redistribution of acetylcholine receptors on cultured muscle cells. *J. Physiol. (Lond.)* **268**:757–773.

Anderson, M. J., M. W. Cohen, and E. Zorytcha (1977) Effects of innervation on the distribution of acetylcholine receptors on cultured muscle cells. *J. Physiol. (Lond.)* **268**:731–756.

Anderson, M. J., Y. Kidokoro, and R. Gruener (1979) Correlation between acetylcholine receptor localization and spontaneous synaptic potentials in cultures of nerve and muscle. *Brain Res.* **166**:185–190.

Baccaglini, P. I. (1978) Action potentials of embryonic dorsal root ganglion neurones in *Xenopus* tadpoles. *J. Physiol. (Lond.)* **283**:585–604.

Baccaglini, P. I., and N. C. Spitzer (1977) Developmental changes in the inward current of the action potential of Rohon-Beard neurones. *J. Physiol. (Lond.)* **271**:93–117.

Bergey, G. K., S. C. Fitzgerald, B. K. Schrier, and P. G. Nelson (1981) Neuronal maturation in mammalian cell culture is dependent on spontaneous electrical activity. *Brain Res.* **207**:49–58.

Bernard, C. (1976) Establishment of ionic permeabilities of the myocardial membrane during embryonic development of the rat. In *Developmental and Physiological Correlates of Cardiac Muscle*, M. Lieberman and T. Sano, eds., pp. 169–184, Raven, New York.

Bernard, C., and Y. M. Gargouil (1970) Aquisitions successive, chez l'embryon du rat, des perméabilités spécifiques de la membrane myocardique. *C.R. Seances Acad. Sci. (Paris) (Series D)* **270**:1495–1498.

Bernard, C., G. Raymond, D. Gros, and Y. M. Gargouil (1969) Réponses électriques et contractions initiales du myocard embryonnaire du rat: Présence d'un canal cinétique lent. *J. Physiol. (Paris) (Suppl. 2)* **61**:216.

Bixby, J. L., and N. C. Spitzer (1982) The appearance and development of chemosensitivity in Rohon-Beard neurones of the *Xenopus* spinal cord. *J. Physiol. (Lond.)* **330**:513–536.

Bixby, J. L., and N. C. Spitzer (1984a) The appearance and development of neurotransmitter sensitivity in *Xenopus* embryonic spinal neurones *in vitro*. *J. Physiol. (Lond.)* **353**:143–155.

Bixby, J. L., and N. C. Spitzer (1984b) Early differentiation of vertebrate spinal neurons in the absence of voltage-dependent Ca^{++} and Na^+ influx. *Dev. Biol.* (in press).

Blackshaw, S., and A. Warner (1976) Onset of acetylcholine sensitivity and end-plate activity in developing myotome muscles of *Xenopus*. *Nature* **262**:217–218.

Blair, L. A. C. (1983) The timing of protein synthesis required for the development of the sodium action potential in embryonic spinal neurons. *J. Neurosci.* **3**:1430–1436.

Bloom, E. M., and I. B. Black (1979) Metabolic requirements for differentiation of embryonic sympathetic ganglia cultured in the absence of exogenous nerve growth factor. *Dev. Biol.* **68**:568–578.

Burstein, D. E., and L. A. Greene (1978) Evidence for RNA synthesis-dependent and -independent pathways in stimulation of neurite outgrowth by nerve growth factor. *Proc. Natl. Acad. Sci. USA* **76**:6059–6063.

Cohen, M. W. (1972) The development of neuromuscular connexions in the presence of D-tubocurarine. *Brain Res.* **41**:457–463.

Cohen, M. W., and R. W. Kullberg (1974) Temporal relationship between innervation and appearance of acetylcholine receptors in embryonic amphibian muscle. *Proc. Can. Fed. Biol. Soc.* **17**:176.

DeHaan, R. L., T. F. McDonald, and H. G. Sachs (1976) Development of tetrodotoxin sensitivity of embryonic chick heart cells *in vitro*. In *Developmental and Physiological Correlates of Cardiac Muscle*, M. Lieberman and T. Sano, eds., pp. 155–168, Raven, New York.

del Castillo, J., and B. Katz (1954) Quantal components of the end-plate potential. *J. Physiol. (Lond.)* **124**:560–573.

Donady, J. J., R. L. Seecof, and S. Dewhurst (1975) Actinomycin D-sensitive periods in the differentiation of *Drosophila* neurons and muscle cells *in vitro*. *Differentiation* **4**:9–14.

Fambrough, D., and J. E. Rash (1971) Development of acetylcholine sensitivity during myogenesis. *Dev. Biol.* **26**:55–68.

Fischbach, G. D. (1972) Synapse formation between dissociated nerve and muscle cells in low density cell cultures. *Dev. Biol.* **28**:407–429.

Fukuda, J., and M. Kameyama (1979) Enhancement of Ca spikes in nerve cells of adult mammals during neurite growth in tissue culture. *Nature* **279**:546–548.

Goodman, C. S., and N. C. Spitzer (1979) Embryonic development of identified neurones: Differentiation from neuroblast to neurone. *Nature* **280**:208–214.

Goodman, C. S., and N. C. Spitzer (1980) Embryonic development of neurotransmitter receptors in grasshoppers. In *Receptors for Neurotransmitters, Hormones, and Pheromones in Insects*, D. B. Sattelle and L. M. Hall, eds., pp. 195–207, Elsevier/North-Holland, Amsterdam.

Goodman, C. S., and N. C. Spitzer (1981) The development of electrical properties of identified neurones in grasshopper embryos. *J. Physiol. (Lond.)* **313**:385–403.

Gruener, R., and Y. Kidokoro (1982) Acetylcholine sensitivity of innervated and noninnervated *Xenopus* muscle cells in culture. *Dev. Biol.* **91**:86–92.

Harris, J. B., M. W. Marshall, and P. Wilson (1973) A physiological study of chick myotubes grown in tissue culture. *J. Physiol. (Lond.)* **229**:751–766.

Henderson, L. P., M. A. Smith, and N. C. Spitzer (1984) The absence of calcium blocks impulse-evoked release of acetylcholine but not *de novo* formation of functional neuromuscular synaptic contacts in culture. *J. Neurosci.* (in press).

Hume, R. I., L. W. Role, and G. D. Fischbach (1983) Acetylcholine release from growth cones detected with patches of acetylcholine receptor-rich membranes. *Nature* **305**:632–634.

Kano, M. (1975) Development of excitability in embryonic chick skeletal muscle cells. *J. Cell Physiol.* **86**:503–510.

Kano, M., and Y. Shimada (1973) Tetrodotoxin-resistant electric activity in chick skeletal muscle cells differentiated *in vitro*. *J. Cell Physiol.* **81**:85–90.

Kano, M., and M. Yamamoto (1977) Development of spike potentials in skeletal muscle cells differentiated *in vitro*. *J. Cell Physiol.* **90**:439–444.

Kano, M., Y. Shimada, and K. Ishikawa (1972) Electrogenesis of embryonic chick skeletal muscle cells differentiated *in vitro*. *J. Cell Physiol.* **79**:363–366.

Kidokoro, Y. (1973) Development of action potentials in the clonal rat skeletal muscle cell line. *Nature (New Biol.)* **241**:158–159.

Kidokoro, Y. (1975) Sodium and calcium components of the action potential in a developing skeletal muscle cell line. *J. Physiol. (Lond.)* **244**:145–159.

Kidokoro, Y., and E. Yeh (1982) Initial synaptic transmission at growth cones in *Xenopus* nerve–muscle cultures. *Proc. Natl. Acad. Sci. USA* **79**:6727–6731.

Kidokoro, Y., M. J. Anderson, and R. Gruener (1980) Changes in synaptic potential properties during acetylcholine receptor accumulation and neurospecific interactions in *Xenopus* nerve–muscle cell culture. *Dev. Biol.* **78**:464–483.

Lamborghini, J. E. (1980) Rohon-Beard cells and other large neurons in *Xenopus* embryos originate during gastrulation. *J. Comp. Neurol.* **189**:323–333.

Lamborghini, J. E. (1981) Kinetics of Rohon-Beard neuron disappearance in *Xenopus laevis*. *Soc. Neurosci. Abstr.* **7**:291.

Lamborghini, J. E., M. Revenaugh, and N. C. Spitzer (1979) Ultrastructural development of Rohon-Beard neurons: Loss of intramitochondrial granules parallels loss of calcium action potentials. *J. Comp. Neurol.* **183**:741–752.

Letourneau, P. C., and N. K. Wessells (1974) Migratory cell locomotion versus nerve axon elongation. Differences based on the effects of lanthanum ion. *J. Cell Biol.* **61**:56–69.

Matsuda, Y., S. Yoshida, and T. Yonezawa (1978) Tetrodotoxin sensitivity and Ca component of action potentials of mouse dorsal root ganglion cells cultured *in vitro*. *Brain Res.* **154**:69–82.

McDonald, T. F., and H. G. Sachs (1975) Electrical activity in embryonic heart cell aggregates. Developmental aspects. *Pflugers. Arch.* **354**:151–164.

Meiri, H., M. E. Spira, and I. Parnas (1981) Membrane conductance and action potential of a regenerating axonal tip. *Science* **211**:709–712.

Miyake, M. (1978) The development of action potential mechanism in a mouse neuronal cell line *in vitro*. *Brain Res.* **143**:349–354.

Miyazaki, S., K. Takahashi, and K. Tsuda (1972) Calcium and sodium contributions to regenerative responses in the embryonic excitable cell membrane. *Science* **176**:1441–1443.

Miyazaki, S., K. Takahashi, K. Tsuda, and M. Yoshii (1974) Analysis of nonlinearity observed in the current-voltage relation of the tunicate embryo. *J. Physiol. (Lond.)* **238**:55–77.

Mori-Okamoto, J., H. Ashida, E. Maru, and J. Tatsuno (1983) The development of action potentials in cultures of explanted cortical neurons from chick embryos. *Dev. Biol.* **97**:408–416.

Nicoll, R. A., A. Padjen, and J. L. Barker (1976) Analysis of amino acid responses on frog motoneurones. *Neuropharmacology* **15**:45–53.

Nieuwkoop, P. D., and J. Faber (1956) *Normal table of* Xenopus laevis *(Daudin)*, North-Holland, Amsterdam.

O'Dowd, D. K. (1981) The timing of RNA synthesis necessary for the development of the Na^+-dependent action potential in cultured neurons. *Soc. Neurosci. Abstr.* **7**:245.

O'Dowd, D. K. (1983a) RNA synthesis dependence of action potential development in spinal cord neurons. *Nature* **303**:619–621.

O'Dowd, D. K. (1983b) Development of voltage-dependent conductances in cultured amphibian spinal neurons. *Soc. Neurosci. Abstr.* **9**:506.

Ohmori, H., and S. Sasaki (1977) Development of neuromuscular transmission in a larval tunicate. *J. Physiol. (Lond.)* **269**:221–254.

Oppenheim, R. W. (1981) Neuronal cell death and some related regressive phenomena during neurogenesis: A selective historical review and progress report. In *Studies in Developmental Neurobiology*, W. M. Cowan, ed., pp. 74–133, Oxford Univ. Press, New York.

Pappano, A. J. (1972a) Action potentials in chick atria: Increased susceptibility to blockage by tetrodotoxin during embryonic development. *Circ. Res.* **31**:379–388.

Pappano, A. J. (1972b) Sodium-dependent depolarization of noninnervated embryonic chick heart by acetylcholine. *J. Pharmacol. Exp. Ther.* **180**:340–350.

Ritchie, A. K., and D. M. Fambrough (1975) Ionic properties of the acetylcholine receptor in cultured rat myotubes. *J. Gen. Physiol.* **65**:751–767.

Seecof, R. L. (1977) A genetic approach to the study of neurogenesis and myogenesis. *Am. Zool.* **17**:577–584.

Sperelakis, N., and K. Shigenobu (1972) Changes in membrane properties of chick embryonic hearts during development. *J. Gen. Physiol.* **60**:430–453.

Spitzer, N. C. (1979) Ion channels in development. *Annu. Rev. Neurosci.* **2**:363–397.

Spitzer, N. C. (1981) Development of membrane properties in vertebrates. *Trends Neurosci.* **4**:169–172.

Spitzer, N. C., and P. I. Baccaglini (1976) Development of the action potential in embryonic amphibian neurons *in vivo*. *Brain Res.* **107**:610–617.

Spitzer, N. C., and J. E. Lamborghini (1976) The development of the action potential mechanism of amphibian neurons isolated in culture. *Proc. Natl. Acad. Sci. USA* **73**:1641–1645.

Spitzer, N. C., and J. E. Lamborghini (1981) Programs of early neuronal development. In *Studies in Developmental Neurobiology*, W. M. Cowan, ed., pp. 261–287, Oxford Univ. Press, New York.

Steinbach, J. H. (1975) Acetylcholine responses in clonal myogenic cells *in vitro*. *J. Physiol. (Lond.)* **247**:393–405.

Strichartz, G., R. Small, C. Nicholson, K. H. Pfenninger, and R. Llinas (1980) Ionic mechanisms for impulse propagation in growing nonmyelinated axons: Saxitoxin binding and electrophysiology. *Soc. Neurosci. Abstr.* **6**:660.

Takahashi, K., S. Miyazaki, and Y. Kidokoro (1971) Development of excitability in embryonic muscle cell membranes in certain tunicates. *Science* **171**:415–418.

Taylor, J. S. H., and A. Roberts (1983) The early development of the primary sensory neurones in an amphibian embryo: A scanning electron microscope study. *J. Embryol. Exp. Morphol.* **75**:49–66.

Thompson, W., D. P. Kuffler, and J. K. S. Jansen (1979) The effect of prolonged reversible block of nerve impulses on the elimination of polyneuronal innervation of newborn rat skeletal muscle fibers. *Neuroscience* **4**:271–281.

Young, S. H., and M.-M. Poo (1983) Spontaneous release of transmitter from growth cones of embryonic neurones. *Nature* **305**:634–637.

Walicke, P. A., and P. H. Patterson (1981) On the role of Ca^{++} in the transmitter choice made by cultured sympathetic neurons. *J. Neurosci.* **1**:343–350.

Willard, A. L. (1980) Electrical excitability of outgrowing neurites of embryonic neurones in cultures of dissociated neural plate of *Xenopus laevis*. *J. Physiol. (Lond.)* **301**:115–128.

Section 2

Glial Development and Neuronal Guidance

The role of glial cells in the development and maintenance of the nervous system has yet to be fully defined. There is no doubt about their contribution to neural morphogenesis or their origins in precursor cells that can give rise to various lineages. In this section are considered some of the issues related to the origin of glia, the interaction of neurons and glia particularly with respect to mitogenic signals, and the relative roles of neurons and glia in neuronal migration.

Federoff has used a variety of techniques ranging from classical microscope methods to tissue culture and immunocytochemistry to analyze glial lineages. By the latter means, he has identified epithelial glioblasts that give rise to proastroblasts leading to astroblasts that in turn engender fibrous stellate astrocytes. Similarly precursors to oligodendroblasts have been isolated in culture; these cells form oligodendrocytes that mature in three distinct maturational stages. Perhaps the most revealing observation is of intermediate forms between astrocytes and oligodendrocytes. This may mean that lineages are not fixed hierarchies but have some plasticity.

Glia are subject to mitogenic signals, the nature of which remains to be fully defined. Brockes and his coworkers have isolated and purified a basic protein, glial growth factor or GGF, that is found in the caudate nucleus and the pituitary gland and promotes mitogenesis in fibroblasts, astrocytes, and Schwann cells.

This factor is present in the regenerating blastema of newt limb buds that depend upon a nerve supply for their division, and it decreases in amount following denervation. Using a monoclonal antibody, these workers have identified a blastemal cell population that depends upon nerve for its division. These cells appear to derive from dedifferentiated Schwann cells and myofibers. Inasmuch as Schwann cells depend on GGF for division, the authors are exploring the possibility that the blastemal derivatives also depend on this signal. Further analysis of the relation between mitogenesis and differentiation states should have a major impact on understanding the role of these important primary processes in morphogenesis.

The role of a third process, cell migration, is now beyond question a major contributor to neural form and the establishment of orderly maps and structures such as those of the cerebellum. Rakic and his colleagues have made important observations on the role of guide glia in such processes. In his present contribution, Rakic shows that both neuron–neuron and neuron–glia modes of migration can be adopted by migratory cell processes. Horizontal neurites of granule cells that follow parallel fibers and ignore glia have specific kinds of lamellipodia. Descending vertical processes of the same cells follow glial surfaces, ignore axons, and lack lamellipodia. Rakic proposes a model of deposition of membrane fragments with specific binding properties at the tip of the leading process. The bifurcation of activities and specificities implied here is consistent with that seen in the distinction between N-CAM and Ng-CAM made by Edelman in his chapter. Should a differential distribution of these molecules be observed, this would represent a signal correlation between molecular dynamics and the regulation of the movement of cellular structures.

Chapter 5

Macroglial Cell Lineages

SERGEY FEDOROFF

ABSTRACT
Early studies of neural cell lineages used silver impregnation methods and light microscopy. Later, electron microscopy, together with tritiated thymidine labeling, and more recently, tissue culture and neuroimmunocytochemistry, brought new insights. Stem cells for neural cell lineages have not yet been unequivocally identified, but we do know that cells of the astroglial lineage are already present at the time of formation of the neural tube. The earliest cells from the astroglial lineage isolated so far in tissue culture are glioblasts. They are epithelial cells and give rise to proastroblasts, which contain vimentin intermediate filaments in their cytoplasm. Proastroblasts form astroblasts, identifiable by the presence of two types of intermediate filaments containing vimentin and glial fibrillary acidic protein. Astroblasts form fibrous astrocytes that have a distinctive stellate shape. The astroglial lineage progresses steadily throughout embryogenesis and into postnatal life.

Oligodendroglial precursor cells are also present in the early stages of neurogenesis. The earliest cells in the lineage that have been isolated in tissue culture so far are small, dark, dividing cells that form oligodendroblasts containing galactocerebroside. Oligodendroblasts form oligodendrocytes, identifiable by the presence of galactocerebroside and myelin basic protein. Postmitotic oligodendrocytes undergo three distinct stages of maturation, forming light, medium, and finally, dark oligodendrocytes. The oligodendroglial lineage seems to be biphasic; some oligodendrocytes are produced early in embryogenesis, followed by a quiescent period and then a second stage of active production early in postnatal life.

Intermediate forms between astrocytes and oligodendrocytes have been observed, indicating that the lineages may have more plasticity than classical hierarchical concepts of lineages encompass.

In this chapter I discuss the lineages of astrocytes and oligodendrocytes and briefly describe the cells intermediate between them. The reader is referred to recent excellent reviews on glial cells and their genesis by Niessing et al. (1980), Skoff (1980), Polak et al. (1982a), and Sturrock (1982). Microglial cells and their origin have been reviewed by Polak et al. (1982b) and will not be dealt with here.

Fundamental to the concept of a cell lineage is the arrangement of cells from various stages of differentiation in a sequence based on morphological, functional, and genetic criteria, the ultimate goal being the expression of the lineage in genetic terms. Assembly of such a sequence of cells is obviously based on the premise that a definite hierarchy underlies the lineage. This is a useful premise at the present stage of knowledge, and it will prevail for purposes of discussion in this chapter, even though cell lineages in nature may be much more flexible and their hierarchies less well defined. Moreover, it must be emphasized that tracing a cell lineage does not imply that each step of the lineage is represented by only one type of cell. On the contrary, a step may consist of a number of cellular subcompartments, perhaps varying in the size of their population, but linked by an ordered sequence of differentiation events. Metcalf and Moore (1971), considering lineages of hemopoietic cells, pointed out that "Diagrams [i.e., charts of cell lineages] must be viewed realistically as representing only an outline of the probable process and much more work will be required to firmly establish whether subpopulations exist within the various cell classes and what additional interrelationships exist between the dividing cell classes during differentiation." These provisos should be kept in mind when reading this chapter.

In constructing a cell lineage, it is usual to begin with a stem cell, followed by a sequence of progenitor cells, and to finish with end cells, that is, nondividing cells that exhibit the phenotype consistent with differentiated cells at the end of the lineage. The most difficult aspect is to identify cells in the early stages of the lineage and to determine their differentiation potential, because the younger the cells, the more alike they are.

A century ago, His (1888, 1889) proposed that primitive neuroepithelium in the germinal zone consists of germinal cells, which are stem cells for the neuronal cell lineage, and columnar cells, which give rise to bipolar primitive spongioblasts, the stem cells for the macroglial cell lineages. This implies that primitive neuroepithelium is not homogeneous and that glial and neuronal cell lineages begin to develop concomitantly early in neurogenesis. His's views were supported by Ramón y Cajal (1906) and Kershman (1938). However, Schaper (1894, 1897), and later, Sauer (1935a,b) argued that differences in the cells represented stages of dividing cells of the same type, and that therefore the primitive neuroepithelium is composed of like cells. Subsequently, investigators who used tritiated thymidine ([^{3}H]thymidine) labeling (Sidman et al., 1959; Fujita, 1963) and extensive study of the ultrastructure of the cells (Smart and Leblond, 1961; Fujita, 1963, 1965; Caley and Maxwell, 1968a,b; Stensaas and Stensaas, 1968; Ling and Leblond, 1973) did not negate the proposals of Schaper. Thus, contrary to His's proposal, the germinal zone was considered to comprise a homogeneous cell population (Fujita, 1963, 1966, 1980; Jacobson, 1978), and in 1970 the Boulder Committee proposed a terminology to supersede the terms originally proposed by His. According to the Boulder terminology, His's germinal and spongioblast cells are called ventricular cells, that is, the earliest progenitors of all neuronal and macroglial cells of the central nervous system. The Boulder Committee (1970) noted, however, that they did not exclude the possibility that ventricular cells may include a number of clones. The "subventricular cells" that form from ventricular cells give rise to precursors of the neurons and macroglia of the central nervous system, with the possible exception of ependymal cells. Proliferating cells that come to possess the morphological properties of macroglial cells are called glioblasts. Some of their daughter cells, those that begin to express astrocytic phenotypes, are called astroblasts.

Early work on the differentiation of neural cells was based on the use of classical histological procedures and saw the development of specific stains: the gold sublimate method for astrocytes (Ramón y Cajal, 1913a) and the silver carbonate method for microglia and oligodendroglia (del Rio Hortega, 1919a). Later, the application of [^{3}H]thymidine labeling, followed by autoradiography at light- (Sidman, 1970) and electron-microscope (Mori and Leblond, 1969, 1970) levels, provided a wealth of information about the timing of the last mitoses of neuronal cells and the patterns of cell migration. Unfortunately, this procedure did little to elucidate glial cell lineages because the cells undergo a fairly large number of divisions before entering the terminal stages of differentiation, and some cells continue to divide throughout life. Study of the morphology of glial cell nuclei in conjunction with [^{3}H]thymidine labeling and electron microscopy, however, did lead to identification of sequences in cell lineages, for example, the oligodendroglial lineage during postnatal development of the brain (Imamoto et al., 1978).

The application of immunocytochemical methods and the discovery of a number of immunological markers was a major breakthrough in the identification of glial cells during their differentiation (Raff et al., 1979; Bignami et al., 1980; Schachner et al., 1983). Identification of cells during the earliest stages of gliogenesis, however, remained a perplexing problem. The cells look alike and immunological markers specific to differentiated or differentiating cells later in the lineage have not yet been expressed.

Tissue culture methods, on the other hand, provide a means of studying neural precursor cells during the earliest stages of neurogenesis. The advantage is that individual undifferentiated cells can be observed over a long period of time until their nature is revealed. Such studies facilitate definition of the stages of differentiation a cell must undergo as it progresses through its lineage. Now it is also possible to isolate cells at a specific stage of a lineage by using cell separation techniques and to propagate cells in large numbers so that biochemical, physiological, and pharmacological studies become feasible (McCarthy and de Vellis, 1980; Doering and Fedoroff, 1982, 1984).

ORIGIN OF MACROGLIA

We do not know precisely when neuroectodermal cells become destined to be neuronal or glial precursor cells. Some neurons are determined in the neural plate, during the late gastrula stage, before the neural tube is formed. In *Xenopus laevis*, for example, these include Mauthner's interneurons, the Rohon-Beard cells (sensory neurons in the spinal cord), a few neurons in cranial nerve ganglia, and some large ventral neurons of the brain stem (Vargas-Lizardi and Lyser, 1974; Spitzer and Spitzer, 1975; see also Warner, this volume). In the chick, such cells include neurons of the medullary reticular formation (McConnell and Sechrist, 1980) and, in the mouse, some brain stem neurons (McConnell, 1981). Buse (1983) cultured single cells from mouse neural plate and showed that they are already destined to form typical neurons.

Thus, at the time the neural tube is formed, a small number of postmitotic cells are already present. By using [^{3}H]thymidine labeling and determining the total number of cells, we calculated that in the neural tube of chick embryos at stage 10 of Hamburger and Hamilton (H&H; 1951), which corresponds to embryonic day 1½ (E1½), 3% of the cells are postmitotic neuronal precursor cells (Fedoroff et al., 1982). Explants from the neural tube of mice at stage 14 of Theiler (1972), which corresponds to E9, give rise in culture to neurons of which 38% are postmitotic, indicating that they had completed their DNA replication before we explanted them (Juurlink and Fedoroff, 1982). On the other hand, at later developmental stages, some cells, though still proliferating, are already destined for a particular lineage. For example, in cultures of disaggregated cells of mouse neopallium at Theiler stage 23 (E15), small, round, proliferating cells occur. They are destined to become neurons, as can be demonstrated by transplanting them into the cerebellum of newborn mice (Doering and Fedoroff, 1982).

It is much more difficult to determine the time of origin of macroglia. [^{3}H]thymidine labeling is not very useful because glial cells proliferate profusely. Immunocytochemical methods and tissue culture are more effective. I now review a number of somewhat unrelated observations that, in combination, may give some clues as to when macroglial cells begin to appear.

The optic nerve contains no neurons, only glial cells, and therefore has been used extensively to study gliogenesis. It develops from the optic stalk and is composed of cells morphologically indistinguishable from the ventricular cells found in other parts of the developing central nervous system (Skoff et al., 1976). In culture, cells of the optic stalk of mice at Theiler stages 16–19 (E10–11) give rise to neurons and to flat cells that are glial precursor cells; however, beginning with Theiler stage 20 (E12), the cells form only glial precursor cells (Juurlink and Fedoroff, 1980). This change is associated with a change in the morphology of ventricular cells from pseudostratified to simple cuboid epithelium, with rearrangement of cells to allow elongation of the optic stalk, and with the migration of axons of the neural retina along the marginal layer of the optic stalk (Juurlink and Fedoroff, 1980). The ventricular cells of the optic stalk thus have the potential to form both neurons and glial cells until they are contacted by axons of the neural retina. Environmental cues, therefore, seem to restrict the differentiation capabilities of ventricular cells of the optic stalk so that only glial cells form. Because no extensive cell necrosis in the ventricular cell population of the optic stalk has been observed (Juurlink and Fedoroff, 1980), it seems unlikely that the restriction of ventricular cells is due to selective cell death of a subpopulation of ventricular cells already committed to a neuronal lineage of differentiation.

In the neural tube of mouse embryos at Theiler stages 14–18 (E9–11), several filamentous bundles extend from the ventricular surface to the external limiting membrane. These bundles react positively for vimentin, a protein of intermediate filaments. The bundles have greater immunolabeling at the pial surface where the processes expand (Schnitzer et al., 1981; Houle and Fedoroff, 1983). Because of the pattern of distribution, we believe that the vimentin-positive fibrous bundles belong to processes of the radial glia precursor cells, which are in the astrocyte cell lineage and are the earliest indicators of glial cell formation detected so far.

Disaggregated cells from the neural tube of mouse embryos at Theiler stage 16 (E10) form epithelial-type colonies in cultures. Disaggregated cells from the neural tube of chicks at H&H stage 17 (E2½) also form epithelial-type colonies. We know that these epithelial cells are astrocyte precursors because when we followed them by time-lapse cinematography or when we isolated them from mouse neural tube cultures on a metrizamide density gradient and then transplanted them into newborn-mouse cerebellum, they formed mature astrocytes (Doering et al., 1983). The morphology of astrocytes in the transplants and in cultures resembles that of normal astrocyte *in vivo* (Doering et al., 1983), and they have vimentin and glial fibrillary acidic protein (GFAP) intermediate filaments (Fedoroff et al., 1983).

Based on observations made on the development of glial cells in the optic stalk (Juurlink and Fedoroff, 1980) and the development of radial glia (Levitt and Rakic, 1980; Choi, 1981; Levitt et al., 1981, 1983) and astrocyte precursor cells (Schnitzer et al., 1981; Juurlink and Fedoroff, 1982; Houle and Fedoroff, 1983; Fedoroff et al., 1983, 1984a) in tissue cultures of neural tube (Fedoroff and Doering, 1980; Juurlink and Fedoroff, 1980; Doering and Fedoroff, 1982), it is evident that divergence of the neuronal and astroglial cell lineages occurs very early in neural tube development and that differentiation proceeds concomitantly and interdependently (Rakic, 1981). Furthermore, it is evident that cells in the ventricular zone are not homogeneous. We still do not know, however, whether some of the ventricular cells are pluripotential and able to give rise to a number of cell lineages or are already restricted in potential and committed to a specific lineage.

ASTROGLIAL CELL LINEAGE

Astroglial Cell Lineage *In Vivo*

Studies of gliogenesis in the corpus callosum and optic nerve of postnatal animals by light and electron microscopy and [^{3}H]thymidine labeling with autoradiography coupled with electron microscopy have led to the proposal of a number of schemes for neural cell lineages (Smart, 1961; Lewis, 1968; Vaughn, 1969; Blakemore and Jolly, 1972; Ling et al., 1973; Paterson et al., 1973; Skoff et al., 1976; Sturrock, 1976; Imamoto et al., 1978). All the schemes trace the development of astroblasts (mitotically active astrocyte precursor cells) from glioblasts that are considered immature cells, and, by some, even pluripotential cells that can give rise to oligodendroblasts as well as astroblasts (Vaughn, 1969; Skoff et al., 1976; Sturrock, 1976; Imamoto et al., 1978). An exception probably occurs in the optic stalk where, according to Skoff et al. (1976), astroblasts may arise directly from ventricular cells. Definition of relationships between glioblasts and ventricular or subventricular cells, although indicated in some schemes, is limited by the methods used in these investigations. These schemes imply that ventricular cells comprise a homogeneous population of the earliest precursors for all neural cells. More recent work related to the formation of radial glia and to tissue cultures of cells from ventricular and subventricular zones in particular indicates that astrocyte precursor cells, glioblasts, and radial glia are present in the ventricular and subventricular zones (Fedoroff and Doering, 1980; Juurlink and Fedoroff, 1980; Choi, 1981; Jurrlink et al., 1981; Levitt et al., 1981).

The work on postnatal gliogenesis indicates that by using electron microscopy coupled with [^{3}H]thymidine autoradiography, astrocytes can be distinguished from oligodendrocytes on the basis of morphological criteria. The degree of immaturity of the cells can be judged, up to a point, by considering whether the cells are mitotically active, and by the composition of organelles and organization of the cytoplasm. For detailed reviews of gliogenesis and identification of cells, the reader is referred to recent reports by Peters et al. (1976), Skoff (1980), Polak et al. (1982a), and Sidman and Rakic (1982).

Astroglial Cell Lineage in Culture

In the last few years, the study of the origin of astrocytes took a leap forward with the discovery that GFAP, a component of intermediate filaments, is found exclusively in astrocytes and some related cells in the central nervous system (Bignami et al., 1980; Eng and DeArmond, 1982; Fedoroff et al., 1984b). In addition, tissue culture methods have been devised that facilitate the isolation of astrocytes and their precursor cells and their propagation in culture. This permits the continuous observation of the stepwise process in their differentiation. A review of this work follows.

In vivo, astrocytes develop continuously throughout the embryonic and early postnatal periods (Smart and Leblond, 1961; Imamoto et al., 1978; Skoff, 1980; Sturrock and Smart, 1980; Schultze and Korr, 1981; Paterson, 1983; Korr et al., 1983). A fragment of nervous tissue taken from a fetus or newborn animal, therefore, contains proliferative astrocyte precursor cells at various stages of differentiation. When disaggregated cells from the fragments are planted in cultures in small numbers, the astrocyte precursor cells attach to the substratum, continue to divide, and form colonies. Many such colonies are true clones. Because the cells continue their programmed sequence of differentiation, various types of colonies form, each with a distinct morphology depending on the stage of differentiation of its cells at the time of explantation (Fedoroff, 1978; Fedoroff and Doering, 1980; Fedoroff et al., 1983, 1984a).

In our laboratory, we have been able to follow the astroglial cell lineage in colony cultures all the way from glioblasts isolated from ventricular and subventricular zones (Fedoroff and Doering, 1980; Juurlink et al., 1981), to differentiated astrocytes (Fedoroff et al., 1983, 1984a). Such cultures are optically ideal for studying changes in the organization of the cytoskeleton during astrocyte differentiation. This is important because the composition and organization of the cytoskeleton are related to cell function, and the presence of specific cytoskeletal proteins related to cell types (Brinkley, 1982; Rungger-Brändle and Gabbiani, 1983; Fedoroff et al., 1984b) greatly facilitates the identification of the cells. We were able to correlate some of the changes in the cytoskeleton, especially in microfilaments and intermediate filaments, with the progression of astrocytes through the lineage (Fedoroff et al., 1983, 1984a; Fedoroff et al., 1984b; Kalnins et al., 1984), and this led to the proposal of a terminology for the astroglial cell lineage (Figure 1).

In culture, glioblast colonies are composed of closely apposed epithelial cells. Their nuclei are large and euchromatic and have a thin layer of heterochromatin along the nuclear envelope. They have a relatively small amount of cytoplasm, many free ribosomes, and few cytoplasmic organelles (Fedoroff et al., 1983). Despite the epithelial nature of their organization and growth pattern, specialized intercellular junctions between adjacent glioblasts were not observed in electron micrographs. Glioblasts in culture have only small bundles of microfilaments

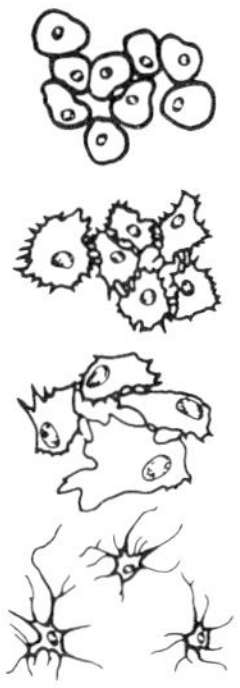

Developmental Stage	Cell Morphology	Microfilaments	Vimentin	GFAP
Glioblasts		0/+	0/+	−
Proastroblasts		++	++	−
Astroblasts		+++	++	+++
Astrocytes		+	++	+++

Figure 1. *Proposed terminology for the astroglial cell lineage.* See text for discussion and definitions.

containing actin. The microfilaments are located along the substratum and also run circumferentially along the cell membrane, outlining each cell. At this stage of development, no intermediate filaments were detected by electron microscopy, and immunocytochemical reactions for GFAP were negative. The glioblasts we observed in culture resemble the glioblasts *in vivo* described by Vaughn (1969) and Sturrock (1976).

The first indication that glioblasts in a colony are beginning to differentiate along the astroglial cell lineage is the commencement of separation of adjacent cells, accomplished by the formation of cytoplasmic projections and interdigitations between neighboring cells. At sites where the cell membranes remain closely apposed to those of neighboring cells, intercellular junctions associated with microfilaments that resemble fascia adherens, appear (Fedoroff et al., 1983). This stage of differentiation is associated with extensive reorganization of the microfilament network. Bundles of microfilaments run perpendicular to the junction and insert into the cell membrane. As the cells move farther apart they are eventually attached to each other only by long slender processes containing a core of microfilaments. Microfilaments are highly concentrated at sites where the processes of adjacent cells meet. The change is also associated with the formation of small, clearly defined focal contacts through which the cytoplasm adheres to the substratum and which can be seen with surface-reflection interference microscopy (Ploem, 1975; Kalnins et al., 1984).

At this stage of differentiation, in addition to the reorganization and expansion of the microfilament network, the cells begin to assemble intermediate filaments that contain vimentin but are negative for GFAP (Fedoroff et al., 1984a). We proposed that cells in the astroglial lineage that have a well-developed actin network and express vimentin intermediate filaments be defined as proastroblasts (Fedoroff et al., 1984b; Figure 1).

Continuous observation of cells in colony cultures by time-lapse cinematography and observation of fixed cells by electron microscopy and immunocytochemistry have indicated that the relatively flat epitheliallike proastroblasts change into cells with more abundant cytoplasm, increased numbers of organelles, and more intermediate filaments, arranged in bundles, which contain both vimentin and

GFAP. Their actin-containing microfilaments and pattern of adhesion foci remain similar to those of proastroblasts. We defined the cells that contain both vimentin and GFAP as astroblasts (Fedoroff et al., 1983, 1984b) to distinguish them from proastroblasts, which contain only vimentin intermediate filaments (Figure 1). The proastroblasts and astroblasts in colony cultures probably correspond to astroblasts (Skoff et al., 1976; Imamoto et al., 1978), intermediate astrocytes (Vaughn, 1969), and young astrocytes (Vaughn, 1969; Sturrock, 1976) *in vivo*.

Abney et al. (1981) described astrocyte precursor cells that were Ran-2^+ (Bartlett et al., 1981) and GFAP$^-$. When they exposed these cells to Ran-2 antibodies and then cultured them for an additional three days before processing for identification of the antibody, they found that many cells having residual Ran-2 antibody were now also GFAP$^+$. Also, on culturing, the Ran-2^+/GFAP$^-$ cells gradually disappeared and Ran-2^+/GFAP$^+$ cells increased; by two weeks, most of the Ran-2^+ cells were also GFAP$^+$. This indicates that Ran-2^+/GFAP$^-$ cells are precursors for Ran-2^+/GFAP$^+$ cells. In our classification, the Ran-2^+/GFAP$^-$ cells are proastroblasts and the Ran-2^+/GFAP$^+$ cells are astroblasts.

In our cultures, the astroblasts form small stellate cells, fibrous astrocytes with long, slender processes that eventually become interconnected, forming an extensive network (Fedoroff et al., 1983). The oval or irregularly shaped nuclei of these cells are situated toward one side of the cell body and contain numerous patches of condensed chromatin. In contrast to proastroblasts and astroblasts, the fibrous astrocytes have relatively little polymerized actin. The only actin-containing microfilaments are found along the plasma membrane of the cell body and its processes. The fibrous astrocytes have no focal contacts, and the cells adhere to the glass substratum only by a mixture of close and far contacts (Kalnins et al., 1984). During the formation of fibrous astrocytes, the elaborate pattern of microfilaments found in proastroblasts and astroblasts is lost. However, fibrous astrocytes still have many microtubules and intermediate filaments throughout their cytoplasm and in their processes. Fibrous astrocytes contain both vimentin and GFAP intermediate filaments. They differ from astroblasts in overall cell morphology: Astroblasts are flat cells that have no processes or only a few relatively short ones; in contrast, fibrous astrocytes have well-defined long processes and distinct perikarya. Both astroblasts and fibrous astrocytes have vimentin and GFAP intermediate filaments (Fedoroff et al., 1983, 1984a), but fibrous astrocytes do not have the elaborate network of actin-containing microfilaments found in astroblasts (Kalnins et al., 1984).

Although the astroglial cell lineage can progress, in culture, in standardized medium all the way from glioblasts to fibrous astrocytes, there are indications that different stages of the lineage may have special nutritional requirements for growth and other requirements for differentiation. For example, the differentiation of glioblasts can be inhibited by a high concentration (20%) of serum in cultures. Once the concentration of serum is reduced, glioblasts form proastroblasts (Fedoroff and Hall, 1979) that do not require any serum in the medium and proliferate in chemically defined medium. However, in order to proceed to the next stage of differentiation, that is, the formation of astroblasts that express GFAP, the cells do require serum (Fisher et al., 1982). When serum is present in the medium, the astroblasts proliferate and mature. Study of the requirements for cell differentiation in culture is a first step in efforts to elucidate regulatory mechanisms of gliogenesis.

Correlation of the Astroglial Cell Lineage in Culture with That *In Vivo*

As the central nervous system develops, cells progress along the lineages, beginning with the most immature precursor cells. It can be assumed that the cell populations of samples taken from the central nervous system at different stages of development differ and that the trend should shift more and more toward the mature end of the lineage. When we obtained disaggregated cells from the neopallia of fetuses of various ages and assayed their composition by the colony culture method, we found that the frequencies of occurrence of colonies varied in a predictable manner according to the age of the cell donor fetus and that the sequence of variation was the same as the sequence of differentiation observed in the culture, that is, glioblasts, proastroblasts, and finally, astroblasts (Fedoroff, 1978, 1980) which gave rise to fibrous astrocytes. This supported our belief that cells in the astrocyte lineage progress through the same sequence *in vivo* as in culture.

It is well known that astroblasts and fibrous astrocytes contain vimentin and GFAP intermediate filaments (Chiu et al., 1981; Dahl et al., 1981; Schnitzer et al., 1981; Shaw et al., 1981; Tapscott et al., 1981; Yen and Fields, 1981; Fedoroff et al., 1983, 1984a). *In vivo*, vimentin-type intermediate filaments appear in the cells of the astrocyte lineage as early as E9 (Houle and Fedoroff, 1983) and GFAP intermediate filaments at E15 (Abney et al., 1981; Fedoroff et al., 1984b). In culture, vimentin-type intermediate filaments are also assembled first, and GFAP-containing intermediate filaments later, toward the final stages of astrocyte differentiation (Dahl et al., 1981; Fedoroff et al., 1981; Schnitzer et al., 1981; Shaw et al., 1981; Tapscott et al., 1981; Yen and Fields, 1981; Fedoroff et al., 1983, 1984a). Abney et al. (1981) prepared cultures from disaggregated cells of 10- and 13-day rat embryo brains and determined when GFAP, Ran-2 antigen (Bartlett et al., 1981), and beating cilia are expressed in astrocyte and ependymal precursor cells. They found that these type-specific cell markers were expressed in cultures of cells from 13-day embryos three days earlier than in cultures of cells from 10-day embryos. Moreover, the first appearance of GFAP, Ran-2 antigen, and cilia in culture was the time at which these markers could be detected in freshly prepared cell suspensions of embryonic or neonatal rat brains. Abney et al. (1981) stated, ". . . that the timing of glial cell development is remarkably similar in dissociated cell cultures and *in vivo*, suggesting that biological clocks are more important than positional information in glial genesis. . . ." They further concluded: "If the positional information (Wolpert, 1969) plays a role in directing the differentiation of glial cell precursors into astrocytes, ependymal cells, and oligodendrocytes, then the information must be acquired before 10 days gestation and be remembered for up to two weeks. Either the precursors of these glial cells have intrinsic biological clocks, or some other cell types in these cultures have the clocks and induce the differentiation of the appropriate glial precursor cells at the correct time."

To further confirm the differentiation potential of astroglial precursor cells at various developmental stages in cultures, we transplanted cell colonies of astroglial precursor cells that originated from E15 or E18 mouse neopallium into the cerebellum of neonatal mice. The cell colonies gave rise to typical fibrous astrocytes. When, using immunocytochemical procedures and morphometry, these were compared with normal fibrous astrocytes of the host brain and those in cultures, all were similar (Doering and Fedoroff, 1982; Fedoroff et al., 1984b).

In response to injury *in vivo*, astrocytes hypertrophy and form large stellate cells known as reactive astrocytes. When astroblasts in culture are treated with dibutyryl cAMP (dBcAMP), they also hypertrophy and form large stellate cells which, by their morphology and by morphometry, correspond closely to reactive astrocytes *in vivo*. The large stellate cells in culture have increased amounts of GFAP-containing intermediate filaments, as do reactive astrocytes *in vivo* (Fedoroff et al., 1983; 1984a,b), a further indication of their similarity.

During later stages of development (e.g., E19–20 in rats), the plasma membrane in astrocytes contains orthogonal arrays of small, uniform, tightly packed particles that can be seen only in freeze-fracture preparations. Landis and Reese (1974) called aggregates of these particles "assemblies." In the central nervous system, these assemblies are found only in astrocytes and ependymal cells and probably in some of their precursor cells. The assemblies increase in number in reactive astrocytes in gliotic scars (Anders and Brightman, 1979, 1982; Landis and Reese, 1981; Brightman et al., 1983). Recently, Landis and Weinstein (1983) examined the assemblies in astrocyte cultures from brains of neonatal rats and concluded that the constituent particles of assemblies appear to be the same in cultures and *in vivo*.

Privat and Leblond (1972) proposed that subependymal cells in postnatal animals can migrate from the subventricular zone. They also defined the migratory cells as free subependymal cells that, in their view, correspond to His's spongioblasts or to glioblasts as defined by the Boulder Committee (1970). Such migrating cells should possess a well-developed cytoskeleton especially adapted for cell motility. We have observed that the glioblasts do not have such a cytoskeleton; however, proastroblasts (vimentin$^+$/GFAP$^-$) do have an extensive actin-containing microfilament network and readily develop adhesion foci and translocate in culture (Kalnins et al., 1984; R. Cleveland and S. Fedoroff, unpublished observations). Therefore, we propose that the migratory cells of the astroglial lineage *in vivo* probably correspond more closely to proastroblasts and astroblasts, which do have the cytoskeleton required for motility, than to glioblasts, which do not. The fibrous astrocytes in culture have a considerably reduced actin-containing microfilament network and are stationary, a cell behavior befitting mature astrocytes *in vivo* considering their function (Fedoroff, 1983, 1984a; Kalnins et al., 1984; R. Cleveland and S. Fedoroff, unpublished observations).

Schachner and her colleagues (1983) developed monoclonal antibodies that bind to C-1 antigen in mice. C-1 antigen is present on GFAP$^-$ cells; it appears very early, at E10 or earlier, and is detectable in astrocytes until postnatal day 10 (P10). Because cells containing C-1 antigen do not have GFAP, they probably correspond to proastroblasts and, possibly, to glioblasts.

Another monoclonal antibody developed by the same group detected M-1 antigen in astrocytes that are GFAP$^+$ in frozen sections of adult mouse cerebellum. Developmentally, the antigen began to appear early postnatally at P7. M-1 antigen is therefore present mainly in mature astrocytes.

From the discussion above, it is evident that the earliest astrocyte precursor cells appear to be strongly programmed genetically, so that when placed in culture conditions, despite the altered environment, they pursue their destined temporal and sequential differentiation.

Relation of Radial Glia to the Astroglial Cell Lineage

During early stages of development, some astrocyte precursor cells that have their soma in the ventricular, subventricular, or the deep part of the intermediate zone have processes extending from the ventricular to the outer surface of the central nervous system. The cell processes lamellate and expand into broad end feet at the pial surface. These cells are radial glia, originally recognized by Ramón y Cajal (1955) and more recently studied extensively by Rakic and his colleagues (Rakic, 1971a,b; Schmechel and Rakic, 1979a,b; Levitt and Rakic, 1980; Levitt et al., 1981, 1983). The radial glia are arranged in such a way that their processes are directed outward and in a radial pattern. They provide guidance for immature neurons that migrate along the radial processes toward the cortical plate. In addition to the neurons, some of the immature glial cells or even more primitive cells may also migrate (Rakic, 1971a).

The earliest cells we have observed that resemble radial glia are vimentin$^+$/GFAP$^-$. They appear in the cervical neural tube of E9 mice (Houle and Fedoroff, 1983) and correspond to proastroblasts. When radial glia begin to express GFAP, they correspond to astroblasts (Fedoroff et al., 1984b). The radial glia are amitotic, but after the neurons, and possibly other cells, have stopped migration, the radial glia cells resume mitotic division and some may retract their processes and differentiate into astrocytes (Choi and Lapham, 1978; Schmechel and Rakic, 1979b; Levitt and Rakic, 1980; Choi, 1981; Choi et al., 1983). Not all cells in the astroglial cell lineage have to go through the stage of radial glia, nor do all radial glia eventually become astrocytes.

The gestational age at which GFAP first appears in radial glia and astroblasts has been reported for a number of species (Bignami and Dahl, 1975; Choi, 1981; Raju et al., 1981; Levitt et al., 1983; Fedoroff et al., 1984b). To find out whether GFAP-containing radial glia and astroblasts appear in various species at the same or different developmental stages, we determined the Carnegie stage of development (Streeter, 1942) for each species at the gestational age when GFAP was first observed. The Carnegie stages ranged from 20 to 23 (Table 1), indicating that GFAP-containing radial glia and astroblasts begin to appear in all species

Table 1. Earliest Detection of Glial Fibrillary Acidic Protein in the Central Nervous System

Species	Site of Appearance	Gestational Age	Carnegie Stages of Development
Human	Spinal cord[a]	8–9 weeks	22
Rhesus monkey	Occipital lobe[b]	E39	20
Mouse	Spinal cord[c]	E15	21
Rat	Spinal cord[d]	E18	23
Chick	Spinal cord[e]	E12	23

[a] Choi, 1981.
[b] Levitt et al., 1983.
[c] Fedoroff et al., 1984b.
[d] Raju et al., 1981.
[e] Bignami and Dahl, 1975.

rather late in embryological development. If one considers that gestational age is only a rough estimation of the developmental stage (Juurlink and Fedoroff, 1979), variation between species probably is even less than our data indicates.

OLIGODENDROGLIAL CELL LINEAGE

Oligodendroglial Cell Lineage *In Vivo*

Del Rio Hortega (1919b), using his silver carbonate technique, demonstrated that Ramón y Cajal's (1913b) so-called third element (i.e., a third cell type in addition to neurons and astrocytes) was itself composed of two types of cells: oligodendrocytes, originating from neuroectodermal cells, and microglial cells, originating from mesodermal cells.

Present information on oligodendroglial genesis *in vivo* is based on morphological studies using electron microscopy combined with [^{3}H]thymidine labeling of cells and subsequent autoradiography. The terminology and stages defined in the oligodendroglial lineage are based on whether or not cells can be labeled with [^{3}H]thymidine, and on their morphology, especially that of their nuclei. For detailed reviews, see Privat (1975), Skoff (1980), and Polak et al. (1982a).

The major development of oligodendroglia occurs during postnatal life and is related roughly to myelination in various parts of the central nervous system. The subventricular layer, composed of immature cells, persists into the postnatal period. Immature cells from this layer migrate into the corpus callosum and other areas of the brain and differentiate into oligodendrocytes (Allen, 1912; Bryans, 1959; Smart, 1961; Lewis, 1968; Blakemore, 1969; Privat and Leblond, 1972; Paterson et al., 1973; Imamoto et al., 1978; Skoff, 1980; Paterson, 1981, 1983; Sturrock, 1982). The proposed sequence of the lineage is as follows: subventricular cells, glioblasts, oligodendroblasts, light oligodendrocytes, medium oligodendrocytes, and finally, dark oligodendrocytes (Imamoto et al., 1978; Skoff, 1980).

The subventricular layer of the lateral ventricles gradually decreases in volume and extent during the postnatal period, but some mitotic activity occurs even in old age (Smart, 1961; Noetzel, 1965; Altman, 1966; Lewis, 1968; Blakemore, 1969; Sturrock, 1979; Hubbard and Hopewell, 1980; Sturrock and Smart, 1980; Paterson, 1983). Hubbard and Hopewell (1980) estimated that in 14-week-old rats about one-third of the cells in the subventricular layer are proliferative. Autoradiographic and electron-microscope evidence indicates that immature cells migrate from the subventricular layer into the corpus callosum and surrounding regions (Lewis, 1968; Privat and Leblond, 1972; Paterson et al., 1973). The morphology of these immature migratory cells is similar in spinal cord, hippocampus, and optic nerve (Skoff, 1980). The immature cells have small nuclei and scanty cytoplasm. Some have an irregularly shaped or larger, rounded nucleus with a visible nucleolus and patches of condensed chromatin. Some have an abundance of ribosomes but few organelles; some have features of astrocytes, whereas others have some oligodendrocyte characteristics (Privat and Leblond, 1972; Skoff et al., 1976; Imamoto et al., 1978; Paterson, 1981). The migratory cells were variously thought to be free subependymal cells (subventricular), glioblasts, or oligodendroblasts (Paterson et al., 1973; Imamoto et al., 1978; Skoff, 1980). Names were given arbitrarily to these cells, but the intent was to indicate that they are immature and still capable of proliferation.

When mitotically active, immature oligodendroglial precursor cells (oligodendroblasts) in the corpus callosum, and presumably elsewhere in the central nervous system, enter postmitotic stages of differentiation, they undergo three morphologically identifiable sequential stages of maturation (Mori and Leblond, 1970; Ling et al., 1973; Ling and Leblond, 1973; Paterson et al., 1973; Privat, 1975; Imamoto et al., 1978). The final division of the oligodendroblasts gives rise to nonproliferative daughter cells, the light oligodendrocytes. These are the largest cells in the series. They have large, pale, centrally located nuclei with large nucleoli and evenly distributed chromatin. The cytoplasm is rich in organelles, contains free ribosomes arranged in rosettes, scanty rough endoplasmic reticulum, and numerous microtubules, suggesting high metabolic activity. The formation of light oligodendrocytes corresponds roughly to the time of rapid myelination, a process in which they probably are involved.

After 4-7 days, the light oligodendrocytes are transformed into medium oligodendrocytes that are smaller and appear to be less active metabolically. The transition from light to medium oligodendrocytes is not abrupt; there are many intermediate cells varying in size and shade. The typical medium oligodendrocyte is denser and smaller, its spherical nucleus has a less distinct nucleolus, and the chromatin is condensed only along the nuclear membrane. The cells have less, but more dense, cytoplasm that contains rough endoplasmic reticulum and has larger and more regularly stacked cisternae and more prominent Golgi apparatus. Microtubules are present in the perikaryon and processes.

About two weeks later, through a number of intermediate stages, the medium oligodendrocytes transform into dark oligodendrocytes that have much denser nuclei and cytoplasm. The nucleus is frequently eccentrically located, the nucleolus is not very prominent, and the chromatin is condensed along the nuclear membrane. The cytoplasm is scanty, the saccules of Golgi are distended, the rough endoplasmic reticulum appears as stacks of cisternae, and lamellar bodies are present. Microtubules are difficult to see. These cells appear to be even less active metabolically than are the medium oligodendrocytes and probably are involved only in the maintenance of the myelin sheath (Imamoto et al., 1978). The dark oligodendrocytes may persist for a long time and are probably replaced at a low turnover rate throughout life (Korr et al., 1983; Paterson, 1983).

Oligodendroglial Cell Lineage in Culture

In culture, disaggregated cells from various regions of brain at early stages of neurogenesis (E12-15) form a layer of flat epithelial cells (glioblasts) and single or clumped small, round, phase-light and phase-dark refractile cells (when observed by phase-contrast microscopy) that usually are located on top of the flat epithelial cells (Fedoroff, 1978; Doering and Fedoroff, 1982, 1984b). When these small cells are separated and planted on plastic- or poly-D-lysine-coated surfaces they do not survive. They prefer astroglial precursor cells as a substratum. The small cells proliferate and reach their highest density after 15-16 days of culturing;

after this period their growth decreases, and after 27 days in culture they are rarely observed (Rioux et al., 1980; Doering and Fedoroff, 1982).

The population of the small cells is not homogeneous. Rioux et al. (1980) identified two types of cells based on their size. They believed both to be oligodendroglia and speculated on their relationship to the described stages in the postmitotic differentiation of oligodendrocytes *in vivo*. Using metrizamide density step-gradient centrifugation, we were able to separate the small cells from the flat epithelial cells and subsequently transplanted them into cerebellum of newborn mice. Such transplants consisted mainly of neurons (Doering and Fedoroff, 1982). However, when we observed cultures by time-lapse cinematography, clumps of the small cells were visible on top of the astroglial precursor cell layer. In these clumps, cells often enlarged and formed neurites that rapidly elongated and grew on top of the astroglial precursor cell layer. Other small cells separated from the clump and migrated along the neurites. We observed that when neurites approached single small cells that were still some distance apart, the small cells were pulled with some force toward neurites, to which they adhered. They then began to-and-fro movement along the neurites. The attraction of the small cells to neurites appeared to be extremely strong (Fedoroff, 1978).

Occasionally in the cultures we saw small cells by themselves in spaces between epithelial cells, and in this situation the cells formed processes (Fedoroff, 1978). Their morphology was that described as characteristic of oligodendrocytes (Lumsden and Pomerat, 1951; Wolfgram and Rose, 1957). We therefore believe that in the cultures initiated from brain at early stages of neurogenesis, the small cells sitting on top of epithelial cells comprise two types: neuronal precursor cells and oligodendroglial precursor cells. This is in accord with the finding that cultures of early embryonic brain contain cells that express 14-3-2 protein (Secchi et al., 1980) and $\gamma\gamma$-enolase (Bologa et al., 1982b), both specific to neurons, as well as cells that express galactocerebroside (GC), a cell surface marker, and myelin basic protein (MBP), both specific to oligodendrocytes.

By using immunological markers and complement-dependent immunocytolysis, precursor cells for oligodendroglia in cultures were identified as GC^-/MBP^- (Bologa et al., 1982a). Abney et al. (1983) used fluorescence-activated cell sorting to isolate oligodendroglial precursor cells from disaggregated cells of newborn-rat corpus callosum, and followed their differentiation in cultures. They found that the oligodendroglial precursor cells were $A4^+$ (a monoclonal antibody that binds to as yet uncharacterized molecules of neurons; Cohen and Selvendran, 1981), were $A2B5^+$ (monoclonal antibodies to polysialoganglioside GQ; Eisenbarth et al., 1979), and bound tetanus toxin. Schachner et al. (1981) developed four monoclonal antibodies, O1, O2, O3, and O4, and showed that oligodendroglial precursor cells are $O3^+$ and $O4^+$. These cells, however, did not have receptors for tetanus toxin. The fact that A4 and A2B5 antigens, and probably receptors for tetanus toxin, are present on neurons as well as oligodendroglial precursor cells implies a common stem cell. The same notion was proposed previously by Fedoroff (1978), based on observation of the high affinity of oligodendroglial precursor cells for neurites.

In seven-day cultures of E14 mouse embryo brain, cells appear that express galactocerebroside and myelin basic protein. *In vivo*, this timing would correspond to that of the newborn mouse. In cultures of brain from newborn animals, $GC^+/$

MBP$^-$ cells appear after seven days and GC$^+$/MBP$^+$ cells appear after 14 days, corresponding to 14 days postnatally *in vivo*. In culture, all cells that were MBP$^+$ were always also GC$^+$. The GC$^+$/MBP$^+$ cells in cultures of newborn (P0) animals were usually larger than GC$^+$/MBP$^-$ cells or GC$^+$/MBP$^+$ cells in cultures of E14 embryonic brain (Bologa-Sandru et al., 1981; Bologa et al., 1982a,b, 1983), indicating some difference in development in the two types of cultures.

It has been suggested that the normal development of oligodendrocytes may depend on the presence of neurons. In E14 embryo cultures, oligodendrocytes develop in the presence of many neurons, but in P0 cultures, oligodendrocytes develop in neuron-free environments (Bologa et al., 1982b). From seven-day cultures of P0 mouse brains, we isolated oligodendrocyte precursor cells (the small round cells) before they expressed GC. When these cells were transplanted into cerebellum of newborn animals, the resulting transplants were composed of typical oligodendroglia that myelinated axons of the host neurons as they sprouted into the transplant (Doering and Fedoroff, 1984). This suggests that oligodendroglial precursor cells in cultures of brain from newborn animals, at least up to seven days of incubation, are fully capable of forming normal oligodendrocytes, even though they are not in the presence of neurons or their precursors.

In culture, the GC$^+$ cells can incorporate [^{3}H]thymidine. The percentage of GC$^+$ cells that becomes labeled with the isotope decreases with increasing age of the culture until, in cultures of 14–35 days of incubation, the labeled cells remain at a constant rate of about 11–12% (Bologa et al., 1982a, 1983). Other investigators found a much smaller percentage (1–5%) of dividing GC$^+$ cells (Abney et al., 1981; Pruss et al., 1981). This may be because of different environmental conditions, as the oligodendroblast proliferation can be influenced by the composition of tissue culture medium (Roussel et al., 1983).

Only a very few GC$^+$/MBP$^+$ cells incorporated [^{3}H]thymidine, and cells that were very strongly positive for MBP did not incorporate [^{3}H]thymidine (Bologa et al., 1982a, 1983; Roussel et al., 1983). These observations indicate that the GC$^+$/MBP$^-$ cells that are mitotically active correspond to oligodendroblasts, and that when the cells which become GC$^+$/MBP$^+$ begin to enter postmitotic differentiation they correspond to oligodendrocytes. The fact that GC$^+$/MBP$^+$ oligodendrocytes, in cultures initiated from brain of 14-day-old mouse embryos, did not incorporate [^{3}H]thymidine indicates that the last division of their precursor oligodendroblasts occurred *in utero* before the fourteenth day postconception.

When the cells are allowed to mature in culture for longer periods, oligodendroblasts, after completing division, enter a postmitotic stage of differentiation. Bradel and Prince (1983) examined such cells by electron microscopy. They found that the descriptions of the ultrastructure of oligodendroblasts and oligodendrocytes *in situ* (Bunge, 1961; Caley and Maxwell, 1968a,b; Mori and Leblond, 1970; Peters et al., 1976; Raine, 1977) also applied to cells in culture (McCarthy and de Vellis, 1980; Bhat et al., 1981; Wollmann et al., 1981; Poduslo et al., 1982; Bradel and Prince, 1983). With increasing time in culture, cells increased in size. The cells in 26-day cultures of P1 mouse brain were larger than those in 14-day cultures; such cells had more cytoplasmic organelles and numerous cell processes. The cell nuclei were eccentric and had prominent nucleoli with some condensed chromatin along the nuclear membranes. The cells had prominent Golgi apparatuses, many poly-ribosomes, prominent microtubules, and numerous mitochondria. The rough endoplasmic reticulum was more developed than in cells of 14-day cultures and was stacked. The cells had an increase in lysosomes and a variable number of cytoplasmic processes. The most important finding was that, in many instances, the processes were continuous with membranous material that formed multilaminar whorls often resembling myelin. In such structures, layers exhibited a periodicity of 13–16 nm and contained a distinctive intraperiod line. Bradel and Prince feel that oligodendrocytes in the 26-day cultures meet all the criteria for light and medium oligodendrocytes *in situ* as originally described by Mori and Leblond (1970). Bradel and Prince did not observe cells resembling dark oligodendrocytes, but Bhat et al. (1981) observed such cells in similar cultures.

Correlation between Oligodendroglial Lineage in Cultures with That *In Vivo*

It seems that, in tissue culture, cells of the oligodendroglial cell lineage progress timewise in a manner similar to that *in vivo* (Abney et al., 1981). The oligodendroglial cell lineage begins early in neurogenesis; in E14 mouse fetuses some postmitotic GC^+/MBC^+ oligodendrocytes are already present. It is not known whether the original stem cell for the oligodendroglial lineage is pluripotential and can give rise to oligodendrocytes and astrocytes or to oligodendrocytes and neurons.

The earliest oligodendroglial precursor cells observed in culture are small, dark cells that proliferate extensively and give rise to GC^+/MBP^- cells. These correspond morphologically to "immature" or free subependymal cells *in situ* (Imamoto et al., 1978; Paterson, 1981). The GC^+/MBP^- cells are proliferative and should be referred to as oligodendroblasts. These cells form GC^+/MBP^+ cells that are postmitotic and on culturing undergo maturation. They correspond to immature oligodendrocytes. The morphology of GC^+/MBP^+ cells as they mature corresponds closely to that of light, medium, and, it seems likely, dark oligodendrocytes observed *in vivo* (Mori and Leblond, 1970). Light and medium oligodendrocytes in culture are capable of forming myelin even in the absence of neurons, a fact that distinguishes them from Schwann cells in the peripheral nervous system, which depend on the presence of neurons for myelination.

It is curious that small oligodendroglial precursor cells are found at different frequencies in cultures initiated from brains of E12–P7 animals. The largest numbers of them were found in cultures of brains of embryos at E14. Cultures of older (E18) brains contain only a few, if any, small cells. In dense cultures of P0 or P7 brains, however, the small cells reappear in large numbers (McCarthy and de Vellis, 1980; Labourdette et al., 1980). It seems that there is a stage of development when small oligodendroglial precursor cells cannot be detected in culture. Barbarese et al. (1983), by use of limiting dilution analysis of fetal rat brain cultures, were able to calculate that perinatal rat brain contains one oligodendroglial precursor cell per 1.3×10^5 brain cells, or a total population of 300–500 precursor cells per brain. This calculation suggests that, in order to detect oligodendrocyte precursor cells in cultures of perinatal brains, considerable selection must occur. It also implies that many more small oligodendroglial precursor cells must have been available during early stages of development, because at that time they were readily observed in large numbers in culture. During early stages, the oligodendroglial precursor cells probably did not replenish themselves at a sufficient rate and most were used for the formation of oligodendrocytes (GC$^+$/MBP$^+$ cells). Because of the depletion in numbers, only a few immature oligodendrocyte precursor cells persisted until postnatal life when, upon receiving the right signals, they began to proliferate vigorously and could again be detected in large numbers in culture.

Oligodendrocytes seem to have two stages of genesis, the first during early neurogenesis and the second postnatally. Oligodendrocyte precursor cells must undergo many divisions postnatally to form the number of oligodendrocytes required for the adult brain.

TRANSITIONAL CELLS

Cells Intermediate between Astrocytes and Oligodendrocytes

Cells that express characteristics between those of astrocytes and oligodendrocytes are often referred to as transitional neuroglia. Such cells have been reported for a long time (Penfield 1924, 1932a,b; del Rio Hortega, 1928; D'Agata, 1950; Klatzo, 1952; Farquhar and Hartmann, 1957; Ramón-Moliner, 1958; Smart and Leblond, 1961; Hartmann, 1962; Koening et al., 1962; Skoff, 1980). This raises the question of whether astroglial and oligodendroglial cell lineages are closely related, and whether they may have part of their lineage in common.

Choi et al. (1983) recently described transitional cells occurring in 11-week-old human spinal cord at the time when myelination begins in that area. They observed cells that had astrocyte characteristics—glycogen and intermediate filaments—and the oligodendrocyte characteristic of electron-dense cytoplasm. In addition, cells with the morphology of astrocytes expressed MBP, which is specific for oligodendrocytes, and cells with the morphology of oligodendrocytes expressed GFAP, which is specific for astrocytes. Choi et al. suggested that radial glia, which always have astrocyte characteristics, may give rise to both astrocytes and oligodendrocytes. Phillips (1973) argued that astrocytes develop before oligodendrocytes do and are already present in large numbers when the number of oligodendrocytes is only beginning to increase. Therefore, it does not seem likely that oligodendrocytes differentiate into astrocytes. The converse, however—that astrocytes might differentiate into oligodendrocytes—seems feasible (Sturrock, 1976).

Recently, Raff et al. (1983) reported that astrocyte precursor cells, when grown in serum-containing medium, develop into fibrous astrocytes; if, however, they are grown in chemically defined medium without serum, they transform into oligodendrocytes with typical morphology, as seen by light microscopy, and, in addition, express GC. Such observations argue strongly for a common progenitor cell, or cells, for the two lineages. At the same time, Abney et al. (1983) report from the same laboratory that oligodendrocyte precursor cells have antigens in common with neurons. These are A4, A2B5, and tetanus toxin binding sites. These findings, contrary to those of Raff et al. (1983), argue for common progenitor cells for oligodendrocytes and neurons. There is some controversy about the tetanus toxin binding sites, because other investigators (Asou and Brunngrabber, 1983; Koulokoff et al., 1983) have not found such sites on cells of the astroglial cell lineage, perhaps because the cells were at different stages of maturity.

Cells Intermediate between Macroglia and Microglia

There are some indications that oligodendrocytes and microglia have common precursor cells. Reyners et al. (1982) described β-astrocytes present in adult brain with ultrastructural characteristics of both protoplasmic astrocytes and light oligodendrocytes. These cells are radio-sensitive and apparently can give rise to oligodendroglia and microglia. Fujita (1963) observed cells with small, dark, round nuclei that formed from ventricular cells and migrated to the subpial region, where some of them matured into astrocytes, some into oligodendrocytes, and some remained as "resting microglia" throughout adult life. Similar cells were found subsequently in other parts of the brain (Fujita et al., 1981).

Resting microglia stain with del Rio Hortega's silver carbonate and have elongated, somewhat irregularly shaped nuclei, clumped chromatin with some clumps adjacent to the nuclear membrane, and occasional nucleoli. Perikaryal cytoplasm is electronlucent and contains a few microtubules but no intermediate filaments or glycogen particles. Fujita and his colleagues (1981) could distinguish resting microglia from phagocytic microglia of hematogenic (monocytic) origin mainly on the basis of differences in cytoplasmic features. In response to injuries, resting microglia hypertrophied; on the one hand, they formed reactive astrocytes and, on the other hand, normal fibrous astrocytes. Fujita et al. concluded that resting microglia form astrocytes after observing intermediate forms between resting microglia and astrocytes under the electron microscope and on the basis of a [^{3}H]thymidine pulse-labeling experiment. In the latter, resting microglia were already labeled three hours after injection of [^{3}H]thymidine; the astrocytes became labeled only after 2½ days and increased in number by 6½ days. Fujita et al. reasoned that it is unlikely that [^{3}H]thymidine would still be present in the tissue 2–6 days after injection and therefore assumed that the microglia must have given rise to the astrocytes.

Cells similar to Fujita's resting microglia were observed by Vaughn and Peters (1968) in the optic nerve. They described small, dark cells that they called the "third neuroglial cell type"; these cells resembled microglia when seen under the light microscope. Electron-microscope observations of developing optic nerve led them to conclude that these cells, like those seen by Fujita, develop from ventricular (matrix) cells. Their nuclei are round or elongated and contain clumps of chromatin adjacent to the nuclear membrane. The cytoplasm is light, but slightly darker than that of astrocytes. It does not contain intermediate filaments, glycogen particles, or microtubules. The rough endoplasmic reticulum consists of long cisternae, and dense laminar bodies are found in the cytoplasm. Thus these cells do not have the cytoplasmic characteristics of either astrocytes or oligodendrocytes, but their ultrastructure is similar to that of Fujita's resting microglia. The small neuroglial cells are probably precursors for both oligodendrocytes and astrocytes. Vaughn and Peters (1968) suggest that they may also "represent a potential source of phagocytic cells," based on the observation that small neuroglial cells are numerous in degenerating tissues, occur near degenerating myelin, contain lipid inclusions, and may acquire the appearance of "Gitter" cells.

The many reports of the existence of cells intermediate between astrocytes and oligodendrocytes, and between macroglia and microglia, suggest that there may indeed be common multipotential cells which, depending on the signals received, give rise to cells in the astrocyte or oligodendrocyte lineages. However, the question as to what is the common progenitor cell for all neural cell lineages remains unanswered.

CONCLUSIONS

The last decade has seen the study of neural cells advance tremendously because of the application of tissue culture methods and of neuroimmunocytochemistry; the continuous addition of new cell markers makes possible the identification of specific cell populations. Remarkable findings were that the destiny of neural cells is already programmed at the earliest stages of neural tube formation and that the cells continue to differentiate sequentially and temporally in tissue culture as they do *in vivo*. Thus a unique model system is available for the study of cell lineages of neural cells under controlled environmental conditions.

The work reviewed in this chapter has so far only scratched the surface of the concept of cell lineages. We have established morphological criteria for definition of cells in the lineages, an important and necessary first step before functional criteria of cells at various stages of the lineages and genetic regulatory mechanisms can be determined.

The studies presented here lead us to a few tentative conclusions. It is established that astroglial and oligodendroglial cells begin to develop early in neurogenesis, rather than after neurons have completed their development, and that the development of the various neural lineages is interdependent. Neurons depend on radial glia for guidance, and some early neural and oligodendroglial precursor cells depend on astroglial precursor cells as a substratum, at least in culture. We are just beginning to explore the nature of this interaction and to recognize the signals that induce cells to proceed to the next stage of differentiation. So far, stem cells for the various neural cells have not been determined.

Cell differentiation in the astroglial cell lineage occurs throughout embryogenesis and in postnatal life. The lineage proceeds in a linear fashion, with sequential development in which immature cells, glioblasts, and proastroblasts form early in embryogenesis, and the more mature cells, the astroblasts, form late. The astroblasts first appear in the spinal cord and later in the brain. There is a regional sequence in addition to the temporal sequence of development.

The oligodendroglial lineage differs from the astroglial lineage in being biphasic. There seems to be a production of oligodendrocytes in the early stages of embryogenesis and then a period of quiescence, followed in postnatal life by another peak of oligodendrocyte production and maturation. Oligodendrocytes also have regional sequences of development, which may be related to the sequence of myelination.

Quantification of cell subpopulations within the lineages at different stages of development has just begun. Present views of cell lineages may be somewhat distorted because of the lack of quantitative perspective. For example, certain cells in tissue culture or cells being examined by electron microscopy may be selected preferentially. In order to have a better perspective, it would be helpful to know the size of various cell subpopulations during development, the rate at which cells exist from the proliferative cell pool into the nonproliferative pool at various times in development, and the actual number of cells entering the mature cell pool found in the adult.

In the study of cell lineages, there is a tendency to look at the forward progression of cells from the undifferentiated state to the state of complete differentiation. The possibility that cell differentiation moves in a backward direction, even for a short time, or, indeed, moves sideways toward another closely related lineage, may be overlooked. Moreover, in this review astroglial and oligodendroglial cell lineages were considered as though all mature astrocytes and all mature oligodendrocytes were alike. In reality, the concept of cell lineage is more complex; glia may specialize for specific functions in various regions of the central nervous system, and there are cells closely related to astrocytes that have their own identity, for example, ependymal cells, Müller cells in the retina, and Bergmann glia in the cerebellum. Cell differentiation may include a certain degree of cell modulation around the classical stem of the lineage, as well as more drastic deviations that represent side branches of differentiation. This aspect of neural cell lineages remains to be investigated.

ACKNOWLEDGMENTS

This work was made possible by grant no. MT-4235 from the Medical Research Council of Canada and a grant from the Saskatchewan Health Research Board. I am most grateful to M. E. Fedoroff for help in preparation of the manuscript and to B. Bell and I. Karaloff for secretarial assistance.

REFERENCES

Abney, E. R., D. P. Bartlett, and M. C. Raff (1981) Astrocytes, ependymal cells, and oligodendrocytes develop on schedule in dissociated cell cultures of embryonic rat brain. *Dev. Biol.* **83**:301–310.

Abney, E. R., B. P. Williams, and M. C. Raff (1983) Tracing the development of oligodendrocytes from precursor cells using monoclonal antibodies, fluorescence-activated cell sorting, and cell culture. *Dev. Biol.* **100**:166–171.

Allen, E. (1912) Cessation of mitosis in central nervous system of the albino rat. *J. Comp. Neurol.* **22**:547–568.

Altman, J. (1966) Proliferation and migration of undifferentiated precursor cells in the rat during postnatal gliogenesis. *Exp. Neurol.* **16**:263–278.

Anders, J. J., and M. W. Brightman (1979) Assemblies of particles in the cell membrane of developing, mature and reactive astrocytes. *J. Neurocytol.* **8**:777–795.

Anders, J. J., and M. W. Brightman (1981) Orthogonal assemblies of intramembranous particles—an attribute of the astrocyte. In *Glial and Neuronal Cell Biology*, S. Fedoroff, ed., pp. 21–35, Alan R. Liss, New York.

Anders, J. J., and M. W. Brightman (1982) Particle assemblies in astrocytic plasma membranes are rearranged by various agents *in vitro* and cold injury *in vivo*. *J. Neurocytol.* **11**:1009–1029.

Asou, H., and E. G. Brunngraber (1983) Absence of ganglioside GM1 in astroglial cells from 21-day-old rat brain: Immunohistochemical, histochemical, and biochemical studies. *Neurochem. Res.* **8**:1045–1057.

Barbarese, E., J. E. Pfeiffer, and J. H. Carson (1983) Progenitors of oligodendrocytes: Limiting dilution analysis in fetal rat brain culture. *Dev. Biol.* **96**:84–88.

Bartlett, P. F., M. D. Noble, R. M. Pruss, M. C. Raff, S. Rathray, and C. A. Williams (1981) Rat neural antigen -2 (Ran-2): A cell surface antigen on astrocytes, ependymal cells, Müller cells and leptomeninges defined by a monoclonal antibody. *Brain Res.* **204**:339–351.

Bhat, S., E. Barbarese, and S. E. Pfeiffer (1981) Requirements for nonoligodendrocyte cell signals for enhanced myelinegenic gene expression in long-term cultures of purified rat oligodendrocytes. *Proc. Natl. Acad. Sci. USA* **78**:1283–1287.

Bignami, A., and D. Dahl (1975) Astroglial protein in the developing spinal cord of the chick embryo. *Dev. Biol.* **44**:204–209.

Bignami, A., D. Dahl, and C. Rueger (1980) Glial fibrillary acidic (GFA) protein in normal neural cells and in pathological conditions. In *Advances in Cellular Neurobiology*, Vol. 1, S. Fedoroff and L. Hertz, eds., pp. 286–310, Academic, New York.

Blakemore, W. F. (1969) The ultrastructure of the subependymal plate in the rat. *J. Anat.* **104**:423–433.

Blakemore, W. F., and R. D. Jolly (1972) The subependymal plate and associated ependyma in the dog. An ultrastructural study. *J. Neurocytol.* **1**:69–84.

Bologa, L., J. C. Bisconte, R. Joubert, P. J. Marangos, C. Debrim, F. Rioux, and N. Herschkowitz (1982a) Accelerated differentiation of oligodendrocytes in neuron-rich embryonic mouse brain cell cultures. *Brain Res.* **252**:129–136.

Bologa, L., A. Z'Graggen, E. Rossi, and N. Herschkowitz (1982b) Differentiation and proliferation: Two possible mechanisms for the regeneration of oligodendrocytes in culture. *J. Neurol. Sci.* **57**:419–434.

Bologa, L., J. C. Bisconte, R. Joubert, S. Margules, and N. Herschkowitz (1983) Proliferation activity and characteristics of immunocytochemically identified oligodendrocytes in embryonic mouse brain cell cultures. *Exp. Brain Res.* **50**:84–90.

Bologa-Sandru, L., H. P. Leigrist, A. Z'Graggen, K. Hoffmann, U. Urismann, D. Dable, and N. Herschkowitz (1981) Expression of antigenic markers during the development of oligodendrocytes in mouse brain cell cultures. *Brain Res.* **210**:217–229.

Boulder Committee (1970) Embryonic vertebrate central nervous system: Revised terminology. *Anat. Rec.* **166**:257–262.

Bradel, E. J., and F. P. Prince (1983) Cultured neonatal rat oligodendrocytes elaborate myelin membrane in the absence of neurons. *J. Neurosci. Res.* **9**:381–392.

Brightman, M. W., W. Zis, and J. Andres (1983) Morphology of cerebral endothelium and astrocytes as determinants of the neuronal microenvironment *Acta Neuropathol. (Berl.) (Suppl.)* **8**:21–33.

Brinkley, B. R. (1982) Summary: Organization of the cytoplasm. *Cold Spring Harbor Symp. Quant. Biol.* **46**:1029–1040.

Bryans, W. A. (1959) Mitotic activity in the brain of the adult rat. *Anat. Rec.* **133**:65–73.

Bunge, R. P. (1961) Glial cells and the central myelin sheath. *Physiol. Rev.* **48**:197–251.

Buse, E. (1983) CNS neurons develop *in vitro* from neuronal precursor cells of mouse embryos, Thieler stage 13. *Proc. Int. Symp. Tissue Culture in Neurobiol. (Saskatoon) Abstr.*, p. B8.

Caley, D. W., and D. S. Maxwell (1968a) An electron microscopic study of neurons during postnatal development of the rat cerebral cortex. *J. Comp. Neurol.* **133**:17–44.

Caley, D. W., and D. S. Maxwell (1968b) An electron microscopic study of the neuroglia during postnatal development of the rat cerebrum. *J. Comp. Neurol.* **133**:45–70.

Chiu, F. C., W. T. Norton, and K. L. Fields (1981) The cytoskeleton of primary astrocytes in culture contains actin, glial fibrillary acidic protein, and the fibroblast-type filament protein, vimentin. *J. Neurochem.* **37**:147–155.

Choi, B. H. (1981) Radial glia of developing human fetal spinal cord: Golgi immunohistochemical and electron microscopic study. *Dev. Brain Res.* **1**:249–267.

Choi, B. H., and L. W. Lapham (1978) Radial glia in the human fetal cerebrum: A combined Golgi immunofluorescent and electron microscopic study. *Brain Res.* **148**:295–311.

Choi, B. H., R. C. Kim, and L. W. Lapham (1983) Do radial glia give rise to both astrocytes and oligodendroglial cells? *Dev. Brain Res.* **8**:119–130.

Cohen, J., and S. Y. Selvendran (1981) A neuron-specific cell-surface antigen in the central nervous system not shared by peripheral neurons. *Nature* **291**:4421–4423.

D'Agata, P. M. (1950) Ricerche sulla morphologia è classificazione dell'oligodendroglia. *Riv. Neurol.* **20**:81–127.

Dahl, D., D. C. Rueger, and A. Bignami (1981) Vimentin, the 57,000 molecular weight protein of fibroblast filaments, is the major cytoskeletal component in immature glia. *Eur. J. Cell Biol.* **24**:191–196.

del Rio Hortega, P. (1919a) Coloración rápida de tejidos normales y patológicos con carbonato de plata amoniacal. *Trab. Lab. Invest. Biol. (Madrid)* **17**:229–235.

del Rio Hortega, P. (1919b) El Tercer elemento de las centros nerviosos. I. La microglia en estado normal. II. Intervención de la microglia en los procesos patológicas. III. Naturaleza probable de la microglia. *Boll. Soc. Esp. Biol. (Madrid)* **9**:69–120.

del Rio Hortega, P. (1928) Tercera aportación al conocimiento morphológico e interpretación functional de la oligodendroglia. *Mem. Soc. Esp. Hist. Nat.* **14**:5. Reprinted in *Arch. Hist. (Buenos Aires)* **6**:132–139.

Doering, L. C., and S. Fedoroff (1982) Isolation and identification of neuroblast precursor cells from mouse neopallium. *Dev. Brain Res.* **5**:229–233.

Doering, L. C., and S. Fedoroff (1984) Isolation and transplantation of oligodendrocyte precursor cells. *J. Neurol. Sci.* **63**:183–196.

Doering, L. C., S. Fedoroff, and R. M. Devon (1983) Fibrous astrocytes and reactive astrocyte-like cells in transplants of cultured astrocyte precursor cells. *Dev. Brain Res.* **6**:183–198.

Eng, L. F., and S. J. DeArmond (1982) Immunochemical studies of astrocytes in normal development and disease. In *Advances in Cellular Neurobiology*, S. Fedoroff and L. Hertz, eds., pp. 145–171, Academic, New York.

Eisenbarth, G. S., F. S. Walsh, and M. Nirenberg (1979) Monoclonal antibody to a plasma membrane antigen of neurons. *Proc. Natl. Acad. Sci. USA* **76**:4913–4917.

Farquhar, M. G., and J. F. Hartmann (1957) Neuroglial structure and relationships as revealed by electron microscopy. *J. Neuropathol. Exp. Neurol.* **16**:18–39.

Fedoroff, S. (1978) The development of glial cells in primary cultures. In *Dynamic Properties of Glial Cells*, E. Schoffeniels, G. Franke, L. Hertz, and D. B. Tower, eds., pp. 83–92, Pergamon, New York.

Fedoroff, S. (1980) Tracing the astrocyte cell lineage in mouse neopallium *in vitro* and *in vivo*. In *Tissue Culture in Neurobiology*, E. Giacobini, A. Vernadakis, and A. Shahar, eds., pp. 349–372, Raven, New York.

Fedoroff, S., and L. C. Doering (1980) Colony culture of neural cells as a model for the study of cell lineages in the developing CNS: The astrocyte cell lineage. *Curr. Top. Dev. Biol.* **16**:283–304.

Fedoroff, S., and C. Hall (1979) Effect of horse serum on neural cell differentiation in tissue culture. *In Vitro* **15**:641–648.

Fedoroff, S., R. V. White, L. Subrahmanyan, and V. I. Kalnins (1981) Properties of putative astrocytes in colony cultures of mouse neopallium. In *Glial and Neuronal Cell Biology*, S. Fedoroff, ed., pp. 1–19, Alan R. Liss, New York.

Fedoroff, S., T. L. Krukoff, and K. R. S. Fisher (1982) The development of chick spinal cord in tissue culture. III. Neuronal precursor cells in culture. *In Vitro* **18**:183–195.

Fedoroff, S., R. V. White, J. Neal, L. Subrahmanyan, and V. I. Kalnins (1983) Astrocyte cell lineage. II. Mouse fibrous astrocytes and reactive astrocytes in cultures have vimentin- and GFP-containing intermediate filaments. *Dev. Brain Res.* **7**:303–315.

Fedoroff, S., J. Neal, M. Opas, and V. I. Kalnins (1984a) Astrocyte cell lineage. III. The morphology of differentiating mouse astrocytes in colony culture. *J. Neurocytol.* **13**:1–20.

Fedoroff, S., J. D. Houle, and V. I. Kalnins (1984b) Intermediate filaments and neural cell differentiation. *Int. J. Neurol.* (in press).

Fisher, G., A. Lentz, and M. Schachner (1982) Cultivation of immature astrocytes of mouse cerebellum in a serum-free, hormonally defined medium. Appearance of the mature astrocytic phenotype after addition of serum. *Neurosci. Lett.* **29**:297–302.

Fujita, S. (1963) The matrix cells and cytogenesis in the developing central nervous system. *J. Comp. Neurol.* **120**:37–42.

Fujita, S. (1965) An autoradiographic study on the origin and fate of the subpial glioblasts in the embryonic chick spinal cord. *J. Comp. Neurol.* **124**:51–60.

Fujita, S. (1966) Application of light and electron microscopic autoradiography to the study of cytogenesis of the forebrain. In *Evolution of the Forebrain*, R. Hassler and H. Stephan, eds., pp. 180–196, Plenum, New York.

Fujita, S. (1980) Cytogenesis and pathology of neuroglia and microglia. *Pathol. Res. Pract.* **168**:271–278.

Fujita, S., Y. Tsuchihashi, and T. Kitamura (1981) Origin, morphology and function of the microglia. In *Glial and Neuronal Cell Biology*, S. Fedoroff, ed., pp. 141–169, Alan R. Liss, New York.

Hamburger, V., and H. L. Hamilton (1951) A series of normal stages in the development of the chick embryo. *J. Morphol.* **88**:49–92.

Hartmann, J. F. (1962) Identification of neuroglia in electron micrographs of normal nerve tissue. In *Proceedings, IVth International Congress of Neuropathology, Vol. 2, Electron Microscopy and Biology*, (Munich), H. Jacob, ed., pp. 32–35, Thieme, Stuttgart.

His, W. (1888) Zur Geschichte des Gehirns sowie der centralen und peripherischen Nervenbahnen beim Menschlichen Embryo. *Abt. Math. Phys. Cl. Kgl. Sachs. Ges. Wiss.* **14**:341–392.

His, W. (1889) Die Neuroblasten und deren Entstehungen im embryonalen Marke. *Abt. Math. Phys. Cl. Kgl. Sach. Ges. Wiss.* **15**:313–372.

Houle, J. D., and S. Fedoroff (1983) Temporal relationship between the appearance of vimentin and neural tube development. *Dev. Brain Res.* **9**:189–195.

Hubbard, B. M., and J. W. Hopewell (1980) Quantitative changes in the cellularity of the rat subependymal plate after X-irradiation. *Cell Tissue Kinet.* **13**:403–413.

Imamoto, K., J. A. Paterson, and C. P. Leblond (1978) Radioautographic investigation of gliogenesis in the corpus callosum of young rats. I. Sequential changes in oligodendrocytes. *J. Comp. Neurol.* **180**:115–137.

Jacobson, M. (1978) *Developmental Neurobiology*, 2nd ed., Plenum, New York.

Juurlink, B. H. J., and S. Fedoroff (1979) The development of mouse spinal cord in tissue culture. I. Cultures of whole mouse embryos and spinal-cord primordia. *In Vitro* **15**:86–94.

Juurlink, B. H. J., and S. Fedoroff (1980) Differentiation capabilities of mouse optic stalk in isolation of its immediate in vivo environment. *Dev. Biol.* **78**:215–221.

Juurlink, B. H. J., and S. Fedoroff (1982) The development of mouse spinal cord in tissue culture. II. Development of neuronal precursor cells. *In Vitro* **18**:179–182.

Juurlink, B. H. J., S. Fedoroff, C. Hall, and E. J. H. Nathaniel (1981) Astrocyte cell lineage. I. Astrocyte progenitor cells in mouse neopallium. *J. Comp. Neurol.* **200**:375–391.

Kalnins, V. I., M. Opas, I. Ahmed, and S. Fedoroff (1984) Astrocyte cell lineage. IV. Distribution of microfilaments and adhesion patterns in astrocytes and their precursor cells in culture. *J. Neurocytol.* (in press).

Kershman, J. (1938) The medulloblast and the medulloblastoma. *Arch. Neurol. Psychiatry* **40**:937–967.

Klatzo, I. (1952) A study of the glia by the Golgi method. *Lab. Invest.* **1**:343–350.

Koening, H., M. B. Bunge, and R. P. Bunge (1962) Nucleic acid and protein metabolism in white matter. Observations during experimental demyelination and remyelination: A histochemical and autoradiographic study of spinal cord of the adult cat. *Arch. Neurol.* **6**:177–193.

Korr, H., W. D. Schilling, B. Schultz, and W. Maurer (1983) Autoradiographic studies of glial proliferation in different areas of the brain of the 14-day-old rat. *Cell Tissue Kinet.* **16**:393–413.

Koulakoff, A., B. Bizzini, and Y. Berwald-Netter (1983) Neuronal acquisition of tetanus toxin binding sites: Relationship with the last mitotic cycle. *Dev. Biol.* **100**:350–357.

Labourdette, G., G. Roussel, and J. L. Nussbaum (1980) Oligodendroglia content of glial cell primary cultures, from newborn rat brain hemispheres, depends on the initial plating density. *Neurosci. Lett.* **18**:203–209.

Landis, D. M. D., and T. J. Reese (1974) Arrays of particles in freeze-fracture astrocytic membranes. *J. Cell Biol.* **60**:316–320.

Landis, D. M. D., and T. J. Reese (1981) Membrane structure in mammalian astrocytes: A review of freeze-fracture studies in adult, developing, reactive and cultured astrocytes. *J. Exp. Biol.* **95**:35–48.

Landis, D. M. D., and L. A. Weinstein (1983) Membrane structure in cultured astrocytes. *Brain Res.* **276**:31–41.

Levitt, P., and P. Rakic (1980) Immunoperoxidase localization of glial fibrillary acidic protein in radial glial cells and astrocytes of the developing rhesus monkey brain. *J. Comp. Neurol.* **193**:815–840.

Levitt, P., M. L. Cooper, and P. Rakic (1981) Coexistence of neuronal and glial precursor cells in the cerebral ventricular zone of the fetal monkey: An ultrastructural immunoperoxidase analysis. *J. Neurosci.* **1**:27–39.

Levitt, P., M. L. Cooper, and P. Rakic (1983) Early divergence and changing proportions of neurons and glial precursor cells in the primate cerebral ventricular zone. *Dev. Biol.* **96**:472–484.

Lewis, P. D. (1968) The fate of subependymal cells in the adult rat brain, with a note on the origin of microglia. *Brain* **91**:721–738.

Ling, E. A., and C. P. Leblond (1973) Investigation of glial cells in semithin sections. II. Variation with age in the numbers of the various glial cell types in rat cortex and corpus callosum. *J. Comp. Neurol.* **149**:73–82.

Ling, E. A., J. A. Paterson, A. Privat, S. Mori, and C. P. Leblond (1973) Investigation of glial cells in semithin sections. I. Identification of glial cells in the brain of young rats. *J. Comp. Neurol.* **149**:43–72.

Lumsden, C. E., and C. M. Pomerat (1951) Normal oligodendrocytes in tissue culture. A preliminary report on the pulsatile glial cells in tissue cultures from the corpus callosum of the normal adult rat brain. *Exp. Cell Res.* **2**:103–114.

McCarthy, K. D., and J. de Vellis (1980) Preparation of separate astroglial and oligodendroglial cell cultures from rat cerebral tissue. *J. Cell Biol.* **85**:890–902.

McConnell, J. A. (1981) Identification of early neurons in the brain stem and spinal cord. II. An autoradiographic study in the mouse. *J. Comp. Neurol.* **200**:273–288.

McConnell, J. A., and J. W. Sechrist (1980) Identification of early neurons in the brain stem and spinal cord. I. An autoradiographic study in the chick. *J. Comp. Neurol.* **192**:769–783.

Metcalf, D., and M. A. S. Moore (1971) *Haemopoietic Cells*, North-Holland, Amsterdam.

Mori, S., and C. P. Leblond (1969) Electron microscopic features and proliferation of astrocytes in the corpus callosum of the rat. *J. Comp. Neurol.* **137**:197–226.

Moris, S., and C. P. Leblond (1970) Electron microscopic identification of three classes of oligodendrocytes and a preliminary study of their proliferative activity in the corpus callosum of young rats. *J. Comp. Neurol.* **139**:1–30.

Niessing, K., E. Scharrer, B. Scharrer, and A. Oksche (1980) Die Neuroglia: Historischer Überblick. In *Neuroglia. I. Handbuch der Mikroskopischen Anatomie des Menschen*, A. Oksche and L. Vollrath, eds., pp. 1–113, Springer-Verlag, Berlin.

Noetzel, H. (1965) Autoradiographische Untersuchungen zur Entwicklung und Differenzierung des Glia. In *Proceedings, Vth International Congress of Neuropathology*, (Amsterdam), H. Jacob, ed., pp. 802–806, Thieme, Stuttgart.

Paterson, J. A. (1981) Postnatal development of oligodendrocytes. In *Glial and Neuronal Cell Biology*, S. Fedoroff, ed., pp. 83–92, Alan R. Liss, New York.

Paterson, J. A. (1983) Dividing and newly produced cells in the corpus callosum of adult mouse cerebrum as detected by light microscopic radioautography. *Anat. Anz.* **153**:149–168.

Paterson, J. A., A. Privat, E. A. Ling, and C. P. Leblond (1973) Investigation of glial cells in semithin sections. III. Transformation of subependymal cells into glial cells, as shown by radioautography after ^{3}H-thymidine injection into the lateral ventricle of the brain of young rats. *J. Comp. Neurol.* **149**:83–102.

Penfield, W. (1924) Oligodendroglia and its relation to classical neuroglia. *Brain* **47**:430–452.

Penfield, W. (1932a) Neuroglia: Normal and pathological. In *Cytology and Cellular Pathology of the Central Nervous System*, Vol. II, E. V. Cowdry, ed., pp. 423–479, Hoeber, New York.

Penfield, W. (1932b) Neuroglia and microglia. The interstitial tissue of the central nervous system. In *Special Cytology*, Vol. III, E. V. Cowdry, ed., pp. 1445–1482, Hoeber, New York.

Peters, A., S. L. Palay, and H. deF. Webster, eds. (1976) *The Fine Structure of the Nervous System: The Neurons and Supporting Cells*, W. B. Saunders, Philadelphia.

Phillips, D. E. (1973) An electron microscopic study of macroglia and microglia in the lateral funiculus of the developing spinal cord in the fetal monkey. *Z. Zellforsch. Mikrosk. Anat. Abt. Histochemie* **140**:145–167.

Ploem, J. S. (1975) General introduction. In *Fifth International Conference on Immunofluorescence and Related Staining Techniques, Ann. NY Acad. Sci.* **254**:4–20.

Poduslo, S. E., K. Miller, and J. S. Walinsky (1982) The production of a membrane by purified oligodendroglia maintained in culture. *Exp. Cell Res.* **137**:203–215.

Polak, M., W. Haymaker, J. E. Johnson, Jr., and F. D'Amelio (1982a) Neuroglia and their reactions. In *Histology and Histopathology of the Nervous System*, W. Haymaker and R. D. Adams, eds., pp. 363–480, Charles C. Thomas, Springfield, Ill.

Polak, M., F. D'Amelio, J. E. Johnson, Jr., and W. Haymaker (1982b) Microglial cells: Origins and reactions. In *Histology and Histopathology of the Nervous System*, W. Haymaker and R. D. Adams, eds., pp. 481–559, Charles C. Thomas, Springfield, Ill.

Privat, A. (1975) Postnatal gliogenesis in the mammalian brain. *Int. Rev. Cytol.* **40**:281–323.

Privat, A., and C. P. Leblond (1972) The subependymal layer and neighboring region in the brain of the young rat. *J. Comp. Neurol.* **146**:277–302.

Pruss, R. M., P. F. Bartlett, J. Gavrilovic, R. P. Lisak, and S. Rattray (1981) Mitogens for glial cells. A comparison of the response of cultured astrocytes, oligodendrocytes and Schwann cells. *Dev. Brain Res.* **2**:19–35.

Raff, M. C., K. L. Fields, S. I. Hakomori, R. Mirsky, R. M. Pruss, and J. Winter (1979) Cell type specific markers for distinguishing and studying neurons and the major classes of glial cells in culture. *Brain Res.* **174**:283–308.

Raff, M. C., R. H. Miller, and M. Noble (1983) A glial progenitor cell that develops *in vitro* into an astrocyte or an oligodendrocyte depending on culture medium. *Nature* **303**:390–396.

Raine, C. S. (1977) Morphological aspects of myelin and myelination. In *Myelin*, P. Morell, ed., pp. 1–49, Plenum, New York.

Raju, T., A. Bignami, and D. Dahl (1981) *In vivo* and *in vitro* differentiation of neurons and astrocytes in the rat embryo. *Dev. Biol.* **85**:344–357.

Rakic, P. (1971a) Guidance of neurons migrating to the fetal monkey neocortex. *Brain Res.* **33**:471–476.

Rakic, P. (1971b) Neuron-glia relationship during granule cell migration in developing cerebellar cortex. A Golgi and electron microscopic study in *Macacus rhesus*. *J. Comp. Neurol.* **141**:283–312.

Rakic, P. (1981) Developmental events leading to laminar and areal organization of the neocortex. In *The Organization of the Cerebral Cortex*, F. O. Schmitt, F. G. Worden, G. Adelman, and S. G. Dennis, eds., pp. 7–28, MIT Press, Cambridge, Mass.

Ramón y Cajal, S. (1906) Génesis de las fibras nerviosas del embrión y observaciones contrarias a la teoría catenaria. *Trab. Lab. Invest. Biol. (Madrid)* **4**:227–294.

Ramón y Cajal, S. (1913a) Sobre un nuevo proceder de impregnación de la neuroglia. *Trab. Lab. Invest. Biol. (Madrid)* **11**:219–237.

Ramón y Cajal, S. (1913b) Contribución al conocimiento de la neuroglia del cerebro humano. *Trab. Lab. Invest. Biol. (Madrid)* **11**:255–315.

Ramón y Cajal, S. (1955) *Histologie du Système Nerveux de l'Homme et des Vertébrés*, Vol. 2, pp. 986–993, trans. by L. Azoulay, Consejo Superior de Investigaciones Científicas, Instituto Ramón y Cajal, Madrid.

Ramón-Moliner, E. (1958) A study of neuroglia. The problem of transitional forms. *J. Comp. Neurol.* **110**:157–172.

Reyners, H., E. G. de Reyners, and J. R. Maisin (1982) The beta astrocyte: A newly recognized radiosensitive glial cell type in the cerebral cortex. *J. Neurocytol.* **11**:967–983.

Rioux, F., C. Derbin, S. Margules, R. Joubert, and J. C. Bisconte (1980) Kinetics of oligodendrocyte-like cells in primary culture of mouse embryonic brain. *Dev. Biol.* **76**:87–99.

Roussel, G., M. Sensenbrenner, G. Labourdette, E. Wittendorp-Rechenmann, B. Pettmann, and J. L. Nussbaum (1983) An immunohistochemical study of two myelin-specific proteins in enriched oligodendroglial cell culture combined with an autoradiographic investigation using [^{3}H]thymidine. *Dev. Brain Res.* **8**:193–204.

Rungger-Brändle, E., and G. Gabbiani (1983) The role of cytoskeletal and cytocontractile elements in pathologic processes. *Am. J. Pathol.* **110**:360–392.

Sauer, F. C. (1935a) The cellular structure of the neural tube. *J. Comp. Neurol.* **63**:13–23.

Sauer, F. C. (1935b) Mitosis in the neural tube. *J. Comp. Neurol.* **62**:377–405.

Schachner, M., S. K. Kim, and R. Zehnle (1981) Developmental expression in central and peripheral nervous system of oligodendrocyte cell surface antigens (O antigens) recognized by monoclonal antibodies. *Dev. Biol.* **83**:328–338.

Schachner, M., I. Sommer, C. Lagenaur, J. Schnitzer, and G. Berg (1983) Autogenic markers of glia and glial subclasses. In *Neuroscience Approached through Cell Culture*, Vol. 2, S. E. Pfeiffer, ed., pp. 115–139, CRC Press, Boca Raton, Fla.

Schaper, A. (1894) Die morphologische und histologische Entwicklung des Kleinhirns der Teleostier. *Anat. Anz.* **9**:489–501.

Schaper, A. (1897) The earliest differentiation in the central nervous system of vertebrates. *Science* **5**:430–431.

Schmechel, D. E., and P. Rakic (1979a) A Golgi study of radial glial cells in developing monkey telencephalon: Morphogenesis and transformation into astrocytes. *Anat. Embryol.* **156**:115–152.

Schmechel, D. E., and P. Rakic (1979b) Arrested proliferation of radial glial cells during midgestation in rhesus monkey. *Nature* **277**:303–305.

Schnitzer, J., W. W. Franke, and M. Schachner (1981) Immunochemical demonstration of vimentin in astrocytes and ependymal cells of developing and adult mouse nervous system. *J. Cell Biol.* **90**:435–447.

Schultze, B., and H. Korr (1981) Cell kinetic studies of different cell types in the developing and adult brain of the rat and the mouse: A review. *Cell Tissue Kinet.* **14**:309–325.

Secchi, J., D. Lecaque, M. A. Cousin, D. Lando, L. Legault-Demarc, and J. P. Raynaud (1980) Detection and localization of 14-3-2 protein in primary cultures of embryonic rat brain. *Brain Res.* **184**:455–466.

Shaw, G., M. Osborn, and K. Weber (1981) An immunofluorescence microscopical study of the neurofilament triplet proteins, vimentin and glial fibrillary acidic protein within the adult rat brain. *Eur. J. Cell Biol.* **26**:68–82.

Sidman, R. L. (1970) Autoradiographic methods and principles for study of the nervous system with thymidine-H^3. In *Contemporary Research Methods in Neuroanatomy*, W. J. H. Nauta and S. O. E. Ebbesson, eds., pp. 252–274, Springer, New York.

Sidman, R. L., and P. Rakic (1982) Development of the human central nervous system. In *Histology and Histopathology of the Nervous System*, W. Haymaker and R. D. Adams, eds., pp. 3–145, Charles C. Thomas, Springfield, Ill.

Sidman, R. L., I. L. Miale, N. Feder (1959) Cell proliferation and migration in the primitive ependymal zone. An autoradiographic study of histogenesis in the nervous system. *J. Exp. Neurol.* **1**:322–333.

Skoff, R. P. (1980) Neuroglia: A reevaluation of their origin and development. *Pathol. Res. Pract.* **168**:279–300.

Skoff, R. P., D. L. Price, and A. Stocks (1976) Electron microscopic autoradiographic studies of gliogenesis in rat optic nerve. I. Cell proliferation. *J. Comp. Neurol.* **169**:291–312.

Smart, I. (1961) The subependymal layer of the mouse brain and its cellular production as shown by radioautography after thymidine-H^3 injection. *J. Comp. Neurol.* **116**:325–349.

Smart, I., and C. P. Leblond (1961) Evidence for division and transformation of neuroglia cells in the mouse brain, as derived from radioautography after injection of thymidine-H^3. *J. Comp. Neurol.* **116**:349–367.

Spitzer, J. L., and N. C. Spitzer (1975) Time of origin of Rohon-Beard cells in spinal cord of *Xenopus laevis*. *Am. Zool.* **15**:781.

Stensaas, L. J., and S. S. Stensaas (1968) An electron microscope study of cells in the matrix and intermediate laminae of the cerebral hemispheres of the 45 mm rabbit embryo. *Z. Zellforsch. Mikrosk. Anat. Abt. Histochem.* **91**:341–365.

Streeter, G. L. (1942) Developmental horizons in human embryos. *Contrib. Embryol. Carneg. Instn.* **30**:213–230.

Sturrock, R. R. (1976) Light microscopic identification of immature glial cells in semithin sections of the developing mouse corpus callosum. *J. Anat.* **122**:521–537.

Sturrock, R. R. (1979) A quantitative lifespan study of changes in cell number, cell division and cell death in various regions of the mouse forebrain. *Neuropathol. Appl. Neurobiol.* **5**:433–456.

Sturrock, R. R. (1982) Gliogenesis in the prenatal rabbit spinal cord. *J. Anat.* **134**:771–793.

Sturrock, R. R., and I. H. M. Smart (1980) A morphological study of the mouse subependymal layer from embryonic life to old age. *J. Anat.* **130**:391–415.

Tapscott, S. J., G. D. Bennett, Y. Toyama, F. Keinbart, and H. Holtzer (1981) Intermediate filament protein in the developing chick spinal cord. *Dev. Biol.* **86**:40–54.

Theiler, K. (1972) Development and normal stages from fertilization to 4 weeks of age. In *The House Mouse*, pp. 168, Springer-Verlag, Berlin.

Vargas-Lizardi, P., and K. M. Lyser (1974) Time of origin of Mauthner's neurons in *Xenopus laevis* embryos. *Dev. Biol.* **38**:220–228.

Vaughn, J. E. (1969) An electron microscopic analysis of gliogenesis in rat optic nerves. *Z. Zellforsch. Mikrosk. Anat. Abt. Histochem.* **94**:293–324.

Vaughn, J. E., and A. Peters (1968) A third neuroglial cell type. An electron microscopic study. *J. Comp. Neurol.* **133**:269–288.

Wolfgram, F., and A. S. Rose (1957) The morphology of neuroglia in tissue culture with comparison to histological preparations. *J. Neuropathol. Exp. Neurol.* **16**:514–531.

Wollmann, R. L., S. Szuchet, J. Barlow, and M. Jerkovic (1981) Ultrastructural changes accompanying the growth of isolated oligodendrocytes. *J. Neurosci. Res.* **6**:757–769.

Wolpert, L. (1969) Positional information and the spatial pattern of differentiation. *J. Theor. Biol.* **25**:1–47.

Yen, S. H., and K. L. Fields (1981) Antibodies to neurofilament, glial filament and fibroblast intermediate filament proteins bind to different cell types of the nervous system. *J. Cell Biol.* **88**:115–126.

Chapter 6

Glial Growth Factor and the Neuronal Control of Cell Division in Amphibian Limb Regeneration

CHRIS R. KINTNER
GREG ERWIN LEMKE
JEREMY P. BROCKES

ABSTRACT

Despite the importance of peptide factors in promoting the division of cultured cells, little is known about their role in the control of cell division in vivo. This chapter is concerned with glial growth factor (GGF), a mitogen found in the nervous system, and its possible role in limb regeneration in Urodele amphibians. GGF has been identified by its mitogenic activity on cultures of dissociated rat Schwann cells; these cells do not respond to a wide variety of other growth factors. GGF activity is present at high levels in the pituitary and caudate nucleus and is active on fibroblasts and astrocytes in addition to Schwann cells. The activity has been purified 10^5-fold to apparent homogeneity from bovine anterior lobes and has been shown by several criteria to reside in a basic protein with a molecular weight of 31 kD.

One possible function of GGF in vivo is to mediate the mitogenic activity of nerves on nonneuronal cells. This possibility is being investigated in the context of limb regeneration, during which the progenitor cells of the regenerating blastema are dependent on the nerve supply for their division. GGF activity is detected in the brains of newts and axolotols and has similar chemical properties to that in bovine pituitary. It is also present in the regenerating blastema, where it decreases in level after denervation.

In order to approach questions about the origin, identity, and proliferation of newt blastemal cells, we have isolated monoclonal antibodies to use as cell markers. One antibody, called 22/18, has provided new evidence about the origin of blastemal cells by dedifferentiation of myofibers and Schwann cells; 22/18 identifies a predominant population of blastemal cells at early stages when Schwann cells appear to make a contribution to the blastema. Only a fraction of the blastemal cell population is $22/18^+$ at later stages of regeneration. The relation of the $22/18^+$ population to the dependency of regeneration on nerve supply is discussed in light of the findings that $22/18^+$ cells are dependent on the nerve for division at both early and late stages of regeneration.

The identification and characterization of various elements that act on cells to promote or retard their division is essential for our understanding of developing and regenerating systems. The diversity of cell types and the difficulties of manipulating cellular environments make the study of cell proliferation *in vivo* rather intractable. For this reason, the study of proliferation in the last 20 years has been dominated by studies in cell culture. There are still remarkably few examples of pure cultures of primary cells that are readily available, so much of this work has been done on cell lines and, in particular, the murine 3T3 cell (Todaro and Green, 1963). A clear outcome of these studies has been to emphasize the importance of mitogenic growth factors. These are generally, though not always, protein molecules that act in a hormonal concentration range ($< 10^{-9}$ M) to promote cell division. One of the earliest of these molecules to be characterized was epidermal growth factor (EGF), originally identified in extracts of mouse submaxillary gland as an activity that promoted premature eye eruption when injected into newborn mice (Cohen, 1962). The identification of platelet-derived growth factor (PDGF) depended upon the recognition that platelet-poor plasma is only weakly mitogenic for a variety of cell types (Balk et al., 1973; Ross et al., 1979) whereas serum is strongly so, due to the release of a powerful mitogen from the platelets during coagulation. The molecular identities of both EGF (Taylor et al., 1972) and PDGF (Waterfield et al., 1983) are well established and their mechanisms of action on cultured cells are under intensive investigation. Studies of receptor-mediated events such as endocytosis, phosphorylation, ion fluxes, and changes in the level of second messengers are vital areas of investigation that have many implications for cell and tumor biology.

In contrast to their effects on cultured cells, the roles played by mitogenic growth factors *in vivo* are unclear. EGF is by now one of the most intensively investigated of all polypeptide hormones in terms of its actions on cultured cells, yet its role in normal development and regeneration has not been defined. One could legitimately ask if protein growth factors have any role in the control of cell division *in vivo*. If so, where are they synthesized and how are they delivered to their target cells? Do cells respond selectively to different growth factors and how is selectivity modulated? Why does cell division stop, and how is it regulated in relation to pattern formation?

These are formidable questions, and it may be important at this stage to identify accessible systems in which the proliferation of a defined population of cells is dependent on an identified source. The early development of the nervous system involves the proliferation of neuronal and glial precursor cells in the neuroepithelium. It is not clear, however, that this process requires the action of mitogenic growth factors or what the source of such factors might be. After the generation of postmitotic neurons, a variety of interactions appear to reflect the mitogenic effect of neurons on nonneuronal cells. The proliferation of central and peripheral glial cells during development and regeneration is often considered to reflect the action of neurons by contact-mediated or diffusable signals (see Brockes and Lemke, 1981, for a review). Perhaps the most dramatic interaction of this sort, however, is the neuronal dependence of limb regeneration in urodele amphibians (Singer, 1952); the innervation of the limb appears to stimulate division of the mesenchymal blastema cells that subsequently give rise to the regenerate.

This chapter is concerned with a new growth factor of nervous tissue that has been identified and assayed by its effect on cultured rat Schwann cells and has been called glial growth factor (GGF). Although GGF acts on other cell types, the rat Schwann cell, as discussed below, does not respond to a variety of other growth factors and therefore has been crucial for the identification of GGF. In the first part of this chapter we give an abbreviated account of the isolation and characterization of GGF and of its distribution and specificity of action. The second part introduces the problem of nervous control of limb regeneration, reviews the circumstantial evidence that GGF may play a role in the phenomenon, and outlines the recent derivation in our laboratory of monoclonal antibodies to blastemal cells. Such reagents promise to be valuable for studies of the origin, identity, and proliferation of these remarkable cells.

CONTROL OF PROLIFERATION OF CULTURED RAT SCHWANN CELLS

In dissociated cultures of the neonatal rat sciatic nerve, all of the cells can be identified by antisera to two cell surface antigens (Brockes et al., 1977). Schwann cells carry on their surface Ran-1, an antigen originally defined by an absorbed mouse antiserum raised against a rat glial cell line (Fields et al., 1975). The fibroblasts of nerve connective tissue express the well-characterized Thy-1 antigen on their surface, as do other cultured fibroblasts (Stern, 1973). All of the cells in sciatic nerve cultures can be identified as Ran-1^+/Thy-1^- Schwann cells or Ran-1^-/Thy-1^+ fibroblasts by indirect immunofluorescence using antibodies conjugated to rhodamine or fluorescein (Brockes et al., 1977). The distinction between the two populations is further supported by use of antisera to the S100 antigen, which is expressed by Schwann cells (Brockes et al., 1979); to fibronectin, which is expressed by fibroblasts (Brockes et al., 1979); and to laminin, which has recently been shown to be made by Schwann cells (Cornbrooks et al., 1983). For studies of Schwann cell proliferation, it is important to remove the fibroblasts because they respond to a variety of mitogens and hence raise the background of the assays. Fibroblasts are removed by treating the cells in suspension with monoclonal antibody to Thy-1 and rabbit complement. This produces populations of Schwann cells that are routinely greater than 99.5% pure and often greater than 99.9% pure (Brockes et al., 1979). Other methods of removing fibroblasts have involved the use of mitotic inhibitors (Wood, 1976) or selections based on adhesion (Kreider et al., 1981).

In a conventional tissue culture medium containing 10% fetal calf serum, purified Schwann cells divide very slowly, taking 7–10 days to double even in the most mitogenic batches of serum. This rate gives a low background in proliferation assays based on counts of cell number, on tritiated thymidine autoradiography, or on the incorporation of ^{125}I-labeled iododeoxyuridine into the DNA of cells growing in microwells (Raff et al., 1978a,b). The microwell assay gives a signal that is linearly related to the logarithm of protein concentration over most of the range (Brockes and Lemke, 1981; Lemke and Brockes, 1984) and is particularly convenient for processing multiple samples. In all of these assays, Schwann cells are strongly stimulated by a protein growth factor of the brain and pituitary that has been purified and characterized. Before describing this analysis, it is important to consider certain unusual aspects of rat Schwann cell proliferation in culture.

As indicated in Table 1, these cells do not proliferate in response to a variety of agents that stimulate division of other cell types (Raff et al., 1978a; Salzer et

Table 1. Agents Tested for Their Effect on Schwann Cell Proliferation[a]

Tissue extracts	Investigators
Bovine liver	Raff et al. (1978a)
Bovine kidney	Raff et al. (1978a)
Bovine muscle	J. P. Brockes (unpublished)
Growth factors	
FGF (brain and pituitary)	Raff et al. (1978a)
	Salzer et al. (1980)
EGF	Salzer et al. (1980)
NGF	Salzer et al. (1980)
PDGF	Lemke and Brockes (1984)
Protein and peptide hormones	
ACTH	Raff et al. (1978a)
TSH	Raff et al. (1978a)
LTH	Raff et al. (1978a)
GH	Raff et al. (1978a)
Vasopressin	Raff et al. (1978a)
Oxytocin	Raff et al. (1978a)
Prolactin	Raff et al. (1978a)
Insulin	Raff et al. (1978a)
	Salzer et al. (1980)
Proinsulin	Salzer et al. (1980)
Others	
Dexamethasone	Salzer et al. (1980)
Prostaglandins	Raff et al. (1978a)
	Salzer et al. (1980)
A23187	Salzer et al. (1980)
Arachidonic acid	Salzer et al. (1980)
Concanavalin A	Salzer et al. (1980)
Phytohemagglutinin	Salzer et al. (1980)
Trypsin	Salzer et al. (1980)
Ouabain	Salzer et al. (1980)

[a] These agents gave no detectable effect on Schwann cell proliferation. Details of the concentrations are given in the original papers.

al., 1980). These include EGF and the fibroblast growth factor preparations derived from bovine brain and pituitary (Gospodarowicz, 1975). Purified human PDGF is also inactive on these cells (Lemke and Brockes, 1984), and this may account in part for the lack of mitogenicity of 10% serum. This lack of responsiveness is a rather unusual feature of Schwann cells and will be referred to later in connection with the neuronal dependence of proliferation in the newt blastema. It appears from recent studies that mouse Schwann cells do proliferate more rapidly in 10% serum (Skaper et al., 1980; White et al., 1983).

The two sources of tissue extracts that promote division are soluble extracts of the brain and pituitary (Raff et al., 1978a; Brockes et al., 1979) and axolemmal membrane preparations from cultured neurons, brain, and peripheral nerve (Salzer et al., 1980; De Vries et al., 1982, 1983). A critical comparison of these two activities must await chemical characterization of the latter. Specific activity in the pituitary is comparable for extracts from the anterior and posterior lobes and is greater than that from brain extracts (Raff et al., 1978a). A variety of tissues have been tested as a source of soluble activity, but only neural tissue has given detectable activity (Table 1). The limit of detection in assays on crude extracts is difficult to estimate accurately, but it is probably less than 2% of that in bovine pituitary extract. Peripheral nerve yields soluble extracts that have an inhibitory activity (Raff et al., 1978a) and thus cannot be reliably assayed.

Schwann cells are also stimulated to divide by raising the intracellular concentration of cyclic AMP (Raff et al., 1978b). A variety of pharmacological treatments, such as cholera toxin or phosphodiesterase inhibitors, act in this way. Although at the time of its discovery this effect of cyclic AMP seemed to be somewhat unusual, several cell types, including 3T3 cells (Rozengurt, 1981), are now considered to be stimulated in the same way.

IDENTIFICATION OF THE ACTIVITY IN BOVINE PITUITARY EXTRACTS

Three lines of evidence indicate that the activity in extracts of the bovine pituitary resides in a protein with a subunit molecular weight of 31 kD on sodium dodecyl sulfate (SDS) polyacrylamide gels.

First, the activity was purified approximately 4400-fold from large quantities of bovine pituitary tissue by a combination of cation-exchange chromatography and gel filtration (Brockes et al., 1980). The most purified fraction (from phosphocellulose columns) displayed significant heterogeneity, but further purification by native gel electrophoresis at pH 4.5 indicated that the activity was associated with a component of molecular weight of 30 kD when analyzed by a second dimension of SDS-gel electrophoresis (Brockes et al., 1980).

Second, after immunizing mice with side fractions from the phosphocellulose column, we derived four monoclonal antibodies that complex with the activity (Lemke and Brockes, 1981, 1984). None of the four inhibits activity, and the immune complexes must therefore be precipitated with a second antibody in order to analyze their reaction. All four antibodies react with a band of molecular weight of 31 kD in SDS gels of partially purified preparations from the pituitary (Lemke and Brockes, 1984). The immunoreactivity in these experiments is stable to heating the preparations in SDS with β-mercaptoethanol prior to electrophoresis.

It could still be argued that the 31-kD species is a carrier molecule that is inactive in stimulating proliferation but is associated with the active species. This possibility is made very unlikely by the third line of evidence: It is possible to recover the biological activity from SDS gels run in the absence of disulfide reducing agents. When various preparations are analyzed in this way, the activity is recovered as a peak of molecular weight of 31 kD (Lemke and Brockes, 1984). Minor peaks of activity at 56 and 110 kD are occasionally observed, suggesting the existence of aggregates. No activity is recovered if the gels are run in the presence of reducing agents.

Taken together, these results provide strong evidence that GGF activity resides in a protein of molecular weight of 31 kD.

LARGE-SCALE PURIFICATION OF BOVINE PITUITARY GGF

The 31-kD species has been purified to apparent homogeneity by combining column chromatography procedures with a final step of preparative SDS-gel electrophoresis (Lemke and Brockes, 1984). A summary of such a purification from 20,000 bovine anterior lobes is shown in Table 2. The estimated purification factor is approximately 100,000-fold (see Lemke and Brockes, 1984, for details), and the 31-kD species is recovered at about 4 µg per 1000 anterior lobes. When corrected for recovery, this corresponds to approximately 10^{-13} moles per lobe. The estimated dissociation constant for the putative high-affinity receptor on Schwann cells is approximately 10^{-10} M, a value comparable to that obtained with EGF (Carpenter et al., 1975) and PDGF (Heldin et al., 1981b; Bowen-Pope and Ross, 1982). A detectable effect on proliferation of cultured Schwann cells is observed at a concentration of 5×10^{-12} M.

A rigorous comparison with other growth factors must await amino-acid-sequencing studies, but the localization of GGF in the nervous system and its action on Schwann cells appear to make it distinct, although there are some striking similarities with PDGF. Both are basic, heat-stable molecules with a molecular weight of approximately 31 kD whose activity can be recovered from SDS gels run in the absence, but not in the presence, of reducing agents (Ross et al., 1979; Antoniades, 1981; Deuel et al., 1981; Heldin et al., 1981a). GGF does not, however, give rise to smaller peptides when treated with reducing agent, and PDGF is not active on cultured rat Schwann cells (Lemke and Brockes, 1984). These differences suffice to distinguish them at present, but it may be that the two growth factors, and others with similar properties such as fibroblast-derived growth factor (Dicker et al., 1981), belong to the same family.

Table 2. Large-Scale Purification of GGF from Bovine Pituitary [a]

Step	Total Protein (mg)	Recovery of Activity (%)	Purification (X-fold)
Crude extract	400,000	—	—
Ammonium sulfate fraction	202,000	100	2
CM-cellulose (batch elution)	1,200	30	100
CM-cellulose (gradient elution)	210	15	250
AcA 44 Ultrogel-gel filtration	19	6.8	1,250
Phosphocellulose	1.1	3.1	10,000
SDS-gel electrophoresis	0.08	0.45	100,000

[a] The activity was purified from 20,000 lyophilized anterior lobes. The calculation of purification and other estimates are given in Lemke and Brockes (1984).

Table 3. Comparison of GGF in Bovine Pituitary and Caudate Nucleus[a]

Property	Investigators
Recovery of activity at $M_r = 31$ kD after SDS-gel electrophoresis	Lemke and Brockes (1984)
Immunoreactivity with monoclonal anti-pituitary GGF at $M_r = 31$ kD	Lemke and Brockes (1984)
Elution of activity from phosphocellulose at pH 6.0	Brockes and Lemke (1981)
Migration of activity on electrophoresis at pH 4.5	Brockes and Lemke (1981)

[a] Bovine pituitary and caudate GGF are indistinguishable with respect to these four properties.

REGIONAL DISTRIBUTION AND CELL TYPE SPECIFICITY

Extracts of brain contain significant activity in stimulating proliferation of rat Schwann cells, although the specific activity of such extracts is approximately one-fifth of those from the pituitary. When bovine brain is dissected into different areas, all areas yield extracts with significant activity on Schwann cells, but there is a five- to 10-fold variation in specific activity that is quite reproducible. The caudate nucleus is the only area with a higher specific activity (1.2-fold) than the pituitary (Brockes et al., 1981), and it provides a convenient source for comparison of brain and pituitary molecules. The activity in partially purified caudate extracts is compared to that in the pituitary in Table 3. The properties of the activities from the two sources are indistinguishable thus far, and the same GGF molecule may well be present in both. The localization of GGF to distinct cell types in the brain is not established, as attempts to localize it by immunohistochemistry have not been successful, perhaps due to its low level and widespread distribution. Nonetheless, its high level in the pituitary and regional distribution in brain are indicative of a neuronal, rather than a glial, localization.

The presence of GGF in the brain raises the question of its action on central glial cells. In mixed cultures of cells from five-day-old rat corpus callosum, partially purified bovine pituitary GGF stimulated the division of astrocytes (Brockes et al., 1980), a result confirmed by Kim et al. (1983), who have successfully used GGF to grow these cells in serum-free medium. On further analysis by native gel electrophoresis at pH 4.5, the activities against astrocytes and Schwann cells migrated together (Brockes et al., 1980), providing strong evidence that the same molecule acts on both peripheral and central glial cells. In parallel experiments, the GGF did not stimulate division of oligodendrocytes or microglia in corpus callosum cultures (Brockes et al., 1980). Its action is not confined to neural cells, since it was capable of stimulating fibroblasts derived from primary rat muscle cultures (Brockes and Lemke, 1981). Therefore, cell specificity may be quite broad, a feature shared with EGF and PDGF, which also apparently act on glial cells (Leutz and Schachner, 1981) as well as on mesenchymal derivatives. As emphasized earlier, the singular feature in this analysis is that the rat Schwann cell responds to GGF but not to other growth factors.

POSSIBLE ROLES OF GGF

Purified Schwann cells have allowed the isolation of a new and potent mitogenic growth factor from the nervous system. It is a basic protein of molecular weight of 31 kD that is found at highest levels in the pituitary and caudate nucleus, and it acts on astrocytes and fibroblasts as well as Schwann cells. What is the biological role of such a molecule both in the nervous system and outside it? The answer to this question is not known, but we can consider several possibilities.

1. GGF may be released from the pituitary as a circulating growth factor/hormone to act on a variety of cell types. It is clearly important to determine if it is present in the plasma of normal and hypophysectomized animals and to identify the cell type(s) in the anterior lobe that make it.

2. GGF may act to stimulate astrocyte or Schwann cell proliferation during development or after injury to the central or peripheral nervous system. It may be significant that GGF acts on the type I astrocyte, according to a recent classification (Raff et al., 1983), and that it is this cell type which is found after lesioning the corpus callosum (Raff et al., 1983).

3. GGF may be important in the pathogenesis of glial tumors. It is particularly interesting that neurofibromatosis involves abnormal Schwann cell proliferation together with astrocytoma in some patients. In view of the recent remarkable association between PDGF and the transforming gene of Simian sarcoma virus (Robbins et al., 1983; Waterfield et al., 1983), the possibility exists that GGF may play a role in some gliomas.

4. GGF may be involved in that set of neurotrophic phenomena that reflects the mitogenic effect of nerve cells. It is this possibility that we wish to consider further in relation to amphibian limb regeneration. We believe that it offers a favorable situation for investigating the issues of growth control *in vivo* that were considered at the beginning of this chapter.

NEURONAL DEPENDENCE OF URODELE LIMB REGENERATION

After limb amputation in Urodele amphibians, the surface of the wound is healed during the first 3–4 days by migrating epidermal cells that form the apical cap. Blastemal cells arise locally at the plane of amputation, divide several times during the next two weeks to form a protruding conical bud at the end of the stump, and undergo cytodifferentiation and morphogenesis to give rise to the internal mesenchymal tissues of the regenerate (see Wallace, 1981, for a review). These events have been most widely studied in newts (*Triturus* and *Notophthalmus* sp.), salamanders (*Ambystoma maculatum*), and axolotls (*Ambystoma mexicanum*). Two key phenomena underlie the early events of regeneration: first, the appearance of blastema cells at the site of amputation; and second, their division, stimulated by the innervation of the limb. While this account is principally concerned with the latter aspect, the first will be considered in connection with cellular identification by monoclonal antibodies.

Although the importance of the nerve in limb regeneration has been recognized since the nineteenth century, the modern analysis of its role is particularly associated with Singer (1952), who studied the dependence of regeneration on innervation of the newt forelimb with quantitative methods. The forelimb is supplied by spinal nerves 3, 4, and 5, which merge at the base of the limb to form the brachial plexus. By interrupting the nerves at different levels, it is possible to vary the density of innervation to the regenerate over a wide range. While approximately 70% of the normal innervation is adequate to ensure regeneration of all members of a group, a reduction below this level leads to a reduction in the proportion of regenerates (Singer, 1946, 1947). There is thus a need for an adequate quantity of innervation. This quantitative requirement can, however, be met by both motor and sensory fibers—with the appropriate surgery, it is possible to derive limbs with a purely sensory (Singer, 1943) or purely motor (Sidman and Singer, 1960) innervation, and both are fully capable of supporting regeneration.

Although the nerve exerts a profound effect on many metabolic activities in the blastema, it is appropriate to concentrate on its effect on cell division. A variety of investigations have shown that denervation prior to amputation or during the early period after amputation leads to a profound drop in the mitotic index in the blastema (Singer and Craven, 1948; Mescher and Tassava, 1975; Maden, 1978) and usually in the tritiated thymidine labeling index (although see Mescher and Tassava, 1975). Denervation is followed by an immediate transient increase in the mitotic index (Singer and Craven, 1948), perhaps due to a release of components from the degenerating nerve, but this rapidly subsides below the control values. It should be noted that the effect of denervation on regeneration is dependent on the stage at which the operation is performed (Singer and Craven, 1948; Grim and Carlson, 1974). If the adult newt is denervated up to approximately 13 days after amputation, the mitotic index falls well below the control and regeneration does not proceed (Singer and Craven, 1948). After this time, the effect of denervation is to produce a much smaller decrease in the index, and regeneration produces a limb that is smaller but well formed. One interpretation of these data is that a critical number of blastemal cells are required to initiate those morphogenetic processes that give rise to form and cytodifferentiation. The difference in the effect of denervation on mitosis at different stages after amputation is an issue that we shall consider later.

The effect of the nerve is usually attributed to the direct action of a mitogenic growth factor on the blastemal cells. Other possibilities have been suggested, for example that the nerve exerts its effect by stimulating the local vascular supply to the blastema (Smith and Wolpert, 1975), but this seems less likely in view of the rapid metabolic effects of denervation. Several procedures have been employed to assay for the putative activity, including the infusion or injection of material into the denervated blastema (Singer et al., 1976) or the stimulation of division in cultured explants of the blastema (Choo et al., 1978) by extracts, growth factors, or cultured dorsal root ganglia (Globus and Globus, 1977). Few explicit suggestions have been made about the molecular identity of the endogenous mitogen. It has been suggested that substance P or other neuropeptides might be responsible (Globus et al., 1983), although this is not obviously consistent with the comparable potency of motor and sensory neurons in sustaining regeneration. Another suggestion has been fibroblast growth factor (Gospodarowicz and Mescher, 1980), although a molecule with these properties has not yet been demonstrated in the blastema.

At this point, it is worth considering what criteria must be fulfilled to establish the molecular identity of the putative neurotrophic factor. It is probably impractical to isolate such a molecule from the blastema itself, so it is important to consider growth factors isolated and characterized in other systems. It might be difficult to obtain a convincing demonstration that one growth factor is important, particularly if there are, in fact, several neurotrophic factors that act synergistically on overlapping populations of blastemal cells. As in the case of neurotransmitter identification, a candidate factor ideally should satisfy a number of criteria that together would constitute an acceptable case. A first criterion is that the candidate factor must be found in the blastema and, second, that the amount of the factor must decrease after denervation.

PRESENCE OF GGF ACTIVITY IN URODELES

GGF has been characterized in detail only in bovine pituitary and brain, but it is possible to detect an activity that stimulates rat Schwann cells in rat pituitary (Raff et al., 1978a) and in chicken, frog, newt, and axolotl brain (G. E. Lemke and J. P. Brockes, unpublished observations). Although it is difficult to obtain adequate amounts of amphibian brain for significant purification of the activity, some progress has been made in its chemical characterization. The activity in newt and axolotl brain elutes from the cation-exchange resins phosphocellulose and mono-S at the same ionic strength as the activity in bovine pituitary and brain. In addition, newt brain activity has been recovered from SDS-gel electrophoresis at a molecular weight of 31 kD. Taken together with the restrictive nature of the rat Schwann cell proliferation assay discussed earlier, it seems likely that a molecule closely related to bovine GGF is present in the urodele nervous system.

Activity is also detectable in the blastema, as shown in Figure 1. The forelimbs of adult axolotls were bilaterally amputated and blastemas were allowed to form. The blastemas, segments of the intact hindlimbs, and the brains were extracted separately, and the extracts were analyzed by phosphocellulose ion-exchange chromatography and the rat Schwann cell proliferation assay. A distinct peak of activity was detected in brain and blastema extracts migrating at a position close to that of bovine GGF. In contrast, activity in control limb segments was barely detectable.

GGF activity in the blastema is apparently decreased by denervation. A group of newts were bilaterally amputated to allow early blastema formation and then unilaterally denervated at the level of the brachial plexus. One week later the right-denervated and left-innervated blastemas were pooled, extracted, and analyzed by ion-exchange chromatography on high-performance mono-S FPLC columns with a volatile buffer. After correction for the logarithmic dose–response relationship, it appears that the activity in the central GGF peak is decreased 80–90% after one week of denervation (Brockes, 1984). Similar results have been obtained in comparable experiments on axolotls. These results indicate that GGF fulfills the first two criteria for the putative factor that is released by nerves to stimulate proliferation in limb regeneration. It is a potent mitogen, present in

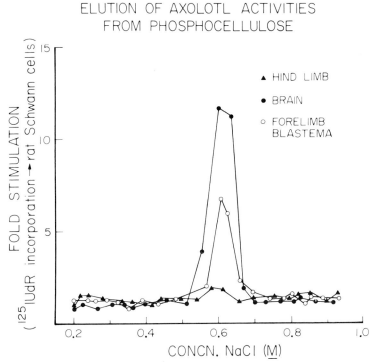

Figure 1. *Elution of axolotl GGF activity from phosphocellulose.* Eight axolotls were bilaterally amputated between elbow and wrist. The blastemas arising after three weeks were removed and a second set of blastemas allowed to form. After three more weeks the animals were sacrificed and the blastemas, brains, and control hindlimb segments were separately pooled and stored at −90°C. Tissue was extracted by homogenization in 0.15 M $(NH_4)_2SO_4$ (Brockes et al., 1980), followed by acidification to pH 4.5, centrifugation, and dialysis into 0.1 M sodium phosphate pH 6.0, 0.2 M NaCl. Samples (3 mg protein) were applied to 0.5 ml phosphocellulose columns and eluted with a 10 ml 0.2–1.0 M NaCl gradient which was monitored by pumping the column effluent through a conductivity cell. Aliquots of the resulting fractions were diluted and assayed in the Schwann cell proliferation assay described previously (Brockes et al., 1980; Lemke and Brockes, 1984).

the amphibian nervous system and in the regenerating blastema, and its level in the blastema decreases on denervation.

A third test would be to introduce blocking antibodies to GGF into the innervated blastema and determine if this mimics the effects of denervation. This approach is the basis of the celebrated immunosympathectomy experiment that used antibodies to NGF (Levi-Montalcini and Booker, 1960)—an experiment that has not yet been successfully accomplished with mitogenic growth factors. None of the four monoclonal antibodies to bovine pituitary GGF is able to block mitotic activity (Lemke and Brockes, 1984), and therefore they may not be appropriate for this type of experiment. It seems important to obtain a polyclonal antiserum to purified GGF or a panel of antibodies raised against synthetic peptides representing different regions of the molecule.

A fourth criterion is that GGF should stimulate the division of the appropriate population of blastemal cells. During the first two weeks after amputation, at the stage when the nerve dependence of regeneration in the adult newt is defined, it is clearly the histologically undifferentiated blastema cells that are stimulated to divide. Although blastema cells may look alike, they probably originate from different tissue sources (Chalkley, 1954), and it is quite possible that there are distinct populations responding to different mitogenic influences. In order to approach questions about the origin, identity, and proliferation of these cells, we recently have derived certain monoclonal antibodies that promise to assist in this task.

MONOCLONAL ANTIBODIES TO THE BLASTEMA

We have immunized Balb/c mice with newt forelimb blastemas, generally from the midbud stage (approximately 16 days after amputation). The blastemas were prepared in a variety of ways, including crude particulate or somewhat purer membrane preparations from whole tissue or dissociated cell homogenates. About 2500 antibody-secreting clones (resulting from 20 fusions) have been screened by indirect immunofluorescence on sections of the regenerating limb. Clones that appeared to give specific staining of the blastema were cloned at least twice and grown as ascites tumors to allow testing of antibody specificity at high immunoglobulin concentrations. In our experience, it is relatively easy to obtain reagents that are specific for the differentiated derivatives such as skin, muscle, or cartilage, but it is much more difficult to obtain ones that are specific for blastemal cells. Many clones react with extracellular matrix materials in the blastema, and these have been discarded as cell markers. Other reagents that appeared in initial screenings to be specific for blastemal cells were found to stain other cell types when tested at high concentration in ascites fluid.

Two monoclonal antibodies that have been useful in dissecting the blastema are 22/18 and 22/31 (see Table 4). The general staining pattern of these reagents in the limb will be described, and three examples will be given of how they have provided new information about the issues considered above. Antibody 22/18 gives no detectable staining of cells in the normal limb. It first appears 4–5 days after amputation in Schwann cells of the nerve sheath at the plane of amputation (Figure 2A,B). Cells positive for 22/18 constitute the majority of cells in the blastema during the first two weeks after amputation (Figure 2C,D); during the third week, the number of $22/18^+$ cells declines and becomes a small minority (Figure 2G,I), although it apparently includes that population differentiating into striated muscle (see below). In the later regenerate, 22/18 staining is lost from all cell types in the limb.

Monoclonal antibody 22/31 stains mesenchymal fibroblasts of the normal limb in the nerve sheath, in muscle and connective tissue, and, most prominently, in the dermis (Figure 2E,F). After amputation it is expressed strongly by a small minority of cells in the early blastema, although the most sensitive immunofluorescence procedures employing antibody amplification do reveal low levels in all blastema cells. During the third week it is expressed strongly by a majority of blastemal cells (Figure 2G,H) but is lost in the later regenerate, at least in terms of expression at higher levels, from all cell types except fibroblasts. It is apparent that during the first three weeks the levels of expression of 22/18 and 22/31 in

Table 4. Properties of Monoclonal Antibodies to the Newt Blastema

Antibody Designation	Specificity in the Mature Limb[a]	Additional Specificity in the Regenerating Limb[a]	Subclass[b]
22/31	Fibroblasts in dermis, muscle, and perichondrium	Small population of blastemal cells at early stages and most, if not all, blastemal cells at later stages	IgM
22/18	No detectable binding	A majority of blastemal cells at early stages and a minority of blastemal cells at later stages	IgM
12/101	Myofibers	A small population of blastemal cells at the dedifferentiation stage of regeneration	IgG

[a] Specificity is determined by staining tissue sections of regenerating and mature limb by immunofluorescence as described in the legend of Figure 2 and, in more detail, in Kintner and Brockes (1984).
[b] Determined as described in Kintner and Brockes (1984).

the blastema are in some sense "reciprocal," and this has been examined in more detail by using a double-label immunofluorescence procedure: Antibody 22/31 is biotinylated and detected with fluorescein-avidin; 22/18 is detected with a heteroantiserum to mouse IgM followed by a rhodamine anti-immunoglobulin to detect the heteroantiserum. These studies reveal that nonoverlapping cell populations show high-level staining with 22/31 and with 22/18 in the nerve sheath and the early blastema. During the third week, some double-labeled cells are detected in the blastema.

It is important to note that it is not possible to infer rigorously the relationships of cell types bearing the two antigens by observing sections at different times. It is quite possible, and indeed we favor the idea, based on the occurrence of double-labeled cells, that many 22/18$^+$ cells in the early blastema convert to 22/31$^+$ cells in the late blastema. This is a point that could only be established by other methods, for example, by injecting a pure population of 22/18$^+$ cells into an irradiated blastema and observing their fate. Nonetheless it appears that particularly 22/18, but also 22/31, "dissect" the blastema, as indicated by the following three results.

1. The origin of blastemal cells after amputation is an interesting and sometimes controversial problem. Anatomists who have studied sections of the regenerating limb with both light- (Thornton, 1938; Chalkley, 1954) and electron-microscope (Hay, 1959; Lentz, 1969) techniques have generally agreed that blastemal cells arise at the plane of amputation by dedifferentiation from the mature tissues, including muscle, cartilage, Schwann cells, and connective tissue. A more experimental approach (see Wallace, 1981, for a review) has been to graft marked tissue into normal or irradiated blastemas and observe its fate. The purity of the grafted cell population is often unclear in such studies, and this leads to an

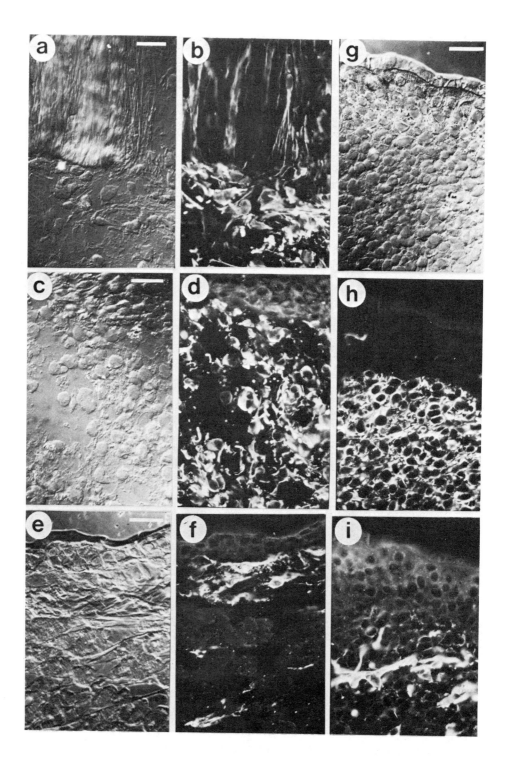

uncertainty in establishing what cell types can dedifferentiate into blastemal cells. The putative dedifferentiation of the multinucleated myofibril, although described in great detail by anatomical methods (Thornton, 1938; Hay, 1959; Lentz, 1969), has always been controversial (see Hay, 1970) in view of the general acceptance of reserve satellite cells as the source of muscle regeneration in frogs and higher vertebrates (Konigsberg et al., 1975; Bischoff, 1979). If dedifferentiation of myofibrils occurs, a prediction is that cells with properties of both myofibrils and blastemal cells should exist at an early stage of regeneration. The use of monoclonal antibodies as cell markers allows one to detect such cells, thus providing a test of the dedifferentiation hypothesis (Kintner and Brockes, 1984).

To obtain a marker for myofibrils, we isolated a monoclonal antibody called 12/101 (Table 4), which gives detectable staining only of striated myofibrils in sections of normal tissue. The staining often has a striated pattern, and the antigen is probably some muscle-specific determinant in the contractile apparatus. After staining sections of the limb 10–12 days after amputation, it is possible to detect both mononucleate cells and occasional multinucleate myofibers staining strongly for both 12/101 and 22/18 (Kintner and Brockes, 1984). The mononucleate cells are situated at the base of the blastema, just proximal to the fragmenting myofibers, and are detected at the stage after wound healing but before differentiation. Double-labeled cells are also detected later in regeneration in the blastema where blastemal cells differentiate into muscle. In both contexts—dedifferentiation and differentiation—12/101 is found only in conjunction with 22/18 and not with 22/31. Although the existence of reserve cells bearing both antigens cannot be ruled out entirely (see Kintner and Brockes, 1984, for discussion), these data provide support that blastemal cells may arise, at least in part, by dedifferentiation of myofibrils.

2. As noted above, 22/18 staining first appears some four days after amputation in the Schwann cells of the nerve sheath (Figure 2A,B). The staining is not in the axons, as evidenced by examination of cross sections, and is not present in

Figure 2. *Regenerating and mature limbs stained with 22/18 and 22/31 antibodies.* Tissue sections were cut from mature and regenerating limbs after fixation with periodate-lysine and then stained with 22/18 and 22/31 antibodies according to two protocols. In the first protocol, 22/18 or 22/31 binding to the section was detected by using a rabbit antibody to mouse immunoglobulin followed by a rhodamine-conjugated goat antibody specific for rabbit immunoglobulin. In the second protocol, binding of both 22/18 and 22/31 could be viewed on the same section by detecting the binding of a biotinylated 22/31 antibody using fluorescein-conjugated avidin and by detecting 22/18 binding as described in the first protocol. *a* and *b*: Tissue section of a regenerating limb at the early bud stage stained with 22/18 using the first protocol and photographed under Nomarski optics (a) or rhodamine optics (b) to show 22/18 staining. Note that the nerve sheath and the blastemal cells lying below it stain with 22/18. *c* and *d*: Tissue section of a regenerating limb at the middle bud stage stained with 22/18 using the first protocol and photographed under Nomarski optics (c) or rhodamine optics (d) to show 22/18 staining. Note that most blastemal cells (middle field) stain with 22/18 while the epidermis (upper field) shows only background staining. *e* and *f*: Tissue section from a mature limb stained with 22/31 using the first protocol and photographed under Nomarski optics (e) or rhodamine optics (f) to show 22/31 staining. Note that fibroblasts in dermis and muscle stain with 22/31. *g–i*: Tissue section from a regenerating limb at the late bud stage, stained with 22/31 and 22/18 using the second protocol and photographed under Nomarski optics (g), fluorescein optics to show 22/31 staining (h), or rhodamine optics to show 22/18 staining (i). Note that the epidermis (upper field) does not stain with either 22/31 or 22/18 and that 22/31 stains most if not all blastemal cells while 22/18 stains a subpopulation. Calibration bars = 50 μm.

the fibroblasts of the sheath as revealed by double-label staining with 22/31. Schwann cells have been noted in previous anatomical descriptions as contributing to the blastema via dedifferentiation (Thornton, 1938; Chalkley, 1954), and in some experimental studies (Wallace, 1972; Maden, 1977) it has been suggested that they may give rise to all the internal tissue of the regenerate. The 22/18 staining supports the idea that Schwann cells contribute because, in sections, $22/18^+$ cells appear to be migrating out of the nerve and into the blastema. It would be desirable to quantitate the incidence of such cells by using a double-label method similar to that used with muscle—for example, by staining for a myelin protein in conjunction with 22/18—but our impression is that Schwann cells make a major contribution to the early blastema.

3. The antibodies show that, irrespective of the onset of cytodifferentiation, there is a difference in the cellular composition of the early and late blastemas. The "reciprocal" expression of 22/18 and 22/31 correlates in time with the transition of the blastema from a full to a partial dependence on the nerve. This correlation might reflect a difference in the dependence of the 22/18 and 22/31 populations on the neuronal mitogen.

In order to investigate this possibility, newts were amputated bilaterally and then denervated unilaterally at six and 14 days after amputation. After allowing six days for the effect of the nerve to diminish, tritiated thymidine was injected into the animals for 24 hours. The limbs were sectioned, stained with the two antibodies, and coated with emulsion for light-microscope autoradiography. The results suggest that proliferation of the $22/18^+$ population is substantially dependent on the nerve at both times after amputation: the proliferation of the $22/31^+$ population is substantially independent at the later time but is somewhat dependent earlier (Brockes, 1984).

These results indicate that 22/18 marks a population whose division is substantially dependent on the nerve, whereas $22/31^+$ cells are substantially independent in the late blastema, where they presumably proliferate in response to other factors. These data illustrate the importance of cell markers in such studies. It is quite possible that injection of a growth factor into the denervated blastema would stimulate division of $22/31^+$ but not $22/18^+$ cells. Clearly, this would not be a candidate for the endogenous neuronal mitogen. Our current aim is to test the role of GGF by injecting either purified bovine GGF or partially purified amphibian GGF into the newt blastema to determine if it stimulates division of $22/18^+$ cells. This seems to be the most pointed test available at present.

CONCLUSIONS

It is interesting that Schwann cells may make an important contribution to the early blastema, in view of the fact that rat Schwann cells do not respond to many mitogenic agents (Table 1). If $22/18^+$ blastema cells initially retain a dependence for division on the neuronal mitogen characteristic of their differentiated progenitors, and if such dependence is lost later because of their progressive dedifferentiation (much as 12/101 is lost by the cells derived from muscle) or conversion to $22/31^+$ cells, this would account broadly for the initial dependence on, and later independence from, the nerve.

The use of these markers will surely assist in understanding the more complicated developmental modulation of the relationship between nerve and limb (Brockes, 1984). It has been recognized for some time that if the limb of an axolotl develops in the absence of a nerve supply, such aneurogenic limbs are deficient in muscle (Popiela, 1976) but otherwise appear normal. The limbs will regenerate in the absence of a nerve supply (Yntema, 1959), but if they are innervated by transplantation to a normal embryo they become nerve-dependent for regeneration (Thornton and Thornton, 1970). The cellular or molecular mechanisms underlying the "addiction" (Singer, 1965) to the nerve are not known. They could involve differences in the cellular composition of the blastema in normal and aneurogenic limbs, influences exerted by the neuronal mitogen on cellular susceptibility to other mitogens, and other as yet unrecognized factors. The availability of markers for the blastemal cells should allow these possibilities to be investigated. We suspect that an understanding of these phenomena would shed light on the more complex questions of cell division *in vivo* that were considered at the beginning of this account.

REFERENCES

Antoniades, H. N. (1981) Human platelet-derived growth factor (PDGF): Purification of PDGF-I and PDGF-II and separation of their reduced subunits. *Proc. Natl. Acad. Sci. USA* **78**:7314–7317.

Balk, S. D., J. F. Whitefield, T. Youdalo, and A. D. Braun (1973) Roles of calcium, serum, plasma and folic acid in the control of proliferation of normal and Rous sarcoma virus-infected chicken fibroblasts. *Proc. Natl. Acad. Sci. USA* **70**:675.

Bischoff, R. (1979) Tissue culture studies on the origin of myogenic cells during muscle regeneration in the rat. In *Muscle Regeneration*, A. Mauro, ed., pp. 13–20, Raven, New York.

Bowen-Pope, D. F., and R. Ross (1982) Platelet-derived growth factor. II. Specific binding to cultured cells. *J. Biol. Chem.* **257**:5161–5171.

Brockes, J. P. (1984) Mitogenic growth factors and nerve dependence of limb regeneration. *Science* **225**:1280–1287.

Brockes, J. P., and G. E. Lemke (1981) The neuron as a source of mitogen. In *Development in the Nervous System*, D. R. Garrod and J. D. Feldman, eds., pp. 309–327, Cambridge Univ. Press, Cambridge.

Brockes, J. P., K. L. Fields, and M. C. Raff (1977) A surface antigenic marker for rat Schwann cells. *Nature* **266**:364–366.

Brockes, J. P., K. L. Fields, and M. C. Raff (1979) Studies on cultured rat Schwann cells. I. Establishment of purified populations from cultures of peripheral nerve. *Brain Res.* **165**:105–118.

Brockes, J. P., G. E. Lemke, and D. R. Balzer (1980) Purification and preliminary characterization of a glial growth factor from the bovine pituitary. *J. Biol. Chem.* **255**:8374–8377.

Brockes, J. P., K. J. Fryxell, and G. E. Lemke (1981) Studies on cultured rat Schwann cells. *J. Exp. Biol.* **95**:215–230.

Carpenter, G., K. J. Lembach, M. M. Morrison, and S. Cohen (1975) Characterization of the binding of ^{125}I-labeled epidermal growth factor to human fibroblasts. *J. Biol. Chem.* **250**:4297–4304.

Chalkley, D. T. (1954) A quantitative histological analysis of forelimb regeneration in *Triturus viridescens*. *J. Morphol.* **94**:21–70.

Choo, A. F., D. M. Logan, and M. P. Rathbone (1978) Nerve trophic effects: An *in vitro* assay for factors involved in regulation of protein synthesis in regenerating amphibian limbs. *J. Exp. Zool.* **206**:347–354.

Cohen, S. (1962) Isolation of a submaxillary gland protein accelerating incisor eruption and eyelid opening in the new-born animal. *J. Biol. Chem.* **237**:1555–1562.

Cornbrooks, C., D. J. Carey, J. A. McDonald, R. Timple, and R. P. Bunge (1983) *In vivo* and *in vitro* observations on laminin production by Schwann cells. *Proc. Natl. Acad. Sci. USA* **80**:3850–3854.

Deuel, T. F., J. S. Huang, R. T. Proffitt, J. V. Baezinger, D. Chang, and B. B. Kennedy (1981) Human platelet-derived growth factor: Purification and resolution into two active protein fractions. *J. Biol. Chem.* **256**:8896–8899.

De Vries, G. H., J. L. Salzer, and R. P. Bunge (1982) Axolemma enriched fractions from the peripheral nervous system and central nervous system are mitogenic for cultured Schwann cells. *Dev. Brain. Res.* **3**:295–299.

De Vries, G. H., L. Minier, and B. L. Lewis (1983) Further studies on the mitogenic response of cultured Schwann cells to rat CNS axolemma-enriched fractions. *Dev. Brain Res.* **9**:87–93.

Dicker, P., P. Pohjanpelto, P. Pettican, and E. Rozengurt (1981) Similarities between fibroblast-derived growth factor and platelet-derived growth factor. *Exp. Cell Res.* **135**:221–227.

Fields, K. L., C. Gosling, M. Megson, and P. L. Stern (1975) New cell surface antigens in rat defined by tumors of the nervous system. *Proc. Natl. Acad. Sci. USA* **72**:1296–1300.

Globus, M., and S. V. Globus (1977) Transfilter mitogenic effect of dorsal root ganglia on cultured regeneration blastemata in the newt, *Notophthalmus viridescens*. *Dev. Biol.* **56**:316–328.

Globus, M., S. V. Globus, A. Kesik, and G. Milton (1983) Roles of substance P and calcium in blastema cell proliferation in the newt *Notophthalmus viridescens*. In *Limb Development and Regeneration*, J. F. Fallon and A. I. Caplan, eds., pp. 513–524, Alan R. Liss, New York.

Gospodarowicz, D. (1975) Purification of fibroblast growth factor from bovine pituitary. *J. Biol. Chem.* **250**:2515–2520.

Gospodarowicz, D., and A. Mescher (1980) Fibroblast growth factor and the control of vertebrate regeneration and repair. *Ann. N.Y. Acad. Sci.* **339**:151–174.

Grim, M., and B. M. Carlson (1974) The formation of muscles in regenerating limbs of the newt after denervation of the blastema. *J. Embryol. Exp. Morphol.* **54**:99–111.

Hay, E. D. (1959) Electron microscopic observation of muscle dedifferentiation in regenerating *Amblystoma* limbs. *Dev. Biol.* **1**:555–585.

Hay, E. D. (1970) Regeneration of muscle in the amputated amphibian limb. In *Regeneration of Striated Muscle and Myogenesis*, A. Mauro, S. A. Shafiq, and A. T. Milhorat, eds., pp. 3–24, Excerpta Medica, Amsterdam.

Heldin, C.-H., B. Westermark, and A. Wasteson (1981a) Platelet-derived growth factor: Isolation by a large-scale procedure and analysis of subunit composition. *Biochem. J.* **193**:907–913.

Heldin, C.-H., B. Westermark, and A. Wasteson (1981b) Specific receptors for platelet-derived growth factor on cells derived from connective tissue and glia. *Proc. Natl. Acad. Sci. USA* **78**:3664–3668.

Kim, S. U., J. Stern, M. W. Kim, and D. E. Pleasure (1983) Culture of purified rat astrocytes in serum-free medium supplemented with mitogen. *Brain Res.* **274**:79–86.

Kintner, C. R., and J. P. Brockes (1984) Monoclonal antibodies identify blastemal cells derived from muscle dedifferentiation in limb regeneration of the newt. *Nature* **308**:67–69.

Konigsberg, U. R., B. H. Lipton, and I. R. Konigsberg (1975) The regenerative response of a single mature muscle fiber isolated *in vitro*. *Dev. Biol.* **45**:260–275.

Kreider, B., A. Mesing, H. Doan, S. Kim, R. Lisak, and D. Pleasure (1981) Enrichment of Schwann cell cultures from neonatal rat sciatic nerve by differential adhesion. *Brain Res.* **207**:433–444.

Lemke, G. E., and J. P. Brockes (1981) An immunochemical approach to the purification and characterization of glial growth factor. In *Monoclonal Antibodies to Neural Antigens*, R. McKay, M. C. Raff, and L. Reichardt, eds., pp. 133–140, Cold Spring Harbor Laboratory, Cold Spring Harbor, N.Y.

Lemke, G. E., and J. P. Brockes (1984) Identification and purification of glial growth factor. *J. Neurosci.* **4**:75–83.

Lentz, T. L. (1969) Cytological studies of muscle de-differentiation and differentiation during limb regeneration of the newt *Triturus*. *Am. J. Anat.* **124**:447–480.

Leutz, A., and M. Schachner (1981) Epidermal growth factor stimulates DNA synthesis of astrocytes in primary cerebellar cultures. *Cell Tissue Res.* **220**:393–404.

Levi-Montalcini, R., and B. Booker (1960) Destruction of the sympathetic ganglia in mammals by an antiserum to a nerve growth protein. *Proc. Natl. Acad. Sci. USA* **46**:384–391.

Maden, M. (1977) The role of Schwann cells in paradoxical regeneration in the axolotl. *J. Embryol. Exp. Morphol.* **41**:1–13.

Maden, M. (1978) Neurotrophic control of the cell cycle during amphibian limb regeneration. *J. Embryol. Exp. Morphol.* **48**:169–175.

Mescher, A. L., and R. A. Tassava (1975) Denervation effects on DNA replication and mitosis during the initiation of limb regeneration in adult newts. *Dev. Biol.* **44**:187–197.

Popiela, H. (1976) In vivo limb tissue development in the absence of nerves: A quantitative study. *Exp. Neurol.* **53**:214–226.

Raff, M. C., E. R. Abney, A. Hornby-Smith, and J. P. Brockes (1978a) Schwann cell growth factors. *Cell* **15**:813–822.

Raff, M. C., A. Hornby-Smith, and J. P. Brockes (1978b) Cyclic AMP as a mitogenic signal for cultured rat Schwann cells. *Nature* **273**:672–673.

Raff, M. C., E. R. Abney, J. Cohen, R. Lindsay, and M. Noble (1983) Two types of astrocytes in cultures of developing rat white matter: Differences in morphology, surface gangliosides, and growth characteristics. *J. Neurosci.* **3**:1289–1300.

Robbins, K. C., H. N. Antoniades, S. G. Devare, M. W. Hunkapiller, and S. A. Aaronson (1983) Structural and immunological similarities between simian sarcoma virus gene product(s) and human platelet-derived growth factor. *Nature* **305**:605–608.

Ross, R., A. Vogel, P. Davies, E. Raines, B. Karuja, M. J. Rivest, C. Gustafson, and J. Glomset (1979) The platelet-derived growth factor. In *Hormones and Cell Culture*, Vol. B., R. Ross and G. H. Sato, eds., pp. 965–971, Cold Spring Harbor Laboratory, Cold Spring Harbor, N.Y.

Rozengurt, E. (1981) Cyclic AMP: A growth-promoting agent for mouse 3T3 cells. *Adv. Cyclic Nucleotide Res.* **14**:429–442.

Salzer, J. L., A. K. Williams, L. Glaser, R. P. Bunge (1980) Studies of Schwann cell proliferation. II. Characterization of the stimulation and specificity of the response to a neurite membrane fraction. *J. Cell. Biol.* **84**:753–766.

Sidman, R. L., and M. Singer (1960) Limb regeneration without innervation of the apical epidermis in the adult newt, *Triturus*. *J. Exp. Zool.* **144**:105–110.

Singer, M. (1943) Regeneration of the forelimb of adult *Triturus*. II. The role of the sensory supply. *J. Exp. Zool.* **92**:297–315.

Singer, M. (1946) Regeneration of the forelimb of adult *Triturus*. V. The influence of number of nerve fibers, including a quantitative study of limb innervation. *J. Exp. Zool.* **101**:299–337.

Singer, M. (1947) Regeneration of the forelimb of adult *Triturus*. VI. A further study of the importance of nerve number, including quantitative measurements of limb innervation. *J. Exp. Zool.* **104**:223–249.

Singer, M. (1952) The influence of the nerve in regeneration of the amphibian extremity. *Quart. Rev. Biol.* **27**:169–200.

Singer, M. (1965) A theory of the trophic nervous control of amphibian limb regeneration, including a re-evaluation of quantitative nerve requirements. In *Regeneration in Animals and Related Problems*, V. Kiortsis and H. A. L. Trampusch, eds., pp. 20–30, North-Holland, Amsterdam.

Singer, M., and L. Craven (1948) The growth and morphogenesis of the regenerating forelimb of adult *Triturus* following denervation at various stages of development. *J. Exp. Zool.* **108**:279–308.

Singer, M., C. E. Maier, and W. S. McNutt (1976) Neurotrophic activity of brain extracts in forelimb regeneration of the urodele, *Triturus*. *J. Exp. Zool.* **196**:131–150.

Skaper, S., M. Manthorpe, R. Adler, and S. Varon (1980) Survival, proliferation and morphological specialization of mouse Schwann cells in a serum-free, fully defined medium. *J. Neurocytol.* **9**:683–697.

Smith, A. R., and L. Wolpert (1975) Nerves and angiogenesis in amphibian limb regeneration. *Nature* **257**:224–225.

Stern, P. L. (1973) Theta alloantigen on mouse and rat fibroblasts. *Nature New Biol.* **246**:76–78.

Taylor, J. M., W. M. Mitchell, and S. Cohen (1972) Epidermal growth factor: Physical and chemical properties. *J. Biol. Chem.* **247**:5928–5934.

Thornton, C. S. (1938) The histogenesis of muscle in the regenerating forelimb of larval *Amblystoma punctatum*. *J. Morphol.* **62**:17–47.

Thornton, C. S., and M. T. Thornton (1970) Recuperation of regeneration in denervated limbs of *Ambystoma* larvae. *J. Exp. Zool.* **173**:293–301.

Todaro, G. J., and H. Green (1963) Quantitative studies of the growth of mouse embryo cells in culture and their development into established lines. *J. Cell Biol.* **17**:299–313.

Wallace, H. (1972) The components of regrowing nerves which support the regeneration of irradiated salamander limbs. *J. Embryol. Exp. Morphol.* **28**:419–435.

Wallace, H. (1981) *Vertebrate Limb Regeneration*, Wiley, Chichester.

Waterfield, M., G. T. Scrace, N. Whittle, P. Stroobant, A. Johnson, A. Wasteson, B. Westermark, C.-H. Heldin, J. S. Huang, and T. Deuel (1983) Platelet-derived growth factor is structurally related to the putative transforming protein p28sis of simian sarcoma virus. *Nature* **304**:35–39.

White, F.U., C. Ceccarini, I. Georgiett, J. M. Matthews, and E. Costantino-Ceccarini (1983) Growth properties and biochemical characterization of mouse Schwann cells cultured *in vitro*. *Exp. Cell Res.* **148**:183–194.

Wood, P. (1976) Separation of functional Schwann cells and neurons from peripheral nerve tissue. *Brain Res.* **115**:361–375.

Yntema, C. L. (1959) Regeneration in sparsely innervated and aneurogenic forelimbs of *Amblystoma* larvae. *J. Exp. Zool.* **140**:101–123

Chapter 7

Mechanisms of Neuronal Migration in Developing Cerebellar Cortex

PASKO RAKIC

ABSTRACT
Light- and electron-microscope studies of the developing cerebellum indicate that the interaction of postmitotic granule cells with Bergmann glial fibers and parallel fibers is crucial for the morphogenetic transformation and migration of the granule cell across the developing molecular layer. Subsequent immunocytochemical analysis has confirmed the glial and neuronal nature of interacting cell processes. More recent studies have shown that the two horizontal neurites of the granule cell, which follow parallel fibers but ignore glial fibers, have lamellipodial endings characterized by specific ultrastructural features. In contrast, the descending, vertical process, which follows a glial surface but ignores nearby axons, lacks lamellipodia and has different cytological and membrane characteristics. A model of differential cell adhesion may account for both the dramatic change in granule cell shape and for the displacement of its soma even if the membranes of adjacent cells remain fixed at any point of their interface. According to this model, membrane fragments with specific binding properties are deposited near the tip of the leading process. The leading tip is highly motile and extends preferentially along the surface with the highest binding affinity. Several lines of evidence support this model, including the pattern and sequence of cell growth in vivo and in vitro, the mode of deposition and molecular composition of membranes, the rate of migration, and the consequences of genetic and experimental alteration of the developmental events involved in neuronal migration.

The most remarkable feature of the developing central nervous system in vertebrates is that individual neurons are constantly in motion although the tissue as a whole maintains independent stability of form. The term "migration" is used here to denote a displacement of postmitotic neurons from their places of origin in the proliferative zones to their final destinations in given structures. Neuronal migration may not be prominent in invertebrates or even in vertebrates with smaller brains, but it is prominent in the brains of larger mammals, including humans (Sidman and Rakic, 1973). Although in several respects the mechanisms of neuronal migration may be similar to the mechanisms of axonal growth, these two developmental processes are distinct entities with specific characteristics and need to be treated separately (Rakic, 1972a, 1985).

Although neurons have to move the greatest distances during formation of the primate telencephalon, perhaps the most instructive example of neuronal migration is the translocation of postmitotic granule cells in the developing cerebellar cortex. The main advantage of using this system is that both the displacement of the soma and the growth of the axon can be analyzed simultaneously in a single cell. Other assets of the cerebellum as a model system are that all cell types and the relationships between them are perhaps better understood than they are in other regions of the central nervous system, and that its structure is remarkably similar in different vertebrate species (Ramón y Cajal, 1955; Llinás, 1969; Palay and Chan-Palay, 1974). Finally, the basic developmental processes and the timing of major cellular events that can serve as a background have been determined in several vertebrate species, including primates (Rakic, 1971a, 1972a, 1973). Much of the following account is taken from work done in my laboratory on cerebellar development in primates and in rodents with single gene mutations affecting the formation of cerebellar structure.

BASIC CYTOLOGICAL ORGANIZATION

The basic cellular organization, synaptic connectivity, and geometric relationship of the normal cerebellar cortex are depicted in Figure 1. An understanding of

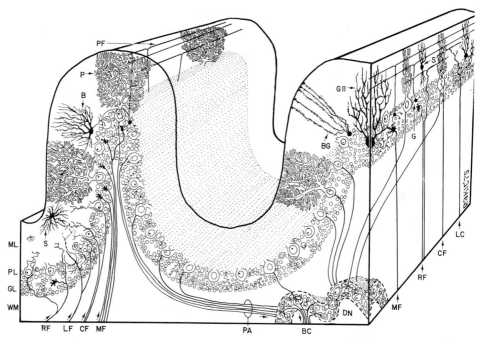

Figure 1. *Design of the cerebellar cortex reconstructed in three dimensions to demonstrate the relative positions, orientations, and connections of different classes of neurons and afferent fibers.* One and one-half cerebellar folia are dissected so that the surface facing the reader is oriented transverse to the folium and the surface on the right side lies longitudinal to the folium. Part of the convoluted dentate nucleus (DN) is drawn in the lower right corner of the diagram. B, basket cell; BC, brachium conjuctivum; BG, Bergmann glial cells; CF, climbing fibers; DN, dentate nucleus; G, granule cells; G II, Golgi type II neurons; GL, granular layer; LF, locus coeruleus fibers; MF, mossy fibers; ML, molecular layer; P, Purkinje cells; PA, Purkinje cell afferents; PF, parallel fibers; PL, Purkinje cell layer; RF, rafe fiber afferents; S, stellate cells; WM, white matter.

these relationships is essential to the analysis and discussion of granule cell migration. There are only three well-distinguished cellular layers: the molecular layer (ML), which contains stellate (S) and basket (B) cells, as well as Purkinje cell dendrites (P) penetrated by myriads of parallel fibers (PF); the Purkinje cell layer (PL), consisting of somata of Purkinje cells and Bergmann glial cells (BG); and the granule cell layer (GL), which in addition to granule cells contains Golgi type II cells. The basic connectivity and synaptic organization is essentially identical throughout the cerebellar hemispheres and vermis and has been precisely delineated (Palay and Chan-Palay, 1974). The Purkinje cell is the principal neuron, and its axons (PA) provide the only efferent outflow from the cortex to deep cerebellar nuclei, including the dentate nucleus (DN). There are two main extrinsic cerebellar afferents: climbing fibers (CF), which form asymmetrical synapses with specific types of spines situated on the large dendrites of Purkinje cells; and mossy fibers (MF), which form asymmetrical synaptic contacts with dendrites of granule cells (G) and Golgi type II cells (GII) in so-called mossy fiber glomeruli. A synaptic link between the mossy fiber input and Purkinje cells is provided by a network of parallel fibers (PF)—composed of the horizontal segments of the T-shaped granule cell axons—which form symmetrical synapses with the small spines of Purkinje cell dendritic branches. The element that contributes the most to the beautiful geometrical organization of the cerebellar molecular layer is the unusually precise parallel orientation of their fibers. The present account is concerned with the role of neuronal migration in building this extraordinarily precise cellular structure.

MAJOR CELLULAR EVENTS

As established at the turn of the century, the cerebellar cortex is built of neurons from two separate sources (Ramón y Cajal, 1955). The earliest neuronal population is generated from the proliferative zone lining the fourth cerebral ventricle, termed the ventricular zone. More recent [^{3}H]thymidine autoradiographic analysis in the monkey showed that during the period between E36 (embryonic day 36) and E63, successive generations of postmitotic neurons leave the ventricular zone and enter the territory of the intermediate zone. These cells, most of which are generated within the first third of the 165-day gestational period, become Purkinje cells, Golgi type II neurons, and neurons of the deep cerebellar nuclei (Rakic, 1979; Gould and Rakic, 1981). However, while the neurons of the deep nuclei remain relatively close to the ventricular surface, Purkinje cells migrate further toward the pial surface (upward-directed arrow in Figure 2). At early fetal ages, Purkinje cells are still small and closely packed and form a multilayered band below the acellular marginal zone. In the rhesus monkey, all Purkinje cells are generated by E43 (Rakic, 1979) and are packed within this multilayered band. During the subsequent growth of the cerebellar cortex, Purkinje cells spread gradually to form a one-cell-thick layer. The extraordinary expansion of the size of the cerebellar cortex from E43 onward is caused by an increase in Purkinje cell somatic and dendritic volume, by the growth and multiplication of glial cells, and, most significantly, by the generation of an enormous number of granule cells from a newly formed proliferative zone, the transient external granular layer (EG in Figure 2).

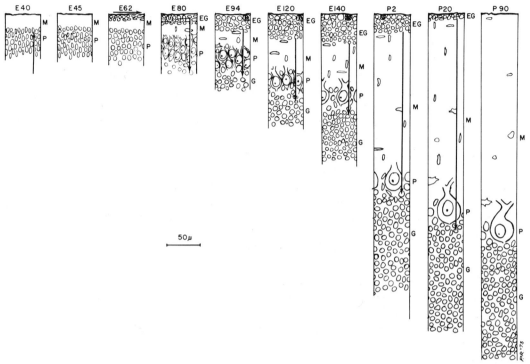

Figure 2. *Semidiagrammatic drawing of the cerebellar cortex at various embryonic (E) and postnatal (P) ages.* This drawing summarizes the pattern and directions of cell migration and the timing of major developmental events in the developing monkey cerebellum. The diagram is based on autoradiographic data from animals injected with [^{3}H]thymidine at various pre- and postnatal ages and killed at different intervals. Arrows indicate the direction of neuronal migration. EG, external granular layer; G, granular layer; M, molecular layer; P, Purkinje cell layer. See Rakic 1971a, 1972a, and 1973 for more details and experimental evidence.

The external granular layer has a unique developmental history and merits close examination. Its progeny provide an example of cell migration unmatched elsewhere in the developing central nervous system. The external granular layer, which was first described in human embryos, begins its ontogenesis as a portion of the cerebellar subventricular zone—the so-called rhombic lip of His (His, 1904). Beginning at roughly E50 in the rhesus monkey, a group of small germinal cells moves across the narrow bank of tissue comprising the rhombic lip and, having attained a subpial position, extends around the sharply angled external groove that separates the protrusion of rhombic lip from the main cerebellar mass. Thus, this population gains access to the free external surface of the cerebellum. At the end of the second and during the first half of the third fetal month, these cells proliferate at an accelerated rate and eventually form a coat over the entire surface of the cerebellum (horizontal arrow in Figure 2). Cell proliferation continues at a rate that exceeds the expansion of the surface area, so that by the fourth fetal month, the external granular layer is thickened to a maximum of 40 µm, the equivalent of 5–8 rows of cells.

The thickness of the external granular layer remains constant for more than a month, but shortly before parturition it gradually begins to decrease and disappears altogether during the third postnatal month (Rakic, 1971a). The point of particular interest here is that the external granular layer generates neurons continuously throughout the final three months of gestation and for an approximately equal postnatal period. Although the external granular layer also produces stellate and basket cells (Rakic, 1972b), their numbers are relatively insignificant in comparison to the enormous production of the granule cells. The granule cells in the monkey begin to be generated between E70 and E80 and are continuously produced during the second half of gestation and well into the postnatal period. The focus of the remainder of this chapter is on how these postmitotic granule cells migrate inward from the external granular layer to their destination below the Purkinje cell layer.

MORPHOGENETIC TRANSFORMATION OF GRANULE CELLS

The developmental history of the postmitotic granule cell provides a unique opportunity to address both the problem of cell migration and that of axonal growth, because this cell class elaborates two distinct types of neurites—one having the cytological and functional properties of the axon and the other having the characteristics of the leading tip of a migrating cell. The morphogenetic transformation and migration of the granule cell is illustrated in Figure 3.

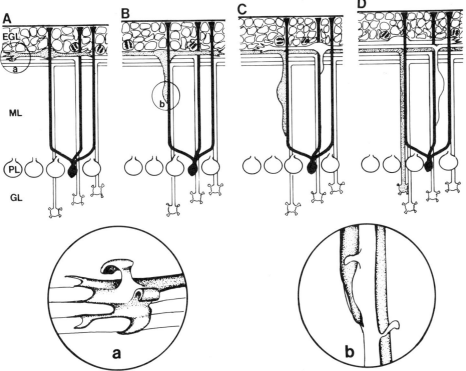

Figure 3. *Semidiagrammatic drawing of the morphogenetic transformation of the cerebellar granule cell (stippled) during its translocation across the molecular layer.* The morphological difference between the leading tips of the two horizontal process and the descending process are illustrated in enlargements *a* and *b*, respectively. EGL, external granular layer; GL, granular layer; ML, molecular layer; PL, Purkinje cell layer. See text for further explanation of the model.

The granule cell of the cerebellar cortex is first recognized after its final mitosis in the external granular layer as a round or oval cell with few cytoplasmic filopodia at its surface. At the time of final division, it lies in the deep stratum of the external granular layer, bordering the developing molecular layer (EGL in Figure 3). Within hours of its origin, the granular cell takes on a bipolar shape by extending two horizontal processes at opposite poles of its body (stippled cell in Figure 3A). As described originally by Ramón y Cajal (1955), these processes lie parallel to the external surface of the cerebellum and longitudinal to individual cerebellar folia (Figure 4). Electron-microscope examination of the developing cerebellum in the monkey fetus (Rakic, 1971a) reveals that the growing tips of these two processes become aligned along previously generated parallel fibers (Figure 5). The tips of these horizontal processes terminate as growth cones that have an elaborate form composed of flat lamellipodia that are visible in rapid Golgi preparations (Figure 4). Although their elaborate form is not obvious in the electron micrographs prepared from a section cut longitudinal to the axonal orientation (Figure 5), their form is revealed in the transverse plane of section (Figure 6). Reconstructed in three dimensions, these growth cones, which resemble a duck's foot with webs (enlargement *a* in Figure 3), contain a relatively light cytoplasmic matrix and few cellular organelles besides occasional clusters of empty membrane-bound vesicles of variable size (Figures 5, 6). The size, the lamellate and elaborate form, and the basic ultrastructural characteristics of the growing axonal tips are remarkably similar to those of growth cones reconstructed from serial sections in the developing optic nerve in the same species at earlier embryonic stages (Williams and Rakic, 1984). As a rule, growth cones of granule cell parallel fibers do not contain ribosomes.

Within less than one day after development of the two horizontal processes, cells elaborate a third cytoplasmic process that penetrates the vertical territory of the developing molecular layer (stippled cell in Figure 3B). The one-day delay from the emergence of the horizontal processes and the development of the descending process is implied from results of the Golgi analysis (Rakic, 1971a). I observed that the horizontal processes are several hundred micrometers longer before the third process appears at the bottom of the cell soma (Figure 4A–C), and, from what is known about growth rate of neurites of granule cells *in vivo* (Sidman et al., 1983), the time needed to achieve that length must be over 24 hours. After the third process develops, all three neurites continue to grow simultaneously; in the adult monkey, parallel fibers reach a final length of about 6 mm (Mugnaini, 1983). As is evident from the three-dimensional electron-

Figure 4. *Photomicrographs of Golgi-stained images of migrating granule cells in the molecular layer of the cerebellar cortex of a newborn monkey. A:* Horizontal bipolar cell in the plane longitudinal to the folium. Note a long (more than 100 μm) horizontal process ending with a lamellate growth cone (arrow). *B* and *C:* Examples of bipolar cells with horizontal processes of over 250 μm that still lack the third descending process. *D:* Development of the third vertical process (arrow) in a section cut longitudinal to the folium. *E:* Cell soma that has descended into the vertical process. Plane longitudinal to the folium. *F:* The silhouette of a vertical bipolar cell in a plane transverse to the folium. A few micrometers away is the vertical shaft of a Bergmann fiber which has progressively more numerous bushy lateral expansions in the deeper zone of the molecular layer. *G:* Another bipolar cell, with a trailing process and a T-shaped attachment to parallel fibers oriented in a plane longitudinal to the folium. Magnifications approximately ×270.

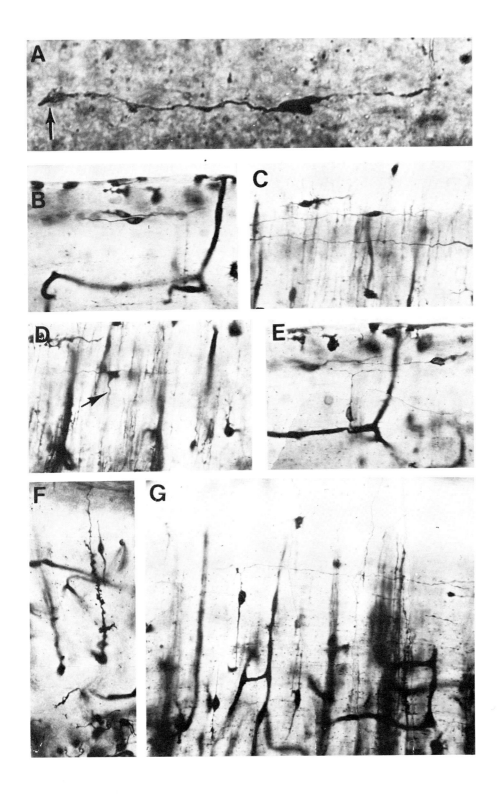

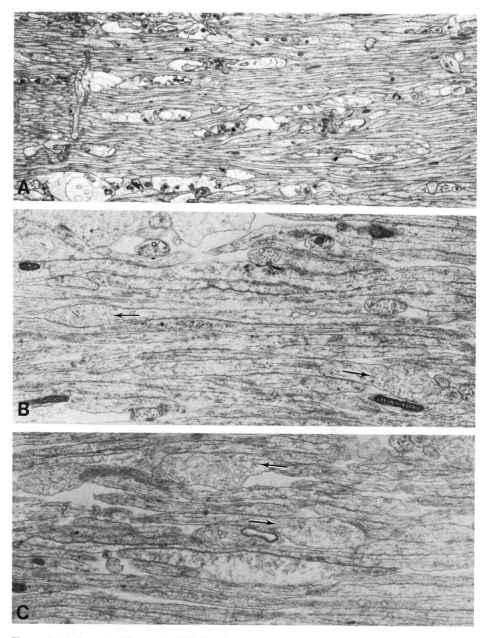

Figure 5. *A: Low-magnification (×6500) electron micrograph of the granule cell axons in the molecular layer of the developing cerebellar cortex cut parallel to the pial surface.* Note transversely cut shafts of Purkinje cell dendrites and radial glial fibers among the perfectly parallel arrangement of granule cell axons. *B: Higher-magnification (×18,000) electron micrograph of the developing molecular layer cut parallel to the pial surface displays profiles of two growth cones (arrow) growing in opposite directions. C: Another field in the same specimen shows several growth cone enlargements filled with empty membrane-bound vesicles.*

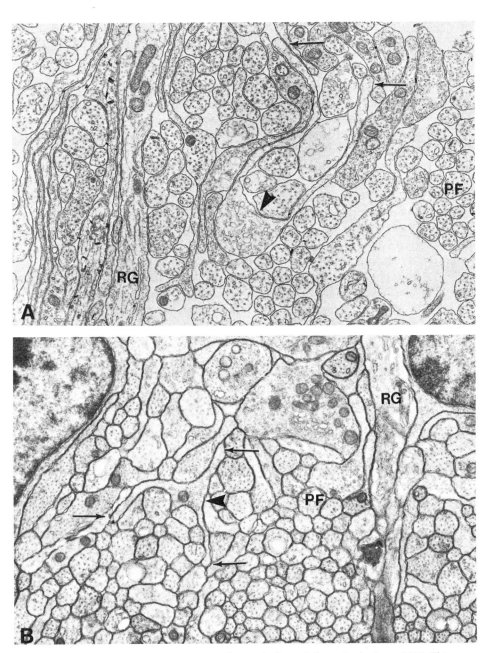

Figure 6. *A: Examples of transversely cut growth cones in the cerebellar molecular layer at E73.* This area, situated just below the external granular layer, contains numerous flattened lamellipodia that form at the tip of the growing parallel fibers. The flattened sheet of the growth cone (arrows) is often in direct continuity with a bulbous enlargement that is filled with vesicles (arrowhead). *B: A similar area of the molecular layer below the external granular layer in a 48-day-old monkey infant.* Note fewer and more densely packed growth cones at this late developmental stage.

microscope reconstructions in fetal monkey, the third process grows at a right angle to the orientation of axons of previously generated parallel fibers and remains attached to nearby Bergmann glial fibers (Rakic, 1971a). The smoothly tapered leading tip of this process (enlargement b in Figure 3) contains a dark cytoplasmic matrix and parallel microtubules, and, in contrast to the horizontal processes, is distinctly rich in membrane-bound and free ribosomes (Figure 7). One can observe various forms of membranous organelles, including rough endoplasmic reticulum, Golgi apparatus cisternae, and transitional vesicles. These organelles are primarily implicated in the transfer of membrane materials. However, although transitional vesicles can be seen in the leading process, they are never abundant or associated in aggregates as they are in the axonal growth cones.

When the descending process attains appropriate length, the nuclear portion of the cell becomes transposed inward within this cytoplasmic cylinder (Figure 3C). Over the course of the next 2–7 days (depending on the stage of development), the somatic portion of the granule cell traverses the full width of the molecular layer in the rhesus monkey (Rakic, 1973). The soma bypasses the row of Purkinje cells, detaches from the surfaces of Bergmann glial cells, and enters the developing granular layer (Figure 3D). There, the granule cell soma develops four or five dendrites which form synapses with mossy fiber terminals that arrived earlier from the brain stem. As a result of this remarkable morphological transformation, the two bipolar horizontal processes and the vertical process that trailed out behind the migrating granule cell are transformed into a T-shaped axon, the horizontal portion of which is termed a parallel fiber (stippled cell in Figure 3D).

The striking difference between the growth behavior of the vertical process and that of the horizontal processes, all of which are formed by the single granule cell, implies a regional difference in the structural composition of both the cell membrane and cytoplasm. The pair of horizontal processes begins to grow first, and from the onset they are oriented in the longitudinal axis of the folium, parallel to the bed of axons derived from previously generated granule cells (Figure 5). On the other hand, the third vertical process, which descends along Bergmann glial fibers, does not have typical growth cones (Figure 7). Not only do these two types of processes have different ultrastructural composition *in situ*, but even when taken out of their cellular environment postmitotic granule cells develop only three processes. Two of the processes assume the typical growth cone morphology of the parallel fiber type, while the third process looks like the descending process (Mangold et al., 1984). The T-shaped divisions, however, do not necessarily form right angles as do their counterparts *in situ*. Although these three processes may take a somewhat irregular course, they nevertheless retain appropriate cytological characteristics *in vitro* (Mangold et al., 1984). Thus, the number, the sequence of development, and the composition of these two classes of processes may be under intrinsic genetic control, whereas their precise geometry appears to depend on the local cellular environment.

Figure 7. *The tips of the leading processes of migrating granule cells at E105 (A), at E138 (B), and in neonatal (C) monkey.* The leading processes project downward along Bergmann glial fibers and occasionally become enveloped by a sheet of glial cytoplasm. There is an abundance of cytoplasmic organelles, including Golgi apparatus, microtubules, and membrane-attached and free ribosomes dispersed in the electron-dense matrix. Note, however, the absence of lamellipodia and of aggregates of clear growth cone vesicles that characterize axonal tips, as illustrated in Figures 5 and 6.

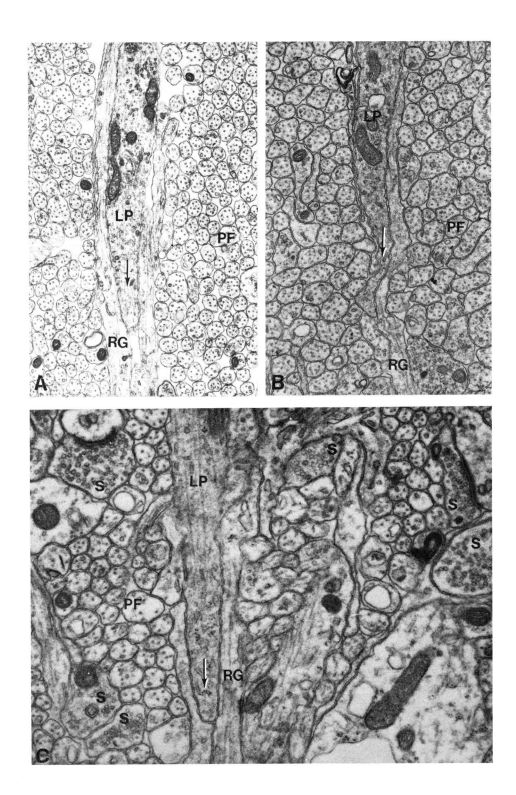

Based on the analysis *in vivo*, it appears that the two types of neurites differ in at least three ways: (1) time of initial outgrowth from the cell soma; (2) ultrastructural characteristics; and (3) behavior with respect to their cellular milieu. The difference in cytological properties between horizontal and descending processes may be determined while the granule cell is still in its round form at the interface between the external granular and molecular layers. It is possible, for example, that one patch of its membrane may already be specialized, induced by early contact with a Bergmann fiber during the last cell division. Alternatively, the difference in surface property may become established at a later time, only after the cell develops two horizontal processes.

If the orientation of the horizontal processes depends on the presence of parallel fibers, one needs to know which forces determine the orientation of the very first of these fibers. This is not well understood, but one possibility is that the orientation of early parallel fibers depends on the direction of arrival of proliferative cells from the rhombic lip to form the external granular layer on the cerebellar surface. There is some supporting evidence for this hypothesis from X-irradiated cerebellar primordia in which axonal orientation is altered (Altman, 1973). Another possibility is that the initial orientation of the first row of parallel fibers is laid down by the cues obtained from the Bergmann glial fibers and their end feet. These fibers are present sufficiently early in cerebellar development to exert such an influence (Rakic, 1971a; Levitt and Rakic, 1980). Furthermore, glial fibers are remarkably well distributed in straight rows forming regular palisades (DeBlas, 1984). The orientation of the early parallel fibers, however, may be based on interaction with many glial fibers arranged in rows in a manner similar to the "guidepost cell" hypothesis proposed for the growth of axons in invertebrates (Bently and Keshishian, 1982), rather than a longitudinal growth along a single glial fiber that is followed by the descending process.

SIGNIFICANCE AND INCIDENCE OF NEURON–GLIA INTERACTION

One can ask whether the associations between the leading descending tips of postmitotic granule cells and the Bergmann glial fibers are specific, developmentally meaningful interactions essential for neuronal translocation across the growing molecular and Purkinje cell layers. One can take the position that it would be difficult for the leading tip of postmitotic granule cells to avoid contact with Bergmann glial fibers, since these two cell classes are intermixed from early developmental stages. However, several lines of evidence provide support for the notion that those relationships are not fortuitous.

Perhaps the most compelling evidence comes from electron-microscope analysis of *in vivo* granule cell migration. Reconstruction from semiserial sections in three cardinal geometrical planes shows constancy of apposition of the leading descending process and Bergmann glial fibers (Rakic, 1971a). The most significant finding from this study is that an individual leading tip may be in contact with several thousand cell processes of various types, but it nevertheless continues to grow only along one glial process (Figure 8). Since the leading tip encounters processes of a wide range of diameters and orientations, the most plausible explanation for its behavior is that it has a preferential affinity to glial surface.

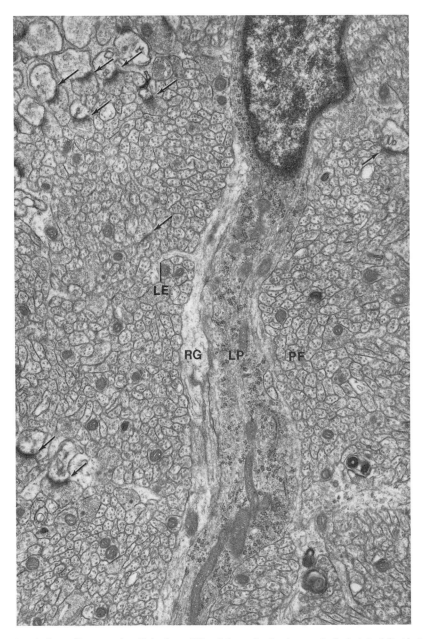

Figure 8. *A descending granule cell in the middle of the molecular layer, attached at its left side to the electronlucent shaft of a Bergmann glial fiber.* The Bergmann fiber has a lamellate cytoplasmic extension (arrow), but the surface shared with the migrating granule cell is relatively smooth. The electron-dense cytoplasm of the granule cell is rich in organelles, including rough endoplasmic reticulum, smooth endoplasmic reticulum, microtubules, mitochrondria, and ribosomes. ×9700.

The same constant phenomenon has been observed in the developing cerebellum of birds, amphibians, rodents (Mugnaini and Forströnen, 1967; Gona, 1978; Landis and Sidman, 1978; Pouwels, 1978) and humans (Zecevic and Rakic, 1976).

Another convincing example of neuron–glia interaction during granule cell migration can be found in the developing cerebellum of the neurological mutant mouse—the *reeler*. In this mutant, most of the cerebellar granule cells are in an abnormal position, situated above the Purkinje layer, rather than in the normal position below it. The point relevant to the present discussion is that many cells that are normally polarized, including Bergmann glial cells, may be disoriented and occasionally inverted (Rakic and Sidman, 1972). In sites where Bergmann fibers lie in an oblique orientation across the attenuated molecular layer, migrating granule cells nonetheless follow them closely and are not seen to take a more direct, but unguided, radial route (Rakic and Sidman, 1972; Caviness and Rakic, 1978). This observation provides supporting evidence that granule cells selectively follow the contours of the Bergmann glial fibers even when their orientation is not orthogonal to the direction of parallel fibers.

Another neurological mutant—the *weaver* mouse—has also provided a useful model for the analysis of granule cell migration. Although studies of this mutant are not without controversy, they are nevertheless highly instructive. First, Rezai and Yoon (1972) found that granule cell migration was slowed down in the cerebellum of heterozygous animals and virtually absent in homozygous mutants. Their study was followed by our discovery that the Bergmann glial fibers are morphologically abnormal in both homozygous and heterozygous animals at developmental stages coinciding with, and even antedating, the expected granule cell migration (Rakic and Sidman, 1973). Postmitotic granule cells in the homozygous *weaver* failed to generate or sustain the two horizontal axonal processes and the third vertical process, and they also were unable to migrate across the molecular layer. We speculated that the glial abnormality might be responsible in part for defective neuronal migration, which eventually causes the death of granule cells that fail to move from their site of origin (Rakic and Sidman, 1973).

The argument that glial change is the indirect cause of granule cell death, rather than the reverse, was based on several additional observations. First, glial abnormality can be detected *in vivo* before cell migration starts and before the wave of granule cell death. Furthermore, the electronlucent, vacuolated appearance of abnormal glial processes in young mice was not indicative of reactive astrogliosis, which becomes prominent only later, in those mutant animals that survive the first two to three weeks of life. Although a Bergmann fiber abnormality has been consistently observed in the cerebellar cortex of young *weaver* mice in several other laboratories (e.g., Hirano and Dembitzer, 1973, 1974; Sotelo and Changeux, 1974), the interpretation of this finding has not been uniform because it has not been possible to determine its relationship to the mode of action of the *weaver* genetic locus (see Caviness and Rakic, 1978, for a review). The most recent studies still fail to provide definitive evidence for any single hypothesis. Thus, the analysis of chimeras made between *weaver* and wild-type mice (Goldowitz and Mullen, 1982), as well as the study of neurite elongation *in vitro* (Sidman et al., 1983), support the notion that the *weaver* locus may be primarily affecting the granule cells. Conversely, some other *in vitro* studies provide evidence that either glial abnormality or the interaction between neuron and glia may play the crucial role (Willinger et al., 1981; Cohen, 1983; Hatten et al., 1984). In spite of these uncertainties, this mutant provides an instructive example of the role of neuron–glia interaction in the formation of normal cellular patterns.

Whatever the molecular basis of the attraction between the two types of granule cell processes and their local environment may be, it enables the growing tips of the horizontal processes to grow exclusively along axons of previously generated granule cells while the descending process makes a specific choice to extend preferentially along a Bergmann glial fiber. The neuron–glia relationship may remain preserved even when Bergmann fibers display severe necrotic changes due to the toxic effect of 6-aminonicotinamide, as long as the integrity of the glial surface is preserved (Sotelo and Rio, 1980). Electron-microscope studies conducted in the normal developing cerebellum of primates (Rakic, 1971a) as well as in other species (Mugnaini and Forströnen, 1967; Gona, 1978; Landis and Sidman, 1978; Pouwels, 1978), including human fetal cerebellum (Zecevic and Rakic, 1976), show the constancy and regularity of the neuron–glia relationship during granule cell migration. Nevertheless, all of these observations, including our own, fail to show obvious morphological specialization between membranes of the leading tip and adjacent glial surfaces. The 20–30 µm intermembrane space contains extracellular material of moderate electron density.

It should be emphasized that the absence of distinct membrane specializations between migrating neuron and glial fibers in specimens prepared for examination by transmission electron microscopy does not rule out either structural or molecular surface differences. Although a systematic freeze-fracture analysis of this process is still lacking, there is some indication that the surface of migrating cells, the plasma membrane of growing parallel fibers, and the somas of immature granule cells have intermembrane particles of different density, size, and distribution pattern during the first two weeks of postnatal life in rodents (Gracia-Segura and Perrelet, 1981). This difference may account for selective outgrowth along the two types of surfaces that postmitotic granule cells encounter at the interface of the external granular and molecular layers. Each type of process makes a specific choice when confronted by different substrate surfaces, and this choice is reflected in the differential behavior of their leading tips.

The close apposition between leading tip of migrating granule cell and Bergmann glial fibers is maintained for the full distance between the place of granule cell origin in the external granular layer and their final destination in the Purkinje cell layer. However, one new property must emerge from these interactions toward the end of the journey as the migrating neurons arrive at the interface between the Purkinje and granular layers. This new property induces cessation of further movement, detachment from the glial surface, and the establishment of the final position of each neuron. One possibility, which lacks experimental evidence, is that the initial contact between granule cells and their synaptic target—the mossy fibers—serves as a signal which inhibits further movement by inducing dendritic growth.

Behavior of cells *in vitro* provides another line of evidence that direct contacts between neurons and glial cells may not be a fortuitous or inconsequential relationship. An apposition between neurons and glial cells has been described in dissociated cerebellar cells in tissue culture condition (Hatten and Liem, 1981; Cohen, 1983). Although these two studies differ in their assessment of the timing and other details of glial fiber outgrowth *in vivo*, they nevertheless agree that Bergmann glial fibers are necessary for the movement, differentiation, and very survival of postmitotic granule cells. These studies, however, did not provide new insight into the nature of the process that determines how far individual cells can move in the radial axis. Studies *in vivo* indicate that this process is likely to be mediated by interaction with processes of other, previously generated neurons and their processes in the local milieu.

POSSIBLE MOLECULAR MECHANISMS OF NEURONAL MIGRATION

Although various neuronal classes may use different mechanisms for the translocation of their bodies relative to neighboring cells, surface-mediated interaction between heterogeneous cells is generally considered to play an essential role. At present, the biophysical and molecular mechanisms of migration are not understood, but the available evidence is compatible with our working hypothesis that glial fibers and migrating neurons possess unique cell surface molecules that allow them to recognize and/or adhere selectively to each other (Rakic, 1981). It is likely that these hypothetical molecules are different from molecules that would bind neurons to each other or molecules that might be responsible for the establishment and maintenance of synaptic junctions. The existence of such "adhesion" molecules was predicted on the basis of electron-microscope observation that the surfaces of migrating neurons and glial cells remain attached to each other even when, due to either mechanical damage or poor fixation, cells in the intermediate zone become separated from other processes by large extracellular spaces (Rakic, 1972a). Transmission electron microscopy displays a high incidence of surface apposition between migrating neurons and radial glial fibers with intermembrane spaces of about 20–300 nm (Rakic, 1971a, 1972a). Although these attachment areas contain low-density fibrillary material, nothing is known about their biochemical makeup. Ultrastructurally similar appositions may be observed between migrating neurons and other classes of adjacent cells, which are not followed by the migrating neurons.

On the basis of observed ultrastructural and growth behavior of granule cell processes during translocation across the molecular layer, as illustrated in Figure 3, several conclusions can be drawn: (1) If we accept a differential adhesion hypothesis, there must be at least two sets of adhesion molecules—one binding neuron to neuron, which may be responsible for the alignment of parallel fibers, and the other binding neurons to glial cells, which may be responsible for the alignment of leading processes and Bergmann glial fibers; (2) both sets of molecules must be present but they may be differentially distributed on the granule cell surface—one set predominantly on the horizontal and the other predominantly on the vertical process; (3) the structural and biochemical difference between the horizontal and vertical processes is intrinsic to the granule cell. These conclusions and their significance for understanding the molecular mechanisms involved in neuronal migration are discussed below.

The crucial observation relevant to understanding granule cell migration is that its leading process extends selectively along the glial fibers and seems to ignore the other cells in its environment (Rakic, 1971a,b). The leading processes follow the glial fibers even when their shape curves and does not follow the often shorter, straight routes of other cells that it encounters during migration. This simple observation indicates selective affinity. The binding affinity between neuron and glial cell surfaces may depend on a single molecule in every structure of the brain and, therefore, all migrating neurons may have equal affinity for all radial glial cells (Rakic, 1981). Nevertheless, the spatiotemporal order of the cell migration could still be achieved because each postmitotic neuron becomes attached only to its nearest glial fibers. Once attached, the leading process simply grows along this guide from the outset of migration in the proliferative zone to an often distant destination.

How can one reconcile the apparent contradiction that migrating neurons and radial glial fibers form a strong bond and yet move along each other? Among several hypotheses that could explain the phenomenon of migration, one is that new membrane is deposited mostly at the tip of the leading process along the surface of radial glial fibers (Figure 9A). According to this hypothesis, the apposing sites of two cells become permanently fixed, and yet the migrating cell moves by the selective growth of the leading process at its tip (Rakic, 1981). The nucleus and its surrounding cytoplasm are translocated within the leading process simultaneously or at a later time. As additional membrane components are added to a growing tip, the leading process progressively extends along the radial glial fiber and the nucleus moves further to a new position within the perikaryal cytoplasm (Figures 9B,C). According to this model, propelling a migrating cell does not require contractile capacity mediated by actin- or myosinlike molecules. However, there must exist some sort of contractile system that is capable of translocating the nucleus and surrounding cytoplasm to the newly formed space created within the leading process.

Several observations give credence to this model. First, there is clearly an increase in the surface area of migrating cells (Rakic et al., 1974). This finding is in harmony with the longstanding conviction that axonal outgrowth is achieved predominantly by membrane deposition at the tips of neurites (Hughes, 1953; Bray, 1973; Pfenninger, 1982). On the basis of calculations from short-survival [^{3}H]thymidine autoradiographic experiments, we found out that the rate of cell movement in the hippocampal formation and neocortex of the rhesus monkey varies between two and five micrometers per hour (Rakic, 1975b; Nowakowski and Rakic, 1979). The translocation of granule cells across the molecular layer at late stages of cerebellar mitogenesis is within the same order of magnitude (Rakic, 1973). Although the speed of migration has not been precisely determined for other brain regions, this approximate rate is compatible with the capacity for generation and insertion of new membrane along the leading process (Rakic et al., 1974; Rakic, 1981). Theoretically, the preferential attachment of the leading tip to the glial surface could be achieved by adhesive forces between the two cells, which interlock in a fashion similar to antibody–antigen complexes. This may involve direct interlocking of heterophilic (Figure 9, *a* in insert) or homophilic (Figure 9, *b* in insert) molecules in the membrane. Alternatively, an additional extracellular molecule that binds heterophilic (Figure 9, *c* in insert) or homophilic (Figure 9, *d* in insert) membrane molecules may be required. Finally, the binding may be achieved by surface electrostatic charges or by modulations of membrane fluidity (Figure 9, *e* in insert).

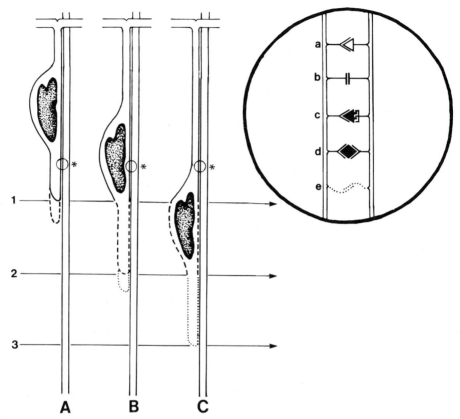

Figure 9. *Diagram showing a possible mechanism for the downward displacement of migratory cells along the surface of radial glial fibers (A, B, and C).* To traverse the distance between levels 1 and 2, new membrane may be inserted along the interface of the two cells (dashed line in A) while the nucleus moves within the cytoplasm of the leading process (B). As the leading tip grows to reach level 3, additional membrane is inserted (dotted line in C) along the glial surface and the nucleus moves to a higher position between levels 1 and 2, resulting in an overall displacement of the cell body. This model does not require movement of the neuronal surface along glial fibers, and the binding sites (small circles marked by asterisks) between the two apposing membranes may remain constant for some time. As indicated in the cartoon within the circle at the right, this attachment may be accomplished by (a) two complementary heterophilic molecules; (b) two homophilic molecules present on the surface of each cell; (c) a third extramembranous molecule that fits between heterophilic molecules; (d) a third extramembranous molecule that fits between two homologous molecules; and (e) by physical forces such as electrostatic charge.

At present it appears that some sort of binding molecules on the plasma membrane are the most likely to play a crucial role in granule cell interaction with its environment. One candidate for such a molecule, termed the neural cell adhesion molecule (N-CAM), has been isolated and characterized by Edelman and his colleagues (Edelman, 1983; Edelman, this volume). This molecule, present on the surface of all neurons but not on the glia, is distributed ubiquitously in the vertebrate central nervous system, including fetal human brain (McClain and Edelman, 1982). It is homophilic (Figure 9, *b* in insert), selectively binds neurons to neurons, and has both embryonic and adult forms. Since anti-N-

CAM antibodies abolish neuron–neuron adhesion *in vitro*, N-CAM may be sufficient to explain the binding and fasciculation of axonal processes such as parallel fibers. However, a recent finding that neuron–glia but not neuron–neuron adhesion can be inhibited by prior incubation of neurons with fragments of specific antibodies prepared against embryonic neuronal membrane indicates the existence of another type of molecule on the neuronal surface (Grumet et al., 1983). This molecule, which apparently has a capacity to bind neurons to glia (Ng-CAM), has been isolated in Edelman's laboratory and is currently being tested (Edelman, this volume). It is expected that application of antiserum against such complementary binding sites to an *in vitro* system may interfere with neuron–glia attachment and ultimately with neuronal migration (e.g., Lindener et al., 1983). On the basis of the behavior of migratory granule cells which move selectively along glial fibers, it is expected that this binding will be heterophilic (Figure 9, *a* in insert) rather than homophilic, since a homophilic molecule would not allow distinction between the surface of adjacent neurons and glial cells (Grumet et al., 1983; Edelman, this volume). The molecule present on the glial cell surface, however, has not been identified as yet.

Although the nature of the adhesion between the leading tip of migrating neurons and the glial surface is not fully understood, it appears that the set of heterophilic molecules should be sufficient to explain all known aspects of glial guidance. Thus, although all migrating neurons may contain the source molecule, a selective and highly patterned migration could be achieved because each postmitotic neuron would follow only the adjacent glial process. However, successive neurons that move along the same radial path may be guided either by the glial fiber, by the trailing processes of their predecessors, or even by the processes of other neurons that have become incorporated into the same radial fascicle. In this manner, a single migratory pathway may attract different populations of neurons at different times in development. Cells that are generated at the same site in the proliferative zone would follow only nearby glial fascicles and, although attached by the same neuron–glia adhesion molecules as their neighbors, they should nevertheless arrive at their unique and appropriate positions. Nor should neurons be attracted to more distant glial fibers, because they are already occupied by other migrating cells or because their unapposed surface lacks a sufficient number of adhesion molecules. The attractiveness of this model is that the existence of glial fibers, which stretch across the developing cerebral wall, minimizes the requirement for a large amount of genetic information and obviates the necessity of producing myriad molecular types for the regulation of neuronal migration and for the acquisition of proper cell positions.

Another aspect of neuronal migration that awaits an explanation is how migrating cells penetrate a terrain densely packed with previously generated cells and their processes. This problem comes into focus particularly during the late stages of migration, when extracellular spaces in the migratory pathway become reduced to the usual 20–30-nm wide intermembranous clefts. Furthermore, neurons generated over the previous several months not only grow in size, but also develop elaborate synaptic connectivity (e.g., Figure 8). One wonders how the tips of migrating neurons loosen the bonds between neuronal processes located within the molecular layer. One possibility is that migrating granule cells release plasminogen or plasminogen activator (Kristocek and Seeds, 1981; Soreq and

Miskin, 1981). This, or a similar enzyme, could be present on the outer surface of the growing tip or it could be released preferentially into the extracellular spaces along radial glial fibers. The study by Moonen et al. (1982) indicates that plasminogen-activator serum protease is indeed present in the developing cerebellar cortex during the migratory period and that the migration of granule cells can be curtailed by inhibitors of the plasminogen activator–plasmin system. Supporting evidence for this hypothesis is that the terrain throughout which neuronal migration is accompanied by massive ingrowth of fibers becomes loosely arranged and separated by large extracellular spaces. The large extracellular spaces are observed even when nearby structures that are not involved in cell migration are densely packed with only 20–30 nm of space between cellular elements (Rakic, 1972a).

This short survey attempts to expose both the complexity and the remarkable selectivity of the migrating behavior of granule cells. It is clear from the results of studies carried out during the last decade that the morphogenetic transformation and translocation of the neuron requires the interaction of heterogeneous classes of cells. It is also evident that defective neuronal migration caused by either genetic or environmental factors may result in severe cerebral malformation in both animals and humans (Rakic and Sidman, 1973; Rakic, 1975a, 1976, 1984). This chapter illustrates that recent advances in molecular neurobiology can provide new challenges and new opportunities to enhance our understanding of these remarkable phenomena.

REFERENCES

Altman, J. (1973) Experimental reorganization of the cerebellar cortex. III. Regeneration of the external germinal layer and granule cell ectopia. *J. Comp. Neurol.* **149**:153–180.

Bently, D., and H. Keshishian (1982) Pathfinding by peripheral pioneer neurons in grasshoppers. *Science* **218**:1081–1088.

Bray, D. (1973) Model for membrane movements in the neural growth cone. *Nature* **244**:93–96.

Caviness, V. S., Jr., and P. Rakic (1978) Mechanisms of cortical development: A view from mutations in mice. *Annu. Rev. Neurosci.* **1**:297–326.

Cohen, J. (1983) In vitro studies of the adhesive interaction between neurons and glial cells. *Proc. Int. Union Physiol. Sci.* **15**:475.

DeBlas, A. L. (1984) Monoclonal antibodies to specific astroglial and neuronal antigens reveal the cytoarchitecture of the Bergmann glia fibers in the cerebellum. *J. Neurosci.* **4**:265–273.

Edelman, G. M. (1983) Cell adhesion molecules. *Science* **219**:450–457.

Goldowitz, D., and R. J. Mullen (1982) Granule cell as a site of gene action in the *weaver* mouse cerebellum: Evidence from heterozygous mutant chimeras. *J. Neurosci.* **2**:1474–1485.

Gona, A. G. (1978) Ultrastructural studies on cerebellar histogenesis in the frog: The external granular layer and the molecular layer. *Brain Res.* **153**:435–447.

Gould, B. B., and P. Rakic (1981) The total number, time of origin, and kinetics of proliferation of neurons comprising the deep cerebellar nuclei in the rhesus monkey. *Exp. Brain Res.* **44**:195–206.

Gracia-Segura, L. M., and A. Perrelet (1981) Freeze-fracture of developing neuronal plasma membrane in postnatal cerebellum. *Brain Res.* **208**:19–33.

Grumet, M., U. Rutishauser, and G. M. Edelman (1983) Neuron–glia adhesion is inhibited by antibodies to neural determinants. *Science* **222**:60–62.

Hatten, M. E., and R. K. H. Liem (1981) Astroglia provide a template for the positioning of cerebellar neurons in vitro. *J. Cell Biol.* **90**:622–630.

Hatten, M. E., R. K. H. Liem, and C. Mason (1984) Defects in specific associations between astroglia and neurons occur in microcultures of *weaver* mouse cerebellar cells. *J. Neurosci.* **4**:1163–1172.

Hirano, A., and H. Dembitzer (1973) Cerebellar alteration in the *weaver* mouse. *J. Cell Biol.* **56**:478–486.

Hirano, A., and H. Dembitzer (1974) Observations on the development of the *weaver* mouse cerebellum. *J. Neuropathol. Exp. Neurol.* **33**:354–364.

His, W. (1904) *Die Entwicklung des menschlichen Gehirns während der ersten Monate*, Hirzel, Leipzig.

Hughes, A. F. (1953) The growth of embryonic neurites. A study of cultures of chick neural tissue. *J. Anat.* **87**:150–162.

Kristocek, A., and N. W. Seeds (1981) Plasminogen activator release at the neuronal growth cone. *Science* **213**:1532–1534.

Landis, D. M. D., and R. L. Sidman (1978) Electron microscopic analysis of histogenesis in the cerebellar cortex of *staggerer* mutant mice. *J. Comp. Neurol.* **179**:831–863.

Levitt, P. R., and P. Rakic (1980) Immunoperoxidase localization of glial fibrillary acidic protein in radial glial cells and astrocytes of the developing rhesus monkey brain. *J. Comp. Neurol.* **193**:815–840.

Lindner, T., F. G. Rathjan, and M. Schachner (1983) L1 mono- and polyclonal antibodies modify cell migration in early postnatal mouse cerebellum. *Nature* **305**:427–430.

Llinás, R., ed. (1969) *Neurobiology of Cerebellar Evolution and Development*, American Medical Association Education and Research Foundation, Chicago, Ill.

Mangold, U., J. Sievers, and M. Berry (1984) 6-Hydroxydopamine-induced ectopia of external granule cells in the subarachnoid space covering the cerebellum. II. Differentiation of granule cells: A scanning and transmission electron microscopic study. *J. Comp. Neurol.* **227**:267–284.

McClain, D. A., and G. M. Edelman (1982) A neural cell adhesion molecule from human brain. *Proc. Natl. Acad. Sci. USA* **79**:6380–6384.

Moonen, G., M. P. Grau-Wagemans, and I. Selar (1982) Plasminogen activator-plasmin system and neuronal migration. *Nature* **298**:753–755.

Mugnaini, E. (1983) The length of cerebellar parallel fibers in chicken and rhesus monkey. *J. Comp. Neurol.* **220**:7–15.

Mugnaini, E., and P. F. Forströnen (1967) Ultrastructural studies on the cerebellar histogenesis. I. Differentiation of granule cells and development of glomeruli in chick embryo. *Z. Zellforsch.* **77**:115–143.

Nowakowski, R. S., and P. Rakic (1979) The mode of migration of neurons to the hippocampus: A Golgi and electron microscopic analysis in fetal rhesus monkey. *J. Neurocytol.* **8**:697–718.

Palay, S. L., and V. Chan-Palay (1974) *Cerebellar Cortex: Cytology and Organization*, Springer, New York.

Pfenninger, K. H. (1982) Axonal transport in the sprouting neuron: Transfer of newly synthesized membrane components to the cell surface. In *Axoplasmic Transport in Physiology and Pathology*, D. G. Weiss and A. Gorio, eds., pp. 52–60, Springer-Verlag, Berlin.

Pouwels, E. (1978) On the development of the cerebellum of the trout, *Salmo gairdeneri*. I. Patterns of cell migration. *Anat. Embryol.* **152**:291–308.

Rakic, P. (1971a) Neuron-glia relationship during granule cell migration in developing cerebellar cortex. A Golgi and electron microscopic study in macaque rhesus. *J. Comp. Neurol.* **141**:283–312.

Rakic, P. (1971b) Guidance of neurons migrating to the fetal monkey neocortex. *Brain Res.* **33**:471–476.

Rakic, P. (1972a) Model of cell migration to the superficial layers of fetal monkey neocortex. *J. Comp. Neurol.* **145**:61–84.

Rakic, P. (1972b) Extrinsic cytological determinants of basket and stellate cell dendritic pattern in the cerebellar molecular layer. *J. Comp. Neurol.* **146**:335–354.

Rakic, P. (1973) Kinetics of proliferation and latency between final division and onset of differentiation of the cerebellar stellate and basket neurons. *J. Comp. Neurol.* **147**:523–546.

Rakic, P. (1975a) Cell migration and neuronal ectopias in the brain. In *Morphogenesis and Malformation of the Face and Brain*, D. Bergsma, ed., pp. 95–129, Alan R. Liss, New York.

Rakic, P. (1975b) Timing of major ontogenetic events in the visual cortex of the rhesus monkey. In *Brain Mechanisms in Mental Retardation*, N. A. Buchwald and M. Brazier, eds., pp. 3–40, Academic, New York.

Rakic, P. (1976) Synaptic specificity in the cerebellar cortex: Study of anomalous circuits induced by single gene mutation in mice. *Cold Spring Harbor Symp. Quant. Biol.* **40**:333–346.

Rakic, P. (1979) Genetic and epigenetic determinants of local neuronal circuits in the mammalian central nervous system. In *The Neurosciences: Fourth Study Program*, F. O. Schmitt and F. G. Worden, eds., pp. 109–127, MIT Press, Cambridge.

Rakic, P. (1981) Neuronal-glial interaction during brain development. *Trends Neurosci.* **4**:184–187.

Rakic, P. (1984) Defective cell-to-cell interactions as causations of brain malformations. In *Malformations of Developments: Biological and Psychological Sources and Consequences*, E. S. Gollin, ed., pp. 239–285, Academic, New York (in press).

Rakic, P. (1985) Principles of neuronal migration. In *Handbook of Physiology: Developmental Neurobiology*, W. M. Cowan, ed., American Physiological Society, Bethesda, Md. (in press).

Rakic, P., and R. L. Sidman (1972) Synaptic organization of displaced and disoriented cerebellar cortical neurons in *reeler* mice. *J. Neuropathol. Exp. Neurol.* **31**:192.

Rakic, P., and R. L. Sidman (1973) Sequence of developmental abnormalities leading to granule cell deficit in cerebellar cortex of *weaver* mutant mice. *J. Comp. Neurol.* **152**:102–132.

Rakic, P., L. J. Stensaas, E. P. Sayre, and R. L. Sidman (1974) Computer aided three-dimensional reconstruction and quantitative analysis of cells from serial electron microscopic montages of fetal monkey brain. *Nature* **250**:31–34.

Ramón y Cajal, S. (1955) *Histologie du Système Nerveux de l'Homme et des Vertébrés*, Vol. 2, trans. by L. Azoulay, Consejo Superior de Investigaciones Científicas, Instituto Ramón y Cajal, Madrid.

Rezai, A., and C. H. Yoon (1972) Abnormal rate of granule cell migration in the cerebellum of "weaver" mutant mice. *Dev. Biol.* **29**:17–26.

Sidman, R. L., and P. Rakic (1973) Neuronal migration, with special reference to developing human brain: A review. *Brain Res.* **62**:1–35.

Sidman, R. L., M. Willinger, and D. M. Margolis (1983) Mouse *weaver* maturation affects granule cell neurite growth *in vitro*. *Birth Defects*: **19**:189–200.

Soreq, H., and R. Miskin (1981) Plasminogen activator in the rodent brain. *Brain Res.* **216**:361–374.

Sotelo, C., and J.-P. Changeux (1974) Bergmann fibers and granular cell migration in the cerebellum of homozygous *weaver* mutant mouse. *Brain Res.* **77**:484–491.

Sotelo, C., and J. P. Rio (1980) Cerebellar malformation obtained in rats by early postnatal treatment with 6-aminonicotinamide. Role of neuron-glia interaction in cerebellar development. *Neuroscience* **5**:1737–1759.

Williams, R. W., and P. Rakic (1984) Form, ultrastructure, and selectivity of growth cones in the developing primate optic nerve: 3-dimensional reconstructions from serial electron micrographs. *Soc. Neurosci. Abstr.* **10**:373.

Willinger, M., D. M. Margolis, and R. L. Sidman (1981) Neuronal differentiation in cultures of *weaver* (*wv*) mutant mouse cerebellum. *J. Supramol. Struct.* **17**:79–86.

Zecevic, N., and P. Rakic (1976) Differentiation of Purkinje cells and their relationship to other components of developing cerebellar cortex in man. *J. Comp. Neurol.* **167**:27–48.

Section 3

The Neural Crest and Development of the Periphery

Studies of the development of the peripheral nervous system (PNS) provide an excellent opportunity to analyze specific local factors influencing migration, differentiation, and mapping of neural precursors in nonneural environments. Most of the PNS arises from the neural crest and the remainder from ectobranchial placodes.

In her chapter, Le Douarin addresses several major issues connected with these themes. She discusses evidence that any level of the crest can give rise to the whole spectrum of sensory and autonomic structures. Given this homogeneous distribution of PNS developmental capabilities in the crest, when does the choice of potentialities occur during ontogeny? It appears likely that two types of precursors are committed early: "S" (sensory) precursors, which are rapidly postmitotic cells that are absent in autonomic ganglia but differentiate into sensory neurons and glia; and "A" (autonomic) precursors, which remain available many days after neuronal birthdates are reached in all types of PNS ganglia. It is clear that positional relationships between PNS ganglia and the central nervous system influence the differentiation of type S precursors, and that the presence of the neural tube is essential for the development of dorsal root ganglia. Similarly,

Le Douarin's studies show a major role of nonneuronal cells in "directing" PNS ontogeny.

Thiery and his colleagues, in the next contribution, study the equally important question of the factors governing the migration of crest cells. He shows that transient but essential routes of migration form between and along fibronectin-rich basement membranes. These narrow pathways lead the migrating cells to sites of arrest, obstacles, and cul-de-sacs when they encounter a milieu relatively depleted of fibronectin. This molecule is shown to be an essential component for the attachment and migration of the crest cells. In a complementary fashion, N-CAM is lost from crest cell surfaces as they migrate and reappears during their aggregation into ganglion rudiments. Cytodifferentiation then occurs in a progressive fashion after these morphogenetic events are enacted. These studies suggest that both cell–substrate and cell–cell adhesion molecules are precisely regulated during such events.

The enteric nervous system, which contains as many as 10^8 intrinsic neurons, has, as Gershon and Rothman show in their chapter, a number of unique features. It lacks internal collagen and has a specialized internal environment that determines the final phenotype of colonizing neural crest precursors that were committed to neuronal lineages before arrival. In some cases, neuronal phenotypes are transient, as indicated by the neurotransmitter-defined subset to which they belong. An example is the transient catecholamine cells that appear in the gut. The enteric microenvironment determines in part the ability of enteric neurons to colonize segments of the bowel. Unlike other ganglionic cells, intrinsic neurons are relatively independent of the influence of nerve growth factor. It remains to be determined what nonneuronal cells and components of the extracellular matrix play a major role in determining colonization.

Hollyday, in her account of motor innervation of vertebrate limbs, addresses the question of whether connection specificity requires specific intercellular recognition or is rather a consequence of a stereotyped cellular production and axonal outgrowth. The bulk of the evidence favors the former: Axonal growth cones of neurons in different motor pools initially differ in their response to limb tissues. Motor axons respond in addition to nonspecific growth cues that are related to mechanical properties of somites and limb tissues. Specific cues appear reflected in the ability of axons to project to dorsal or ventral muscle tissue in an appropriate fashion after experimental perturbation. Another constraint appears to be axonal interactions: In the absence of neighbors axons will expand into "foreign" territory. These studies suggest that a heterarchy of factors developing epigenetically are responsible for neural patterns in the limb and emphasize the role of discretely distributed local cues in the limb bud as a major one in this remarkable process. This example leads naturally to the considerations of microscopic factors involved in formation of neurites and synapses, taken up in the next section.

Chapter 8

In Vivo and *In Vitro* Analysis of the Differentiation of the Peripheral Nervous System in the Avian Embryo

NICOLE M. LE DOUARIN

ABSTRACT

The development of the peripheral nervous system (PNS) has been investigated by using two different methodologies. One is based on in vivo *grafting experiments, taking advantage of the quail–chick marking technique that allows a definite pool of grafted cells to be distinguished from those of the host embryo irrespective of the duration and site of the graft. The second consists of explanting* in vitro *the progenitor cells of the PNS (i.e., the neural crest cells) and, by monitoring the culture conditions, investigating their potentialities for differentiation. It appears that, although the neural crest is regionalized in several distinct areas, yielding different PNS structures in normal development, spatial disturbances of this preexisting order do not result in major abnormalities in PNS ontogeny. This means that one neural crest area can be substituted for another to provide the embryo with sensory, sympathetic, and parasympathetic ganglia. Thus there exists at all levels of the neural crest a larger range of developmental potentialities than those that will actually be expressed during development. The question then arises of when, during ontogeny, selection operates on these potencies. Back-transplantation of developing quail PNS ganglia into the neural crest migration pathway of younger chick hosts revealed that all types of PNS ganglia (sensory and autonomic as well) contain undifferentiated cells able to divide and provide the host with a variety of autonomic ganglion neurons and glia. In contrast, the capacity to yield sensory neurons in the host exists only in quail spinal ganglia containing mitotic sensory neuroblasts. These results and* in vitro *culture experiments revealed that a certain degree of heterogeneity already exists in the neural crest itself. The neural crest is in fact composed of subpopulations of cells in which developmental capacities are restricted to a certain spectrum of definite phenotypes whose development is selectively triggered by factors arising from the tissue environment of the embryo.*

EMBRYONIC FATE MAPS OF PERIPHERAL NERVOUS SYSTEM STRUCTURES

The peripheral nervous system (PNS) arises mainly from a transitory embryonic structure, the neural crest. However, in the head, ectobranchial placodes along with cephalic neural crest cells contribute to the formation of the sensory ganglia of certain cranial nerves.

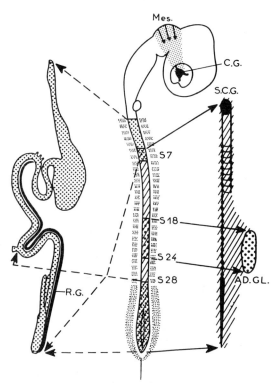

Figure 1. *Levels of origin of adrenomedullary cells and autonomic ganglion cells.* The spinal neural crest caudal to the level of the fifth somite gives rise to the ganglia of the orthosympathetic chain. The adrenomedullary cells originate from the spinal neural crest between the levels of somites (S) 18 and 24. The vagal neural crest (somites 1–7) gives rise to the enteric ganglia of the preumbilical region, the ganglia of the postumbilical gut originating from both the vagal and lumbosacral neural crest. The ganglion of Remak (R. G.) is derived from the lumbosacral neural crest (posterior to the somite-28 level). The ciliary ganglion (C. G.) is derived from the mesencephalic crest (Mes.). AD. GL., adrenal gland; S. C. G., superior cervical ganglion.

The ontogeny of PNS ganglia has been studied extensively during the last decade by several groups using a cell marking technique based on stable structural differences between the nuclei of quail and chick cells (Le Douarin, 1969, 1973). The site of origin along the neural axis of different peripheral ganglia can then be determined by constructing appropriate quail–chick chimeras. By microsurgery, defined regions of the neural primordium were removed from a chick (or quail) embryo and replaced by the equivalent primordium from a quail (or chick) at the same developmental stage. Analysis of the resulting chimeras allowed construction of a "fate map" of the neural crest and of the placode derivatives (see Le Douarin, 1982; D'Amico-Martel and Noden, 1983; Le Douarin et al., 1984, for reviews).

A craniocaudal regionalization of the neural crest could be recognized with respect to its participation in the development of the autonomic nervous system (ANS; Figure 1). The sympathetic chain derives from the entire length of the neural crest from the level of the fifth somite caudad, with the chromaffin cells of the adrenal medulla originating specifically from the level of somites 18–24. The great majority of enteric ganglia arise from the "vagal" neural crest, opposite somites 1–7. Neural crest cells from this region start migrating in a ventral direction at around the eight- to 10-somite stage and become localized in the area of the branchial arches (Le Douarin and Teillet, 1973; Le Lièvre and Le Douarin, 1975).

The precursors of the enteric ganglia become incorporated in the developing wall of the foregut, which is of mesodermal origin. Thereafter they migrate caudally along the gut, colonizing it down to the cloacal end and giving rise to the myenteric and submucosal plexuses. An additional, although minor, contribution to these structures in the postumbilical gut is made by the lumbosacral level of the crest, which gives rise essentially to the parasympathetic ganglion of Remak (Teillet, 1978).

Thus, it is evident that the cervicodorsal crest, situated between somites seven and 28, does not provide the developing gut with ganglionic cells. The migration of crest cells from this area is limited to the dorsal trunk structures, and, apart from the Schwann cells lining the nerves, these crest cells do not penetrate the dorsal mesentery.

The origin of the dorsal root ganglia (DRG) at the cervicotruncal level of the neural crest can be traced easily in quail–chick chimeras; the ganglia arise at the level of each somite from the corresponding transverse region of the neural crest from somite six downward (see Le Douarin, 1982 and Le Douarin et al., 1984, for more details).

Our knowledge of the ontogeny of the sensory ganglia located along the cranial nerves first came from the pioneering work of embryologists at the turn of the century, who noticed that cells break off from the cephalic neurogenic placodes and form ganglia in conjunction with neural crest cells. However, their conclusions long remained tentative, because cells of placodal and crest origin rapidly become indistinguishable from each other during the ontogenetic process. Application of the quail–chick chimera system to this problem enabled the respective roles of placodal and crest cells to be more accurately defined (Noden, 1978b; Narayanan and Narayanan, 1980; Ayer-Le Lièvre and Le Douarin, 1982; D'Amico-Martel and Noden, 1983), and the current state of our knowledge regarding this question is summarized in Figure 2.

HOMOGENEOUS DISTRIBUTION OF PNS DEVELOPMENTAL CAPABILITIES IN THE NEURAL CREST

It has long been known in developmental biology that the fate of a given anlage does not necessarily reflect the totality of the developmental potential of its component cells. Isolation *in vitro*, or *in vivo* transplantation in heterotopic situations, often revealed broader differentiation capacities in such anlagen than those actually expressed in normal development. This means that certain potentialities are repressed in some way during the normal course of ontogeny. Whether they are repressed through an active inhibitory mechanism or because certain favorable conditions or factors are lacking is an open question.

Although a marked regionalization of the neural crest has been demonstrated in normal development, particularly with respect to the precursors of the ANS (see Figure 1), further experiments by our group have demonstrated that the developmental potentials of various crest cell populations (e.g., vagal, cervicodorsal, or lumbosacral) are not restricted to those expressed during normal ontogeny. In fact, heterotopic transplantations of pieces of the neural primordium, for

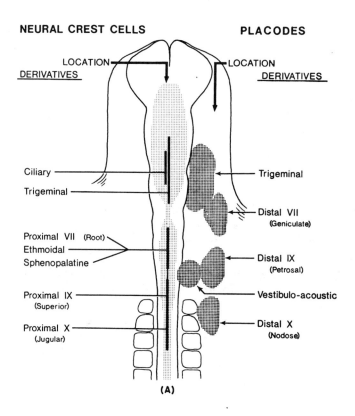

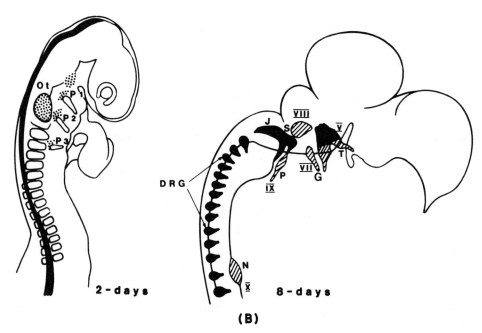

example transfer of the vagal primordium to the adrenomedullary region of the neuraxis and vice versa, have demonstrated that any level of the crest can give rise to the whole spectrum of sensory and autonomic structures (Le Douarin and Teillet, 1974). This was confirmed more recently by another approach in which pieces of neural crest from various levels (including the prosencephalon) of the neural axis of a quail were transplanted into the crest migratory pathway of a chick embryo at the adrenomedullary level (Figure 3). This supernumerary crest dissociates and migrates rapidly, finally stopping in the normal sites of arrest of the trunk crest cells of the host, that is, in the DRG, in the sympathetic ganglion chains and plexuses, and in the adrenal medulla. In each of these sites the grafted crest cells differentiated according to their final location regardless of their axial origin in the donor embryo. In one striking example, the prosencephalic crest, from which no PNS ganglia are known to originate, gave rise to DRG neurons and to sympathetic and adrenomedullary cells. Other levels of donor crest behaved similarly. Only the relative proportion of neurons and glia arising from the various crest populations, initially similar in size, were found to be variable (Le Lièvre et al., 1980; Schweizer, 1980).

Therefore although certain regional differences were observed in the crest cell population—particularly in its mesectodermal derivatives, which can arise only from the cephalic area (Le Lièvre and Le Douarin, 1975; Noden, 1978a, 1983)—its developing potentialities with respect to PNS cell components are distributed rather homogeneously throughout the neural axis. Some kind of selection must then take place which elicits some of these potentialities and represses others during the ontogeny of each particular type of PNS ganglion. One of the crucial problems is to determine whether the totality of these presumptive phenotypes is contained in a single multipotential cell or in several partly or completely committed cells, and if so, when the commitment of these different progenitor cells will occur.

WHEN DOES THE CHOICE AMONG THE VARIOUS POTENTIALITIES OF NEURAL CREST CELLS OCCUR DURING PNS ONTOGENY?

This choice can occur during the migration process or at the site where PNS ganglia are formed. If the process of migration does not involve some kind of selectivity, the question is raised as to whether the potentialities that will never be expressed in a particular type of ganglion are ever definitely switched off. If

---

Figure 2. *A: Schematic drawing of a stage-9.5 chick embryo indicating positions of neural crest and placodal anlagen for cranial sensory and autonomic ganglia.* The limits of each anlage were ascertained by comparing measurements taken at the time of surgery with the distribution of neurons containing the quail marker. Although the actual placodal neurogenic areas are undoubtedly smaller than drawn here, it is not possible to be more precise, due to the regulative properties of surface ectoderm. (Reproduced with permission from D'Amico-Martel and Noden, 1983.) *B*: Drawing on the left shows the distribution of placodes in a two-day-old embryo. ot, otic placode; P1, P2, P3, epibranchial placodes. Drawing on the right shows the sensory spinal (DRG) and cranial ganglia at eight days, the neurons of which are either of crest (black) or placodal (stripes) origin. V, VII, VIII, IX, and X indicate the number of cranial nerves bearing sensory ganglia. T, trigeminal; G, geniculate; J. and S., jugular and superior; P. and N., petrosal and nodose.

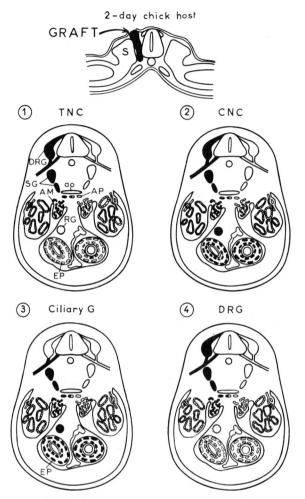

Figure 3. *Diagram summarizing the experiments of back-transplantation of quail neural crest cells or crest derivatives at the adrenomedullary level of a two-day-old chick host.* The top center drawing illustrates the experimental procedure; the grafted tissue is inserted between the somite (S) and the neural tube of the host at the level of somites 18–24 and at a stage preceding the migration of host crest cells. 1 and 2: Supernumerary crest cells from (1) the truncal level (TNC) or (2) the cephalic level (CNC) become localized in the normal sites of arrest of the host trunk crest cells; quail cells participate in the formation of the dorsal root ganglia (DRG), the sympathetic ganglia (SG), the aortic plexuses (AP), and the adrenal medulla (AM). In addition, cephalic crest cells (2) penetrate the dorsal mesentery and populate the ganglion of Remak (RG) and the enteric plexuses (EP). 3 and 4: Results from the graft of a fragment of five- to seven-day ciliary ganglion (3) or DRG (4). In both types of grafts, quail cells are found as Schwann cells along the rachidian nerves, as supportive cells of the sympathetic structures, and as catecholamine-containing cells in the sympathetic ganglia, aortic plexuses, and adrenal medulla. However, two major differences are observed between the two kinds of grafts; ciliary ganglion cells (3) penetrate the dorsal mesentery and colonize the cholinergic Remak and enteric ganglia, as already seen for the graft of supernumerary cephalic crest cells (2), but they never give rise to DRG cells. In marked contrast, grafted DRG cells (4) participate in the formation of host sensory ganglia but do not penetrate the dorsal mesentery. The distribution of the grafted ganglion cells varies with the age of the ganglion at grafting time (see text).

they are not, it would mean that undifferentiated (crestlike) cells would remain present in PNS ganglia even after they had developed into functional structures.

Some answers to these questions were provided by experiments in which developing PNS ganglia, removed from quail embryos at various embryonic stages, were implanted into the neural crest migration pathway of two-day-old chick hosts (Figure 3). The rationale underlying this experimental paradigm was that if certain cells in the ganglion possessed differentiating capacities that could be triggered by the environment of a younger embryo then this might be a means of revealing them (Le Douarin et al., 1978, 1979; Ziller et al., 1979; Le Lièvre et al., 1980; Ayer-Le Lièvre and Le Douarin, 1982; Schweizer et al., 1983).

Various types of ganglia—sensory (spinal ganglia) and autonomic (ciliary, sympathetic ganglia)—were selected and taken from the quail donor from days four to 15 of incubation. No experiments involving postnatal ganglia have been done so far. Except for the early stages, the ganglia were fragmented in order to implant two to three thousand cells. The ganglion was always inserted as an undissociated, compact graft as shown in Figure 3. Irrespective of its nature, the peripheral cells of the implant detached soon after grafting, and this phase of ganglion disruption resulted in the complete (or nearly complete) dispersion of the ganglion cells within three days. Simultaneously, most grafted quail cells divided, as shown by incorporation of tritiated thymidine (Dupin, 1984), and pycnotic figures also appeared at the graft site. At days 2–3 postgrafting, the cells originating from the graft could be seen dispersed among the host somitic cells. This phase of "dispersion" was followed by a phase of "homing," during which the progeny of the grafted ganglion cells became localized in the neural crest derivatives of the host. This process involved a mechanism of "sorting-out," in which some kind of cell–cell recognition mechanism presumably accounts for the differential localizations of the grafted ganglion cells, which varied according to both the nature of the ganglion and its developmental stage when grafted.

The quail ganglionic tissues were always implanted at the adrenomedullary (somites 18–24) level of a 25-somite-stage host, and subsequent observations all conformed to one of the following three patterns of quail cell distribution:

1. The grafted ganglion cells or their progeny homed to the host DRG, to the sympathetic chain ganglia and plexuses, and to the suprarenal gland, where they differentiated respectively into sensory and sympathetic neurons, adrenomedullary cells, and the corresponding satellite cells. In all cases, Schwann cells of quail type were present in peripheral nerves at the level of the graft. This pattern was found only with DRG grafts, and even then only when the DRG were removed from the donor before seven days of incubation. This stage coincides with the time when all DRG neurons become postmitotic in the quail (our unpublished observations). In birds, DRG contain two populations of neurons, the large lateroventral (LV) neurons, whose birthdates extend from four to six days in the quail, and the smaller mediodorsal (MD) neurons, whose birthdates cover a longer period of time, lasting until seven days of incubation. It was possible to implant selectively fragments of either LV or MD areas of quail DRG into the chick neural crest migration pathway. It turned out

that the pattern of derivatives described above was obtained with both LV and MD fragments, but only when they still contained cycling sensory neuroblasts, that is, until six days with the LV region and seven days with the MD region (Schweizer et al., 1983).

2. In the second pattern, the quail cells were distributed in the sympathetic ganglia and plexuses, in the adrenal medulla, and as Schwann cells along the nerves. In this case, the derivatives yielded by the graft were all of the autonomic type, and the migration of the grafted cells was restricted to the dorsal mesenchyme of the host. The most noticeable fact was that no cells (except, in a few cases, a small number of nonneuronal cells) of graft origin were found within the host DRG. This distribution was observed either when DRG (taken from the quail donor at eight days onward) or sympathetic ganglia (4.5–6 days) or ciliary ganglia (taken at 10–15 days) were grafted.

3. In the third pattern, in addition to the locations described in (2), quail cells were also found in the enteric ganglia. This was the result obtained with young ciliary ganglia (4.5–6 days) and with distal sensory cranial ganglia of the vagus nerve (nodose ganglia) removed from five- to 10-day-old quail embryos. It is thus clear that, at least until 10–15 days of incubation, autonomic ganglion cells and adrenomedullary cells can arise from all types of PNS ganglia investigated so far. Their differentiation into either adrenergic, peptidergic, or cholinergic neurons and paraganglion cells depends on the environment of the host embryo within which they become localized rather than on the nature (sensory, sympathetic, or parasympathetic) of the ganglion implanted. The exact developmental stage at which this capacity disappears for each type of PNS ganglion has not yet been determined. It appears, however, that the proliferation rate and the extent of migration of the ganglion cells decrease with the age of the ganglion (Dupin, 1984).

A striking fact was that the capacity to provide the host with cells able to home to the DRG and to differentiate into sensory neurons and glia existed only in the quail spinal ganglia and not in the other ganglia tested. Furthermore, a correlation could be established between the presence of cycling sensory neuroblasts and the potentiality of DRG fragments to yield sensory neurons in the host. This led us to devise an experiment in which the fate of neuronal and nonneuronal cells could be followed in the host in a selective manner. Such an opportunity was provided by cranial nerve sensory ganglia, such as the nodose (nerve X) and petrosal (nerve IX) ganglia, which develop distally with respect to the central nervous system (CNS). As mentioned above, it could be shown by a variety of methods, and especially by using the quail–chick marker system, that these ganglia have a mixed neural crest–placodal origin (see Le Douarin, 1982; D'Amico-Martel and Noden, 1983). In both the nodose and the petrosal ganglia, all neurons are derived from ectodermal placodes, whereas the nonneuronal cell population, destined to differentiate into satellite cells, arises entirely from the neural crest (Ayer-Le Lièvre and Le Douarin, 1982; D'Amico-Martel and Noden, 1983). Therefore, by virtue of exchanges of neural primordia between quail and chick embryos, it was possible to construct nodose and petrosal ganglia in which either the

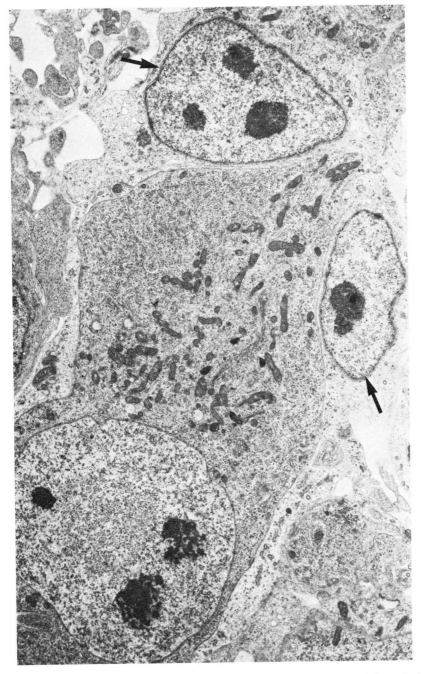

Figure 4. Nodose ganglion of a seven-day-old chimeric chick embryo that has received the graft of a quail rhombencephalon (see Figure 5). The neurons belong to the chick species; the nonneuronal cells (arrows) have the quail nuclear marker. ×7400.

neurons or the nonneuronal cells selectively carried the quail nuclear marker (Figure 4).

The following two questions could then be answered:

1. Do the postmitotic neurons of the nodose and petrosal ganglia die after back-transplantation within the neural crest migration pathway of the chick embryo?
2. Does the nonneuronal population of the ganglion possess neuronal (sensory, autonomic, or both) potentialities that can be revealed in back-transplantation experiments?

The experimental design is represented in Figure 5. The answer to the first question was positive. When pieces of nodose ganglia in which the neurons

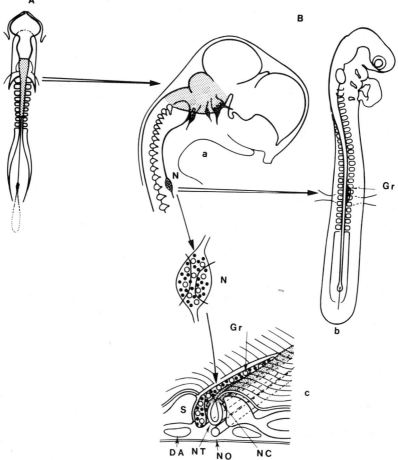

Figure 5. *Making a chimeric nodose ganglion, and its graft into a two-day-old chick host.* A: A quail–chick (or chick–quail) chimera is made by replacing the rhombencephalo–vagal–neural primordium of the host with the same piece of neural primordium from the donor. The grade is made in embryos at the six- to 11-somite stage. B: Grafting pieces of the chimeric nodose (N) ganglion from five- to nine-day-old chimeras (a) between neural tube (NT) and somites (S) at the brachial level of a two-day-old chick host embryo (b); transverse section of the host at the graft level (c). DA, dorsal aorta; Gr, graft; NC, host neural crest; NO, notochord. (Reproduced with permission from Ayer-Le Lièvre and Le Douarin, 1982.)

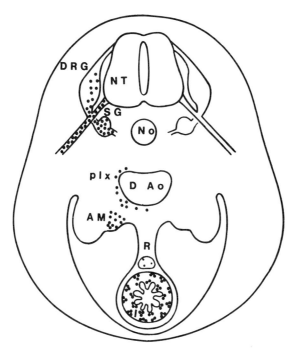

Figure 6. *Diagram showing the results of the experiment represented in Figure 5.* Section through an eight-day-old host embryo indicating the localization of the quail cells derived from the grafted quail–chick chimeric nodose ganglion: Quail cells are found in the dorsal root ganglion (DRG), the rachidian nerves, the sympathetic ganglia (SG), plexuses (plx), adrenomedulla (AM), the ganglia of the enteric plexuses (I), and Remak's ganglion (R). In the DRG, nerves, and Remak's ganglion the quail cells are of the glial type. In the other structures they also form neurons and paraganglion cells. (Reproduced with permission from Ayer-Le Lièvre and Le Douarin, 1982.)

were carrying the quail marker were transplanted into a chick embryo, no quail neurons of any type were found in the host several days later.

As for the second question, no sensory neurons with the quail marker developed in the host DRG. However, besides Schwann and satellite cells located in the expected structures of the host, many autonomic neurons and paraganglion cells developed from the graft and were distributed not only in sympathetic ganglia and adrenergic paraganglia but also in the intramural enteric plexuses (Figure 6; Ayer-Le Lièvre and Le Douarin, 1982, and unpublished observations concerning the petrosal ganglion).

From these experiments a certain number of conclusions can be drawn: (1) In back-transplantation experiments, sensory neurons arise only from spinal ganglia and not from the distal sensory ganglia of cranial nerves IX and X. The proximal sensory ganglia of these nerves, whose neurons are of neural crest origin (jugular and superior ganglia), have not been investigated so far in this respect. Moreover, this capacity of spinal ganglia is realized only if they still contain cycling sensory neuroblasts, and it disappears when all sensory neurons are postmitotic. (2) The nonneuronal cell population of all types of PNS ganglia so far investigated contains progenitor cells inducible toward the autonomic neuronal and the adrenomedullary phenotypes if they are subjected to the appropriate embryonic microenvironments between days two and six of chick embryo development. This period coincides with the period during which the neural crest progenitors normally reach their destination, undergo gangliogenesis, and begin to differentiate into the large array of cell types that constitute the PNS. Importantly, the phenotypes that will be commonly expressed by all types of PNS ganglia in these experiments correspond to Schwann cells, adrenomedullary cells, and autonomic ganglion cells. In contrast, the sensory neuronal and glial phenotypes formed in DRG are only provided by early spinal ganglia.

A quantitative analysis carried out by Dupin (1984) on ciliary ganglion grafts revealed that the numbers of these inducible autonomic progenitors decrease with time during embryonic life, although they still exist at 15 days of incubation, that is, one day before hatching.

THE "AUTONOMIC–SENSORY CELL LINES SEGREGATION" HYPOTHESIS

In view of these experimental facts, I have formulated the hypothesis that, early in neural crest ontogeny, two types of precursors become committed and therefore diverge from a possible common ancestor. I call those that differentiate into sensory neurons and glia and home to the host DRG in the back-transplantation experiment type S (for sensory) precursors. The second is type A (for autonomic), present in all PNS ganglia tested so far. All type S cells become rapidly postmitotic in spinal ganglia and apparently are absent in the nonneuronal population of both the distal cranial nerve sensory ganglia and the autonomic ganglia. Whether they never reach these sites of gangliogenesis or whether they die there very early on cannot, at present, be ascertained. Type A progenitors remain available many days after neuronal birthdates are reached in all types of PNS ganglia (see Le Douarin, 1984a,b for detailed discussion).

INFLUENCE OF THE POSITIONAL RELATIONSHIPS BETWEEN PNS GANGLIA AND THE CNS ON THE DIFFERENTIATION OF NEURAL CREST-DERIVED SENSORY NEURONS

One striking fact is that neural crest-derived sensory neurons develop in the embryo in ganglia that form in close vicinity to the CNS. The sensory neurons differentiating in these ganglia (i.e., spinal ganglia and proximal ganglia of certain cranial nerves such as the jugular and superior) establish connections with the CNS very early in development by sending processes to the spinal cord and hindbrain as soon as the neural crest cells aggregate to form a ganglion. This can be seen very clearly by applying antibodies against neurofilament proteins on early mouse or chick embryos (Figure 7). One can hypothesize that such an early relationship may involve some trophic activity displayed by the CNS with regard to the type S precursors, which could survive only if supplied with a growth factor of neural tube origin. Although no direct evidence is so far available to support this hypothesis, certain data make it plausible. In a recent study, the role of the neural tube and the notochord on gangliogenesis by neural crest-derived cells has been investigated (Teillet and Le Douarin, 1983). The total excision of both these axial organs resulted in the total absence of sensory and sympathetic ganglia at the level considered. The fate of the neural crest cells

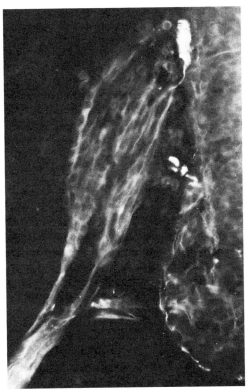

Figure 7. *Demonstration of neurofilament protein immunoreactivity in an early dorsal root ganglion.* Transverse section in the trunk region of a 10.5-day-old mouse embryo. Antibodies directed against the 150-kD protein subunit of neurofilaments selectively decorate neurons and their processes in the dorsal root ganglion primordium and in the ventral horn of the spinal cord (lower right). Dorsal projections of sensory neurons very rapidly reach the spinal cord (as indicated by the well-developed, intensely labeled, dorsal white column, upper right), as the aggregation of neural crest cells to form the DRG takes place around embryonic day 10.

that had migrated in the somitic area prior to the excision could be followed through the quail–chick marker. It appeared that in the absence of neural tube and notochord, the neural crest and the somitic cells were sites of intense cell death, resulting in complete failure of development of vertebrae, striated muscles, and spinal and sympathetic chain ganglia. If the notochord was left *in situ*, vertebrae (although abnormal), muscles (some, but not the complete set), and sympathetic adrenergic ganglia developed, but spinal ganglia did not form at all. If, in contrast, the neural tube (but not the notochord) was left *in situ*, DRG developed normally, as did the sympathetic chain ganglia.

It appears, therefore, that the presence of the neural tube is necessary for DRG development.

HETEROGENEITY OF THE NEURAL CREST CELL POPULATION DEMONSTRATED IN TISSUE CULTURE

If heterogeneity is established early in the neural crest cell population, it might be apparent in tissue culture where environmental conditions can be monitored to trigger selectively the expression of one or another of their developmental potentialities. This is why a series of studies have been undertaken in several laboratories with a view to addressing these questions.

Expression of Neuronal Phenotypes by Cultured Crest Cells

The cephalic and truncal neural crest can be obtained in a pure state by surgery, either when it is still in the neural fold or, for the mesencephalic area, when it is already in the process of migration (see Fauquet et al., 1981; Ziller et al., 1983).

When cultured *in vitro* in serum-containing medium (DMEM with 15% fetal calf or horse serum), isolated mesencephalic or trunk neural crest cells grew rapidly but did not exhibit a morphologically identifiable neuronal phenotype. However, acetylcholine-synthesis activity was significantly higher in these cultures than in the freshly removed crest (Smith et al., 1979). Moreover, catecholamine synthesis became detectable, particularly when fetal calf serum was used instead of horse serum. A modulatory effect of serum was actually observed on catecholamine- and acetylcholine-synthesis activities, with horse serum favoring the latter and fetal calf serum the former (Fauquet et al., 1981). In these cultures, none of the neuronal markers tested, such as tetanus toxin or neurofilament protein immunoreactivity, revealed any neuronal phenotypic expression in culture or in the crest prior to cultivation. The intermediate filament protein, vimentin, was present in 100% of crest cells both prior to cultivation and in culture, and an antibody to desmin labeled a small proportion of cells in mesencephalic cultures (Ziller et al., 1983).

A strikingly contrasting result was found when equivalent neural crest explants were cultured in a fully defined medium totally devoid of serum but containing hormones, growth factors, and transferrin—the "Basic Brazeau Medium" (BBM; Ziller et al., 1981, 1983). In a few hours, a subpopulation of crest cells from both trunk and mesencephalic regions readily differentiated into neurons without dividing. Expression of tetanus-toxin binding sites and neurofilament protein synthesis could be detected very early in these cells. Coexpression of neurofilament protein and vimentin was a general rule in the neurons while only vimentin was detectable in the still cycling, nonneuronal, flat cells (Figure 8).

Application of a pulse of depolarizing current to cells with typical neurite outgrowths revealed their aptitude to generate action potentials from day four in culture (Bader et al., 1983), thus confirming their neuronal nature. Biochemical or histochemical analysis failed to reveal catecholamine synthesis and storage in these cells, and only low levels of acetylcholine synthesis could be detected.

A strong inhibitory effect of serum factor(s) on neurite outgrowth was demonstrated, and it appeared clear that conditions stimulating proliferation of crest cells, such as the presence of serum, were incompatible with promotion of the expression of a neuronal phenotype by the precursors thus revealed in the crest. In fact it became apparent that culture in serum-free medium stops proliferation and triggers neurite extension in cells which, *in vivo*, would normally have gone through several divisions before differentiating.

Culture of neural crest cells with or without the neural tube (Cohen, 1977; our unpublished observations) can, however, yield neurons even in the presence of serum if 10–15% of chick embryo extract is added to the medium. The neurons

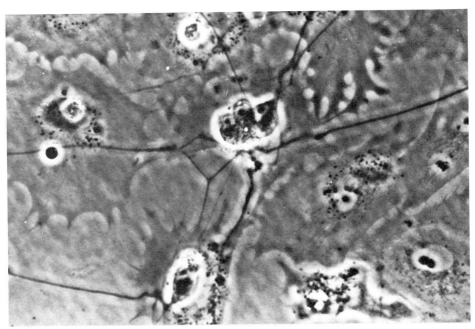

Figure 8. *A three-day culture of mesencephalic neural crest cells in a fully defined medium (BBM), observed under phase-contrast microscopy.* Differentiated neurons with refringent cell bodies and long neurites form a network on an underlayer of flat cells.

appear in the culture only after several days, and at least some of them exhibit the catecholaminergic phenotype. Inhibition of cell proliferation by cytosine arabinoside prevents their differentiation (Kahn and Sieber-Blum, 1983), suggesting that they arise from different progenitors than those which differentiate within 24 hours of culture in BBM (Ziller et al., 1981, 1983). If such a contention is confirmed by further experiments now in progress in our laboratory, it would support the hypothesis that several types of precursors for PNS neurons are segregated early in the crest cell population. One type can differentiate into neurons in culture without dividing but is prevented from doing so by some serum component(s). A second type develops into neurons in the presence of serum once it has gone through a number of cell cycles, but only if some growth or differentiation factor is provided by the embryo extract. Cell division would then be necessary before this neural crest precursor can extend neurites.

It is tempting to establish a parallel between the conclusions drawn from the *in vivo* experiments described above and the results of *in vitro* culture. The progenitor cells of neurons that differentiate in the absence of serum and do not synthesize significant amounts of either acetylcholine or catecholamine could correspond to type S precursors, which would find appropriate conditions to develop in BBM. The precursors that divide in the presence of high levels of chick embryo extract and later synthesize catecholamine and acetylcholine would then be of type A. However, only cloning experiments could allow one to find out whether these two distinct neuronal phenotypes really arise from different, irreversibly committed neuronal precursors, thus confirming the hypothesis of an early segregation of sensory and autonomic cell lines in neural crest ontogeny.

CONCLUSIONS

The results reported above might appear at first glance somewhat contradictory. Through *in vivo* analysis, two notions have emerged.

If, by using the quail–chick chimera system, one follows the fate of the neural crest progenitors of the PNS without disturbing (or doing so as little as possible) the normal course of development, the neural crest is divided into craniocaudally distributed regions, each devoted to yielding a definite array of differentiated cell types of the PNS. It seems, therefore, as though a restricted role is attributed to each region of the neural crest in the building up of complex PNS structures.

If, by contrast, the natural arrangement of the craniocaudal neural crest territories is artificially disturbed, no major alterations take place in the arrangement of PNS structures, one of the presumptive territories roughly being able to substitute for another even if their respective roles in normal development are very different.

Therefore the initially suggested notion of heterogeneity should be replaced by that of homogeneity of the neural crest along the neuraxis as far as the potentialities of its component cells are concerned. The role of the nonneuronal tissues of the embryo thus becomes preponderant in "directing" PNS ontogeny.

It is clear that, at the cellular level, this role can be exerted either by inducing multipotential cells toward a particular array of differentiation pathways in each type of PNS ganglia or by selecting from among partly or completely committed cells those which, in each category of structures, will be allowed to develop.

The observations resulting both from back-transplantation of developing PNS ganglia into the neural crest cell migration pathway of younger hosts and from *in vitro* cultures of neural crest cells have shown that a certain degree of heterogeneity already exists in the neural crest itself. The neural crest appears to be composed of subpopulations of cells. In some of them at least, developmental capacities are restricted to a certain spectrum of definite phenotypes whose development is selectively triggered by factors arising from the tissue environment of the embryo. Although a certain degree of commitment exists in such crest cells, developmental choices can be imposed by environmental factors on certain of them. Such is the case, for example, for sympathetic neurons, in which the nature of the transmitter synthesized (acetylcholine or catecholamine) and/or the cellular morphology exhibited (sympathetic neurons or "small intense fluorescent" cells) can be modulated by extrinsic cues (see Patterson, 1978; Landis, 1983, for reviews).

ACKNOWLEDGMENTS

This work was supported by the Centre National de la Recherche Scientifique, the Délégation Générale à la Recherche Scientifique et Technique, the Ministère de l'Industrie et de la Recherche, the National Institutes of Health, and the Association pour le Développement de la Recherche contre le Cancer.

REFERENCES

Ayer-Le Lièvre, C., and N. M. Le Douarin (1982) The early development of cranial sensory ganglia and the potentialities of their component cells studied in quail–chick chimeras. *Dev. Biol.* 94:291–310.

Bader, C. R., D. Bertrand, E. Dupin, and A. C. Kato (1983) Development of electrical membrane properties in cultured avian neural crest. *Nature* **305**:808–810.

Cohen, A. M. (1977) Independent expression of the adrenergic phenotype by neural crest cells *in vitro*. *Proc. Natl. Acad. Sci. USA* **74**:2899–2903.

D'Amico-Martel, A., and D. M. Noden (1983) Contributions of placodal and neural crest cells to avian cranial peripheral ganglia. *Am. J. Anat.* **166**:445–468.

Dupin, E. (1984) Cell division in the ciliary ganglion of quail embryos *in situ* and after back-transplantation into the neural crest migration pathways of chick embryos. *Dev. Biol.* **105**:288–299.

Fauquet, M., J. Smith, C. Ziller, and N. M. Le Douarin (1981) Differentiation of autonomic neuron precursors *in vitro*: Cholinergic and adrenergic traits in cultured neural crest cells. *J. Neurosci.* **1**:478–492.

Kahn, C. R., and M. Sieber-Blum (1983) Cultured quail neural crest cells attain competence for terminal differentiation into melanocytes before competence for terminal differentiation into adrenergic neurons. *Dev. Biol.* **95**:232–238.

Landis, S. C. (1983) Factors which influence the transmitter functions of sympathetic ganglion cells. In *Autonomic Ganglia*, L. G. Elfvin, ed., pp. 453–473, Wiley, Chichester.

Le Douarin, N. M. (1969) Particularités du noyau interphasique chez la caille japonaise (*Coturnix coturnix japonica*). Utilisation de ces particularités comme "marquage biologique" dans les recherches sur les interactions tissulaires et les migrations cellulaires au cours de l'ontogenèse. *Bull. Biol. Fr. Belg.* **103**:435–452.

Le Douarin, N. M. (1973) A biological cell labeling technique and its use in experimental embryology. *Dev. Biol.* **30**:217–222.

Le Douarin, N. M. (1982) *The Neural Crest*, Cambridge Univ. Press, Cambridge.

Le Douarin, N. M. (1984a) A model for cell line divergence in the ontogeny of the peripheral nervous system. In *Cellular and Molecular Biology of Neuronal Development*, I. Black, ed., pp. 3–28, Plenum, New York.

Le Douarin, N. M. (1984b) The neural crest and the development of the peripheral nervous system. In *Handbook of Physiology*, W. M. Cowan, ed., American Physiological Society, Bethesda, Md. (in press).

Le Douarin, N. M., C. S. Le Lièvre, G. Schweizer, and C. M. Ziller (1979) An analysis of cell line segregation in the neural crest. In *Cell Lineage, Stem Cells and Cell Determination*, N. Le Douarin, ed., pp. 353–365, Elsevier/North Holland, Amsterdam.

Le Douarin, N. M., and M. A. Teillet (1973) The migration of neural crest cells to the wall of the digestive tract in avian embryo. *J. Embryol. Exp. Morph.* **30**:31–48.

Le Douarin, N. M., and M. A. Teillet (1974) Experimental analysis of the migration and differentiation of neuroblasts of the autonomic nervous system and of neurectodermal mesenchymal derivatives, using a biological cell marking technique. *Dev. Biol.* **41**:162–184.

Le Douarin, N. M., M. A. Teillet, and J. Fontaine-Perus (1984) Chimaeras in the study of the peripheral nervous system of birds. In *Chimaeras in Developmental Biology*, N. M. Le Douarin and A. McLaren, eds., Academic, London (in press).

Le Douarin, N. M., M. A. Teillet, C. Ziller, and J. Smith (1978) Adrenergic differentiation of cells of the cholinergic ciliary and Remak ganglia in avian embryo after *in vivo* transplantation. *Proc. Natl. Acad. Sci. USA* **75**:2030–2034.

Le Lièvre, C. S., and N. M. Le Douarin (1975) Mesenchymal derivatives of the neural crest: Analysis of chimaeric quail and chick embryos. *J. Embryol. Exp. Morphol.* **34**:125–154.

Le Lièvre, C. S., G. G. Schweizer, C. M. Ziller, and N. M. Le Douarin (1980) Restrictions of developmental capabilities in neural crest cell derivatives as tested by *in vivo* transplantation experiments. *Dev. Biol.* **77**:362–378.

Narayanan, C. H., and Y. Narayanan (1980) Neural crest and placodal contributions in the development of the glossopharyngeal–vagal complex in the chick. *Anat. Rec.* **196**:71–82.

Noden, D. M. (1978a) The control of avian cephalic neural crest cytodifferentiation. I. Skeletal and connective tissues. *Dev. Biol.* **67**:296–312.

Noden, D. M. (1978b). The control of avian cephalic neural crest cytodifferentiation. II. Neural tissues. *Dev. Biol.* **67**:313–329.

Noden, D. M. (1983) The role of the neural crest in patterning of avian cranial skeletal, connective and muscle tissues. *Dev. Biol.* **96**:144–165.

Patterson, P. H. (1978) Environmental determination of autonomic neurotransmitter functions. *Annu. Rev. Neurosci.* **1**:1–77.

Schweizer, G. (1980) Recherches sur la ségrégation des lignées cellulaires dans la crête neurale de l'embryon d'oiseau. Thèse de 3e cycle, Université de Paris VI.

Schweizer, G., C. Ayer-Le Lièvre, and N. M. Le Douarin (1983) Restrictions of developmental capacities in the dorsal root ganglia during the course of development. *Cell Differ.* **13**:191–200.

Smith, J., M. Fauquet, C. Ziller, and N. M. Le Douarin (1979) Acetylcholine synthesis by mesencephalic neural crest cells in the process of migration *in vivo*. *Nature* **282**:853–855.

Teillet, M. A. (1978) Evolution of the lumbo-sacral neural crest in the avian embryo: Origin and differentiation of the ganglionated nerve of Remak studied in interspecific quail-chick chimaerae. *W. Roux's Arch. Dev. Biol.* **184**:251–268.

Teillet, M. A., and N. M. Le Douarin (1983) Consequences of neural tube and notochord excision on the development of the peripheral nervous system in the chick embryo. *Dev. Biol.* **98**:192–211.

Ziller, C., E. Dupin, P. Brazeau, D. Paulin, and N. M. Le Douarin (1983) Early segregation of a neuronal precursor cell line in the neural crest as revealed by culture in a chemically defined medium. *Cell* **32**:627–638.

Ziller, C., N. M. Le Douarin, and P. Brazeau (1981) Différenciation neuronale de cellules de la crête neurale cultivée dans un milieu défini. *C.R. Acad. Sci.* **292**:1215–1219.

Ziller, C., J. Smith, M. Fauquet, and N. M. Le Douarin (1979) Environmentally directed nerve cell differentiation: *In vivo* and *in vitro* studies. *Prog. Brain Res.* **51**:59–74.

Chapter 9

Gangliogenesis in the Avian Embryo: Migration and Adhesion Properties of Neural Crest Cells

JEAN-PAUL THIERY
GORDON C. TUCKER
HIROHIKO AOYAMA

ABSTRACT

The peripheral nervous system derives almost entirely from the neural crest, a transient structure from which ectodermal cells emigrate and subsequently localize in different territories. The detailed patterns of crest cell migration were reconstructed after immunolabeling crest cells and fibronectin, a suitable marker of basement membranes. Routes of migration specific to each axial level form transiently between or along fibronectin-rich basement membranes. Most crest cells do not exhibit invasive properties, but instead remain confined as a compact multicellular layer in the pathways. The formation of the different types of ganglia and plexuses can be readily understood by following the progressive remodeling of the pathways concomitantly with the development of the surrounding tissues. In many cases, narrow and transient pathways lead crest cells to their site of arrest, where they encounter physical obstacles as well as a local milieu rapidly depleted of fibronectin.

In vitro studies reveal that fibronectin is an essential component of the extracellular matrix for both the attachment and migration of crest cells. Directional migration is likely to be provided by population pressure, unique motility properties, and the ability to respond to exogenous fibronectin. The neural cell adhesion molecule (N-CAM) is shown to be directly involved in the aggregation of crest cells into ganglion rudiments. Its early appearance together with the liver cell adhesion molecule (L-CAM) in all the cells of the blastoderm prior to gastrulation, its subsequent enrichment in the neural epithelium, which becomes devoid of L-CAM, its transient disappearance during migration of crest cells, and finally, its de novo expression at the surface of aggregating crest cells strongly support the hypothesis put forward by Edelman (1983) that cell surface modulation of adhesive molecules by prevalence is of paramount importance in early developmental stages. In contrast, cytodifferentiation, as evidenced by the expression of neurofilaments, occurs progressively after the morphogenetic processes. Following neuronal cell death, a definitive axonal network becomes progressively established while N-CAM converts from an embryonic to an adult form; this conversion, a second case of modulation by chemical modification, strengthens the binding between individual axons within fascicles.

The different phases of neural crest development into mature ganglia apparently involve only a limited number of specific cell surface and extracellular matrix molecules, the expression of which is, however, precisely regulated both in time and space.

Studies on the fate of the vertebrate neural crest have the potential to address key questions on the mechanisms controlling morphogenesis and organogenesis. Indeed, crest cells can be followed throughout their phases of migration, proliferation, adhesion, differentiation, and death. The development of neural crest derivatives has already been described as involving both tissues and extracellular matrices. However, this statement does not account for the apparently extremely sophisticated mechanisms through which crest cells develop into defined structures such as craniofacial elements and the sensory and autonomic ganglia of the nervous system.

Let us consider a very young embryo. The ectoderm, that is, the upper of the three layers constituting the primitive embryo, appears thicker at its center; this area delimits the neural plate that will soon invaginate in its middle region as its lateral borders begin to fold up. As shown in Figure 1, cells situated at the apex of these folds are the presumptive neural crest cells. At this stage, nothing distinguishes them from the adjacent cells, presumptive epidermal ectoderm and epithelial cells of the neural plate. However, at about the time when the neural folds of the ectoderm contact each other, the very cells that initiate this contact begin losing their ectodermal–epithelial structure and "round up." These cells, now referred to as neural crest cells, proliferate, separate from the ectoderm and the recently formed neural tube, and migrate in a newly created space between these epithelia. From then on, neural crest cells are confronted by a rapidly changing environment that contributes to their directed migration and final location.

Small and restricted to a well-defined region as it might be at the beginning, this cell population settles finally in a wide variety of structures and tissues, including muscle, bone, cartilage, dermis and connective tissues in the head and the neck, pigment cells, endocrine and paraendocrine cells, and neuronal and glial lineages of the peripheral nervous system. The latter give rise to sensory and sympathetic ganglia, plexuses around the aorta, and numerous blood vessels, enteric ganglia, and species-specific ganglia such as Remak's ganglion in the avian embryo.

The most immediate questions regarding the development of crest cells into ganglia can be outlined as follows: How are the migration properties of crest cells acquired? What mechanism allows them to reach their specific sites of localization? What prevents crest cells from migrating in certain territories? Do the surrounding tissues and the extracellular material play an active or a passive role in the process of migration and localization? How may adhesive properties be modulated to account for the release, migration, and aggregation of crest cells? How do the different cell lineages segregate? What controls the formation of the peripheral neuronal network?

Because the chapters by Le Douarin and by Gershon in this volume deal with many aspects of crest cell differentiation, we focus here mostly on the problem of cell migration and cell adhesion. One of the major prerequisites to understanding the mechanism underlying crest cell migration and aggregation is a detailed knowledge of the pathways of migration. Recently, several different approaches have allowed us to delineate accurately these pathways and to design *in vitro* and *in vivo* experiments to assess the respective roles and properties of the extracellular matrix, of the surrounding tissues, and most importantly, of the

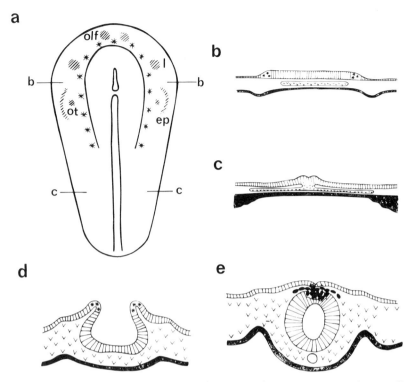

Figure 1. *Neural crest cell individualization.* a: The positions of presumptive neural crest cells and of other ectodermal specializations have been localized in a head-process–stage chick embryo. Inside the inner curve are prospective cells of the brain and spinal cord; external cells correspond to future ectodermal cells. Presumptive placodal structures intervene in the latter: olfactory placodes (olf), lens (l), and ear (ot) are surrounded by epibranchial placodes (ep) (hatched areas). Prospective neural crest cells (stars) are located approximately at the periphery of the future central nervous system (drawn after Rudnick, 1948; Rosenquist, 1981). b: Transverse section at the level of the head primordium indicates the position of crest cells with respect to ectoderm and neural plate. c: More caudally, cells ingress at the level of the primitive streak to form the mesodermal layer between the upper and deep layer; the upper layer will develop further only into ectoderm. d: Formation of the neural tube. e: Closure of the neural tube and migration of crest cells. The ectoderm will progressively form a separate layer over the neural tube while, centrally, the notochord individualizes (see also Figure 7). In the chick embryo, only one day elapses from egg laying to neural plate formation, encompassing the stage of gastrulation and neural induction.

neural crest cells. In the first part of this chapter, we describe the migration pathways, taking the opportunity to define the dynamic aspects of tissue morphogenesis associated with crest cell dispersal. Then we examine the adhesive properties of crest cells throughout their development.

MIGRATION PATHWAYS OF NEURAL CREST CELLS

Early Phase of Migration

In the early avian embryo, neural crest cells individualize from the dorsal border of the closing neural tube. After detachment from the neural tube, crest cells encounter an obstacle formed by the two partially fused basement membranes of the ectoderm and the neural tube. Migration is permitted through a local expansion of this space. How might it be achieved? It might be the consequence of extensive secretion—by surrounding tissues and by neural crest cells themselves—of hyaluronic acid, a polymer whose hydration properties favor the formation of new space (Toole, 1976). On the other hand, local proliferation of crest cells might induce the separation of the basement membranes. Later migration is greatly influenced by the nature of adjacent tissues near the neural tube. Therefore it varies according to the level of the embryo that is being considered. Moreover, as the neural tube closes in a craniocaudal direction, neural crest cells migrate according to changing temporal and spatial patterns at different levels, because organogenesis of adjacent structures, such as the presumptive vertebrae, occurs craniocaudally with a different time schedule. Figures 2 and 3 sum up the different opportunities of migration for neural crest cells at various levels of the embryo, the main sections corresponding to the principal obstacles or available spaces with which the cells are confronted. These routes of migration have been established through the work of many investigators (see Weston, 1970; Noden, 1975; Tosney, 1978; Duband and Thiery, 1982a; Thiery et al., 1982a; Le Douarin et al., 1983; Tucker and Thiery, 1984).

Cephalic Levels. At the level of the prosencephalon, the optic vesicle, a local expansion of the neural tube whose lateral borders remain fused to the ectoderm, prevents crest cells from migrating farther laterally (Duband and Thiery, 1982a). Trapped between the neural tube and the optic vesicle, these cells will later migrate between the ectoderm and the optic cup to form the endothelial cells

---

Figure 2. *Routes of migration at cephalic and vagal levels.* This three-dimensional reconstitution of a 15-somite-old chick embryo shows the location of neural crest cells along the length of the cephalic and vagal parts. The results presented here were obtained through the use of several techniques, including grafting experiments between chick and quail embryos (Le Douarin, 1982), fibronectin immunolabeling (Duband and Thiery, 1982a; Thiery et al., 1982a), acetylcholinesterase histochemical detection (Cochard and Coltey, 1983), and staining with a monoclonal antibody that recognizes migrating crest cells (Vincent and Thiery, 1983, 1984). In this drawing, the embryo is sectioned at various transverse levels, and the ectoderm is partially removed on the right side to reveal the main obstacles, either epithelial or mesenchymal, that prevent crest cells (solid ovals) from invading adjacent structures. The first slice presents the optic vesicle and constriction, and the surrounding and accumulated neural crest cells. As represented in the second and third slices, mesencephalic cells locate above the mesenchyme under the ectoderm, a situation occurring down to the otic placode (shown between the fourth and fifth slices), except for a small acellular space rostral to the auditory pit. On the third slice, a ventral migration arises followed by neuronal and glial precursor cells of the trigeminal ganglion, whose mesencephalic nucleus is built with a particular class of crest cells that remain at the site of closure of the neural tube. The auditory pit, cut along its middle between the fourth and fifth slices, is the second of the main obstacles encountered by neural crest cells. Two dense flows are clearly evidenced rostral and caudal to this structure in continuity with trapped cells shown on the transverse section and with anterior rhombencephalic and anterior vagal crest cells. The latter occupy the subectodermic space with respect to the first five somites and, more caudally, they are able to colonize the ventral pathway that lines the neural tube. The last slice shows the seventh somite level, where both ventral and lateral pathways exist. Thus neural crest cells remain confluent and never enter adjacent tissues; instead, they follow developing spaces and are prevented from invading the whole embryo because dense epithelia form important obstacles that include blood vessels, the walls of which are often used subsequently as a pathway.

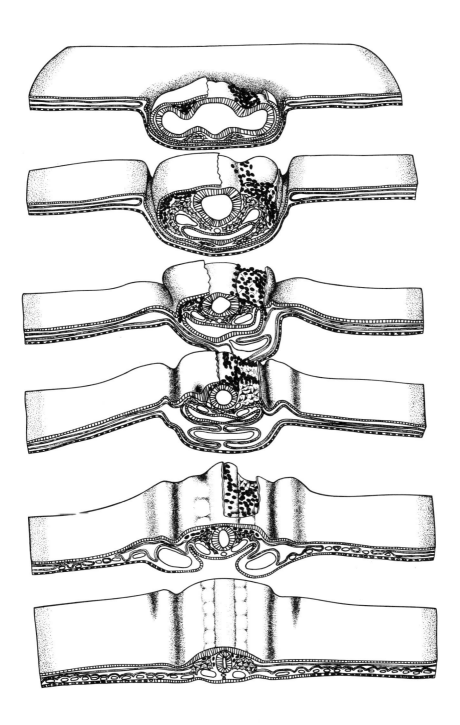

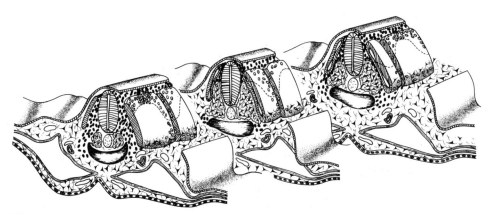

Figure 3. *Migration at trunk level.* After the dissociation of the somite, its sclerotomal part rapidly acquires a characteristic shape that will condition the routes of migration of neural crest cells, as they do not invade this structure. Trunk levels were studied extensively and new data were collected concerning the location of crest cells around the sclerotome. At the level of its middle, the sclerotome appears as a high and wide mass of cells joining the myotome and neural tube, allowing no space except at its dorsal apex, where crest cells accumulate and form the rudiment of the dorsal root ganglion. Note that some crest cells distribute ventrally near the aorta; these come from the adjacent intersomitic flow that is already evidenced in the intermediate transverse section between the middle and the end of the somite. There, the sclerotome appears less important and its diameter is much reduced, so that crest cells follow their route between sclerotome and myotome laterally. Such a pattern is more obvious at the intersomitic level where the sclerotome is on the verge of joining the next one and therefore fills only a small area. Crest cells, no longer restricted in their ventral migration, occupy the whole space between myotome and neural tube, soon reaching the ventral parts of the embryo. In older embryos, each sclerotome expands rapidly in all directions and fuses with its adjacent homologue. Neural crest cells are then divided into two subpopulations. The first follow the intersomitic pathway and colonize the ventral parts while inserting themselves longitudinally under the sclerotome before differentiating into autonomic ganglia and plexuses. This particular pathway is responsible for the presence of groups of crest cells separated from the second subpopulation—namely, the dorsal root ganglion precursors—by a huge sclerotome. Note also the shape of the dorsal aorta: It looks like a cylindrical tube with extensions on both sides dorsally between two adjacent sclerotomes. Neural crest cells also follow these developing aortic branches. Solid ovals, visible crest cells; open ovals, crest cells hidden under the myotome; solid triangular shapes, mesenchyme.

of the cornea (Hay, 1980). Proliferating rapidly, they also join the flow of crest cells from a more rostral level (these cells give rise to mesenchymal components around the telencephalon, the optic vesicle itself, and olfactory placodes) and the flow of crest cells at the anterior mesencephalic level (which will form craniofacial components) (Noden, 1975, 1978).

Mesencephalic neural crest cells first meet a wide acellular space between the ectoderm and the neural tube. This absence of a physical barrier enables them to migrate much farther laterally under the ectoderm basement membrane until the local mesenchyme fills the space. Crest cells continue their migration ventrally while remaining between the ectoderm and mesodermal cells (Pratt et al., 1975; Tosney, 1982). Finally, they reach the oropharyngeal region and the first visceral arch, to form the greater part of the maxillary and mandibular mesenchyme (Le Lièvre and Le Douarin, 1975; Noden, 1975). Some crest cells also localize near the choroid fissure to form the ciliary ganglion, while others, accumulating more ventrally, give rise to other parasympathetic ganglia and numerous small paraganglia.

More caudally, anterior rhombencephalic crest cells follow the same pathway between the ectoderm and the mesenchyme, although the pattern of migration is quite different at the onset as there is no acellular space. The ectoderm fuses partially with the endoderm and therefore acts as a physical barrier. Another difference becomes evident; a second migratory pathway appears ventrally along the neural tube. Crest cells following this path will accumulate near the aorta and the developing cardinal vein and contribute to the trigeminal ganglion (Duband and Thiery, 1982a). In fact, some crest cells remain at the site of closure of the neural tube and later will incorporate into the roof of the brain to form the mesencephalic nucleus of the trigeminal nerve (Narayanan and Narayanan, 1978).

Otic placodes appearing at the median rhombencephalic level as a result of a local thickening of the lateral ectoderm constitute another obstacle to the migration of neural crest cells. Though numerous, crest cells remain at this level for several hours and finally progress ventrally and surround the newly invaginated otic vesicle. No further migration is observed, and the cells contribute to the mesenchyme of the internal ear (Le Lièvre and Le Douarin, 1975).

As revealed by their final location and differentiation, cephalic neural crest cells play an important part in the ontogenesis of craniofacial components, whereas the dorsal part of the head arises from mesodermal cells only. The bone and cartilage of intermediate regions originate from the contribution of both mesodermal- and neural crest-derived cells; the latter are consequently called mesectodermal cells. The major part of connective and muscular tissues, together with the walls of vessels and glands in the ventrolateral part of the head and neck, are built up from mesencephalic and rhombencephalic neural crest cells (see Johnston, 1966; Le Lièvre and Le Douarin, 1975; Le Lièvre, 1978; Noden, 1978, for reviews).

Vagal and Trunk Levels. In these regions, the pattern of migration is correlated with the development of the adjacent mesoderm. This tissue soon compartmentalizes into small entities called somites, which represent the anlagen of vertebrae; these structures line the neural tube, and appear as a row of dense aggregates of cells individualizing from mesoderm in a craniocaudal direction all the way from the otic placode down to the developing caudal parts of the embryo. The morphogenesis of the somite is fairly similar from one level to another; typically, several hours after its condensation from mesoderm, it disrupts into two components, the dermomyotome and the sclerotome. The dermomyotome, composed of two epitheliallike rows, will further develop into dermis and striated muscle, whereas the sclerotome will progressively invade the neighborhood, joining its symmetrical homologue to include ventrally the notochord, another transient structure, and envelop the whole neural tube, thus generating a vertebra. Interestingly, a vertebra does not arise from two opposite somites; it derives from two consecutive pairs, the posterior part of the first pair of sclerotomes associating with the anterior half of the second. Sensory ganglia, although first trapped between the neural tube and the middle of the somite as shown below, actually develop between two adjacent vertebrae. Meanwhile, dermomyotomes retain their original configuration but expand considerably.

Anterior vagal crest cells migrate exclusively between the basement membranes of ectoderm and somitic mesoderm, already disrupted into dermomyotome and sclerotome. Posterior vagal crest cells migrate simultaneously under the ectoderm and between the somitic mesoderm and the neural tube; at that level, migrating crest cells first encounter a nondissociated somite and are soon after confronted by its dissociation (Thiery et al., 1982a).

Trunk crest cells detach during a much longer period of time so that two groups, whose respective behaviors are utterly different, can be evidenced: A first flow runs in the ventral pathway between the neural tube and the somitic mesenchyme; a second flow, corresponding to prospective melanocytes, invades the epidermis and follows the lateral pathway between ectodermal and somitic tissues (Teillet and Le Douarin, 1970). The second flow occurs much later than the first. The disruption of the somite at this level takes place several hours after the onset of migration, so that trunk crest cells are confronted with a nondissociated somite much longer than are posterior vagal ones.

Ontogenesis of Peripheral Nervous System from Vagal and Trunk Levels

Enteric Plexuses. Having migrated between the ectoderm and the somite, neural crest cells from somites one to seven soon reach the apex of the pharynx and its local expansions, the branchial pouches alternating with visceral arches. Mesodermal tissue—still separated laterally into two epithelial layers, the somato- and splanchnopleural mesoderms, that enclose the pleuroperitoneal space—no longer remains attached to the somite. Instead, these two layers, running parallel, meet each other at the lateral border of the endoderm, which is in the process of constituting the digestive tract. Therefore, one group of crest cells will enter the narrow space between the endoderm and the splanchnopleural-epithelium lining, thus providing the developing gut with prospective enteric neuroblasts, while at the same time the other group will insert between the somatopleure and the ectoderm (Thiery et al., 1982a). This population will later give birth to the mesectodermal derivatives of visceral arches and to the glial cells of the nodose ganglion (Ayer-Le Lièvre and Le Douarin, 1982; D'Amico-Martel and Noden, 1983).

Crest cells accumulate at the junction between the endoderm and the splanchnopleural mesoderm until a space becomes available between the two epithelia (Figure 4A) when the splanchnopleural mesoderm starts to delaminate (Tucker and Thiery, 1984). Initially in close contact, the endoderm and splanchnopleural mesoderm first separate dorsolaterally, allowing loose cells from the splanchnopleural mesoderm and crest cells to fill the resulting space. The later localization of crest cells is closely linked with this delamination of cells from the splanchnopleural wall; as the embryo gets older, the splanchnopleural wall gives rise to a thin epithelium that envelops the gut and lines the newly created mesenchyme, which contains smooth muscle cell precursors. Delamination and mesenchymal differentiation into muscle follow the same rostrocaudal pattern, governing the closure of the gut in the anterior part of the digestive tract.

Concomitant with their rostrocaudal progression in the delaminating dorsolateral space, crest cells are soon also able to migrate dorsoventrally in a given transverse

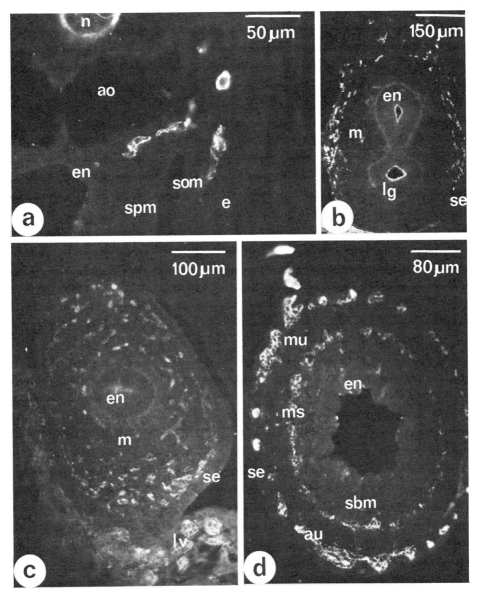

Figure 4. *Final location of crest cells: ontogeny of enteric plexuses. a*: The colonization of the digestive tract by neural crest cells reveals the powerful migratory capabilities of this population. Vagal crest cells enter the space between the endoderm (en) and the splanchnopleural mesoderm (spm) as soon as it is available, this event taking place at the 24th somite stage. A second flow runs parallel between the somatopleural mesoderm (som) and endoderm. *b*: Crest cells progress as a confluent monolayer and prefer the proximity of the delaminating splanchnopleural epithelium during their early concomitant dorsoventral and rostrocaudal migration. Thus they are found in a 3½-day-old chick embryo at the periphery of the mesenchyme (m) in the region of the developing pulmonary tract (lg); this configuration can also be seen in much older embryos. *c*: Probably as a result of intense delamination in the enlarging portion of the gut corresponding to the proventriculus, the crest cells appear dispersed in the gut mesenchyme (m). The smooth muscle layer (mu) forms rapidly after 4½ days, and crest cells regroup only at the periphery to organize into the external plexus. *d*: As colonization proceeds caudally, neural crest cells will form two plexuses, probably because of the opportunity to migrate first in the already formed mesenchyme and then on both sides of the differentiated circular smooth muscle layer. Therefore two plexuses with their fiber connections can be visualized in a 15-day-old chick embryo. ao, aorta; au, Auerbach's plexus; e, ectoderm; lg, lungs; ms, Meissner's plexus; mu, circular muscle layer; sbm, submucosa; se, serosa.

plane, again as a result of this delamination. During this rapid remodeling phase, neural crest cells remain preferentially near the thinning splanchnopleural epithelium and adopt a one-cell-thick confluent layer configuration (Figure 4B). Isolated crest cells appearing in the splanchnopleural mesenchyme probably withdraw from this layer when mesenchymal cells detach from the epithelium, and they frequently bear neurites, suggestive of their neuronal differentiation.

As migrating crest cells follow the closure of the gut, they are slightly delayed by the presence of developing obstacles, namely, evaginations of the endoderm such as the hepatic, biliary, and pancreatic buds. When nearing the umbilical endoderm, crest cells no longer retain a confluent configuration; rather, the front of migration appears dispersed through the already established mesenchyme soon after passing these buds (a configuration similar to that illustrated in Figure 4C). The mesenchyme seems to provide an adequate substrate for the migration of crest cells. Crest cells reach the mesenchyme while progressing, whereas at more rostral levels other crest cells have already stopped, unable to migrate farther in a possibly modified environment. Such a scheme could explain the presence of one enteric plexus down to the gizzard and two main concentric plexuses that extend from the duodenum down to the rectum before regrouping into two plexuses (Figure 4D); at that caudal level, prospective neuroblasts and supporting cells distribute into the mesenchyme.

A second contribution to the innervation of the gut takes place at caudal levels from somite 28 onward. For the most part, however, caudal neural crest cells participate in the formation of another peripheral nervous structure—the ganglion of Remak. These cells progress caudocranially in the dorsal mesentery and contribute to the intrinsic innervation of the gut to a limited extent only (Teillet, 1978).

Only intense proliferation can explain this colonization from a very small number of neural crest cells. This is all the more so because the digestive tract elongates enormously during organogenesis, thus reducing the real speed of migration by keeping the front of migration in step with the general growth of the gut.

Sensory Ganglia and Sympathetic Chains. At the level corresponding to somites six and seven (the posterior vagal region) and more caudally all through the trunk region, a ventral pathway opens up to neural crest cells, which accumulate there after the somites dissociate. The description of migration routes would be incomplete if we omitted the intersomitic pathway, a major one for the ontogeny of the peripheral nervous system (Figure 5). Cells facing the middle of a somite resume their progression after it disrupts, but they are soon stopped by the expanding sclerotome and form the anlagen of sensory ganglia. Meanwhile, cells arising between two somites continue migrating ventrally between the basement membranes of intersomitic grooves (where the aorta expands dorsally), follow the curvature of the aorta, and reach the lateral and ventral aspect of this blood vessel. From now on, these cells no longer migrate only in a transverse plane; they distribute themselves along the aorta. A continuous layer of neural crest cells connecting the sympathetic ganglion rudiments, now condensed from migrating intersomitic cells, is clearly visible, and together with several layers around the aorta announce the developing aortic plexuses. Sympathetic structures, at first continuous, clearly metamerize at later stages of the embryo. Trunk neural

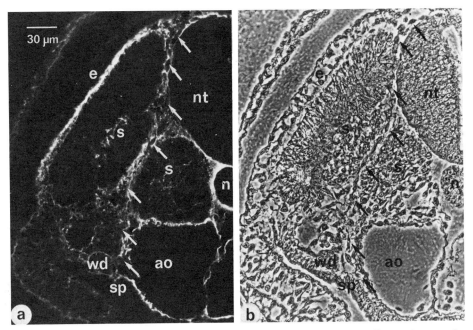

Figure 5. *Intersomitic pathway in trunk region. a*: Antibodies directed against fibronectin proved to be invaluable in finding the intersomitic route of migration. A well-delimited pathway between the fibronectin-rich basement membranes of neural tube (nt), ectoderm (e), and somite (s) can be visualized on an oblique section from a 25-somite-stage embryo. *b*: Phase-contrast microscopy; an almost continuous string of crest cells can be seen from the dorsal aspect of the neural tube to the aorta (ao). n, notochord; sp, splanchnopleural mesoderm; wd, Wolffian duct.

crest cells from the intersomitic pathway, still present along the kidney tubules, form the adrenal medulla and sympathetic plexuses. Crest cells can be evidenced between myotome and sclerotome, as long as a space is available, in continuity with the intersomitic flow (Figure 6).

Specificity of Localization. These studies point to several important features of crest cells and their environment. The latter seems to dictate migration routes by forming transient compartments in the embryo, and also seems to determine their final location, possibly by physical trapping, such as is seen in the dorsal root ganglia. Crest cells, together with a few other cell types in the developing embryo, are endowed with unique properties of motility; nevertheless, it should be emphasized that in several cases intrinsic migration is difficult to evaluate because the embryo grows very rapidly at those stages.

The description of neural crest cell pathways of migration provides some clues as to how their astonishingly precise dispersal might be achieved. However, it is still very difficult to give definitive answers to several questions: Why do some mesencephalic crest cells regroup just behind the eye to form the ciliary ganglion? Why do trunk crest cells first follow a ventral pathway when a potentially available pathway exists laterally, as in more rostral levels? What prevents crest cells from invading the somites and, later, the dermomyotome and the sclerotome? Why do crest cells accumulate around the aorta and not continue in the mesentery

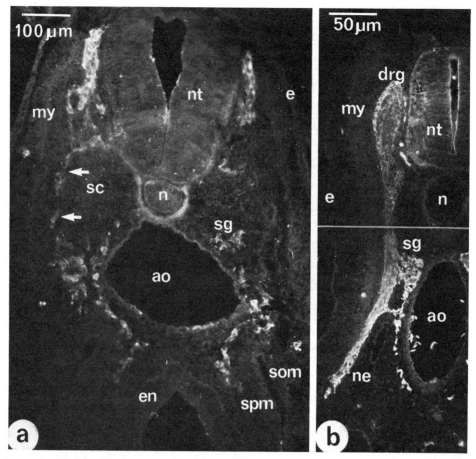

Figure 6. *Final location of crest cells: sensory and sympathetic nervous system. a*: Level of the sixth somite of a 28-somite-stage embryo. Neural crest cells can be seen migrating around the sclerotome (sc) near the intersomitic space (arrows); others are trapped above the sclerotome and will form the rudiment of the dorsal root ganglion (right part). Others, reaching the ventral parts, will give rise to sympathetic ganglia (sg), to aortic plexuses, and, at this level, to precursors of enteric ganglion cells just entering between the endoderm (en) and the splanchnopleural mesodermal epithelium (spm). *b*: The duration of migration may not last more than 20 hours; parts of the peripheral nervous system elements are well organized at as early as four days of incubation. The dorsal root (drg) and sympathetic (sg) ganglion can be easily distinguished, as can nerve fibers (ne) following the myotome (my) toward the limb bud. Few crest cells line the ventrolateral part of the aorta (ao) at this level, corresponding to a transversal section through the eighth somite. e, ectoderm; nt, neural tube.

to invade the gut? The following sections will try to throw some light upon the possible regulatory mechanisms that allow crest cells to choose their environment.

THE EXTRACELLULAR MATRIX AS A CRITICAL SUBSTRATE FOR NEURAL CREST CELL MIGRATION

Molecular Organization of the Extracellular Matrix

The extracellular matrix corresponds to an acellular space filled with a dense meshwork of fibrils. Such a space appears between epithelial tissues, and in young embryos it is usually indistinguishable from the basement membranes of these tissues (Hay, 1981).

Histochemical and immunohistological analyses have revealed the presence of several distinct components, including glycosaminoglycans, collagens, laminin, and fibronectin, which are already deposited and assembled in the early embryo. Laminin and collagen type IV have a restricted distribution in the basal lamina, a 100-Å-thick structure juxtaposed to the basal cell surface of epithelial cells. In contrast, fibronectin, collagen types I and III, and glycosaminoglycans are found mostly in the less organized extracellular material of the basement membranes. All these high-molecular-weight components assemble into fibrillar structures as a result of homologous and heterologous interactions. Interestingly, functional domains have been defined in fibronectin, in collagens, and to a lesser degree, in laminin.

Fibronectin

Structure and Function. Fibronectin is a high-molecular-weight glycoprotein composed of two almost identical subunits of 240 kilodaltons each. By an unknown process, this dimer forms polymers that appear as 10-nm fibrils in the embryo. For a long time these fibrils were actually considered to be collagens. Indeed, collagen and fibronectin may be closely associated in the fibrils. Fibronectin contains several distinct binding domains, and a region proximal to the amino terminus that interacts specifically with collagens is particularly well characterized. Two regions also interact with glycosaminoglycans, and at least one domain is involved in binding to cell surfaces (Hynes and Yamada, 1982). Although a well-defined cell surface receptor has not yet been characterized, it is likely that direct interaction between fibronectin and the cell surface can occur. Whether or not there is a unique receptor or multiple receptors remains to be defined. In addition, the affinity for such receptors may vary with different cell types or at different stages of cell differentiation. Indeed, fibronectin has two apparently contradictory effects: first, it increases cell-to-substrate adhesion and is considered to be an excellent cell-spreading factor; and second, it favors cell motility, offering suitable anchorage sites to cell filopodia, for which it provides efficient traction. Some cells will respond to fibronectin by intensive spreading, whereas others will migrate. The precise mechanisms responsible for these two states have not yet been elucidated, but cells will migrate on fibronectin deposited by an exogenous source, whereas, in the presence of their own fibronectin, a fibrillar meshwork is progressively organized around the cell surface, contributing to stable adhesion to the substratum (Couchman et al., 1982). Where is fibronectin with respect to these anchorage sites (also called adhesive plaques or focal contacts)? Some reports have shown that fibronectin is organized at these sites as parallel fibers in alignment with microfilament bundles (Hynes and Destree, 1978). Others have shown that fibronectin may participate in the formation of the plaque (Singer, 1982), but that it is subsequently removed from this region (Avnur and Geiger, 1981). At any rate, during migration transient binding may be expected to occur in order to allow translocation of the cell.

Interaction between Crest Cells and Fibronectin. Fibronectin was found to be associated with all basement membranes, including those used as migration

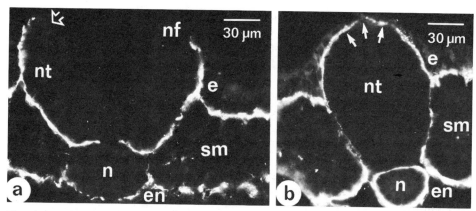

Figure 7. *Distribution of fibronectin before individualization of neural crest cells.* Immunofluorescence labeling for fibronectin. *a*: During neurulation in this eight-somite-old chick embryo, immediately caudal to the last somite, fibronectin appearing in basement membranes separates the neural tube (nt) from adjacent ectoderm (e) and somitic mesenchyme (sm). Note that notochord (n) remains closely associated with the neural tube. *b*: A few hours later, in the 16-somite stage at the level of the 15th somite, just prior to crest cell individualization, fibronectin is found around the tightly bound, dense epithelial structures. Crest cell presumptive territory is indicated by arrows. en, endoderm; nf, neural folds.

pathways of crest cells. However, it appears long before the individualization of crest cells from the neural tube in the presumptive pathways as well as in other regions not concerned with crest cells (Figure 7). Therefore fibronectin per se cannot trigger migration of crest cells and its presence cannot completely explain their preference for the pathways described above (Duband and Thiery, 1982b).

Most interestingly, apart from a small population originating from cephalic, vagal, and extreme caudal levels, neural crest cells are unable to synthesize and deposit fibronectin in the extracellular network, in contrast to all surrounding embryonic structures (Newgreen and Thiery, 1980). This inability might well sensitize them to the presence of fibronectin in the neighborhood and prevent them from being entrapped in their own fibronectin network. As to fibronectin-secreting crest cells, it may be that these cells, which are "pioneers" at the levels we have discussed, undergo an early differentiation into mesectodermal derivatives. Nevertheless, another simple explanation can be made: Their secretions would make up for the absence of fibronectin at these levels in the early phase of migration.

The large body of data accumulating on the structure and functions of fibronectin, and the fact that crest cells may not interact directly *in vivo* with basal lamina components, including laminin (Newgreen et al., 1982), prompted us to investigate whether fibronectin has a pivotal role in crest cell behavior.

Roles of Different Extracellular Matrix Components in Adhesion and Migration of Crest Cells

In Vitro *Studies (Rovasio et al., 1983).* In an appropriate medium and on a defined substrate, cultures of neural tube explants taken from the trunk level at the onset of crest cell individualization develop a halo of cells that originate from the dorsal border of the neural tube, thus mimicking the migration of crest cells. Reminiscent of the *in vivo* configuration, the developing halo exhibits a confluent morphology, with numerous cells spreading and extending fine filopodia. In fibronectin-free medium, very few crest cells could migrate from a neural tube explanted on a three-dimensional type I collagen substrate; in a 1:1 mixture of fibronectin and collagen type I, crest cells stop blebbing and form confluent, two-dimensional aggregates behind the front of the migration. The migration is initiated more rapidly and the crest cells are flattened, therefore suggesting a higher adhesiveness, when the substrate is pure fibronectin. Indeed, the level of binding of crest cells to pure fibronectin or fibronectin-containing extracellular matrices is much higher than it is to pure collagen. Adhesion on a laminin-coated substrate varies according to the time of culture: Thus, crest cells adhere more slowly to laminin in the early phase of migration and progressively show a higher binding as the age of the culture increases, so that binding to laminin appears slightly higher than binding to fibronectin after 48 hours; crest cell binding to the latter decreases during the same period (Figure 8).

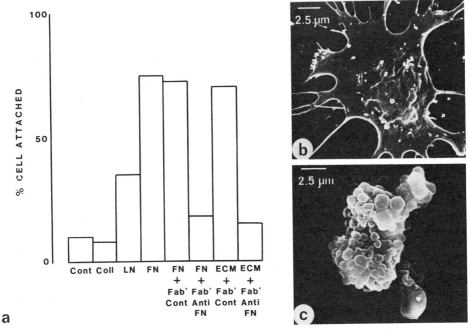

Figure 8. *Neural crest cell adhesion* in vitro. Adhesive properties of crest cells as a function of extracellular matrix components. *a*: Crest cells were cultured for 24 hours and dissociated before incubation on plastic dishes previously uncoated or coated with bovine serum albumin as a control (Cont), with type I collagen (Coll), with laminin (LN), with plasma fibronectin (FN), or with chick fibroblast extracellular matrix (ECM). In the presence of fibronectin either alone or together with collagens I and III and glycosaminoglycans in the extracellular matrix, more than 70% of the crest cells adhere to the substrate in 60 min. In both cases, binding is strongly reduced when monovalent antifibronectin antibodies are added prior to or with the crest cells. Concomitantly, changes in shape occur in the presence or absence of antifibronectin. *b*: A flattened neural crest cell adhering to plasma fibronectin and emitting numerous filopodia. *c*: Cells cannot spread in the presence of antifibronectin antibodies and exhibit numerous blebs.

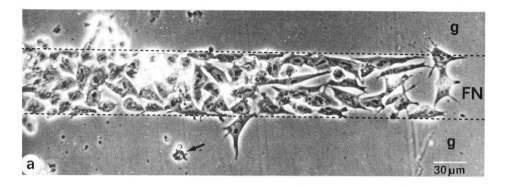

Figure 9. In vitro *migration on fibronectin substrate. a*: Neural crest cells were followed with time-lapse video equipment on a fibronectin–glass alternating substrate. After 24 hours in culture with the neural tube segment placed perpendicular to the fibronectin-coated region (FN), crest cells locate exclusively on fibronectin. Crest cells that have left this zone round up and soon become paralyzed. *b*: Track of the pioneer cell (1) together with track of a cell (2) located within the dense population, the latter track demonstrating a much higher degree of persistence in the direction of movement than does the former. In fact, isolated crest cells are unable to migrate unidirectionally; only when they are maintained as a confluent layer can the cell population as a whole translocate. g, glass substrate.

When confronted with alternating fibronectin- and laminin-serum-coated substrates, crest cells migrate almost exclusively on fibronectin stripes, whereas they aggregate on laminin. A model of the *in vivo* migration can be achieved by depositing a neural tube explant perpendicularly to fibronectin stripes (Figure 9). Crest cells respect very precisely the fibronectin-glass boundary and remain confluent at all times in the fibronectin-rich area. It is only in such a configuration that crest cells can maintain a persistent direction of movement. Isolated crest cells, such as pioneer cells at the front of migration, exhibit a random migration because no adjacent cell prevents them from changing their orientation.

The highest apparent speed of locomotion is observed when crest cells migrate in a fibronectin-rich matrix deposited by confluent mesenchymal cells. In this case, crest cells expanding in a halo *in vitro* reorganize the fibronectin-containing matrix deposited by fibroblasts. The three-dimensional network progressively becomes oriented by the pioneer crest cells and arranges in thicker bundles, making an alveolar pattern above crest cell surfaces.

It is likely that factors other than fibronectin are involved in increasing the ability to migrate. Perhaps the three-dimensional nature of this fibrillar matrix is more favorable than is a two-dimensional amorphous substrate. In particular, we cannot exclude some contact guidance effect and possible reorganization of the matrix in the direction of migration *in vivo*. However, a specific role for fibronectin is further evidenced as monovalent antibodies directed against the fragment containing the cell-binding region of fibronectin inhibit crest cell adhesion and migration on fibronectin-coated substrates (Figure 10). In the presence of these antibodies, most crest cells round up and start a very limited random migration; they regain a directional movement after the addition of plasma fibronectin. Inhibition of adhesion and migration is also observed in the presence of complex extracellular matrices, strongly suggesting that fibronectin, but not other components, is directly involved in the binding of crest cells.

In Vivo Studies. The fact that the extracellular matrix deposited *in vivo* by adjacent tissues such as the somite, ectoderm, and neural tube has a chemical composition similar to that deposited by fibroblasts cultured *in vitro*, together with the results of these experiments, suggest that neural crest cells also bind *in vivo* to the fibronectin present in all migration pathways. An attempt has been made to determine the role of fibronectin *in vivo*; in a series of preliminary experiments, microinjections of antibodies directed against the fragment containing the cell-binding component were performed. The first level chosen was the mesencephalon, where an important acellular space is present, thus allowing a sufficient quantity of antibody to be injected without causing severe damage to adjacent tissues. However, antibodies were also introduced at the vagal level, as many crest cells are found to accumulate between the ectoderm and the somite. In the mesencephalon, crest cells migrate under the ectoderm as a multicellular confluent layer that is clearly visible at low magnification in living

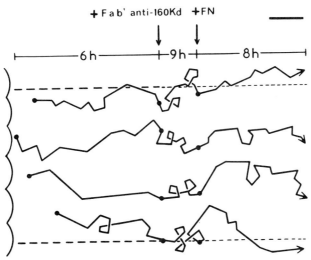

Figure 10. *Requirement for fibronectin in vitro: migration and directionality.* Addition of antibodies directed against the cell-binding domain of fibronectin (anti-160 kD) in a culture of neural crest cells migrating on fibronectin stripes for six hours results in almost complete arrest (loss of directional persistence and reduced speed of locomotion). Subsequent addition of plasma fibronectin in the medium enables crest cells to resume their oriented migration, although they can leave the stripes because the substrate is now uniformly coated with fibronectin. Undulating line, edge of neural tube in culture; dashed line, limits of the fibronectin stripe.

embryos. Injecting antibodies into one side of the embryo in this region results in a translucent space where the crest cell number is greatly reduced compared to the other (control) half, which contains its normal quantity of crest cells; analysis of transverse sections confirms this phenomenon (Thiery et al., 1982c). The same is true for the vagal level; the injected part shows very few crest cells, but the other half contains a normal accumulation of cells between dermomyotome and ectoderm that nearly reaches the apex of the pharyngeal endoderm (Thiery, 1985).

Essential as it may be *in vivo*, fibronectin is not sufficient in certain cases to promote migration. For instance, as already noted in the trunk region for the lateral pathways, crest cells do not first enter this fibronectin-rich space. Such a phenomenon could be accounted for by the presence of other components, such as chondroitin sulfate, that prevent adhesion or that delay the progression of crest cells. In fact, undersulfated chondroitin sulfate can inhibit migration, an observation that can be correlated with the decrease of synthesis and export of chondroitin sulfate in the trunk lateral pathway just before lateral migration can take place (Newgreen and Thiery, 1980; Newgreen et al., 1982). Hyaluronic acid, the most abundant glycosaminoglycan in crest cell pathways, was also shown to retard crest cell migration *in vitro*. However, its removal by hyaluronidase treatment *in vivo* prevents migration of crest cells (Pratt et al., 1976). It is now accepted that hyaluronic acid does not serve as a substrate for migration but allows the formation of transient spaces when hydrating. So far, all the studies on the role of glycosaminoglycans have indicated that these components may modulate the behavior of crest cells. As crest cells themselves produce large amounts of the components (Pintar, 1978), it is not yet understood what regulates the local concentrations of the components and at what relative concentration they become inhibitory (Derby, 1978). It is even possible that very high local concentrations of fibronectin can inhibit migration because crest cell filopodia adhere more frequently under this condition, becoming unable to explore the environment and initiate an efficient translocation.

GANGLIOGENESIS

Crest Cell Aggregation

By varying the conditions of culture *in vitro*, particularly the substrate for neural crest cells, it is possible to obtain the formation of two- and three-dimensional clusters of closely juxtaposed cells or aggregates. Such clusters form on collagen- or bovine serum–albumin-coated substrates when heat-inactivated fetal calf serum is added to the medium or when fibronectin is heated nearly to its thermal denaturation temperature. Denatured fibronectin allows some cell-to-substratum adhesion, but favors mostly cell-to-cell adhesion (Rovasio et al., 1983). *In vivo*, neural crest cells accumulate in small defined spaces rapidly depleted of fibronectin (Thiery et al., 1982a). As soon as they form small clusters they organize themselves into epitheliallike structures. However, only a few gap junctions can be detected in the rudiment of the dorsal root ganglia at four days of development (Figure 11). At the same time many other junctions, called terminal bar junctions, can

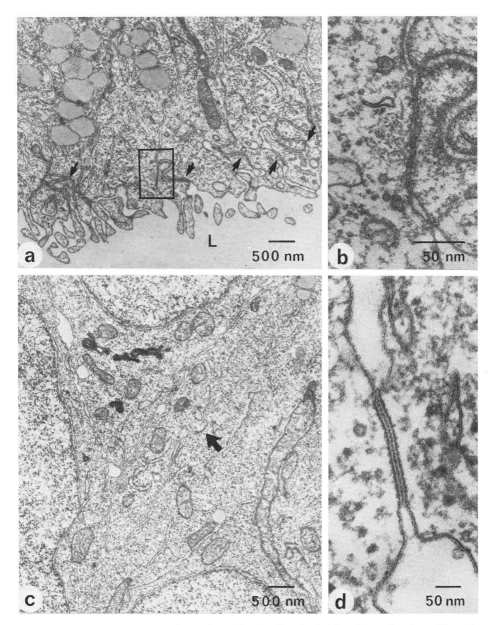

Figure 11. *Junctions in early gangliogenesis. a*: Neural tube of a 3½-day-old quail embryo. The side facing the lumen (L) is shown. Specialized intercellular junctions (tight and gap junctions) can be seen in every juxtaposed cell surface near the lumen. *b*: Higher magnification of one of the terminal bar junctions indicated by box in a. *c*: Rudiment of a dorsal root ganglion of a 3½-day-old quail embryo. Neighboring cells form epitheliallike tissues. The arrow indicates one of the very few gap junctions found in the ganglionic anlage. *d*: At a higher magnification, a gap junction can be identified easily. All the samples were prepared according to method of Ginzberg and Gilula (1979).

be evidenced in the neural tube. The surrounding tissues, including the ectoderm, the somite, and the notochord, also contain differentiated junctions (Revel and Brown, 1975). Thereafter, in most of these tissues, including ganglia (Pannese, 1974), junctions disappear; the only exception is the ectoderm in which desmosomes become predominant (Overton, 1975). It is therefore unlikely that the formation of ganglion anlagen results from the appearance of differentiated junctions, especially since these are gap junctions.

The role of cell adhesion molecules (CAMs) (Edelman, 1983) was investigated during gangliogenesis; the neural cell adhesion molecule (N-CAM) can be seen in very early stages of the embryo (Thiery et al., 1982b), even in the blastoderm before gastrulation. At this stage the liver cell adhesion molecule (L-CAM) also appears at the surface of the epithelial cells of the blastoderm (Edelman et al., 1983). During gastrulation, while superficial cells lateral to the primitive streak ingress, a central area begins to appear as a zone destined to give rise to the central nervous system. Ancestors of crest cells have been found to localize at the periphery of the hemicyclelike shape of the presumptive central nervous system (Rosenquist, 1981) and are therefore likely to express both L-CAM and N-CAM. As the neural plate starts to fold and thicken, however, cells at the boundary between the presumptive ectoderm and neural plate lose L-CAM while, like all the cells of the neural plate, they stain more with anti-N-CAM antibodies (Figure 14). After this period of determination, crest cells enter a phase of individualization corresponding to the already described separation from the neural tube epithelium, during which they progressively lose N-CAM. As migration occurs, the N-CAM level stabilizes at a small value and finally reaches an intermediate higher plateau during the process of aggregation.

Maturation of Ganglia

Very rapidly after aggregation, neural crest cells begin differentiating into neurons and glial cells. Although the timing of neuron and glial cell differentiation and cell death is not yet defined for all types of ganglia, it is already known that the first neurons are born very rapidly after the aggregation of crest cells and that cell death occurs while the last neurons appear. Such is the case in the dorsal root ganglia (Carr and Simpson, 1978). This early differentiation can be evidenced by the silver impregnation technique (Ramón y Cajal, 1955) and by antibodies to intermediate filaments (Holtzer et al., 1982). Crest cells contain vimentin during their migration both *in vivo* and *in vitro*, but start to express neurofilament proteins during neurite outgrowth; the presence of both classes of intermediate filaments is detectable during the early stages of neural differentiation (Figure 12). Crest cells express a much greater amount of N-CAM during differentiation than during their migration; this is particularly true for autonomic ganglia, which form compact aggregates both *in vitro* and *in vivo* (Figure 13). This phenomenon occurs prior to or at the onset of neuronal differentiation at the surface of both neuron and glial cell precursors. Slightly later, N-CAM is found on neuron cell bodies and their processes but, at least *in vitro*, is absent from differentiated glial cells. In the neurons, the N-CAM level increases even further as soon as neurite outgrowth occurs. However, these observations are only qualitative, as they are based on immunofluorescence-labeling studies alone.

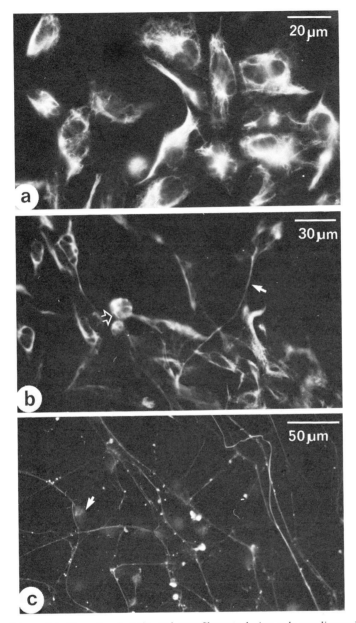

Figure 12. *Intermediate filaments, vimentin, and neurofilaments during early gangliogenesis. a: In vitro,* vimentin is found in all migrating neural crest cells as a dense meshwork of cytoplasmic fibrils. *b:* Ganglion rudiments from 4½-day-old quail embryos cultured for two days. Both morphologically undifferentiated crest cells and neurons (arrows) contain vimentin. *c:* In a similar culture, only neurons (arrow) contain the neurofilament proteins. A transient coexpression of vimentin and neurofilament protein is found in these young neurons. Antibodies were kindly given to us by D. Paulin (Institut Pasteur, Paris).

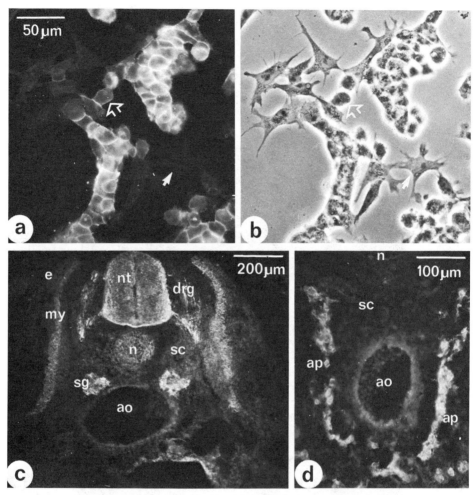

Figure 13. *N-CAM expression* in vitro *and* in vivo. *a* and *b*: Neural crest cell clusters are easily obtained after 48 hours of culture in heat-inactivated fetal calf serum on a glass substrate (Rovasio et al., 1983). As revealed by antibodies directed against N-CAM, these cultures exhibit small aggregates together with flattened cells (solid arrow) still migrating and rounded ones (open arrow) in the process of attaching or detaching from the aggregates, suggesting reversibility of the aggregation process. *c*: *In vivo* N-CAM staining of a four-day-old chick embryo, at the level of the heart, reveals the presence of sensory (drg) and sympathetic (sg) ganglia. Myotome (my) and neural tube (nt) also stain heavily. *d*: More caudally, in a five-day-old chick embryo, aortic plexuses (ap) are heavily labeled by anti-N-CAM antibodies. ao, aorta, e, ectoderm; n, notochord; sc, sclerotome.

Aggregation studies carried out on crest cells *in vitro* show that, according to their dissociation conditions, these cells express a calcium-dependent or a calcium-independent mechanism of adhesion. A comparative analysis of the aggregation of crest cells and of neural retina cells (Tables 1, 2) in our microaggregation assay reveals very similar behavior. Particularly interesting is that anti-N-CAM monovalent antibodies inhibit neural crest cell aggregation.

Recent studies have strongly suggested that cell–cell adhesion is mediated by direct interactions between N-CAM on opposed cell surfaces (Edelman, 1983). Furthermore, N-CAM has been shown to undergo a transition from an embryonic form with a very high sialic acid content to an adult form with a much reduced sialic acid content (Rothbard et al., 1982). Precise measurement of the strength of binding between N-CAMs showed that the efficacy of binding depends on the concentration of molecules but, more importantly, on the state of the molecules; a much higher binding is obtained between adult N-CAM molecules than between embryonic N-CAMs, at least in part as a result of the decrease in negative charges (Hoffman and Edelman, 1983). It was interesting to observe that, in the central nervous system, the transition between embryonic and adult N-CAM occurs at different times in different areas of the brain, correlating with a number of decisive events in the final maturation of this tissue (Edelman and Chuong, 1982; Chuong and Edelman, 1984). The simplest hypothesis is that the expression of adult N-CAM corresponds to the phase of final stabilization of the neuronal network. It was therefore of interest to determine how early such a transition would occur in neurons derived from both crest cells and neurogenic placodes, since both types contribute to the formation of cranial sensory ganglia.

Although an exhaustive analysis has not yet been performed (Figure 15), it appears that ciliary and placodal ganglia are the first to express the adult form of N-CAM. In the ciliary ganglion it is found to occur just after the period of massive neuronal death (Landmesser and Pilar, 1974). Therefore, a definitive neuronal network may be established in these cranial ganglia before hatching, as the early expression of adult N-CAM may contribute to the network's final stabilization. Indeed, in these ganglia neuronal death is mostly the result of competition at the target level within the fully developed axonal network. In contrast, in the dorsal root ganglion, adult N-CAM appears progressively just before hatching, and the transition is only completed a long time after hatching, an adult form being observed at three months. In this case, the period of cell death occurs long before the completion of innervation, possibly because of the

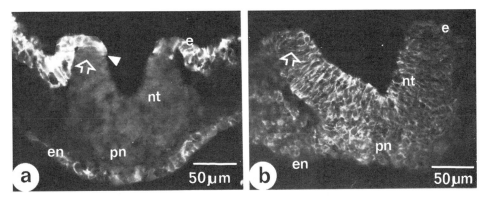

Figure 14. *L-CAM and N-CAM expression in early stages. a*: Every ectodermal cell (e) of a 10-somite-old chick embryo is stained with anti-L-CAM antibodies, whereas cells of the neural fold are not. The last cells stained (arrow head) probably correspond to the boundary between presumptive neural crest cells (open arrow) and true ectoderm. Neural tube (nt) and adjacent mesenchyme, which will soon condense into a somite, are not stained. Faint but effective staining appears at the surface of endodermal cells (en). *b*: At the same time and level, while neurulation and ectoderm formation occur, N-CAM was found in most tissues with, nevertheless, an increased intensity in the neural tube and presumptive notochord area (pn).

Table 1. Aggregation of Cultured Quail Neural Crest Cells[a]

Cells	Condition of Aggregation	Aggregation (%)
TC	$-Ca^{2+}$ $-Fab'$	8
	$+Ca^{2+}$ $-Fab'$	49
	$+Ca^{2+}$ $+Fab'$	51
LTE	$-Ca^{2+}$ $-Fab'$	41
	$-Ca^{2+}$ $+Fab'$	18
	$+Ca^{2+}$ $-Fab'$	40
	$+Ca^{2+}$ $+Fab'$	18

[a] Neural crest cells cultured for 48 hours with heat-inactivated fetal calf serum (Rovasio et al., 1983) were harvested by one of two different treatments to recover cells bearing the activity of one of the two types of adhesion systems. LTE treatment [0.0001% trypsin and 1 mM in EDTA in HCMF (hepes-buffered calcium–magnesium-free saline, pH 7.4) for 20 min at 37°C] makes cells with a calcium-independent system of cell adhesion (CIDS); TC treatment (0.04% trypsin and 1 mM CaCl$_2$ in HCMF for 20 min at 37°C) makes cells with a calcium-dependent system of cell adhesion (CDS) (Brackenbury et al., 1981; Ogon et al., 1982).

Aggregation of dissociated cells was carried out in 96-well (each U-bottom well with a diameter of 6 mm) tissue culture seroclusters, which were precoated by incubation overnight at 10°C with 0.5% bovine serum albumin in HCMF to prevent cells from adhering to the bottom during the aggregation assay. In each well, 4×10^3 cells were inoculated in 40 µl of HCMF with or without 1 mM CaCl$_2$. After 30 min of incubation at 37°C in a gyratory shaker at a speed of 150 rpm, 30 µl of 2% glutaraldehyde solution in HCMF was added to each well and the total particle numbers were counted with a hemocytometer.

The degree of cell aggregation was measured by the decrease of total particle number, that is, the percentage of aggregation = (1.0 − total particle number at 30 min/initial total particle number) × 100. Aggregation of LTE-treated neural crest cells with or without Ca^{2+} was inhibited more than 50% by the Fab' fragment of anti-N-CAM, which suggests that N-CAM corresponds to CIDS in neural crest cells. On the other hand, aggregation of TC-treated neural crest cells was not inhibited at all by the Fab' fragment of anti-N-CAM, which suggests that CDS is unrelated to N-CAM.

low level of diffusible trophic factors (Hamburger and Oppenheim, 1982). It is likely that sensory targets are progressively receiving their afferent innervation, even after hatching.

This type of innervation involves extensive arborizations. Furthermore, most sensory fibers are mingled with motor root fibers still in the process of reorganization, as the number of myotubes increases enormously. Thus an embryonic

Table 2. Aggregation of Chick Neural Retina Cells[a]

Method	Condition of Aggregation	Aggregation (%)	Inhibition (%)
Previous	$-Fab'$	90	—
	$+Fab'$	14	84
Present	$-Fab'$	79	—
	$+Fab'$	34	57

[a] Because the present assay method of cell aggregation is slightly different from the previous method (Brackenbury et al., 1981), we compared the two methods using neural retina cells from seven-day-old quail embryos. This table shows that the inhibition of neural retina cell aggregation by Fab' fragments of anti-N-CAM antibodies appears to be less effective in the present method than in the previous one. The aggregation of LTE-treated neural retina cells, which have N-CAM as a calcium-independent system and have no calcium-dependent system, is inhibited almost completely (84%) in the previous method, but it is inhibited partially (57%) in the present method.

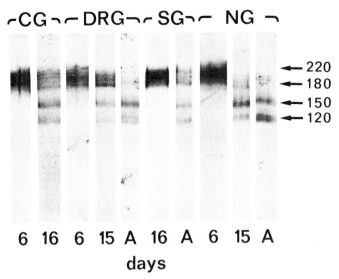

Figure 15. *Conversion of N-CAM from embryonic (E) to adult (A) form in crest- and placode-derived ganglia.* Quail ciliary ganglia (CG), dorsal root ganglia (DRG) and sympathetic ganglia (SG) from the brachial level, and nodose ganglia (NG) were taken from embryos at different days of incubation and from 3-month-old adult (A) quail. According to the size of ganglia, 1–10 ganglia were dissolved in 10 μl of sample buffer. They were boiled for 3 min and then loaded on SDS-polyacrylamide gel (6.5%) for electrophoresis (Laemmli, 1970). Both forms of N-CAM were detected by immunoblotting (Towbin et al., 1979) using anti-N-CAM polyclonal antibody and ^{125}I-labeled protein A. The E-form of N-CAM migrates as a fuzzy band of M_r = 180–240 kD as shown in CG6. The A-form of N-CAM migrates as three discrete bands of M_r = 120, 140, and 180 kD as shown in CG16. E-to-A conversion occurs progressively in each ganglion, but at different ratios. The gross rates appear to be CG > NG > DRG > SG. In SG, E-to-A conversion does not occur during embryonic age, and even in 3-month-old quail considerable amounts of E-CAM remain.

N-CAM may be necessary all through that period to allow plasticity and fiber displacement in bundles. Finally, it was surprising to find that, in sympathetic ganglia, N-CAM never fully converts to an adult form. Indeed, these ganglia do not make precise connections with their targets; multiple, ill-defined, synaptic contacts can only be recognized because of the presence of vesicles (Elfvin, 1983). A similar situation is also known to exist for the enteric plexuses, where a relative independence of the smooth muscle and the neuron processes is observed (Gershon, 1981). The incomplete conversion of N-CAM from the embryonic to the adult form may be required to maintain the neuronal network in the very diffuse state necessary for the proper function of autonomically innervated viscera.

AN OVERALL VIEW

The history of development of the neural crest from its earliest stages in the blastula to its derivatives in the adult offers a remarkable diversity of phenotypic expressions. Particularly interesting is the adhesive behavior of crest cells. Successive phases of cell–cell, cell–extracellular matrix, and cell–cell interactions have been characterized. Very early on they may involve both L-CAM and N-CAM, then fibronectin, and later, N-CAM (Figure 16). Aggregation of crest cells and formation of neurite bundles may be controlled by N-CAM. Interestingly,

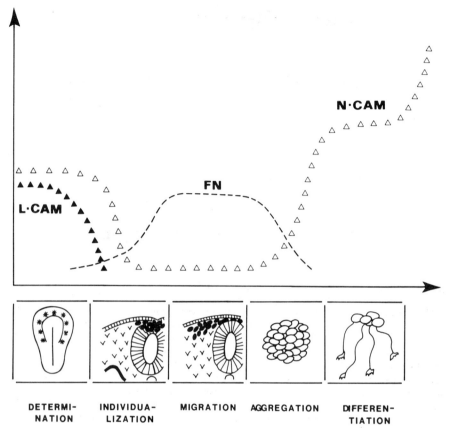

Figure 16. *Correlation between adhesive properties of crest cells, the expression of the cell adhesion molecules, and the presence of fibronectin.* Five epochs were chosen to correspond to the main changes of behavior of this population: determination; individualization from the neural epithelium (see Figure 1); migration through the embryo; final aggregation; and onset of differentiation into ganglia. The loss of L-CAM may occur during the stage of determination, and possibly at the onset of desolidarization from the neural epithelium. N-CAM increases until the crest cells depart from the neural tube. During migration, the N-CAM level falls while adhesivity to fibronectin in the extracellular environment reaches a higher plateau. This is reversed during aggregation; the binding to fibronectin is much reduced, this molecule disappearing from the ganglion rudiment but remaining abundant in the immediate environment. N-CAM is likely to increase at the surface of differentiating neurons and may contribute to early formation of fascicles. Later, the conversion from embryonic to adult N-CAM will further reinforce this type of interaction. Symbols as in previous figures.

a new CAM that mediates adhesion between neurons and glial cells (Ng-CAM) (Grumet et al., 1983) has been found to appear on neurons of the ganglia (Thiery et al., 1985) at the time when satellite cells start to appear (Holton and Weston, 1982). Furthermore, the transition from embryonic N-CAM to the adult form occurs at different stages of development in the different types of ganglia, possibly in correlation with the timing and with their respective modes of innervation.

The transitory period during which neural crest cells lose their epithelial structure and migrate as mesenchymal cells is accompanied by a rapid increase in binding to the extracellular matrix. Both *in vivo* and *in vitro* experiments indicate that fibronectin is directly involved in the adhesion of crest cells to the extracellular matrix. In addition, cell movement is facilitated in the presence of this component. One of the most interesting properties of crest cells is their inability to produce fibronectin in their immediate environment, combined with an ability (in contrast to other mesenchymal cells) to migrate in response to exogenous fibronectin. This intrinsic ability to migrate in the young embryonic environment is limited to a small number of embryonic cells, including primordial germ cells (Wylie and Heasman, 1982) and lymphoid precursor cells (F. Dieterlen-Lièvre, personal communication), which together with crest cells and tumor cells such as sarcoma cells (Erickson et al., 1980) do not deposit fibronectin. When introduced into trunk crest cell pathways all these cells are able to migrate, whereas mesenchymal cells or latex beads covered with fibronectin are unable to do so (Bronner-Fraser, 1982). Interestingly, crest cells *in vitro* progressively lose their ability to bind to fibronectin and become more adhesive to laminin; meanwhile, the duration of contact increases markedly between crest cells (Newgreen et al., 1979), which start to form two- and three-dimensional aggregates. These crest cells are likely to be equivalent to those found in ganglion rudiments during an early stage of differentiation into neurons and glial cells. Glial cells also have been reported to bind to laminin (Rogers et al., 1983), and they may have acquired laminin receptors just after their stabilization into ganglion rudiments.

One of the major issues is to decipher the mechanisms that control the transitory expression of adhesive properties to both cells and extracellular matrix components. Cell surface modulation of adhesive molecules is clearly involved throughout the life of crest cells and their derivatives. Transient or stable expression of surface receptors for fibronectin and laminin, as well as of L-CAM and N-CAM, has been described as being involved in the diverse behavior of crest cells.

The analysis of the development of neural crest cell derivatives, particularly of the peripheral nervous system, leads to the idea that a limited number of time-modulated, specific cell surface molecules, such as CAMs and specific receptors for the extracellular matrix, may provide enough constraints in a rapidly developing system to ensure normal morphogenesis. It has been found, so far, that each crest cell is not predetermined to reach a defined site, but rather, that the environment provides transient spaces through which confluent crest cells divide and migrate. Directionality may be provided in great part by the transient pathways and by the ability of crest cells to proliferate rapidly, to respond to a fibronectin-rich environment, and to exhibit persistence of movement if maintained as a confluent layer. Final localization of crest cells must also be partly a consequence of the environment imposed by the epithelia, which act as physical obstacles, and by sites that are unfavorable for migration, either because they are devoid of fibronectin or because they contain suprathreshold levels of other inhibitory components.

It is tempting to state that once the primary anlagen of the embryo, and particularly the neural plate, are disposed of after gastrulation, subsequent development can be considered to be a succession of simple steps which intimately interplay and are precisely regulated in both time and space. Plasticity is nevertheless maintained throughout development, not only by the number of cell divisions, but also because the interactive processes, rather than requiring a vast repertoire of specific molecules, need only a small set of molecules that can vary in number and structure as well as in their ontogeny and localization.

ACKNOWLEDGMENTS

We thank our colleagues Bruce Cunningham, Annie Delouvée, Jean-Loup Duband, Gerald Edelman, Warren Gallin, Roberto Rovasio, and Michel Vincent for their fruitful collaboration. Gerald Edelman, Françoise Dieterlen-Lièvre, and Nicole Le Douarin contributed critical readings of the manuscript. Excellent technical assistance was provided by Monique Denoyelle and Louis Addade.

This research was supported by grants from the Centre National de la Recherche Scientifique (ATP 82-3701), the Ministère de l'Industrie et de la Recherche (81E1082), the Institut National de la Santé et de la Recherche Médicale (CRL82-4018), the Ligue Nationale Française contre le Cancer, the Fondation pour la Recherche Médicale, and the Association pour le Développement de la Recherche contre le Cancer.

G. C. Tucker was supported by a fellowship from Ecole Polytechnique, and H. Aoyama by fellowships from Institut National de la Santé et de la Recherche Médicale and the Fondation Fyssen.

REFERENCES

Avnur, Z., and B. Geiger (1981) The removal of extracellular fibronectin from areas of cell–substrate contact. *Cell* **25**:121–132.

Ayer-Le Lièvre, C. S., and N. M. Le Douarin (1982) The early development of cranial sensory ganglia and the potentialities of their component cells studied in quail–chick chimeras. *Dev. Biol.* **94**:291–310.

Brackenbury, R., U. Rutishauser, and G. M. Edelman (1981) Distinct calcium-independent and calcium-dependent adhesion systems of chicken embryo cells. *Proc. Natl. Acad. Sci. USA* **78**:387–391.

Bronner-Fraser, M. (1982) Distribution of latex beads and retinal pigment epithelial cells along the ventral neural crest pathway. *Dev. Biol.* **91**:50–63.

Carr, V. McM., and S. B. Simpson (1978) Proliferative and degenerative events in the early development of chick dorsal root ganglia. I. Normal development. *J. Comp. Neurol.* **182**:727–740.

Chuong, C.-M., and G. M. Edelman (1984) Alterations in neural cell adhesion molecules during development of different regions of the nervous system. *J. Neurosci.* **4**:2354–2368.

Cochard, P., and P. Coltey (1983) Cholinergic traits in the neural crest: acetylcholinesterase in crest cells of the chick embryo. *Dev. Biol.* **98**:221–138.

Couchman, J. R., D. A. Rees, M. R. Green, and C. G. Smith (1982) Fibronectin has a dual role in locomotion and anchorage of primary chick fibroblasts and can promote entry into the division cycle. *J. Cell Biol.* **93**:402–410.

D'Amico-Martel, A., and D. M. Noden (1983) Contributions of placodes and neural crest cells to avian cranial peripheral ganglia. *Am. J. Anat.* **166**:445–468.

Derby, M. A. (1978) Analysis of glycosaminoglycans within the extracellular environments encountered by migrating neural crest cells. *Dev. Biol.* **66**:321–336.

Duband, J. L., and J. P. Thiery (1982a) Distribution of fibronectin in the early phase of avian cephalic neural crest cell migration. *Dev. Biol.* **93**:308–323.

Duband, J. L., and J. P. Thiery (1982b) Appearance and distribution of fibronectin during chick embryo gastrulation and neurulation. *Dev. Biol.* **94**:337–350.

Edelman, G. M. (1983) Cell-adhesion molecules. *Science* **219**:450–457.

Edelman, G. M., and C. M. Chuong (1982) Embryonic to adult conversion of neural cell adhesion molecules in normal and staggerer mice. *Proc. Natl. Acad. Sci. USA* **79**:7036–7040.

Edelman, G. M., W. J. Gallin, A. Delouvée, B. A. Cunningham, and J. P. Thiery (1983) Early epochal maps of two different cell adhesion molecules. *Proc. Natl. Acad. Sci. USA* **80**:4384–4388.

Elfvin, L. G. (1983) *Autonomic Ganglia*, Wiley, Chichester.

Erickson, C. A., K. W. Tosney, and J. A. Weston (1980) Analysis of migratory behavior of neural crest and fibroblastic cells in embryonic tissues. *Dev. Biol.* **77**:142–156.

Gershon, M. D. (1981) The enteric nervous system. *Annu. Rev. Neurosci.* **4**:227–272.

Ginzberg, R. D., and N. B. Gilula (1979) Modulation of cell junctions during differentiation of the chicken otocyst sensory epithelium. *Dev. Biol.* **68**:110–129.

Grumet, M., U. Rutishauser, and G. M. Edelman (1983) Neuron–glia adhesion is inhibited by antibodies to neural determinants. *Science* **222**:60–62.

Hamburger, V., and R. W. Oppenheim (1982) Naturally occurring neuronal death in vertebrates. *Neurosci. Comment.* **1**:39–55.

Hay, E. D. (1980) Development of vertebrate cornea. *Int. Rev. Cytol.* **63**:263–322.

Hay, E. D. (1981) *Cell Biology of Extracellular Matrix*, Plenum, New York.

Hoffman, S., and G. M. Edelman (1983) Kinetics of homophilic binding by E and A forms of the neural cell adhesion molecule. *Proc. Natl. Acad. Sci. USA* **80**:5762–5766.

Holton, B., and J. A. Weston (1982) Analysis of glial cell differentiation in peripheral nervous tissue. I. S100 accumulation in quail embryo spinal cultures. *Dev. Biol.* **89**:64–71.

Holtzer, H., G. S. Bennett, S. J. Tapscott, J. M. Croop, and Y. Toyama (1982) Intermediate-size filaments: Changes in synthesis and distribution in cells of the myogenic and neurogenic lineages. *Cold Spring Harbor Symp. Quant. Biol.* **46**:317–329.

Hynes, R. O., and A. T. Destree (1978) Relationship between fibronectin (LETS protein) and actin. *Cell* **15**:875–886.

Hynes, R. O., and K. M. Yamada (1982) Fibronectins: Multifunctional modular glycoproteins. *J. Cell Biol.* **95**:369–377.

Johnston, M. C. (1966). A radioautographic study of the migration and fate of cranial neural crest cells in the chick embryo. *Anat. Rec.* **156**:143–156.

Laemmli, U. K. (1970) Cleavage of structural proteins during the assembly of the head of bacteriophage. *Nature* **227**:680–685.

Landmesser, L., and G. Pilar (1974) Synaptic transmission and cell death during normal ganglionic development. *J. Physiol. (Lond.)* **241**:737–749.

Le Douarin, N. M. (1982) *The Neural Crest*, Cambridge Univ. Press, Cambridge.

Le Douarin, N. M., P. Cochard, M. Vincent, J. L. Duband, G. C. Tucker, M. A. Teillet, and J. P. Thiery (1983) Nuclear, cytoplasmic and membrane markers to follow neural crest cell migration: A comparative study. In *The Role of Extracellular Matrix in Development*, R. L. Trelstad, ed., pp. 373–398, Alan R. Liss, New York.

Le Lièvre, C. S. (1978) Participation of neural crest-derived cells in the genesis of the skull in birds. *J. Embryol. Exp. Morphol.* **47**:17–37.

Le Lièvre, C. S., and N. M. Le Douarin (1975) Mesenchymal derivatives of the neural crest: Analysis of chimaeric quail and chick embryos. *J. Embryol. Exp. Morphol.* **34**:125–154.

Narayanan, C. H., and Y. Narayanan (1978) Determination of the embryonic origin of the mesencephalic nucleus of the trigeminal nerve in birds. *J. Embryol. Exp. Morphol.* **43**:85–105.

Newgreen, D. F., and J. P. Thiery (1980) Fibronectin in early avian embryos: Synthesis and distribution along the migration pathways of neural crest cells. *Cell Tissue Res.* **211**:269–291.

Newgreen, D. F., M. Ritterman, and E. A. Peters (1979) Morphology and behavior of neural crest cells of the chick embryo *in vitro*. *Cell Tissue Res.* **203**:115–140.

Newgreen, D. F., I. L. Gibbons, J. Sauter, B. Wallenfels, and R. Wutz (1982) Ultrastructural and tissue culture studies on the role of fibronectin, collagen and glycosaminoglycans in the migration on neural crest cells in the fowl embryo. *Cell Tissue Res.* **221**:521–549.

Noden, D. M. (1975) An analysis of the migratory behavior of avian cephalic neural crest cells. *Dev. Biol.* **42**:106–130.

Noden, D. M. (1978) The control of avian cephalic neural crest cytodifferentiation. I. Skeletal and connective tissues. *Dev. Biol.* **67**:396–312.

Ogon, S., T. S. Okada, M. Takeichi (1982) Cleavage stage mouse embryos share a common cell adhesion system with teratocarcinoma cells. *Dev. Biol.* **92**:521–528.

Overton, J. (1975) Development of cell junctions of the adhaerens type. *Curr. Top. Dev. Biol.* **10**:1–32.

Pannese, E. (1974) The histogenesis of the spinal ganglia. *Adv. Anat. Cell Biol.* **47(5)**:1–97.

Pintar, J. E. (1978) Distribution and synthesis of glycosaminoglycans during quail neural crest morphogenesis. *Dev. Biol.* **67**:444–464.

Pratt, R. M., M. A. Larsen, and M. C. Johnston (1975) Migration of cranial neural crest cells in a cell-free hyaluronate-rich matrix. *Dev. Biol.* **44**:298–305.

Pratt, R. M., G. Morriss, and M. C. Johnston (1976) The source, distribution and possible role of hyaluronate in the migration of chick cranial neural crest cells. *J. Gen. Physiol.* **68**:15–16.

Ramón y Cajal, S. (1955) *Histologie du Système Nerveux de l'Homme et des Vertébrés*, Vol. 1, trans. by L. Azoulay, Consejo Superior de Investigaciones Científicas, Instituto Ramón y Cajal, Madrid.

Revel, J. P., and S. S. Brown (1975) Cell junctions in development, with particular reference to the neural tube. *Cold Spring Harbor Symp. Quant. Biol.* **40**:443–455.

Rogers, S. L., P. C. Letourneau, S. L. Palm, J. McCarthy, L. T. Furcht (1983) Neurite extension by peripheral and central nervous system neurons in response to substratum-bound fibronectin and laminin. *Dev. Biol.* **98**:212–220.

Rosenquist, G. C. (1981) Epiblast origin and early migration of neural crest cells in the chick embryo. *Dev. Biol.* **87**:201–211.

Rothbard, J. B., R. Brackenbury, B. A. Cunningham, and G. M. Edelman (1982) Differences in carbohydrate structures of neural cell-adhesion molecules from adult and embryonic chicken brains. *J. Biol. Chem.* **257**:11064–11069.

Rovasio, R. A., A. Delouvée, K. M. Yamada, R. Timpl, and J. P. Thiery (1983) Neural crest cell migration: Requirements for exogenous fibronectin and high cell density. *J. Cell Biol.* **96**:462–473.

Rudnick, D. (1948) Prospective areas and differentiation potencies in the chick blastoderm. *Ann. N.Y. Acad. Sci.* **49**:761–772.

Singer, I. J. (1982) Association of fibronectin and vinculin with focal contacts and stress fibers in stationary hamster fibroblasts. *J. Cell Biol.* **92**:398–408.

Teillet, M. A. (1978) Evolution of the lumbo-sacral neural crest in the avian embryo: Origin and differentiation of the ganglionated nerve of Remak studied in interspecific quail–chick chimaeras. *W. Roux' Arch. Dev. Biol.* **184**:251–268.

Teillet, M. A., and N. M. Le Douarin (1970) La migration des cellules pigmentaires étudiée par la méthode des greffes hétérospécifiques de tube nerveux chez l'embryon d'oiseau. *C. R. Acad. Sci.* **270**:3095–3098.

Thiery, J. P. (1985) Roles of fibronectin in embryogenesis. In *Fibronectin*, D. F. Mosher, ed., Academic (in press).

Thiery, J. P., J. L. Duband, and A. Delouvée (1982a) Pathways and mechanism of avian trunk neural crest cell migration and localization. *Dev. Biol.* **93**:324–343.

Thiery, J. P., J. L. Duband, U. Rutishauser, and G. M. Edelman (1982b) Cell adhesion molecules in early chick embryogenesis. *Proc. Natl. Acad. Sci. USA* **79**:6737–6741.

Thiery, J. P., R. A. Rovasio, J. L. Duband, and A. Delouvée (1982c) The extracellular matrix in early embryogenesis. In *Hormones and Cell Regulation*, Vol. 7, J. N. Dumont and J. Nunez, eds., pp. 319–334, Elsevier/North Holland, Amsterdam.

Thiery, J. P., A. Delouvée, M. Grumet, and G. M. Edelman (1985) Initial appearance and regional distribution of the neuron-glia cell adhesion molecule (Ng-CAM) in the chick embryo. *J. Cell Biol.* (in press).

Toole, B. P. (1976) Morphogenetic role of glycosaminoglycans (acid mucopolysaccharides) in brain and other tissues. *Curr. Top. Neurobiol.* **6**:275–329.

Tosney, K. W. (1978) The early migration of neural crest cells in the trunk region of the avian embryo: An electron microscopic study. *Dev. Biol.* **62**:317–333.

Tosney, K. W. (1982) The segregation and early migration of cranial neural crest cells in the avian embryo. *Dev. Biol.* **89**:13–24.

Towbin, H., T. Staehelin, and J. Gordon (1979) Electrophoretic transfer of proteins from polyacrylamide gels to nitrocellulose sheets: Procedure and some applications. *Proc. Natl. Acad. Sci. USA* **76**:4350–4354.

Tucker, G. C., and J. P. Thiery (1984) Mechanism of avian neural crest cell migration in the gut. *Dev. Biol.* (submitted).

Vincent, M., and J. P. Thiery (1983) A cell surface determinant expressed early on migrating avian neural crest cells. *Dev. Brain Res.* **9**:235–238.

Vincent, M., and J. P. Thiery (1984) A cell surface marker for neural crest and placodal cells: Further evolution in peripheral and central nervous system. *Dev. Biol.* **103**:468–481.

Weston, J. A. (1970) The migration and differentiation of neural crest cells. *Adv. Morphogenet.* **8**:41–114.

Wylie, C. C., and J. Heasman (1982) Effects of the substratum on the migration of primordial germ cells. *Philos. Trans. R. Soc. Lond. [Biol.]* **299**:177–183.

Chapter 10

Neuronal Determination in the Enteric Nervous System

MICHAEL D. GERSHON
TAUBE P. ROTHMAN

ABSTRACT

The mature enteric nervous system is a large autonomic division that is able to function independently of the brain and spinal cord. It shows great phenotypic diversity among its component neurons and has unique characteristics that include the absence of internal collagen, support from enteric glial cells, and a specialized internal environment. The neural crest-derived neurons and glia of the enteric nervous system have been found to colonize the bowel as precursors that are morphologically unrecognizable. Their final phenotype is acquired within the gut itself. For neurons, the microenvironment of the bowel has been shown to influence this expression with regard to neurotransmitter-defined subset; however, the early expression of the neuronal marker, neurofilament protein, suggests that these precursor cells become committed to the neuronal lineage before they reach the bowel, possibly within the premigratory or early migratory neural crest. Initial expression of a neuronal phenotype may be transient, as in the case of catecholaminergic cells that appear in the gut during development. The intrinsic enteric neurons also seem to be dependent on the enteric microenvironment for their ability to colonize segments of bowel. An abnormal microenvironment in the terminal colon of the lethal spotted mutant mouse may lead to failure of colonization and subsequent aganglionosis of the abnormal segment. The intrinsic neurons of the ENS seem to be relatively independent of nerve growth factor for their survival and development, but they may be influenced transynaptically by an extrinsic innervation.

As Langley (1921) first defined it, the autonomic nervous system has three divisions. These are the sympathetic, characterized by a thoracicolumbar outflow of preganglionic axons from the spinal cord; the parasympathetic, characterized by an outflow of preganglionic axons from the brain and the sacral spinal cord; and the enteric, characterized by the presence of many intrinsic neurons that receive no direct input at all from the central nervous system (CNS). The enteric division, or enteric nervous system (ENS), has been until recently less studied than the other two autonomic divisions. This neglect occurred in part because Langley's original distinction of the system as a separate autonomic unit became obscured, and for a long time investigators tended to regard the neural plexuses of the bowel as parasympathetic relay ganglia (Kuntz, 1953). It is now apparent that the ENS is a unique region of the peripheral nervous system (PNS; Gershon, 1981a; Gershon and Erde, 1981) that merits a separate classification. The ENS also has a number of intriguing features, not found elsewhere, that make it suitable for neurobiological research, not the least of which is the ability to mediate reflex activity *in vitro* in the complete absence of CNS control. In order to do this, the ENS contains sensory receptors, intrinsic primary afferent neurons, interneurons, and at least two types of motor neuron, one that excites and one that inhibits the enteric smooth muscle layers.

THE MATURE ENS

Neuronal cell bodies of the ENS are confined to two ganglionated plexuses, the submucosal (Meissner's) and the myenteric (Auerbach's) (see Schofield, 1968; Gershon and Erde, 1981, for references). Of these, the myenteric plexus, situated between the circular and longitudinal muscle layers of the muscularis externa, is by far the larger. The mature myenteric plexus exists as a glia-enclosed tube that lacks the endoneurial and perineurial sheaths that characterize peripheral nerve elsewhere in the body (Gabella, 1971, 1972). At the ultrastructural level (Figures 1, 2), there appears to be little extracellular space within the plexus, and no collagen at all is found inside the boundary established by the basal

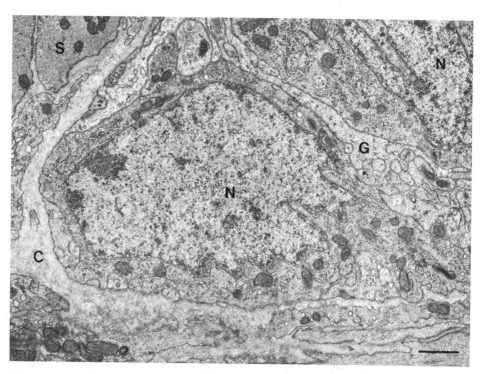

Figure 1. *Electron micrograph of the myenteric plexus of a mature guinea pig.* A neuron fills the center of the field. Note the exclusion of collagen from the interior and the restriction of basal lamina material to the perimeter of the plexus. N, neuron; C, collagen; S, smooth muscle; G, glia. Calibration bar = 1 μm.

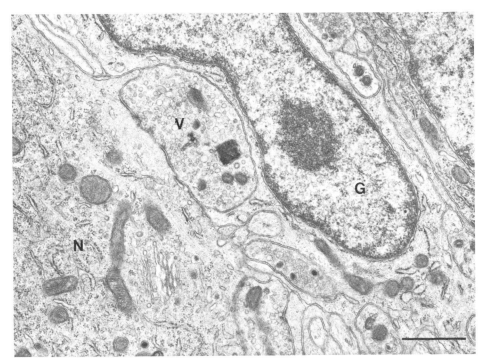

Figure 2. *Enteric glial cell (G) in the myenteric plexus of a guinea pig.* Note the glial filaments in the cytoplasm, the irregular contour of the cell supporting the neuronal elements. N, neuron; V, varicosity making a synapse. Calibration bar = 0.5 μm.

lamina that surrounds and lies external to the glial sheath of the plexus. Support of the neural elements is derived from cells that, in their shape and intermediate filament content, resemble the astroglial cells of the CNS (Figure 2; Gabella, 1971; Cook and Burnstock, 1976b). These supporting cells seem, in fact, to be a highly specialized form of glial cell that differs from the typical Schwann cell of the PNS. Not only is it particularly rich in intermediate filaments, but it is also equally rich in the astrocytic (Yen and Fields, 1981) intermediate filament protein, glial fibrillary acidic protein (GFAP; Jessen, 1981). In addition, unlike Schwann cells (Carey et al., 1983), the enteric glial cell fails to form a basal lamina.

The interior of the enteric ganglia differs from that of other peripheral ganglia in an additional respect. Enteric ganglia are totally avascular (Gabella, 1972); the capillaries that nurture these ganglia lie outside of them. The enteric capillaries that supply the submucosal plexus are fenestrated and similar to those that make up the bulk of the wall of the alimentary canal (Simionescu et al., 1974). These vessels are extremely permeable, even to protein, so that the enteric lymph, and presumably extracellular tissue fluid, are rich in protein that has emerged from the vessels (Mayerson et al., 1960). In contrast, the capillaries that lie just outside the myenteric plexus have thicker walls than those of the capillaries of the rest of the gut and are not fenestrated (Gershon and Bursztajn, 1978). These myenteric capillaries resemble cerebral capillaries (Brightman and Reese, 1969). In common with the cerebral vessels, the myenteric capillaries are relatively impermeable to such tracer proteins as Evans blue-labeled albumin or horseradish peroxidase (Gershon and Bursztajn, 1978). Tight junctions between endothelial cells prevent these proteins from leaking out of vessels via an interendothelial cell route. As a result, there is a functional barrier between the blood and the myenteric layer that includes the plexus. As is true of the blood–brain barrier (Brightman and Reese, 1969), specialized endothelial cells are a major component of the impermeable layer. The enteric glia, like the astrocytes of the brain, do not seem to prevent entry of macromolecules into the neural parenchyma (Bursztajn and Gershon, 1977); nevertheless, the intrinsic myenteric neurons appear to require a specialized environment for their function.

Two distinct neural components make up the ENS. In addition to the intrinsic ganglionated plexuses and their intraalimentary projections, the bowel also receives an extrinsic innervation (see Gershon and Erde, 1981), which consists of the enteric projections of dorsal root and cranial nerve (sensory) ganglion cells, and the sympathetic (postganglionic) and parasympathetic (preganglionic) neurons. Some intrinsic neurons of the myenteric plexus also have extraalimentary projections and provide a reciprocal centripetal input to the prevertebral sympathetic ganglia (Szurszewski and Weems, 1976). The number of intrinsic enteric neurons is extremely large, approximating that of the spinal cord (Furness and Costa, 1980); in humans, the number of enteric neurons has been estimated to be over 10^8. The huge size of the intrinsic component of the ENS dwarfs that of the extrinsic component. The number of preganglionic parasympathetic vagal fibers reaching the bowel in humans, for example, has been estimated at less than 2000 (Hoffman and Schnitzlein, 1969), and the sympathetic postganglionic axons have been found to comprise less than 1% of the neurites of the myenteric plexus (Manber and Gershon, 1979).

Phenotypic diversity of intrinsic enteric neurons is very great (Furness and Costa, 1980, 1982; Gershon, 1981a; Gershon and Erde, 1981). This diversity is evident morphologically when myenteric neurons are categorized on the basis of shape (Gunn, 1959, 1968; Schofield, 1968; Feher and Vajda, 1972), cholinesterase reactivity (see Figure 12; Coupland and Holmes, 1958; Leaming and Cauna, 1961; Lassman, 1962; Sutherland, 1967), monoamine oxidase activity (Furness and Costa, 1971), or affinity for silver stains (Schofield, 1968). The apparent diversity of myenteric neurons has been confirmed and extended by the use of electron microscopy. Myenteric neuronal cell bodies have been classified into nine different sets (Cook and Burnstock, 1976a); their terminals have been divided into eight (Cook and Burnstock, 1976a) and 10 (Furness and Costa, 1980) different subsets by using ultrastructural criteria. Even more striking than the morphological differences between enteric neurons is the phenotypic diversity defined by neurotransmitter type. A still-expanding list of established or putative enteric neurotransmitters includes: acetylcholine (ACh), norepinephrine (NE), serotonin (5-HT), adenosine triphosphate (ATP), gamma aminobutyric acid (GABA), Met- and Leu-enkephalin, vasoactive intestinal polypeptide (VIP), somatostatin, substance P (SP), neurotensin, cholecytokinin (octapeptide), neuropeptide Y, and other peptides (Furness and Costa, 1980, 1982; Gershon, 1981a; Gershon and Erde, 1981; Sundler et al., 1983).

The unique features of the mature ENS are advantageous for studies of neuronal phenotypic expression and, in particular, for studies of the generation of phenotypic diversity. The ENS resembles the CNS but is less complex and is more accessible for experimental manipulation. The ENS has a greater number of types of neuron than do sympathetic or parasympathetic ganglia, but a smaller and more manageable number than the brain. Moreover, many of these types are known; assayable neurotransmitter-related properties are available for all of the known neurons, and can be used in developmental studies as phenotypic markers.

DERIVATION OF THE ENS

The ganglion cells of the ENS are neuroectodermal derivatives that descend from those cells of the neural crest that migrate to and colonize the bowel (Yntema and Hammond, 1954; Andrew, 1971; Le Douarin, 1982). In chicks, these neurons have been shown to arise from the rhombencephalic (somites 1–7) and sacral (caudal to somite 28) levels of the neuraxis (Le Douarin and Teillet, 1973; Le Douarin, 1982), although the exact pathways traversed by the neural crest precursor cells en route to the bowel have not been mapped. The levels of neural crest that colonize the mammalian gut have not been identified precisely; nevertheless, the colonization of the mammalian bowel probably also involves the migration of neural crest-derived precursor cells to an appropriate enteric location.

In common with other ganglionated regions of the PNS, the precursor cells that colonize the bowel and there give rise to the ENS are derivatives of the neural crest. The precursor cells of the enteric neurons differ from those of other regions of the PNS in their levels of origin within the neuraxis, the massive expansion of their ultimate number, the extraordinary phenotypic diversity of their neuronal expression, and the independence and complexity of the neural circuits to which these cells give rise.

The Microenvironment of the Gut Influences Enteric Neuronal Phenotypic Expression

Le Douarin and her colleagues have shown that neural crest removed from regions of quail embryos that do not ordinarily give rise to enteric neurons will do so when transplanted to the rhombencephalic level of chick embryo hosts (Le Douarin and Teillet, 1974; Smith et al., 1977; Le Douarin, 1982). These experiments indicate that neural crest cells do not themselves bear directional or "homing" information that enables them to reach their correct destination in the gut. Instead, they appear to migrate along defined pathways (Newgreen and Thiery, 1980; Thiery et al., 1982); thus, the heterotopic substitution of quail cells for chick cells into the "vagal" level of the neural crest does not result in the failure of these cells to "find" the gut (Le Douarin, 1980, 1982). Once in the bowel, the neural crest primordia appear to develop into at least some of the same types of neuron as would normally be found in the ENS.

For example, when truncal neural crest of quail donor is grafted to replace the vagal neural crest of a chick recipient, cholinergic neurons (Smith et al., 1977), peptidergic neurons (Fontaine-Perus et al., 1982), and serotonergic neurons (Figure 3; Gershon et al., 1983a) develop in the gut, whereas catecholaminergic neurons do not. These neurons of the chimeric bowel are all of quail origin and bear the distinctive quail nuclear marker of nucleolar-associated heterochromatin.

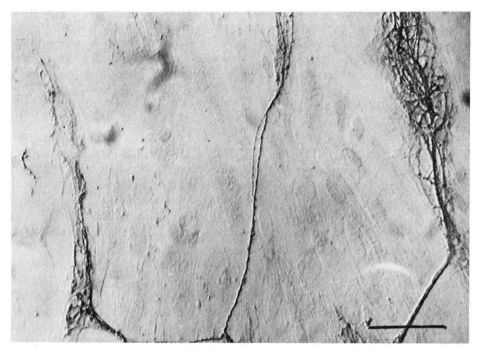

Figure 3. *SP-immunoreactivity in varicose neurites in the outgrowth zone of an organotypic tissue culture of an explant of fetal mouse gut.* The tissue was explanted on day E11, prior to the appearance of any neurons in the bowel and grown in culture for two weeks. The precursors of SP-immunoreactive neurons, therefore, were present in the bowel at the age of explantation. Calibration bar = 50 μm.

If the truncal neural crest is left in or grafted onto its appropriate location in an embryo, it will give rise to the peptidergic neurons and noradrenergic neurons of dorsal root and sympathetic ganglia (Le Douarin, 1982). Truncal neural crest also forms the cells of the adrenal medulla, which, like sympathetic ganglion cells, store catecholamines and peptides. Although peptidergic neurons are derivatives of both grafted and nongrafted truncal neural crest, the fate of truncal neural crest is clearly changed by grafting it in such a way that it colonizes the bowel. Within the gut, its emigrés give rise to neuronal types (serotonergic neurons; Figure 3) that they would not ordinarily produce, and they fail to develop types (catecholaminergic cells) that they would produce if left *in situ*. The enteric microenvironment thus appears to play an important role in influencing neuronal phenotypic expression. This influence could be mediated through selection or induction. The enteric microenvironment, for example, might permit already committed serotonergic cells (if these are present in the truncal neural crest) to survive in the bowel; they would not do so if they reached one of the normal targets of the truncal neural crest. Alternatively, the cells of the neural crest may not yet be committed with respect to choice of a neurotransmitter. Uncommitted cells migrating from the truncal neural crest might, if this is the case, be induced by the appropriate enteric microenvironment to express the genes that determine a serotonergic phenotype.

There are limits to the flexibility of neural crest derivatives. Although there is evidence that premigratory neural crest cells may be pluripotent (Sieber-Blum and Cohen, 1980; Kahn and Sieber-Blum, 1983), restriction of the developmental options open to them occurs rapidly once these cells have begun to migrate (Kahn et al., 1980). Le Douarin and her colleagues have proposed that migrating neural crest cells follow a pattern of binary choices (Le Douarin, 1980, 1982; Le Lièvre et al., 1980). They apparently become committed to a neuronal lineage, then to an autonomic or a sensory lineage, and finally to a neurotransmitter-defined subset. Actually, it could be argued that the initial choice of a neurotransmitter is tentative or at least not irreversible. Autonomic ganglia, for example, remain plastic and susceptible to a change in neurotransmitter characteristics when grown under differing culture conditions even after these cells have become postmitotic (Bunge et al., 1978; Patterson, 1978; Black and Patterson, 1980; Potter et al., 1981). A change in initial phenotypic expression may also occur *in vivo*. The sympathetic innervation of the sweat glands, which appears to change from noradrenergic to cholinergic and VIP-ergic, seems to be an example of such a phenotypic change (Landis and Keefe, 1980; Landis, 1981, 1983).

In order to evaluate the hypothesis that the enteric microenvironment critically influences the phenotypic expression of enteric neurons, it is necessary to know when and where the final phenotype of enteric neurons is first expressed. If the precursors of enteric neurons were found, for example, to express neurotransmitter properties before they reach the bowel, a role for the enteric microenvironment, at least in the initial expression of these characteristics, would be ruled out. Similarly, it is important to try to define the timing of the commitment of precursor cells to the neuronal lineage, the stability of initial expression of a given phenotype, the factors involved in enabling neural crest-derived precursor cells to colonize given segments of bowel, and ultimately, to try to ascertain what types of interaction (between enteric neuronal precursors and nonneuronal cells, or both) are involved in the mediation of microenvironmental influences.

Phenotypic Expression by Enteric Neurons Begins in the Gut Itself

Many studies have attempted to follow the colonization of the gut by neuronal progenitor cells by using techniques such as acetylcholinesterase (AChE) histochemistry (a property of virtually all enteric neurons; Gabella, 1981; see Figure 12), or silver stains to identify the developing neurons (Okamoto and Ueda, 1967; Cantino, 1970; Webster, 1973; Okamoto et al., 1982). These studies rely on neuronal characteristics that are not displayed by the progenitor cells. The markers thus demonstrate neuronal phenotypic expression or maturation, not enteric colonization by the precursors of neurons. There is no reason to assume that the timing of phenotypic expression will necessarily reflect the timing of the colonization of a given segment of bowel by the precursor cells.

In order to investigate the appearance of enteric neurons, we selected neurotransmitter-related properties as markers of the presence of committed enteric neurons (Gershon et al., 1982, 1984; Rothman and Gershon, 1982; Rothman et al., 1984). Assays were devised to detect the presence in mouse gut of cholinergic, serotonergic, peptidergic (SP and VIP), and noradrenergic elements. For cholinergic neurons, a two-stage assay was developed that required the hemicholinium-sensitive, high-affinity uptake of [^{3}H]choline and its subsequent conversion, mediated by choline acetyltransferase, to [^{3}H]acetylcholine by intact tissue (Rothman and Gershon, 1982). Serotonergic neurons were detected by the specific, fluoxetine-sensitive, radioautographic demonstration of uptake of [^{3}H]serotonin and the immunocytochemical detection of 5-HT-like immunoreactivity in the gut (Figure 4). Peptidergic neurites were detected immunocytochemically (Rothman et al., 1984). Noradrenergic elements were detected radioautographically by using the specific, desmethylimipramine-sensitive uptake of [^{3}H]norepinephrine, and immunocytochemically by using antisera directed against tyrosine hydroxylase (TH) (Figure 5; Gershon et al., 1982, 1984). In these assays, neuronal appearance and phenotypic expression displayed a proximodistal order in development. Cholinergic and serotonergic elements were found for the first time in the foregut on embryonic (E) day 12 and appeared in the terminal hindgut on day E14. Peptidergic neurons were detected in the foregut two days later than were cholinergic and serotonergic neurons on day E14, and were present in the terminal hindgut by the time of birth. The noradrenergic elements were detected in the most anterior portions of the foregut in small numbers on day E14, but did not fill out the foregut until day E15 (Figure 5). The entire gut was noradrenergically innervated at the time of birth.

These observations correlated well with ultrastructural morphological observations on the appearance of enteric neurons. The neurotransmitter-related phenotypic markers were present when neurons could first be recognized in the gut, and neurons could not be recognized before the markers were detected. An interesting finding was the presence in the foregut of a very low level of

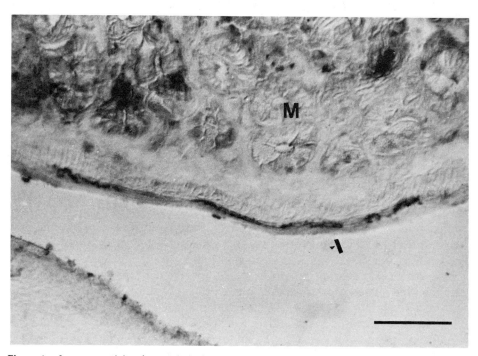

Figure 4. *Immunoreactivity of serotonin in the murine small intestine.* Serotonin-containing enteroendocrine cells in the mucosa (M) and nerve fibers in the myenteric plexus (arrow) are immunoreactive. Calibration bar = 50 μm.

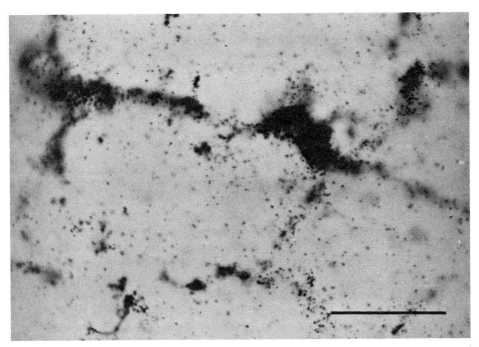

Figure 5. *Immunoreactivity of TH and simultaneous radioautographic localization of [^{3}H]norepinephrine in the fetal rat gut at day E15.* The gut has become innervated by ingrowing (extrinsic) sympathetic axons. All of the fibers, including the expanded growth cones, are doubly labeled. Peroxidase reaction product diffusely darkens the processes, which are also overlain by black radioautographic silver grains. Calibration bar = 30 μm.

production of [^{3}H]acetylcholine from [^{3}H]choline, as early as day E11, prior to both the appearance of recognizable neurons and to the massive rise in [^{3}H]acetylcholine biosynthesis associated with their development. This last observation suggested that otherwise unrecognizable neuronal precursors might be present in the gut prior to enteric neuronal phenotypic expression. In support of this idea, production of ACh by neural crest cells in the process of migration has been reported (Smith et al., 1979).

Enteric Neuronal Precursors Colonize the Whole Gut Before They Can Be Recognized as Neurons

In order to determine whether cells that have yet to express any known phenotypic characteristics of neurons are present in the gut, segments of gut were explanted at various gestational ages and grown in culture for two weeks. At the end of this time the cultures were examined for the presence of neurons by using the neurotransmitter-related techniques outlined above for examination of the embryonic bowel. Because cells cannot migrate into the explanted gut from other embryonic locations, any neurons arising in the cultures had to have been derived from cells that were present in the initial explants. The culture technique therefore represents an assay system for the presence of the progenitors of enteric neurons that does not depend on the manifestation of neuronal marker properties by

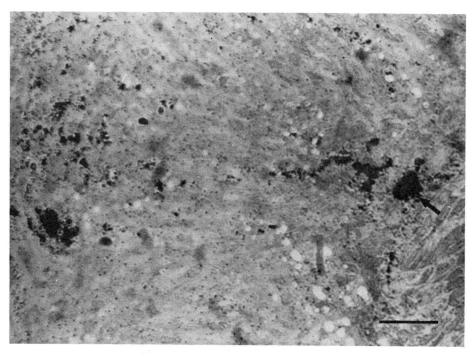

Figure 6. *Specific uptake of [³H]serotonin into cell bodies (arrow) and neurites in an organotypic tissue culture of the duodenum of an ls/ls mouse.* The proximal bowel receives the precursors of serotonergic neurons in mutant as well as normal mice, and these develop and survive in organotypic tissue culture. Calibration bar = 30 μm.

the cells. A similar assay that involves growing explants of bowel on the chorioallantoic membrane (CAM) of chick embryonic hosts has been used to detect the presence of enteric neuronal precursors in the developing alimentary canal of the chick (Smith et al., 1977; Allan and Newgreen, 1980).

In using the culture system to assay the developing murine gut for the presence of neuronal precursor cells, we found that they were present in both foregut and hindgut as early as day E9, that is, only half a day after the gut is fully formed. Moreover, the cultures contained all four of the phenotypic markers that were examined, ACh, 5-HT (Figure 6), SP (Figure 7), and VIP prior to the time these phenotypes are expressed *in situ*. These experiments establish the following four points.

1. The precursors of enteric neurons migrate to and colonize the bowel before they definitively express enteric neuronal marker properties.
2. The precursors of enteric neurons reside in the mammalian gut for a substantial time before phenotypic expression is evident. This apparent latent period is at least three days in the foregut and even longer in the hindgut. In this respect, mammalian enteric development differs from that of the chick, in which the equivalent lag period has been estimated at 12 hours (Allan and Newgreen, 1980).
3. In contrast to neuronal phenotypic expression, no proximodistal sequence could be demonstrated in colonization of the bowel with precursor cells. Precursors

were detectable at the same early period in both the hindgut and the foregut. This observation makes it unlikely that the demonstrated proximodistal sequence that occurs in neuronal phenotypic expression (Okamoto and Ueda, 1967; Webster, 1973; Okamoto et al., 1982; Rothman and Gershon, 1982a) is due to a migration of precursors in a slow proximodistal progression down the bowel. The proximodistal order thus might be due to the interaction of precursor cells already in place with proximodistally maturing nonneuronal elements of the gut. The proximodistal pattern of neuronal phenotypic expression in rats and rabbits is replaced by one in which neuronal appearance first begins at opposite ends of the developing gut and then moves toward the middle (Cantino, 1970). In the early colonization of the hindgut, the mammalian bowel again differs from the avian, in which the hindgut is not colonized by enteric neuronal precursor cells until 3–4 days after the foregut (Allan and Newgreen, 1980; Le Douarin, 1982). This difference might be due to the presence of the avian ganglion of Remak, which does not occur in mammals. Alternatively, a wider source from within the neuraxis of enteric neuronal precursor cells (or an earlier migration of sacral progenitors) might occur in mammals than in birds.

It is not clear from earlier work in either birds or mammals how much distal migration, or movement, of neural crest cells through the enteric tube actually

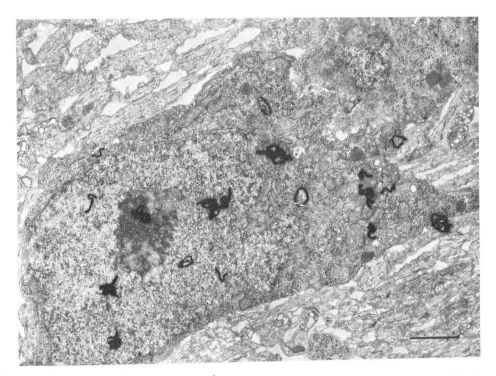

Figure 7. *A neuron specifically labeled by [³H]serotonin in the small intestine of a 15-day-old quail–chick chimeric embryo.* The truncal neural crest of a quail embryo (13 somites) was grafted onto the vagal region of a chick embryonic recipient (8 somites). The [³H]serotonin-labeled cell bears the large quail nucleolar marker. Stain has been extracted from DNA with EDTA. Serotonergic neurons, therefore, are of neural crest origin and develop from heterotropic grafts. This experiment was done in collaboration with Phillip Cochard in the laboratory of Nicole Le Douarin. Calibration bar = 1 μm.

takes place. Transplantation of fragments of splanchnopleure in quail–chick chimeras has demonstrated that the entire foregut and a significant portion of the midgut is situated just beneath the vagal level of the neural crest at the time when the enteric ganglion precursor cells leave this region (Noden, 1984). Thus the neural crest cells move ventrally and become associated with what eventually will become a very long piece of the gut but which, at the time of colonization, is only a short, rostrally situated primordium. This is the region of the chick gut that first displays neuronal precursors (Le Douarin, 1982; Thiery et al., 1982). If proliferation of progenitors takes place, a considerable distal spreading of neuronal precursor cells could occur through the massive elongation of the bowel rather than through a specific caudal translocation of the neural crest cells relative to other cells of the developing gut. Such a distal movement might take place, but is not essential and need not be postulated. In fact, the degree to which neural crest cells actually ambulate has been questioned (Noden, 1983). They are clearly displaced along preferential pathways that lead them to their final destination in embryos (Newgreen et al., 1979; Newgreen and Thiery, 1980; Thiery et al., 1982); however, their relative movement with respect to other cells must be evaluated as one part of an integrated series of dorsoventral movements of mesenchymal and epithelial tissues (Noden, 1983, 1984). Factors involved in migration and the nature of the preferential pathways remain to be elucidated.

4. All of the markers looked for, including peptidergic characteristics, were present in the cultures explanted prior to the time when neuronal phenotypic expression was evident. This suggests that the early wave of precursor cells that colonizes the gut has within it the potential to form a variety, if not all, of the types of enteric neuron that ultimately appear in the ENS. It is not necessary to postulate, and the evidence does not support, sequential waves of precursors of the various types. The sequential order in phenotypic expression, in which peptidergic neurons follow ACh and 5-HT, may thus be due to a sequential change that occurs in the enteric microenvironment as a function of ontogenetic time. Alternatively, peptidergic neurons may appear later because methods used to detect the peptidergic phenotype are less sensitive than those used for the small-molecule transmitters.

The Gut Is Colonized by the Precursors of Enteric Glia Prior to the Onset of Neuronal Phenotypic Expression

The immunoreactivity of GFAP (Uyeda et al., 1972; Yen and Fields, 1981) was used to detect enteric glia (Jessen, 1981; Rothman et al., 1982b). In adult mice, the antiserum (obtained from Dr. G. Nilaver) brilliantly demonstrated glial cells of the enteric plexuses but failed to react with peripheral Schwann cells in myelinated or unmyelinated nerves in the spinal roots, and sphincter, and mesentery of the gut. The antiserum thus distinguishes between enteric glia and Schwann cells. The organotypic tissue culture assay was again used to detect the presence of precursors in the bowel prior to the onset of phenotypic expression. As with neuronal precursors, both foregut and hindgut were found to be colonized by the precursors of enteric glia, at least as early as day E10. It is thus likely that the early neural crest population that colonizes the alimentary tract contains glial, as well as neuronal, precursor cells.

Enteric Neuronal Precursors May Make Neurofilaments Before They Colonize the Bowel

If it is true, as the experiments described above (Rothman and Gershon, 1982) suggest, that neuronal progenitor cells are present in the gut and are responsible for the synthesis of [^{3}H]acetylcholine from [^{3}H]choline before neurons can be recognized, these cells might have become committed to the neuronal lineage before they reached the bowel. A second approach to the detection of cells committed to the neuronal lineage has been to use neurofilament immunoreactivity as an indicator (Tapscott et al., 1981a,b). In mature tissues, neurofilament immunoreactivity is a useful neuronal marker (Bennett et al., 1978; Yen and Fields, 1981). In the developing CNS, it has been shown to be an early sign of neuronal determination (Tapscott et al., 1981a,b). Our observations on neurofilament immunoreactivity in the developing chick have made use of an antiserum (obtained from Dr. G. S. Bennett) that was produced in rabbits immunized with the 160-kD neurofilament polypeptide eluted from SDS-polyacrylamide gels. The antiserum, on nitrocellulose immunoblots, reacts with chick 160-kD neurofilament protein but not with the 180- or 70-kD neurofilament peptides, vimentin, desmin, prekeratin, actin, or tubulin (Bennett et al., 1981). In adult animals, the antiserum reacts only with neurons, although neurofilament-immunoreactive cells appear in some embryonic sites where neurons are not expected to develop. In the developing chick gut, neurofilamentlike immunoreactivity (Figure 8) was detected

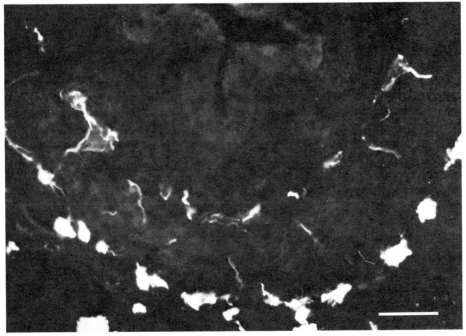

Figure 8. *Neurofilament immunoreactivity in the developing small intestine of a 10-day-old chick embryo.* Neurites (white) show neurofilament immunoreactivity in both the myenteric (outer) and submucosal plexus (inner). This experiment was done in collaboration with Robert Payette. Calibration bar = 50 μm.

in neuritic processes in the foregut by the fourth day of incubation, but did not appear in the wall of the hindgut until the seventh day, although neurofilament-immunoreactive neurites abound in the ganglion of Remak after day five (Gershon et al., 1983b).

Within the postumbilical bowel, development of neurofilament immunoreactivity was more rapid in the colorectum than it was in the more proximally located terminal ileum and caecal appendages. This distal-proximal appearance of neurofilament immunoreactivity probably reflects the distal-proximal migration of sacral neural crest cells in the postumbilical gut. This is particularly evident when the long caecal appendages are considered. These appendages are connected to the terminal but not to the proximal ileum or to the ganglion of Remak. Even the tips of the appendages display neurofilament immunoreactivity at the same time as the ileum, so it seems likely that they are not colonized only by vagal precursors growing distally down the length of the ileum and then proximally up the caecal appendages from the ileocaecal junction. When enteric explants were removed and grown in culture or as grafts on the CAM, their ability to become neuralized *in vitro* correlated with whether neurofilament immunoreactivity could be found in the given segment of gut at the age of explantation.

These observations support the hypothesis that enteric neuroblasts are already committed to the neuronal lineage and thus are expressing a neuronal protein (neurofilament protein) when they colonize the bowel. Masses of cells (rather than the neuritic processes seen in the intestine) that express neurofilament immunoreactivity were found transiently in the caudal branchial arches in proximity to the pharynx before neurofilament immunoreactivity was seen in the wall of the gut. These cells disappeared from the branchial arches by developmental day six. During the time that cells which displayed neurofilament immunoreactivity were present in the branchial arches, these structures had neurogenic potential. This was demonstrated by the correlation of the neuralization of branchial-arch explants *in vitro* with the presence, in the arches, of cells that manifest neurofilament immunoreactivity. No similar neurogenic potential could be demonstrated in the more rostral branchial arches that lack neurofilament-immunoreactive cell bodies. The caudal branchial arches are situated below the ventrally migrating vagal neural crest and are known to be invaded by migratory neural crest cells (D'Amico-Martel and Noden, 1983; Noden, 1983, 1984). The location of the caudal branchial arches, the transience of their population of cells showing neurofilament immunoreactivity, and the correlation of that transient presence with neurogenic potential are consistent with the view that the masses of neurofilament-immunoreactive cells in the caudal branchial arches might, at least in part, be enteric neuronal precursors, derived from the vagal neural crest, migrating through this region of the embryo on their way to the gut. Support for this view comes from parallel observations made by Ciment and Weston (1982), who used a monoclonal antibody that was initially prepared from dorsal root ganglia. Their nerve-specific antibody recognizes a transient population of cells in the branchial arches that is in exactly the same location as are the neurofilament immunoreactive cells. Ciment and Weston (1983) also showed that the branchial arches could colonize aneural explants of chick hindgut grown as CAM grafts. Even prior to the appearance of cells with neurofilament immunoreactivity in the branchial arches, we were able to find, at stages 10–11, a subset of cells that express neurofilament immunoreactivity in the premigratory and early migratory neural crest only in the vagal regions of the neuraxis. These cells are likely precursors of the branchial-arch population. Mitoses were detected in both the neural crest- and branchial arch-immunoreactive cells; proliferative expansion of the population probably takes place while the cells migrate. A considerable increase in cell number has, in fact, occurred between the neural crest and the branchial arches if, as proposed, the vagal neural crest is the source of the immunoreactive cells. These data suggest that the vagal neural crest cells that populate the ENS may become committed to the neuronal lineage while still in the vagal region of the neuraxis. If so, it would be unlikely that the enteric microenvironment plays a role in neuronal determination, although a role of the enteric microenvironment, in the determination of neuronal *type*, as indicated by the work with interspecies chimeras, still seems likely.

Catecholaminergic Cells Appear Transiently in the Developing Gut and May Represent a Natural Change in Phenotypic Expression by Developing Neuroblasts

Experiments done on catecholamine-containing cells in the developing gut indicate that the initial expression of a cell's phenotype may not be stable during the development of enteric neurons. A population of proliferating cells in the fetal gut has been found to express transiently the aspects of a catecholaminergic phenotype (TC cells) during development in both rats (Cochard et al., 1978; Teitelman et al., 1978; Figure 9) and mice (Teitelman et al., 1981b). In rats, these

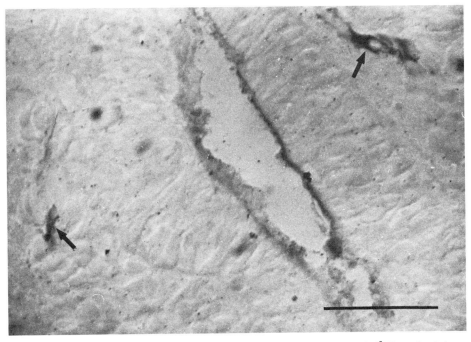

Figure 9. *TH-immunoreactive cells (arrows) do not radioautographically label with [^{3}H]norepinephrine at day E12 in the developing fetal rat gut.* The tissue has been incubated with [^{3}H]norepinephrine and prepared for the simultaneous demonstration of TH by immunocytochemistry and [^{3}H]norepinephrine by radioautography. Calibration bar = 50 μm.

cells appear to be noradrenergic and to contain both TH (Figure 9) and dopamine beta hydroxylase (DBH); in mice, they are dopaminergic and contain TH but no DBH (Teitelman et al., 1981b). These TC cells have been reported to take up NE as well as to synthesize the amine, and the uptake has been said to persist beyond the age when rat TC cells normally disappear (Jonakait et al., 1979). However, these observations were made by using the histofluorescence of NE to detect the amine; therefore, NE taken up by TC cells could not be distinguished in these experiments from endogenous NE.

We have recently reassessed the ability of TC cells to take up NE. TC cells were identified by the immunocytochemical demonstration of TH, and uptake was evaluated radioautographically with [^{3}H]norepinephrine (Gershon et al., 1982, 1984). When TC cells were most numerous in the bowel of both rats and mice, no cells were labeled by [^{3}H]norepinephrine (days E12–13; Figure 9). In rats, but not in mice, labeling of cell bodies by [^{3}H]norepinephrine was found on days E14–15 (Figure 10). No cells showed TH immunoreactivity on day E15 (see Figure 5), although a few were doubly labeled by [^{3}H]norepinephrine and TH immunoreactivity on day E14. In rats, therefore, TC cells contain TH immunoreactivity but *do not* take up [^{3}H]norepinephrine prior to day E14, and their disappearance is followed by the appearance of a second cell population that lacks TH immunoreactivity but *does* take up [^{3}H]norepinephrine. The transient appearance of some cells that express *both markers* on day E14 suggests, but does not prove, that TC cells change their phenotype and are the precursors of the

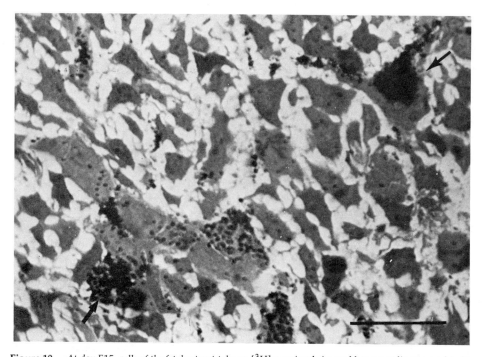

Figure 10. *At day E15, cells of the fetal rat gut take up [^{3}H]norepinephrine and become radioautographically labeled even though TH immunoreactivity has disappeared.* This is a 1.0 μm thick epon section. Radioautographic labeling of several cells is evident (arrows). Calibration bar = 20 μm.

cells found later in development that lack TH but take up [^{3}H]norepinephrine. The TC cells in both rats and mice have been demonstrated to be members of a proliferating population (Teitelman et al., 1981b). It is therefore probable that cells of the E15 rat gut that take up [^{3}H]norepinephrine are not persisting cells that have changed their properties, but are the progeny of the earlier-appearing cells that could synthesize, but not take up the amine. Cells that take up [^{3}H]norepinephrine are rare or absent in the gut of newborn rat, indicating that they may also be transient. These results show that genes responsible for different aspects of the noradrenergic phenotype need not necessarily be coupled in their expression. They also show that initial phenotypic expression of a given cell may not be permanent. Although uptake of [^{3}H]norepinephrine into cell bodies does not occur *in vivo* in mouse gut at E13, it is found in mouse bowel explanted and grown in culture. The potential for development of cells able to take up [^{3}H]norepinephrine thus exists in mice as well as in rats. Why the genes responsible for the uptake of NE and the synthesis of DBH are expressed differently in the rat and murine bowels remains to be determined. The expression of the uptake property *in vitro* in mouse indicates that the cellular microenvironment has an important influence in determining this characteristic.

TC cells appear in many locations in the embryo other than the bowel (Teitelman et al., 1981a,b; Jonakait et al., 1982). In the rat pancreas, such cells are the precursors of cells that ultimately produce glucagon (Teitelman et al., 1981a). The glucagon-producing A cells of the pancreas probably are not neural crest derivatives (Le Douarin, 1982). Transient expression of some aspects of a catecholaminergic phenotype may thus be found in the precursors of cells in a number of locations that ultimately come to contain an apparently unrelated secretory product. The TC characteristic does not appear to be limited to neural crest derivatives. Enteric TC cells have not yet been shown to be of neural crest origin, although they do appear to give rise to a cell of the myenteric plexus (the cell that takes up [^{3}H]norepinephrine). The disappearance of TC cells seems to be coincident with the innervation of the gut by an extrinsic, noradrenergic sympathetic innervation (Figure 5). A causal relationship between these two events is possible, but has not been established.

The Enteric Microenvironment Changes as a Function of Time during Ontogeny

The sequential expression of different enteric neuronal phenotypes or the change in expression of one (such as the TC cells discussed above) could be due to sequential changes occurring in the enteric microenvironment. Such changes, reflected in an increasing compartmentation of developing enteric neurons, were found by examining the developing bowel ultrastructurally (Epstein et al., 1980; Gershon et al., 1980, 1981a,b; Gintzler et al., 1980). Neurons first appear in the enteric mesenchyme prior to the development of either of the smooth muscle coats of the bowel. The circular layer of smooth muscle forms next, dividing the mesenchyme into inner submucosal and outer myenteric compartments. Neurons always move to the outer compartment, and a subset migrates through the circular muscle to form the submucosal plexus. At this time, no basal lamina separates the developing myenteric plexus from either the muscle or the mesenchyme, and extensive interdigitations and contacts form between muscle and nerve. Finally, a basal lamina surrounds the myenteric plexus as glia appear within it and the longitudinal muscle develops outside it. The neural elements are separated from their surroundings, and the mature, specialized interior of the plexus can be discerned. Thus, development is associated with a progressive restriction, or walling off, of neuronal elements. The developmental significance of these changes remains to be established but, especially in view of the evident need to maintain a restricted environment (see discussion of the mature myenteric plexus, above; Gershon and Bursztajn, 1978) for proper functioning of the myenteric plexus, they are likely to be significant.

Nerve Growth Factor (NGF) Probably Does Not Affect Development of Intrinsic Enteric Neurons

In order to determine if NGF is one microenvironmental factor that may play a role in the development of enteric neurons, ENS development was examined in the offspring of maternal guinea pigs that had been actively immunized to mouse NGF prior to and during pregnancy (Gershon et al., 1983c; immunogen and animals obtained from Dr. E. M. Johnson). This active immunization of animals induces a transplacental deprivation of NGF in developing fetuses (Gorin and Johnson, 1979, 1980). It is especially effective in guinea pigs and produces a profound destruction of both sympathetic and dorsal root ganglion neurons (Johnson et al., 1980). Adult sensory neurons are affected by the treatment and show a loss of SP, but they are not destroyed (Schwartz et al., 1982). NGF thus appears to be necessary for the survival of the developing sensory ganglia of the dorsal roots, and the function, but not the survival, of adult neurons of this class. The ENS contains both autonomic projections and intrinsic primary afferent neurons. It is reasonable to postulate that NGF may be an important factor in the development of some, if not all, enteric neurons. Serotonergic neuronal development was used as a marker for exclusively intrinsic enteric neurons; SP and VIP immunoreactivity were used as markers for neurons that are both intrinsic and extrinsic to the gut; noradrenergic neuronal integrity marked the effectiveness of the transplacental treatment of the embryos with maternal anti-NGF. As expected, uptake of [^{3}H]norepinephrine by enteric neurites, as well as the histofluorescence of NE in enteric nerves, was markedly reduced in the sympathectomized animals; however, enteric noradrenergic neurites were less affected than were other sympathetic axons or dorsal root ganglia. Despite this, no change was found in SP- or VIP-immunoreactive elements of the bowel and there was a significant *increase* in uptake of [^{3}H]serotonin, which was found to be a result of increased axonal uptake. There was no increase in the number of serotonergic cell bodies. It is possible that the loss of noradrenergic innervation, which occurs as a result of immunosympathectomy, causes the observed proliferation of serotonergic neurites. Apparently all enteric serotonergic neurons in the adult myenteric plexus receive a noradrenergic innervation, and they may be the only neurons that do so (Figure 11; Gershon, 1981b; Erde et al., 1982; Gershon and Sherman, 1982a,b). The destruction of the noradrenergic input may remove an inhibitory noradrenergic transsynaptic influence on serotonergic axonal sprouting. This remains to be proved, but despite severe damage to

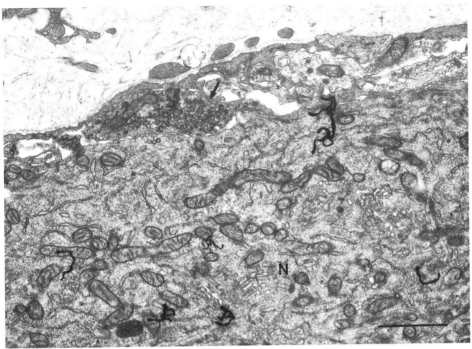

Figure 11. *Simultaneous visualization of noradrenergic and serotonergic elements of the myenteric plexus.* Aldehyde–permanganate fixation after incubation with [^{3}H]serotonin. A noradrenergic terminal (arrow) is identified by its content of dense cored vesicles (about 40 nm diameter) and a serotonergic neuronal cell body (N) is identified by radioautographic labeling. Most, if not all, serotonergic cell bodies receive many noradrenergic contacts. Calibration bar = 0.5 μm.

sympathetic and sensory neurons, immunosympathectomy has no demonstrable adverse effect on the development of intrinsic enteric neurons, so it is unlikely that they are NGF-independent.

Aganglionosis Develops in Human and Murine Bowel

As might be expected from the complexity of enteric neuronal development, involving as it does differentiation, migration, population expansion, and selection of an appropriate neurotransmitter, congenital malformations that involve the ENS are not rare and cause a considerable amount of perinatal morbidity and mortality. One condition of interest, both as a medical problem and for the insight it can provide into enteric neuronal development, is Hirschsprung's disease. This condition is found in 1:2000–1:10,000 live human births and is associated with the total absence of intrinsic enteric ganglion cells (but not the projections of axons from cells outside the gut) from a portion of the bowel (Smith, 1967; Wood, 1979). Consequently, the gut dilates proximal to the aganglionic, effectively nonpropulsive region. The lethal spotted mutant mouse (*ls/ls*) provides an experimental model for the investigation of this condition. The mice, like their human counterparts with Hirschsprung's disease, develop megacolon proximal to a terminal segment of aganglionic gut (cf. Figures 12, 13; Lane, 1966; Bolande, 1975). In both the human (Smith, 1967; Wood, 1979) and

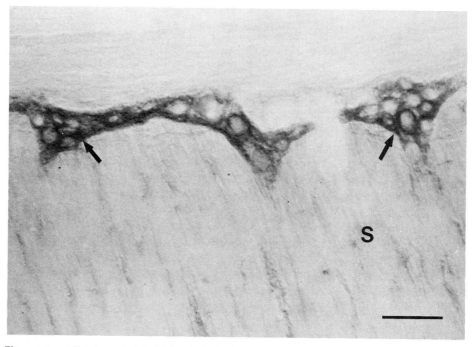

Figure 12. *AChE demonstrated histochemically in the terminal colon of a normal mouse.* AChE reactivity can be seen in cell bodies and neurites of the myenteric plexus (arrows) and in nerve fibers within the circular layer of smooth muscle (S). Calibration bar = 50 μm.

murine (Lane, 1966; Bolande, 1975) conditions of congenital megacolon, the segment of bowel that becomes aganglionic is usually the terminal rectum or rectosigmoid. The neurites of adrenergic and cholinergic neurons, extrinsic to the gut, grow into aganglionic segments (Howard, 1972; Rawdon and Dockray, 1983; Figure 13); serotonergic neurites are excluded from the aganglionic segments in both humans (Rogawski et al., 1978) and mice (Rothman et al., 1982a, 1984). It has also been found that VIP-containing axons and neurites with SP immunoreactivity are present in the aganglionic segments of *ls/ls* (see below) and piebald mice, another strain that develops megacolon (Vaillant et al., 1982), although there are smaller numbers of these types of neurites in aganglionic regions than are in the terminal and distal colon of normal mice. 5-HT is found entirely in intrinsic enteric neurons (Gershon, 1981b); VIP, SP, and ACh can be present in either intrinsic neurons or axons of extrinsic origin (SP is present in peripheral processes of dorsal root neurons), and NE is an exclusive marker of extrinsic axons. Therefore it is possible that the microenvironment of the gut in aganglionic zones is not conducive to the growth and/or development of intrinsic neural elements; it might permit ingrowth of only extrinsic autonomic or sensory axons.

One view of the origin of congenital megacolon ascribes the condition to a slowing of the normal migration of neural crest cells down the gut (Okamoto and Ueda, 1967; Webster, 1973). This assumes that all enteric ganglion cells, even those of the postumbilical bowel, are derived from the vagal neural crest.

It is proposed that if vagal precursors do not reach the end of the gut within a limited time frame, the muscle closes over or some other change occurs, and the late emigrés cannot enter. Evidence supporting this theory is that the normal gradient of proximodistal appearance of AChE-containing ganglion cells is retarded in humans and mutant mice. The theory of slow migration, however, assumes that mammals are different from birds, in which there is a sacral, as well as a vagal, contribution to the colonization of the colon with neuronal precursors (Le Douarin and Teillet, 1973). In addition, the proximodistal gradient in neuronal appearance is not related to the downward migration of precursor cells (Rothman and Gershon, 1982). Finally, the slow migration of precursor cells cannot by itself account for the ability of some types of neurites (noradrenergic) to enter aganglionic regions whereas others (serotonergic) cannot. We have investigated enteric neuronal development in *ls/ls* mice in order to determine the role of the enteric microenvironment in the genesis of congenital megacolon, and to gain insight into mechanisms of colonization of the gut.

Aganglionosis in Developing and Adult *ls/ls* Mice Is Limited to the Terminal 2 mm of Bowel

Neurites within the ganglionic and aganglionic zones of the gut of developing and adult mutant mice were studied with respect to their content of AChE,

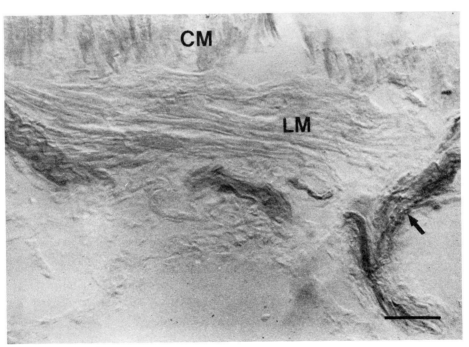

Figure 13. *AChE demonstrated histochemically in the terminal colon of lethal spotted* (ls/ls) *mutant mouse.* Extrinsic nerves are entering the colon (arrow); some contain AChE. However, no neuronal cell bodies are visible in this segment of gut. CM, circular muscle; LM, longitudinal muscle. Calibration bar = 50 μm.

immunoreactive SP, VIP, and 5-HT, and their ability to take up [^{3}H]serotonin (Rothman et al., 1982a, 1984). In both the fetal gut of developing mutant mice and in the mature bowel of adult animals abnormalities were limited to the terminal 2 mm of colon. The ENS in the proximal alimentary tract was indistinguishable from that of control animals for all of the parameters examined. In the terminal bowel, the normal plexiform pattern of the innervation and the ganglion cell bodies was replaced by a coarse reticulum of nerve fibers that stained for AChE and could be traced from extrinsic nerves running to the colon from the pelvic wall. These coarse nerve bundles contained greatly reduced numbers of fibers that displayed SP- and VIP-like immunoreactivity, but a serotonergic innervation was totally missing. During development, AChE and uptake of [^{3}H]serotonin appeared in neural elements in the foregut of mutant mice on day E12, at about the same time that these markers appeared in the foregut of normal mice. By day E14, neurons expressing one or the other marker were recognizable as far distally as about 2 mm from the anus. The timing of neuronal development, therefore, is the same in normal and *ls/ls* mice. This suggests that the defect may not arise from a slowed migration of neuronal precursor cells down the gut but from a failure of these precursors to enter the terminal bowel.

Neuronal Precursor Cells Are Not Found in the Terminal Bowel of *ls/ls* Mice

The appearance of neurons in segments of gut grown for two weeks as explants in culture was used as an assay for the presence of neuronal progenitor cells in segments of fetal bowel at the time of explantation. Both AChE activity and uptake of [^{3}H]serotonin developed between days E10 and E20 in neurons of proximal bowel explants *in vitro* (see Figure 6). At all times, the terminal 2 mm of mutant, but not normal, fetal gut gave rise to aneuronal cultures (Figure 14; cf. Figure 6). In some mutant mice, rare, small, ectopically situated pelvic ganglia were found just outside aganglionic segments of fetal colon. Uptake of [^{3}H]serotonin, normally a marker for intrinsic enteric neurites, was detected in these ganglia.

These experiments can be interpreted to support the hypothesis that the terminal 2 mm of the gut in *ls/ls* mice is intrinsically abnormal, and thus cannot be colonized by the precursors of enteric neurons. The defect seems to be specific, in that both cells and processes of intrinsic enteric neurons, including all serotonergic and most peptidergic neurites, apparently are excluded from the abnormal region, although extrinsic nerve fibers, including sympathetic and sensory axons, are able to enter the aganglionic zones. Examination has failed to reveal a significant proximodistal displacement of neural progenitor cells through the enteric tube during development, even of the normal murine bowel. Therefore, a defect in the migration of precursor cells down the alimentary tract to the terminal gut probably is not substantially involved in the pathogenesis of aganglionosis. This conclusion is supported by the normality of the enteric nervous system in proximal regions of the mutant gut, and the presence of enteric types of neuron outside of the aganglionic region, but at the same level.

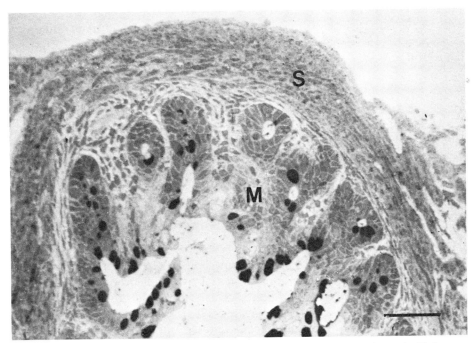

Figure 14. *Radioautography following incubation of an explant of terminal colon from a fetal* ls/ls *mouse with [³H]serotonin fails to reveal specific labeling.* This tissue was explanted on day E13 and grown for two weeks in organotypic tissue culture. The mucosa (M) has been maintained. Smooth muscle (S) has developed in the absence of nerve and no neuronal cell bodies are visible. Calibration bar = 50 μm.

Normal but Not Presumptive Aganglionic Bowel Can Be Colonized by Sources of Neural Crest-Derived Cells in Cocultures

The failure of enteric neurons to develop in cultures of the terminal 2 mm of bowel provides an ideal target for coculture experiments. Experiments using cocultures were designed to answer the question: Can enteric neuronal precursor cells enter and survive in these zones of bowel if presented to them, thereby bypassing the migratory pathway? To examine this, the terminal gut from *ls/ls* mice was grown in combination with a source of neuronal precursor cells (Jacobs-Cohen et al., 1983). Potential donors included foregut from normal and *ls/ls* mice (E10), mouse neural tube and crest (somites 1–7, E8), chick and quail foregut (E5), and ganglion of Remak (E6). The chick hindgut, which in contrast to that of mice (see above; Le Douarin, 1982) can be obtained prior to its colonization by precursor cells, served as a control. AChE histochemistry was used as a neuronal marker. All potential sources of enteric ganglia were able to migrate into and colonize the chick hindgut. Species differences proved no barrier to the process. In contrast, none of the potential sources, from mouse or bird, could colonize the presumptive aneuronal hindgut of *ls/ls* mice. These experiments should be repeated using normal, aganglionic mouse colon (if it can be obtained) or marked sources of neuronal precursor cells; nevertheless, they presently support the idea that the terminal bowel of ls/ls mice becomes aganglionic because precursor cells fail to enter it, and so justify a search for the abnormality involved.

Extracellular Matrix and Neural Crest Migration

Neural crest cells have been shown to migrate passively along defined pathways in a cell-free space that is filled with extracellular matrix proteins (ECM), and that is bounded by basal laminae (Weston and Butler, 1966; Pratt et al., 1975; Bancroft and Bellairs, 1976; von der Mark et al., 1976; Derby, 1978; Derby and Pintar, 1978; Pintar, 1978; Weston et al., 1978; Bolender et al., 1980; Lofberg et al., 1980; Bronner-Fraser, 1982; Weston, 1982). The ECM has been found to be especially rich in glycosaminoglycans (GAGs) wherever neural crest cells are migrating. The migratory pathways are also intensely marked by fibronectin immunoreactivity (Newgreen and Thiery, 1980; Thiery et al., 1982). Fibronectin is a major glycoprotein (Hay, 1978, 1981; Yamada, 1981) of the extracellular matrix. Most neural crest cells are unable to synthesize fibronectin themselves (Newgreen and Thiery, 1980; Le Douarin, 1982). It has been postulated that fibronectin may direct the migration of neural crest cells precisely because they lack either fibronectin or its binding sites, and thus are not retarded in their movement (Bronner-Fraser, 1982; Weston, 1982). Collagen has been found in the crest migration routes of amphibians (Lofberg and Ahlfors, 1978), and a role, especially for type I collagen, in determining the final destination of neural crest precursors or in influencing their phenotypic expression has not been excluded. Basal laminae contain type IV collagen and laminin, and these ECM proteins may influence neural crest migration (Weston et al., 1978; Newgreen and Thiery, 1980). Microfibrils also are present in the loose ECM of cell-free regions of early embryos (Hay, 1978, 1981; Newgreen et al., 1979). Neural crest cells surely come into contact with these elements, which could be collagenous or procollagenous (Newgreen and Thiery, 1980). It seems clear that ECM proteins and GAGs are factors that, in part, control and/or define the pathways of neural crest migration. Changes in ECM have been shown to be associated with, and may determine the cessation of, migration, localization, and subsequent development of neural crest cells (Bunge et al., 1978; Weston et al., 1978; Weston, 1982). Therefore, in considering the failure of ganglion cells to develop in a segment of bowel, one of the possible abnormalities we are currently evaluating is a localized defect in one or another of the constituents of the ECM.

CONCLUSION

We have framed a number of working hypotheses based on the observations and literature presented above. These are meant to direct the progress of future work and are not intended to be taken as statements that definitively explain the development of the ENS. We propose that the neural crest precursors of enteric neurons become committed to the neuronal lineage and begin making neuron-specific intermediate filaments while in the vicinity of the neural tube (within the premigratory neural crest or at the initiation of migration). These cells are still proliferating. The migrating precursors remain plastic with respect to choice of neurotransmitter and may remain so even after an initial transmitter is expressed. Both the final neurotransmitter-defined phenotype and the ability of precursor cells to colonize segments of gut are determined by an interaction between the precursor cells and their microenvironment. Microenvironmental factors that are important may include components of the extracellular matrix and interactions with nonneuronal cells, but probably not NGF.

ACKNOWLEDGMENTS

The authors wish to thank Ms. Diane Sherman for her invaluable technical assistance, especially for the electron microscopy. We also thank Mrs. Carol Haaksma for her help with the photography. The work outlined in this chapter was supported, in part, by grants NS15547 and NS12969 from the National Institutes of Health, grants BNS8204904 and BNS8309156 from the National Science Foundation, the National Foundation March of Dimes, The Dysautonomia Foundation, and the Alfred Jurzykowski Foundation.

REFERENCES

Allan, I. J., and D. F. Newgreen (1980) The origin and differentiation of enteric neurons of the intestine of the fowl embryo. Am. J. Anat. 157:137–154.

Andrew, A. (1971) The origin of intramural ganglia. IV. The origin of enteric ganglia: A critical review and discussion of the present state of the problem. J. Anat. 108:169–184.

Bancroft, M., and R. Bellairs (1976) The neural crest cells of the trunk region of the chick embryo studied by SEM and TEM. ZOON 4:73–85.

Bennett, G. S., S. A. Fellini, J. M. Croop, J. J. Otto, J. Bryan, and H. Holtzer (1978) Differences among 100 Å filament subunits from different cell types. Proc. Natl. Acad. Sci. USA 75:4364–4368.

Bennett, G. S., S. J. Tapscott, J. A. Kleinbart, P. B. Antin, and H. Holtzer (1981) Different 10-nm filament proteins in cultured chick neurons and non-neuronal cells. Science 212:567–569.

Black, I. B., and P. H. Patterson (1980) Developmental regulation of neurotransmitter phenotype. In Current Topics in Developmental Biology, Vol. 15, Neural Development, Part I, A. A. Moscona and A. Monroy, eds., pp. 27–40, Academic, New York.

Bolande, R. P. (1975) Animal model of human disease. Hirschsprung's disease, aganglionic or hypoganglionic megacolon; animal model: Aganglionic megacolon in piebald and spotted mutant mouse strains. Am. J. Pathol. 79:189–192.

Bolender, D. L., W. G. Seliger, and R. R. Markwald (1980) A histochemical analysis of polyanionic compounds found in the extracellular matrix encountered by migrating cephalic neural crest cells. Anat. Rec. 196:401–412.

Brightman, M. W., and T. S. Reese (1969) Junction between intimately apposed cell membranes in the vertebrate brain. J. Cell Biol. 40:648–677.

Bronner-Fraser, M. (1982) Distribution of latex beads and retinal pigment epithelial cells along the ventral neural crest pathway. Dev. Biol. 91:50–63.

Bunge, R., M. Johnson, and C. D. Ross (1978) Nature and nurture in development of the autonomic neuron. Science 199:1409–1416.

Bursztajn, S., and M. D. Gershon (1977) Discrimination between nicotinic receptors in vertebrate ganglia and skeletal muscle by alpha-bungarotoxin and cobra venoms. J. Physiol. (Lond.) 269:17–31.

Cantino, D. (1970) A histochemical study of the nerve supply to the developing alimentary tract. *Experientia* **15**:766–767.

Carey, D. J., C. F. Eldridge, J. Carson, R. T. Cornbrooks, and R. P. Bunge (1983) Biosynthesis of type IV collagen by cultured rat Schwann cell. *J. Cell Biol.* **97**:473–479.

Ciment, G., and J. A. Weston (1982) Early appearance in neural crest and crest-derived cells of an antigenic determinant present in avian neurons. *Dev. Biol.* **93**:355–367.

Ciment, G., and J. A. Weston (1983) Enteric neurogenesis by branchial and mesenchymal cells of birds. *Neurosci. Abstr.* **9**:969.

Cochard, P., M. Goldstein, and I. B. Black (1978) Ontogenetic appearance and disappearance of tyrosine hydroxylase and catecholamines in the rat embryo. *Proc. Natl. Acad. Sci. USA* **75**:2986–2990.

Cook, R. D., and G. Burnstock (1976a) The ultrastructure of Auerbach's plexus in the guinea pig. I. Neuronal elements. *J. Neurocytol.* **5**:171–194.

Cook, R. D., and G. Burnstock (1976b) The ultrastructure of Auerbach's plexus in the guinea pig. II. Non-neuronal elements. *J. Neurocytol.* **5**:195–206.

Coupland, R. E., and R. L. Holmes (1958) Auerbach's plexus in the rabbit. *J. Anat. (Lond.)* **92**:651.

D'Amico-Martel, A., and D. Noden (1983) Contributions of placodal and neural crest cells to avian cranial peripheral ganglia. *Am. J. Anat.* **166**:445–468.

Derby, M. (1978) Analysis of glycosaminoglycans within the extracellular environments encountered by migrating neural crest cells. *Dev. Biol.* **66**:331–336.

Derby, M. A., and J. E. Pintar (1978) Histochemical specificity to streptomyces hyaluronidase and chondroitinase ABC. *Histochem. J.* **10**:529–547.

Erde, S. M., M. D. Gershon, and D. L. Sherman (1982) Serotonergic neurons of the small intestine: Distribution and interactions with other neurons. *Neurosci. Abstr.* **8**:551.

Epstein, M. L., D. L. Sherman, and M. D. Gershon (1980) Development of serotonergic neurons in the chick duodenum. *Dev. Biol.* **77**:22–40.

Feher, E., and J. Vajda (1972) Cell types in the nerve plexus of the small intestine. *Acta Morphol. Acad. Sci. Hung.* **20**:13–25.

Fontaine-Perus, J. C., M. Chanconie, and N. Le Douarin (1982) Differentiation of peptidergic neurons in quail–chick chimaeric embryos. *Cell Differ.* **11**:183–193.

Furness, J. B., and M. Costa (1971) Monoamine oxidase histochemistry of enteric neurons in the guinea-pig. *Histochemie* **28**:324–326.

Furness, J. B., and M. Costa (1980) Types of nerves in the enteric nervous system. *Neuroscience* **5**:1–20.

Furness, J. B., and M. Costa (1982) Identification of gastrointestinal neurotransmitters. In *Handbook of Experimental Pharmacology*, G. Bertaccini, ed., pp. 383–460, Springer-Verlag, Berlin.

Gabella, G. (1971) Glial cells in the myenteric plexus. *Z. Naturforsch.* **26B**:244–245.

Gabella, G. (1972) Fine structure of the myenteric plexus. *J. Anat. (Lond.)* **111**:69–97.

Gabella, G. (1981) Structure of muscles and nerves in the gastrointestinal tract. In *Physiology of the Gastrointestinal Tract*, L. R. Johnson, ed., pp. 197–241, Raven, New York.

Gershon, M. D. (1981a) The enteric nervous system. *Annu. Rev. Neurosci.* **4**:227–272.

Gershon, M. D. (1981b) Storage and release of serotonin and serotonin-binding protein by serotonergic neurons. In *Cellular Basis of Chemical Messengers in the Digestive System*, M. I. Grossman, M. A. B. Brazier, and J. Lechago, eds., pp. 285–298, Academic, New York.

Gershon, M. D., and S. Bursztajn (1978) Properties of the enteric nervous system: Limitation of access of intravascular macromolecules to the myenteric plexus and muscularis externa. *J. Comp. Neurol.* **180**:467–488.

Gershon, M. D., and S. M. Erde (1981) The nervous system of the gut. *Gastroenterology* **80**:1571–1594.

Gershon, M. D., and D. L. Sherman (1982a) Identification of and interactions between noradrenergic and serotonergic neurites in the myenteric plexus. *J. Comp. Neurol.* **204**:407–421.

Gershon, M. D., and D. L. Sherman (1982b) Selective demonstration of serotonergic neurons and terminals in electron micrographs: Loading with 5,7-dihydroxytryptamine and fixation with $NaMnO_4$. *J. Histochem. Cytochem.* **30**:769–773.

Gershon, M. D., M. L. Epstein, and L. Hegstrand (1980) Colonization of the chick gut by progenitors of enteric serotonergic neurons: Distribution, differentiation and maturation within the gut. *Dev. Biol.* **77**:41–51.

Gershon, M. D., D. L. Sherman, and A. R. Gintzler (1981a) An ultrastructural analysis of the developing enteric nervous system of the guinea-pig small intestine. *J. Neurocytol.* **10**:271–296.

Gershon, M. D., G. N. Teitelman, and T. P. Rothman (1981b) Development of enteric neurons from non-recognizable precursor cells. *Ciba Found. Symp.* **83**:51–61.

Gershon, M. D., T. P. Rothman, G. N. Teitelman, T. Joh, and D. Reis (1982) Incomplete expression of a catecholaminergic phenotype in cells that transiently appear during development in the fetal rat gut. *Neurosci. Abstr.* **8**:189.

Gershon, M. D., P. Cochard, D. Sherman, and T. P. Rothman (1983a) Origin of enteric serotonergic neurons from the neural crest and determination of the serotonergic phenotype by the enteric microenvironment: A study using quail–chick interspecies chimeras. *Neurosci. Abstr.* **9**:306.

Gershon, M. D., R. F. Payette, and T. P. Rothman (1983b) Development of the enteric nervous system. *Fed. Proc.* **42**:1620–1625.

Gershon, M. D., T. P. Rothman, D. L. Sherman, and E. M. Johnson (1983c) Effect of prenatal exposure to anti-NGF on the enteric nervous system (ENS) of the guinea pig. *Anat. Rec.* **205**:62A.

Gershon, M. D., T. P. Rothman, T. H. Joh, and G. N. Teitelman (1984) Transient and differential expression of aspects of the catecholaminergic phenotype during development of the fetal bowel of rats and mice. *J. Neurosci.* (in press).

Gintzler, A. R., T. P. Rothman, and M. D. Gershon (1980) Ontogeny of opiate mechanisms in relation to the sequential development of neurons known to be components of guinea pig's enteric nervous system. *Brain Res.* **189**:31–48.

Gorin, P. D., and E. M. Johnson (1979) Experimental auto-immune model of nerve growth factor deprivation: Effects on developing peripheral sympathetic and sensory neurons. *Proc. Natl. Acad. Sci. USA* **76**:5382–5386.

Gorin, P. D., and E. M. Johnson (1980) Effects of long-term nerve growth factor deprivation on the nervous system of the adult rat: An experimental autoimmune approach. *Brain Res.* **198**:27–42.

Gunn, M. (1959) Cell types in the myenteric plexus of the cat. *J. Comp. Neurol.* **111**:83–93.

Gunn, M. (1968) Histological and histochemical observations on the myenteric and submucosal plexuses of mammals. *J. Anat. (Lond.)* **102**:223–239.

Hay, E. D. (1978) Fine structure of embryonic matrices and their relation to the cell surface in ruthenium red-fixed tissues. *Growth* **42**:399–423.

Hay, E. D. (1981) Collagen and embryonic development. In *Cell Biology of Extracellular Matrix*, E. D. Hay, ed., pp. 379–409, Plenum, New York.

Hoffman, H. H., and H. N. Schnitzlein (1969) The number of vagus nerves in man. *Anat. Rec.* **139**:429–435.

Howard, E. R. (1972) Hirschsprung's disease: A review of the morphology and physiology. *Postgrad. Med. J.* **48**:471–477.

Jacobs-Cohen, R. J., M. D. Gershon, and T. P. Rothman (1983) The influence of the local microenvironment on the colonization of the gut by neural crest derived neuronal precursor cells. *Anat. Rec.* **205**:90A.

Jessen, K. R. (1981) Removal of the ganglionated plexuses from the gut wall: Advantages for studies of the enteric nervous system. *Scand. J. Gastroenterol.* **71**:91–102.

Johnson, E. M., P. D. Gorin, L. D. Brandeis, and J. Pearson (1980) Dorsal root ganglion neurons are destroyed by exposure *in utero* to maternal antibody to nerve growth factor. *Science* **210**:916–918.

Jonakait, G. M., J. Wolff, P. Cochard, M. Goldstein, and I. B. Black (1979) Selective loss of noradrenergic phenotypic characters in neuroblasts of the rat embryo. *Proc. Natl. Acad. Sci. USA* **76**:4683–4686.

Jonakait, G. M., K. A. Markey, M. Goldstein, and I. B. Black (1982) Transient expression of catecholaminergic traits in cranial nerve ganglia of the embryonic rat. *Neurosci. Abstr.* **8**:754.

Kahn, C. R., and M. Sieber-Blum (1983) Cultured quail neural crest cells attain competence for terminal differentiation into melanocytes before competence to terminal differentiation into adrenergic neurons. *Dev. Biol.* **95**:232–238.

Kahn, C. R., J. T. Coyle, and A. M. Cohen (1980) Head and trunk neural crest *in vitro*: Autonomic neuron differentiation. *Dev. Biol.* **77**:340–348.

Kuntz, A. (1953) *The Autonomic Nervous System*, Bailliere, Tindall, and Cox, London.

Landis, S. C. (1981) Environmental influences on the postnatal development of rat sympathetic neurons. In *Development in the Nervous System*, D. R. Garrod and J. D. Feldman, eds., pp. 147–160, Cambridge Univ. Press, Cambridge.

Landis, S. C. (1983) Development of cholinergic sympathetic neurons: Evidence for transmitter plasticity *in vivo*. *Fed. Proc.* **42**:1633–1638.

Landis, S. C., and D. Keefe (1980) Development of cholinergic sympathetic innervation of eccrine sweat glands in rat footpad. *Neurosci. Abstr.* **6**:379.

Lane, P. W. (1966) Association of megacolon with two recessive spotting genes in the mouse. *J. Hered.* **57**:29–31.

Langley, J. N. (1921) *The Autonomic Nervous System*, Part I. W. Heffer, Cambridge.

Lassmann, G. (1962) The demonstration of specific cholinesterase in the nervous formations of the human appendix. *Acta Histochem.* **13**:113–122.

Leaming, D. B., and N. Cauna (1961) A qualitative and quantitative study of the myenteric plexus of the small intestine of the cat. *J. Anat. (Lond.)* **95**:160–169.

Le Douarin, N. M. (1980) The ontogeny of the neural crest in avian embryo chimeras. *Nature* **286**:663–669.

Le Douarin, N. M. (1982) *The Neural Crest*, Cambridge Univ. Press, Cambridge.

Le Douarin, N. M., and M. A. Teillet (1973) The migration of neural crest cells to the wall of the digestive tract in avian embryo. *J. Embryol. Exp. Morphol.* **30**:31–48.

Le Douarin, N. M., and M. A. Teillet (1974) Experimental analysis of the migration and differentiation of neuroblasts of the autonomic nervous system and of neuroectodermal mesenchymal derivatives, using a biological cell marking technique. *Dev. Biol.* **41**:162–184.

Le Lièvre, C. S., G. G. Schweizer, C. M. Ziller, and N. M. Le Douarin (1980) Restrictions of developmental capabilities in neural crest cell derivatives as tested by *in vivo* transplantation experiments. *Dev. Biol.* **77**:362–378.

Lofberg, J., and K. Ahlfors (1978) Extracellular matrix organization and early neural crest migration in the axolotyl embryo. *ZOON* **6**:87–101.

Lofberg, J., K. Ahlfors, and C. Fallstrom (1980) Neural cell migration in relation to extracellular matrix organization in the embryonic axolotyl trunk. *Dev. Biol.* **75**:148–167.

Manber, L. M., and M. D. Gershon (1979) A reciprocal adrenergic–cholinergic axoaxonic synapse in the mammalian gut. *Am. J. Physiol.* **236**:E738–E745.

Mayerson, H. S., C. G. Wolfram, H. H. Shirley, Jr., and K. Wasserman (1960) Regional differences in capillary permeability. *Am. J. Physiol.* **198**:155–160.

Newgreen, D., and J. P. Thiery (1980) Fibronectin in early avian embryos: Synthesis and distribution along the migration pathways of neural crest cells. *Cell Tissue Res.* **211**:269–291.

Newgreen, D. F., M. Ritterman, and E. A. Peters (1979) Morphology and behavior of neural crest cells of chick embryo *in vitro*. *Cell Tissue Res.* **203**:115–140.

Noden, D. M. (1983) The embryonic origins of avian cephalic and cervical muscles and associated connective tissues. *Am. J. Anat.* **168**:257–276.

Noden, D. M. (1984) Craniofacial development: New views on old problems. *Anat. Rec.* **208**:1–13.

Okamoto, E., and T. Ueda (1967) Embryogenesis of intramural ganglia of the gut and its relation to Hirschsprung's disease. *J. Pediatr. Surg.* **2**:437–443.

Okamoto, E., M. Satani, and K. Kuwata (1982) Histologic and embryonic studies on the innervation of the pelvic viscera in patients with Hirschsprung's disease. *Surg. Gynecol. Obstet.* **155**:823–828.

Patterson, P. H. (1978) Environmental determination of autonomic neurotransmitter functions. *Annu. Rev. Neurosci.* **1**:1–17.

Pintar, J. E. (1978) Distribution and synthesis of glycosaminoglycans during quail neural crest morphogenesis. *Dev. Biol.* **67**:444–464.

Potter, D. D., S. C. Landis, and E. J. Furshpan (1981) Chemical differentiation of sympathetic neurons. In *Neurosecretion and Brain Peptides*, J. B. Martin, S. Reichlin, and K. L. Bick, eds., pp. 275-285, Raven, New York.

Pratt, R. M., M. A. Larsen, and M. C. Johnston (1975) Migration of cranial neural crest cells in a cell-free hyaluronate-rich matrix. *Dev. Biol.* **44**:298-305.

Rawdon, B. B., and G. J. Dockray (1983) Directional growth of sympathetic nerve fibres *in vitro* towards enteric smooth muscle and heart from mice with congenital aganglionic colon and their normal littermates. *Dev. Brain Res.* **7**:53-59.

Rogawski, M. A., J. T. Goodrich, M. D. Gershon, and R. J. Touloukian (1978) Hirschsprung's disease: Absence of serotonergic neurons in the aganglionic colon. *J. Pediatr. Surg.* **13**:608-615.

Rothman, T. P., and M. D. Gershon (1982) Phenotypic expression in the developing murine enteric nervous system. *J. Neurosci.* **2**:381-393.

Rothman, T. P., and M. D. Gershon (1984) Regionally defective colonization of the terminal bowel by the precursors of enteric neurons in lethal spotted ls/ls mutant mice. *Neuroscience* (in press).

Rothman, T. P., G. Nilaver, and M. D. Gershon (1982a) Neural crest-microenvironment interactions in the formation of enteric ganglia: An analysis of normal and lethal spotted mutant mice. *Neurosci. Abstr.* **8**:6.

Rothman, T. P., G. Nilaver, and M. D. Gershon (1982b) Colonization of normal and congenitally aganglionic bowel by cells that contain glial fibrillary acidic protein. *J. Cell Biol.* **95**:54a.

Rothman, T. P., G. Nilaver, and M. D. Gershon (1984) Colonization of the developing murine enteric nervous system and subsequent phenotypic expression by the precursors of peptidergic neurons. *J. Comp. Neurol.* **225**:13-23.

Schofield, G. C. (1968) Anatomy of muscular and neural tissues in the alimentary canal. In *Handbook of Physiology*, Vol. 4, Sect. 6, *Alimentary Canal*, C. F. Code, ed., pp. 1579-1627, American Physiological Society, Washington, D.C.

Schwartz, J. P., J. Pearson, and E. M. Johnson (1982) Effect of exposure to anti-NGF on sensory neurons of adult rats and guinea pigs. *Brain Res.* **244**:378-381.

Sieber-Blum, M., and A. M. Cohen (1980) Clonal analysis of quail neural crest cells. *Dev. Biol.* **80**:96-106.

Simionescu, M., N. Simionescu, and G. E. Palade (1974) Morphometric data on the endothelium of blood capillaries. *J. Cell Biol.* **60**:128-152.

Smith, B. (1967) Myenteric plexus in Hirschsprung's disease. *Gut* **8**:308-312.

Smith, J., P. Cochard, and N. M. Le Douarin (1977) Development of choline acetyltransferase activities in enteric ganglia derived from presumptive adrenergic and cholinergic levels of neural crest. *Cell Differ.* **6**:199-216.

Smith, J., M. Fauquet, C. Ziller, and N. M. Le Douarin (1979) Acetylcholine synthesis by mesencephalic neural crest cells in the process of migration *in vivo*. *Nature* **282**:853-855.

Sundler, F., E. Moghimzadeh, R. Hakanson, M. Ekelund, and P. Emson (1983) Nerve fibers in the gut and pancreas of the rat displaying neuropeptide immunoreactivity. Intrinsic and extrinsic origin. *Cell Tissue Res.* **230**:487-493.

Sutherland, S. D. (1967) The neurons of the gall bladder and gut. *J.Anat. (Lond.)* **10**:701-709.

Szurszewski, J. H., and W. A. Weems (1976) A study of peripheral input to and its control by postganglionic neurons of the inferior mesenteric ganglion. *J. Physiol.* **256**:541-556.

Tapscott, S. J., G. S. Bennett, and H. Holtzer (1981a) Neuronal precursor cells in the chick neural tube express neurofilament proteins. *Nature* **292**:836-838.

Tapscott, S. J., G. S. Bennett, Y. Toyama, F. Kleinbart, and H. Holtzer (1981b) Intermediate filament proteins in the developing chick spinal cord. *Dev. Biol.* **86**:40-54.

Teitelman, G. N., T. H. Joh, and D. J. Reis (1978) Transient expression of a noradrenergic phenotype in cells of the rat embryonic gut. *Brain Res.* **158**:229-234.

Teitelman, G. N., T. H. Joh, and D. J. Reis (1981a) Transformation of catecholaminergic precursors into glucagon (A) cells in mouse embryonic pancreas. *Proc. Natl. Acad. Sci. USA* **78**:5225-5229.

Teitelman, G. N., M. D. Gershon, T. P. Rothman, T. H. Joh, and D. J. Reis (1981b) Proliferation and distribution of cells that transiently express a catecholaminergic phenotype during development in mice and rats. *Dev. Biol.* **86**:348-355.

Thiery, J. P., J. L. Duband, and A. Delouvée (1982) Pathways and mechanisms of avian trunk neural crest cell migration and localization. *Dev. Biol.* **93**:324–343.

Uyeda, C. T., L. F. Eng, and A. Bignami (1972) Immunological study of the glial fibrillary acidic protein. *Brain Res.* **37**:81–89.

Vaillant, C., A. Bullock, R. Dimaline, and G. J. Dockray (1982) Distribution and development of peptidergic nerves and gut endocrine cells in mice with congenital aganglionic colon and their normal littermates. *Gastroenterol.* **82**:291–300.

von der Mark, H., K. von der Mark, and S. Gay (1976) Study of differential collagen synthesis during development of the chick embryo by immunofluorescence. I. Preparation of collagen type I and type II specific antibodies and their application to early stages of the chick embryo. *Dev. Biol.* **48**:237–249.

Webster, W. (1973) Embryogenesis of the enteric ganglia in normal mice and in mice that develop congenital aganglionic megacolon. *J. Embryol. Exp. Morphol.* **30**:573–585.

Weston, J. A. (1982) Motile and social behavior of neural crest cells. In *Cell Behavior*, R. Bellairs, A. Curtis, and G. Dunn, eds., pp. 429–470, Cambridge Univ. Press, Cambridge.

Weston, J. A., and S. L. Butler (1966) Temporal factors affecting localization of neural crest cells in the chick embryo. *Dev. Biol.* **14**:246–266.

Weston, J. A., M. A. Derby, and J. E. Pintar (1978) Changes in the extracellular environment of neural crest cells during their early migration. *ZOON* **6**:103–113.

Wood, J. D. (1979) Congenital megacolon: Hirschsprung's disease. (1979) In *Spontaneous Animal Models of Human Diseases*, Vol. I, E. J. Andrews, B. C. Ward, and N. H. Altman, eds., pp. 29–34, Academic, New York.

Yamada, K. M. (1981) Fibronectin and other structural proteins In *Cell Biology of Extracellular Matrix*, E. D. Hay, ed., pp. 95–114, Plenum, New York.

Yen, S.-H., and K. L. Fields (1981) Antibodies to neurofilament, glial filament, and fibroblast intermediate filament proteins bind to different cell types of the nervous system. *J. Cell Biol.* **88**:115–126.

Yntema, C. L., and W. S. Hammond (1954) The origin of intrinsic ganglia of trunk viscera from vagal neural crest in the chick embryo. *J. Comp. Neurol.* **101**:515–541.

Chapter 11

Development of Motor Innervation in Vertebrate Limbs

MARGARET HOLLYDAY

ABSTRACT

The developmental processes underlying the formation of specific connections between spinal motoneurons and muscles of the vertebrate limb have been studied intensively in recent years. A major issue has been whether connection specificity requires specific intercellular recognition processes or is instead a simple consequence of a stereotyped pattern of cell production and axon outgrowth. The results of a variety of experimental perturbations clearly favor the existence of recognition processes. Motoneurons belonging to different motor pools differ in how their axonal growth cones respond to limb tissue. These differences are expressed during initial axon outgrowth and not through a subsequent selection mechanism such as cell death.

At least three factors influence the outgrowth process: nonspecific growth cues, specific growth cues, and axonal interactions. The existence of nonspecific growth cues is indicated by the ability of either foreign or normal axons to form typical limb-specific branching patterns. In addition, both the spinal segments contributing to a limb plexus and the gross patterning of nerves distal to the plexus are constrained by nonspecific growth cues, probably mechanical properties of the somites and limb tissue. It is important to recognize these constraints on axonal outgrowth, especially when interpreting motoneuron projections in experimentally perturbed situations.

Motor axons also respond to specific growth cues, which probably are situated at several discrete loci or choice points within the limb. Observations of axonal trajectories and changes in neighbor relationships in normal embryos provide indirect evidence for the existence of sequential choice points. The best experimental evidence for guidance cues at a choice point involves the ability of axons to project appropriately to either dorsal or ventral muscle tissue after experimental manipulations of the limb at or proximal to the plexus, the first choice point that axons encounter when growing to the limb.

Axonal interactions also influence projection patterns. In the absence of neighboring axons, some axons expand their projections into foreign nerves, either at the plexus or farther distally along common nerve pathways; the majority of the axons present appear to project along their normal pathway. Hence, while the process of limb innervation appears dominated by axons responding to specific growth cues, axonal interactions may also play a role in normal development, perhaps in fine-tuning the projection patterns, especially along common nerve pathways.

These and ongoing studies of the organizational principles underlying axonal guidance are bringing us close to the point where it should be feasible to characterize growth cues at cellular and molecular levels.

A major challenge in studies of nervous system development is understanding how neurons establish highly patterned connections with target cells, either other neurons or such peripheral tissues as muscle or skin. Of possible explanations, one extreme model is Sperry's chemoaffinity hypothesis (Sperry, 1963), which assumes significant differentiation of neuron types in terms of the presence of distinctive molecules on the surfaces of the cells and/or their growth cones. In such models, neurons form selective contacts with the appropriate target cells because they recognize or preferentially bind to similarly distinctive molecules on the surfaces of the targets. Such models also presume the existence of distinctive molecules along the pathway through which axons grow. At the other extreme of models to explain connectivity patterns are those that attribute to neurons no intrinsic differentiating characteristics; the only relevant factors are presumed to be position and neuronal birth order. It is imagined that a carefully orchestrated differentiation sequence of physical properties of the environment into which the axons grow could yield an appropriately ordered projection pattern (Horder, 1978). These models explicitly exclude any active recognition or selective adhesion between growth cones and cues, either along the pathway or at the postsynaptic target.

The vertebrate limb has proved to be a useful system in which to study the developmental mechanisms involved in establishing patterns of neuronal connections. Each muscle in the adult limb is innervated by a group of motoneurons whose cell bodies tend to be clustered together in well-defined positions within the ventral horn of the spinal cord. This group is called the motor pool. The motoneurons for limb muscles are present in enlargements of the ventral horn called the lateral motor columns. It is possible to identify the motor pool for individual muscles by retrograde transport of intramuscularly injected horseradish peroxidase (HRP). Such experiments have now been conducted in a variety of species (*Xenopus:* Lamb, 1976. Chick: Landmesser, 1978; Hollyday, 1980; Straznicky and Tay, 1983. Rat: Rootman et al., 1981; Smith and Hollyday, 1983), in most cases corroborating and extending previous data on motor-pool position based on retrograde degeneration or chromatolysis after nerve injury (Goering, 1928; Elliott, 1942, 1944; Romanes, 1946, 1951, 1964; Sprague, 1951; Sharrard, 1955; Cruce, 1974).

One of the most important generalizations to emerge from these studies on the organization of motor pools in the adult animal is that motor-pool position is perfectly correlated with the embryonic position of the muscles. Neither adult position nor function is as well correlated with the position of motor pools. The limb muscles of all vertebrates are formed from two sheets of muscle precursor tissue, a dorsal and a ventral premuscle mass that surrounds the centrally condensing skeletal elements of the limb. All muscles are derived exclusively from one of these two masses. Muscles derived from the ventral mass are innervated by medially or dorsomedially positioned motoneurons. Dorsally derived muscles are innervated by laterally or ventrolaterally positioned motoneurons. These consistent relationships suggest that the underlying developmental processes are shared by all vertebrate species.

In the chick, muscles formed from neighboring portions of each dividing muscle mass are innervated by adjacent motor pools (Hollyday, 1980). This arrangement would be consistent with the possibility that, within each premuscle mass, the normal orderly innervation pattern is achieved by growing axons simply maintaining the same relative neighbor relationships exhibited by their cell bodies in the lateral motor column. There are at least two ways in which axons could maintain neighbor relations during growth to the periphery. Axons might actively recognize axons from neighboring cells because of homophilic affinity, or they might be passively guided to neighboring regions of the muscle masses by a network of parallel channels or other mechanical guides (Horder, 1978; Singer et al., 1979). An additional feature of the organization of motor pools that is less easy to understand is that neighboring limb muscles derived from *different* muscle masses are *not* innervated by adjacent motor pools. To account for this, nonrecognition models would have to presume one of the following: (1) A difference in the time of axonal growth to the dorsal and ventral muscle masses; or (2) a segregation of axons to dorsal and ventral masses to different regions of the spinal nerve, perhaps as far proximally as the site of axon emergence from the spinal cord. The alternative is to admit some form of active recognition of the dividing point between dorsal and ventral axon pathways at the position where the spinal nerves come together and form the limb plexus. As discussed in the following, it seems most likely that recognition processes are critical not only in accounting for the disposition of motor axons to dorsal and ventral premuscle masses, but also for most other aspects of nerve patterning. In this chapter I attempt to document this general point and to review what is and is not known about the organization of such recognition processes.

MOTONEURONS IN DIFFERENT MOTOR POOLS ARE DIFFERENTIATED

Several types of experiments have shown that motoneurons belonging to different motor pools are, in fact, different from one another in a way that fits Sperry's chemoaffinity hypothesis, although no data are yet available on the inferred molecular basis for these differences. This conclusion was reached by designing experiments to test the alternative hypotheses and eliminating them, based on the results of such experiments. A summary of these experiments is described below.

Lance-Jones and Landmesser (1980b) tested to see whether the initial position of outgrowing axons determined their targets by rotating a few segments of spinal cord about the rostrocaudal axis. These rotations were made after neural tube closure, at the time when the firstborn motoneurons were completing their terminal mitoses and before any axons had emerged from the ventral root. Axonal projections to the limb were studied electrophysiologically and by orthograde HRP transport at stages after the mature innervation pattern had formed. Lance-Jones and Landmesser found that the relative anteroposterior position in which axons arrived at the plexus did not determine the distal projections of those axons. Rather, axons would alter their growth trajectory to project to their normal targets.

A second indication of intrinsic motoneuron differentiation comes from experiments involving rearrangement of the relative positions of muscle precursors within the limb rather than from manipulations of spinal cord position (Whitelaw

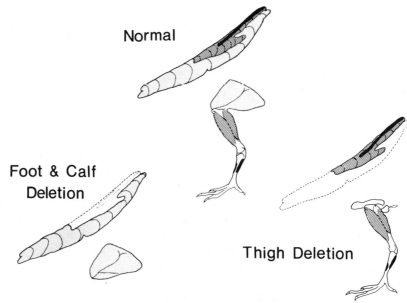

Figure 1. *Summary diagram illustrating positions of motor pools and the groups of muscles they innervate in normal chicks and in experimental animals with partial hindlimbs.* In all pictures, thigh muscles and motoneurons are labeled with stippling, calf muscles and motoneurons with diagonal lines, intrinsic foot muscles and motoneurons with black. Bottom left: After embryonic surgery resulting in the absence of segments distal to the knee, the remaining thigh muscles were innervated by thigh motoneurons; calf and intrinsic foot motoneurons died. Bottom right: After surgery resulting in absence of the thigh muscles, the remaining calf and intrinsic foot muscles were appropriately innervated; thigh motoneurons died. These results indicate that motoneurons do not innervate muscles based on the order of axon ingrowth and the proximity of uninnervated tissue, but rather that they can discriminate among various limb tissues to innervate their appropriate targets. (Data from Hollyday, 1980; Whitelaw and Hollyday, 1983a.)

and Hollyday, 1983a; see Figure 1). The rationale for this manipulation is similar to the spinal cord rotation experiment, except that the possible significance of both axon position and order of arrival could be tested simultaneously. A set of preliminary experiments showed that surgical manipulations of the limb bud did not alter the birthdates or the settling patterns of motoneurons within the lateral motor columns. Hence, the timing of arrival ingrowth was likewise not perturbed. We reasoned that if motoneurons did not differ from one another except by position in the lateral motor column, and cell body position determines the growth trajectory within the limb, perhaps by the sequence of arrival at different muscles, then by rearranging the relative positions of limb muscles it should be possible to alter the innervation pattern in a predictable way. If, for example, axons were making connections based on order of axon ingrowth and the proximity of uninnervated targets, one would expect motoneurons that normally innervate thigh muscles to innervate whatever set of muscles was closest to the spinal cord. This should have been true, even when thigh muscles were absent and when only calf and foot muscles were present in a surgically created partial limb. Contrary to this prediction, thigh motoneurons did not innervate calf muscles in a thighless limb; calf motoneurons did. What happened to the thigh motoneurons? They died. When the innervation of hindlimbs amputated at the knee was studied, analogous results were obtained. The thigh muscles were innervated by motoneurons from the normal motor pools, and calf and intrinsic foot motoneurons were absent from the lateral motor columns. In neither situation did the entire lateral motor column project to a partial limb; instead, the remaining muscles were selectively innervated by their normal motor pools. It is not possible to attribute these findings to developmental processes such as the time of axonal arrival and the proximity of undifferentiated targets. Rather, one is led to the conclusion that motoneurons in different motor pools must differ from one another in characteristic ways. One is forced to conclude that limb tissue is also differentiated in such a way as to permit motor axons from different pools to innervate their correct target.

While significant for establishing the existence of intrinsic differences between motoneurons, these sorts of experimental manipulations do not allow us to specify the nature of the differences. It is not clear, for example, whether each motor pool has a unique identity or specifier, presumably in the form of a chemical label, or whether motor pools might share certain specifiers in a way that is analogous to the sequence of pathway choices made by axons as they grow from the spinal cord into the limb.

WHEN ARE MOTONEURON DIFFERENCES EXPRESSED—DURING INITIAL AXONAL OUTGROWTH OR VIA A SUBSEQUENT SELECTION MECHANISM SUCH AS CELL DEATH?

Differences among motoneurons might yield the adult pattern either because they influence the process of axonal growth or because they act at or after the time of synapse formation to adjust an initial pattern. If either of the latter, there should be a mechanism to select which of the possible synaptic contacts will be made or stabilized. In the case of the vertebrate limb, an obvious candidate for a selection mechanism is selective cell death. It is known that at least 40% of the motoneurons initially present within the lateral motor column die during the time of initial synapse formation (Harris-Flannagan, 1969; Hamburger, 1975; Oppenheim and Majors-Willard, 1978; Lance-Jones, 1982). In the chick, at least 95% of that initial population can be labeled by a peripheral injection of HRP, indicating that those cells have sent an axon into the limb bud (Oppenheim and Chu-Wang, 1977). The critical question is the degree to which the pattern resulting from initial axonal growth differs from that observed after synapse formation and cell death.

This issue has been examined in the axolotl, frog, chick, and rat, and somewhat differing results have been obtained in the four species. In the chick, Lance-Jones and Landmesser (1981a) and Hollyday (1983a) have used orthograde transport of HRP injected into various regions of the spinal cord to map the projections of motoneurons whose axons exit via identified spinal segments. The labeled axons are solidly filled, as in Golgi- or silver-stained preparations, and can be followed with certainty in serial sections. This method has the technical advantage over retrograde labeling studies in that both the cell bodies and axons are solidly filled, and HRP leakage to unintended locations should not go undetected. The disadvantage of this technique is that it may be biased toward finding selective outgrowth. One needs to be certain that all or a large majority of the fibers are labeled so that any existing aberrantly projecting axons do not escape detection. As currently applied, this technique has been limited to analysis of projections that vary along the rostrocaudal axis of the spinal cord. Because at any rostrocaudal level there are normally several different motor pools present that are expected to project to a subset of the entire pattern, it is not possible to detect projection errors within this subset. This leaves, however, a number of peripheral nerves that would not be expected to have labeled axons after injections into particular segments, because the relevant motor pools are located only in uninjected segments. These are the sorts of errors that one should detect if they are present.

Using the method of orthograde HRP transport, both Lance-Jones and Landmesser (1981a) and Hollyday (1983a) have found that by the time recognizable nerves have formed, axons emerging from a particular spinal segment project to the subset of nerves in the limb that one would expect, given the known segmental locations of various motor pools in the adult. The experiments have been done in both the hindlimb and the wing before the onset of the major period of cell death, and no significant differences have been described. While it remains possible that motor axons explore some range of their environment as they grow, and that a small population of aberrantly projecting axons remains undetected by this technique, there is no evidence that large numbers of incorrectly projecting remnants of such explorations become incorporated into the definitive nerve branches of the pre-cell-death embryo of the chick. Although it has not yet been technically possible to study the accuracy of initial projections of motoneurons belonging to a single motor pool, all of the available evidence points to a high degree of accuracy, at least at the level of a single spinal segment. If projection errors are commonly made that can account for the loss of at least 40% of all motoneurons produced, they must be limited to muscles that normally received innervation from the segment in question. Resolution of this issue will require techniques with greater precision than those now employed to investigate this question.

These findings contrast with those reported by Pettigrew et al. (1979) in the chick wing and by Nurcombe et al. (1981) in the rat. In both of these studies, HRP was injected into muscle or muscle precursor tissue and the location of labeled motoneurons plotted at different stages of development. Comparison of the distributions of labeled cells from animals of different ages showed a considerable reduction in the numbers of labeled cells from all segments in older animals, as well as the complete loss of labeled neurons from segments both rostral and caudal to the post-cell-death motor-pool position. It seems likely that the differences between these findings and those described above can be best explained by technical artifacts. As shown by Rootman et al. (1981), it is difficult to prevent HRP leakage, especially in undifferentiated muscle precursor tissue, which is where evidence for large numbers of erroneous projections has been obtained.

Present evidence suggests that the precision of initial motor outgrowth in amphibian species may be less than that in chicks and mammals. In both *Xenopus* and the axolotl, evidence for a change in the projections of certain spinal segments to regions of the developing limb has been obtained. At the earliest stages of axonal ingrowth to the *Xenopus* hindlimb bud, Lamb (1976) found that when he injected a particular region of the presumptive thigh muscle, approximately one-third of the motoneurons labeled with HRP were located in the medialmost third of the lateral motor column. At later stages, and after motoneuron death, only the lateral and central two-thirds of the lateral motor column contained HRP-labeled cells when this particular region of the thigh was injected. These results suggest that some of the motoneurons originally projecting to the hindlimb were removed by cell death. In this case, the possibility that early widespread labeling was an artifact related to leakage of HRP from the injection site seems to have been excluded by the finding that, subsequent to cell death, the abnormally located labeled cells that resulted from an early injection disappeared, whereas normally located labeled cells remained.

As both synaptic endings, as well as growth cones, can pick up HRP and transport it back to the cell body, these anatomical experiments cannot tell us whether the inappropriately projecting synapses actually made synapses. They do suggest, however, that axonal growth is not so tightly controlled as to prevent axons from growing into muscle regions where they will not eventually terminate. Evidence for elimination of incorrect synapses during initial limb innervation has been described in the axolotl. McGrath and Bennett (1979) reported that before the flexor (ventral) muscle sheet cleaved to form the individual muscles of the limb, muscle fibers all across the sheet were innervated by two spinal nerves in approximately equal percentages. In one portion of the muscle sheet examined, nearly half of the muscle cells impaled with recording microelectrodes received synaptic input from one spinal nerve that normally never supplies the fibers found in that region in the adult. Approximately half of the muscle fibers receiving innervation from the unexpected nerve were dually innervated from the expected nerve as well. After the individual muscles separated from the flexor sheet, the percentages of muscle fibers innervated by each of the nerves examined closely approximated the percentages found in the adult. The explanation for the observed changes is not clear because embryonic cell death has not been described for axolotl motoneurons, probably because of the difficulty in unequivocally identifying them by use of conventional Nissl stains. Electrical coupling between muscle fibers or artifactual current spread also has not been discounted in this system.

It is conceivable that motor innervation plays a generalized trophic role in limb differentiation and growth. If such a role exists, it is possible that the smaller numbers of motoneurons present in the amphibian at any given time during limb innervation may have to be more widely distributed to yield the required trophic effect.

In conclusion, while there may be significant species differences in the precision of initial motor innervation, and hence in the time when intrinsic motoneuron differences are expressed, it seems clear that, at least in the chick, such differences are expressed at the time of initial axonal ingrowth to the limb. This implies that differentiated motor axons normally respond to growth cues present within the limb bud at that time, rather than distributing themselves nonspecifically to a variety of muscles only to be eliminated if the match is inappropriate. The rest of this chapter focuses on our current understanding of the nature of the growth cues used by axons to form their normal stereotyped pattern of projections to limb muscles.

NONSPECIFIC GROWTH CUES

It has been recognized for many years that the gross pattern of nerve branches is determined by limb tissue rather than by properties of the innervating fibers (Harrison, 1907, 1935; Detwiler, 1936; Piatt, 1956, 1957; Stirling and Summerbell, 1977; Morris, 1978; Hollyday, 1981). Axons from both appropriate and inappropriate spinal segments can form the same branching patterns, so it is likely that at least some of the growth cues in the limb can be recognized by all classes of axons. I consider some nonspecific growth cues in this section. The character of the cues necessary for establishing the specific match between motor pool and muscle are considered in the next section.

Cues for Plexus Formation

The most proximal element of the gross nerve pattern is the limb plexus. This is where axons emerging from several spinal segments converge and intermingle before forming the various peripheral nerves that innervate specific muscles or groups of muscles. Available evidence suggests the site of plexus formation is correlated with the position of the limb girdle relative to the adjacent spinal segments (Hamburger, 1939; Morris, 1978; Hollyday, 1981; Lance-Jones and Landmesser, 1981b). In experimentally manipulated limbs, where the girdle has been shifted along the rostrocaudal axis of the spinal cord or rotated about its anteroposterior axis, the adjacent spinal segments are commonly observed to form a plexus in a characteristic position relative to the rotated or shifted limb girdle.

Axons have not been observed avoiding an adjacent but inappropriate plexus to join their normal but distant plexus if the limb has been rotated or shifted by more than two or three segments. These observations indicate that the essential cues for plexus formation are not a property of the somitic tissue in which the individual spinal nerves are found, but appear to be associated with the limb-specific tissues where the spinal nerves form the plexus instead.

Are the cues responsible for leading axons to the plexus-forming region analogous to tunnels or mechanical guides, or does the point of axonal convergence reflect an active axonal response to growth cues in the environment? The best evidence addressing this issue comes from experimental manipulation of the hindlimb. In contrast to the forelimb (or wing) which is supplied by a single plexus, the hindlimb is normally innervated via two plexuses, the crural and the sciatic. Axons from segments 23 to 25 (and occasionally from 22) form the crural plexus, whereas axons from segments 25 to 29 or 30 form the sciatic plexus. The contribution of segment 25 ("bifurcalis") to both plexuses is characteristic of hindlimb innervation. Nerves emerging from the crural plexus supply muscles of the anterior thigh; the posterior thigh, calf, and intrinsic foot muscles are innervated from the sciatic plexus.

Hindlimb-bud shifts or rotations, and grafts of supernumerary hindlimbs, consistently result in a redistribution of the adjacent spinal nerves into each plexus. After a supernumerary limb is added, fewer than normal numbers of lumbar segments are observed forming a plexus, and segmental nerves other than nerve 25 often divide to join both the crural and sciatic plexuses, even though they never do so in the normal situation. These latter observations, in particular, suggest that the decision as to which plexus to join need not represent an active response to specific growth cues, but may instead simply reflect the simple proximity of pathways leading to each plexus-forming region. None of the experimental observations to date require the conclusion that axons respond actively to cues proximal to the normal plexus-forming region.

Are the cues which are essential for plexus formation specific properties of the limb girdle tissue, or are they properties of limb tissue more generally? A preliminary answer to this question is provided by observations on plexus formation in partial hindlimbs that lack proximal limb tissues (Whitelaw and Hollyday, 1983a; M. Hollyday, unpublished observations). Normal crural and sciatic plexuses have been observed despite the apparent absence of precartilage condensations associated with the hindlimb girdle. Although these observations cannot exclude the involvement of other proximal tissues in plexus formation, they suggest that once having reached limb premuscle tissue, axons respond to specific growth cues which result in normal plexus patterns despite the absence of girdle tissues. Hence, there may be no obligatory role for the limb-girdle tissues in providing essential growth cues to the limb-innervating axons. The previously observed correlations between limb-girdle position and plexus formation may actually reflect a combination of developmental events, the tendency of the segmental nerves to grow peripherally without much exploration along the rostrocaudal axis of the embryo, and the presence of specific growth cues in the proximal limb tissues distal to the girdle elements. When combined, stereotyped limb plexuses result despite the absence of growth cues specific for plexus formation.

This explanation of plexus formation is consistent with the observed sequence of limb innervation and differentiation of various limb tissues, and of the growth behavior of the population of axons from a single spinal segment. At the time of plexus formation, the only morphological sign of tissue differentiation within the limb is the precartilage condensation of the femur or humerus and the incipient predorsal and preventral muscle masses on either side of the future skeletal element. The precartilage condensations of the girdle elements become morphologically distinct only at stage 28 or later (stages according to Hamburger and Hamilton, 1951), well after the plexuses and individual distal nerve branches have formed (Romer, 1927).

Some additional observations imply that the nonspecific cues leading axons to one plexus-forming region or another may be separated by as little as 250 μm. These observations also suggest an explanation for why axons are not observed growing long distances to join their appropriate plexus region if experimentally displaced. Motor axons emerge from the neural tube and collect at the end of the dermomyotome from stage 17 to stage 20 (Hollyday, 1983a, and unpublished observations). The population of axons comprising the different spinal nerves grow ventral to the dermomyotome at stages 21–22, separated by approximately 250 μm center-to-center or by only 100–150 μm from the posterior border of one nerve to the anterior border of the adjacent one. Before converging at the limb plexus-forming region, axons from a single segment do not overlap with those from neighboring segments; the population of growth cones at the end of the nerve normally extends in a radius of 150 μm or less. Thus, the normal range of exploration proximal to the plexus is insufficient to detect the presence of cues leading to another plexus-forming region.

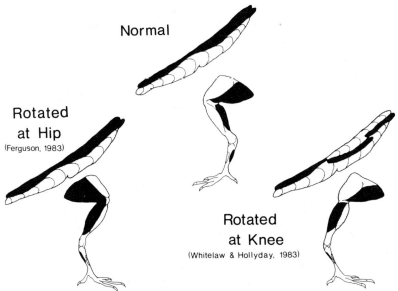

Figure 2. *Summary diagram of motoneuron position in relation to muscle position in normal chicks and after two types of limb rotations.* Top: Normal chick. Limb muscles derived from the ventral premuscle mass are innervated by medially positioned motoneurons (black). Limb muscles derived from the dorsal premuscle mass are innervated by laterally positioned motoneurons (stippled). (Data from Romer, 1927; Hollyday, 1980.) Bottom left: After rotation of the entire hindlimb about its dorsoventral axis, dorsally and ventrally derived muscles continue to be innervated by the appropriate motoneurons as a consequence of axons selecting their normal pathways distal to the limb plexus. (Data from Ferguson, 1983.) Bottom right: After dorsoventral rotation of the hindlimb at the level of the knee, muscles distal to the rotation are innervated by motoneurons that normally innervate muscles derived from the opposite muscle mass. Once having selected pathways to either dorsal or ventral tissue at the limb plexus, the first choice point in the pathway, axons are constrained by the mechanical properties of the limb tissue to innervate the subset of either dorsally or ventrally derived limb muscles. (Data from Whitelaw and Hollyday, 1983c.)

Axons converge at the site of plexus formation at stages 23–24. At the point of convergence, axons from a single segment spread out within a region approximately 300 μm along the rostrocaudal axis. Whether this increased range of potential axonal explorations reflects the removal of a mechanical barrier or the presence of a different growth-promoting substrate is not at all clear. Within the plexus-forming region, axons from different segments overlap with each other; in general, the degree of overlap is such that growth cones from one segment extend as far rostral or caudal as those from the next neighboring segment. This enlarged range of growth-cone exploration from a single nerve may be sufficient to explain the ability of a reduced number of spinal segments to form the full complement of peripheral nerves emerging distal to the plexus. This property is likely to be essential for understanding the abilities of motor axons to selectively innervate their appropriate muscles after reversals of a few spinal cord segments (Lance-Jones and Landmesser, 1980b) and to expand into inappropriate pathways in the absence of the normal axonal population in the plexus (Hollyday, 1983b).

Growth Cues for Peripheral Nerves

Once having entered the limb via a particular plexus, the subsequent pathway choices appear to be restricted by mechanical properties of the limb tissue. With the exception of a few muscles in the proximal thigh that can receive innervation from either the crural or sciatic plexus, most muscles receive innervation from axons that follow stereotyped pathways distal to the plexus. At the plexus, axons diverge from one another to form nerves leading to either dorsal or ventral muscle mass tissue. Dorsoventral rotation of the limb tissue distal to the plexus yields systematic patterns of misinnervation of distal muscles (Whitelaw and Hollyday, 1983c; see Figure 2). Ventrally derived muscles are then innervated by laterally positioned motoneurons that normally supply dorsal muscles; the dorsally derived muscles are innervated by medially situated motoneurons. These results suggest that once axons grow into nerves leading them to either dorsal or ventral muscles, they are constrained to innervate muscles derived from that premuscle mass exclusively. Furthermore, irrespective of the appropriateness or inappropriateness of the axons involved, they tend to form a complete pattern of peripheral nerves appropriate for the available muscles. These observations indicate that there are nonspecific growth cues within the peripheral limb tissues that both restrict axons to stereotyped nerve pathways and permit such growth by any class of motor axon. While nonspecific growth cues are clearly present, they cannot, given the observations described earlier, account for the development of the normal projection pattern. They do, however, play a role. Recognition of their presence is additionally important for the interpretation of other experimental results as discussed further below.

SPECIFIC GROWTH CUES

Growth Cues Are Not Chemotrophic Agents Released from Distant Targets

One general model of specific growth cues involves release from the target tissues of attracting substances, perhaps analogous to nerve growth factor, which diffuse along the prospective nerve pathways and guide axons to their appropriate muscles. One need only admit the possibility that axons can respond to more than one attracting substance to account for experimental findings of growth of inappropriate axons along particular pathways. In the normal situation, however, axons would respond to their preferred attractant, producing the precise pattern of muscle innervation observed.

Observations on the innervation patterns of partial hindlimbs make it appear extremely unlikely that specific growth-promoting cues depend on the presence of the targets themselves. In partial hindlimbs amputated at the calf by removal of the distal tip of the limb bud before axonal ingrowth to the limb, the nerves leading to the calf and intrinsic foot muscles form normally. Axons within these nerves grow through the thigh along their normal pathway until they reach the level of amputation (Whitelaw and Hollyday, 1983a). The vast majority of the axons appeared to dead-end just beneath the dermal layer of the healed stump; a few axons were seen projecting to distal thigh muscles. Unless one were to

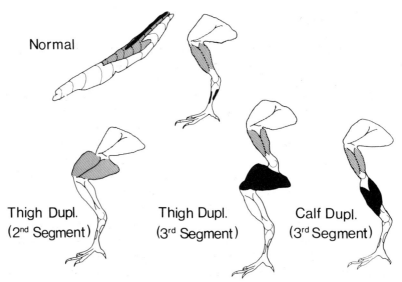

Figure 3. *Summary diagram illustrating motor pool position and the muscles innervated in normal chicks and in chicks having serially duplicated limb segments.* Top: Normal chick. Thigh muscles and motoneurons labeled with stippling, calf muscles and motoneurons with diagonal lines, intrinsic foot muscles and motoneurons with black. Bottom row: The pattern of muscle innervation after embryonic surgery resulting in addition of limb segments along the proximodistal axis of the limb. The first or most proximal segment, which was always a thigh, was appropriately innervated by thigh motoneurons. The second segment, whether a second thigh (left) or a calf (center and right) was innervated by calf motoneurons. The third segment, whether a thigh (center) or a calf (left and right), was always innervated by motoneurons that normally supply intrinsic foot muscles. Muscles in segments further distal than the third were never innervated. These results indicate that the presence of additional distal muscles did not stimulate axonal growth further than normal and suggest that local growth cues are the dominant factor in axonal guidance.

postulate that under these experimentally perturbed conditions the calf-specific attracting cues were then produced by distal thigh tissue, a possibility we cannot exclude but think unlikely, these observations suggest that the growth cues are local ones, independent of the presence or absence of distant target tissues. It appears that the axonal growth responsible for establishing the nerve pattern is more analogous to that of a bloodhound following local scents along the ground than it is to a bird dog spotting prey at a distance and pointing a hunter in the appropriate direction.

Additional data consistent with the view that the cues affecting axonal growth are local properties of the substrate, not chemical attractants released from a distant target, are results from experiments on serially duplicated limbs (Whitelaw and Hollyday, 1983b; see Figure 3). In these experiments, surgical addition of a second thigh in place of a calf failed to stimulate growth of thigh motoneurons into the second thigh. Similarly, addition of a second calf distal to the normal calf failed to attract calf motoneurons to innervate it. Whereas negative findings are never as convincing as positive ones, these results are consistent with the view that axonal growth is controlled by local interactions between growth cones and the substrate, not by cues acting from a distance. These observations do not, however, provide information about the distribution of growth cues within the limb itself.

Growth Cues Are Located at Several Places in the Limb Bud

An unresolved question about specific growth cues is their distribution. Does all of the axonal sorting into muscle-specific axon bundles take place at or close to the limb plexus (Lance-Jones and Landmesser, 1981a), or are additional growth cues found farther distally so that the process of limb innervation involves axons making a sequence of choices at successive positions along the outgrowth pathway? Our own observations favor the latter alternative. Our studies, and those of Lance-Jones and Landmesser, consist of observations of axons solidly filled with HRP reaction product as a result of injections of the tracer into the vicinity of the motoneuronal somas or directly into the various peripheral nerves. In both studies, the positions of labeled axons relative to other labeled and unlabeled ones are interpreted as reflecting their growth trajectories. While distances between nerve branching points clearly increase as the limb grows, it seems unlikely that the relative positions of individual axons within nerve trunks change much during growth. Hence, observations of axonal trajectories in normal embryos provide several clues as to the location and distribution of the local growth cues. Places in the limb bud where axons change their direction of growth and neighbor relations with other axons are inferred to be sites of active responses to local growth cues. Lance-Jones and Landmesser's study focused primarily on thigh muscles innervated from the crural plexus and probably involved fewer labeled axons than our own, which has included tracing axonal projections to the distal limb muscles in both the calf and wing. It is likely that these two differences account for our differing interpretations.

Axons tend to maintain fixed neighbor relationships while growing along common nerve trunks and abruptly change neighbor relations at several well-defined places called choice points. The first obvious choice point along the outgrowth path is at the plexus, where axons from the segmental nerves converge and subsequently diverge toward the dorsal or ventral premuscle masses. There is no strict correlation between the relative dorsal–ventral position of an axon proximal to the plexus and the route via which it exits. At stages 23–24, when axons first grow into the plexus (Roncali, 1970; Fouvet, 1973), the distance over which they change their position relative to other axons extends over approximately 50–75 μm (M. Hollyday, unpublished observations). Some axons appear to change direction with a single turn; others appear to make one or more detours, wandering perhaps 100 μm or more before exiting from the plexus along the dorsal or ventral nerve branch. Individual axons have not been observed to have growth cones extended toward both branches, but the possibility that they exist cannot be dismissed. How these axonal trajectories correspond to the exploration distance of a single growth cone *in vivo* is not clear from observations made in fixed specimens. In our material, the maximum width of a lamellipodial sheet seen at the end of a growth cone has been approximately 30 μm, and most range from 5 to 10 μm across. Filopodial extensions visible in the light microscope extend no farther than 15–20 μm from the growth cone. Any error in these measured distances should be underestimates of the actual values, due either to the inability of the light microscope to resolve the distalmost extensions of the filopodia or to the retraction of either filopodia or lamellipodia when living cells are filled with the HRP tracer. Neither possible source of error should affect our major conclusion, however, which is that individual axons alter their growth trajectories over relatively small distances.

Once having made this first choice, axons from the same segment tend to grow together, maintaining their relative neighbor relationships until they reach additional pathway choice points located distal to the plexus. These distal choice points become visible in embryos at stages 26½–29 as the individual nerves separate from the dorsal and ventral sheets. As has been observed in the plexus region, some axons change their courses and neighbor relationships rather abruptly over distances of 50 μm or less, whereas others do this over distances extending up to 150 μm or more. The explanation for these differences is unknown. One possibility is that the growth cues are distributed in a noncontinuous or patchy array at the various choice points. Another possibility is that axons differ in their strength of response to continuously distributed cues. Until more is known about the cellular and molecular nature of the growth cues *in vivo* and the ways in which growth cones respond to them, these questions will remain unanswered. They do not affect our major conclusion, however, which is that the process of limb innervation involves a sequence of active pathway choices at several places along their growth trajectory.

How many choice points do growth cones encounter on their way to the periphery? Clearly, the answer depends on the nerve under consideration. For some motor pools, such as supracoracoideus in the brachial region and adductor in the lumbosacral region, a single pathway choice at the plexus region could suffice to lead axons directly to their target muscle. The more distal a muscle is located, the greater the number of branch points a growth cone must navigate successfully to reach its target muscle. For example, to reach the muscle adductor indicis situated in the dorsal autopod or hand of the chick wing, a growth cone need execute only a series of seven or eight choices successfully.

It will be interesting to learn whether the growth cues for accurate axonal guidance are coded in such a way as to specify only the appropriate series of choice points in the pathway or whether the entire pathway from plexus to peripheral destination is marked in a continuous fashion.

Experimental Evidence for Choice Points in the Nerve Pathways

The best experimental evidence for the presence of growth cues in the nerve pathways that serve as choice points for outgrowing axons comes from experiments in which the hindlimb bud was rotated about its dorsoventral axis (Ferguson, 1983). In these embryos, muscles were innervated from their normal motor pools despite the reversed orientation of the limb. Although the precise level of the rotation with respect to the limb girdle and plexus-forming regions was not well described, the sciatic plexus and the nerves forming distal to the plexus were normally patterned, implying that the limb was rotated proximal to or just at the level of the plexus. Assuming this to be true, the results indicate that an axon's decision to grow along either dorsal or ventral nerve pathways within the limb does not depend on the orientation of the limb bud with respect to the body, but rather can be attributed to its ability to respond to growth cues in the plexus region that direct the axon to its preferred dorsal or ventral pathway.

This interpretation is consistent with results obtained on the innervation of grafted supernumerary limbs (Hollyday, 1981). In these experiments, the limb girdles were included in the graft, and the plexuses that were formed by the various spinal nerves were correlated with the positions of the girdle elements and their relationship to the spinal cord. Both supernumerary hindlimbs and wings were grafted adjacent to the lumbar spinal cord. Although the rostrocaudal position of the motor pools supplying a given muscle varied, the consistent finding was that muscles derived from either dorsal or ventral muscle mass were innervated by lateral or medially positioned motoneurons, respectively. This is the normal pattern of motor-pool organization and implies that even when confronted with a foreign plexus region, axons recognize cues leading to the preferred dorsal or ventral pathway.

These results were originally interpreted as reflecting a "hierarchy of neuronal specificities" between motoneurons and the muscles of the limb (Hollyday et al., 1977). It now seems unlikely that a hierarchy exists in the sense of a motoneuron with an ordered series of preferences for different muscles (Lance-Jones and Landmesser, 1981b). An alternative way of interpreting the systematic misinnervation in supernumerary limbs compatible with our current understanding of the developmental process of limb innervation is that the cues in each plexus region leading to the divergence of axons for dorsal and ventral pathways are common ones that can be interpreted correctly by any lateral motor column neuron. This interpretation is also consistent with the innervation patterns after rostrocaudal rotations of a large piece of spinal cord following shifts of the limb bud along the rostrocaudal axis of the embryo. In both of these situations, ventrally derived muscles were innervated by appropriately located medial motoneurons, and dorsally derived muscles were supplied by laterally positioned ones (Lance-Jones and Landmesser, 1981b).

Experimental evidence for the existence of choice points distal to the plexus is at present inconclusive. One such test is to determine whether axons seek out appropriate nerves when the limb is rotated about its anteroposterior axis. When a wing is rotated about its dorsoventral and anteroposterior axes simultaneously, distal to the plexus, axons were observed projecting in accord with their relative anterior or posterior position in the spinal cord (Stirling and Summerbell, 1979). However, after the wing was rotated about its anteroposterior axis only, axons were observed to compensate for the rotation and project to their appropriate muscles (Stirling and Summerbell, 1983). One possible explanation for the differing findings is that if both limb axes are rotated distal to the plexus, thus forcing dorsal-preferring axons to enter ventral tissue and vice versa, axons in such abnormal pathways fail to respond to specific growth cues along the distal pathways. Their projected pattern would then be dominated by the nonspecific growth cues in the limb tissue discussed previously. A satisfactory resolution of these issues awaits a more complete description of the level of the rotations in relationship to the nerves being studied and the position(s) along the nerve pathway at which axons alter their trajectory. We think it likely, however, that a careful study of axonal trajectories in rotated limbs will provide experimental evidence for choice points located distal to the plexus.

Proximal Pathway Choices Constrain Subsequent Distal Ones

Once having selected a dorsal or ventral pathway into the limb, an axon's choice of subsequent pathways is restricted to those normally accessed from the two major subdivisions of each plexus. Rotations of the hindlimb about the dorsoventral axis at the level of the knee produced systematic patterns of misinnervation of the calf and intrinsic foot muscles (Whitelaw and Hollyday, 1983c). Ventrally derived muscles situated adjacent to dorsal thigh muscles as a result of the rotation were innervated by motoneurons that normally would supply dorsal muscles. The pathways leading to the muscles in the rotated segments were entirely normal proximal to the rotation; once axons entered the rotated limb segments, they formed nerve branching patterns appropriate for the foreign muscles they encountered. Hence, unlike the findings following rotations of either the spinal cord (Lance-Jones and Landmesser, 1981b) or the limb tissue at or proximal to the limb plexus region (Ferguson, 1983), axons did not alter their growth trajectory to reach their appropriate muscles. Rather, axons grew as if they were constrained in their choices of distal nerve pathways as a result of their previous choice at the plexus (see Figure 2).

This interpretation can also explain the apparently conflicting results of Summerbell and Stirling (1981) and Ferguson (1983). The first investigators rotated wing buds about their dorsoventral axes at the level of the shoulder or upper arm and found that, in the majority of cases, dorsally and ventrally derived muscles situated distal to the rotation were innervated by inappropriate motoneurons. These findings are similar to our own after dorsoventral rotations at the knee (Whitelaw and Hollyday, 1983c). Although Summerbell and Stirling did not explicitly state the level of their rotations with respect to the brachial plexus, their description of their experimental embryos indicates that it was distal to where the axons diverged to form dorsal and ventral nerve pathways, that is, the plexus region. Their findings, like our own, then indicate that pathway choices are progressively restricted as axons grow into the limb. While this may not be surprising, it allows for the possibility that the amount of information needed to guide axons to their muscles is considerably less than the number of muscles in the limb, because only the behavior at branch points along the pathway need be controlled. One implication of this hypothesis is that motor pools whose axons traverse common pathways within the limb would share both some specifying labels and responses to the growth cues. Other labels and cues would be expected to be unique in order to account for their divergence at the branch points along the major nerve pathways or at the target muscles themselves.

This view of axonal pathfinding in the limb bud is also consistent with observations on the innervation of limbs in which additional limb tissue has been added along the proximodistal axis of the limb (Whitelaw and Hollyday, 1983b). In these experiments, a normal thigh was the limb segment most proximal to the spinal cord. When a second thigh was grafted in place of the normal calf, its muscles were innervated by motoneurons that would normally supply calf muscles; the first, or most proximal, thigh was innervated by normal thigh motor pools. When a second calf was grafted in the position of the intrinsic foot muscles, its muscles were innervated by motoneurons that normally supply intrinsic foot muscles. Both of the more proximal limb segments, the thigh and the first calf, were normally innervated. The generalization that emerges from these experiments is that muscles are innervated in accordance with their position along the proximodistal axis. These results further support a view of axonal pathfinding in which the choices made proximally affect the innervation patterns of the distal limb tissue. Once having responded to proximal growth cues, some portion of the axonal population is no longer available for distal innervation.

The sequential pattern of innervation along the proximodistal axis also warrants comment. We think it unlikely that this pattern reflects an intrinsic program to grow to a specific distance and then to stop. Were this the case, we should have found that the limb segment most proximal to the spinal cord was always innervated by thigh motoneurons. We found instead that when a calf replaced a thigh as the most proximal limb segment, this proximally located calf was appropriately innervated by calf motoneurons and not by thigh motoneurons (Whitelaw and Hollyday, 1983a). It seems more likely that the results of both the deletion and the duplication experiments depend on motor axons recognizing and responding specifically to a preferred set of growth cues. However, when the preferred set is absent, the presence or absence of other axons influences the behavior of the remaining ones.

Aberrant Nerve Pathways and Local Specific Growth Cues

The bloodhound model of axonal pathfinding seems to account for most, but not all, of the published experimental observations on limb innervation. The one potential difficulty for this interpretation is that after adding supernumerary hindlimbs or shifting hindlimbs along the rostrocaudal axis of the spinal cord, occasional axons have been observed taking abnormal routes from the wrong plexus to reach muscles in the proximal thigh. In some of the cases described, it has been shown directly—or it is highly likely—that the axons taking these aberrant or cross-country pathways were from the appropriate motor pool; in others, foreign axons were observed taking unusual routes in the limb. In none of the experiments did large numbers of axons appear to be involved; the nerves were consistently smaller in size than their normal counterparts, but their presence was unequivocal.

One possible explanation for these findings is that attractant chemical cues act over relatively short distances only and could therefore stimulate axonal growth over unusual terrain in these particular circumstances. In normal development they might play a role in fine-tuning the pathfinding abilities of axons along major nerve pathways normally shared by axons from several motor pools. An alternative explanation for the observed aberrant pathways takes into account the initial pattern of axonal ingrowth to the limb bud. When axons first grow into the limb, they spread out as two sheets of axons on opposite sides of the humerus or femur, one for dorsal premuscle tissue, the other for ventral. The anteroposterior extent of the sheets at stage 25 varies in size, depending on the plexus, but ranges between approximately 300 and 500 µm. The stereotyped pattern of peripheral nerves, characteristic of wing or hindlimb tissue, emerges from these two sheets of axons between stages 25 and 28 or 29. Within a single plexus, axons from individual segments distribute within the dorsal and ventral sheets, maintaining their relative position along the anteroposterior axis of the limb while overlapping considerably with axons from neighboring segments. The extent of overlap is such that axons from the segments forming the anterior and posterior borders of the axonal sheets spread out, reaching as far as the edge of axons extending from the opposite border. Axons entering the plexus in middle positions overlap extensively with axons from the border segments.

Is there normally any overlap of axons entering the limb via different plexuses? In the stage-25 hindlimb, ventrally growing axons from the crural plexus extend as far posteriorly as the anteriormost axons from the sciatic plexus. Dorsally growing axons from the two plexuses are separated by only 100–150 μm. Were this exploration distance expanded in certain experimental situations, such as in the absence of axons from adjacent segments or when foreign nerves entered the limb tissue, one might see the results of such explorations maintained as aberrant nerve branches.

AXONAL INTERACTIONS ALSO INFLUENCE PROJECTION PATTERNS

If axonal projections were determined solely by the specific response of growth cones to limb-specific cues in the substrate, one would expect axons to project along their preferred pathways irrespective of the presence or absence of other axons in the limb. The available data are not in agreement with regard to this issue. Lance-Jones and Landmesser (1980a) found that after removal of either segments 23 and 24, or 24 and 25, the axons in the remaining segments, either 23 or 25, failed to expand their projections to all of the muscles of the anterior thigh normally supplied by the crural plexus. We have not repeated these experiments, but instead did comparable experiments in the wing (Hollyday, 1983b). In normal embryos, segments 13–16 consistently innervate the wing; segment 17 supplies the wing in approximately half the animals we used. After removal of segments 13–15, the peripheral projections of axons from segment 16 alone, or 16 and 17 together, were traced by using orthograde transport of HRP. The nerve patterns were studied at stage 28, after individual nerves had formed but before cell death, just as in the Lance-Jones and Landmesser study. In all of the cases examined in which the intensity of staining was adequate, axons from the remaining segments were observed projecting to unexpected muscles. The numbers of axons involved appeared to be only a small fraction of the total number innervating the wing, suggesting that axons did not expand evenly to fill all available nerve pathways. The most commonly observed expanded projection was to the biceps brachii muscle, which is normally innervated from a branch of the brachialis inferior nerve trunk that supplies distal forearm muscles. The expanded projections were not, however, restricted to muscles found along major nerve trunks; a few of the expanded projections were to proximal nerves emerging directly from the wing plexus. These results suggest that axons might increase their exploration range at a number of places along the nerve pathways (at both the plexus and at branching points along common nerve pathways) in the absence of neighboring axons.

The apparent size of nerve 16 or nerves 16 and 17, and also presumably the number of axons in the nerves, varied in different cases and was generally correlated with the number of unexpected nerves in which labeled axons were found. In cases in which segment 17 alone innervated the wing, even fewer unexpected nerves had unlabeled axons. These observations suggest that the number of axons within a nerve pathway affects either the degree to which axons explore their environment during initial outgrowth, or the withdrawal of inappropriately projecting axons.

The innervation patterns observed after a few spinal cord segments were deleted did not, however, display any tendency for the remaining axons to project to limb muscles following a proximal-to-distal sequence. The motor pools in segments 16 and 17 normally project to muscles in the distal wing; a number of more proximal muscles are normally supplied by motor pools in segments 13–15 (Straznicky and Tay, 1983; M. Hollyday and R. D. Jacobson, unpublished observations). In a number of our experimental cases, no labeled axons were seen to project to some of the proximal limb muscles, whereas the nerves projecting to distal segments contained many labeled axons.

These results lead us to suggest that, although the dominant determinant of axonal growth to the limb is the recognition of specific, preferred, growth cues, this is not the only determinant. Interactions between axons also seem to be significant. The most commonly observed type of expanded projection was to muscles along major nerve trunks leading to more than one distal target, so it seems likely that axonal interactions normally play a role in fine-tuning the projection patterns of axons traveling along common nerve pathways. Growth-cue recognition may be the dominant influence for guiding axons to the major nerve trunks within the limb. The more subtle aspects of axonal sorting appear to depend on axonal interactions, either along the growth pathways or at the target itself.

CONCLUSION

It no longer seems possible to entertain the idea that limb innervation is a simple consequence of highly ordered spatiotemporal patterns of axonal outgrowth. Motoneurons must differ in some intrinsic way that causes them to respond differently during the outgrowth process to cues in the limb. This conclusion seems well established. More recent work has turned to the character of the processes involved in axon guidance. The observations are not consistent with guidance by cues from the ultimate target sensed at a distance. They instead suggest responses to more local cues, probably distributed at a number of perhaps discrete loci in the developing limb bud. These observations provide the basis for further characterization of the organization of the processes underlying axonal guidance, as well as for efforts to characterize the molecular bases of such processes.

ACKNOWLEDGMENTS

The work cited from the author's laboratory was supported by grants from the Block Fund, the Spencer Foundation, and the National Institutes of Health (NS-14066). I gratefully acknowledge the helpful comments of Paul Grobstein during the preparation of this manuscript.

REFERENCES

Cruce, W. L. R. (1974) The anatomical organization of hindlimb motoneurons in the lumbar spinal cord of the frog, *Rana catesbiana*. *J. Comp. Neurol.* **153**:59–76.

Detwiler, S. R. (1936) *Neuroembryology: An Experimental Study*, Macmillan, New York (reprint, Hafner, New York, 1964).

Elliott, H. C. (1942) Studies on the motor cells of the spinal cord. I. Distribution in the normal human cord. *Am. J. Anat.* **70**:95–117.

Elliott, H. C. (1944) Studies on the motor cells of the spinal cord. IV. Distribution in experimental animals. *J. Comp. Neurol.* **81**:97–103.

Ferguson, B. (1983) Development of motor innervation of the chick following dorsal-ventral limb bud rotations. *J. Neurosci.* **3**:1760–1772.

Fouvet, B. (1973) Innervation et morphogenèse de la patte chez l'embryo de poulet. I. Mise en place de l'innervation normale. *Arch. d'Anat. Microscop. Exp.* **62**:269–280.

Goering, J. H. (1928) An experimental analysis of the motor-cell columns in the cervical enlargement of the spinal cord in the albino rat. *J. Comp. Neurol.* **46**:125–151.

Hamburger, V. (1939) The development and innervation of transplanted limb primordia of chick embryos. *J. Exp. Zool.* **80**:347–389.

Hamburger, V. (1975) Cell death in the development of the lateral motor column of the chick embryo. *J. Comp. Neurol.* **160**:535–546.

Hamburger, V., and H. Hamilton (1951) A series of normal stages in the development of the chick embryo. *J. Morphol.* **88**:49–92.

Harris-Flannagan, A. E. (1969) Differentiation and degeneration in the motor horn of the foetal mouse. *J. Morphol.* **129**:281–306.

Harrison, R. G. (1907) Experiments in transplanting limbs and their bearing upon the problems of the development of nerves. *J. Exp. Zool.* **4**:239–281.

Harrison, R. G. (1935) On the origin and development of the nervous system: Studies by the methods of experimental embryology. *Proc. R. Soc. Lond. (Biol.)* **118**:117–165.

Hollyday, M. (1980) Organization of motor pools in chick lumbar lateral motor column. *J. Comp. Neurol.* **194**:143–170.

Hollyday, M. (1981) Rules of motor innervation in chick embryos with supernumerary limbs. *J. Comp. Neurol.* **202**:439–465.

Hollyday, M. (1983a) The development of motor innervation of chick limbs. In *Limb Development and Regeneration*, Part A, J. F. Fallon and A. I. Caplan, eds., pp. 183–194, Alan R. Liss, New York.

Hollyday, M. (1983b) Patterns of initial wing innervation in normal chick embryos and in embryos lacking some brachial spinal cord segments. *Neurosci. Abstr.* **9**:210.

Hollyday, M., V. Hamburger, and J. M. G. Farris (1977) Localization of motor neuron pools supplying identified muscles in normal and supernumerary legs of chick embryo. *Proc. Natl. Acad. Sci. USA* **74**:3582–3586.

Horder, T. J. (1978) Functional adaptability and morphogenetic opportunism, the only rules for limb development? *ZOON* **6**:181–192.

Lamb, A. H. (1976) The projection patterns of the ventral horn to the hind limb during development. *Dev. Biol.* **54**:82–99.

Lance-Jones, C. (1982) Motoneuron cell death in the developing lumbar spinal cord of the mouse. *Dev. Brain Res.* **4**:473–479.

Lance-Jones, C., and L. Landmesser (1980a) Motoneurone projection patterns in embryonic chick limbs following partial deletions of the spinal cord. *J. Physiol. (Lond.)* **302**:559–580.

Lance-Jones, C., and L. Landmesser (1980b) Motoneurone projection patterns in the chick hind limb following early partial reversals of the spinal cord. *J. Physiol. (Lond.)* **302**:581–602.

Lance-Jones, C., and L. Landmesser (1981a) Pathway selection by chick lumbosacral motoneurons during normal development. *Proc. R. Soc. Lond. (Biol.)* **214**:1–18.

Lance-Jones, C., and L. Landmesser (1981b) Pathway selection by embryonic chick motoneurons in an experimentally altered environment. *Proc. R. Soc. Lond. (Biol.)* **214**:19–52.

Landmesser, L. (1978) The distribution of motoneurones supplying chick hindlimb muscles. *J. Physiol. (Lond.)* **284**:371–389.

McGrath, P. A., and M. R. Bennett (1979) Development of the synaptic connections between different segmental motoneurons and striated muscles in an axolotl limb. *Dev. Biol.* **69**:133–145.

Morris, D. G. (1978) The functional motor innervation of supernumerary hind limbs in the chick embryo. *J. Neurophysiol.* **41**:1450–1465.

Nurcombe, V., P. A. McGrath, and M. R. Bennett (1981) Postnatal death of motor neurons during the development of brachial spinal cord of the rat. *Neurosci. Lett.* **27**:249–254.

Oppenheim, R. W., and I.-W. Chu-Wang (1977) Spontaneous cell death of spinal motoneurons following peripheral innervation in the chick embryo. *Brain Res.* **125**:154–160.

Oppenheim, R. W., and C. Majors-Willard (1978) Neuronal cell death in the brachial spinal cord of the chick is unrelated to the loss of polyneuronal innervation in wing muscle. *Brain Res.* **154**:148–152.

Pettigrew, A., R. Lindeman, and M. R. Bennett (1979) Development of the segmental innervation of the chick forelimb. *J. Embryol. Exp. Morphol.* **49**:115–137.

Piatt, J. (1956) Studies on the problem of nerve pattern. I. Transplantation of the forelimb primordium to ectopic sites in *Amblystoma*. *J. Exp. Zool.* **131**:173–201.

Piatt, J. (1957) Studies on the problem of nerve pattern. II. Innervation of the intact forelimb by different parts of the central nervous system in *Amblystoma*. *J. Exp. Zool.* **134**:103–125.

Romanes, G. J. (1946) Motor localization and the effects of nerve injury on the ventral horn cells of the spinal cord. *J. Anat. (Lond.)* **80**:117–131.

Romanes, G. J. (1951) The motor cell columns of the lumbo-sacral spinal cord of the cat. *J. Comp. Neurol.* **94**:313–364.

Romanes, G. J. (1964) The motor pools of the spinal cord. In *Organization of the Spinal Cord: Progress in Brain Research*, Vol. 11, J. C. Eccles and J. P. Schade, eds., pp. 93–119, Elsevier, Amsterdam.

Romer, A. S. (1927) The development of the thigh musculature of the chick. *J. Morphol. Physiol.* **43**:347–385.

Roncali, L. (1970) The brachial plexus and the wing nerve pattern during early developmental phases in chicken embryos. *Monitore Zool. Ital. N.S.* **4**:81–98.

Rootman, D. S., W. G. Tatton, and M. Hay (1981) Postnatal histogenic death of rat forelimb motoneurons. *J. Comp. Neurol.* **199**:17–27.

Sharrard, W. J. W. (1955) The distribution of permanent paralysis in the lower limb in poliomyelitis. *J. Bone Joint Surg. (Br.)* **37B**:540–558.

Singer, M., R.H. Nordlander, and M. Egar (1979) Axonal guidance during embryogenesis and regeneration in the spinal cord of the newt: The blueprint hypothesis of neuronal pathway patterning. *J. Comp. Neurol.* **185**:1–22.

Smith, C. L., and M. Hollyday (1983) The development and postnatal organization of motor nuclei in the rat thoracic spinal cord. *J. Comp. Neurol.* **220**:16–28.

Sperry, R. W. (1963) Chemoaffinity in the orderly growth of nerve fiber patterns and connections. *Proc. Natl. Acad. Sci. USA* **50**:703–710.

Sprague, J. M. (1951) Motor and propriospinal cells in the thoracic and lumbar ventral horn of the rhesus monkey. *J. Comp. Neurol.* **95**:103–124.

Stirling, R. V., and D. Summerbell (1977) The development of functional innervation in the chick wing-bud following truncations and deletions of the proximal-distal axis. *J. Embryol. Exp. Morphol.* **61**:233–247.

Stirling, R. V., and D. Summerbell (1979) The segmentation of axons from the segmental nerve roots to the chick wing. *Nature* **278**:640–643.

Stirling, R. V., and D. Summerbell (1983) Familiarity breeds contempt: The behavior of axons in foreign and familiar environments. In *Limb Development and Regeneration*, Part A, J. Fallon and A. Caplan, eds., pp. 217–226, Alan R. Liss, New York.

Straznicky, C., and C. Tay (1983) The localization of motoneuron pool innervating wing muscles in the chick. *Anat. Embryol.* **166**:209–218.

Summerbell, D., and R. V. Stirling (1981) The innervation of dorso-ventrally reversed chick wings: Evidence that motor axons do not actively seek out their appropriate targets. *J. Embryol. Exp. Morphol.* **61**:233–247.

Whitelaw, V., and M. Hollyday (1983a) Thigh and calf discrimination in the motor innervation of the chick hindlimb following deletions of limb segments. *J. Neurosci.* **3**:1119–1215.

Whitelaw, V., and M. Hollyday (1983b) Position dependent motor innervation of the chick hindlimb following serial and parallel duplications of limb segments. *J. Neurosci.* **3**:1216–1225.

Whitelaw, V., and M. Hollyday (1983c) Neural pathway constraints in the motor innervation of the chick hindlimb following dorso-ventral rotations of distal limb segments. *J. Neurosci.* **3**:1226–1233.

Section 4

Formation of Neurites and Synapses

In this section, the themes developed in previous sections are extended to the issues of neurite formation, axon growth, and neural guidance, as well as to synapse formation. In addition, the metabolic decisions made by developing neurons are also considered.

Letourneau discusses the activities essential to neurite growth: locomotory movements at the neurite tip driven by growth cones and organization of structural components into the neurite itself. The elongation of the neurite is a complex process consisting of expansion of the plasmalemma at the tip and the accumulation of cytoskeletal components. The main external element in guiding neurite growth is the exertion of tension at local, firmly adherent contacts. At the same time, it is clear that differential adhesivity at the neurite tip influences the direction of growth-cone activity; neurites are directed to more adhesive surfaces. The present issues related to directivity that are being actively investigated concern the effects of chemotactic gradients, those of electrical fields, and the existence of discontinuities in adhesivity.

Goodman and his colleagues explore the interconnection of developing neurons in grasshoppers and fruit flies. In these insects, specific filopodial contacts occur at early stages of neuronal contact, resulting in more or less stereotyped arrangements of selective fasciculation. Goodman proposes that many different molecules are likely to be involved in this process. This is expressed in the labeled pathways hypothesis: (1) Pioneering neurons establish stereotyped axonal pathways; (2) these axonal pathways are differentially labeled on their cell surfaces; and (3) later growth cones are differentially determined in their ability to make specific choices of such labeled pathways. These workers have been searching for monoclonal antibodies that might uniquely label or correspond to such pathways. Antibodies have been found that show individual antigenic distinction among different fascicles, but they have not yet been related functionally to actual choice or cellular adhesion. The availability of *Drosophila* as a model opens up the possibility of using molecular genetic approaches to test this hypothesis.

Edwards and Meyer take up a complementary issue that contrasts with the apparent stereotypy seen by the authors of the previous chapter. Their purpose is not, however, to address the underlying mechanisms of specificity in functional connections. It is, rather, to address the capability of the insect nervous system to respond actively and dynamically to changed milieux. Insect neurons are capable of localized growth in embryonic and postembryonic periods in response to axotomy, deafferentation, and metamorphosis. Edwards and Meyer infer a means for intercellular communication between neurons to account for these events. They describe studies of the cercal–sensory giant interneuron pathways of the terminal ganglion of the cricket central nervous system to provide evidence for this notion. Their data suggest that presynaptic terminals on target dendrites serve a regulatory role, signaling the perikaryon to synthesize and transport proteins needed for dendritic growth and function. These authors also review the evidence for integrative flexibility in insect nervous systems after damage or loss of parts and point out many aspects of insect postembryonic development that do not reach finality at metamorphosis.

Willard and his colleagues, in their chapter, take up the changes in state that must occur during neuronal development, particularly in the regulation of axonal elongation. They propose the hypothesis that certain proteins act as environmentally sensitive switches to decide between developmental pathways, for example, growth of an axon or termination of growth. Investigating molecular differences between pregrowth, growth, and stationary states of neurons, they have directed their attention to growth-associated proteins in regenerating neurons of fish and amphibians. The changes observed include increased new synthesis of polypeptides, decreased synthesis of others, and a very large increase (100-fold or more) in a small number of growth-associated proteins (GAPs). Three of these GAPs have properties of integral membrane proteins and turn over and are transported at different rates. Similar proteins have been seen in rabbits where regeneration does not occur. One of the GAPs (GAP-43) appears to be greatly enriched in a growth-cone particle fraction, raising the possibility of its association with axonal growth.

An additional example of other gene products expressed in a regulated and segmental fashion in development is the neurofilament proteins. Willard and his coworkers have demonstrated a definite sequence of appearance of neurofilament polypeptides: M and L are present in rabbit optic nerve from the time of birth, whereas polypeptide *H* does not appear until 12 days postnatally. The authors suggest that this may be related to the later need for a cross-linking function of *H*. They further raise the hypothesis that this is connected with the entry of the axon into a stable state.

How can such metabolic decisions be made and what factors regulate the long-term growth of neurons? Berg and his colleagues provide a specific example in their studies of cholinergic development using chick ciliary ganglion neurons in cell culture. Such neurons form cholinergic synapses on eye muscles and receive cholinergic inputs as well. Two distinct factors have been isolated with different functions in promoting the growth of such neurons in culture. One, with a molecular weight of 20 kD, stimulates overall growth of the neurons but does not change choline acetyltransferase activity. The other, with a molecular weight of 50 kD, has no effect on growth but does stimulate choline acetyltransferase. The ciliary ganglion neurons can be shown to develop acetylcholine receptors in culture as well as develop functional cholinergic synapses on each other. Berg and his coworkers provide evidence that such receptors are distinct membrane components from analysis of α-bungarotoxin binding sites. The synaptic component was shown to share an antigenic determinant with acetylcholine receptors of muscle and electric organs.

The studies described in this section indicate that it is now possible to dissect both the metabolic-regulatory and mechanochemical aspects of neuronal growth and interaction at the molecular level. The combination of tissue culture techniques, monoclonal antibody analysis, and gene cloning with appropriate functional assays provides the necessary ingredients for analysis of the mechanisms behind neuronal pattern formation. An extensively studied example of such pattern formation, the retinotectal map, is the subject of the next section.

Chapter 12

Axonal Growth and Guidance

PAUL C. LETOURNEAU

ABSTRACT

Neurite growth is the expression of two distinct activities: locomotory movements at the neurite tip, and the synthesis and organization of structural components into a neurite. Growth-cone motility is a cyclic activity in which filopodia and lamellipodia are protruded, adhesive contacts are made with other surfaces, and mechanical forces are exerted within the protrusions by actomyosin-mediated events. This process produces work that can be used to direct neurite growth. The neurite elongates by expansion of the cell surface and by accumulation of cytoskeletal components. The neurite plasmalemma is expanded predominantly at the neurite tip, possibly by exocytotic addition of anterogradely transported vesicles to the cell surface. The cytoskeleton also advances at the neurite tip, but it is unclear whether individual cytoskeletal fibers move or whether net assembly of the cytoskeleton from monomers occurs in the neurite.

Neurite growth can be induced pharmacologically to occur without the protrusive and tensile activities of growth cones. Rather than providing a constant pull to sustain neurite growth, the exertion of tension at the neurite tip may simply promote neurite growth in a localized manner. This is the key element in the guidance of neurites by extrinsic factors. Differential adhesivity of the neurite tip to available surfaces will direct neurites onto the more adhesive surface. Gradients of soluble chemicals and electrical fields are potential directive factors as well, if they act on the locomotory apparatus to produce local differences in force exertion at adhesive contacts.

The shapes of nerve cells are truly fantastic: profusely branched dendrites in the neuropil and long axons that enter virtually every tissue and organ. It is all the more remarkable to consider that neuronal morphologies and patterns of synaptic connections are extremely characteristic from individual to individual. In large part these specific connections arise by the highly regulated growth of axons from neuronal perikarya to target cells during embryogenesis. A number of factors influence the routes taken by growing axons. This chapter will attempt to explain mechanisms of axonal growth in ways that can be related to regulatory factors that operate during axonal morphogenesis.

The cellular events that are immediately responsible for nerve fiber growth can be separated into two simultaneous but distinguishable activities. One is surface protrusion and related locomotory events at the neurite tip, which has been known as the growth cone since so named by Ramón y Cajal (1890). These movements of the cell surface and the associated intracellular lattice of microfilaments probably involve mechanisms that operate in other migratory tissue cells. The other activity is assembly and stabilization of the cytoskeletal fibers and membranous components of the nerve fiber. This may be more specialized than motility, for no other cell type produces structures with the dimensions of axons. However, cell elongation, on a smaller scale, is a general phenomenon, and general mechanisms for plasmalemmal and cytoskeletal growth should be examined. These two activities, motile behavior and neurite construction, involve different organelles and are largely restricted to closely related but separate cytoplasmic regions. This discussion of neurite growth will try to integrate these processes. This is important because guidance of axonal growth often occurs via extrinsic influences on motile behavior at the neurite tip (for other recent perspectives on neurite growth, see Johnston and Wessells, 1980; Bray, 1982; Carbonetto and Muller, 1982; Lasek, 1982; Letourneau, 1982b; Wessells, 1982; Landis, 1983).

MOTILE ACTIVITY AT THE NEURITE TIP IS A GENERAL PROCESS

The locomotory movements at the tips of growing neurites are similar to what is seen at the leading edge of other migratory cells. This behavior can be subdivided into three activities that seem to operate in a cyclic manner: protrusion of thin cytoplasmic processes, adhesion of these extensions to other surfaces, and exertion of mechanical forces within the extended processes. These forces create tensions that can be used to move cellular structures. Successful repetition of each phase of this cycle advances the cell margins, but if adhesions break during exertion of force, the protrusion is withdrawn and no advancement occurs. This cycle of behavior comprises an exploration of the cell's local environment for surfaces that are suitable for cell migration (Albrecht-Buhler, 1976).

Motile behavior at the tips of growing neurites was predicted by Santiago Ramón y Cajal (1890) and confirmed by Ross Harrison's (1910) pioneering *in vitro* studies of living neurons (Figure 1). Many highly resolved observations of the movements of growth-cone margins *in vitro* have been made. Detailed interactions of filopodial and lamellipodial protrusions with cells and other surfaces have been reported, and the exertion of tensions within filopodia extended at neurite tips documented. Even adhesive contacts to glass can be examined. These observations reinforce the notion that protrusive activity explores the environment around the neurite tip and produces intracellular tensions that pull the neurite tip across sufficiently adhesive surfaces. Protrusive activity of neurite tips has been demonstrated in several *in vivo* situations, perhaps most elegantly in grasshopper embryos, where filopodia extend 50 μm or more to interact with specific cells and axons (Goodman, this volume). This work emphasizes that selection of axonal pathways may involve filopodial contacts that occur at significant distances from the neurite tip.

Growth-Cone Structure Is Dynamic during Motility

The highly motile margins of neurite tips contain a limited variety of structural components (Figures 2, 3). Actin filaments fill the broad, flat, growth-cone pe-

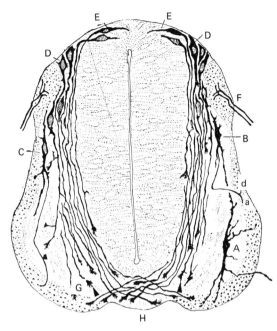

Figure 1. *The first published drawing of growth cones, reported by Ramón y Cajal (1890).* This cross section of the spinal cord from a four-day-old chick embryo was stained by the rapid Golgi method and shows the expanded tips of neurites growing ventrally from cell bodies in the dorsal cord. Thin processes projecting from these growth cones may be filopodia (arrowheads).

riphery in a latticelike arrangement of long overlapping filaments, although in filopodia and lamellipodia the filaments are often tightly packed into bundles (Kuczmarski and Rosenbaum, 1979; Letourneau, 1983). This is a localized, highly developed elaboration of the network of short actin filaments seen within the cortex of neurites and the perikaryon (Hirokawa, 1982; Schnapp and Reese, 1982). Smooth-membraned vesicles are scattered within the motile region, and vesicles frequently cluster at the bases of filopodia and lamellipodia. Microtubules penetrate forward from the base of the growth cone and terminate in the growth-cone margin, but neurofilaments stop farther back and are absent from the motile areas (Letourneau, 1983). It is not clear, but all these structures and the surrounding plasma membrane seem to be connected to the matrix of actin filaments, either directly or via intermediary filaments and other associated proteins.

During protrusion, adhesion, and exertion of tension, these structural elements probably undergo considerable rearrangement, as well as assembly and disassembly of labile components (Pollard, 1981; Alberts et al., 1983). Actin filaments are in a dynamic equilibrium, with perhaps 50% or more of the actin at the neurite tip in the unassembled state (Bray and Gilbert, 1980; Blikstad and Carlsson, 1982; Lasek, 1982; Wang et al., 1982). It seems reasonable to assume that protrusion of filopodia or lamellipodia involves locally regulated polymerization of actin, whereas withdrawal of protrusions is accompanied by actin disassembly. However, many movements of the growth-cone margins may also result from rearrangements of the assembled filament lattice. Filament organization can be influenced by a variety of actin-binding proteins, including fimbrin, filamin, and

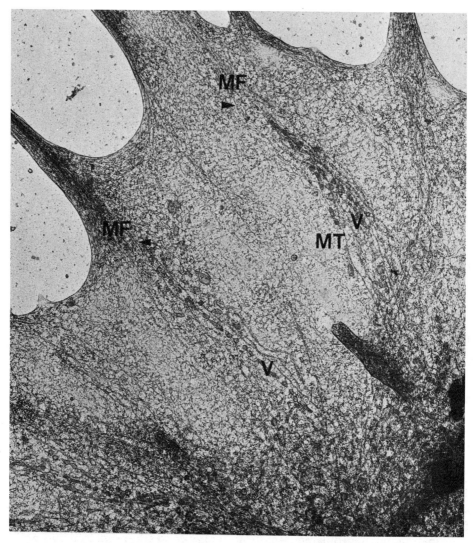

Figure 2. *Portion of the motile anterior margin of a whole-mounted growth cone.* Network and bundles of actin filaments (MF) fill the cell margin. Smooth membraned vesicles (V) and microtubules (MT) penetrate forward from base of growth cone. Note how these structures are aligned with bundles of actin filaments (arrowheads) in filopodia and lamellipodia. ×24,500. (From Letourneau, 1979.)

alpha-actinin, all of which cross-link actin filaments, vinculin, or fodrin (Willard, this volume), which may connect actin filaments to the plasma membrane, and gelsolin, which fragments actin filaments to break up a filament network. And of course, when actin and myosin filaments interact, mechanical energy is produced to slide the filaments past each other or to exert force on structures linked to the interacting actomyosin complex (Letourneau, 1981; Figure 4). The tensions exerted by filopodia on the surfaces they contact and the rapid withdrawal of nonadherent filopodia are probably based on actomyosin (Figure 5). In addition, the extensive cross-linking of the actin network with other structures means that actomyosin-derived forces can be transmitted in complex ways within the growth cone. Although the biochemistry of actin is becoming better known, the way in which motile activity is localized and regulated at the neurite tip is still a mystery.

The plasma membrane is also in a dynamic state during protrusive activity at the neurite tip. Protrusion involves expansion of the cell surface, either by addition from internal sources, perhaps by exocytosis, or by lateral flow of the plasmalemma from a local reservoir of folded cell surface. Pulse-chase experiments indicate that filopodia and lamellipodia are covered by "new" membrane (Pfenninger and Maylie-Pfenninger, 1981a,b). Endocytic vesicles are rapidly labeled with tracers at the neurite tip (Wessells et al., 1974; Bunge, 1977), and regions

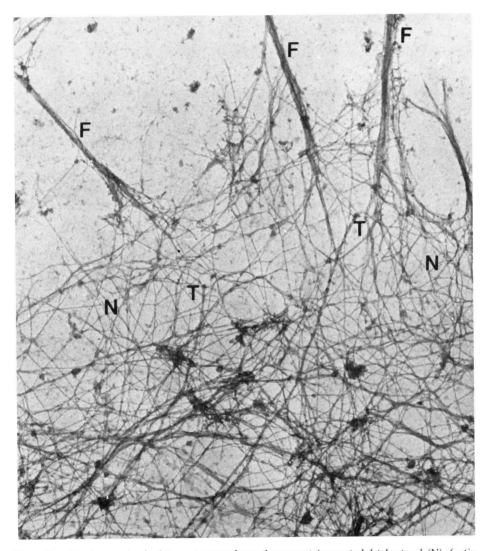

Figure 3. *Anterior margin of a detergent-extracted growth cone contains a cytoskeletal network (N) of actin filaments within which the ends of two microtubules (T) are seen.* The microtubule ends interact with actin filaments that are associated with bundles of actin filaments in filopodia (F). ×35,000. (From Letourneau, 1983.)

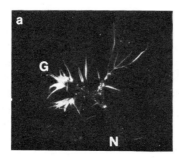 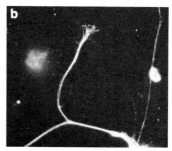

Figure 4. *Immunocytochemical localization of actin (a) and myosin (b) in detergent-extracted neurites*. Actin is concentrated in a detergent-stable form in the growth-cone margin (G), but is not detectable in the neurite (N). Myosin, distributed differently, is present in the growth cone and along the neurite. ×900.

of the growth-cone margin that were retracting at the moment of fixation contain aggregates of vesicles that may have been withdrawn from the surface by endocytosis (Tosney and Wessells, 1983). The cyclic nature of motile behavior would seem an appropriate place to find recycling of plasma membrane components (Heuser, 1978). Yet the structural features of membrane dynamics at the growth cone remain unclear because it is so difficult to fix the plasma membrane in motile areas. Rapid freezing may be capable of capturing membrane events, although Tosney and Wessells (1983) have shown that growth-cone behavior should be visually monitored until the instant of fixation.

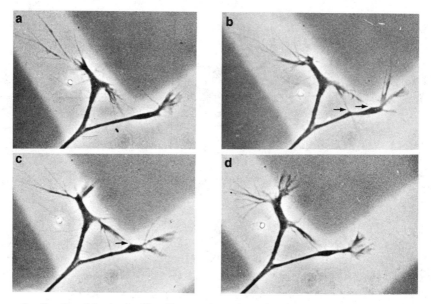

Figure 5. *Exertion of tension by filopodia* on points of attachment (arrows) to an adjacent neurite branch. Elapsed time between successive frames: *a* and *b*, 230 sec; *b* and *c*, 113 sec; *c* and *d*, 524 sec. ×650. (From Letourneau, 1982b.)

Adhesion Is an Element of Motile Behavior

In general, migrating cells must adhere to a surface in order to move over it. Using interference reflection microscopy to see the cell–substratum contacts, it appears that adhesive sites frequently correspond to portions of the growth-cone margin that contain actin filament bundles (Letourneau, 1979, 1981; Figure 6). The growth cone is the major adhesive site of neurites on many surfaces. The temporal or causal relationships among adhesion, filament bundles, and the production of forces are not known. Nevertheless, the actin filaments in the bundles have a uniform polarity (Kuczmarski and Rosenbaum, 1979; Alberts et al., 1983) and will interact with myosin: They will either pull backward on their attachments at the filopodial tip, or the myosin and associated components will be pulled forward if the filopodial tip is firmly anchored to a surface.

Adhesive bonds are formed by specific membrane molecules that act as adhesive ligands. A large glycoprotein called N-CAM (Edelman, this volume), and a complex containing proteoglycans (Schubert et al., 1983) are two cell surface components implicated in cell adhesion between neurons. N-CAM and several lectin receptors are distributed on both growth cones and neurites, although topographic differences indicate that the membrane composition of growth cones is different from that of neurites (Pfenninger and Maylie-Pfenninger, 1981a,b). A growth cone may carry several adhesive ligands that recognize different components on other cells or extracellular surfaces. Neurons from both the central and peripheral nervous systems will extend neurites on the basal laminar glycoprotein laminin, but only peripheral nervous system neurons extend neurites on the extracellular glycoprotein fibronectin (Rogers et al., 1983). This difference is probably due to cellular differences in the presence of binding sites for fibronectin.

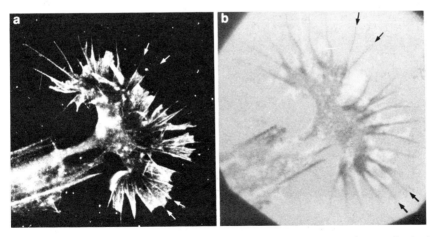

Figure 6. *Immunocytochemical localization of actin (a) in a growth cone, and interference reflection image (b) of the same growth cone.* Many linear actin arrays (white arrows) coincide with linear adhesions (black arrows) of the growth-cone margin to the substratum. ×1075. (From Letoureau, 1981.)

NEURITE GROWTH IS A LOCALIZED ACCUMULATION OF MATERIALS

The motility just described is a generalized cell behavior, associated with a specific histotypic process—the elongation of a neurite. A simple calculation shows that a 15-μm neuron will double its volume and multiply its surface area fivefold upon extension of a 700-μm neurite. Obviously, neurite growth is a regional accumulation of microtubules, neurofilaments, and plasma membrane (Figure 7) and is the product of complex metabolic events involving the synthesis, assembly, disassembly, and degradation of each molecular component (Lasek, 1982). Because neurites are so highly elongated, transport processes are also important in the relationships of these metabolic activities to neurite growth. For example, the rates of neurite growth *in vivo* and *in vitro* are close to the rates of transport of cytoskeletal components in mature axons (Lasek, 1982; Willard, this volume). Thus, transport of the cytoskeleton may be a limiting process that is critical to the mechanisms of neurite elongation.

An extremely significant finding that contributes to the understanding of neurite growth is that neurites severed from their perikaryon with microneedles will continue to elongate for several hours, even in the presence of cycloheximide and even after contracting into a ball before growth recommences (Shaw and Bray, 1977; Bray et al., 1978; Wessells et al., 1978). Thus neurites are not pushed or telescoped from the perikaryon as stable structures, but rather neurite structure

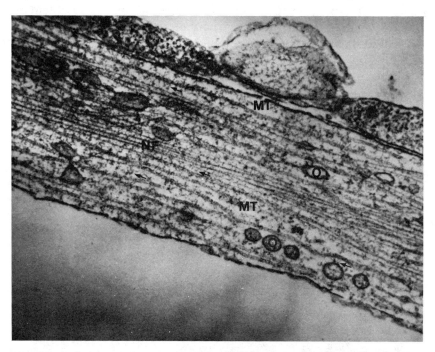

Figure 7. *Longitudinal section along a neurite in tissue culture.* Neurofilaments (NF) and microtubules (MT) are the major cytoskeletal elements, running longitudinally in the neurite. An indistinct network of material (arrows) lies centrally among the neurofilaments and microtubules, as well as beneath the plasmalemma. Membranous organelles (O) may be in transit along the neurite. ×31,000. (From Yamada et al., 1971.)

is capable of internal rearrangements and may be subject to extensive disassembly and reassembly of components. Although morphometric analyses of different categories of neurite structure have not been applied to the growth of surgically isolated neurites, the results indicate that smaller precursors of neurite structures may be exported from the soma and be destined for assembly within the neurites. The lability and dynamic character of neurite growth can be seen without resorting to microsurgery, as neurites undergo variable periods of advance, halt, and, on occasion, retraction.

The plasmalemma and microtubules are particularly relevant to this discussion because they comprise a basic framework for the neurite. Distinct cellular mechanisms regulate the assembly, disassembly, and stability of microtubules and membranes (Alberts et al., 1983). The following discussion emphasizes that the dynamic nature of microtubules and the plasma membrane is a key element in the mechanisms of neurite growth.

Membrane Expansion Occurs at the Distal Neurite

The continued elongation of surgically isolated neurite segments shows that the neurite need not be connected to the perikaryon for its surface to be expanded. Because the surface of the neurite is not highly folded to provide a reservoir of surface membrane, growth of the neurite plasmalemma probably occurs by the addition of material from internal sources. Growth by intercalation at many sites along the neurite seems unlikely, because particles placed as markers along a neurite do not move apart from one another or move farther from the perikaryon as a neurite grows (Bray, 1970, 1973). In addition, the lower surface of a neurite can remain anchored to an adhesive substratum as the tip advances without stopping neurite growth (Letourneau, 1979). Metabolic labeling of membrane precursors indicates that newly synthesized components are rapidly added to the distal neurite without initially being incorporated into the surface of the perikaryon or the proximal neurite (Pfenninger and Johnson, 1983).

When labeled and unlabeled lectins were used in combination to mark either new or old cell surface, new membrane seemed to be added to the surfaces of filopodia and lamellipodia (Pfenninger and Maylie-Pfenninger, 1981a,b). The abundant smooth vesicles in the growth cone and neurite are the probable sources of this new surface (Yamada et al., 1971). In extruded axoplasm from invertebrate neurons, such vesicles move anterogradely at rapid rates, which correspond to the rates at which metabolically labeled membrane precursors appear in distal axons (Lasek, 1982). A simple assumption is that vesicles fuse with the surface of protruding processes, as during exocytosis, and displace the existing membrane backward. Surprisingly, Tosney and Wessells (1983) found very few vesicles in growth-cone regions that were expanding just prior to fixation. However, glutaraldehyde may not fix membranous structures well, especially in regions of membrane expansion.

Thus, protrusive activity at the neurite tip may be the principal source of new surface for neurite growth. Yet newly incorporated membrane does not flow back in bulk to become part of the neurite surface, because much endocytotic activity at the growth cone removes plasmalemma (Wessells et al., 1974; Bunge, 1977). Some of these components may be recycled to the surfaces of new protrusions, but other endocytic vesicles enter lysosomes and move retrogradely to the soma. Consequently, only a portion of the membrane surface added to growth cones may be available for incorporation into a neurite.

The considerable heterogeneity of the cell surface components on neurites is not easily explained in terms of a single source for new membrane derived from filopodial expansion. Some components are concentrated more on growth cones than on neurites, others are more prevalent on neurites, and still others are uniformly distributed. Perhaps immature or incomplete membrane is added via protrusion at the tip, and the neurite membrane is further differentiated by additions or deletions of components by other means and at other locations (Carbonetto and Fambrough, 1979). Membrane heterogeneity may also be based on local restrictions of the mobility of specific membrane components on neurites such as NGF receptors or Ca^{2+} channels.

Microtubules Are a Dynamic Cytoskeletal Framework

There is general agreement that intact microtubules are required for the growth and maintenance of nerve fibers (Yamada et al., 1971; Daniels, 1975). However, the dynamics of microtubule assembly and movement in growing neurites are not completely understood. Cytoplasmic microtubules are subject to assembly and disassembly according to a variety of conditions (Margolis and Wilson, 1981; Alberts et al., 1983; Oliver et al., 1983). In particular, certain microtubule-associated proteins (MAPS) may block the less stable "minus" ends of microtubules and prevent microtubule disassembly. In mature axons, there is a subset of microtubules that are very stable to depletion of tubulin monomers (Lasek, 1982). An early event in neurite initiation may be stabilization of microtubules in neurites by the activation or production of MAPS that prevent microtubule depolymerization. Direct posttranslational modification of tubulin can also change the dynamics of microtubules.

It has not been established whether microtubules grow and shrink in elongating neurites. They might be stable elements that are formed at microtubule-organizing centers in the perikaryon and transported distally in the neurites (Spiegelman et al., 1979). This issue is significant in view of such diverse morphological changes of growing neurites as retraction of neurite segments or the branching of neurites. In order for microtubule assembly to occur within neurites, a supply of tubulin monomers is needed; these must be transported distally either as monomers or as labile polymerized microtubules. An indirect indication that free tubulin is present at neurite tips comes from studies with taxol, a drug that stimulates tubulin polymerization. When taxol is added to growing neurites, they stop growing immediately, and microtubules at the neurite tip are thrown abnormally into loops that turn back toward the perikaryon (Letourneau and Ressler, 1984). Considering what is known about taxol, this response may be due to a rapid, uncontrolled assembly of tubulin onto microtubule ends projecting into the growth cone. This explosive polymerization may exhaust a pool of tubulin monomers that normally add to the ends of these microtubules as the growth cone advances.

Another characteristic of microtubules that may be related to neurite growth is length; microtubules in neurites of cultured neurons are surprisingly long

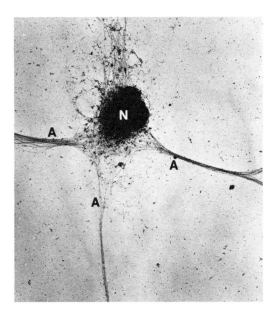

Figure 8. *The perikaryon and proximal neurite segments of the cytoskeleton of a detergent-extracted sensory neuron.* Microtubules converge within the perikaryon and pass down the neurites (A). Individual microtubules can be traced for long distances along the neurites. N, nucleus. ×1500. (From Letourneau, 1982a.)

(Bray and Bunge, 1981; Letourneau, 1982a). Perhaps many extend continuously from the perikaryon to the growth cone, just as if the microtubules had continued to polymerize as they extended down the neurite (Figures 8, 9). In addition, many more microtubule ends are present at the neurite tip than elsewhere along neurites. Thus if free tubulin is present in neurites, the neurite tip would be the principal site of microtubule assembly. Free tubulin has been quantitatively demonstrated in mature axons. This is not feasible in cultures of dispersed neurons, but with techniques for injection of labeled tubulin into the somata of individual neurons, the question of free tubulin and microtubule assembly in growing neurites can be resolved.

Neurofilaments Add Stability

What neurofilaments do in growing neurites is not clear, for they are much less frequent in embryonic or regenerating axons and in some cases may even be absent from growing neurites (Shaw et al., 1981; Lasek, 1982). General roles ascribed to neurofilaments are supported in neurite structure and axoplasmic transport, although no drug studies confirm their specific functions (Lazarides, 1980; Lasek, 1982; Willard, this volume). They are far more stable structurally than other components of neurites, and probably they do not disassemble once polymerized in the perikaryon until they are degraded by specific axonal proteases. When present, neurofilaments are associated with microtubules in the core of growing neurites (Hirokawa, 1982; Leterrier et al., 1982; Schnapp and Reese, 1982), but they always terminate short of the protrusive regions of growth cones,

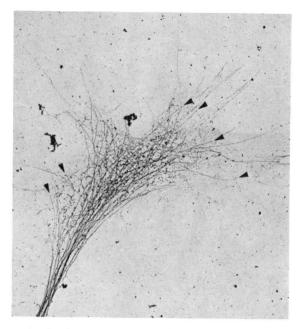

Figure 9. *Growth cone of a detergent-extracted neurite* showing many microtubule ends (arrowheads) that project into the anterior margins of the growth cone. ×4500. (From Letourneau, 1983.)

whereas microtubules penetrate forward into the network of actin filaments. This may reflect differences between neurofilaments and microtubules in their interactions with actin or differences in transport and/or assembly within the neurite tip. A reasonable proposal is that neurofilaments have no role in neurite elongation per se, but they are transported down neurites and participate in consolidation of the neurite structure laid down by the dynamic activities of the microfilaments, microtubules, and plasmalemma.

THE RELATION OF GROWTH-CONE MOTILITY TO NEURITE GROWTH

Growth-Cone Motility Is Not Required for Neurite Growth

What is the relationship of motility at the neurite tip to the metabolic activities and transport processes that comprise neurite elongation? Because the movements of filopodia and lamellipodia capture one's attention, and because the presence of actin and myosin implies generation of mechanical force, growth-cone motility has been considered necessary for neurite growth. This idea was supported by findings that neurite growth is rapidly and reversibly inhibited by cytochalasin B (CB), which disrupts the actin filament network at the neurite tip (Yamada et al., 1971). Recently, Lois Marsh, working in my laboratory, has shown that neurons plated in the presence of CB extend neurites that grow for long periods but do not exhibit filopodial activity (Marsh and Letourneau, 1984). Networks and bundles of actin filaments are absent from these CB-treated neurites; instead, actin is aggregated irregularly along neurites into densities from which short actin filaments project (Figure 10). It is not known whether actin has any functions in these CB-treated neurons, but the results show that the cyclic protrusive behavior and tensions usually seen at neurite tips are not required for neurite growth.

Other groups have reported that nonneuronal cells plated in medium containing CB extend cylindrical processes that are filled with microtubules or intermediate filaments (Vasiliev, 1982). This suggests that a general ability exists among cells to produce elongated cytoplasmic processes by mechanisms that are separable from actin-mediated spreading of the cell cortex and actomyosin-linked translocation of the cell. In neurons, this tendency to elongate is highly developed to continue during the long periods of neurite growth.

Axoplasmic transport and the movements of pigment granules are also insensitive to CB (Lasek, 1982; Beckerle and Porter, 1983). Perhaps these activities, as well as events underlying neurite growth, use an intracellular system of motility that is based on microtubules and intermediate filaments. After growth of a neurite ceases, axoplasmic transport continues to be critical to axonal maintenance. What additional events and materials distinguish neurite growth from axonal transport and mere axonal maintenance? Several proteins appear at higher levels within regenerating axons, and their name—growth-associated proteins—signifies their potential roles in neurite growth (Willard, this volume).

Growth-Cone Motility Promotes Localized Neurite Growth—Key to Guidance

Having removed neurite growth from the shadow of growth-cone motility and established it as a separate process, the two activities will be resynthesized to examine how the pathways of axonal growth are determined. The next section attempts to relate growth-cone activity to neurite growth; finally, cell–substratum adhesion, chemical gradients, and electrical fields will be discussed as primary cues for the guidance of growing axons.

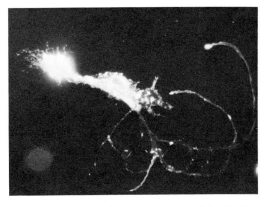

Figure 10. *Immunocytochemical localization of actin in a neuron cultured in the presence of cytochalasin B. Actin-rich filopodia are absent from the neurite tip, and actin is concentrated into unusual aggregates along the lengths of neurites.* ×880.

Protrusion of lamellipodia and filopodia has already been proposed as a source of cell surface for the growing neurite. In the presence of CB, the plasmalemma at the neurite tip seems to expand via limited undulations and bulging movements that produce much less surface area than does normal protrusive behavior during an equivalent time period. It seems an inescapable conclusion that normally a vast excess of surface area is added during growth-cone protrusion. Much membrane may be lost during retraction and lysosomal degradation unless an efficient system for membrane recycling is present at the neurite tip. It may be important to stabilize the newly added membrane of protrusions against removal by endocytosis and thus to permit the assembly and transport of neuritic cytoskeletal structures into the growth-cone margin. Adhesions of protrusions to substrata seem to stabilize the cell surface, as growth cones are quite spread when on a highly adhesive surface (Figure 11). The length and number of filopodia are also greater on adhesive substrata, and this seems due to a reduced retraction of the firmly attached protrusions (Letourneau, 1979). Thus adhesion, together with protrusion, may provide stabilized cell surface for incorporation onto the elongating neurite.

The actin filaments within the growth cone and cortical layer of neurites exert tensions on neurite structure that are visible in several ways. When filopodia or growth cones are released from adhesive contacts, they frequently snap, as if under tension. Growth cones respond to tensions produced by lateral displacement during micromanipulation with a change in direction of growth that minimizes the lateral tensile forces (Wessells and Nuttall, 1978; Bray, 1982). Short incubation with 40 µg/ml trypsin induces a rapid retraction of neurites, but if

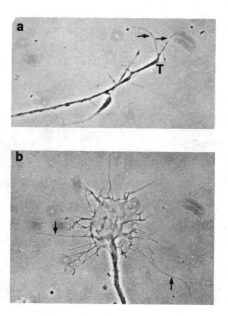

Figure 11. *Neurite tips of sensory neurons cultured on untreated glass (a) or polyornithine-treated glass (b). Note the extensive spreading of the growth cone on polyornithine, as well as the greater number and length of filopodia (arrows). ×1150. (From Letourneau, 1979.)*

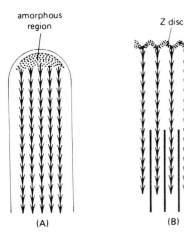

Figure 12. *Schematic drawing showing the polarity of actin filaments in a microvillus (A) and a muscle sarcomere (B).* The arrowheads indicate the polarity of decoration with S1 fragments of myosin. (Adapted from Alberts et al., 1983.)

CB is added with the trypsin, the neurites do not retract for as long as two hours of continuous exposure to trypsin. Finally, neurites extended by cells plated in CB are divided profusely into fine branches, as if the cell cortex is weakened without the actin network, and the cytoskeletal core of microtubules and neurofilaments splays out to establish many thin neurites.

The tensions developed in networks of actin filaments may have restrictive effects on neurite growth. Tensions developed in the cortical layer of neurites may be primarily exerted tangential to the plasmalemma because of the interconnections between the short, oblique actin filaments. In the transverse plane, these tensions might account for the cylindrical shapes of neurites. Along the longitudinal axis, these tensions pull neurites straight on nonadhesive surfaces and retract them if their tips are detached from the substratum.

At the neurite tip, actin is elaborated into networks of long filaments in the spread growth-cone margin and in bundles of actin filaments in filopodia and lamellipodia (Figures 2, 3). Tensions within nonadherent filopodia will displace them sideways or backward. However, firmly adherent filopodia resist the tensions, and like a tug-of-war, the mechanical energy may be used to pull forward intracellular structures that are associated with the actin network. Tosney and Wessells (1983) found that adherent filopodia and lamellipodia contain bundles of actin filaments (Letourneau, 1981). These probably have the same relative polarity to the filopodia tip as actin filaments do to the Z-line of a sarcomere (Alberts et al., 1983), and so, when mechanical force is generated within a firmly adherent filopodium, myosin and associated structures may slide forward along the immobilized actin filaments (Figure 12).

In whole-mounted specimens it is apparent that microtubules project forward into the growth-cone margin in alignment with the axes of filopodia and the lateral borders of lamellipodia (Figures 2, 3). Vesicles and smooth-membraned sacs are often associated with these microtubules. In cytoskeletons of detergent-extracted growth cones, the interactions of microtubules with the network of actin filaments are more visible (Letourneau, 1983). Microtubules are enmeshed in the actin network, and by virtue of interconnections among the actin filaments and their associations with these microtubules, the microtubule ends may be

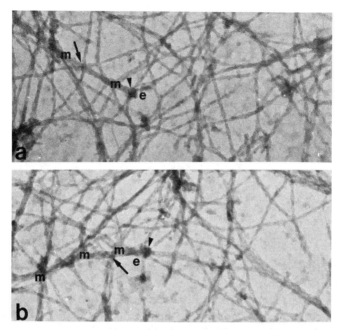

Figure 13. *Two regions of the anterior margin of a detergent-extracted growth cone.* Two microtubule ends (e) are seen, enmeshed in a network of actin filaments that are identified by decoration with the S1 fragment of myosin. Some filaments appear to pass over the microtubule (arrows), but others may contact it (arrowheads). ×65,000. (From Letourneau, 1983.)

subjected to the mechanical forces that are developed in filopodia and lamellipodia (Figure 13).

It is unknown whether these forces actually pull the long microtubules forward from the base of the growth cone, or whether they act only locally to position the ends of microtubules in the growth-cone margin. It may be sufficient for the forces to influence the positions of the microtubule ends if microtubules are translocated by other means or if tubulin is assembling onto these microtubule termini. Once the microtubules are projected forward into the growth-cone margin, neurofilaments and vesicular precursors of the plasmalemma may be channeled forward along them (Figure 2). This hypothesis requires further elucidation of the myosin localization in the growth cone, of the molecular nature of associations between actin filaments, microtubules, and the plasmalemma, and of the events of microtubule translocation or assembly in the growth-cone margin.

Thus, growth-cone motility has profound effects on the elongation process, although it is not required for neurite growth. Protrusion and adhesion of filopodia and lamellipodia provide cell surface that can be incorporated onto the neurite as the cytoskeleton advances. The exertion of tension within firmly adherent filopodia develops forces within the growth-cone margin that influence the movement and assembly of neurite structures. The key to understanding how the three extrinsic factors, discussed below, elicit directed growth of neurites is that these effects of adhesion and tension occur locally in the growth-cone margin.

EXTRINSIC FACTORS THAT GUIDE NEURITE GROWTH

Neurites Are Directed by Differential Adhesivity

Differential adhesivity is the most widely accepted proposal to explain guidance of growth cones. In essence, the model proposes that filopodial protrusion can randomly explore the adhesivity of surfaces around the neurite tip, and when tensions are exerted, again randomly, the mechanical forces will promote neurite elongation in the vicinity of more adherent filopodia while filopodia with weak adhesions will be withdrawn and retracted. Hence, the neurite will grow in the direction of greater adhesivity. This is straightforward and testable in several *in vitro* situations. The demonstration and testing of adhesions *in vivo* is much more challenging.

Neurite guidance by differential adhesivity clearly works on patterned substrata that contain sharp boundaries between areas of different adhesivity (Letourneau, 1975). Filopodia at a boundary explore both available surfaces, but the growth cone remains on the surface of greater adhesivity and will even turn corners around a square of less adhesive substratum (Figure 14). Discontinuities of adhesion similar to these artificial examples may exist in embryos and define pathways for neurite growth (Katz and Lasek, 1980). An example may be the adhesion to and growth of neurites along previously extended fiber bundles in a process called fasciculation (Nakai, 1960; Nakajima, 1965; Bray et al., 1980; Wessells et al., 1980). In the grasshopper embryo, filopodia extend many micrometers from their neurite tip to sample the surfaces of many fascicles, but the overwhelming majority of filopodial attachments are made with the fascicle that the neurite eventually joins (Goodman, this volume). Cinematographic sequences made *in vitro* show filopodia extending from a growth cone to another neurite and then exerting tension on the attachment point (Figure 5). If the attachment holds, the force can draw the neurites together; if the attachment breaks, the growth cone may withdraw (Dunn, 1971). In similar fashion, the filopodia of grasshopper growth cones may find suitably adhesive fascicles and guide neurite growth by way of events occurring far from the neurite tip.

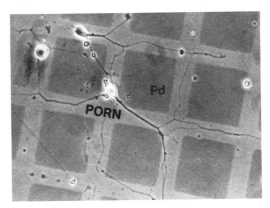

Figure 14. *Preferential extension of neurites on the more adhesive regions of a patterned substratum.* Growth cones turn, branch, or continue straight on the polyornithine-treated surface (PORN) and rarely cross onto the less adhesive palladium (Pd) surface. ×140. (From Letourneau, 1975.)

The term "contact guidance" has a long history in studies of cell motility, and for this purpose it is interpreted as stating that variations in the texture and asymmetry of available surfaces can produce variations in adhesivity that will influence growth-cone behavior (Weiss, 1941; Dunn, 1982). For example, neurite growth is decidedly biased in an aligned matrix of collagen fibrils (Ebendal, 1982). Perhaps adhesions are larger or more frequent for filopodia extended along collagen fibrils than for filopodia extended across the axis of collagen alignment. Consequently, tensions will more effectively orient neurite growth along the fibrils. Whether extracellular matrices in embryos are aligned in situations where cell migration occurs (Trinkaus, 1982) is controversial. Other three-dimensional features such as furrows or ridges could also influence neurite growth, but they remain to be demonstrated *in vivo*.

Contact guidance and discontinuities of adhesion do not explain the specific directions of neurite growth along pathways. Gradients of adhesivity would provide directional cues, although there are no *in vitro* reports of neurite growth along demonstrated adhesive gradients, as in Carter's (1965) haptotactic experiments. Perhaps the range and steepness of adhesive gradients have not been properly designed into *in vitro* models. However, gradients of intercellular adhesion and deposition of extracellular materials may exist along the proximodistal axis of the developing wings of moths (Nardi and Kafatos, 1976; Nardi, 1983). These gradients are in the correct direction to guide the sensory axons that grow centrally along the wing epithelium, and an exciting experiment will be to measure growth cone–substratum adhesion along the wing axis. Several years ago, cell adhesion assays suggested the presence of regional adhesive affinities of retinal cells for the optic tectum (Barbera, 1975; Gottlieb et al., 1976), and more recently, work has been undertaken to analyze whether regional differences in adhesivity actually guide retinal axons across the surface of the tectum (Bonhoeffer and Huf, 1982).

How Many Adhesive Affinities?

It is useful to speculate on how many adhesive affinities are needed to account for the specific morphogenesis of neuronal pathways (Edelman, this volume). Theories of chemoaffinity, such as Sperry's original hypothesis (1963), require the existence of a multiplicity of specific adhesive ligands or highly regulated topographic gradients in the concentrations of several ligands. Other hypotheses, however, propose that the order of axonal pathways and synapses can be developed from the temporal and spatial modulation of general processes such as the initiation of neurite growth and cell adhesion, involving relatively few adhesive ligands (Horder and Martin, 1978; Rager, 1980; Easter, Fraser, Goodman, this volume). A cell surface glycoprotein, N-CAM, is the major ligand for adhesion between neurons (Edelman, this volume). N-CAM is distributed over the surface of neurites and growth cones, and the addition of anti-N-CAM antibodies to culture media disrupts neurite–neurite fasciculation (Rutishauser et al., 1978) as well as neurite–muscle interactions. In my laboratory, Sherry Rogers has evidence suggesting that binding sites for the extracellular glycoproteins fibronectin and laminin are present on embryonic neurons, but they are expressed differently on peripheral versus central neurons. Her analyses of neurite growth on fragments from different regions of fibronectin indicate that neurons can interact with two separate domains of fibronectin (Rogers et al., 1985). Again, peripheral and central neurons differ in their growth on the fibronectin fragments. Cultured nonneuronal cells of several types release proteins into culture media that bind to polyornithine-treated surfaces and promote neurite outgrowth (Collins, 1980; Adler and Varon, 1981). These molecules may be distinct from fibronectin and laminin, indicating an additional adhesive specificity present on neurites.

These studies collectively suggest that growth cones may carry at least four or five different receptors or binding sites for adhesion to other cells and extracellular surfaces. Spatial and temporal differences in the expression of these binding sites on growth cones, and in the deposition of extracellular substances like fibronectin or laminin, could account for many features of neurite growth. Relatively simple modifications of ligands, such as sialylation of N-CAM and the binding of glycosaminoglycans to fibronectin or laminin, could modulate adhesive processes and produce changes in neurite growth without invoking additional adhesive interactions. One may wonder whether these adhesive ligands need be restricted to growth cones. This may be unnecessary, since the more elaborate adhesivity of growth cones, relative to neurites, may be caused by the regional restriction of the locomotory apparatus (i.e., the organization and action of the actin filament network) rather than a regional concentration of adhesive ligands on the growth cone.

Neurites May Be Guided by Soluble Gradients

The orientation of morphogenetic movements by gradients of soluble molecules, or chemotaxis, is an old idea, and recently the chemotactic mechanisms of leukocytes have been analyzed in detail (Zigmond, 1982). Growth cones of dorsal root ganglion neurons turn rapidly toward a micropipette releasing nerve growth factor (NGF) (Gunderson and Barrett, 1980; Figure 15), and over long periods of culture, neurite growth is oriented toward a diffusible source of NGF (Letourneau, 1978). A chemotactic response can be dissected into three parts: (1) sensation of the attractant, (2) determination of direction of a gradient, and (3) modulation of locomotion. Receptors for NGF are present on neurites and growth cones of responsive cells (Rohrer and Barde, 1982), and spatial variations in occupancy of NGF receptors along a growth cone and neurite may be used to sense directionality of a gradient. It is not clear how NGF-receptor interactions are transformed or integrated to provide spatial information and modulate locomotory behavior. Rapid changes in ion fluxes, Ca^{2+} binding, and membrane enzyme activity are reported to follow NGF binding to responsive neurons, but it is not known whether any of these events is primary in a sequence (Schubert et al., 1978; Greene and Shooter, 1980; Skaper and Varon, 1980; Connolly et al., 1981; Pfenninger and Johnson, 1981).

Locomotory changes that turn a growth cone may involve increases in protrusion, adhesion, or the generation of mechanical force. Gunderson and Barrett (1980) reported no change in growth-cone adhesion, but they did report a greater number of filopodia on the side of growth cones that was closer to the NGF-containing pipette. When NGF is added to sympathetic neurons that have been deprived of NGF for several hours, rapid stimulation of protrusion is seen.

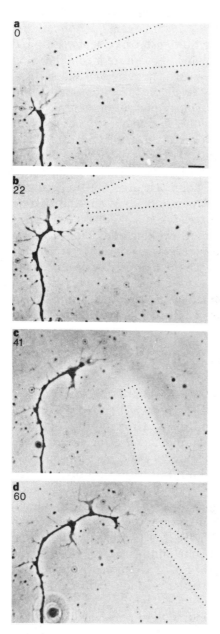

Figure 15. *Turning response of the tip of a sensory neurite* exposed to a gradient of nerve growth factor released from a micropipette. Time in minutes as indicated. (From Gunderson and Barrett, 1979.)

Ciliary neurons, which do not require NGF for survival, do respond to NGF with enhanced spreading of growth cones and more rapid neurite growth. These studies indicate that NGF modulates the locomotory activity of growth cones by mechanisms that are separable from NGF's effects on neuronal metabolism. Demonstrations of the local modulation of specific locomotory events will clarify how gradients of NGF operate to direct neurite growth, and the work with NGF can be a model for understanding other potential chemotactic responses of

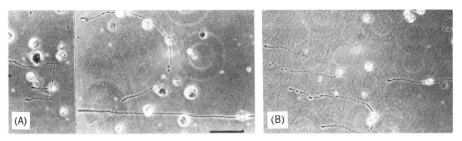

Figure 16. *Oriented growth of neurites toward cathode positioned to the left of the microscopic field.* Strength of the field in A was 120 mV/mm, and in B, 170 mV/mm. ×125. (From Hinkle et al., 1981.)

neurites. A persistent question in this research will be: Do the factors really act in a soluble form or do they exist as bound to substrata?

Do Electrical Fields Influence Neurites?

The presence of endogenous electrical fields in several biological systems raises the prospect that axonal growth is oriented by such fields. Several recent studies show quite convincingly that neurites grow toward a cathode in fields ranging from 7 to 190 mV/mm (Hinkle et al., 1981; Patel and Poo, 1982; Figure 16). When the direction of the field was reversed, neurites changed direction and curved toward the new position of the cathode. Several mechanisms are proposed to explain the directive influences of electrical fields. Steady fields can induce electrophoretic movements of plasmalemmal molecules, and local accumulations of NGF receptors, adhesive ligands, or other components such as Ca^{2+} channels could elicit local changes in protrusive activity. Electrical fields might also influence membrane potential and ion conductance in a local manner. These interesting findings suggest that the nature and cellular effects of electrical fields in embryos deserve further analysis. It will be necessary to sort out direct effects of electrical fields on neurites from effects on all the other cells and materials also subject to the fields, which then influence neurites by other means.

CONCLUSIONS

In vitro studies have demonstrated the potential influences of three extrinsic factors on neurite growth. Cellular and molecular approaches are being applied to elucidate how each factor might modulate neurite growth. *In vivo* work must test whether these influences operate in embryos. Extrinsic factors may act synergistically to direct neurites. For example, discontinuities in adhesivity may be coupled with chemotactic gradients to produce directed growth of axons along specific pathways. Experimental manipulations should test whether particular extrinsic factors are required for the development of pathways, and new methods of observation should provide increased resolution of growth-cone activity in embryos.

ACKNOWLEDGMENTS

The author thanks Dr. Sherry Rogers for a critical reading of the manuscript. The unpublished work of Lois Marsh, Alice Ressler, and Dr. Sherry Rogers has been included in this chapter. The author's research has been supported by grants from the Minnesota Medical Foundation, the Graduate School of the University of Minnesota, the Muscular Dystrophy Association, the National Institutes of Health, and the National Science Foundation.

REFERENCES

Adler, R., and S. Varon (1981) Neurite guidance by polyornithine-attached materials of ganglionic origin. *Dev. Biol.* **81**:1–11.

Alberts, B., D. Bray, J. Lewis, M. Raff, K. Roberts, and J. D. Watson (1983) *Molecular Biology of The Cell*, Garland, New York.

Albrecht-Buhler, G. (1976) Filopodia of spreading 3T3 cells. *J. Cell Biol.* **69**:275–286.

Barbera, A. (1975) Adhesive recognition between developing retinal cells and the optic tecta of the chick embryo. *Dev. Biol.* **46**:167–191.

Beckerle, M. C., and K. R. Porter (1983) Analysis of the role of microtubules and actin in erythrophore intracellular motility. *J. Cell Biol.* **96**:354–362.

Blikstad, I., and L. Carlsson (1982) On the dynamics of the microfilament system in Hela cells. *J. Cell Biol.* **93**:122–128.

Bonhoeffer, F., and J. Huf (1982) *In vitro* experiments on axon guidance demonstrating an anterior-posterior gradient on the tectum. *EMBO J.* **1**:427–431.

Bray, D. (1970) Surface movements during the growth of single explanted neurons. *Proc. Natl. Acad. Sci. USA* **69**:905–910.

Bray, D. (1973) Branching patterns of individual sympathetic neurons in culture. *J. Cell Biol.* **56**:702–712.

Bray, D. (1982) Filopodial contraction and growth cone guidance. In *Cell Behavior*, R. Bellairs, A. Curtis, and G. Dunn, eds., pp. 299–318, Cambridge Univ. Press, Cambridge.

Bray, D., and M. B. Bunge (1981) Serial analysis of microtubules in cultured rat sensory axons. *J. Neurocytol.* **10**:589–605.

Bray, D., and D. Gilbert (1980) Cytoskeletal elements in neurons. *Annu. Rev. Neurosci.* **4**:505–523.

Bray, D., C. Thomas, and G. Shaw (1978) Growth cone formation in cultures of sensory neurons. *Proc. Natl. Acad. Sci. USA* **75**:5226–5229.

Bray, D., P. Wood, and R. P. Bunge (1980) Selective fasciculation of nerve fibers in culture. *Exp. Cell Res.* **130**:241–250.

Bunge, M. B. (1977) Initial endocytosis of peroxidase or ferritin by growth cones of cultured cells. *J. Neurocytol.* **6**:407–439.

Carbonetto, S., and D. M. Fambrough (1979) Synthesis, insertion into the plasma membrane and turnover of alpha-bungarotoxin receptors in chick sympathetic neurons. *J. Cell Biol.* **81**:555–569.

Carbonetto, S., and K. J. Muller. (1982) Nerve fiber growth and the cellular response to axotomy. *Curr. Top. Dev. Biol.* **17**:33–76.

Carter, S. B. (1965) Principles of cell motility. The directionality of cell movement and cancer invasion. *Nature* **208**:1183–1187.

Collins, F. (1980) Neurite outgrowth induced by substrate associated material from nonneuronal cells. *Dev. Biol.* **79**:247–252.

Connolly, J. L., S. A. Green, and L. A. Greene (1981) Pit formation and rapid changes in surface morphology of sympathetic neurons in response to nerve growth factor. *J. Cell Biol.* **90**:176–180.

Daniels, M. (1975) Role of microtubules in growth and stabilization of nerve fibers. *Ann. NY Acad. Sci.* **253**:535–544.

Dunn, G. A. (1971) Mutual contact inhibition of extension of chick sensory nerve fibers *in vitro. J. Comp. Neurol.* **143**:491–508.

Dunn, G. A. (1982) Contact guidance of cultured tissue cells: A survey of potentially relevant properties of the substratum. In *Cell Behavior*, R. Bellairs, A. Curtis, and G. Dunn, eds., pp. 247–280, Cambridge Univ. Press, Cambridge.

Ebendal, T. (1982) Orientational behavior of extending neurites. In *Cell Behavior*, R. Bellairs, A. Curtis, and G. Dunn, eds., pp. 281–299, Cambridge Univ. Press, Cambridge.

Gottlieb, D. I., K. Rock, and L. Glaser (1976) A gradient of adhesive specificity in developing avian retina. *Proc. Natl. Acad. Sci. USA* **73**:410–414.

Greene, L. A., and E. M. Shooter (1980) The nerve growth factor. *Annu. Rev. Neurosci.* **3**:353–402.

Gunderson, R. W., and J. N. Barrett (1979) Neuronal chemotaxis. Chick dorsal root axons turn toward high concentrations of nerve growth factor. *Science* **206**:1079–1081.

Gunderson, R. W., and J. N. Barrett (1980) Characterization of the turning response of dorsal root neurites toward nerve growth factor. *J. Cell Biol.* **87**:546–555.

Harrison, R. G. (1910) The outgrowth of the nerve fiber as a mode of protoplasmic movement. *J. Exp. Zool.* **9**:787–848.

Heuser, J. E. (1978) Synaptic vesicle exocytosis and recycling during transmitter discharge from the neuromuscular junction. In *Transport of Macromolecules in Cellular Systems*, S. C. Silverstein, ed., pp. 455–464, Dahlem Konferenzen, Berlin.

Hinkle, L., C. D. McCaig, and K. R. Robinson (1981) The direction of growth of differentiating neurones and myoblasts from frog embryos in an applied electric field. *J. Physiol. (Lond.)* **314**:121–135.

Hirokawa, N. (1982) Cross-linker system between neurofilaments, microtubules and membranous organelles in frog axons revealed by the quick-freeze, deep-etching method. *J. Cell Biol.* **94**:129–142.

Horder, T. J., and K. A. C. Martin (1978) Morphogenetics as an alternative to chemospecificity in the formation of nerve connections. *Symp. Soc. Exp. Biol.* **32**:275–358.

Johnston, R., and N. K. Wessells (1980) Regulation of the elongating nerve fiber. *Curr. Top. Dev. Biol.* **16**:165–206.

Katz, M. J., and R. J. Lasek (1980) Guidance cue patterns and cell migration in multicellular organisms. *Cell Motil.* **1**:141–158.

Kuczmarski, E. R., and J. L. Rosenbaum (1979) Studies on the organization and localization of actin and myosin in neurons. *J. Cell Biol.* **80**:356–371.

Landis, S. C. (1983) Neuronal growth cones. *Annu. Rev. Physiol.* **45**:567–580.

Lasek, R. J. (1982) Translocation of the neuronal cytoskeleton and axonal locomotion. *Philos. Trans. R. Soc. Lond. (Biol.)* **299**:313–327.

Lazarides, E. (1980) Intermediate filaments as mechanical integrators of cellular space. *Nature* **283**:249–256.

Leterrier, J. F., R. K. H. Liem, and M. Shelanski (1982) Interactions between neurofilaments and microtubule-associated proteins: A possible mechanism for intraorganellar bridging. *J. Cell Biol.* **95**:982–986.

Letourneau, P. C. (1975) Cell-to-substratum adhesion and guidance of axonal elongation. *Dev. Biol.* **44**:92–101.

Letourneau, P. C. (1978) Chemotactic response of nerve fiber elongation to nerve growth factor. *Dev. Biol.* **66**:183–196.

Letourneau, P. C. (1979) Cell-substratum adhesion of neurite growth cones and its role in neurite elongation. *Exp. Cell Res.* **124**:127–138.

Letourneau, P. C. (1981) Immunocytochemical evidence for colocalization in neurite growth cones of actin and myosin and their relationship to cell-substratum adhesions. *Dev. Biol.* **85**:113–122.

Letourneau, P. C. (1982a) Analysis of microtubule number and length in cytoskeletons of cultured chick sensory neurons. *J. Neurosci.* **2**:806–814.

Letourneau, P. C. (1982b) Nerve fiber growth and its regulation by extrinsic factors. In *Neuronal Development*, N. C. Spitzer, ed., pp. 213–254, Plenum, New York.

Letourneau, P. C. (1983) Differences in the organization of actin in the growth cones compared with the neurites of cultured neurons from chick embryos. *J. Cell Biol.* **97**:963–973.

Letourneau, P. C., and A. H. Ressler (1984) Inhibition of neurite initiation and growth by taxol. *J. Cell Biol.* **98(4)**:1355–1362.

Margolis, R. L., and L. Wilson (1981) Microtubule treadmills–possible molecular machinery. *Nature* **293**:705–711.

Marsh, L., and P. C. Letourneau (1984) Growth of neurites without filopodial or lamellipodial activity in the presence of cytochalasin B. *J. Cell Biol.* (in press).

Nakai, J. (1960) Studies on the mechanism determining the course of nerve fibers in tissue culture. II. The mechanism of fasciculation. *Z. Zellforsch.* **52**:427–499.

Nakajima, S. (1965). Selectivity in fasciculation of nerve fibers *in vitro*. *J. Comp. Neurol.* **125**:193–204.

Nardi, J. B. (1983) Neuronal path finding in developing wings of the moth *Manduca sexta*. *Dev. Biol.* **95**:163–174.

Nardi, J., and F. Kafatos (1976) Polarity and gradients in lepidopteran wing epidermis. *J. Embryol. Exp. Morphol.* **36**:489–512.

Oliver, J. M., J. M. Caron, and R. D. Berlin (1983) New concepts of the control of cell surface structure and function. In *Muscle and Nonmuscle Motility*, Vol. 2, A. Stracher, ed., pp. 153–201, Academic, New York.

Patel, N., and M. M. Poo (1982) Orientation of neurite growth by extracellular electric fields. *J. Neurosci.* **2**:483–496.

Pfenninger, K. H., and M. P. Johnson (1981) Nerve growth factor stimulates phospholipid methylation in growing neurites. *Proc. Natl. Acad. Sci. USA* **78**:7797–7800.

Pfenninger, K.H., and M. P. Johnson (1983) Membrane biogenesis in the sprouting neuron. I. Selective transfer of newly synthesized phospholipid into the growing neurite. *J. Cell Biol.* **97**:1038–1042.

Pfenninger, K. H., and M. F. Maylie-Pfenninger (1981a) Lectin labeling of sprouting neurons. I. Regional distribution of surface glycoconjugates. *J. Cell Biol.* **89**:536–546.

Pfenninger, K. H., and M. F. Maylie-Pfenninger (1981b) Lectin labeling of sprouting neurons. II. Relative movement and appearance of glycoconjugates during plasmalemmal expansion. *J. Cell Biol.* **89**:547–559.

Pollard, T. D. (1981) Cytoplasmic contractile proteins. *J. Cell Biol.* **91(2)**:156–165.

Rager, G. (1980) Specificity of nerve connections by unspecific mechanisms. *Trends Neurosci.* **3**:43–44.

Ramón y Cajal, S. (1890) Sur l'origine et les ramifications des fibres nerveuses de la moelle embryonnaire. *Anat. Anz.* **5**:609–613; 631–639.

Rogers, S. L., P. C. Letourneau, S. C. Palm, J. McCarthy, and L. T. Furcht (1983) Neurite extension by peripheral and central nervous system neurons in response to substratum-bound fibronectin and laminin. *Dev. Biol.* **98**:212–220.

Rogers, S. L., J. B. McCarthy, S. L. Palm, L. T. Furcht, and P. C. Letourneau (1985) Neuron-specific interactions with proteolytic fragments of fibronectin. *J. Neurosci.* (in press).

Rohrer, H., and Y. A. Barde (1982) Presence and disappearance of nerve growth factor receptors on sensory neurons in culture. *Dev. Biol.* **89**:309–315.

Rutishauser, U., W. E. Gall, and G. M. Edelman (1978) Adhesion among neural cells of the chick embryo. IV. Role of the cell surface molecule CAM in the formation of neurite bundles in cultures of spinal ganglia. *J. Cell Biol.* **79**:382–393.

Schnapp, B. J., and T. S. Reese (1982) Cytoplasmic structure in rapid-frozen axons. *J. Cell Biol.* **94**:667–679.

Schubert, D., M. Lacorbiere, C. Whitlock, and W. Stallcup (1978) Alterations in the surface properties of cells responsive to nerve growth factor. *Nature* **273**:718–723.

Schubert, D., M. Lacorbiere, F. G. Klier, and C. Birdwell (1983) A role for adherons in neural retina cell adhesion. *J. Cell Biol.* **96**:990–998.

Shaw, G., and D. Bray (1977) Movement and extension of isolated growth cones. *Exp. Cell Res.* **104**:55–62.

Shaw, G., M. Osborn, and K. Weber (1981) Arrangement of neurofilaments, microtubules and microfilament-associated proteins in cultured dorsal root ganglion cells. *Eur. J. Cell Biol.* **24**:20–27.

Skaper, S. D., and S. Varon (1980) Properties of the sodium extrusion mechanism controlled by nerve growth factor in chick embryo dorsal root ganglionic cells. *J. Neurochem.* **34**:1650–1660.

Sperry, R. W. (1963) Chemoaffinity in the orderly growth of nerve fiber patterns and connections. *Proc. Natl. Acad. Sci. USA* **50**:703–710.

Spiegelman, B. M., M. A. Lopata, and M. W. Kirschner (1979) Aggregation of microtubule initiation sites preceding neurite outgrowth in mouse neuroblastoma cells. *Cell* **16**:253–263.

Tosney, K. W., and N. K. Wessells (1983) Neuronal motility: The ultrastructure of veils and microspikes correlates with their motile activities. *J. Cell Sci.* **61**:389–411.

Trinkaus, J. P. (1982) Some thought on directional cell movement during morphogenesis. In *Cell Behavior*, R. Bellairs, A. Curtis, and G. Dunn, eds., pp. 471–498, Cambridge Univ. Press, Cambridge.

Vasiliev, J. M. (1982) Pseudopodial attachment reactions. In *Cell Behavior*, R. Bellairs, A. Curtis, and G. Dunn, eds., pp. 135–158, Cambridge Univ. Press, Cambridge.

Wang, Yu-Li, F. Lanni, P. L. McNeil, B. R. Ware, and D. L. Taylor (1982) Mobility of cytoplasmic and membrane-associated actin in living cells. *Proc. Natl. Acad. Sci. USA* **79**:4660–4664.

Weiss, P. (1941) Nerve patterns: The mechanics of nerve growth. *Growth* (Suppl.) **5**:163–203.

Wessells, N. K. (1982) Axon elongation: A special case of cell locomotion. In *Cell Behavior*, R. Bellairs, A. Curtis, and G. Dunn, eds., pp. 225–246, Cambridge Univ. Press, Cambridge.

Wessells, N. K., and R. P. Nuttall (1978) Normal branching, induced branching, and steering of cultured parasympathetic motor neurons. *Exp. Cell Res.* **115**:111–122.

Wessells, N. K., M. A. Luduena, P. C. Letourneau, J. T. Wrenn, and B. S. Spooner (1974) Thorotrast uptake and transit in embryonic glia, heart fibroblasts and neurons *in vitro*. *Tissue Cell* **6**:757–776.

Wessells, N. K., S. R. Johnson, and R. P. Nuttall (1978) Axon initiation and growth cone regeneration in cultured motor neurons. *Exp. Cell Res.* **117**:335–344.

Wessells, N. K., P. C. Letourneau, R. P. Nuttall, M. Luduena-Anderson, and J. M. Geiduschek (1980) Responses to cell contacts between growth cones, neurites and ganglionic non-neuronal cells. *J. Neurocytol.* **9**:647–664.

Yamada, K. M., B. S. Spooner, and N. K. Wessells (1971) Ultrastructure and function of growth cones and axons of cultured nerve cells. *J. Cell Biol.* **49**:614–635.

Zigmond, S. H. (1982) Polymorphonuclear leukocyte response to chemotactic gradients. In *Cell Behavior*, R. Bellairs, A. Curtis, and G. Dunn, eds., pp. 183–202, Cambridge Univ. Press, Cambridge.

Chapter 13

Cell Recognition during Neuronal Development in Grasshopper and *Drosophila*

COREY S. GOODMAN
MICHAEL J. BASTIANI
JONATHAN A. RAPER
JOHN B. THOMAS

ABSTRACT

During development, neurons find and interconnect with other neurons in a remarkably precise way. The unfolding of neuronal specificity undoubtedly involves a series of highly specific recognition events between individual neurons. What cellular interactions underlie this specific neuronal recognition? What molecules underlie these specific cellular interactions? To answer these questions, we began several years ago to study cell recognition during neuronal development in the grasshopper embryo. In this chapter, we review what we have learned from cellular and immunological studies of the grasshopper embryo. Cell recognition at early stages of neuronal development is mediated largely by specific filopodial contacts and leads to the stereotyped patterns of selective fasciculation. Our results suggest that such recognition is likely to involve the temporal and spatial expression of many different molecules, and our monoclonal antibody studies reveal cell surface antigens whose distribution correlates with these predictions. We end the chapter by reviewing our recent studies on the same cellular interactions in the Drosophila *embryo, which leads to a consideration of the future prospects for a molecular genetic solution to this problem using* Drosophila.

CELL RECOGNITION DURING NEURONAL DEVELOPMENT

Specific cell recognition occurs throughout much of neuronal development, from cell migration to the outgrowth of axonal processes and the formation of specific synaptic connections. The initial question is: During which period of development, and in which species, can these events best be analyzed and manipulated at the cellular and molecular level? Insects have a relatively simple nervous system, which, in addition to a complex brain (10^5 neurons), includes a chain of relatively simple segmental ganglia, each containing about 1000 pairs of neurons. Unfortunately, if we wait until these neurons have formed many of their specific synaptic connections late in embryonic development, even this "simple" nervous system appears hopelessly complex. However, if we look early enough, when only a handful of neurons have sent out processes, then the system is indeed simple and accessible enough to allow us to ask how individual neurons distinguish their appropriate targets among a limited population of embryonic neurons.

At these early stages of neuronal development, cell recognition occurs most dramatically at the tips of growing axons, called growth cones, and at their fingerlike extensions, called filopodia. Growth cones radiate many filopodia (approximately 0.1 μm in diameter, up to 50 μm in length) that transiently explore the environment, many contacting the surfaces of other cells. Filopodia strongly adhere to some of these surfaces, and to others their adhesion is much weaker. If adhesion is weak during the contractile cycle, filopodia retract; if, however, adhesion is strong, then tension in that direction is increased during the contractile cycle and the leading tip of the growth cone advances toward the point of attachment (Bray, 1982; Letourneau, 1982).

Our results in the grasshopper embryo suggest that cell recognition at these early stages of neuronal development is mediated largely by specific filopodial interactions. The high affinity that growth cones and their filopodia show for particular neuronal surfaces gives rise to the stereotyped patterns of selective fasciculation in which growth cones, confronted with a scaffold of axon fascicles, choose particular axon bundles along which to extend.

We begin here by focusing on the analysis and manipulation of a single neuron, called G, at a single choice point at which it fasciculates on a pair of axons, called P1 and P2, in a particular axon bundle, called the A/P fascicle (Raper et al., 1983a,b,c, 1984; Bastiani et al., 1984b). The lessons learned from the G growth cone concerning cell recognition, filopodial adhesion, selective fasciculation, and filopodial insertion apply equally well to other growth cones and selective recognition events throughout grasshopper development (Bastiani and Goodman, 1984a,b).

Our results suggest that cell recognition during neuronal development is likely to involve the temporal and spatial expression of many different molecules. In this chapter we describe monoclonal antibodies that reveal cell surface antigens whose temporal and spatial distribution in the embryo correlates with the prediction of cellular studies that neurons whose axons fasciculate together share common surface antigens (Kotrla and Goodman, 1983, 1984). In order to isolate and characterize the surface molecules implicated by our cellular and immunological studies of the grasshopper embryo, we have shifted our emphasis to cell recognition in the central nervous system (CNS) of the *Drosophila* embryo. We end by discussing our recent cellular studies, which show that the early embryonic development of the fly CNS is largely identical, albeit in a miniature form, to the grasshopper CNS in terms of the identified neurons (e.g., G) and their selective fasciculation (e.g., the A/P fascicle) (Thomas et al., 1984), thus opening the way for future molecular genetic approaches.

SELECTIVE FASCICULATION IN THE GRASSHOPPER EMBRYO

The grasshopper's CNS is segmentally arranged with a chain of cephalic, thoracic, and abdominal ganglia. Each segmental ganglion is generated by a precise pattern of precursor cells, containing two bilaterally symmetric plates of 30 neuroblasts

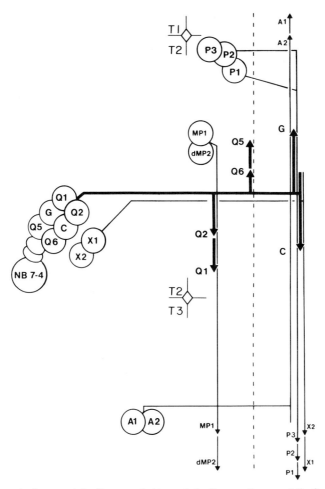

Figure 1. *Schematic diagram of the divergent choices made by the growth cones of the first six progeny of NB 7–4 as they fasciculate on different longitudinal axon fascicles.* All six growth cones choose the same axon fascicle to extend across the posterior commissure, yet make divergent choices in the contralateral neuropil. Q1 and Q2 turn posteriorly on the MP1/dMP2 fascicle. G extends anteriorly on the A/P fascicle. C extends posteriorly in this same axon bundle once other axons (including X1 and X2) have joined the bundle. Q5 and Q6 extend anteriorly in a different, more medial pathway (dashed line). T1, T2, and T3, first, second, and third thoracic ganglion. (From Bastiani et al., 1984b.)

(NBs), an unpaired median neuroblast (MNB), and seven midline precursors (MPs). Each NB is a stem cell, maintaining its large size as it divides repeatedly to produce a chain of smaller ganglion mother cells. Each ganglion mother cell in the chain divides once more, thus producing a chain of paired ganglion cells that subsequently differentiate into neurons. The second thoracic ganglion (T2) contains about 2000 neurons largely arranged as 1000 pairs of symmetric bilateral homologues. In the example discussed here, NB 7–4 in the T2 segment generates a string of progeny that are accessible throughout their development. NB 7–4 gives rise to about 100 progeny. With our emphasis on the early events of cell recognition, we have been concerned only with the first six progeny of NB 7–4, called Q1 and Q2, G and C, and Q5 and Q6 (Figure 1).

Q1's growth cone meets the growth cone of its contralateral homologue Q1 at the midline, and there they fasciculate on one another to form one of the first axon fascicles in the posterior commissure (Goodman et al., 1982; Harris and Goodman, 1983; Raper et al., 1983a,b). Although several different axon fascicles subsequently develop in the posterior commissure, the next five siblings of Q1 (namely Q2, G, C, Q5, and Q6) fasciculate on the Q1 axon bundle, even though other bundles are within filopodial grasp.

However, upon reaching the contralateral side of the developing neuropil, the growth cones of these clonally related neurons diverge as they make cell-specific choices about which way to grow (Raper et al., 1983a). These growth cones, like most other embryonic growth cones, find themselves in an environment surrounded by the axons of other previously differentiating neurons. These axons run in fascicles which take the form of a scaffold of nearly orthogonal axon bundles. There are already about 20 different longitudinal axon fascicles when the G growth cone makes its cell-specific choice (Bastiani et al., 1984b). As shown schematically in Figure 1, the growth cones of Q1 and Q2 turn posteriorly along a medial axon pathway (the MP1/dMP2 fascicle; MP1 is one of two progeny of midline precursor 1; dMP2 is one of two sibling progeny of MP2), G and C turn in opposite directions along a lateral axon pathway (the A/P fascicle), and Q5 and Q6 turn anteriorly along a different intermediate pathway (unidentified fascicle) (Raper et al., 1983b).

As the G growth cone turns anteriorly, it always fasciculates on the A/P fascicle in preference to other nearby bundles. At this stage, the A/P fascicle

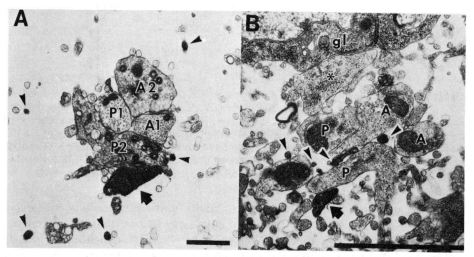

Figure 2. *Selective fasciculation in grasshopper and* Drosophila. The A/P fascicle and the tip of the G growth cone from a grasshopper embryo at 40% of development (*A*) and a 12-hour-old *Drosophila* embryo (*B*) are visible in semiserial TEMs of HRP-filled G neurons. In both species the A/P fascicle contains the G growth cone and the A1, A2, P1, and P2 axons at this stage. Note that in both species the tip of the G growth cone (large arrow) is in contact with the P axons rather than the A axons. Arrowheads show G's filopodia. In *Drosophila* the axons are smaller in diameter and the fascicles are closer together than in grasshopper. Asterisk in B is another bundle, probably homologous to the D fascicle in the grasshopper. Calibration bar = 1 μm.

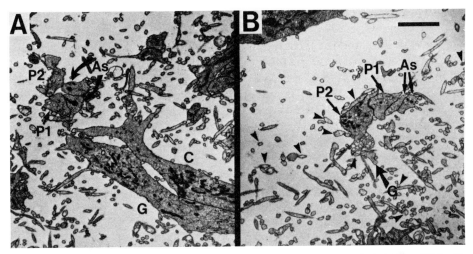

Figure 3. *Ultrastructure of the G and C growth cones and their filopodia in relation to the A/P fascicle.* TEMs taken from serial thin-sections. Two sections showing the growth cones (A) and terminal lamellipodia and filopodia (B) of the G and C growth cones just before the G growth cone gets onto the A/P fascicle. Note the A/P fascicle contains the axons of the A1, A2, P1, and P2 neurons at this time. The G and C growth cones at this stage are suspended, most likely by their filopodia, just medial and ventral to the A/P fascicle. The small black arrowheads in B are G's filopodia, and the large black arrow is the tip of G's growth cone, all of which appear to prefer and wrap around the P axons. The reconstruction shown in Figure 4 is based on serial sections from the preparation shown in this figure. Calibration bar = 1.5 μm.

contains the axons of the A1, A2, P1, and P2 neurons (Figures 2A, 3A). The axons of the A1 and A2 neurons extend anteriorly through the dorsal lateral neuropil. They meet and fasciculate on the two posteriorly growing axons of the P1 and P2 neurons. The axon of the P3 neuron joins the fascicle shortly after P1 and P2. C's growth cone extends posteriorly and fasciculates on the same axon bundle as G's, except 10 or so hours later in development. By the time C begins to extend predominantly in the posterior direction, there are several additional axon profiles in the bundle, including X1, X2, and others (Figure 1).

This pattern of selective fasciculation, coupled with our knowledge of cell lineages, leads to two simple yet interesting observations. First, cells from the same precursor fasciculate in different axon bundles. Second, axon bundles contain cells from different precursors. For example, just after G joins the A/P fascicle, it contains two axons (the As) from one precursor, two axons (the Ps) from a second precursor, and one axon (G) from a third precursor. Shortly thereafter, they are joined by two other axons (the Xs) from yet another precursor. Other progeny from these same precursors join other fascicles, as demonstrated by G's siblings. Thus, the axons of unrelated neurons of diverse embryonic origin demonstrate a high affinity for one another, giving rise to their specific patterns of selective fasciculation and quite possibly to their selective connectivity. These observations will have predictive value later, when we consider our search for cell surface molecules mediating this cell recognition.

SELECTIVE FILOPODIAL ADHESION

In order to understand the recognition events leading to the patterns of selective fasciculation, we considered the ultrastructure of the G growth cone, its filopodia, and its filopodial contacts just before, and shortly after, it climbs onto the A/P fascicle (Raper et al., 1983c; Bastiani et al., 1984b). Semiserial and serial reconstructions from the transmission electron microscope (TEM) were made from either horseradish peroxidase (HRP)-filled neurons (e.g., Figure 2A) or from serial thin-sections in which axons, growth cones, and filopodia were individually identified (e.g., Figure 3).

There is a particular suborganization to the A/P fascicle. The relative apposition of the axons we observed in many different embryos suggests that A axons adhere tightly to other A axons, P axons to P axons, but A axons not so tightly to P axons and vice versa. Within the fascicle, the two tightly apposed P axons can be located anywhere around the two A axons. Interestingly, in the four cases examined in which the tip of G's growth cone had just climbed onto the A/P fascicle, G was always closely associated with the P and not the A axons (e.g., Figure 2A), irrespective of the relative orientation of the four axons within the A/P fascicle. This finding suggests that not only is G able to distinguish the A/P fascicle from other axon bundles, but also that it is able to distinguish the P axons from the A axons within this fascicle (Raper et al., 1983c; Bastiani et al., 1984b).

We observed the filopodia of the G growth cone in the light microscope by visualizing cells after intracellular injection of either HRP or Lucifer yellow followed by HRP-immunocytochemistry with a serum antibody to Lucifer yellow (Taghert et al., 1982). The shape of the G growth cone, its number of filopodia, and its rate of extension depend upon its location in the neuropil. As the G growth cone reaches its choice point and slows down, it usually becomes quite broad and complex in shape, with numerous filopodia radiating in many directions. It sometimes has several anteriorly directed bumps with filopodia extending in tufts from each of these bumps, called "active sites." These tufts radiate filopodia that appear to contact the surfaces of many different axon fascicles. But once the G growth cone climbs onto the A/P fascicle, its shape and rate of extension change dramatically. The faster-extending G growth cone becomes long and tapered and has fewer filopodia, most of which extend from the tip of the growth cone and contact the A/P fascicle.

The most illuminating observations, however, come from TEM reconstructions of G's filopodia and their contacts just before the growth cone climbs onto the A/P fascicle (Figures 2, 3, and 4). There are several other longitudinal axon bundles within 10 μm of the A/P fascicle. For example, one of these bundles, the D fascicle, runs about 5–10 μm dorsal and slightly medial to the A/P fascicle and consists of two axons (the D neurons) just before G extends onto the A/P fascicle. The filopodia of the G growth cone extend for over 30 μm and are within grasp of the D fascicle, the A/P fascicle, and many other axon bundles. Thus it is of interest to examine the extent of filopodial contact with these different axon fascicles before G's growth cone makes its cell-specific choice to extend anteriorly upon the A/P fascicle.

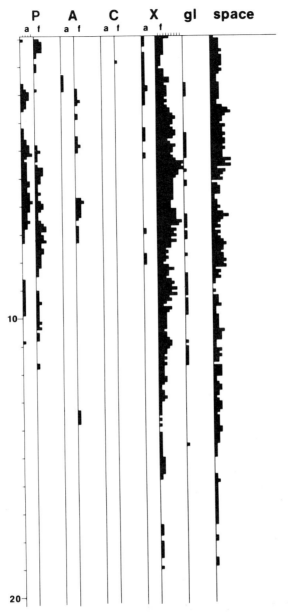

Figure 4. *Selective filopodial adhesion.* TEM serial-section reconstruction (approximately 20 μm, 200 sections) of the filopodia of the G growth cone (see Figure 3). The filopodia from the leading 5 μm of the G growth cone were reconstructed for 20 μm anteriorly (scale shown on left). The G filopodia were often out in space (space), touching unidentified filopodia (X,f) from 40 or more different axons, or occasionally touching other unidentified axons (X,a), glial cells (gl), or the filopodia from the C cell (C,f). Most interesting was the specificity they showed among the axons (a) and filopodia (f) of the P cells (P) and the A cells (A). G's filopodia show a highly selective preference for the axons and filopodia of the two P cells as compared to the two A cells.

According to Bray's model (1982), filopodia are actively extended and retracted over a several-minute cycle. Those filopodia that contact a particularly adhesive surface are retained as their contractile cycle produces tension rather than retraction. At any one moment, one might expect to find more filopodia contacting particularly adhesive cell surfaces. Furthermore, the filopodia are likely to have longer and more extensive contacts with the more adhesive surfaces than with less adhesive ones.

In one example from a TEM reconstruction at a time before the G and C growth cones extend onto the A/P fascicle, the filopodia from the G and C growth cones were in extensive contact with the A/P fascicle, whereas they made only brief contact with the D fascicle or other nearby fascicles. Moreover, they ran along and contacted the axons of the A/P fascicle for many micrometers. This suggests that the filopodia of the G and C growth cones adhere more strongly to the A/P fascicle than to others nearby (Raper et al., 1983c; Bastiani et al., 1984b).

Before the G and C growth cones extend onto the A/P fascicle, they leave the Q1 commissural fascicle containing their sibling's axons and extend dorsally and laterally toward the A/P fascicle, appearing to be guided by strong filopodial and lamellipodial adhesion with the P axons (Figure 3). In fact, the leading filopodia and lamellipodia from G's growth cone appear to wrap around the P axons and show striking preference for them over the A axons at this time (Figure 3B). When the G filopodia were serially reconstructed in this embryo, remarkably specific patterns of filopodial contact were revealed (Figure 4) (Bastiani et al., 1984b). Not only do the G filopodia show a preference for the P axons, but this preference also extends to the P filopodia over the A filopodia (in Figure 4, compare $P_{a,f}$ to $A_{a,f}$). In addition, some of G's filopodia contacted glia (gl in Figure 4), some were in space, and some contacted the axons or filopodia of unidentified neurons (X, representing approximately 40 different axons within filopodial grasp).

Thus, it appears that G's filopodia can distinguish the A/P fascicle from other axon bundles, and, within this fascicle, they can distinguish the P axons from the A axons. Although the electron micrographs present a static picture of a dynamic process, the results strongly suggest that this selective fasciculation is likely to be mediated by differential adhesion of the filopodia of the G growth cone to the A/P fascicle and, in particular, to the P axons.

SELECTIVE FILOPODIAL INSERTION

In the process of studying the selective adhesion of the filopodia from the G growth cone onto the P axons, we discovered another novel and highly specific interaction between these cells, as revealed by TEM serial-section reconstructions. Two filopodia were inserted into the G growth cone, one quite deeply (Figure 5), and both induced the formation of coated pits and coated vesicles at the filopodial tips. Interestingly, both of these filopodia were from the P neurons. This interaction of filopodial insertion is highly specific, since filopodia from many other nearby growth cones and axons (approximately 40) contact the

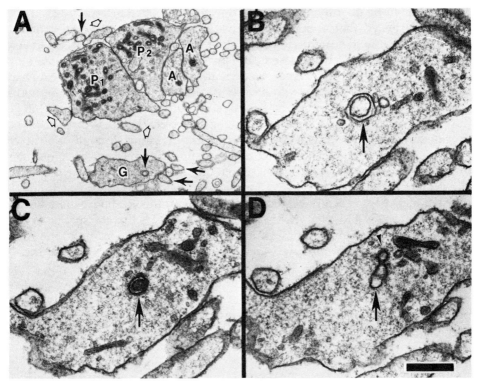

Figure 5. *Four consecutive sections showing selective filopodial insertion.* TEM serial-section reconstruction of the G growth cone and the A/P fascicle (see Figures 3, 4) reveals the insertion of P filopodia into the G growth cone (closed arrows mark P filopodia, open arrows mark G filopodia). In this preparation, two filopodia were penetrating the G growth cone, and both came from the P axons. In both cases, coated pits and coated vesicles were being induced at the tips of the inserted filopodia. *A*: Relationship of G growth cone, A/P fascicle, and filopodia. *B*: Shank of inserted P filopodium in G growth cone. *C*: Tip of P filopodium (note dense staining at tip of filopodium). *D*: Induction of coated pit and coated vesicle (arrowhead) just beyond filopodial tip. Calibration bar = 1 μm (A) and 0.2 μm (B–D).

surface of the G growth cone, yet none penetrate it or induce coated vesicles (Bastiani et al., 1984b).

We first discovered specific filopodial insertions while studying the first growth cones in the CNS of the grasshopper embryo (Bastiani and Goodman, 1984a, b). Here too, the interaction is highly specific and induces coated pits and vesicles. In both cases, coated pits and vesicles are induced only at the tips of the penetrating filopodia.

The specificity of the filopodial insertions and induction of coated vesicles has several interesting implications. Only filopodia from particular cells (in this case the P neurons) are observed inserting into other identified growth cones (in this case the G growth cone), even though filopodia from many other cells contact their surfaces. Just as the data on selective filopodial adhesion suggests that axons in different bundles have different cell surfaces among which filopodia can distinguish, so the data on selective filopodial insertion suggests that the filopodia from different neurons have different cell surfaces among which growth cones can distinguish.

What might be the function of these selective filopodial insertions? We have already suggested that selective filopodial adhesion is the mechanism of cell recognition that guides a growth cone onto a particular axon fascicle. However, growth cones typically make a series of such selective fasciculation choices on the way to their appropriate target, sequentially getting on and then off particular axon bundles. One attractive hypothesis is that the selective adhesion of filopodia is a dynamic process such that the adhesive properties of a growth cone change during the course of these navigations along the scaffold of axon fascicles. One speculation is that the interactions involved in navigating through one choice point might induce the cell to change its expression of the cell surface molecules involved either in the filopodial adhesion at that choice point or in the filopodial adhesion at a subsequent choice point. How, though, does the transcriptional, translational, and/or posttranslational machinery in the cell body know where the growth cone is and what it is doing? One possibility is that the specific events described here might mediate such inductive changes by signaling to the cell's machinery via receptor-mediated endocytosis. We propose that when the G growth cone encounters the P axons, the selective adhesion of the G filopodia onto the P axons guides the G growth cone onto the A/P fascicle. The selective insertion of the P filopodia into the G growth cone induces changes in the G growth cone that either enhance this adhesive interaction with the P axons or prepare it for subsequent encounters.

LABELED PATHWAYS HYPOTHESIS

Our descriptive studies show that the G growth cone and those of its sibling neurons extend in very close apposition to other specific axons in the developing neuropil (Raper et al., 1983b; Bastiani et al., 1984b). Similarly, observations at earlier stages of development show selective fasciculation patterns among the very first axons in the grasshopper CNS (Goodman et al., 1982; Bastiani et al., 1984a). These observations led us in 1982 to propose the "labeled pathways" hypothesis (Goodman et al., 1982; Raper et al., 1983b). The hypothesis proposes (1) that pioneering neurons establish stereotyped axonal pathways; (2) that these axonal pathways are differentially labeled on their cell surfaces; and (3) that later growth cones are differentially determined in their ability to make specific choices of which labeled pathways to follow. It includes the notions that filopodia are actively involved in sampling the surfaces of axon bundles within their grasp and that differential filopodial adhesion mediates the selective fasciculation.

This hypothesis predicts that G's growth cone traverses its very precise route in the developing neuropil by first recognizing, and then extending along, the Q1 and Q2 axons (its earlier siblings) in the Q1 fascicle across the posterior commissure, and subsequently, the P1 and P2 axons in the A/P fascicle anteriorly across the contralateral neuropil. Alternatively, of course, G's growth cone might be guided through the neuropil by cues extrinsic to these specific axons, and it might therefore use the axons merely as convenient mechanical substrates. On the basis of this hypothesis, we anticipated that ablating the A and P axons (and more specifically, the P axons alone) should prevent G from locating and extending anteriorly in its proper location in the contralateral neuropil.

EXPERIMENTAL TEST OF THE HYPOTHESIS

We tested the labeled pathways hypothesis by examining the effects of ablating the A1, A2, P1, P2, and P3 axons on the behavior of the G growth cone (Raper et al., 1983c, 1984). If the A/P fascicle specifically guides G's growth cone through the neuropil, its ablation should prevent G's normal behavior. If G's growth cone is determined to elongate upon particular axons within the fascicle, only the ablation of those particular axons should affect G's behavior. Our results suggest that the A/P fascicle plays an important role in guiding G's growth cone and that it is specifically the P axons that appear to be most active in this role. These results, as described in detail below, strongly support the hypothesis.

For these experiments, the embryos were cultured outside of their eggshells, embryonic membranes, and yolks in a supplemented RPMI culture medium. The ablation of neuronal cell bodies and axons was accomplished under Nomarski optics with a sharp microelectrode. Each manipulated embryo had its own internal control, since one G neuron faced a perturbed environment whereas the contralateral G neuron in the same segment faced a control environment.

The experimental manipulations were performed during a relatively narrow time window after the growth cones of both the A1 and A2 neurons had turned into the ganglionic connectives, but before the growth cones of the G, A1, or P1 neurons reached the location at which G turns anteriorly—before the A/P fascicle had formed in the T2 segment (approximately 37% of embryogenesis; see Figure 6A). If the embryo is removed from its egg and cultured at 29°C for 40–48 hours, G's growth cone continues extending to a lateral position in the neuropil and then turns onto the A/P fascicle and extends in the anterior–posterior axis (Figure 6B). In favorable cultures, the distalmost tip of G's growth cone advances anteriorly into the ganglionic connective joining the second and first thoracic segments (equivalent to 42% of embryogenesis). Thus we are able to attain sufficient development in culture to examine the cues which guide G's growth cone through its choice point and then anteriorly through the neuropil of the second thoracic ganglion.

The axons of the A1, A2, P1, P2, and P3 neurons were prevented from making the A/P fascicle in the second thoracic ganglion by cutting the portion of the ganglion connectives in which the A1 and A2 axons run and by killing the P1, P2, and P3 cell bodies and separating them from their axons. Manipulations were performed on one side of the embryo only, so that the opposite side served as an internal control.

Figure 6 shows the morphology of the control and experimental G cells from camera lucida drawings of Lucifer yellow-filled neurons viewed with Nomarski optics. The morphologies of a control G and sham-manipulated G are shown in Figure 6B. The axons of both neurons extend well anteriorly through the normal, lateral portion of the neuropil.

Figure 6C shows a preparation in which the A1, A2, P1, P2, and P3 axons were ablated on one side. The control G extends anteriorly past the anterior commissure; the experimental G extends anteriorly for only a short distance. The multiple, posteriorly directed processes of the experimental G appear to wander and branch anomalously.

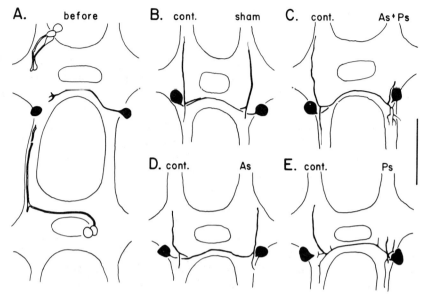

Figure 6. *Experimental test of labeled pathways hypothesis.* Effects of axon ablations on the behavior of the G growth cone. The morphology of the G growth cone (two-dimensional camera lucida drawings) in examples of manipulated embryos subsequently cultured for 40–48 hours. All manipulations were performed on the right sides, leaving the left sides as internal controls. *A:* Relative positions of the A1, A2, P1, P2, P3, and G growth cones at the time of manipulation. *B:* Sham-manipulated. *C:* A1, A2, P1, P2, and P3 axons ablated; G growth cone behaves abnormally. *D:* A1 and A2 axons ablated; G growth cone behaves normally. *E:* P1, P2, and P3 axons ablated; G growth cone behaves abnormally. G growth cone requires the P axons for normal extension. Calibration bar = 100 μm.

No effect upon G was detected if only the A1 and A2 axons were prevented from joining the A/P fascicle (Figure 6D). However, ablation of only the P cells does affect G's morphology. Figure 6E shows a preparation in which the P1, P2, and P3 cells were killed on the right side. The control G has made considerable progress anteriorly, while several processes on the experimental side made considerably less anterior progress. The two more medial of these processes grew in abnormally medial positions.

These light-level observations show the normal versus abnormal branching patterns of the G growth cone in only two dimensions and without the perspective of other axons. In order to observe the three-dimensional behavior of the G growth cone and its possible interaction and/or affinity with other axons, we examined several experimental embryos in semiserial TEM reconstructions (Raper et al., 1984). For example, when only the P axons were ablated, light-level observations showed that the G growth cone behaved abnormally and extended many branches in the contralateral neuropil. Upon closer examination, the G growth cone appeared to be "confused," with no clear affinity for any particular longitudinal axon fascicle. On the control side, the A/P fascicle developed normally (Figure 7A). On the experimental side, the A axons formed the A/P fascicle (albeit without the P axons), and several other axons subsequently joined the fascicle (Figure 7B). Presumably these unidentified axons normally have a high affinity for the A axons and thus behave normally in the absence of the P axons.

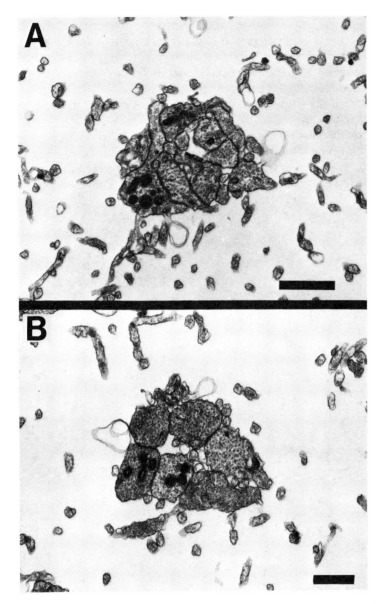

Figure 7. *TEM sections of the A/P fascicle on the control (A) and the experimental (B) sides of a cultured embryo in which the P axons were ablated before the A/P fascicle had formed.* The A/P fascicle develops normally in culture. On the control side, the A/P fascicle is likely to consist of the two A axons, the three P axons, and three other axons, including G, which have joined the bundle. On the experimental side, the G axon has not joined the bundle (see Figure 8); rather, the fascicle is likely to consist of the two A axons and four other axons that have joined the bundle. Calibration bars = 1 μm.

However, although some other axons joined the abnormal A/P fascicle, G did not. Rather, in the absence of the P axons, G's growth cone extended directly below the A/P fascicle and branched abnormally many times in both the ventromedial and ventrolateral neuropil (Figure 8). None of these multiple branches showed a high affinity for any particular axon fascicle, although the branches and filopodia of the G growth cone appeared to contact or to be near more than 20 different axon fascicles.

These experimental results confirm observations based on TEM reconstructions of normal embryos. The filopodia of the G growth cone have a high affinity for the P axons, and in their absence appear to be uninterested in any other fascicle within a wide region of the contralateral neuropil (Figure 8). Not only do these

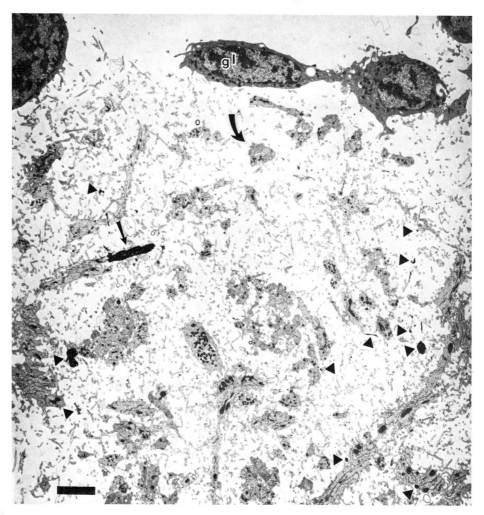

Figure 8. *Abnormal behavior of G growth cone in embryo in which the P axons were ablated.* Same preparation as Figure 7. The A/P fascicle forms around the A axons (large curved arrow). The G growth cone extends past the A/P fascicle (straight arrow) and has branched extensively throughout the ventral and lateral neuropil (arrowheads) without tightly associating with any other axon bundle. Calibration bar = 5 μm.

results strongly support the labeled pathways hypothesis, but also they suggest that the neuropil is subdivided into more labeled pathways than previously thought. For example, the A/P fascicle appears to be subdivided into separately labeled P axons and A axons. Furthermore, no other longitudinal axon fascicle appears to express the same label as the P axons.

MONOCLONAL ANTIBODY CORRELATE OF THE HYPOTHESIS

Most embryonic growth cones thus appear to use the surface labels on previously differentiated axonal pathways for selective fasciculation. Given the specificity that the G growth cone shows for the A/P fascicle as compared to all other longitudinal fascicles, and for the P axons as compared to the A axons within the A/P fascicle, it seems reasonable to hypothesize that each longitudinal axon fascicle, and perhaps subsets of axon bundles within the fascicles, are labeled by different molecules. Can we uncover this postulated code of molecular surface labels? One way is to make monoclonal antibodies that recognize cell surface molecules specifically expressed on small subsets of axons early in development (Kotrla and Goodman, 1983, 1984). Such antibodies might reveal the cell surface molecules used in selective fasciculation. The function of these surface molecules could be tested by applying the antibodies to embryos growing in culture in an attempt to block growth-cone guidance.

Mice were immunized with a variety of tissues, including cells and/or membrane fractions from embryonic neuroepithelium. Thus far we have examined about 2000 monoclonal antibodies (screened on whole-mount 40% neuroepithelium) and have obtained five that label cell surface antigens on small subsets of embryonic axons in the grasshopper embryo at 40% of development (Kotrla and Goodman, 1983, 1984); the antigen recognized by a sixth monoclonal antibody (Mes-2) is not expressed in the CNS until 48.5% of development and is described in the next section (Figure 9). In the CNS of the embryo at 40% of development, each segment contains an axonal scaffold consisting of about 100 different axonal pathways: approximately 20 longitudinal fascicles, approximately 20 fascicles in each of the three commissures, and approximately 20 fascicles exiting the peripheral nerves. At any one time and place, an individual growth cone might have to distinguish among the 20-odd bundles within filopodial grasp. Of the 100 or so different axon bundles, the five monoclonal antibodies recognize antigens on the surface of 1–9 different bundles, depending upon the antibody.

The most specific, and to us the most interesting, of these are Mes-3 and Mes-4, both of which recognize the same single longitudinal axon bundle—the MP1/dMP2 fascicle. Other axon bundles at 40% of development are not stained by either Mes-3 or Mes-4, including, for example, the vMP2 fascicle, the D fascicle, and the A/P fascicle. Fortunately, we know much about the neurons that establish the MP1/dMP2 fascicle (MP1, dMP2, and pCC) and about several of the identified neurons whose growth cones choose to fasciculate with this bundle (e.g., the H-cell sibling). We are presently testing to see if either or both of these monoclonal antibodies, when applied to embryo growing in culture, block the selective fasciculation of this specific axon bundle (S. Kuwada and S. Helfand, work in progress).

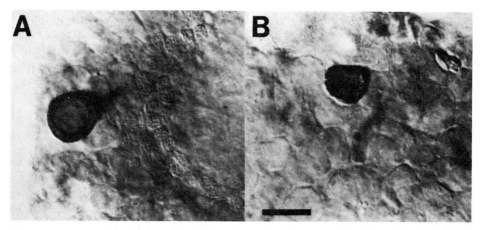

Figure 9. *Staining of two identified motoneurons in the metathoracic (T3) ganglion by the Mes-2 monoclonal antibody at 55% of development in the grasshopper embryo.* Mes-2 transiently stains a surface antigen on four out of 1000 neurons in each T3 hemiganglion (the other two neurons are unidentified interneurons X and Y). *A*: Staining of the fast extensor tibiae motoneuron (FETi), revealed with an HRP-labeled second antibody. *B*: Staining of the slow extensor tibiae motoneuron (SETi). For implications, see text. Calibration bar = 20 μm.

In the grasshopper, Mes-3 and Mes-4 appear to recognize molecular labels that distinguish a single axon fascicle from all other axon bundles in the embryonic axonal scaffold at 40% of development. Thus they provide a strong correlate for the labeled pathways hypothesis by demonstrating that individual axon fascicles are antigenically distinct while growth cones are choosing among them.

TRANSIENT EXPRESSION OF A CELL SURFACE ANTIGEN

We discussed earlier the dynamic aspects of growth-cone guidance and selective filopodial adhesion. Growth cones typically make a series of selective fasciculation choices on the way to their appropriate target, sequentially getting onto and then off particular axon bundles. Thus the selective adhesion of their filopodia is likely to be a dynamic process such that their adhesive properties change during the course of development, possibly as a result of selective filopodial insertion. These adhesive changes suggest changes in surface molecules.

Within this context, it is interesting to consider Mes-2, whose antigen is transiently expressed on only four neurons (out of approximately 1000) in each metathoracic (T3) hemisegment during a narrow time window of development (Table 1) (Kotrla and Goodman, 1983, 1984). Included among these four are the two excitatory motoneurons that innervate the extensor tibiae (ETi) muscle in the leg (Figure 9)—FETi (from NB 4–4) and SETi (from NB 2–3). The other two neurons are unidentified interganglionic interneurons called IN X and IN Y; nothing is known about their axonal pathways. The Mes-2 antigen is on the cell surface, as indicated by antibody staining of the cells either in highly dissected living preparations or in fixed but unpermeabilized preparations (cytoplasmic antigens require detergent treatment in these whole-mount preparations).

The growth cones of FETi and SETi follow the same final axonal pathway and innervate the same muscle. Of the 50 or so motoneurons in each T3 hemisegment whose growth cones innervate muscles at the base or in the limb bud, only FETi and SETi express the Mes-2 antigen. Although these two motoneurons arise from different neuroblasts, they alone among their two families share a final pathway and target, and they alone among their families express the Mes-2 antigen. Thus it seems reasonable to suppose that the Mes-2 antigen might be involved in their common pathway and/or target choice.

Further support for this notion comes from an examination of the temporal pattern of expression of the Mes-2 antigen. The antigen is not expressed on either cell while their two growth cones extend in different central and peripheral axonal pathways. SETi's growth cone pioneers nerve 3 (Ho and Goodman, 1982) and extends distally into the femur of the metathoracic limb bud, where it extends along the muscle pioneers for the ETi muscle (Ho et al., 1983). During this period, the FETi growth cone reaches the exit point of the CNS, where the axon bundles for nerves 3, 4, and 5 diverge. Although it is within short filopodial grasp (2 μm) of SETi's axon, FETi's growth cone extends out to nerve 5 (Figure 10). At this time (about 40% of development), neither cell expresses the Mes-2 antigen. FETi's growth cone extends past the muscle pioneer for coxal muscle 133A and then makes another choice at a pathway trifurcation.

At 49% of development, FETi's growth cone once again comes within contact of SETi's axon at a choice point in the limb bud. However, the second time they meet they behave quite differently toward one another; FETi's growth cone chooses to fasciculate with SETi's axon and follow it distally to the ETi muscle pioneer. Clearly, one or both of the cells has changed since their last encounter over 24 hours earlier.

Interestingly, expression of the Mes-2 antigen begins just as FETi's growth cone reaches the SETi axon at 48.5% of development. Their expression of the Mes-2 antigen is transient during this period of axon outgrowth and innervation (Table 1). After reaching and innervating their common muscle, expression of the Mes-2 antigen dramatically decreases and eventually disappears, never to be seen again on their surface during larval and adult life. The best hypothesis is that the Mes-2 antigen is involved in the selective fasciculation of FETi's growth cone onto SETi's axon. When neither cell expresses the antigen, they do not fasciculate; when both cells express the antigen, they do. At the very least, the Mes-2 antibody demonstrates the remarkable specificity of a cell surface antigen for a very small subset of neurons (four out of 1000, two of which fasciculate in the periphery), and the transient expression of this antigen during the narrow

Table 1. Transient Expression of Mes-2 Cell Surface Antigen

Cell	Staining							
	40%	45%	50%	55%	60%	65%	70%	80%
SETi	−	−	+ +	+ +	+	+	−	−
FETi	−	−	+	+ +	+ +	+ +	−	−
IN X	−	−	+	+	+ +	+ +	+ +	+ +
IN Y	−	−	−	−	+	+ +	+	+

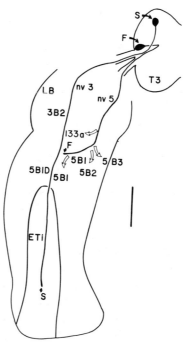

Figure 10. Camera lucida drawing of the axons of the SETi and FETi motoneurons at 48.5% of development, when both FETi and SETi begin expressing the Mes-2 antigen on their surfaces. At this time, FETi's growth cone has just arrived at SETi's axon in the femur of the limb bud (LB). The cell bodies of both motoneurons are in the T3 ganglion, and SETi's growth cone at this stage is already extending across the extensor tibiae (ETi) muscle pioneers (Ho et al., 1983). On its way to its target, FETi's growth cone comes within 2 μm of SETi's axon as it exits the CNS; neither cell expressed the Mes-2 antigen. On its way out to nerve 5, FETi's growth cone extends past the muscle pioneer for coxal muscle 133a, where the growth cone of motoneuron Df gets off. It then reaches a pathway trifurcation in the limb bud, in which it extends along nerve 5B1 instead of nerves 5B3 or 5B2. It then extends off nerve 5B1 towards SETi's axon in nerve 3B2 (which becomes 5B1D). After 48.5% of development, FETi fasciculates on SETi's axon and extends distally towards the ETi muscle. For implications, see text. Calibration bar = 50 μm.

time window of development in which axon outgrowth and specific cell recognition occur.

SELECTIVE FASCICULATION MODEL

Cellular and immunological studies on cell recognition and selective fasciculation in the grasshopper embryo suggest the model presented in Figure 11. Individual axon fascicles (a and c), or subsets of axons within a fascicle (d), residing in a particular region of neuropil are labeled by unique cell surface molecules. A growth cone extends along the axonal scaffold (1, 2, and 3) according to its adhesive affinity for these surface molecules. We favor a hypothesis based on dynamic, rather than static, processes in which the surface molecules on the growth cone change during the course of these navigations as a result of cellular interactions at either past or present intersections. By a series of sequential changes in surface affinities, an individual growth cone switches from one axon fascicle to another as it extends toward its final target.

SELECTIVE FASCICULATION IN THE *DROSOPHILA* EMBRYO

Whereas the grasshopper embryo is an ideal system for cellular and immunological studies of neuronal cell recognition, the *Drosophila* embryo has obvious attributes

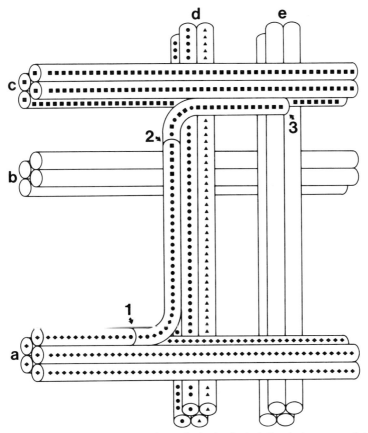

Figure 11. *Schematic model for cell recognition and selective fasciculation during neuronal development.* This model is suggested by our cellular and immunological studies on cell recognition and selective fasciculation in the grasshopper embryo. Individual axon fascicles (a and c), or subsets of axons within a fascicle (d), residing in a particular region of neuropil are labeled by unique cell surface molecules (diamonds, circles, triangles, squares). An individual growth cone extends along the axonal scaffold (positions 1, 2, and 3 for the same growth cone) according to its adhesive affinity for these surface molecules. We favor a hypothesis based on dynamic, rather than static, processes in which the surface molecules on the growth cone change during the course of these navigations, as a result of cellular interactions at either past or present intersections. For example, by the time the growth cone gets to position 2, it is already starting to express an affinity for "squares." By a series of sequential changes in surface affinities, an individual growth cone switches from one axon fascicle to another as it extends towards its final target. In a mixed fascicle, such as d, the fasciculation of two different subsets of axons on each other must represent some common surface affinity in addition to their different specificities.

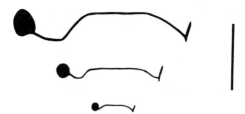

Figure 12. *The homologous G neuron at comparable stages in the development of the grasshopper (top), moth (middle), and fly (bottom) embryo.* These were revealed by camera lucida drawings of Lucifer yellow injections. The grasshopper (*Schistocerca americana*) is at 40% of development, the moth (*Manduca sexta*) is of unknown age, and the fly (*Drosophila melanogaster*) is at 12 hours of development. Calibration bar = 50 µm.

for a potential molecular genetic approach. The problem with studying cell recognition in the CNS of *Drosophila* has always been the small size, inaccessibility, and lack of knowledge about its embryonic neurons. However, we recently discovered that we could scale down the same cellular approaches used so successfully with the grasshopper to study the developing CNS of the *Drosophila* embryo. Much to our delight, we found that the CNS of the early fly embryo is a miniature replica of the grasshopper embryo in terms of its identified neurons, their growth cones, and their selective fasciculation choices (Thomas et al., 1984). For example, *Drosophila* has a neuron homologous to the grasshopper G neuron (Figure 12); its growth cone turns anteriorly in a lateral bundle of axons. When this fascicle was reconstructed in the TEM, it was found to consist of two P axons and two A axons. Moreover, the tip of the G growth cone associates with the P and not the A axons, just as in the grasshopper (Figure 2). These and other results (Thomas et al., 1984) indicate that the patterns of selective fasciculation are largely identical in both insect species.

FUTURE PROSPECTS

In order to isolate and characterize the surface molecules implicated by cellular and immunological studies of the grasshopper embryo, we have turned our attention to the *Drosophila* embryo for molecular genetic approaches. The early cell-recognition events involving selective filopodial contacts and fasciculation, so well characterized in the grasshopper, occur between hours 10 and 13 of fly development (Thomas et al., 1984). Fortunately, this is before the appearance of neurotransmitters, synapses, and excitability. We can now isolate 100,000 *Drosophila* CNSs from this three-hour period in order to construct specific cDNA libraries and probes and to immunize mice for monoclonal antibody production.

We are isolating specific cDNA probes and monoclonal antibodies to the embryonic fly CNS in an attempt to isolate genes expressed in small subsets of embryonic neurons. We hope that characterization of these genes, the proteins they encode, and the neurons that express them (with immunocytochemical and *in situ* hybridization approaches) will lead us to molecules expressed on the surface of neurons whose axons fasciculate in the neuropil. The functional role of these molecules in cell recognition can then readily be tested by a genetic analysis in *Drosophila*.

ACKNOWLEDGMENTS

We thank Michael Bate, Kathryn Kotrla, and Francis Thomas for their contributions to the work described in this review. This work was supported by grants from the National Science Foundation, the March of Dimes, the National Institutes of Health, and the McKnight Foundation (to C.S.G.); by an NICHHD training grant (to M.J.B.); and by a Helen Hay Whitney postdoctoral fellowship (to J.B.T.).

REFERENCES

Bastiani, M. J., and C. S. Goodman (1984a) Neuronal growth cones: Specific interactions mediated by filopodial insertion and induction of coated vesicles. *Proc. Natl. Acad. Sci. USA* **81**:1849–1853.

Bastiani, M. J., and C. S. Goodman (1984b) The first growth cones in the central nervous system of the grasshopper embryo. In *Cellular and Molecular Approaches to Neuronal Development*, I. Black, ed., pp. 63–84, Plenum, New York.

Bastiani, M.J., S. Dulac, and C. S. Goodman (1984a) The first growth cones in insect embryos: Model system for studying the development of neuronal specificity. In *Model Neural Networks and Behavior*, A. Selverstone, ed., Plenum, New York (in press).

Bastiani, M. J., J. A. Raper, and C. S. Goodman (1984b) Pathfinding by neuronal growth cones in grasshopper embryos. III. Selective affinity of the G growth cone for the P cells within the A/P fascicle. *J. Neurosci.* **4**:2311–2328.

Bate, C. M. (1976) Embryogenesis of an insect nervous system. I. A map of the thoracic and abdominal neuroblasts in *Locusta migratoria*. *J. Embryol. Exp. Morphol.* **35**:107–123.

Bate, C. M., and E. B. Grunewald (1981) Embryogenesis of an insect nervous system. II. A second class of precursor cells and the origin of the intersegmental connectives. *J. Embryol. Exp. Morphol.* **61**:317–330.

Bray, D. (1982) Filopodial contraction and growth cone guidance. In *Cell Behavior*, R. Bellairs, A. Curtis, and G. Dunn, eds., pp. 299–318, Cambridge Univ. Press, Cambridge.

Goodman, C. S., J. A. Raper, R. Ho, and S. Chang (1982) Pathfinding by neuronal growth cones in grasshopper embryos. *Symp. Soc. Dev. Biol.* **40**:275–316.

Harris, A. L., and C. S. Goodman (1983) Guidance of a growth cone to its contralateral pathway in the absence of the axon it normally follows during grasshopper embryogenesis. *Soc. Neurosci. Abstr.* **9**:1044.

Ho, R. K., and C. S. Goodman (1982) Peripheral pathways are pioneered by an array of central and peripheral neurones in grasshopper embryos. *Nature* **297**:404–406.

Ho, R. K., E. E. Ball, and C. S. Goodman (1983) Muscle pioneers: Large mesodermal cells that erect a scaffold for developing muscles and motoneurones in grasshopper embryos. *Nature* **301**:66–69.

Kotrla, K. J., and C. S. Goodman (1983) Transient expression of cell surface antigen on two neurons that share a common final pathway and target in the grasshopper embryo. *Soc. Neurosci. Abstr.* **9**:1045.

Kotrla, K. J., and C. S. Goodman (1984) Transient expression of cell surface antigen on a small subset of neurones during embryonic development. *Nature* **311**:151–153.

Letourneau, P. C. (1982) Nerve fiber growth and its regulation by extrinsic factors. In *Neuronal Development*, N. C. Spitzer, ed., pp. 213–254, Plenum, New York.

Raper, J. A., M. J. Bastiani, and C. S. Goodman (1983a) Pathfinding by neuronal growth cones in grasshopper embryos. I. Divergent choices made by the growth cones of sibling neurons. *J. Neurosci.* **3**:20–30.

Raper, J. A., M. J. Bastiani, and C. S. Goodman (1983b) Pathfinding by neuronal growth cones in grasshopper embryos. II. Selective fasciculation onto specific axonal pathways. *J. Neurosci.* **3**:31–41.

Raper, J. A., M. J. Bastiani, and C. S. Goodman (1983c) Guidance of neuronal growth cones: Selective fasciculation in the grasshopper embryo. *Cold Spring Harbor Symp. Quant. Biol.* **48**:587–598.

Raper, J. A., M. J. Bastiani, and C. S. Goodman (1984) Pathfinding by neuronal growth cones in grasshopper embryos. IV. The effects of ablating the A and P axons upon the behavior of the G growth cone. *J. Neurosci.* **4**:2329–2345.

Taghert, P. H., M. J. Bastiani, R. K. Ho, and C. S. Goodman (1982) Guidance of pioneer growth cones: Filopodial contacts and coupling revealed with an antibody to Lucifer Yellow. *Dev. Biol.* **94**:391–399.

Thomas, J. B., M. J. Bastiani, C. M. Bate, and C. S. Goodman (1984) From grasshopper to *Drosophila*: A common plan for neuronal development. *Nature* **310**:203–207.

Chapter 14

Developmental Interactions between Neurons in Insects

JOHN S. EDWARDS
MARK R. MEYER

ABSTRACT

The insect nervous system, which traditionally has been considered to mediate rigidly stereotyped behavior and thus, by implication, to be limited in the flexibility of its growth and injury responses, is proving to be capable of responding actively to a changed milieu.

Selected examples of embryonic and postembryonic neuronal growth patterns illustrate the capacity of neurons for localized growth and for response to changes in their neural connections after deafferentation and axotomy or during metamorphosis. These interactions between neurons imply a means for intercellular communication in addition to those which serve integrative behavioral functions.

Our studies of the cercal sensory pathway and its relationship to the metabolism of identified giant interneurons in the cricket Acheta domesticus *provide a window to possible mechanisms underlying the growth and maintenance of neurons.*

Fundamental changes in our perceptions of the insect nervous system have accelerated in the last decade. Once seen to be a rigidly wired vehicle for rather strictly stereotyped behavior and, as such, remote in principle and practice from that of vertebrates, the insect nervous system is now undergoing a reevaluation that admits more flexibility and plasticity. Neurons and neuropil once considered to be rigidly specified in terms of form and function are now known to be capable of a variety of reactive responses to their milieu.

It may be no coincidence that, at the same time, earlier emphasis on a rigid repertoire of innate behavior has yielded to capacities for learning and flexibility of behavior. The honeybee, with a brain less than a millionth the mass of our own, has long been known to learn skillfully, albeit within circumscribed modalities. Isolated grasshopper ganglia have been shown to "learn," and we now know that flies as seemingly dopey as *Drosophila* are capable of learning. Recently, crayfish have been shown to modulate their escape response in complex ways in relation to the size of their current morsel of food (Bellman and Krasne, 1983), and such flexibility will doubtless be shown for insects as well.

Our objective in this chapter is to present selected examples of interactions between insect neurons that demonstrate their capacity for developmental and metabolic responses to their milieu. Aspects of neuronal interactions underlying mechanisms of specificity in the formation of functional connections are touched on but are not a major theme. Rather, we are concerned with the changes in form and metabolic function that may serve to underlie the mechanisms of recognition and the formation and maintenance of functional connectivity, but do not necessarily constitute the immediate mechanism for neural specificity.

We first consider structural aspects of embryonic and postembryonic neuronal growth patterns and examine some general themes concerning the independence of the periphery in development, the dependence of central development on peripheral input, and neuronal responses to damage and sensory deprivation. We then present new data relating metabolic responses to changing peripheral input and their relations to the phenomena considered at the beginning of this chapter.

EMBRYONIC NEURAL GROWTH PATTERNS

During embryogenesis, axons are generated as outgrowths from central ganglion cells (the progeny of rigidly specified nests of neuroblasts) or from epidermal cells that differentiate at specific loci in the periphery. The growing axons may produce few temporary branches in the course of growth toward their targets, as with sensory axons, whose growth cones seem generally to "know" where they should go. Alternatively, they may produce supernumerary branches in both the embryonic neuropil and the periphery, as is typical of motor neurons.

Sensory Systems

The first neurons of the sensory system invariably arise in highly specified positions and forge pioneering connections with the center by means of a relay in which distalmost axons growing centripetally contact successively more proximal cells (Goodman et al., 1982). Pathfinding by these cells is mediated by filopodial extrusions of the growth cones, which explore the cellular landscape ahead and make choices that imply recognition of surface characteristics. Experimental deletion of pioneer fibers in embryonic sensory appendages of crickets (Edwards et al., 1981) supports this notion, but Keshishian (1980) questions the necessity of pioneer tracts in locust legs for the formation of directive pathways to the center.

In the locust cercal sensory nerve, the pathways taken by pioneer afferents through the neuropil are strikingly linear. It is the later-arriving axons that diverge from the primary pathway to establish the characteristic cercal sensory projection (Shankland and Goodman, 1982).

Motor Systems

In contrast to sensory development, the generation of first motor connections between center and periphery is characterized by the formation of many supernumerary branches in both the neuropil and the periphery. Supernumerary branches disappear from the neuropil as the growing axon leaves, whereas peripheral branches to inappropriate targets regress at the time contacts are established with the correct muscle. The appearance of transmitter in these axons coincides with the formation of axonal branches over the surface of the muscle. These events, as elucidated in the locust, have close parallels with comparable developmental stages in vertebrates (Spitzer, 1979).

POSTEMBRYONIC NEURONAL GROWTH PATTERNS

This topic has been reviewed many times (Edwards, 1969; Bate, 1978; Anderson et al., 1980) and it is sufficient here to outline major features. New sensory neurons arise in the epidermis, and their sensilla are added to the sensory array at each molt during postembryonic development. The new axons of these sensory neurons associate with the arbor of existing sensory bundles, following twigs to branches and branches to major sensory nerves by mechanisms generally described as contact guidance. After entry to the ganglia via peripheral nerve trunks, sensory neurons normally follow preexisting tracts to regions of termination where postsynaptic fibers must grow appropriately to accommodate new synapses. Pathfinding within the neuropil lies outside our topic, but it should be noted that the question is under active experimental investigation through techniques of transplantation (Murphey et al., 1983) and genetic modification (e.g., Palka and Schubiger, 1980; Stocker, 1982).

The postembryonic growth of motor neurons and interneurons involves an increase in cell dimensions but, with localized exceptions, there is little or no increase in the size of cell populations formed during embryogenesis. Most exceptions to this rule arise during the transformation of the nervous system during metamorphosis of holometabolous insects.

INTERDEPENDENCE OF CENTER AND PERIPHERY DURING DEVELOPMENT

In short, the bulk of evidence at both embryonic and postembryonic stages of development speaks for an independence of the sensory system from the necessity for central contact, but conversely, a dependence of the center on the periphery for full development. Some typical examples at the tissue and single-cell level from embryonic and postembryonic stages serve to illustrate this generalization.

Sensory Systems

The independence of the periphery was first established at the tissue level with the visual system (Meinertzhagen, 1973). The arthropod eye, with its precise spatial geometry, has provided important data at the level of cell patterning in the receptor (see Anderson et al., 1980, for a review). The locust retina, for example, continues to proliferate and generate new ommatidia after connections with the brain have been severed (Anderson, 1978). Axons from sensilla in isolated imaginal disks of *Drosophila* differentiating *in vitro* form organized nerve bundles (Edwards et al., 1978), and antennal chemoreceptors differentiate normally in the absence of the brain in the moth *Manduca sexta* (Sanes et al., 1976). The integumental milieu does exercise control of sensillar identity; the surrounding epidermis determines the pattern of central projection in locusts (Anderson and Bacon, 1979).

In contrast to peripheral autonomy, central neurons depend on contact with afferent fibers for their differentiation. In flies the process of laminar differentiation proceeds rapidly after contact with retinular cells. For example, patches of wild-type retina on the eye of *Drosophila* overlie islands of normal lamina surrounded by disorganized neuropil that lack retinal contacts (Meyerowitz and Kankel, 1978).

Macagno's (1979) work with *Daphnia* took such analyses of the visual system to the single-cell level. The arrival of the first axons from a bundle of ommatidial neurons, the so-called lead fibers that make contact with laminar cells throughout their extensive growth cones, triggers the differentiation of laminar cells.

A compelling example of the role of afferent fibers in evoking central differentiation comes from work on the brain of *Manduca sexta* (Schneiderman et al., 1982), in which the olfactory lobe of the brain is sexually dimorphic. Transplants, at larval stages, of male antennal imaginal disks to females give rise to gynandromorphic moths in which the brain on the side bearing the male antenna contains the characteristic male macroglomerular complex.

Motor Systems

The embryonic development of motor nerves has been examined in grasshopper embryos, in which both motor neurons and mesodermal cells destined to become muscles are large and directly observable under Nomarski optics (Ho et al., 1983). Muscle pioneer cells arise early in embryonic development when distances between them and the neurons destined to provide their innervation are short and the neural environment is relatively simple. They evidently serve as a scaffold that provides the cues for neuronal outgrowth. The general picture arising from this study implies a capacity for recognition and specificity, and yet it is also known that some motor neurons do make supernumerary branches during embryogenesis, as is the case with the dorsal unpaired median extensor tibiae neuron of the locust that has been observed by Goodman et al. (1979). Motor-nerve outgrowth in the embryo shows some flexibility; neurons normally destined to innervate a limb that has been excised early in embryogenesis will leave the ganglion variably by other routes (Whitington and Seifert, 1982).

The capacity of insect motor neurons to regenerate peripheral parts has long been known (Bodenstein, 1953), but the first approaches to the structure of central arborizations, by use of cobalt filling, emphasized their stability. Motor neurons and their central processes were found to persist unchanged in the adult long after the leg they supplied had been amputated (Pitman et al., 1973), and the removal from the locust wing base of developing sensilla that provide input to flight motor neurons was found to have no perceptible effect on the development of motor innervation (Kutsch, 1974).

Motor neurons may persist even after prolonged severance from the cell body (Clark, 1976a,b). In the cricket *Teleogryllus oceanicus*, axons and arborizations of an identified thoracic motor nerve persisted long after the neurite connecting the cell body was severed. Slow degeneration sets in after about 50 days, but before that time some supernumerary branches or sprouts may be formed. The extensive glial wrapping within the neuropil was evidently necessary for metabolic support; distal segments of the same neuron severed outside the neuropil degenerated quickly.

Dendrites of motor neurons were found to sprout profusely in the neuropil of cockroach ganglia after nerve roots entering the ganglia were cut (Pitman and Rand, 1982). The sprouting proved to be independent of direct surgical damage to the identified neuron (the metathoracic fast coxal depressor). The sprouts reach regions not normally occupied by the neuron, but grow preferentially in regions containing degenerating neurons. It is not known whether the sprouts make functional contacts with other neurons. Similar, though less prolific, sprouting of supernumerary dendrites was also observed in crickets by Clark (1976a,b) after severing the cell body of an identified neuron from its central arborization. Again the relationships established with surrounding cells by these new processes is unknown.

Collateral sprouting of neurons in response to muscle denervation, a well-known phenomenon in the vertebrates, also occurs in insects. A particular case has been analyzed by Donaldson and Josephson (1981), who followed the consequences of severing motor nerves to the extensor tibiae muscles of *Teleogryllus oceanicus*. A fast fiber innervates the middle region of the muscle and a slow axon innervates the ends. After the fast axon is severed, slow-axon collaterals advance into the fast-muscle field from each end. The new connections are stable and remain functional after the fast muscle has regenerated, as in goldfish ocular muscles in which foreign innervation persists in a functional form after regeneration of the native nerve (Scott, 1977).

Clearly, insect motor neurons retain the capacity for growth and sprouting during postembryonic and adult stages. During the major reorganization of the body at metamorphosis, many neurons are lost and new ones generated, but some undergo a restructuring of their arborizations, reflecting the transition during the pupal stage from the relatively sedentary, wormlike larva to the mobile adult insect. Axons in longitudinal tracts may shorten dramatically (Pipa, 1978), and surviving central neurons may be radically restructured. The arborization of identified larval motor neurons has been shown to collapse and withdraw partially during metamorphosis, after which more extensive branches are rebuilt (Truman and Reiss, 1976).

GROWTH AND FORM OF IDENTIFIED INTERNEURONS

The advent of the cobalt- and other axon-staining techniques such as horseradish peroxidase opened the way to analysis of motor and sensory arborizations in ganglia through the relatively simple method of filling through a peripheral stump. The arborization of interneurons presented greater difficulty. They could be filled en masse by cutting connectives between segmental ganglia, a seemingly nonselective technique that is useful in practice because the interneurons most commonly studied are giant plurisegmental fibers and, because they fill at a faster rate than the major population of small fibers, they can be distinguished. The alternate approach, based on the constancy of cell body and axon position, is the intracellular injection of marker, sometimes after physiological recordings have been made.

In the remainder of this chapter we concentrate on the properties of a set of giant interneurons of orthopteroid insects (e.g., cockroach, cricket, grasshopper) that receive mechanosensory input from posterior sensory appendages, the abdominal cerci. The abdominal cerci are covered with sensory hairs arrayed in remarkably constant patterns. Two sets of sensilla have proved to be most amenable to study: the filiform sensilla, which are long, delicately inserted hairs whose associated neurons respond to mechanical displacement by air movements, including sound; and a second set of clavate sensilla, situated on the basal ventromedial surface, whose role in gravity reception is now established. The activation of a set of giant interneurons by filiform sensilla provides for a startle response (Edwards and Palka, 1974). Cerci regenerate after removal in immature stages, and the new sensory neurons of the regenerate are able to reestablish the startle reflex even after prolonged deafferentation (Palka and Edwards, 1974). The system has now been quite thoroughly characterized (see Anderson et al., 1980, for a review) and has been explored in considerable detail by Murphey and his colleagues (e.g., Murphey et al., 1983).

Effects of deafferentation on giant interneurons were examined in one of the earliest applications of the cobalt filling technique. Tweedle et al. (1973) reported that the arborizations of an identified neuron in the adult cockroach *Periplaneta americana* remained unchanged after severance. This conclusion was consistent with the general concepts of that time concerning the fixity of insect neurons. However, subsequent studies of both locust embryos and postembryonic development in crickets revealed significant effects of peripheral connections on the growth and form of giant interneurons. In the locust embryo, cercal innervation influences the number of dendritic branches and their size (Shankland et al., 1982). The earliest phases of development of the identified medial giant interneuron (MGI) proceed normally, but at about 70% of full development, when afferents would normally arrive, their absence causes stunting of arborizations within the area of neuropil to which they would normally project.

MGI dendrites arise as secondary processes that sprout at characteristic points on the axon after the growth cone has passed on. They arborize to form a miniature version of the adult interneuron. Other than ephemeral filopodia, the embryonic interneuron forms no supernumerary branches and there is no secondary pruning, but rather a progressive branching with little variability between individuals. The majority of these branches remain within the cercal domain. In the absence of cercal input there are 30% fewer dendritic branches, and growth is stunted in regions to which cercal afferents normally project, but the effects are restricted to that part of the interneuron within the cercal domain; dendritic growth in other regions is enhanced. The localized response to afferent contact found in embryos also holds during postembryonic development. When the comparable experiment was performed in crickets during postembryonic development by removing cerci from hatchlings and regenerates from subsequent instars throughout postembryonic development (Murphey et al., 1975), the suppression of dendritic growth was consistent with the overall diminution of neuropil volume (30%) in deafferented regions. However, the major branching pattern was not altered and the axonal process of the giant interneuron was virtually unaffected outside the cercal domain. If cerci were permitted to regenerate after a period of several instars with no cercal input, afferents from the regenerate compensated for the intervening loss, producing dendrites of near normal size. Nonetheless, persistent imbalance in contralateral inhibitory input from the intact cerci prevented the restoration of normal sensory function, and deprivation of sensory stimuli during postembryonic development in this system appears to have no effect on the growth of dendrites (Murphey et al., 1977); it seems that cell contact, rather than the sensory input, is sufficient for normal growth.

A further example of localized response of identified neurons to deafferentation comes from the auditory system of *Teleogryllus* (Hoy et al., 1978), in which an identified auditory interneuron in the prothoracic ganglion was deprived of normal auditory input by removing the ipsilateral foreleg (on which the ear is situated) immediately after the animal hatched. The sensory neurons of the ear are unusual in that they do not make contact with their central targets until after the completion of embryonic development, so that amputation of the foreleg at hatching prevents access of any sensory neurons to the auditory interneuron. The form of the interneuron in adults unilaterally deprived of auditory input was altered significantly. As might be expected, dendrites that normally lie among auditory afferents were reduced in size, but the medially directed dendrites that normally would arborize in the ipsilateral ventral acoustic neuropil crossed the midline to terminate in the contralateral counterpart.

The capacity for exuberant sprouting of motor neurons in cockroach thoracic ganglia (Pitman and Rand, 1982) is met, indeed surpassed, by the giant interneurons of the cricket. When the axons of these well-known cells are cut where they leave the terminal abdominal ganglion to traverse the length of the ventral nerve cord, their dendritic arborization within the neuropil of the ganglion undergoes extensive sprouting, particularly from the ends of primary dendrites (Roederer and Cohen, 1983). The same cells respond to sectioning or crushing at more distal points along the axon by producing neurites from the axon stump, and under certain treatments the exceptional emergence of processes from the cell body may occur.

Taken together, the examples given above support the view that the growth of arborizations of insect interneurons is a regionally autonomous response to local conditions related principally to the presence or absence of presynaptic elements. The embryonic response of deprived interneurons, which give evidence of some hypertrophy outside the domain of deafferentation, and the response of the cricket auditory interneurons seem to be exceptions to the general observation that growth of interneurons is simply related to the number of afferents for which it is a target. Metabolic studies to be discussed next demonstrate dynamic interaction between afferent axons and their central targets and may provide insight into the mechanism of interaction between center and periphery.

METABOLIC INTERACTIONS IN THE CRICKET CERCAL SENSORY–GIANT INTERNEURON PATHWAY

Characterizing the role of peripheral (i.e., sensory) neurons in the growth and maintenance of central target neurons has been an intensely active area of study in neurobiology for more than half a century (see Cowan, 1970; Smith, 1977, for reviews). Despite these efforts, the mechanisms by which sensory neurons induce or promote the growth of their postsynaptic targets remain obscure. The rules governing these mechanisms are, at best, reflected only by alluring paradoxes. For example, we would like to understand why some classes of central neurons are so strongly dependent on intact afferent innervation for proper development—or even survival—whereas others seem indifferent to the presence or absence of sensory synapses.

It is clear from the examples reviewed above that there is a striking contrast between insect peripheral and central neurons in their ability to develop and differentiate in each other's absence. As with vertebrate central neurons, neurons comprising the arthropod central nervous system (CNS) demonstrate a wide variety of responses to deafferentation, from apparent cell death (Macagno, 1979; Maxwell and Hildebrand, 1981) or marked retardation of dendritic growth (Murphey et al., 1975) to little, if any, detectable deviation from normal development (Tweedle et al., 1973; Sanes et al., 1977).

Morphological Correlates of Deafferentation

Identifiable giant interneurons (GI) in the terminal ganglion of the cricket CNS respond to deafferentation carried out throughout postembryonic development with diminished growth of their primary dendritic processes and dendritic spines (Murphey et al., 1975; Murphey and Levine, 1980); the large somata and axonal processes of these GIs, far removed from the dendritic branches (Figure 2), do not appear to be affected by prolonged deafferentation. Even GI dendritic branches contralateral to the deafferented zone of neuropil are unaffected. Thus, the reactivity of GI to loss of presynaptic input is a localized phenomenon, concentrated in glomerular regions of the ganglion neuropil where synapses are made between cercal sensory neurons and GI dendrites.

The cercal sensory–GI pathway has been well characterized morphologically. This feature, together with the findings that postsynaptic target cells survive denervation and that the effects of deafferentation are discrete, makes this pathway a useful system with which to explore the nature of the interaction between sensory and central neurons during development.

When the cerci (posterior sensory appendages) are removed, degeneration of cercal afferent axons follows swiftly. The pace of degeneration in distal (central) segments of sensory neurons is astonishing. The cytological picture in terminals of cercal sensory axons at 3.5 hours after cercectomy is already chaotic, with swollen and burst mitochondria embedded in an amorphous osmiophilic matrix (K.-D. Ernst and J. S. Edwards, in preparation) as shown in Figure 1. The majority of sensory axons and terminals have degenerated within 1–2 weeks of cercectomy. Despite such rapid removal of presynaptic neurons, the marked effects of deafferentation on GI dendrites take longer to manifest themselves,

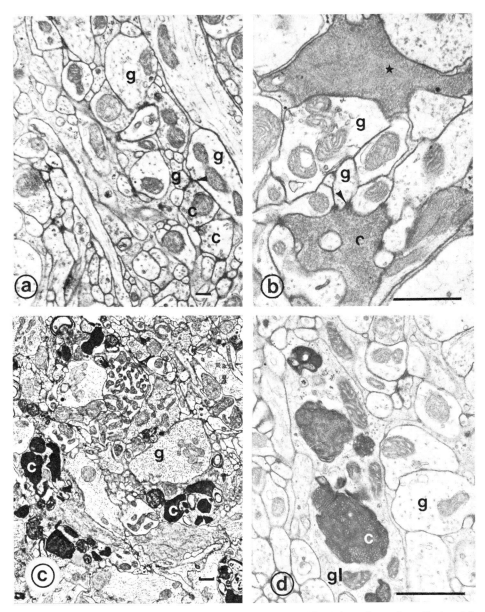

Figure 1. *Neuropil in terminal abdominal ganglion of the house cricket* Acheta domesticus. Region rich in cercal sensory fibers and synapses with giant interneuron arborizations is shown. *a*: Normal animal. Several afferent fibers (c) synapse (arrow) onto giant interneuron processes (g). *b*: 3.5 hours after amputation of cercus. Degeneration is evident in cercal afferent terminals (c). Mitochondria (star) are swollen, and cytoplasm is vesiculate and osmiophilic. (Print courtesy of Dr. K.-D. Ernst.) *c*: 12 hours after amputation of cercus. Dark profiles of degenerating cercal afferents (c) are separated from giant fiber arborizations (g) by glial processes. *d*: 12 hours after amputation. Detail showing vesiculate and densely osmiophilic cercal afferents (c) engulfed by glia (gl) and separated from giant interneuron processes (g). Calibration bars = 1 μm.

maximal retardation of dendritic growth occurring only after a period of deafferentation sustained throughout postembryonic growth. Furthermore, deafferentation during stages of ongoing neural development seems to produce more profound effects than when cercectomy is performed in the adult animal, in which the GIs are fully matured (Meyer and Edwards, 1982). Taken together, these findings parallel results reported for various vertebrate central neurons (see Smith, 1977) and serve to underscore the recurring theme that neurons are most susceptible to changes in their state of innervation during critical periods of active growth.

Biochemical Correlates of Deafferentation

How might sensory neurons influence or modulate the growth of their postsynaptic targets? The many possibilities include what may be considered both passive and active roles played by intact sensory terminals in synaptic areas of ganglion neuropil.

Abundant sensory synapses on GI dendrites might serve as an anchoring substrate for growing dendritic processes and spines. Loss of these terminals after deafferentation might then lead to passive retraction of spines and processes in the absence of the proper neuropilar framework. However, a number of observations argue against this possibility. First, maximal changes in dendritic length are closely correlated with long-term duration of deafferentation. If a retraction mechanism were operating, one would expect reductions in dendritic length to occur in a time course similar to the loss of terminals; yet loss of sensory terminals begins rapidly (within hours) after cercectomy. Second, the finding that the morphological effects of deafferentation on GI dendritic length can largely be reversed by allowing the return of sensory innervation (Murphey et al., 1977) strongly implies that active GI dendritic growth is involved. Though the dense glomerular structure of the neuropil may confer some structural stability to the conformation of the giant interneuron dendritic arbors, it is more likely that local metabolic transneuronal interactions actively participate in their development. Studies on the development of GI in grasshopper embryos also lend support to this notion (Shankland et al., 1982).

If more active processes are in play, what might they be? The word "trophic" is often invoked to describe an active growth influence of a cell upon a target cell, and there are many well-documented examples of such trophic or inductive properties of neurons in the current literature of neurobiology (Purves, 1976; Black, 1978). Clearly, developmental interactions in the growing cricket CNS could be cataloged under this rubric, but we are also interested in looking for the presence of a cellular metabolic mechanism whereby sensory neurons might play a definable role in regulating the growth of the GI dendrites.

When we compared the *in vitro* incorporation of [^{3}H]leucine into control and chronically deafferented terminal ganglia, we found that uptake and incorporation into deafferented ganglia was markedly reduced (Meyer and Edwards, 1982). No such changes were observed when ganglia that had been deafferented after adult development were compared to controls. Thus, changes in amino acid metabolism appeared to correlate with the time and duration of deafferentation, in accord with the magnitude of alterations in neuropil volume and GI dendrite

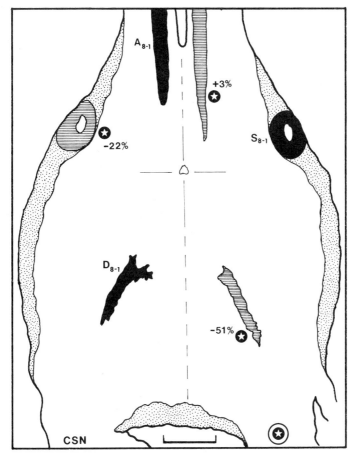

Figure 2. *Protein metabolism in control and chronically deafferented MGI.* Schematic representation of terminal ganglion derived from composite horizontal sections through ganglion to show bilateral organization of elements of the medial giant interneuron (MGI; interneuron 8–1). Somata (S_{8-1}), axon segments (A_{8-1}), and primary dendrites (D_{8-1}) are superposed on tracing of ganglion. Cercal sensory nerves (CSN) enter posteriorly; chronically cercectomized CSN is starred. Deafferented MGI elements are denoted by stars and cross-hatchings. Mean percent decreases in incorporation of [^{3}H]leucine in MGI elements are indicated, as determined from quantitative autoradiographic analysis (see Meyer and Edwards, 1982). Calibration bar = 100 μm.

length observed in previous studies. Clearly, deafferentation had a striking effect on protein synthesis in the ganglion, and this was not the typical injury response observed for most denervated neurons (see Watson, 1974).

In order to look specifically at the metabolic consequences of deafferentation in the GI, we took full advantage of the well-characterized properties of the cercal sensory–GI pathway in the terminal ganglion. The medial giant interneuron (MGI; interneuron 8–1) receives a majority of its sensory innervation from the cercal nerve. It is distinctive in its size, location, and topography. More importantly, the primary dendritic branches, soma, and axon of MGI are large and visible at the light-microscope level, and MGI thus lends itself well to the methods of quantitative autoradiography (Figure 2). The presence of paired MGIs in the ganglion allows comparison of chronically deafferented and normally innervated neurons within the same ganglion. Thus, we could look directly at protein metabolism within discrete cellular compartments (e.g., primary dendrites, somata, axons) of control and deafferented MGI after incubation in [^{3}H]leucine and preparation for autoradiography.

We found that the primary dendrites of deafferented MGIs contained more than 50% less labeled protein than paired contralateral control dendrites and that deafferented somata incorporated 22% less [^{3}H]leucine into protein than did control somata (Figure 2). In contrast, grain densities within the axons of control and deafferented MGI were not significantly different. The results of this set of experiments showed for the first time that removal of presynaptic input from postsynaptic central interneurons can lead to reductions in intracellular protein metabolism within discrete regions of the target cell. It is certainly reasonable to assume that modulation of the status of the protein synthetic machinery of central target MGIs may have profound effects on the overall metabolism and subsequent growth of the MGI (and other target interneuron) dendrites. Although our findings do not shed light on the specific means by which sensory neurons influence MGI protein synthesis, they do invite several hypotheses (see Figure 3 for schematic representation).

Studies in other neuronal systems give evidence that extraneuronal or transneuronal (e.g., presynaptic, glial) sources may participate actively in the intercellular transfer or exchange of metabolites (Gainer et al., 1977; Meyer and Bittner, 1978). Although we have no experimental data to suggest that this process operates in the cricket nervous system, compelling evidence certainly does exist for the metabolic inductive effect of presynaptic neurons on developing postsynaptic neurons in the vertebrate autonomic nervous system (e.g., Black et al., 1971). By such a mechanism, a giant interneuron dendrite, rather far removed from perikaryal sources of protein and lacking the conventional ribosomal machinery for intrinsic synthesis, may be dependent on transneuronal sources for essential metabolites, especially during critical growth periods. However, our finding that changes in protein metabolism also take place in the MGI soma (a nonsynaptic region) after deafferentation—and this in view of the rather short (90 min) incubation of ganglia in [^{3}H]leucine—does not support the transneuronal transfer hypothesis. Rather, a more generalized regulatory mechanism seems likely.

Our data better support a scheme in which the presence of presynaptic terminals on target dendrites would serve a more regulatory role. In this case, the presence of increasing numbers of terminals during development would signal the perikaryon to synthesize and transport greater amounts of the proteins needed for dendritic growth and function. This mechanism might involve the retrograde transport of a factor or metabolic cue from terminal to dendrite to soma, followed by the increased synthesis and somatofugal shipment of macromolecules to the dendrites. This circuit of signals and shipments would rely strongly on retrograde intraneuronal transport, which is known to occur in a wide variety of neurons. Retrograde transport has been invoked as a likely means for regulation of neuron growth and differentiation (Stöckel and Thoenen, 1975; Schwartz, 1979) and for communication of metabolic information between the soma and distant neuronal outposts. Another requisite of such a mechanism would include a means by which metabolites (e.g., structural proteins) could be selectively routed or gated

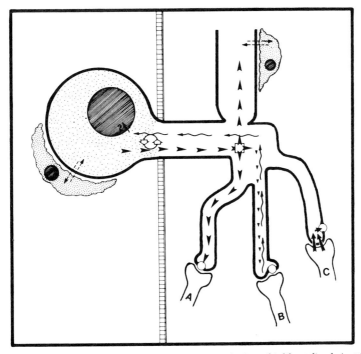

Figure 3. *Metabolic interactions in a giant interneuron.* Drawing depicts a highly stylized giant interneuron (e.g., MGI) composed of soma (in cortex compartment) and neurite, giant axon segment, and primary dendrite arbor in neuropil. Cercal sensory afferents synapse onto dendrites at *A*, *B*, and *C*. Stippled areas in soma and surrounding glial cells indicate regions known to possess ribosomes and hence, capacity for protein synthesis. Arrows between neurons and glia indicate possible exchange of metabolites. Three possible modes of metabolic interactions between afferent neurons and target GI are depicted. *A*: Soma supplies dendrite with protein by orthograde transport (large arrowheads) on supply–demand basis. Absence of normal complement of sensory terminals requires less dendritic area and soma accordingly transport fewer metabolites to this region. *B*: Orthograde transport is regulated in accord with retrograde cues (wavy arrows) that are provided by presence of sensory axon terminals. Coupling mechanisms are depicted by cycling arrows. Boxed region in neurite indicates presence of hypothetical gate whereby routing of metabolites to discrete compartments of the neuron can be directed and regulated. *C*: Active transneuronal communication whereby regional dendritic growth is modulated by local metabolic factors elaborated by synaptic contacts. Such factors could regulate synthesis or turnover of dendritic membrane components (embedded circles) independent of perikaryal influences.

by the neuron and targeted to discrete cellular areas in proportion to their biosynthetic or maintenance needs.

Finally, it is conceivable that the reduced intraneuronal protein synthesis observed in deafferented MGIs is a secondary consequence of decreased dendritic growth. Thus, less elaboration of dendritic apparatus would require less protein synthesis by the soma on a supply and demand basis. Such an indirect mechanism still would not explain the obvious localized influence of presynaptic elements on growth of the dendritic processes. Rather, as borne out in studies of developing MGI in the embryonic grasshopper (Shankland et al., 1982), specific localized interactions do appear to take place between sensory neurons and growing target interneuron processes. Results from analysis of the receptor pharmacology of the cricket terminal ganglion, discussed below, further tend to reinforce this view.

Effects of Deafferentation on Central Cholinergic Receptors in the CNS

Because protein metabolism in the terminal ganglion and, more specifically, in dendritic regions of MGI is markedly altered as a consequence of deafferentation, we suppose that proteins important to neuronal structure and function may undergo changes in their rate of synthesis or turnover. It is therefore of interest to attempt to determine the identity of at least some of these macromolecules in order to study more precisely the ways in which presynaptic input can influence the properties of postsynaptic targets. Because the regions of synaptic contact between pre- and postsynaptic elements are so important in nerve function and developmental interactions, it would be especially useful to focus our attention on those proteins involved in synaptic function. Such macromolecules would include structural proteins, enzymes, and receptor proteins residing in synaptic regions of the postsynaptic target cells. Receptor proteins, because of their deposition at the surface of the synaptic membrane, might be particularly susceptible to regulation, as is certainly the case at the vertebrate neuromuscular junction (e.g., see Changeux, 1979).

The known pharmacology and physiology of neurotransmission in the insect nervous system points strongly to the likelihood that transmission across sensory pathways is mediated by cholinergic mechanisms (Gerschenfeld, 1973; Callec, 1974). More to the point, detailed investigations on the pharmacology of the cercal sensory–GI pathway in a variety of orthopteroid insects closely related to crickets corroborate the impression that this pathway is predominantly cholinergic insofar as excitatory transmission is concerned (Shankland et al., 1971; Harrow et al., 1979). Therefore, in all likelihood, acetylcholine receptors (AChRs) are an integral component of cricket cercal sensory–GI synapses, although the biochemical properties of such putative AChRs have not been characterized. If AChRs are, in fact, situated within the cercal sensory–GI pathway of the cricket, it follows that assay of these protein macromolecules might provide an indication of their possible regulation by the state of afferent innervation. Our strategy was first to characterize the AChRs in the terminal ganglion and then to determine how they might be influenced when presynaptic input was eliminated by deafferentation.

A number of earlier studies on the neuropharmacology of insects concluded that insect AChRs, unlike most vertebrate AChRs, possessed hybrid or mixed nicotinic–muscarinic properties (but see review by Harrow et al., 1982). We thought it likely that cholinergic binding sites in the cricket terminal ganglion could be probed with either muscarinic or nicotinic labeled ligands. Our initial studies, therefore, were carried out with the potent, highly specific muscarinic antagonist [^{3}H]quinuclidinyl benzilate (Meyer and Edwards, 1980; Meyer et al., 1983; M. R. Meyer, in preparation).

Studies with Labeled Muscarinic Ligands. We developed a modified filter binding assay to characterize the binding of [^{3}H]quinuclidinyl benzilate to terminal ganglion homogenates. The ligand was shown to bind reversibly and with high specificity

Table 1. Comparative Pharmacology of Cholinergic Binding Sites in Terminal Ganglion Homogenates

	[^{3}H]Quinuclidinyl Benzilate	[^{125}I]α-Bungarotoxin Sites
B_{max}	1900 fmol/mg Protein	2129 fmol/mg Protein
K_D (app)	9.9×10^{-9} M	2.9×10^{-9} M
K_I^a(M)		
Antagonists		
Atropine	1.2×10^{-7}	3.9×10^{-6}
D-tubocurarine	1.2×10^{-6}	8.8×10^{-7}
α-bungarotoxin	$>3.0 \times 10^{-5}$	1.5×10^{-8}
Agonists		
Oxotremorine	6.0×10^{-5}	1.6×10^{-4}
Nicotine	1.3×10^{-4}	7.9×10^{-7}
Carbamylcholine	1.5×10^{-3}	1.1×10^{-4}
Deafferentation Response^b		
Short term	−	+
Long term	+	+

^{a} Inhibition constants (K_I) determined by competitive drug displacement experiments at K_D concentrations of labeled ligands. K_I values estimated from the equation $K_I = IC_{50}/\{1 + ([L]/K_D)\}$. IC_{50} values were determined graphically.
^{b} Change in either B_{max} or K_D values from kinetic analysis of binding isotherms.

to homogenates. Kinetic analysis of the binding isotherms disclosed a single class of sites of high affinity ($K_D \simeq 10 \times 10^{-9}$ M) that were present at a very high density ($B_{max} \simeq 2000$ fmol/mg protein) in the ganglion. Competitive ligand displacement experiments showed that although the pharmacology of [^{3}H]quinuclidinyl benzilate binding was predominantly muscarinic, the labeled ligand was nonetheless also displaceable with characteristically nicotinic agents (i.e., nicotine, curare) (Table 1). Thus, we were able to detect and characterize reliably in the terminal ganglion a unique class of AChRs that were present in sufficient concentration to monitor the possible effects of deafferentation.

When adult ganglia were subjected to bilateral cercal deafferentation, we observed no loss of binding sites (or change in affinity) after one week; in two out of every four experiments, we detected substantial increases in the concentration of AChRs (Figure 4). Two weeks after deafferentation, only slight decreases (12%) in AChR density (B_{max}) were noted. In sharp contrast, however, chronic deafferentation carried on throughout postembryonic development resulted in a 40% decrease in binding-site density without a corresponding change in binding-site affinity.

These results closely parallel changes observed for neuropil volume, MGI dendritic growth, and protein metabolism discussed earlier. We conclude from this that receptor macromolecules which may be localized at cercal sensory–GI synapses, and which may be intimately involved in synaptic transmission, are developmentally regulated by the state of afferent innervation. This certainly provides further evidence that dynamic metabolic interactions take place between pre- and postsynaptic neurons during development and that these interactions

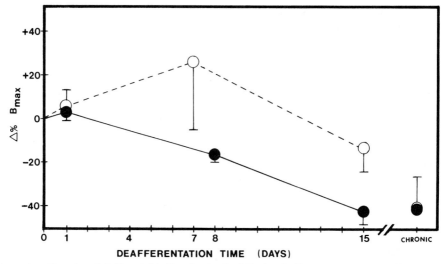

Figure 4. *Alteration of cholinergic binding-site density following deafferentation.* Terminal ganglia were bilaterally deafferented either throughout postembryonic development (chronic) or in adult crickets. Pooled ganglia from deafferented and matched-control animals were homogenized and incubated in either [^{3}H]quinuclidinyl benzilate (open circles) or [^{125}I]α-bungarotoxin (closed circles), and duplicate measurements of specific binding were calculated for at least five ligand concentrations. The resulting binding isotherms were analyzed by Scatchard analysis to determine binding-site densities (B_{max}). Each point represents mean percent difference (± SEM) in B_{max} values between control and deafferented groups derived from at least two separate experiments performed at each time point (except for chronic group incubated with [^{125}I]α-bungarotoxin, which was performed once).

may occur via mechanisms involving the biosynthesis and/or turnover of proteins localized on the neuron cell surface.

Studies with Labeled Nicotinic Ligands. In order to characterize more fully the pharmacological properties of AChR populations in the terminal ganglion and to explore the possibility of labeling AChRs for autoradiographic localization within ganglia, we next developed a binding assay to probe for cholinergic sites of the nicotinic class. The labeled active neurotoxic polypeptide of banded-krait venom, [^{125}I]α-bungarotoxin, has been used extensively to detect nicotinic AChRs at the vertebrate neuromuscular junction (Changeux, 1979), and has been especially useful for biochemical and autoradiographic studies of receptor metabolism and regulation (Fertuck and Salpeter, 1976; Fambrough et al., 1979). Similarly, it has been employed as a highly specific ligand to characterize distinct classes of binding sites in a variety of other preparations, including the insect CNS (see Harrow et al., 1982, for review of insect AChRs).

In the cricket terminal ganglion, [^{125}I]α-bungarotoxin binds with high affinity ($K_D \simeq 3 \times 10^{-9}$ M) to a single class of saturable sites that are present at high concentration (~2100 fmol/mg protein; ca. 40 fmol/ganglion) (Table 1). The pharmacological profile of the toxin binding sites obtained from drug displacement experiments demonstrated that, in a manner analogous to [^{3}H]quinuclidinyl benzilate binding, [^{125}I]α-bungarotoxin was displaceable by both nicotinic and muscarinic (e.g., atropine, oxotremorine) agents. In all cases, however, nicotinic agents were more potent displacers than were muscarinic agents. Our conclusion from these data was that, in all likelihood, we were detecting a single class of putative AChRs with predominantly nicotinic properties, or perhaps heterogeneous populations of distinct AChR subtypes.

In accord with this conclusion, results from kinetic analysis of [^{125}I]α-bungarotoxin binding isotherms after short-term deafferentation differed strikingly from those of parallel experiments performed with [^{3}H]quinuclidinyl benzilate. A loss of [^{125}I]α-bungarotoxin sites was noted within one week (Figure 4) of bilateral cercectomy. Within 15 days, a loss of more than 40% of the binding sites was seen; this was comparable in magnitude to the maximal decreases found after chronic deafferentation maintained throughout postembryonic development. Measurement of the rate of receptor loss indicated that approximately 50% of the binding sites were lost within nine days of deafferentation. Furthermore, and also in contrast to results from the [^{3}H]quinuclidinyl benzilate binding studies, changes in receptor number were accompanied by changes in the affinity of the receptor for the labeled ligand (M. R. Meyer et al., in preparation).

In the only other study to date on the effects of deafferentation on the properties of AChRs in the developing insect CNS, Hildebrand and his colleagues (Sanes et al., 1977; Hildebrand et al., 1979) concluded that prolonged absence of antennal sensory afferents during development of antennal lobes in the brain of *Manduca sexta* did not substantially alter the detectable levels of [^{125}I]α-bungarotoxin binding in those regions. Our results with both labeled α-bungarotoxin and quinuclidinyl benzilate stand clearly in contrast to these findings, and once more they illustrate the fact that various classes of invertebrate central neurons can display widely different developmental responses to loss of presynaptic terminals.

Despite some similarities in the pharmacological profiles and binding data for [^{3}H]quinuclidinyl benzilate and [^{3}H]α-bungarotoxin in the cricket terminal ganglion, the finding of distinct differences for the regulation of binding-site densities following short-term deafferentation suggests possibilities that warrant further investigation. The finding that toxin binding sites are rapidly lost after removal of cerci suggests that the sites are situated directly, perhaps monosynaptically, in cercal sensory–GI circuitry. Certainly, evidence from physiological recording supports this notion, as does the concept that this class of AChR is a postsynaptic component of the pathway. In contrast, the slow time course for loss of [^{3}H]quinuclidinyl benzilate sites, together with the lack of obvious physiological effects of muscarinic agents on cercal sensory–GI transmission, favors the likelihood that these sites may not be intrinsically situated in the pathway but may, in fact, be polysynaptically influenced by the state of afferent innervation. Such long-distance metabolic communication across several synapses recently has been shown to occur in mammalian brain (Dietrich et al., 1982).

Because the deafferentation data seem compatible with the presence of two AChR populations or subtypes rather than one class of AChRs with mixed pharmacology, we would like to know more about the nature and distribution of the cholinergic binding sites in the ganglion. As a first step toward localizing AChRs, we have begun autoradiographic studies of ligand binding. Frozen sections of terminal ganglia, after incubation with [^{125}I]α-bungarotoxin, show heavy labeling in regions of neuropil (cercal glomerulus) where cercal afferents synapse directly onto target GI dendrites (Figure 5). In particular, the label

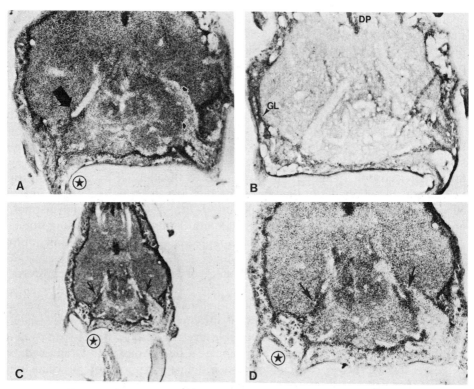

Figure 5. *Autoradiographic localization of [^{125}I]α-bungarotoxin binding sites.* Terminal ganglia from adult crickets that had been unilaterally deafferented (starred side) for 15 days were isolated and fixed, and 10-μm horizontal frozen sections were incubated with 20 nM [^{125}I]α-bungarotoxin for one hour at 22°C. Sections were rinsed and prepared for autoradiography (exposure time, six days). *A*: Section shows heavy accumulation of toxin labeling around large primary dendritic process of a GI on left side (large arrow). Deafferented (starred) side cannot be compared with control side due to asymmetry of section orientation. Grain density is higher in ganglion neuropil than in cortical regions containing somata or over sensory axon projection (small arrows). ×110. *B*: Adjacent section, incubated in medium containing 10^{-3} M nicotine (to displace specific binding) in addition to labeled toxin. Note that labeling in neuropil has been substantially reduced. Heavy labeling remaining in glial (GL) and dorsal pit (DP) regions represents nonspecific binding of toxin to nonneuronal elements within the terminal ganglion. ×110. *C*: Section taken through another unilaterally deafferented (starred side) terminal ganglion. Degeneration is apparent in cercal sensory nerve on deafferented side when compared to control nerve. Large processes of GI dendrites can be seen (arrows) on both sides of the ganglion. Note low labeling densities over cercal nerves. ×55. *D*: Higher-magnification view. The distal tips of GI dendrites are heavily invested with labeled toxin (arrows). These are regions where synapses between afferent axons and target GIs are likely to occur. Cercal glomerular neuropil shows high density of label. Grain density seems somewhat lower over deafferented regions of neuropil (also apparent in C). ×110.

appears to be concentrated in regions immediately surrounding primary dendritic processes of MGI, and thus the localization of [^{125}I]α-bungarotoxin seems to coincide with key anatomical landmarks that are associated with the cercal sensory–GI pathway. Curiously, in unilaterally deafferented ganglia, we have been unable in some cases to detect significant differences in grain densities between intact and deafferented regions of neuropil. This could well be due to the rather diffuse cercal projections into the ganglion, as well as to the overall high density of sites distributed throughout the ganglion. Nonetheless, our studies with labeled cholinergic ligands have shown that cell surface macromolecules that may be intimately involved in synaptic function in the insect CNS are metabolically regulated by the presence of peripheral neurons. These results, taken together with previous findings for changes in morphology and protein metabolism in deafferented postsynaptic central neurons, demonstrate the dynamic metabolic interactions that exist during development of the insect nervous system.

CONCLUSION

Our objective in focusing on selected cellular interactions in the insect nervous system has been to emphasize the flexibility of growth and metabolic activity in neurons that until recently have been regarded as rigidly programmed and unresponsive to a changing milieu. We have not addressed the implications of this flexibility for integrative activities, nor for pathfinding and specificity in development; rather, we have sought to show that communication between neurons is not restricted to integrative functions.

Nonetheless, we should note the known range of integrative flexibility, beginning with the virtually instantaneous, such as the alteration of gait pattern following limb loss at the short end of the spectrum. Here metabolic changes or growth are almost certainly not immediately involved. Longer-term changes encompass adaptive modification of escape responses in cockroaches after unilateral sensory loss (e.g., Vardi and Cambi, 1982) in which synaptic connections may well change. Long-term effects are exemplified by changes in auditory habituation characteristics in crickets exposed to continuous tone pulses throughout development (Murphey and Matsumoto, 1976). Mechanisms underlying these changed functions, other than postulated adjustment of synaptic efficiency, have not been found and correlated changes in neuron form have not been reported. Central and peripheral neurons are certainly capable of changes of form by sprouting; sometimes, as with muscle reinnervation or acoustic interneuron deafferentation, in a seemingly adaptive manner, or in other cases, as in massive denervation of ganglia, where exuberant sprouting of motor and interneurons can be induced. Subtler changes in neuron form, at the level of dendritic spine morphology, have recently been reported in honeybees; interneurons in the corpora pedunculata in the bee brain change in form during adult life (Coss et al., 1980). Samples taken from young (newly emerged), older (nurses), and oldest (foragers) bees show no change in spine density, but do show a progressive shortening of stem length and an increase in spine–head width, expressed as group means. Dendritic spine form is reported to change rapidly in a subpopulation of the same neurons at the time of the first orientation flight (Brandon and Coss, 1982). Changes of dendritic spine form have been associated with changes in the sensory milieu of vertebrates. A counterpart may now have been found in an insect, although some caution is appropriate here, for many aspects of the postembryonic development of insects do not reach finality at metamorphosis and the changes observed may be postmetamorphic maturational changes as well as responses to a progressively more complex environment.

All changes discussed above, with the exception of instantaneous circuit switching, must have implications for the metabolic activity of the neurons and presumably for their metabolic communication with neighboring cells, both glial and neuronal. Our studies with the cercal sensory–giant fiber system of crickets reveal aspects of the metabolic communication between sensory neurons and their central targets that point to possible mechanisms of interaction underlying changes in form and function. To the extent that so far our work with identified cells parallels studies with vertebrates at the cell population level, we consider our findings to have general applicability and hope that further probing will reveal the more intimate details of these sustaining and nurturing relationships between neurons.

REFERENCES

Anderson, H. (1978) Postembryonic development of the visual system of the locust *Schistocerca gregaria*. II. An experimental investigation of the formation of the retina-lamina projection. *J. Embryol. Exp. Morphol.* **46**:147–170.

Anderson, H., and J. Bacon (1979) Developmental determination of neuronal projection patterns from wind-sensitive hairs in the locust, *Schistocerca gregaria*. *Dev. Biol.* **72**:364–373.

Anderson, H., J. S. Edwards, and J. Palka (1980) Developmental neurobiology of invertebrates. *Annu. Rev. Neurosci.* **3**:97–139.

Bate, M. (1978) Development of sensory systems in arthropods. In *Handbook of Sensory Physiology*, Vol. 9, *Development of Sensory Systems*, M. Jacobson, ed., pp. 1–53, Springer-Verlag, New York.

Bellman, K. L., and F. B. Krasne (1983) Adaptive complexity of interactions between feeding and escape in crayfish. *Science* **221**:779–781.

Black, I. B. (1978) Regulation of autonomic development. *Annu. Rev. Neurosci.* **1**:183–214.

Black, I. B., I. A. Hendry, and L. L. Iversen (1971) Transsynaptic regulation of growth and development of adrenergic neurones in a mouse sympathetic ganglion. *Brain Res.* **34**:229–240.

Bodenstein, D. (1953) Regeneration. In *Insect Physiology*, K. D. Roeder, ed., pp. 866–878, Wiley, New York.

Brandon, J. G., and R. G. Coss (1982) Rapid dendritic spine stem shortening during one-trial learning: The honey bee's first orientation flight. *Brain Res.* **252**:51–61.

Callec, J.-J. (1974) Synaptic transmission in the central nervous system of insects. In *Insect Neurobiology*, J. E. Treherne, ed., pp. 119–178, Elsevier/North Holland, Amsterdam.

Changeux, J.-P. (1979) Molecular interactions in adult and developing neuromuscular junction. In *The Neurosciences: Fourth Study Program*, F. O. Schmitt and F. G. Worden, eds., pp. 749–778, MIT Press, Cambridge.

Clark, R. (1976a) Structural and functional changes in an identified cricket neuron after separation from the soma. I. Structural changes. *J. Comp. Neurol.* **170**:253–266.

Clark, R. (1976b) Structural and functional changes in an identified cricket neuron after separation from the soma. II. Functional changes. *J. Comp. Neurol.* **170**:267–278.

Coss, R. G., J. G. Brandon, and A. Globus (1980) Changes in morphology of dendritic spines on honeybee calycal interneurons associated with cumulative nursing and foraging experiences. *Brain Res.* **192**:49–59.

Cowan, W. M. (1970) Anterograde and retrograde transneuronal degeneration in the central and peripheral nervous system. In *Contemporary Research Methods in Neuroanatomy*, J. H. Nauta and S. O. E. Ebbesson, eds., pp. 217–251, Springer-Verlag, New York.

Dietrich, W. D., D. Durham, O. H. Lowry, and T. A. Woolsey (1982) "Increased" sensory stimulation leads to changes in energy-related enzymes in the brain. *J. Neurosci.* **2**:1608–1613.

Donaldson, P. L., and R. K. Josephson (1981) Collateral sprouting of insect motoneurons. *J. Comp. Neurol.* **196**:317–327.

Edwards, J. S. (1969) Postembryonic development and regeneration of the insect nervous system. *Adv. Insect Physiol.* **6**:97–137.

Edwards, J. S., and J. Palka (1974) The cerci and abdominal giant fibres of the house cricket *Acheta domesticus*. I. Anatomy and physiology of normal adults. *Proc. R. Soc. Lond. (Biol.)* **185**:83–121.

Edwards, J. S., M. Milner, and S. W. Chen (1978) Integument and sensory nerve differentiation of *Drosophila* leg and imaginal disk *in vitro*. *Roux Arch. Dev. Biol.* **185**:59–77.

Edwards, J. S., S. W. Chen, and M. W. Berns (1981) Cercal sensory development following laser microlesions of embryonic apical cells in *Acheta domesticus*. *J. Neurosci.* **1**:250–258.

Fambrough, D. M., P. N. Devreotes, J. M. Gardner, and D. J. Card (1979) The life history of acetylcholine receptors. *Prog. Brain Res.* **49**:325–334.

Fertuck, H. C., and M. M. Salpeter (1976) Quantitation of junctional and extrajunctional acetylcholine receptors by electron microscope autoradiography after [^{125}I]α-bungarotoxin binding at mouse neuromuscular junctions. *J. Cell Biol.* **69**:144–158.

Gainer, H., I. Tasaki, and R. J. Lasek (1977) Evidence for the glia–neuron protein transfer hypothesis from intracellular perfusion studies on squid giant axons. *J. Cell Biol.* **74**:524–530.

Gerschenfeld, H. M. (1973) Chemical transmission in invertebrate central nervous systems and neuromuscular junctions. *Physiol. Rev.* **53**:1–119.

Goodman, C. S., M. O'Shea, R. McCaman, and N. C. Spitzer (1979) Embryonic development of identified neurons: Temporal pattern of morphological and biochemical differentiation. *Science* **204**:1219–1222.

Goodman, C. S., J. A. Raper, R. K. Ho, and S. Chang (1982) Pathfinding by neuronal growth cones in grasshopper embryos. *Symp. Soc. Dev. Biol.* **40**:275–316.

Harrow, I. D., B. Hue, M. Pelhate, and D. B. Sattelle (1979) α-Bungarotoxin blocks excitatory postsynaptic potentials in an identified insect interneurone. *J. Physiol. (Lond.)* **295**:63P–64P.

Harrow, I. D., J. A. David, and D. B. Sattelle (1982) Acetylcholine receptors of identified insect neurons. *Ciba Found. Symp.* **88**:12–31.

Hildebrand, J. G., L. M. Hall, and B. C. Osmond (1979) Distribution of binding sites for [^{125}I]-labeled α-bungarotoxin in normal and deafferented antennal lobes of *Manduca sexta*. *Proc. Natl. Acad. Sci. USA* **76**:499–503.

Ho, R. K., E. E. Ball, and C. S. Goodman (1983) Muscle pioneers: Large mesodermal cells that erect a scaffold for developing muscles and motoneurons in grasshopper embryos. *Nature* **301**:66–69.

Hoy, R. R., G. B. Casaday, and S. Rollins (1978) Absence of auditory afferents alters the growth pattern of an identified auditory neuron. *Soc. Neurosci. Abstr.* **4**:115.

Keshishian, H. (1980) The origin and morphogenesis of pioneer neurons in the grasshopper metathoracic leg. *Dev. Biol.* **80**:388–397.

Kutsch, W. (1974) The influence of the wing sense organs on the flight motor pattern in maturing adult locusts. *J. Comp. Physiol.* **88**:413–424.

Macagno, E. R. (1979) Cellular interactions and pattern formation in the development of the visual system of *Daphnia magna* (Crustacea, Branchiopoda). I. Interactions between embryonic retinular fibers and laminar neurons. *Dev. Biol.* **73**:206–238.

Maxwell, G. D., and J. G. Hildebrand (1981) Anatomical and neurochemical consequences of deafferentation in the development of the visual system of the moth *Manduca sexta*. *J. Comp. Neurol.* **195**:667–680.

Meinertzhagen, I. A. (1973) Development of the compound eye and optic lobe of insects. In *Developmental Neurobiology of Arthropods*, D. Young, ed., pp. 51–104, Cambridge Univ. Press, Cambridge.

Meyer, M. R., and G. D. Bittner (1978) Biochemical studies of trophic dependencies in crayfish giant axons. *Brain Res.* **143**:213–232.

Meyer, M. R., and J. S. Edwards (1980) Muscarinic cholinergic binding sites in an orthopteran central nervous system. *J. Neurobiol.* **11**:215–219.

Meyer, M. R., and J. S. Edwards (1982) Metabolic changes in deafferented central neurons of an insect, *Acheta domesticus*. I. Effects upon amino acid uptake and incorporation. *J. Neurosci.* **2**:1651–1659.

Meyer, M. R., G. R. Reddy, and J. S. Edwards (1983) Sensory innervation regulates the level of cholinergic receptors on insect central neurons during postembryonic development. *Soc. Neurosci. Absr.* **9**:320.

Meyerowitz, E. M. and D. R. Kankel (1978) A genetic analysis of visual system development in *Drosophila melanogaster*. *Dev. Biol.* **62**:112–142.

Murphey, R. K., and R. B. Levine (1980) Mechanisms responsible for changes observed in response properties of partially deafferented insect neurons. *J. Neurophysiol.* **43**:367–382.

Murphey, R. K., and S. G. Matsumoto (1976) Experience modifies the plastic properties of identified neurons. *Science* **191**:564–566.

Murphey, R. K., B. Mendenhall, J. Palka, and J. S. Edwards (1975) Deafferentation slows the growth of specific dendrites on identified giant interneurons. *J. Comp. Neurol.* **159**:407–418.

Murphey, R. K., S. G. Matsumoto, and B. Mendenhall (1977) Recovery from deafferentation by cricket interneurons after reinnervation by their peripheral field. *J. Comp. Neurol.* **169**:335–346.

Murphey, R. K., J. P. Bacon, D. S. Sakaguchi, and S. E. Johnson (1983) Transplantation of cricket sensory neurons to ectopic locations: Arborizations and synaptic connections. *J. Neurosci.* **3**:659–672.

Palka, J., and J. S. Edwards (1974) The cerci and abdominal giant fibres of the house cricket, *Acheta domesticus*. II. Regeneration and effects of chronic deprivation. *Proc. R. Soc. Lond. (Biol.)* **185**:105–121.

Palka, J., and M. Schubiger (1980) Formation of central patterns by receptor cell axons in *Drosophila*. In *Development and Neurobiology of Drosophila*, O. Siddiqi, P. Babu, L. M. Hall, and J. C. Hall, eds., pp. 223–246, Plenum, New York.

Pipa, R. L. (1978) Patterns of neural reorganization during the postembryonic development of insects. *Int. Rev. Cytol. Suppl.* **7**:403–438.

Pitman, R. M., and K. Rand (1982) Neural lesions can cause dendritic sprouting of an undamaged adult insect motoneurone. *J. Exp. Biol.* **96**:125–130.

Pitman, R. M., C. D. Tweedle, and M. J. Cohen (1973) The form of nerve cells: Determination by cobalt impregnation. In *Intracellular Staining in Neurobiology*, S. B. Kater and G. Nicholson, eds., pp. 83–97, Springer-Verlag, New York.

Purves, D. (1976) Long-term regulation in the vertebrate peripheral nervous system. *Int. Rev. Physiol.* **10**:125–177.

Roederer, E., and M. J. Cohen (1983) Regeneration of an identified central neuron in the cricket. I. Control of sprouting from soma, dendrites, and axon. *J. Neurosci.* **3**:1835–1847.

Sanes, J. R., J. G. Hildebrand, and D. J. Prescott (1976) Differentiation of insect sensory neurons in the absence of their normal synaptic targets. *Dev. Biol.* **52**:121–127.

Sanes, J. R., D. J. Prescott, and J. G. Hildebrand (1977) Cholinergic neurochemical development of normal and deafferented antennal lobes during metamorphosis of the moth *Manduca sexta*. *Brain Res.* **119**:389–402.

Schneiderman, A. M., S. G. Matsumoto, and J. G. Hildebrand (1982) Transsexually grafted antennae influence development of sexually dimorphic neurons in moth brain. *Nature* **298**:844–846.

Schwartz, J. H. (1979) Axonal transport: Components, mechanisms and specificity. *Annu. Rev. Neurosci.* **2**:467–504.

Scott, S. A. (1977) Maintained function of foreign and appropriate junctions on reinnervated goldfish extraocular muscles. *J. Physiol.* **268**:87–109.

Shankland, M., and C. S. Goodman (1982) Development of the dendritic branching pattern of the medial giant interneuron in the grasshopper embryo. *Dev. Biol.* **92**:489–506.

Shankland, D. L., J. A. Rose, and C. Donniger (1971) The cholinergic nature of the cercal nerve–giant fiber synapse in the sixth abdominal ganglion of the American cockroach, *Periplaneta americana* (L.). *J. Neurobiol.* **2**:247–262.

Shankland, M., D. Bentley, and C. S. Goodman (1982) Afferent innervation shapes the dendritic branching pattern of the medial giant interneuron in grasshopper embryos raised in culture. *Dev. Biol.* **92**:507–520.

Smith, D. E. (1977). The effect of deafferentation on the development of brain and spinal nuclei. *Prog. Neurobiol.* **8**:349–367.

Spitzer, N. C. (1979) Ion channels in development. *Annu. Rev. Neurosci.* **2**:363–397.

Stöckel, K., and H. Thoenen (1975) Retrograde axonal transport of nerve growth factor: Specificity and biological importance. *Brain Res.* **85**:337–341.

Stocker, R. F. (1982) Genetically displaced sensory neurons in the head of *Drosophila* project via different pathways into the same specific brain regions. *Dev. Biol.* **94**:31–40.

Truman, J. W., and S. E. Reiss (1976) Dendritic reorganization of an identified motoneuron during metamorphosis of the tobacco hornworm moth. *Science* **192**:477–479.

Tweedle, C. D., R. M. Pitman, and M. J. Cohen (1973) Dendritic stability of insect central neurons subjected to axotomy and deafferentation. *Brain Res.* **60**:471–476.

Vardi, N., and J. M. Cambi (1982) Functional recovery from lesions in the escape system of the cockroach. II. Physiological recovery of the giant interneurons. *J. Comp. Physiol.* **146**:299–309.

Watson, W. E. (1974) Cellular response to axotomy and to related procedures. *Br. Med. Bull.* **30**:112–115.

Whitington, P. M., and E. Seifert (1982) Axon growth from limb motoneurons in the locust embryo: The effect of target limb removal on the path taken to the central nervous system. *Dev. Biol.* **93**:206–215.

Chapter 15

Changes of State during Neuronal Development: Regulation of Axon Elongation

MARK B. WILLARD
KARINA MEIRI
MARCIE GLICKSMAN

ABSTRACT

On its way to becoming a functional neuron, a cell must make a series of "decisions," each demarcating the transition from one developmental state to the next. Each state is characterized by a particular molecular configuration that defines the potential of the cells in that state. The process of making decisions between developmental states is much better understood in certain simple systems than it is in neuronal development. For example, the bacteriophage λ employs proteins that function as environmentally sensitive switches that serve to decide between two alternative developmental pathways. Here we consider the possibility that proteins with analogous switching functions may participate in the developmental decision of a neuron to grow an axon and terminate its growth. Furthermore, we consider that the delayed appearance of certain neurofilament polypeptides during development may be involved in a decision of the axon to enter a stable state.

The developmental program of a cell is reflected in the series of states through which the cell passes as it traverses its life cycle. A fundamental problem is to describe the molecular configuration that distinguishes these states and to determine the mechanism for the decision to leave one state and enter another. Cells must pass through many states on their way to becoming functional neurons (e.g., see Figure 1). Here we consider certain differences between the molecular configurations of four of these states: (1) the state prior to axon extension (which we designate the pregrowth state); (2) the state in which a neuron extends an axon (the growth state); (3) the state in which a neuron maintains an axon that has ceased elongating (the stationary state); and (4) a hypothetical state in which an axon is stabilized after it has formed synaptic connections (the stable state). Certain differences between these states raise the possibility that axon growth may be regulated in part by a small number of proteins that can influence the behavior of multiple cellular constituents in such a manner as to switch them to the proper configuration for axon growth. The clearest precedents for such developmental pleiotropic switching functions and their mechanisms are found in nonneuronal systems. The development of a prokaryotic organism, the bac-

Figure 1. *Schematic illustration of certain states through which a cell may need to pass on its way to becoming a functional neuron.* The potential of each state is defined by its particular molecular configuration, and each state is demarcated by a decision (A–D) to enter or leave that state.

teriophage λ, serves to illustrate the utility of such switching functions, as well as certain other developmental strategies that may have analogies in regulating changes of state during neuronal development.

PHAGE λ'S MECHANISMS FOR DECIDING BETWEEN DEVELOPMENTAL STATES

Perhaps more of what there is to know about a particular developmental program has been described for the life cycle of the bacteriophage λ than for any other organism. In the course of this cycle, a branch point is encountered that requires λ to decide which of two quite different developmental pathways will be traversed. The role of certain molecular processes in making this decision has been described, and it is useful to have in mind the principles involved in this decision-making process when considering the changes in state of a cell that is as complicated and incompletely described as is a neuron. Consequently, we begin by summarizing several strategies employed by phage λ to negotiate its life cycle successfully (see Echols, 1980, and Hendrix et al., 1983, for reviews).

Phage λ is a syringe-shaped virus about 50 nm in diameter and 150 nm long. Its icosahedral head contains a double-stranded DNA chromosome, about 50 kb long, that comprises about 60 genes. One phase of the life cycle of λ begins when a freely diffusing virus encounters a bacterium of the appropriate pedigree; the virus binds to specific surface receptors and injects its chromosome, one end of which subsequently attaches covalently to the other, forming a double-stranded DNA circle. The bacterial RNA polymerase attaches to two sites on the chromosome and begins to transcribe the genetic information into messenger RNA; transcription from the two sites proceeds in opposite directions. Some of the protein products of these "early genes" participate in the decision as to which of the two developmental pathways will be followed in that cell. The choice is this: On the one hand, the virus can pursue a vegetative (or lytic) pathway, which results in the production of about 100 new phage particles per cell; these are released when the bacterium is broken open by specific phage gene products. Alternatively, the virus can follow the lysogenic pathway, which leads to the incorporation of the viral chromosome into the bacterial chromosome; in the lysogenic state, the viral chromosome is replicated as part of the bacterial chromosome and is passed to progeny bacteria in a nearly dormant state called a prophage. In response to certain environmental conditions, a prophage can reenter the vegetative state; its chromosome is excised from the bacterial chromosome, phage are produced and released, and the host is killed.

The successful accomplishment of the vegetative pathway requires the expression of a different set of genes than does the lysogenic pathway. To make new phage via the vegetative pathway requires the expression of genes coding for the head and tail proteins as well as for the enzymes that ultimately digest the bacterial cell wall to release the newly formed phage. If the lysogenic pathway is to be successfully executed, the gene coding for the enzyme that integrates the viral chromosome into the bacterial chromosome must be expressed, and the vegetative functions that lead to the death of the cell must be efficiently repressed. This repression is accomplished by the synthesis of a repressor protein that binds to the λ DNA in a position appropriate for blocking the initiation of transcription of genes required for the expression of vegetative functions. The decision as to which of these two developmental pathways will be followed is thus in part a decision as to which genes will be transcribed. A key arbitrator in arriving at this decision is a short-lived protein product of the gene called *cII*; *cII* is transcribed from one of the two promoters that is initially active during phage infection (Figure 2). The *cII* protein functions to favor the lysogenic response in several ways: It promotes transcription from a site on the chromosome that results in the expression of the gene required for integrating the λ chromosome into the bacterial chromosome; in addition, it promotes transcription (at another site) of the repressor gene, whose product not only prevents transcription of vegetative functions, but promotes its own transcription (from yet another site on the DNA). The efficacy of the *cII* protein in accomplishing these functions is influenced by the internal milieu of the bacterium; certain bacterial gene products interact with certain metabolites to determine the stability of the *cII* protein and consequently determine how effectively it will accomplish these lysogeny-promoting functions. The importance of the *cII* protein in the decision to lysogenize is illustrated by mutations in the *cII* gene that destroy its function; phage carrying these mutations lysogenize very poorly.

An additional participant in the decision is the product of the gene named *cro*, which is transcribed from the same promotor as is *cII*; *cro*'s functions favor

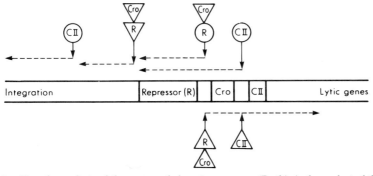

Figure 2. *How the products of three genes of phage* λ—*repressor* (R; *this is the product of the gene called* cI *in traditional* λ *nomenclature*), cro, *and* cII—*act to switch the phage to one developmental state or another—the lytic state or the lysogenic state.* The dashed lines indicate RNA transcripts that start at various points on the λ chromosome, a segment of which is represented by the bar. The influence of the products of the λ regulatory genes upon transcription from these sites is indicated by circles (stimulation of transcription) or triangles (inhibition of transcription).

the vegetative response (Figure 2). It binds to sites on the DNA in such a way that it prevents the transcription of the repressor gene and also prevents binding to the DNA of the repressor protein itself.

cII, the repressor, and *cro* illustrate that the decision to grow vegetatively or to lysogenize can be strongly influenced by a small number of gene products. In addition, the multiple functions of both cII and *cro* serve to lock the system into one or the other pathway, even if initially the decision between the two pathways is equivocal. These gene products appear to act as switches that commit the phage to its choice of developmental pathway.

A second aspect of λ regulation, the timing of the production of gene products relevant to different temporal phases of the vegetative pathway, is also accomplished in part by strategies involving RNA transcription (Figure 3). The RNA transcripts that are initiated at the earliest promoter sites on the λ DNA are aborted at termination signals before the transcription process reaches the genes (e.g., those whose products digest the cell wall) that must be expressed only during the later stages of the vegetative phase of λ's infectious cycle. Among the gene products synthesized from the early transcripts is a protein that functions to block this termination and allows subsequent transcripts originating at the same early promoters to continue, so that additional genes are expressed. Among these additional gene products is a protein that serves as an antiterminator of transcription at termination sites further along the chromosome; antitermination at these latter sites allows the genes for structural proteins and lysis to be finally transcribed and expressed. Thus, the temporal sequence of expression of λ genes is controlled in part by the requirement of the products of early genes for late genes to be transcribed.

A third strategy employed by λ for regulating its gene expression, and hence its course of development, is the rearrangement of the spatial relationships of its genes (Figure 4). Such a mechanism appears to be involved in determining whether the λ chromosome is integrated into the bacterial chromosome (as it must be to establish a lysogenic state) or excised from the chromosome (as it must be if a prophage is to enter the lytic state). Two phage proteins named *int* and *xis* are involved in these reactions. The *int* gene product by itself appears to favor integration of the phage chromosome into the bacterial chromosome, whereas the expression of both gene products appears to favor excision. To ensure that *int* is preferentially produced during the establishment of the lysogenic state, it is transcribed from a cII-dependent promoter that resides within the neighboring *xis* gene. The resulting messenger RNA contains sufficient information for the translation of *int*, but not *xis*, and integration is favored. During the

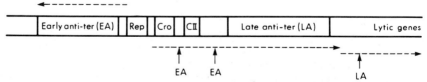

Figure 3. *How the products of a λ gene (early antiterminator) serves to block termination of transcription (broken line) at certain sites (EA), to allow the delayed expression of a gene (late antiterminator). The product of this late antiterminator gene in turn blocks termination of transcription at a site (LA) that allows the expression of lytic genes required only in the vegetative phase of λ development.*

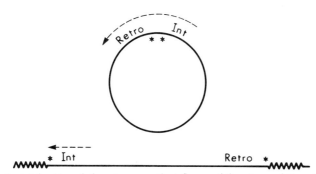

Figure 4. *How the* int *function of phage* λ *escapes the influence of the retroregulation site by means of the separation that occurs when the circular* λ *chromosome is inserted into the bacterial chromosome (squiggled line) by a recombination event that occurs between the two asterisks.* The lower figure illustrates the inserted chromosome in the prophage state; the broken line illustrates transcription of the *int* gene in the vegetative state (upper) or upon induction of the prophage (lower).

vegetative phase of the life cycle, when the *int* function is not required, transcripts (from a different promoter) that include the *int* gene contain additional information downstream from the *int* gene; this information serves to prevent translation of the *int* message, apparently by making the RNA more labile. (This mechanism of control has been called retroregulation.) Paradoxically, *int* and *xis* are transcribed from this same promoter that yields retroregulation when a prophage is induced, when both *int* and *xis* must be expressed in order to accomplish the excision of the phage chromosome from the bacterial chromosome. How does *int* expression escape this downstream retroregulation? The point at which the circular λ DNA is broken when it is integrated into the bacterial chromosome lies between the *int* gene and the retroregulation site (Figure 4). Therefore, in the prophage, the *int* gene is physically separated from this retroregulation site; the transcript containing *int* will lack the retroregulation sequence, and both *int* and *xis* will be expressed, as they must be to achieve efficient excision of the λ chromosome. Thus by altering the spatial relationships of its genetic elements, λ ensures that regulatory elements will influence the expression of a gene only under the appropriate circumstances.

These three mechanisms—pleiotropic switching functions to determine a developmental pathway, sequential requirements for gene products to exert temporal control, and chromosomal spatial relationships that can be altered to regulate gene expression—are strategies that are effectively employed to elaborate and perpetuate the creature λ.

STATES OF NEURONAL DEVELOPMENT

The purpose of describing the regulation of λ development here is to raise the possibility that analogous strategies may be employed in some aspects of neuronal development. A cell must make a sequence of decisions on its way to becoming a functional neuron, and each decision can be considered to demarcate the transition from one cellular state to another (Figure 1). Decision A, the commitment of a multipotential cell to a single developmental pathway, illustrates the potential for analogy between the development of λ and the development of a neuron. Like the commitment of λ to a lysogenic or vegetative pathway, the selection of a neuronal or glial pathway by a stem cell must involve the activation of expression of certain gene modules and the repression of others. Presumably, there are cII and *cro* analogues in neurons that interact with parameters that are sensitive to developmental time and place, and so affect the selection of the particular constellation of genes that will be expressed. Some of the factors involved in this decision are considered in detail elsewhere in this volume.

A specific example of the switching of neuronal gene expression during neuronal differentiation is the decision as to which neurotransmitter will be synthesized by sympathetic neurons. At an early stage in development, these neurons produce and store noradrenaline, but a population later switches to the production and storage of acetylcholine. This transition can be mimicked in cultured sympathetic neurons, where the switch in transmitter type can be influenced by exogenous factors as well as by polarization of the cell membrane (Patterson, 1978; Johnson et al., 1981; Landis and Keefe, 1983). This switch appears to reflect, at least in part, alterations in the amounts of the enzymes that synthesize noradrenaline and acetylcholine (Wolinsky and Patterson, 1983). It is easy to imagine that a variant of λ's lytic–lysogenic or prophage–lytic decision-making mechanism may be at work in selecting the expression of the appropriate set of genes for each transmitter phenotype.

Another set of important decisions encountered during the development of the nervous system is the set involved in the neuron's production of an axon. First, the neuron must decide that the time and place is appropriate for axon growth. To execute this decision the neuron must construct a motile growth cone with the capacity to accomplish a sequence of tasks correctly: The growth cone must translocate in the appropriate direction, responding to cues as to which way to turn and which fascicle of axons to follow; it must branch to achieve an appropriate degree of arborization; after forming transient connections with potential target cells, it must form stable synapses with the correct partners; finally, the neuron must ascertain that an acceptable constellation of synaptic contacts has been established and terminate axon growth.

The decision to grow an axon and the decision to terminate axon growth (decisions B and C, respectively, in Figure 1) demarcate the boundaries of a developmental state, which we refer to as the growth state (Skene and Willard, 1981a,c), in which all of the machinery for growing an axon is present and operable. Previously and subsequently, this machinery does not function to grow an axon, either because some of its elements are missing or because the machinery is turned off or diverted toward other ends. Analogous to the vegetative and prophage states of phage λ, the growth state represents one phase in the life cycle of a neuron. We consider next certain molecular alterations that may be important in establishing and terminating such a growth state.

MOLECULAR DIFFERENCES BETWEEN THE PREGROWTH, GROWTH, AND STATIONARY STATES

The stationary state, in which a neuron has successfully completed the elaboration of an axon, resembles the pregrowth state in that neither state involves axon elongation. However, a consideration of the nature of the stationary axon suggests that these states must be quite different. The major morphological features of a stationary axon that can be seen in the electron microscope are a myriad of membranous organelles, including mitochondria, smooth endoplasmic reticulum, and vesicular elements, embedded in a lattice formed by linear elements—the neurofilaments and microtubules—that are linked to each other, to the membranous organelles, and to the plasma membrane by thin fibrillar cross bridges (e.g., Hirokawa, 1982; Schnapp and Reese, 1982). This electron-microscope representation of an axon reflects the steady state of a system that is actually in rapid flux. Materials synthesized in the cell body enter the axon and move to their destinations by anterograde axonal transport, whereas retrograde axonal transport conveys materials from the terminal regions toward the cell body. Electrophoretic analysis of newly synthesized radiolabeled proteins has indicated that different collections of proteins move down the axon at different velocities (e.g., Baitinger et al., 1982). The most rapidly moving (group I, velocity $\simeq$ 250 mm/day) include plasma membrane proteins such as the Na^+/K^+-ATPase. The moving vehicles that carry these proteins have recently been observed in living axoplasm by means of contrast-enhanced video-microscope techniques (Allen et al., 1982). When viewed in the electron microscope, these vehicles appear to include membranous vesicular and tubular structures (Tsukita and Ishikawa, 1980). Mitochondria move more slowly, and together with a protein called fodrin (a relative of erythrocyte spectrin; Levine and Willard, 1981), as well as certain unidentified organelles, are components of group II (velocity $\simeq$ 40 mm/day). Conventional phase-contrast microscopy has demonstrated that mitochondria and uncharacterized "particles" move in living axons at velocities similar to those of group II proteins (Forman et al., 1983) and presumably represent the transport vehicles for some of these group II proteins. Groups III and IV (velocity $\simeq$ 6 mm/day) include actin and two forms of myosin, as well as many other cytoplasmic proteins (e.g., Willard et al., 1979; Brady et al., 1981); these groups may provide force-generating machinery used to transport other organelles. The transport of proteins that compose the neurofilamentous-microtubular cytoskeleton gives rise to group V, which may represent the progressive displacement down the axon of the intact cytoskeleton (Lasek and Hoffman, 1976). The entry of proteins into the steady-state stationary axon in this highly ordered sequence must be exactly balanced by their exit. The fates of specific proteins are known in only a few instances; proteins may exit from axons by means of their release from axon terminals, their degradation, and their return to the cell body.

Because a neuron in the stationary state must provide for this flux of materials through the axon, whereas a neuron in a pregrowth state does not, it is likely that the molecular configurations of the two states are different, even though neither involves axon growth. Furthermore, a neuron in the pregrowth state would not necessarily express genes whose products are needed only for axonal or synaptic functions.

On the other hand, the possibility that the stationary state may resemble the growth state is raised by the dynamic nature of the steady-state stationary axon. In many respects the transport processes that normally supply the mature axon with new plasma membrane, cytoskeleton, and cytoplasm would appear to have the potential to fabricate new axons if the balance between delivery and exit of materials were altered appropriately (Lasek and Black, 1977). Indeed, neurons are able to carry out certain activities associated with axon growth (e.g., the formation of structures that morphologically resemble growth cones, and the formation of limited, although abortive, sprouts) without the benefit of new information from the cell body.

In an effort to evaluate whether special gene products, unique to the growth state, are required if a neuron is to grow an axon, several laboratories have compared the proteins synthesized and axonally transported by neurons that grow axons with those of neurons with stationary axons (Giulian et al., 1980; Bisby, 1981; Perry and Wilson, 1981; Skene and Willard, 1981a,c; Benowitz et al., 1983; see Forman, 1981, and Willard and Skene, 1982, for reviews). The nervous systems of fishes and amphibians provide good experimental systems for addressing this question, because in these animals the developmental transition from the growth state to the stationary state is readily reversible; when their axons are injured, they typically regrow and form functional synaptic connections. For example, when the optic nerve of the toad *Bufo marinus* is crushed, the retinal ganglion cells regrow their axons and form functional synaptic connections in the optic tectum. Using a variety of electrophoretic techniques, we compared radiolabeled proteins transported into regenerating axons with proteins transported into uninjured control axons and observed the following differences (Skene and Willard, 1981a): (1) Most axonally transported polypeptides were radiolabeled to a higher degree in regenerating than in stationary axons. The average increase in labeling was about threefold in the case of group I polypeptides and tenfold in the case of group IV polypeptides. These changes most likely reflect the increase in synthesis of materials (e.g., components of plasma membrane and cytoskeleton) required to elaborate structures of the growing axon. (2) The labeling of a few polypeptides decreased during regeneration. It seems reasonable to suppose that these polypeptides may perform functions, such as those related to neurotransmitter release, that are not required in the regenerating axon. (3) The labeling of a small number of proteins, designated growth-associated proteins, or GAPs, increased much more than average during regeneration—as much as a 100-fold absolute increase. The increased labeling of GAPs followed a time course that roughly corresponded to the regenerative growth of axons; after lag periods of duration similar to the delay between axon injury and initial axon outgrowth, the labeling of GAPs increased to high levels that were maintained through the period of regeneration and that returned to basal levels at about the time that functional connections were restored. This pattern of growth-associated induction suggests that the GAPs may perform functions that are unique to the growth state of the neuron.

Three prominent GAPs in toad (GAPs-50, -43, and -24, designated by their molecular weight in kilodaltons) were group I polypeptides with properties characteristic of integral proteins of the plasma membranes (Skene and Willard, 1981b). GAP-24 turned over very rapidly ($t_{1/2}$ is several hours), a property that, together with its rapid rate of transport, indicates that the function it performs could be rapidly initiated or terminated by the induction or cessation of its synthesis in the cell body. GAP-50 and GAP-43, which turn over more slowly ($t_{1/2}$ is several days), appear to be transported preferentially to axon tips. GAP-50 is a fucosylated glycoprotein.

In mammals, the developmental decision to leave the growth state and establish a stationary state is readily reversed when axons of the peripheral nervous system (PNS) are injured; these axons are often successfully regenerated. When a rabbit peripheral nerve, the hypoglossal nerve, is crushed, a protein with the same molecular weight and isoelectric point as toad GAP-43 is axonally transported at increased levels as the axons of this nerve regenerate (Skene and Willard, 1981c). On the other hand, neurons of the adult mammalian central nervous system (CNS) generally do not recover from axonal injury. Correspondingly, GAP-43 transport is not increased when the axons of adult rabbit retinal ganglion cells (CNS neurons) are injured by crush of the optic nerve. However, these same retinal ganglion cells transport GAP-43 in neonatal rabbit during a developmental period when the axons of these neurons are elongating, and the level of GAP-43 transport decreases as the animal matures. Thus, elevated levels of GAP-43 correlate with regenerative axon growth in the toad CNS and mammalian PNS, and with developmental growth in the mammalian CNS. A second rabbit polypeptide, GAP-23, bears certain physical similarities to toad GAP-24, although the two differ somewhat in molecular weight and isoelectric point. Like GAP-43, the elevated transport of GAP-23 correlated with axon growth during development of the mammalian CNS (visual system) and was not induced by crush of the adult optic nerve. Unlike GAP-43, it was labeled significantly in the hypoglossal nerve of adult rabbit and was not further induced when that nerve was crushed and axon regeneration was initiated. One explanation for the behavior of GAP-23/24 is that it performs a growth-related function, such as the remodeling of neuromuscular junctions, that is ongoing in mammalian peripheral nerves. If so, the differential expression of GAP-23 and GAP-43 would illustrate that different aspects of the growth state (e.g., major axon elongation and synapse formation) can be regulated independently (Skene and Willard, 1981c). Proteins with certain behaviors and properties similar to toad and rabbit GAPs have also been observed in regenerating goldfish optic nerve and in rat spinal neurons (Giulian et al., 1980; Benowitz et al., 1981; Bishy, 1981).

FUNCTION OF GAPs

The induction of GAP transport during periods of axon growth indicates that the transition from a stationary state to a growth state involves a major alteration in expression of GAP gene products. The potential significance of the GAPs is that their growth-associated increases suggest that they may perform special growth-specific functions; if so, their failure to be induced after injury to axons of the mammalian CNS would explain in part the failure of these neurons to regenerate axons. To evaluate this possibility, it will be necessary to determine the functions of the GAPs. Although no specific GAP function has yet been identified, certain observations provide substrates for speculation. For example, it would be anticipated that if any of the cellular structures that are essential for axon growth are composed of proteins peculiar to the growth state, the supply of these proteins from the cell body would be essential for axon growth. The growth cone is a candidate for such a growth-related structure. Although it is composed of many constitutents (e.g., actin and tubulin) that are shared with the rest of the cell, in some respects its composition is different. For example, lectins, which bind to certain glycoproteins, label the growth cone differently from the way in which they label the rest of the neuron (Pfenninger and Maylie-Pfenninger, 1981). Recently Pfenninger and his coworkers (1983) have partially purified growth-cone particles from embryonic rat brain by means of subcellular fractionation. When we compared the proteins of growth-cone particles (supplied by the Pfenninger laboratory) with GAP-43 (which can be identified in membranes of the neonatal rat superior colliculus) by two-dimensional gel electrophoresis, GAP-43 proved to be greatly enriched in the growth-cone particle fraction and was one of the major polypeptide constituents of the fraction (experiments of K. Meiri, L. Ellis, K. H. Pfenninger, and M. Willard). If GAP-43 should prove to be an essential component of the growth cone, its induction would be a prerequisite for successful axon growth and an essential feature of entry into the growth state.

What function might a GAP perform in the growth cone? A clue may be that only a small number of growth-associated proteins have been observed, which suggests that only a small number of growth-specific functions must be performed to convert an axon from a steady state to a growth state. (This argument is qualified by our ignorance of how many growth-specific proteins have escaped detection by the assays employed, which typically detect most sensitively those proteins that most rapidly incorporate methionine.) This consideration leads to the conjecture that a GAP, like the *cII*, *cro*, or repressor proteins of λ, might perform a switching function with pleiotropic consequences so as to influence multiple components of the cell to reorganize in a mode appropriate for axon extension. Like the λ regulatory proteins, a GAP might function as a switch at a branch point where, in the neuron, the metabolic pathways that lead to the steady state, or axon growth, diverge. There are several specific ways that a protein might perform such a pleiotropic switching function. For example, if a GAP were a component, or a regulator, of an ion channel in the growth cone membrane, many of the elements of the cell (e.g., actin, myosin, neurofilaments, microtubules, and membrane-bounded organelles) that were exposed to the altered cytoplasmic milieu created by the action of the GAP might respond by reorganizing into a growth mode. Alternatively, a GAP could influence the behavior of many other molecules if it modified them. For example, if a GAP were a specific protease or kinase, it might modify a number of proteins (e.g., actin, tubulin) in such a way as to alter their behavior in a manner appropriate for axon growth. Similarly, by modifying a few key proteins, a GAP might initiate a sequence of modifications (e.g., a kinase activating a lipase) that would result in the multiple altered molecular configurations necessary for axon growth. It is interesting to note that the receptors for certain growth factors [epidermal growth factor (EGF) and platelet-derived growth factor (PDF)], and the gene products of certain RNA tumor viruses that are involved in neoplastic transformations (the *src* gene products) are protein kinases that phosphorylate tyrosine residues (see Skene, 1983, for a review). As we anticipate in the case of GAPs, these kinases appear to promote aspects of cell growth by performing pleiotropic switching functions, mediated by the modification of other proteins. We do not yet know whether any of the GAPs are themselves kinases. However, we have recently observed that radioactive phosphate can be incorporated into GAP-43 when a membrane fraction from neonatal rat superior colliculus is incubated with ATP labeled in the gamma position with ^{32}P (experiments of K. Meiri). The degree of phosphorylation is greatly increased when calcium and calmodulin are added to the reaction mixture; GAP-43 is one of the most highly labeled polypeptides under these conditions. The phosphorylation of GAP-43 suggests that its function may be subject to regulation by posttranslational modification, as well as by its synthesis or introduction into the axon. Alternatively, in view of the observation that kinases invariably phosphorylate themselves, the labeling of GAP-43 with ^{32}P could reflect a kinase activity of GAP-43 itself; then GAP-43, like the EGF receptor, PDF receptor, and *src* kinases, might perform a switching function with pleiotropic consequences by phosphorylating other proteins. Because there is more than one GAP, it is likely that several mechanisms, such as those discussed here, may be involved in the conversion of an axon to a growth state.

THE DECISION TO EXPRESS GAPs

If indeed there are proteins that are essential for axon growth but are not transported into steady-state axons, then the decision to express GAPs is a critical element in the decision to grow an axon and to terminate its growth. Like the lysogeny versus vegetative growth decision mediated by the *c*II gene product in λ development, a neuron's decision as to whether to grow an axon must take into account the relevant prevailing conditions. For example, to initiate a growth state in order to recover from axon injury, a steady-state neuron must have access to the information that its axon is no longer intact. Three pathways that could potentially convey information about the integrity of the axon in such a way as to participate in the decision to enter a growth state are reviewed below (Willard and Skene, 1982).

First, stationary-state neurons have the capacity to take up certain substances at their terminals and to transport them retrogradely to the cell body where they can alter the expression of neuronal genes. For example, nerve growth factor that reaches the cell bodies of sympathetic neurons by this pathway can alter the synthesis of enzymes involved in neurotransmission. If a substance that repressed GAP synthesis in the cell body were obtained from the periphery (e.g., the postsynaptic cell) by this route, the supply of GAP repressor to the cell body would be interrupted if the axon were injured; this interruption would lead to the induction of GAP gene expression, which would favor the establishment of a growth state.

A second pathway, which could regulate levels of GAPs in a manner that would assure their induction only when communication between the neuron and target cell had been interrupted, could utilize regulatory molecules synthesized in the neuronal cell body itself. For example, suppose an inactive GAP repressor were synthesized in the neuronal cell body and were activated, by means of posttranslational modification, only after it had been transported anterogradely into a functional axon terminal. The return of the activated repressor to the cell body by means of retrograde axonal transport would serve to prevent the synthesis of the GAP in neurons with functional synapses. If this pathway were to be interrupted by injury to the axon, only inactive repressor would be present in the cell body; consequently, the GAP would be induced in preparation for axon growth. Certain features of such a pathway are illustrated by an 18-kD polypeptide that is a component of the group I axonally transported proteins in retinal ganglion cells and that appears to become modified when it reaches the axon terminals—the optic tectum in the toad, and the superior colliculus and lateral geniculate nucleus in the rabbit (Kelly et al., 1980; Skene et al., 1982). The modification is manifested as an apparent decrease in the net negative charge of the protein and could therefore result from the loss of negatively charged moieties, such as phosphate, or the addition of positively charged moieties. Two observations raise the possibility that this modification requires an intact synaptic terminal. First, the tectal form of the protein was not generated in the ends of axons of a toad optic nerve that had been crushed (Skene et al., 1982). Second, a toad optic tectum that had been treated previously with α-bungarotoxin failed to convert the nerve form of the 18-kD polypeptide to the tectal form (Skene et al., 1982). Retinal-tectal transmission in the toad is mediated by nicotinic acetylcholine receptors (Freeman et al., 1980); apparently, the disruption of the synapse caused by the binding of α-bungarotoxin to these receptors (Freeman, 1977) is sufficient to disrupt the mechanism by which the 18-kD polypeptide is modified in the optic tectum. We do not know whether the tectal form of the 18-kD polypeptide returns to the cell body by retrograde transport; however, many group I proteins appear to do so (Bisby, 1981). Neither do we know whether the altered form of the 18-kD polypeptide could change the expression of GAP genes. However, the behavior of the 18-kD polypeptide illustrates the existence of a pathway whereby the cell body could monitor the state of the synaptic terminals and alter the neuron's state of growth accordingly.

An interesting consequence of regulatory mechanisms (such as the two outlined above) that depend upon intact synapses to supply repressors of GAPs, is that the concentration of repressor in the cell body might be proportional to the number of intact synaptic connections that supply the repressor. The neuron could use such a feature to "count" its synapses and regulate the number formed during the course of developmental or regenerative growth (Willard and Skene, 1982). As increasing numbers of synapses were formed by the growing axon, the level of active repressor in the cell body would increase until the concentration was sufficient to terminate the expression of the relevant growth-related genes. At that point, synapse formation and axon growth would cease.

A third mechanism that could ensure that GAP genes would be expressed when axons are injured would result if the expression of these genes were contingent on GAP-inducing factors produced by nonneuronal cells in response to the injury. Such a factor, or a "messenger" produced as a consequence of the release of such factors, might then be transported retrogradely to the cell body, where it would induce the expression of genes necessary for the growth state. The observation (Skene and Shooter, 1983) that the synthesis and release of a 37-kD polypeptide is induced in nonneuronal cells when a rat sciatic nerve is crushed illustrates the potential of nonneuronal cells to respond to axotomy of neurons. The observation (e.g., Kristensson and Olsson, 1976) that molecules supplied exogenously at a site of axon injury can be retrogradely transported illustrates the availability of a retrograde pathway that could convey such an exogenously supplied inducer to the neuronal cell body.

Of these mechanisms for regulating the growth state, the third, in which growth-associated genes are regulated by exogenously supplied inducers from nonneuronal cells, is supported by the most compelling evidence. In particular, certain neurons of the mammalian CNS that would not normally regenerate an injured axon can accomplish substantial axon elongation when they are provided with an implanted tissue graft from a mammalian peripheral nerve (Aguayo et al., 1982; Benfey and Aguayo, 1982). These experiments demonstrate that a mature mammalian CNS neuron has the potential to regenerate its axon and, moreover, that information supplied by nonneuronal cells of the peripheral nerve is sufficient to allow the expression of this potential. A part of the contribution of the implanted peripheral nerve may be to supply an appropriate substrate suitable for axon growth. If in addition, as considered above, alterations in neuronal gene expression are required for successful axon growth, then all of the factors necessary to accomplish this gene regulation must have been supplied by the nonneuronal cells of the peripheral nerve.

The metabolic state of the CNS axon does in fact appear to influence its ability to regenerate into a peripheral nerve graft, as illustrated by the behavior of dorsal root ganglion (DRG) sensory neurons (Richardson, 1983). The axon of each of these neurons branches, sending one process to the periphery and another process centrally to the spinal cord; when injured, the central process does not normally regenerate within the spinal cord. Yet when a piece of peripheral sciatic nerve is experimentally implanted into the spinal cord, the central processes of the DRG neurons will regenerate into it, but the degree of regeneration is dramatically increased—by a factor of at least 10—if the peripheral processes of these DRG neurons are cut. The observation of preferential central regeneration of peripherally injured DRG neurons illustrates the importance of the metabolic state of the neuron in determining its ability to grow an axon. When the central processes of two DRG neurons from either side of the spinal cord are confronted with the same peripheral implant, one, with an injured peripheral process, will grow a central axon into the implant, whereas the other, with an intact peripheral process, will do so much less frequently. One explanation for the enhanced ability of central DRG axons to regenerate after their peripheral axons have been cut is that an intact peripheral axon supplies the DRG cell body with a repressor of GAPs sufficient to override inductive influences of the implanted peripheral nerve tissue. Severing the peripheral nerve would terminate the supply of repressor and allow the induction of GAPs, and the central process could then regenerate into the implanted peripheral nerve.

The three hypothetical mechanisms discussed here, as well as other pathways for regulating the expression of growth-related genes, are not mutually exclusive. In fact, the interpretation that different modes of axon growth (e.g., limited sprouting and axon elongation) require different combinations of neuronal gene products (e.g., GAP-23/24 and GAP-43, respectively) suggests that several different pathways are required to regulate independently the expression of different growth-related genes (Skene and Willard, 1981c).

THE PREGROWTH STATE

The pregrowth state is separated from the growth state by the initial decision of a developing neuron to extend an axon. Does this transition require a change in the program of gene expression to produce proteins required for axon growth or, alternatively, are all of the gene products necessary for axon growth present in the pregrowth state, awaiting a growth-triggering signal that is encountered only when the cell arrives at the appropriate time and place in the developing organism? It might be anticipated that the pregrowth state would differ from the growth state with respect to gene expression to an even greater extent than the growth state differs from the stationary state. Although both the pregrowth and stationary states might do without GAPs, the pregrowth neuron might also do without materials designed specifically for axonal functions (e.g., neurofilament polypeptides and transmitter-related enzymes) that must be constantly supplied to the axon in both the growth and stationary states. Although it would be difficult to evaluate directly the change in gene expression that accompanies the transition from a pregrowth to a growth state *in vivo*, certain cell lines can be induced to extend neurites synchronously in culture. For example, PC-12 cells, a clonal cell line derived from a rat medullary pheochromocytoma, can be induced to extend neurites when it is supplied with NGF (Greene and Tischler, 1976). When newly synthesized proteins from NGF-treated and naive cultures of PC-12 cells were compared by two-dimensional electrophoresis, the radiolabeling of one protein increased by approximately tenfold after NGF treatment, but 90% of the proteins changed by less than a factor of two (Garrels and Schubert, 1979). However, the relationship of PC-12 neurites to axons, and the relationship of the states of PC-12 cells in culture to developmental states of neurons, is ambiguous. Nonetheless, it is interesting that when PC-12 cells are conditioned by treatment with NGF under conditions (suspension in culture medium) in which they do not produce neurites, they will subsequently produce neurites much more rapidly than will cultures that have not been so primed (Greene et al., 1982). Because this conditioning response requires RNA synthesis, it could reflect the need for synthesis of certain proteins required for neurite growth (Burstein and Greene, 1978).

Immunofluorescence assays of the expression of the products of individual genes during the developmental period that encompasses the transition from the pregrowth to the growth state have revealed changes in gene expression that could be critical for this transition. A monoclonal antibody raised against fractionated growth-cone particles stains neurons in developing rat spinal cord when they begin to extend their axons, but does not appear to react with neuroblasts while they are still dividing (Wallis et al., 1983). If these antigens represent growth-cone components whose function is necessary for axon growth, then their expression some time after the cessation of neuroblast proliferation would be an obligate event in accomplishing the transition from a pregrowth to a growth state. During development of the grasshopper nervous system, certain neurons become reactive with a monoclonal antibody only when their axons have extended to a point where they contact other axons that they will subsequently follow en route to their target (see Goodman et al., this volume). The expression of the gene product recognized by this antibody thus appears to represent the entry of the neuron into a substate of the growth state; this substate is apparently required if the growing axon is to find its way to the appropriate target. This suggests that a correctly timed sequence of alterations in gene expression may be required for a neuron to pass successfully through the growth state, and that the sequence may be different for different neurons. Just as there are mechanisms for assuring the appropriate sequence of expression of the genes involved in the vegetative development of phage λ, there must be mechanisms which ensure that the correct sequence of alterations in gene expression is accomplished during axon growth.

EXPRESSION OF NEUROFILAMENT POLYPEPTIDES DURING DEVELOPMENT

A specific example of the sequential expression of neuronal genes during development is the expression of the neurofilament genes. The nature of the products of these genes is partially understood, and the delayed expression of one of the neurofilament genes raises the speculation that its regulation might be involved in a transition of a developing axon to a more stable state (decision D in Figure 1).

Structure and Function of Neurofilaments

The neurofilaments are long, slender filaments, approximately 100 Å in diameter, which comprise a major fraction of the total protein of many neurons. Within the cell body they tend to run singly or in fascicles; within the axons of some neurons, they are extensively cross-linked to each other, to microtubules, to membranous organelles, and to the plasma membrane (e.g., Hirokawa, 1982; Schnapp and Reese, 1982). The resulting structure resembles a three-dimensional lattice. The neurofilaments belong to a more general class of morphologically similar organelles called intermediate filaments; these include vimentin, keratin, desmin, and glial filaments, all of which have diameters of about 100 Å. Unlike these other intermediate filaments, each of which is composed of polypeptide subunits with molecular weights between 50 and 60 kD, mammalian neurofilaments are composed of three polypeptides with apparent molecular weights of approximately 200, 145, and 73 kD. These will be referred to here as H, M, and L, respectively. Each of these neurofilament polypeptides begins at the N-terminal end with a short, positively charged segment followed by two α-helical regions separated by a short nonhelical region (Geisler et al., 1983). This sequence of structures, which consumes about 40–50 kD of the polypeptide, appears to be common to all intermediate filament polypeptides and to be the region that forms the filamentous structure (Steinert et al., 1980; Geisler and Weber, 1982). In the case of shorter intermediate filament polypeptides such as vimentin, most of the polypeptide is devoted to this filament-forming structure. In contrast, the neurofilaments have large C-terminal regions left over—about 160 kD and 100 kD in H and M, respectively. These C-terminal regions, which can be phosphorylated, are probably able to extend from the core of the filament and mediate interactions with the environment. For example, the H polypeptide has been demonstrated to be a component of the cross bridges between neurofilaments within axons (Willard and Simon, 1981; Hirokawa et al., 1984).

The cross-bridging function performed by a phosphorylated form of H appears to be a specifically axonal function. When neurons (e.g., pyramidal cells of the cerebral cortex and the hippocampus and spinal motor neurons) are immunofluorescently stained with antibodies specific for each of the neurofilament polypeptides, all three antibodies stain axons intensely, whereas only anti-M and anti-L stain the cell bodies and dendrites intensely (Shaw et al., 1981; Hirokawa et al., 1984). Although neurofilaments are often cross-linked to each other in cell bodies and dendrites as well as in axons, these cross bridges tend to be shorter than those of the axons, and therefore may be composed of some protein other than this phosphorylated form of H: for example, the C-terminal region of M (Willard, 1983; Hirokawa et al., 1984), or an unphosphorylated form of H. An explanation for the preferential reactivity of certain anti-H antibodies with the axon, as compared to the cell body and dendrites, is that the axonal form of H is posttranslationally modified by phosphorylation (e.g., Julien and Mushinksi, 1982), such that certain antibodies recognize only the axonal form (Goldstein et al., 1983; Sternberger and Sternberger, 1983). It is consistent with this explanation that our polyclonal anti-H antibody, which reacts with interneurofilament cross bridges in axons but does not react strongly with cell bodies, does not react with dephosphorylated H. The variation of different antibody preparations with respect to their reactivity with the phosphorylated and unphosphorylated forms of the neurofilament polypeptides may also explain the following anomaly: In some species, anti-M (instead of anti-H) antibodies react specifically with axons, whereas anti-H antibodies do not distinguish between axons and cell bodies (Dahl et al., 1981; Tapscott et al., 1981b; Dahl, 1983). It is possible that these anti-M antibodies are specific for a phosphorylated axonal form of M, whereas these anti-H antibodies may react with both the phosphorylated and unphosphorylated forms of H.

The neurofilaments appear to function as structural elements of neurons. However, they may perform additional, more glamorous, roles. Their extensive cross-linking to membrane-bounded organelles of the axoplasm suggests that they are involved either in holding these organelles in place or in moving them down the axon. A particularly exotic property of neurofilaments is that the M polypeptide reacts with antibodies that otherwise are quite specific for α-melanocyte-stimulating hormone (Drager et al., 1983). The function of this resemblance between a region of the neurofilament polypeptide and a region of the hormone remains to be elucidated, but raises the possibility that neurofilaments may perform quite unexpected functions within the cell. With the exception that erythrocytes of developing chickens have been found to contain a polypeptide very similar to the L polypeptide of neurofilaments (Granger and Lazarides, 1983), neurofilament polypeptides have been detected only in neurons, and for this reason have been used as a diagnostic test for neuronal cell type (Osborn and Weber, 1983).

Developmental Expression

In view of the neuronal specificity of neurofilament proteins and of the axonal specificity of a phosphorylated form of H, it is interesting to consider how neurofilament protein genes are expressed in relationship to the states of neuronal development. In the neural tube of the developing chicken, neurofilament antigens have been detected in dividing neuroblasts shortly before they undergo their last mitotic division (Tapscott et al., 1981a,b). Soon after they begin to express neurofilament antigens, these cells lose their reactivity with antibodies against vimentin, the intermediate filament protein previously expressed in these cells. Similarly, mammalian retinal ganglion cells express certain neurofilament antigens shortly after they begin to send out axons (Shaw and Weber, 1983). Prior to this, their major intermediate filament protein has also been vimentin, which disappears soon after the neurofilament antigens appear. An exception is the horizontal cells of some, but not all, mammals in which vimentin continues to be expressed along with the neurofilament proteins in the adult neuron (Shaw and Weber, 1983). Neurofilament polypeptides are thus synthesized for the first time slightly before or at the same time as the transition of a neuron from a pregrowth to a growth state.

Like the sequential expression of the lytic functions during vegetative λ development, the three neurofilament polypeptides appear in a programmed sequence during development of the retinal ganglion cells of rabbit. M and L are present in the rabbit optic nerve from the time of birth, whereas H does not accumulate to detectable levels until 8–12 days after birth (Levine et al., 1982; Willard and Simon, 1983). A similar sequence has been observed in the rat cerebral cortex (Shaw and Weber, 1982) and in the rat optic nerve (Pachter and Liem, 1984). In the latter, the appearance of M has been reported to precede the appearance of L; as in the rabbit, both polypeptides appear before H.

During the period after H first appears in the axon, the apparent velocities of axonal transport of the neurofilament polypeptides change in rabbit retinal ganglion cells (Willard and Simon, 1983). The velocities of M and L decrease from about 9 mm/day in the six-day-old rabbit to about 1 mm/day in the adult. Although the two events are not necessarily related, their temporal correspondence suggests that the reduction in transport velocity could be a consequence of the introduction of H into the axon. For example, if uncross-linked neurofilaments composed of M and L are transported into the axon during the initial stages of neuronal development, the introduction of the H cross-linking function might precipitate a "phase transition" from a state of independent neurofilamentous elements to one characterized by the highly cross-linked matrix that composes the adult cytoskeleton. This matrix could be more difficult to translocate down the axon than its independent elements, and this could account for the ninefold decrease in the transport velocity of the neurofilament proteins.

During approximately the same period that the velocity of the group V neurofilament polypeptides decreases, the velocity of the group IV polypeptides (actin and myosin, etc.) decreases about twofold. This reduction could also be a consequence of the advent of the H cross-linking function. For example, perhaps the additional cross-links provided by H retard the passage of group IV material. Alternatively, if group IV and group V are partially coupled to each other, as they would be if group IV provided the motive force for the transport of group V (Levine et al., 1982; Levine and Willard, 1983), the drag upon the group IV transport machinery produced by the cross-linked cytoskeleton might result in the twofold reduction in the rate of the group IV movement.

What function could be served by the sequential introduction of neurofilament proteins into developing axons? The delayed appearance of the cross-linking polypeptide H, compared to the core polypeptides L and M, suggests that the cross-linking function of H is not needed or cannot be tolerated until late in development. One speculation as to why this might be so is that the appearance of H marks the transition from a state of growth to a state of stability (e.g., decision D in Figure 1). For example, suppose that the cross-linking of the cytoskeleton by H serves to stabilize the axon in the configuration that prevails when H is introduced into the axon. Then it would be beneficial for the neuron to await a sign that the proper configuration had been attained before it initiated the transport of H. If this signal were provided only by a correctly contacted postsynaptic cell, the induction of H by this signal after it was conveyed to the neuronal cell body would ensure that only axons with proper synaptic connections would be stabilized by the cross-linking function of H. It is interesting to note that, if such a hypothetical transition from the plastic state of the axon were irreversible, the induction of the H cross-linking function would provide the molecular basis for a critical period after which the plastic functions that were terminated when H was induced could no longer be accomplished. These observations raise the possibility that the decision to stabilize an axon involves regulation of the expression of the H gene, and that this decision may take into account the suitability of the connections that have been made.

ACKNOWLEDGMENTS

The work from our laboratory on growth-associated proteins that is reviewed here was initiated and developed by Dr. Pate Skene. We collaborated with Dr. Nobutaka Hirokawa and Carolyn Simon on the decoration of neurofilaments with antibodies *in situ* and *in vitro*, respectively. We also thank C. Baitinger, D. Clements, and R. Cheney for discussion, and Jan Hoffman and Caroline White for typing. This work was supported by grant EY02682 and a Neuromuscular Research Center Grant.

REFERENCES

Aguayo, A. J., P. M. Richardson, S. David, and M. Benfy (1982) In *Repair and Regeneration of the Nervous System*, J. G. Nichols, ed., pp. 91–105, Springer-Verlag, Berlin.

Allen, R. D., J. Metuzals, I. Tasaki, S. T. Brady, and S. P. Gilbert (1982) Fast axonal transport in squid giant axon. *Science* **218**:1127–1129.

Baitinger, C., J. Levine, T. Lorenz, C. Simon, P. Skene, and M. Willard (1982) Characteristics of axonally transported proteins. In *Axoplasmic Transport*, D. G. Weiss, ed., pp. 110–120, Springer-Verlag, Berlin.

Benfey, N., and A. J. Aguayo (1982) Extensive elongation of axons from rat brain into peripheral nerve grafts. *Nature* **296**:150–152.

Benowitz, L. I., V. E. Shashoua, and M. G. Yoon (1981) Specific changes in rapidly transported proteins during regeneration of the goldfish optic nerve. *J. Neurosci.* **1**:300–307.

Benowitz, L. I., M. G. Yoon, and E. R. Lewis (1983) Transported proteins in the regenerating optic nerve: Regulation by interactions with optic tectum. *Science* **222**:185–188.

Bisby, M. A. (1981) Reversal of axonal transport: Similarity of proteins transported in anterograde and retrograde directions. *J. Neurochem.* **36(2)**:741–745.

Brady, S. T., M. Tytell, K. Heriot, and R. J. Lasek (1981) Axonal transport of calmodulin: A physiologic approach to identification of long-term associations between proteins. *J. Cell Biol.* **89**:607–614.

Burstein, D. E., and L. A. Greene (1978) Evidence for RNA synthesis-dependent and -independent pathways in stimulation of neurite outgrowth by nerve growth factor. *Proc. Natl. Acad. Sci. USA* **75**:6059–6063.

Dahl, D. (1983) Immunohistochemical differences between neurofilaments in perikarya, dendrites, and axons. *Exp. Cell Res.* **149**:397–408.

Dahl, D., A. Bignami, N. T. Bich, and N. H. Chi (1981) Immunohistochemical localization of the 150K neurofilament protein in the rat and rabbit. *J. Comp. Neurol.* **195**:659.

Drager, U. C., D. L. Edwards, and J. Kleinschmidt (1983) Neurofilaments contain α-melanocyte-stimulating hormone (α-MSH)-like immunoreactivity. *Proc. Natl. Acad. Sci. USA* **80**:6408–6412.

Echols, H. (1980) Bacteriophage λ development. In *The Molecular Genetics of Development*, P. J. Leighton and W. F. Loomis, eds., pp. 1–16, Academic, New York.

Forman, D. S. (1981) Axonal transport and nerve regeneration: A review. In *Spinal Cord Reconstruction*, C. C. Kao, R. P. Bunge, and P. J. Reit, eds., pp. 75–86, Raven, New York.

Forman, D. S., K. J. Brown, and D. R. Livengood (1983) Fast axonal transport in permeabilized lobster giant axons is inhibited by vanadate. *J. Neurosci.* **3**:1279–1288.

Freeman, J. A. (1977) Possible regulatory function of acetylcholine receptor in maintenance of retinotectal synapses. *Nature* **269**:218–222.

Freeman, J. A., J. T. Schmidt, and R. E. Oswald (1980) Effect of alpha-bungarotoxin on retinotectal synaptic transmission in the goldfish and the toad. *Neuroscience* **5**:929–942.

Garrels, J. I., and D. Schubert (1979) Modulation of protein synthesis by nerve growth factor. *J. Biol. Chem.* **254**:7978–7985.

Geisler, N., and K. Weber (1982) The amino acid sequence of chicken muscle desmin provides a common structural model for intermediate filament proteins. *EMBO J.* **1**:1649–1656.

Geisler, N., E. Kaufmann, S. Fischer, U. Plessmann, and K. Weber (1983) Neurofilament architecture combines structural principles of intermediate filaments with carboxy-terminal extensions increasing in size between triplet proteins. *EMBO J.* **2**:1295–1302.

Giulian, D., H. des Ruisseaux, and D. Cowburn (1980) Biosynthesis and intraaxonal transport during neuronal regeneration. *J. Biol. Chem.* **255**:6494–6501.

Goldstein, M. E., L. A. Sternberger, and N. H. Sternberger (1983) Microheterogeneity ("neurotypy") of neurofilament proteins. *Proc. Natl. Acad. Sci. USA* **80**:3101–3105.

Granger, B. L., and E. Lazarides (1983) Expression of the major neurofilament subunit in chicken erythrocytes. *Science* **221**:553–556.

Greene, L. A., and A. S. Tischler (1976) Establishment of a noradrenergic clonal line of rat adrenal pheochromocytoma cells which respond to nerve growth factor. *Proc. Natl. Acad. Sci. USA* **73**:2424–2428.

Greene, L. A., D. E. Burstein, and M. M. Black (1982) The role of transcription-dependent priming in nerve growth factor promoted neurite outgrowth. *Dev. Biol.* **91**:305–316.

Hendrix, R. W., J. Roberts, F. Stahl, and R. Weisberg, eds. (1983) *Lambda II*. Cold Spring Harbor Laboratory, Cold Spring Harbor, N.Y.

Hirokawa, N. (1982) Cross-linker system between neurofilaments, microtubules, and membrane organelles in frog axons revealed by the quick-freeze, deep-etching method. *J. Cell Biol.* **94**:129–142.

Hirokawa, N., M. A. Glicksman, and M. B. Willard (1984) Organization of mammalian neurofilament polypeptides within the neuronal cytoskeleton. *J. Cell Biol.* **98**:1523–1536.

Johnson, M., L. Iacovitti, D. Higgins, R. Bunge, and H. Burton (1981) Growth and development of sympathetic neurons in tissue culture. *CIBA Symp.* **83**:108–122.

Julien, J.-P., and W. F. Mushinski (1982) Multiple phosphorylation sites in mammalian neurofilament polypeptides. *J. Biol. Chem.* **257**:10467–10470.

Kelly, A. S., J. A. Wagner, and R. B. Kelly (1980) Properties of individual nerve terminal proteins identified by two-dimensional gel electrophoresis. *Brain Res.* **185**:192–197.

Kristensson, K., and Y. Olsson (1976) Retrograde transport of horseradish peroxidase into transected axons. 3. Entry into injured axons and subsequent localization in perikaryon. *Brain Res.* **115**:201–213.

Landis, S. C., and D. Keefe (1983) Evidence for neurotransmitter plasticity *in vivo*; developmental changes in properties of cholinergic sympathetic neurons. *Dev. Biol.* **98**:349–372.

Lasek, R. J., and M. M. Black (1977) How do axons stop growing? Some clues from the metabolism of the proteins in the slow component of axonal transport. In *Mechanisms, Regulation and Special Functions of Protein Synthesis in the Brain*, S. Roberts, A. Lajtha, and W. Gispen, eds., pp. 161–169, Elsevier/North Holland, Amsterdam.

Lasek, R. J., and P. N. Hoffman (1976) The neuronal cytoskeleton, axonal transport and axonal growth. *Cold Spring Harbor Symp. Quant. Biol.* **3**:1021–1049.

Levine, J., and M. Willard (1981) Fodrin: Axonally transported polypeptides associated with the internal periphery of many cells. *J. Cell Biol.* **90**:631–743.

Levine, J., and M. Willard (1983) Redistribution of fodrin (a component of the cortical cytoplasm) accompanying capping of cell surface molecules. *Proc. Natl. Acad. Sci. USA* **80**:191–195.

Levine, J., C. Simon, and M. Willard (1982) Mechanistic implications of the behavior of axonally transported proteins. In *Axoplasmic Transport*, D. G. Weiss, ed., pp. 275–278, Springer-Verlag, Berlin.

Osborn, M., and K. Weber (1983) Biology of disease. Tumor diagnosis by intermediate filament typing: A novel tool for surgical pathology. *Lab. Invest.* **48**:372–394.

Pachter, J. S., and R. K. H. Liem (1984) Differential appearance of neurofilament triplet polypeptides in developing rat optic nerve. *Dev. Biol.* **103**:200–210.

Patterson, P. H. (1978) Environmental determination of neurotransmitter functions. *Annu. Rev. Neurosci.* **1**:1–17.

Perry, G. W., and D. L. Wilson (1981) Protein synthesis and axonal transport during nerve regeneration. *J. Neurochem.* **37**:1203–1217.

Pfenninger, K. H., and M. F. Maylie-Pfenninger (1981) Lectin labeling of sprouting neurons. II. Relative movement and appearance of glycoconjugates during plasmalemma expansion. *J. Cell Biol.* **89**:547–559.

Pfenninger, K. H., L. Ellis, M. P. Johnson, L. B. Friedman, and S. Somio (1983) Nerve growth cones isolated from fetal rat brain: Subcellular fractionation and characterization. *Cell* **35**:573–584.

Richardson, P. M. (1983) Effect of peripheral nerve injury on regeneration of axons in the dorsal column of the spinal cord. *Soc. Neurosci. Abstr.* **9**:698.

Schnapp, B. J., and T. S. Reese (1982) Cytoplasmic structure in rapid frozen axons. *J. Cell Biol.* **94**:667–679.

Shaw, G., and K. Weber (1982) Differential expression of neurofilament triplet proteins in brain development. *Nature* **298**:277–279.

Shaw, G., and K. Weber (1983) The structure and development of the rat retina: An immunofluorescence microscopical study using antibodies specific for intermediate filament proteins. *Eur. J. Cell Biol.* **30**:219–232.

Shaw, G., M. Osborn, and K. Weber (1981) An immunofluorescence microscopical study of the neurofilament triplet proteins, vimentin and glial fibrillary acidic protein within the adult rat brain. *Eur. J. Cell Biol.* **26**:68–82.

Skene, J. H. P. (1983) An oncogene abounds in brains. *Trends Neurosci.* **6**:353–354.

Skene, J. H. P., and E. Shooter (1983) Denervated sheath cells secrete a new protein after nerve injury. *Proc. Natl. Acad. Sci. USA* **80**:4119–4173.

Skene, J. H. P., and M. Willard (1981a) Changes in axonally transported proteins during axon regeneration in toad retinal ganglion cells. *J. Cell Biol.* **89**:86–95.

Skene, J. H. P., and M. Willard (1981b) Characteristics of growth-associated proteins (GAPs) in regenerating toad retinal ganglion cells. *J. Neurosci.* **1**:419–426.

Skene, J. H. P., and M. Willard (1981c) Axonally transported proteins associated with axon growth in rabbit central and peripheral nervous systems. *J. Cell Biol.* **89**:96–103.

Skene, J. H. P., M. Willard, and J. A. Freeman (1982) Modification of an axonally transported protein in toad retinotectal terminals. *Soc. Neurosci. Abstr.* **9**:247.

Steinert, P. M., W. W. Idler, and R. D. Goldman (1980) Intermediate filaments of baby hamster kidney (BHK-21) cells and bovine epidermal keratinocytes have similar ultrastructures and subunit domain structures. *Proc. Natl. Acad. Sci. USA* **77**:4534–4538.

Sternberger, L. A., and N. H. Sternberger (1983) Monoclonal antibodies distinguish phosphorylated and nonphosphorylated forms of neurofilaments *in situ*. *Proc. Natl. Acad. Sci. USA* **80**:6126–6130.

Tapscott, S. J., G. S. Bennett, and H. Holtzer (1981a) Neuronal precursor cells in the chick neural tube express neurofilament proteins. *Nature* **292**:836–838.

Tapscott, S. J., G. S. Bennett, Y. Toyama, F. Klienbart, and H. Holtzer (1981b) Intermediate filament proteins in the developing chick spinal cord. *Dev. Biol.* **86**:40–54.

Tsukita, S., and H. Ishikawa (1980) The movement of membranous organelles in axons. Electron microscopic identification of anterogradely and retrogradely transported organelles. *J. Cell Biol.* **84**:513.

Wallis, I., L. Ellis, and K. H. Pfenninger (1983) Neuronal differentiation antigens expressed during sprouting and synaptogenesis. *Soc. Neurosci. Abstr.* **9**:1178.

Willard, M. (1983) Neurofilaments and axonal transport. In *Neurofilaments*, C. A. Marotta, ed., pp. 86–116, Univ. Minnesota Press, Minneapolis.

Willard, M., and C. Simon (1981) Antibody decoration of neurofilaments. *J. Cell Biol.* **89**:198–205.

Willard, M., and C. Simon (1983) Modulations of neurofilament axonal transport during the development of rabbit retinal ganglion cells. *Cell* **35**:551–559.

Willard, M., and J. H. P. Skene (1982) Molecular events in axonal regeneration. In *Repair and Regeneration of the Nervous System*, J. G. Nicholls, ed., pp. 71–89, Springer-Verlag, Berlin.

Willard, M., M. Wiseman, J. Levine, and P. Skene (1979) Axonal transport of actin in rabbit retinal ganglion cells. *J. Cell Biol.* **81**:581–591.

Wolinsky, E., and P. H. Patterson (1983) Tyrosine hydroxylase activity decreases with induction of cholinergic properties in cultured sympathetic neurons. *J. Neurosci.* **3**:1495–1500.

Chapter 16

Cholinergic Development and Identification of Synaptic Components for Chick Ciliary Ganglion Neurons in Cell Culture

DARWIN K. BERG
MICHELE H. JACOB
JOSEPH F. MARGIOTTA
RAE NISHI
JES STOLLBERG
MARTIN A. SMITH
JON M. LINDSTROM

ABSTRACT

Chick ciliary ganglion neurons form cholinergic synapses on smooth and striated muscle in the eye and, in turn, receive cholinergic input from preganglionic terminals. Two components that stimulate long-term growth and development of the neurons in cell culture can be isolated from embryonic eye tissue. One component, with an apparent size of about 20 kD as determined by gel filtration of crude extract, stimulates overall growth of the neurons without changing neuronal levels of choline acetyltransferase activity. The other component, with an apparent size of 50 kD, stimulates the development of choline acetyltransferase in culture without changing the rate of neuronal growth. A rabbit antiserum raised against an impure preparation of the second component specifically blocks the increases it normally causes in choline acetyltransferase activity.

Ciliary ganglion neurons acquire acetylcholine receptors in cell culture and establish functional cholinergic synapses on each other. Synaptic acetylcholine receptors on the neurons appear to be distinct from high-affinity α-bungarotoxin binding sites, since ultrastructural studies with embryonic ganglia reveal that the toxin sites are not present in synaptic membrane. Instead the sites are concentrated predominantly on dendritic surfaces near regions of innervation. Moreover, levels of acetylcholine sensitivity and α-bungarotoxin binding can apparently be regulated independently, suggesting that the two membrane properties represent different membrane components. Monoclonal antibody studies demonstrate that ciliary ganglion neurons share an antigenic determinant with the "main immunogenic region" of acetylcholine receptors from muscle and electric organ. Ultrastructural studies on ganglionic tissue indicate that, in contrast to toxin sites, the cross-reacting component is situated predominantly in synaptic membrane, as expected for a synaptic acetylcholine receptor.

Cell–cell interactions play important roles in guiding neuronal development and synapse formation *in vivo*. In some cases the interactions have been shown to be mediated by diffusible substances. A prominent example comes from studies on neuronal survival, where it has been demonstrated that most vertebrate neurons do not survive beyond a critical period in development if deprived of interaction with appropriate target tissue for innervation. For sympathetic and dorsal root ganglion neurons there is good evidence that the dependence is mediated by nerve growth factor, a protein that promotes survival and continued development of the neurons and is thought to be released by the target tissue. Recently another component that promotes survival of some sensory neuron populations, at least in cell culture, has been purified to homogeneity from pig brain (Barde et al., 1982). This raises hopes that additional factors serving other neuronal populations in a similar manner will soon be isolated (see Berg, 1984, for a review).

Diffusible components have also been shown to influence or instruct the neuron as to the choice of neurotransmitter that it will synthesize. Thus a component produced by nonneuronal cells and released into culture medium can induce superior cervical ganglion neurons of rat to reduce levels of norepinephrine synthesis and to produce substantial levels of acetylcholine (ACh) and form cholinergic synapses instead. Synaptic input can also regulate the capacity of neurons to synthesize neurotransmitters (see Patterson, 1978, and Black, 1982, for reviews).

Cell–cell interactions are clearly important for synapse formation. For example, it has been known for some time that the presynaptic nerve regulates the number and distribution of acetylcholine receptors and acetylcholinesterase molecules in skeletal muscle. Synaptic activity has been shown to be a primary mechanism in mediating this regulation (see Fambrough, 1979, for a review). However, soluble components may also play a role, since factors present in nerve tissue extracts can induce clustering of ACh receptors on muscle cells (Podleski et al., 1978; Jessell et al., 1979; Olek et al., 1983). The regulation of synaptic components on neurons is much less well understood.

CHICK CILIARY GANGLION NEURONS

The chick ciliary ganglion provides a convenient source of parasympathetic neurons for studying neuronal development and synapse formation (see Landmesser and Pilar, 1978; Pilar and Tuttle, 1982, for reviews). The ganglion is composed of two populations of neurons: ciliary neurons that innervate striated muscle in the iris and ciliary body, and choroid neurons that innervate smooth muscle in the choroid layer. Both populations are cholinergic and both receive cholinergic input from preganglionic fibers that arise from the accessory oculomotor nucleus in the midbrain. The preganglionic fibers also form electrical synapses with ciliary neurons, and many of the fibers contacting both kinds of neurons contain substance P and enkephalin (Erichsen et al., 1982). The function of these neuropeptides in the ganglion is not known.

Functional cholinergic transmission to all ganglionic neurons is established before embryonic day seven. Between embryonic days eight and 14, half of the neurons in both the ciliary and choroid populations are lost through naturally ocurring cell death, leaving a total of about 3000 neurons in the ganglion. Information from the target tissue is clearly necessary for the normal survival and development of the neurons, as removal of the optic vesicle results in the death of nearly all ganglionic neurons between embryonic days eight and 14. (Functional innervation of the target muscle by ganglionic neurons normally occurs during this period.) The electrical synapses on ciliary neurons are established at about embryonic day 16 and are maintained throughout adulthood in conjunction with the cholinergic synapses.

Survival in Cell Culture

The demonstration that ciliary ganglion neurons will survive and grow in dissociated cell culture (Helfand et al., 1976) opened the way for regulatory studies on the neurons *in vitro*. Early on it was found that all of the neurons present in the embryonic ganglion, including those destined to die normally *in vivo*, could survive and grow in cell culture (Nishi and Berg, 1977). Moreover, the neurons could develop and form functional synapses (see below) when supplied with appropriate target cells or culture media supplements (Nishi and Berg, 1977, 1979; Tuttle et al., 1980; Margiotta and Berg, 1982). These results suggested that the cell culture system might be useful for identifying mechanisms by which the target tissue normally supports the survival and development of the ganglionic neurons.

Varon and his colleagues first devised a short-term assay for identifying factors that promote the survival of ciliary ganglion neurons in culture. They demonstrated that extracts prepared from embryonic eye tissue, the normal synaptic target for the neurons, contain the highest levels of survival factor activity (Adler et al., 1979). The amount of activity in eye tissue is developmentally regulated and increases substantially during the period in which neuronal survival becomes dependent on the target tissue (Landa et al., 1980). Furthermore, the time course of neuronal dependence in culture on the survival activity appears to parallel that expected from *in vivo* observations. Thus neurons in five-day embryonic ganglia do not require the survival factor for maintenance in explant culture for a three-day period, whereas neurons in eight-day ganglia do (Adler and Varon, 1982). These properties of the survival activity are consistent with those predicted for a target-derived component mediating survival of the neurons *in vivo*.

Current efforts are focused on purifying the active component from extracts of embryonic eye tissue. The activity appears to be associated with a component of about 20 kD (Varon et al., 1983) with an isoelectric point of about five (Manthorpe et al., 1982a). When the partially purified material was examined by gel filtration (Manthorpe et al., 1980), some of the activity could be recovered as a component of about 35 kD. Similar active components have been described in other tissue extracts (Bonyhady et al., 1980; Hill et al., 1981; Bonyhady et al., 1982; Nieto-Sampedro et al., 1982). Preliminary studies suggest that the component can support the survival of sensory and sympathetic neurons in culture as well (Manthorpe et al., 1982b), but definitive tests await the availability of purified factor.

The ciliary ganglion cell culture system has also been very useful for the identification of components that stimulate neurite production; some of these are known to be different from the survival factor described here.

LONG-TERM GROWTH AND DEVELOPMENT

Chick ciliary ganglion neurons can be maintained for at least several weeks in dissociated cell culture (Nishi and Berg, 1977, 1979; Tuttle et al., 1980). In the presence of skeletal myotubes or culture medium containing appropriate supplements, the neurons acquire levels of choline acetyltransferase (CAT) activity characteristic of developmental stages well beyond the period of ganglionic cell death *in vivo*. These features suggested that the culture system also could be used for identifying mechanisms governing long-term growth and development.

To identify components that have long-term stimulatory effects on the neurons, basal culture conditions were needed that would allow for prolonged maintenance of the neurons in the absence of nonneuronal cells and tissue extracts. The basal conditions would then provide a backdrop against which the stimulatory effects of exogenous factors could be recognized. The conditions chosen included a culture substratum coated with collagen and the residue of lysed fibroblasts to enhance neuronal attachment, and a basic culture medium containing horse serum and an elevated K^+ concentration to enhance survival. All of the neurons present in eight-day embryonic ganglia could survive in culture under these control conditions for at least two weeks, and slow growth and cholinergic development occurred (Nishi and Berg, 1981).

Extracts prepared from eye tissue of embryonic chicks were tested for stimulatory effects on neurons maintained under the basal conditions. A two- to fourfold stimulation of CAT levels was observed over a two-week period (Figure 1). Maximal stimulation was obtained with extract concentrations of about 0.02 mg/ml protein. Similar effects were seen on neuronal growth by measuring the relative levels of lactic acid dehydrogenase (LDH), a cytoplasmic marker enzyme. Fractionation of the eye extract by gel filtration revealed that separate com-

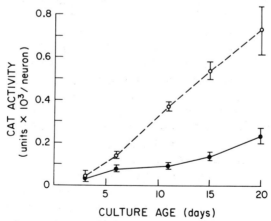

Figure 1. *Stimulation of neuronal CAT levels by embryonic eye extract.* Ciliary ganglion neurons were grown on substrata coated with fibroblast residue and fed with culture medium consisting of Eagle's Minimum Essential Medium, 10% (vol/vol) horse serum, and 25 mM K^+ (basal conditions, filled circles), or the same medium supplemented with 5% (vol/vol) embryonic eye extract (open circles). Cell counts and CAT assays were performed on the same cultures. Values represent the mean of 4–18 cultures pooled from three to nine experiments; bars indicate the standard error. (From Nishi and Berg, 1981.)

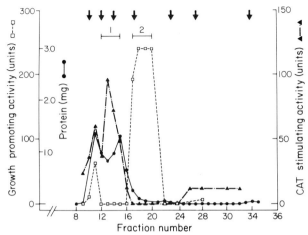

Figure 2. *Fractionation of eye extract by gel filtration.* Embryonic chick extract (3 ml) was applied to a Bio Gel P200 column (105 ml, 1.8 × 35 cm) equilibrated with phosphate-buffered saline solution at 4°C. Fractions (3 ml) were collected, measured for protein, and tested for effects on ciliary ganglion neurons by growing cultures in basal conditions supplemented with diluted aliquots from the fractions and then assaying the cultures for LDH activity, CAT activity, and the number of surviving neurons. To calculate units of stimulatory activity, cultures were also grown in a range of eye extract concentrations and assayed for CAT and LDH activities to construct a dose–response curve. Column fraction dilutions were chosen to yield responses estimated to be in the linear range of the assay; actual volumes assayed were the same in all cases. A unit of GPA was defined as the amount of stimulation necessary to produce half the maximum increment in LDH levels caused by eye extract; a unit of CSA was defined in the same way with respect to CAT levels. Horizontal bars indicate fractions pooled for CSA (pool 1) and GPA (pool 2). The arrows from left to right indicate the elution positions for blue dextran (200 kD), human transferin (80 kD), ovalbumin (40 kD), soybean trypsin inhibitor (26 kD), cytochrome c (13.5 kD), α-bungarotoxin (8 kD), and [^{3}H]choline chloride (140 D). The elution position of blue dextran marks the void volume of the column and that of [^{3}H]choline chloride indicates the completely included volume. (From Nishi and Berg, 1981.)

ponents were responsible for the effects on neuronal growth and on cholinergic development.

Growth-Promoting Activity

Stimulation of neuronal growth was caused by "growth-promoting activity" (GPA), which migrated primarily as a component of 20 kD when the extract was fractionated by gel filtration (Figure 2). Small amounts of activity were recovered in the excluded and completely included portions of the column eluate. Titration experiments demonstrated that the 20-kD GPA material could stimulate the neurons to the same maximal rate of growth as did saturating amounts of eye extract. GPA had no effects on the basal levels of CAT activity associated with the neurons.

Preliminary experiments suggest that the partially purified ciliary survival factor provided by Varon and his colleagues from eye tissue of embryonic chicks has properties similar to those of GPA in the growth assay (J. Fujii and D. Berg, unpublished observations), and that a GPA fraction has effects similar to those of the survival factor in the survival assay (M. Manthorpe and S. Varon, unpublished observations). Further purification will be necessary to demonstrate whether the two activities reside in the same component.

CAT-Stimulating Activity

Stimulation of CAT levels in culture is caused primarily by an activity in eye extract that migrates as a component of 50 kD when fractionated by gel filtration (Figure 2). Minor amounts of stimulatory activity are also recovered in the excluded and completely included fractions of the column eluate. The 50-kD activity, termed CAT-stimulating activity (CSA), is unable to drive neurons to the same maximal rate of CAT development that is achieved by saturating amounts of eye extract unless GPA is also added to the medium. Thus although GPA has no direct effect on the development of CAT levels, it does have a permissive effect with respect to CSA action. CSA has no effect on neuronal growth in culture.

Vitreous humor from embryonic chick eye is a relatively good source of CSA. Although it contains only 20% as much total activity as does whole-eye extract (omitting vitreous humor and the lens), it has about eight times the specific activity. Fractionation of vitreous humor by P200 gel filtration and DEAE cellulose column chromatography achieves a 300-fold enrichment compared to material in eye extract. The partially purified material is active at 50–100 ng/ml. Examination of the material by SDS-polyacrylamide gel electrophoresis still reveals a number of components (J. Stollberg and D. Berg, unpublished observations).

Antiserum to the partially purified CSA can be obtained. Pooled fractions of CSA from a P200 fractionation of vitreous humor were injected into rabbits. The immune serum was tested by mixing dilutions of it with CSA fractions, and then assaying the mixtures for the ability to stimulate CAT development in the

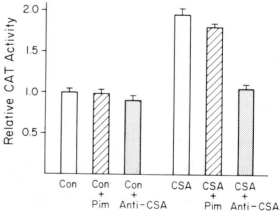

Figure 3. *Effects of anti-CSA antiserum on neurons in cell culture.* An antiserum against partially purified CSA was prepared by immunizing rabbits with material obtained by gel filtration of vitreous humor extract from chick embryos. The serum was tested for anti-CSA activity by mixing aliquots with a CSA fraction, diluting the mixture into basal culture medium, and testing the medium for ability to support CAT development over a 10-day period in culture. Results were normalized to a value of 1 for control cultures and are expressed as the mean (vertical bars) of triplicate cultures plus or minus standard error (horizontal lines). Con, control cultures grown with basal conditions; CSA, test cultures receiving medium supplemented with CSA; Pim, serum obtained from the rabbit prior to immunization; Anti-CSA, rabbit anti-CSA antiserum.

neuronal cultures. The serum contained specific anti-CSA activity in that it blocked CAT induction by CSA without influencing neuronal survival, neuronal growth, or basal levels of CAT activity (Figure 3). The serum was effective at dilutions of 1:1000 and did not require removal of the antigen–antibody complex for blockade. Preimmune serum had no effect. The blockade of CSA activity may have resulted from nonspecific trapping of CSA by antigen–antibody complexes formed in the medium, but it seems more likely that CSA is immunogenic in partially purified form and that specific anti-CSA antibodies were responsible for the blockade. If so, experiments with purified antibodies may eventually permit examination of the role of CSA *in vivo*. When probed with the anti-CSA antiserum, immunoblots of CSA material active at 50–100 ng/ml revealed a number of labeled components. This is not surprising, since the original immunogen was heterogeneous.

CSA is heat labile, elutes from both anion and cation exchange columns at low ionic strength in neutral pH buffers, and migrates as a component of 50 kD. Similar properties have been reported (Weber, 1981; Fukada, 1983) for a component released by nonneuronal cells that induces rat superior cervical ganglion neurons to synthesize ACh and form cholinergic synapses in culture (see Patterson, 1978, for a review).

SYNAPSE FORMATION

Nerve–Muscle Synapses

Most ciliary ganglion neurons form functional synaptic contacts with cocultured skeletal myotubes (Nishi and Berg, 1977). Transmission is reversibly blocked by D-tubocurarine and irreversibly blocked by α-bungarotoxin, as expected for nicotinic innervation of muscle. In some cases synaptic transmission is adequate to initiate muscle action potentials. Functional nerve–muscle synapses can be detected within hours of adding ciliary ganglion neurons to the myotubes.

Nerve–Nerve Synapses

Ciliary ganglion neurons do not normally innervate each other *in vivo*, at least for those times examined. Under appropriate conditions in cell culture, however, the neurons can form functional synaptic contacts with each other whether or not myotubes are present in the culture (Margiotta and Berg, 1982). Intracellular recordings reveal that most of the neurons receive synaptic input in the absence of stimulation. Postsynaptic potentials range in amplitude from 0.5 to 15 mV and occur at frequencies of 2–300 per min. The depolarizations often sum and occasionally trigger action potentials in the neurons. The proportion of neurons receiving spontaneous synaptic input varies greatly among cultures, but it can be as high as 95% both for neurons grown alone and for neurons grown with skeletal myotubes. The synaptic potentials are reversibly blocked by D-tubocurarine and by bungarotoxin-3.1, but not by α-bungarotoxin. This pharmacological profile parallels that found for ACh receptors on neurons in culture (Ravdin and Berg, 1979).

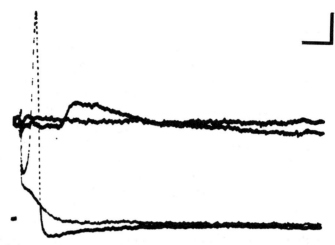

Figure 4. *Evoked synaptic transmission between ciliary ganglion neurons.* Neurons were grown for eight days with skeletal myotubes. Intracellular stimulation of one neuron (lower traces) evoked a synaptic potential recorded intracellularly in a second neuron (upper traces). Two superimposed trials are shown. In the first case presynaptic stimulation was subthreshold and failed to elicit a presynaptic action potential or a postsynaptic response. In the second case a stronger stimulus generated a presynaptic action potential that evoked a short-latency synaptic potential in the second neuron. Repeated trials evoked similar one-for-one responses with the same latency (not shown). Vertical calibration bar: 5 mV for upper traces and 10 mV for lower traces. Horizontal calibration bar: 5 msec.

Much of the spontaneous synaptic activity appears to be driven by spontaneous action potentials in the neurons. Intracellular recording from the neurons reveals the presence of action potentials without electrical stimulation. Addition of 10 μM tetrodotoxin to the culture medium rapidly abolishes most synaptic activity (Margiotta and Berg, 1982).

Intracellular stimulation of a single neuron can also evoke short-latency synaptic potentials in a second nearby neuron when cultures are grown with or without muscle (Figure 4). When tested in this manner, preliminary results indicate that about 20% of the neuron pairs in a microscopic field of view are connected by functional synapses. Another study of synapse formation between ciliary ganglion neurons grown in the absence of muscle failed to detect functional transmission (Crean et al., 1982). Ultrastructural analysis did reveal synaptic structures between the neurons, but levels of ACh sensitivity associated with the neurons apparently were too low to support cholinergic transmission. Differences in culture technique in the two studies most likely account for the observed differences in sensitivity (see below).

The fact that ciliary ganglion neurons can innervate each other in cell culture demonstrates that no inherent feature of the neurons prevents this process *in vivo*. Rather, other conditions, absent in culture, must be responsible. An interesting possibility is that normal preganglionic input is important. Synaptic contacts from preganglionic terminals are established on all ciliary ganglion neurons before the neurons innervate their targets. Studies on frog parasympathetic neurons *in situ* indicate that denervation of the ganglion permits the formation of abnormal intraganglionic synapses, which subsequently disappear when the preganglionic fibers are allowed to reestablish proper synapses (Sargent and

Dennis, 1981). Conceivably, a similar process normally is at work restricting synapse formation by ciliary ganglion neurons, and this constraint is relaxed when the preganglionic input is removed in culture.

ACh RECEPTORS

To understand the steps involved in synapse formation, it is necessary to study the recruitment and regulation of identified synaptic components. A central component for synaptic function on ciliary ganglion neurons is the ganglionic nicotinic ACh receptor. ACh receptors from skeletal muscle and electric organ have been studied in considerable detail, and much is known about their expression and regulation during innervation of muscle. The progress with muscle ACh receptor has been due in large part to specific probes that permit the identification and quantification of the receptors. Useful probes for muscle receptors include such snake α-neurotoxins as α-bungarotoxin and antisera and monoclonal antibodies raised against the receptors. α-Bungarotoxin has also been claimed to be a probe for neuronal ACh receptors, but current evidence from ciliary ganglion neurons suggests that the toxin binding sites on the neurons are different from the synaptic ACh receptors (see below). To date, the only reliable assays for neuronal ACh receptors are physiological.

We have compared the development of ACh sensitivity by ciliary ganglion neurons grown in cell culture under several conditions. An intracellular microelectrode was used to record membrane conductance changes caused by brief pulses of ACh applied to the neurons by pressure ejection from a nearby micropipette. Control experiments indicated that the application immersed the soma in ACh, and the responses probably reflected the "summed" responses of most of the somatic ACh receptors. For these experiments neurons from eight-day-old embryonic ganglia were grown on a substratum of collagen and fibroblast residue in normal medium (MEM and 10% horse serum) supplemented either with 3% (vol/vol) eye extract (eye medium), with 20 mM K$^+$ for a final concentration of 25 mM (K$^+$ medium), or with both (K$^+$/eye medium). After seven days in culture, neurons grown in eye medium displayed high levels of ACh sensitivity, whereas neurons grown in either K$^+$ or K$^+$/eye medium displayed much lower levels when examined in the standard test medium (Figure 5). The high levels obtained with neurons grown in eye medium represented development *in vitro*; time-course studies indicated that ACh sensitivity increased markedly between days one and three in culture and then was maintained for the remainder of the week (Figure 6). ACh sensitivities for neurons grown in K$^+$ medium remained low at all times tested. These results differed radically from those obtained for α-bungarotoxin binding to neurons grown under the same conditions (see below).

The levels of ACh sensitivity obtained with eye extract were adequate to support spontaneous synaptic transmission between neurons (Margiotta and Berg, 1982). In contrast, Crean et al. (1982) found that neurons grown on a collagen substratum, instead of the fibroblast residue used here, lose ACh sensitivity in culture and are unable to support synaptic transmission, even though electron-microscope analysis confirms the presence of morphological synapses on such neurons. Coculture with skeletal myotubes prevents the neurons from losing

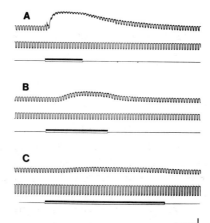

Figure 5. *ACh conductance for ciliary ganglion neurons in culture.* Neurons grown for seven days in eye (A), K⁺/eye (B), or K⁺ medium (C) were tested in a standard recording medium consisting of MEM buffered with 5 mM Hepes (instead of HCO_3^-) and containing 10% (vol/vol) horse serum and 5.4 mM Ca^{2+}. In each case the upper trace indicates the membrane potential, the middle trace shows applied current, and the bottom trace (bar) marks the application of ACh (0.1 mM). Vertical calibration bar: 20 mV and 0.1 nA (A and C) or 0.2 nA (B). Horizontal calibration bar: 1 sec. (From Smith et al., 1983.)

ACh sensitivity on the collagen substratum. These observations suggested that nerve–muscle contact may play a role in sustaining functional ACh receptors on the neurons, and they prompted an analysis of substratum effects on neuronal ACh sensitivity. Tuttle (1983) found that the ACh sensitivity of ciliary ganglion neurons could be supported by myotubes, myotube residue, or residue from primary fibroblasts on a collagen substratum, but not by collagen alone, muscle-conditioned medium on collagen, or residue from fibroblasts that had been serially passaged in cell culture. From these results he concluded that it was likely that the original fibroblast material supported neuronal ACh sensitivity because the primary fibroblast population had been contaminated with muscle cells. Passage of the fibroblasts removed the muscle cells and, in so doing, removed the capacity to support neuronal ACh sensitivity. The interesting im-

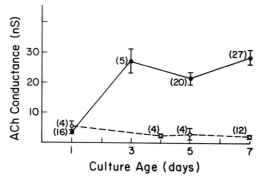

Figure 6. *Development of ACh sensitivity.* Neurons grown in eye medium (solid line) and in K⁺ medium (dashed line) were tested for ACh sensitivity at the indicated times. The values represent the mean plus or minus standard error. The numbers in parentheses indicate the number of neurons tested, taken from a total of 21 cultures from 14 separate platings. (From Smith et al, 1983.)

plication is that nerve–muscle contact is necessary for maintenance of functional ACh receptors on the neurons and that the interaction is specific, because it cannot be mimicked by nerve contact with other cell types. In related studies we have found that neurons maintained for 5–8 days with eye medium on a substratum of collagen and polylysine (Messing, 1982) express more than half of the sensitivity found for neurons on fibroblast residue and that the neurons remain able to form functional synapses with each other. Since neurons at one day in culture, at least on fibroblast residue, have very little ACh sensitivity, it seems that polylysine substrata can support the acquisition of functional receptors. Either polylysine is able to substitute partially for required muscle components in this regard, or direct contact with muscle is not normally required for expression of ACh receptors on the neurons.

α-BUNGAROTOXIN BINDING SITES

For quantitative studies on neuronal ACh receptors it is important to have a method that measures receptor number directly. α-Bungarotoxin binding has been proposed as a quantitative assay for neuronal nicotinic ACh receptors, as it is for ACh receptors in skeletal muscle and electric organ. Chick ciliary ganglion neurons have high-affinity binding sites for α-bungarotoxin (Chiappinelli and Giacobini, 1978; Gangitano et al., 1978; Ravdin and Berg, 1979). Although D-tubocurarine blocks the binding as expected for the nicotinic receptor, and although development of the receptors has been studied both *in vivo* (Chiappinelli and Giacobini, 1978; Gangitano et al., 1978; Gangitano et al., 1981) and in cell culture (Messing and Kim, 1981), the identity of the site remains in question. It has been known for some time that saturation of the site with α-bungarotoxin does not inhibit ACh sensitivity on the neurons (Ravdin and Berg, 1979; Chiappinelli et al., 1981; Ravdin et al., 1981; Margiotta and Berg, 1982), in contrast to the complete blockade of receptor function caused by the toxin in muscle. Two additional lines of evidence now suggest that the α-bungarotoxin binding site on ciliary ganglion neurons is different from the synaptic ACh receptor.

Independent Regulation

Levels of ACh sensitivity and α-bungarotoxin binding appear to vary independently for the neurons during development in cell culture. Thus neurons grown for seven days in eye medium (see above) have high levels of ACh sensitivity, whereas neurons grown in K^+ or K^+/eye medium have relatively low levels. In contrast, neurons grown in eye and K^+/eye medium have low levels of toxin binding; neurons grown in K^+ medium have higher levels (Figure 7). These findings for α-bungarotoxin binding sites are unexpected not only because they fail to parallel changes in the levels of ACh sensitivity, but also because they run counter to the general development of the neurons. Eye extract increases growth and development of the neurons above that obtained with elevated K^+ concentrations alone (Nishi and Berg, 1981), yet the number of toxin binding sites is considerably reduced when eye extract is included in the growth medium. Control experiments demonstrate that eye extract does not inhibit toxin binding

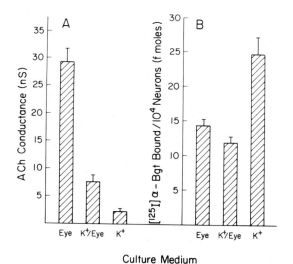

Figure 7. *Differences in ACh conductances and α-bungarotoxin binding levels.* Neurons grown for seven days in the indicated culture medium were tested for sensitivity to pulses of 0.1 mM ACh (*A*) and for levels of α-bungarotoxin binding (*B*). Results are expressed as the mean plus or minus standard error; n = 12–22 neurons from three to six experiments in *A*, and 24–36 cultures from eight to 12 experiments in *B*. (From Smith et al., 1983.)

when applied acutely, and the pharmacology of toxin binding with respect to inhibition by D-tubocurarine and carbamylcholine remain identical for neurons grown with and without eye extract (Smith et al., 1983). Eye extract does not cause a redistribution of toxin sites away from the soma and does not shift the dose–response curve for ACh sensitivity on the neurons.

The most straightforward explanation for the lack of correlation between toxin binding and ACh sensitivities during growth in culture is that the two membrane properties represent different membrane components. A similar conclusion has been reached in a recent study on the rat pheochromocytoma cell line PC-12, where it was found that growth of the cells in serum-free medium dramatically reduces ACh receptor function (as reflected in carbachol-induced Na^+ flux and in catecholamine release) but causes no reduction in the number of α-bungarotoxin binding sites associated with the cells (Mitsuka and Hatanaka, 1983). It should be kept in mind, however, that in both studies the ACh receptor assays measured receptor function rather than receptor number. Accordingly, the changes observed could have reflected a control of receptor function that was independent of receptor number.

Extrasynaptic Distribution

A second line of study arguing against equating the toxin binding site with the synaptic ACh receptor comes from ultrastructural analysis of the distribution of toxin sites on the neurons. Embryonic ganglia (14-, 16-, and 20-day-old) were incubated with horseradish peroxidase-conjugated α-bungarotoxin (HRP–α-bungarotoxin), washed, fixed, stained for HRP reaction product, sectioned, and examined under the transmission electron microscope. Dense reaction product was associated with most of the short processes or "pseudodendrites" emerging from both choroid and ciliary neurons in the regions where the cells receive preganglionic innervation (Figure 8). Less-dense labeling was occasionally associated with adjacent regions of smooth soma membrane on the neurons. Most synapses clearly were not labeled by the conjugate (Figure 9). Similar results were obtained with adult tissue, where only a few synapses (five out of 43) could be described as being even lightly labeled. Synapses on the neurons are characterized by a parallel arrangement and thickening of the pre- and postsynaptic membranes, a widening of the cleft, an enhanced postsynaptic density, and an accumulation of clear synaptic vesicles adjacent to the presynaptic membrane (De Lorenzo, 1960; Hess, 1965; Landmesser and Pilar, 1972). Close juxtaposition of labeled surface membrane and unlabeled synapse was often apparent; the layer of reaction product coating the nonsynaptic portion of the surface membrane

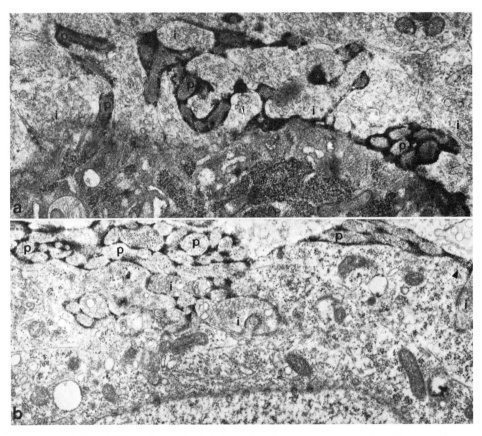

Figure 8. *Binding of HRP–α-bungarotoxin to the embryonic neuron surface.* A choroid neuron from a 14-day-old chick embryo (a) and a ciliary neuron from a 16-day-old chick embryo (b) are heavily labeled on the surface of the numerous short processes (p) emerging from the soma in the region of preganglionic innervation (i) after incubation in 10^{-7} M HRP–α-bungarotoxin and development of the HRP reaction product. Less-dense labeling occasionally is present on the smooth surface membrane in the vicinity of the heavily labeled processes (arrowhead). ×80,100 (a); ×51,000 (b). (From Jacob and Berg, 1983.)

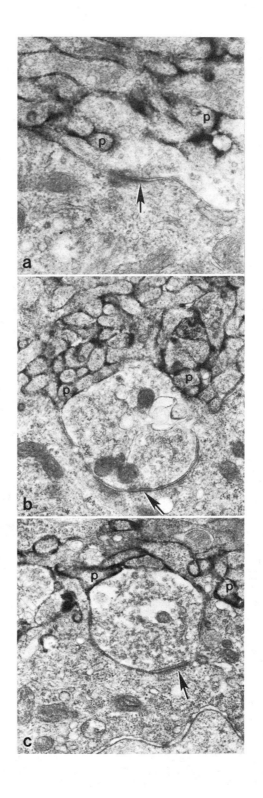

did not extend into the specialized synaptic zone. The labeling was specific, as it was blocked by coincubation with the cholinergic antagonists D-tubocurarine (100 μM) and hexamethonium (1.4 mM), or by native α-bungarotoxin (10 μM) (Jacob and Berg, 1983).

In labeling studies of this kind it is important to demonstrate that the toxin conjugate did have access to the synaptic membrane. Two control experiments indicate that this was the case. First, unconjugated HRP, which is nearly as large as HRP–α-bungarotoxin (40 versus 48 kD), heavily stained all plasma membranes, including synaptic membranes, when incubated with the ganglia and crosslinked with glutaraldehyde before the tissue was washed. Second, an HRP–monoclonal antibody conjugate, which is at least four times larger than HRP–α-bungarotoxin, readily penetrated the synapses and labeled synaptic membrane, as well as the entire surface of the neurons. The antibody conjugate, thought to be specific for the Na^+/K^+ ATPase (Fambrough and Bayne, 1983), was prepared and kindly provided by J. Lippincott-Schwartz and D. Fambrough.

These findings strongly suggest that the α-bungarotoxin binding site on chick ciliary ganglion neurons cannot be the synaptic ACh receptor. The identity of the sites remains unknown, but their proximity to synaptic areas on the neurons implies a synaptic role. Possibly the sites represent membrane components important for the formation or maintenance of synapses. Alternatively the sites may represent receptors, responding to ACh or some other ligand, which influence synaptic function over a relatively slow time course and therefore do not need to be tightly clustered within the postsynaptic membrane. The results at present, of course, do not exclude the possibility that the toxin binding sites have a more obscure relationship to synaptic receptors, such as being their inactive precursors or degradation products.

SYNAPTIC ANTIGENS

Immunological approaches have been valuable for identifying new synaptic antigens at the neuromuscular junction and for further characterizing such known synaptic antigens as the ACh receptor in muscle (see Kelly and Hall, 1982, for a review). Progress is also being made in identifying presynaptic antigens at nerve–nerve synapses (see Reichardt and Kelly, 1983, for a review).

One approach to identifying postsynaptic antigens on neurons is to screen antibodies that recognize known components of the neuromuscular junction for cross-reactivity with neuronal components. Swanson et al. (1983) have found that monoclonal antibodies which recognize determinants on ACh receptors from skeletal muscle and electric organ also recognize determinants in the lateral spiriform nucleus of chick brain. Indirect immunofluorescence microscopy revealed

---

Figure 9. *Absence of HRP–α-bungarotoxin labeling at synapses on embryonic neurons.* The pre- and postsynaptic membranes of synapses on the somata of 14- (*a*), 16- (*b*), and 20-day-old (*c*) chick embryo ciliary neurons are not labeled. The synapses (arrows) are characterized by relatively straight and thickened pre- and postsynaptic membranes, enhanced postsynaptic densities, a widened synaptic cleft, and an accumulation of clear synaptic vesicles adjacent to the presynaptic membrane. Short processes (p) in the immediate vicinity of the unreacted synapses are heavily labeled. ×74,400 (a); ×56,100 (b); ×60,800 (c). (From Jacob and Berg, 1983.)

that determinants characteristic of all four types of ACh-receptor subunits could be identified in the nucleus, and the distribution of antibody binding in the tissue did not overlap with α-bungarotoxin binding. Best cross-reaction was obtained with monoclonal antibodies that recognized determinants on a region of the α subunit of muscle and electric-organ ACh receptor designated the "main immunogenic region" (MIR), which is exposed on the extracellular surface of the membrane (Tzartos and Lindstrom, 1980; Tzartos et al., 1981).

Immunological Cross-Reaction

We have screened anti-MIR monoclonal antibodies for cross-reaction with components on ciliary ganglion neurons in the hope of finding probes for the synaptic ACh receptor on the neurons. Initial tests were carried out at the light-microscope level by using indirect immunofluorescence to detect antibody binding to ciliary ganglion neurons grown in dissociated cell culture. Specific binding to the neurons was observed with three monoclonal antibodies, designated 6, 35, and 203 (Figure 10). All three are thought to bind to the same determinant (or closely positioned determinants) since they compete for binding to muscle and electric-organ ACh receptor (see below). Substitution of nonimmune rat serum at a comparable concentration of IgG for the antibodies did not result in significant labeling of the neurons.

Ultrastructural Distribution

HRP-conjugated monoclonal antibody-35 (HRP-35) was prepared to examine the distribution of antibody binding on neurons in the ganglion. Transmission electron microscopy of labeled ganglia from 16-day-old embryos revealed the dense HRP reaction product associated with synaptic membranes; reaction product often

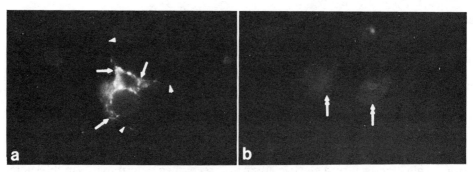

Figure 10. *Binding of anti-MIR monoclonal antibodies to a surface component on ciliary ganglion neurons in culture.* Neurons grown for six days on glass coverslips in eye medium were incubated for one hour in either 10^{-7} M antibody 203 (*a*) or in nonimmune rat serum at an equivalent concentration of IgG (*b*). After washing, the cultures were incubated for one hour at room temperature with fluorescein-conjugated goat anti-rat IgG antiserum, rinsed, fixed, and examined with fluorescence microscopy. Thin patches of fluorescent staining are seen along a portion of the surfaces of the two phase-bright rounded neuron cell bodies in a. The portion of the neuron surface that is stained is often contacted by a neurite (arrows). The neurites, for the most part, appear unlabeled (arrowheads). There is essentially no fluorescent staining on the surfaces of neurons (double arrows) in b. ×210. (From Jacob et al., 1984.)

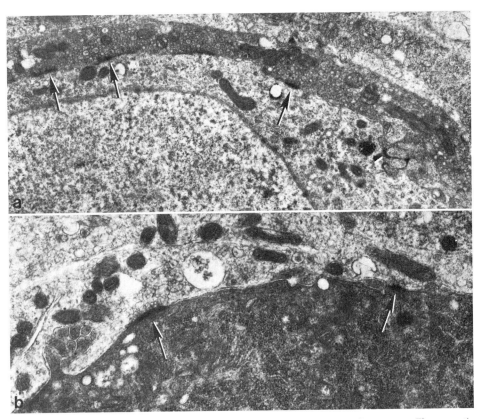

Figure 11. *Binding of anti-MIR antibodies on synaptic membranes of ciliary ganglion neurons.* The synaptic membranes (arrows) of a ciliary (*a*) and a choroid (*b*) neuron from a 16-day-old embryonic chick ganglion are heavily labeled after incubation in 10^{-7} M HRP-35 and development of the HRP reaction product. Labeling does not extend beyond the specialized synaptic zones in areas of pre- and postganglionic cell apposition. Reaction product is also present along a small portion of the surface of the small processes (arrowhead) that project out from the ciliary neuron (a) in the vicinity of the large presynaptic terminal or calyx. The surface of the small processes emerging from this particular choroid neuron (b) are not labeled. ×38,100 (a); ×62,500 (b). (From Jacob et al., 1984.)

filled the synaptic clefts of clearly identifiable synapses on both ciliary and choroid neurons (Figure 11). Labeled synapses were present on both the smooth portion of the cell soma and the short processes emerging from the postsynaptic cell in the region of preganglionic innervation. Reaction product did not extend beyond the specialized synaptic contact zones in areas of pre- and postganglionic cell apposition. On all neurons with at least some reaction product, the vast majority of synapses were labeled. Dense reaction product was also present on a portion of the numerous small processes emerging from the postsynaptic cell in the vicinity of presynaptic terminals, but usually only a small fraction of the surface membrane on the processes was labeled. HRP reaction product was not found elsewhere on neuronal membrane or on the surfaces of other cells in the ganglion.

Ganglionic labeling with HRP-35 was specific, because it was blocked by antibodies known to compete for binding to the MIR. Thus an excess of antibody

6, 35, or 203 all blocked HRP-35 labeling, while nonimmune rat serum had no effect (Jacob et al., 1984).

These results demonstrate that a synaptic component on chick ciliary ganglion neurons shares an antigenic determinant with the MIR of muscle and electric-organ ACh receptor. It is tempting to conclude that the cross-reacting component is the synaptic ACh receptor on the neurons, but it should be noted that these monoclonal antibodies do not block receptor function on the neurons (or on muscle) and therefore a direct demonstration of their association with the receptor is not possible. Accordingly, it will be important to determine whether antigenic determinants associated with other subunits of muscle and electric-organ ACh receptor can be identified in the ganglion as they are in the lateral spiriform nucleus of chick brain (Swanson et al., 1983). It will also be important to examine and understand the relationship between levels of ACh sensitivity on the neurons and numbers of antibody binding sites.

The significance of the binding to neuronal processes is unclear. The binding represents a small amount of the total HRP-35 labeling on the neurons and is much less than the amount of α-bungarotoxin binding associated with the processes (Jacob and Berg, 1983). Antibody precipitation experiments carried out on detergent extracts of embryonic ciliary ganglia should demonstrate whether any of the determinants recognized by antibody 35 are associated with molecules carrying α-bungarotoxin binding sites. From the ultrastructural studies, however, it is clear that most of the antibody-35 binding sites should represent separate molecules.

CONCLUSIONS

The chick ciliary ganglion has been useful in examining the dependence of neuronal development on information from synaptic target tissue. Studies in cell culture have identified components from the tissue that promote survival and stimulate long-term growth and development of the neurons. Such components are candidates for regulatory factors that act on the neurons during normal development *in vivo*.

The identification of a synaptic component on chick ciliary ganglion neurons that shares an antigenic determinant with the MIR of muscle and electric organ ACh receptor raises hopes that a probe for neuronal ACh receptors has been found. The MIR of muscle and electric-organ ACh receptor is associated with the α subunit, as are both the ACh binding site and the α-bungarotoxin binding site. On ciliary ganglion neurons, however, the α-bungarotoxin binding site and the cross-reacting antigenic determinant appear to be on separate membrane components. The relationship of these membrane components and their significance for synaptic function await further study.

ACKNOWLEDGMENTS

This work was supported by National Institutes of Health Grant NS-12601 and by grants from the Muscular Dystrophy Association and the American Heart Association, with funds contributed in part by the California Heart Association.

REFERENCES

Adler, R., and S. Varon (1982) Neuronal survival in intact ciliary ganglia *in vivo* and *in vitro*: Ciliary neuronotrophic factor as a target surrogate. *Dev. Biol.* **92**:470–475.

Adler, R., K. B. Landa, M. Manthorpe, and S. Varon (1979) Cholinergic neuronotrophic factors: Intraocular distribution of trophic activity for ciliary neurons. *Science* **204**:1434–1436.

Barde, Y.-A., D. Edgar, and H. Thoenen (1982) Purification of a new neurotrophic factor from mammalian brain. *EMBO J.* **1**:549–553.

Berg, D. K. (1984) New neuronal growth factors. *Annu. Rev. Neurosci.* **7**:149–170.

Black, I. (1982) Stages of neurotransmitter development in autonomic neurons. *Science* **215**:1198–1204.

Bonyhady, R. E., I. A. Hendry, C. E. Hill, and I. S. McLennan (1980) Characterization of a cardiac muscle factor required for the survival of cultured parasympathetic neurons. *Neurosci. Lett.* **18**:197–201.

Bonyhady, R. E., I. A. Hendry, and C. E. Hill (1982) Reversible dissociation of a bovine cardiac factor that supports survival of avian ciliary ganglionic neurons. *J. Neurosci. Res.* **7**:11–21.

Chiappinelli, V. A., and E. Giacobini (1978) Time course of appearance of α-bungarotoxin binding sites during development of chick ciliary ganglion and iris. *Neurochem. Res.* **3**:465–478.

Chiappinelli, V. A., J. B. Cohen, and R. E. Zigmond (1981) The effects of α- and β-neurotoxins from the venoms of various snakes on transmission in autonomic ganglia. *Brain Res.* **211**:107–126.

Crean, G., G. Pilar, J. B. Tuttle, and K. Vaca (1982) Enhanced chemosensitivity of chick parasympathetic neurons in co-culture with myotubes. *J. Physiol. (Lond.)* **331**:87–104.

De Lorenzo, A. J. (1960) The fine structure of synapses in the ciliary ganglion of the chick. *J. Biophys. Biochem. Cytol.* **7**:31–36.

Erichsen, J. T., H. J. Karten, W. D. Eldred, and N. C. Brecha (1982) Localization of substance P-like and enkephalin-like immunoreactivity within preganglionic terminals of the avian ciliary ganglion: Light and electron microscopy. *J. Neurosci.* **2**:994–1003.

Fambrough, D. M. (1979) Control of acetylcholinergic receptors in skeletal muscle. *Physiol. Rev.* **59**:165–227.

Fambrough, D. M., and E. K. Bayne (1983) Multiple forms of $(Na^+ + K^+)$-ATPase in the chicken. *J. Biol. Chem.* **258**:3926–3935.

Fukada, K. (1983) Studies on the cholinergic differentiation factor for sympathetic neurons. *Int. Soc. Neurochem. Abstr. (Suppl.)* **41**:589.

Gangitano, C., L. Fumagalli, G. de Renzis, and C. O. Sangiacomo (1978) α-Bungarotoxin-acetylcholine receptors in the chick ciliary ganglion during development. *Neuroscience* **3**:1101–1108.

Gangitano, C., L. Fumagalli, A. Del Fa, and C. Olivieri-Sangiacomo (1981) α-Bungarotoxin receptors in the chick ciliary ganglion: Behavior *in vivo* and *in vitro*. *Neuroscience* **6**:273–279.

Helfand, S. L., G. A. Smith, and N. K. Wessells (1976) Survival and development in culture of dissociated parasympathetic neurons from ciliary ganglia. *Dev. Biol.* **50**:541–547.

Hess, A. (1965) Developmental changes in the structure of the synapse on the myelinated cell bodies of the chicken ciliary ganglion. *J. Cell Biol.* **25**:1–19.

Hill, C. E., I. A. Hendry, and R. E. Bonyhady (1981) Avian parasympathetic neurotrophic factors: Age-related increases and lack of regional specificity. *Dev. Biol.* **85**:258–261.

Jacob, M. H., and D. K. Berg (1983) The ultrastructural localization of α-bungarotoxin binding sites in relation to synapses on chick ciliary ganglion neurons. *J. Neurosci.* **3**:260–271.

Jacob, M. H., D. K. Berg, and J. M. Lindstrom (1984) Shared antigenic determinant between *Electrophorus* acetylcholine receptor and a synaptic component on chick ciliary ganglion neurons. *Proc. Natl. Acad. Sci. USA* **81**:3223–3227.

Jessell, T. M., R. E. Siegel, and G. D. Fischbach (1979) Induction of acetylcholine receptors on cultured muscle by a factor extracted from brain and spinal cord. *Proc. Natl. Acad. Sci. USA* **76**:5397–5401.

Kelly, R. B., and Z. W. Hall (1982) Immunology of the neuromuscular junction. In *Neuroimmunology*, J. Brockes, ed., pp. 1–48, Plenum, New York.

Landa, K. B., R. Adler, M. Manthorpe, and S. Varon (1980) Cholinergic neuronotrophic factors. III. Developmental increase of trophic activity for chick embryo ciliary ganglion neurons in their intraocular target tissues. *Dev. Biol.* **74**:401–408.

Landmesser, L., and G. Pilar (1972) The onset and development of transmission in the chick ciliary ganglion. *J. Physiol. (Lond.)* **241**:691–713.

Landmesser, L., and G. Pilar (1978) Interactions between neurons and their targets during *in vivo* synaptogenesis. *Fed. Proc.* **37**:2016–2022.

Manthorpe, M., S. Skaper, R. Adler, K. Landa, and S. Varon (1980) Cholinergic neuronotrophic factors: Fractionation properties of an extract from selected chick embryonic eye tissues. *J. Neurochem.* **34**:69–75.

Manthorpe, M., G. Barbin, and S. Varon (1982a) Isoelectric focusing of the chick eye ciliary neuronotrophic factor. *J. Neurosci. Res.* **8**:233–239.

Manthorpe, M., S. D. Skaper, G. Barbin, and S. Varon (1982b) Cholinergic neuronotrophic factors. Concurrent activities on certain nerve growth factor-responsive neurons. *J. Neurochem.* **38**:415–421.

Margiotta, J. F., and D. K. Berg (1982) Functional synapses are established between ciliary ganglion neurons in dissociated cell culture. *Nature* **296**:152–154.

Messing, A. (1982) Cholinergic agonist-induced down regulation of neuronal α-bungarotoxin receptors in cultured chick ciliary ganglion neurons. *Brain Res.* **232**:479–484.

Messing, A., and S. U. Kim (1981) Development of α-bungarotoxin receptors in cultured chick ciliary ganglion neurons. *Brain Res.* **208**:479–486.

Mitsuka, M., and H. Hatanaka (1983) Selective loss of acetylcholine sensitivity in a nerve cell line cultured in hormone-supplemented serum-free medium. *J. Neurosci.* **3**:1785–1790.

Nieto-Sampedro, M., E. R. Lewis, C. W. Cotman, M. Manthorpe, S. D. Skaper, G. Barbin, F. M. Longo, and S. Varon (1982) Brain injury causes a time-dependent increase in neuronotrophic activity at the lesion site. *Science* **217**:860–861.

Nishi, R., and D. K. Berg (1977) Dissociated ciliary ganglion neurons *in vitro*: Survival and synapse formation. *Proc. Natl. Acad. Sci. USA* **74**:5171–5175.

Nishi, R., and D. K. Berg (1979) Survival and development of ciliary ganglion neurons grown alone in cell culture. *Nature* **277**:232–234.

Nishi, R., and D. K. Berg (1981) Two components from eye tissue that differentially stimulate the growth and development of ciliary ganglion neurons in cell culture. *J. Neurosci.* **1**:505–513.

Olek, A. J., P. A. Pudimat, and M. P. Daniels (1983) Direct observation of the rapid aggregation of acetylcholine receptors on identified cultured myotubes after exposure to embryonic brain extract. *Cell* **34**:255–264.

Patterson, P. H. (1978) Environmental determination of autonomic neurotransmitter functions. *Annu. Rev. Neurosci.* **1**:1–17.

Pilar, G., and J. B. Tuttle (1982) A simple neuronal system with a range of uses: The avian ciliary ganglion. In *Progress in Cholinergic Biology: Model Cholinergic Synapses*, I. Hanin and A. M. Goldberg, eds., pp. 213–247, Raven, New York.

Podleski, T. R., D. Axelron, P. Ravdin, I. Greenberg, M. M. Johnson, and M. M. Salpeter (1978) Nerve extract induces increase and redistribution of acetylcholine receptors on cloned muscle cells. *Proc. Natl. Acad. Sci. USA* **75**:2035–2039.

Ravdin, P. M., and D. K. Berg (1979) Inhibition of neuronal acetylcholine sensitivity by α-toxins from *Bungarus multicinctus* venom. *Proc. Natl. Acad. Sci. USA* **76**:2072–2076.

Ravdin, P. M., R. M. Nitkin, and D. K. Berg (1981) Internalization of α-bungarotoxin on neurons induced by a neurotoxin that blocks neuronal acetylcholine sensitivity. *J. Neurosci.* **1**:849–861.

Reichardt, L. F., and R. B. Kelly (1983) A molecular description of nerve terminal function. *Annu. Rev. Biochem.* **52**:871–926.

Sargent, P. B., and M. J. Dennis (1981) The influence of normal innervation upon abnormal synaptic connections between frog parasympathetic neurons. *Dev. Biol.* **81**:65–73.

Smith, M. A., J. F. Margiotta, and D. K. Berg (1983) Differential regulation of acetylcholine sensitivity and α-bungarotoxin-binding sites on ciliary ganglion neurons in cell culture. *J. Neurosci.* **3**:2395–2402.

Swanson, L. W., J. Lindstrom, S. Tzartos, L. C. Schmued, D. D. M. O'Leary, and W. M. Cowan (1983) Immunohistochemical localization of monoclonal antibodies to the nicotinic acetylcholine receptor in chick midbrain. *Proc. Natl. Acad. Sci. USA* **80**:4532–4536.

Tuttle, J. B. (1983) Interaction with membrane remnants of target myotubes maintains transmitter sensitivity of cultured neurons. *Science* **220**:977–979.

Tuttle, J. B., J. B. Suszkiw, and M. Ard (1980) Long-term survival and development of dissociated parasympathetic neurons in culture. *Brain Res.* **183**:161–180.

Tzartos, S. J., and J. M. Lindstrom (1980) Monoclonal antibodies used to probe acetylcholine receptor structure: Localization of the main immunogenic region and detection of similarities between subunits. *Proc. Natl. Acad. Sci. USA* **77**:755–759.

Tzartos, S., D. E. Rand, B. L. Einarson, and J. M. Lindstrom (1981) Mapping of surface structures of *Electrophorus* acetylcholine receptor using monoclonal antibodies. *J. Biol. Chem.* **256**:8635–8645.

Varon, S., M. Manthorpe, F. M. Longo, and L. R. Williams (1983) Growth factors in regeneration of neural tissues. In *Nerve, Organ, and Tissue Regeneration: Research Perspectives*, F. J. Seil, ed., pp. 127–155, Academic, New York.

Weber, M. J. (1981) A diffusible factor responsible for the determination of cholinergic functions in cultured sympathetic neurons. *J. Biol. Chem.* **256**:3447–3453.

Section 5

Map Formation in the Retinotectal System

The development of the retinotectal projection was chosen as the main subject of an entire section of this book for reasons that Cowan and Hunt clearly outline in their scholarly review: (1) It is a central system subject to manipulation for a long period of time; (2) the projection is topographically ordered; (3) it is accessible to experimental attack at levels ranging from molecular to morphologic description; and (4) in some animals it is capable of regeneration, thus allowing a comparison of this process with original development.

In their extensive, historically based analysis, Cowan and Hunt take up the development of polarity in the retina and tectum, the growth of optic fibers from the retina to the brain, the selection by these fibers of terminal zones and target cells in the optic lobe, and the refinement of the central pattern of retinotectal connections. They suggest a mechanism to account for the selective loss of ganglion cells projecting to the wrong region. In addition, they place their historical review in the context of models and molecular mechanisms, some of which are discussed by succeeding authors in this section.

In his contribution, Easter focuses attention upon the role of the axonal pathway and emphasizes the continuing and dynamic aspect of retinotectal growth. Ages of optic axons in the goldfish span a wide range inasmuch as they are produced continually at the retinal margin. In a particular cohort, axons remain associated with those of similar ages in the nerve tract and exit systematically from their fascicles onto the tectum. Thus they run along the outer edge of the tectum, with ventral retinal axons in the dorsal position and dorsal axons in a ventral position; temporal axons exit the fascicles first and nasal axons last. The eccentric growth of new tectal cells creates a new boundary for apposition of a subsequent generation of axons. These observations leave little doubt that terminal arbors must shift their positions because of the almost circular symmetric growth of retina as contrasted to the eccentric growth of tectum. This has profound implications for models at the molecular level. Continuous relocation puts several constraints on any model, suggesting that fixed molecular addresses on target sites on the tectum are unlikely.

Schmidt takes up the complementary roles of the differential affinity of neural membranes and the activity-dependent stabilization of synapses in forming the map. He emphasizes the studies by himself and his colleagues demonstrating that the latter is an essential stage in refining that map. He argues for two stages: (1) the formation of a crude map by selective fiber–fiber adhesion, selective fiber pathway interactions, and interfiber competition at the target; and (2) the final sharpening of the map by correlated activity of optic fibers from neighboring ganglion cells. Where the arbors of such cells overlap, summation of postsynaptic responses causes a selective stabilization of synapses. Such a position is supported by Schmidt's studies of the effects of a tetrodotoxin block on the developing map and is in good accord with analyses of activity and synapse elimination elsewhere in the nervous system. Schmidt includes an elegant short review of these analyses in the last part of his chapter.

In his chapter, Fraser describes a detailed mechanistic model for the patterning of nerve connections based on adhesive interactions between cells. The model, which is relatively simple, is based on adhesive interactions between cells and can be tested against much but not all of the data in the literature. Fraser then adds activity-dependent competition between terminals (consistent with Schmidt's views in the previous chapter) and provides a simple mechanism for such competition. The hybrid model that ensues can explain the main features of retinotectal patterning as well as the formation of ocular-dominance columns in lower vertebrates. While the reader is urged to study the model in detail, it may be valuable to state here that it consists of: (1) a strong position-independent adhesion between optic nerve fibers and tectum; (2) a strong repulsion between optic nerve fiber terminals; and (3) two weak position-dependent adhesions between optic fiber terminals and tectum—a dorsoventral one and a slightly weaker anteroposterior one. These strength rankings and two different adhesion categories are shown to be sufficient to create a map.

The value of such models is to provide a guide to relating molecular mechanisms and higher level interactions. Following the meeting upon which his book is based, evidence was obtained that N-CAM is a strong candidate for at least one of the components postulated to create the hierarchical and dynamic adhesions helping to form a map [Fraser, S. E, B. A. Murray, C.-M. Chuong, and G. M. Edelman (1984) Proc. Natl. Acad. Sci. USA 81:4222–4226].

The chapters in this section serve to illustrate the progressive analytical depth that has been achieved in understanding one example of neural mapping. While far from solution, the retinotectal projection serves to indicate the multistage, dynamic, and heterarchical series of interactions that are required to form orderly projections. While certain molecular interactions are absolutely required, activity and selection based on the epigenetic development of the whole system are also essential.

Chapter 17

The Development of the Retinotectal Projection: An Overview

W. MAXWELL COWAN
R. KEVIN HUNT

For more than four decades the representation of the retina upon the optic tectum has been one of the most intensively studied model systems in developmental neurobiology. There are several reasons for this. First, it is a central neural system—the retina and optic nerve being outgrowths from the diencephalon—yet one that is readily accessible to experimental manipulation throughout life. Second, the projection formed by the retina upon the optic tectum is topographically ordered in an extremely precise manner. Indeed, there is no other neural system that is more convenient for the study of the topographic patterning of connections. Third, the system lends itself readily to analysis by a variety of experimental procedures that range from simple behavioral testing to physiological recording and a number of neuroanatomical mapping techniques. And fourth, as Matthey (1926) first pointed out, in certain vertebrates the optic nerve is capable of regeneration even in adult life, so it is possible to study not only the embryological development of the retinotectal projection, but also its reestablishment after such manipulative procedures as optic nerve transection and eye rotation.

It was the capacity of the optic nerve for regeneration that led Sperry to select the retinotectal system, first of amphibians and later of fish, for his now classic series of studies on the role of function in the formation of neural circuits. In his earliest experiments, Sperry (1943a) rotated one eye in a series of adult newts and frogs and observed a persistence of misdirected visual striking behavior that was not corrected by experience. Later, when he sectioned the optic nerve before rotating the eye, he found that when vision was restored, the animal's visual behavior was again inverted and was never corrected with practice (Sperry, 1943b, 1944). In yet another series of experiments he transplanted left eyes into right-eye sockets or cross-sutured the optic nerves so that each eye would reinnervate the wrong optic lobe (Sperry, 1944, 1945). In both instances, when the optic nerve had regenerated the animals showed persistent misdirected vision in either the horizontal or the vertical dimension of visual space. Finally, when he made localized lesions in the tectal lobes of such animals, he found that there were localized scotomas in the visual field that persisted throughout the life of the animal (Sperry, 1948).

After these behavioral studies, Attardi and Sperry (1963) went on to show that when selected parts of the goldfish retina were ablated and the optic nerve divided, the regenerating fibers from the surviving parts of the retina grew back only to the corresponding regions of the optic tectum, while the regions related to the ablated sectors of the retina remained uninnervated. In a companion study, Arora and Sperry (1962) showed that optic fibers could reform connections at their appropriate tectal loci even when they were forced to regenerate along an abnormal pathway (see also Sperry, 1965).

Although the initial purpose of these experiments was to determine whether neural circuitry is shaped primarily by function or by prior structural determinants, Sperry was quick to recognize that his findings provided a general framework for a new theory for the formation of neural connections. This theory, which he later termed the "chemoaffinity hypothesis," underwent a slow evolution (see Hunt and Cowan, 1984), but in its final form proposed that during development nerve cells acquire distinctive chemical labels which serve to specify their topographic positions within their fields of origin, the course that their axons normally follow, and the pattern of connections that they can form within their target field (Sperry, 1951, 1963, 1965). The actual establishment of connections was ascribed to the recognition by the growing axons of a matching or complementary set of labels, distributed in an orderly fashion, upon the cells within the target field. Thus, in the case of the retinotectal system, the hypothesis holds that ganglion cells in different parts of the retina acquire distinctive labels, and that these labels determine in some way the ordered pattern of connections that their axons form within the optic tectum. A necessary corollary of this is that neurons in different regions of the tectum similarly acquire position-defining labels that determine with which subset of ganglion cell axons they can form stable connections. Sperry considered that the relevant labels could be generated by a variety of mechanisms, including self-differentiation and embryonic induction; in the specific case of the retina and tectum, he proposed that both structures undergo a polarized, fieldlike differentiation which endows their constituent cells with the necessary positional markers.

In this chapter, we consider five sets of issues that derive from Sperry's chemoaffinity hypothesis. The first of these concerns the establishment of polarity and positional marking of cells in the retina and tectum; that is, the nature of the developmental program which specifies the spatially ordered differentiation of neurons in the retina and tectum. The second set of issues may be thought of as falling under the general heading of axonal pathfinding. They concern the mechanisms that are involved at each step from the initial outgrowth of optic axons from the bodies of the retinal ganglion cells, their course across the vitreal surface of the retina, their coming together at the optic disk to form the optic nerve, their growth through the optic nerve to the chiasm, the decision to cross into the contralateral optic tract or to remain on the same side of the brain, the giving off of collateral branches to other visual relay centers in the diencephalon, and finally their ingrowth into the optic tectum of the midbrain. Viewed broadly, the problem of pathfinding concerns fiber-to-fiber associations, directed and temporarily constrained axonal growth, and the ability of axons to discriminate among the various substrates encountered by their growth cones at each point along their course. The third set of issues concerns the mechanisms of target selection. Once the growing axons have reached the optic tectum, they have to identify the topographically appropriate region; within that region they must select the appropriate class or classes of neurons; finally, they must form synapses with these neurons. In considering target selection, therefore, we have to determine to what extent finding the correct synaptic partners in the tectum is governed by orderly interactions among the growing optic fibers (such as the selective association of optic fibers from neighboring retinal positions), and to what extent it is attributable to the direct recognition by optic axons of the appropriate postsynaptic elements. A fourth set of issues concerns what we may call the refinement of the retinotectal map. Since, a priori, it is unlikely that the connections formed by optic fibers within the tectum would be precise and without error from the very beginning, several questions arise, including: How are the necessary refinements brought about? To what extent do they involve the relocation of optic axons, the death of retinal and tectal neurons, the elimination of axonal collaterals, and the pruning of terminal arbors? The final set of issues obviously concerns the molecular mechanisms that underlie each event in the establishment of the retinotectal projection; the identification of the molecular labels that distinguish retinal and tectal neurons by class and position; the characterization of the differentiation events by which these labels arise during development; and the functional roles they play in axonal pathfinding and target selection.

Before turning to these issues, it may be useful to mention that there are several problems which we do not discuss at length here and to review briefly the essential anatomical features of the retinotectal system in commonly studied vertebrate species. In all vertebrates, the tectum develops from the roof of the mesencephalon, while the retina arises by a rather more complex sequence of outgrowth and later invagination, which transforms the neuroepithelial optic vesicle into a laminated optic cup. The details of retinal and tectal histogenesis vary somewhat among different species. In mammals and birds, there is a spatially graded pattern of neurogenesis and neuronal maturation that is more or less radially organized in the retina, but directed more or less rostrocaudally in the tectum. In fish and frogs, these developmental gradients are more sharply organized, with the cells in the central retina becoming postmitotic at an early stage (when only a few hundred neurons—mainly ganglion cells—are present); subsequently, annuli of new neurons continue to be added to the peripheral margin of the retina throughout the life of the animal (Straznicky and Gaze, 1971; Jacobson, 1976a). Similarly, the first neurons in the tectum are generated near its rostral pole, and successive generations of neurons are added from a roughly linear strip of germinal cells near the caudomedial border of the growing tectal lobe.

Space does not permit us to consider in detail the histogenesis of the retina and tectum, except to note that the various cell types in both structures are generated within the ventricular zone in well-defined sequences, that they migrate to their definitive layers in an orderly manner, and that they then take on their distinctive morphologies, acquire their physiological properties, neurotransmitters, receptors, and so forth, and begin to form both local and more distant connections. We do not deal with the mechanics of synaptogenesis, although obviously this is one of the key events in the establishment of connections between the retina and tectum. And, since relatively little work has been done on the subject of how retinal ganglion cells of different classes establish connections with their counterparts in the optic tectum, this topic also receives scant attention. However, it is worth noting that all retinal ganglion cells respond to particular stimulus features (e.g., direction of movement, color, etc.) and that each such class of ganglion cell must contact a subset of tectal neurons that will have comparable features (Arora and Sperry, 1963). The establishment of these connections occurs within (but apparently independent of) the general topographic ordering of the retinotectal projection (Chung et al., 1973); the latter is the principal concern of this chapter.

The essential topographic features of the retinotectal projection are well known and can be summarized in four broad statements. (1) To a first approximation, the retina and tectum can both be considered as two-dimensional sheets with each point on the retina projecting to a corresponding point on the surface of the tectum. (2) The topographic arrangement of the projection in most vertebrates is such that the nasal (or anterior) parts of the retina are connected to posterior (caudal) parts of the tectum, while the superior (upper or dorsal) retinal quadrants project to the lateral (or ventral) half of the tectum. (3) Within this broad topographical arrangement there is considerable orderliness, which is best described by saying that in the normal, fully developed retinotectal projection, neighboring ganglion cells in the retina project to neighboring cells in the tectum. (4) The terminal arbors of the optic axons end in certain well-defined layers in the tectum, and each contacts a specific subset of tectal neurons in those layers. There are, of course, a number of differences in the morphology of the projection in different species. (In birds, for example, the optic fibers run in a fiber layer on the surface of the tectum and only plunge into its cellular layers close to their point of termination, whereas in fish and frogs the fibers approach the tectum in distinct bundles, called brachia, and then often run for considerable distances within the tectal gray matter before terminating.) These differences are relatively minor and should not obscure the main features outlined above, which are common to all vertebrates.

In considering in sequence the issues involved in the establishment of polarity and the acquisition of positional labels by cells in the retina and tectum, of pathfinding by the axons of the retinal ganglion cells and their selection of an appropriate site for termination in the optic tectum, and the subsequent refinement and validation of the connections that are initially formed, we are more or less following the natural chronology of events in the developing visual system. However, two important points should be noted. The first is that in amphibians and fish, from which so much of our knowledge of this system is derived, these four phases overlap considerably. New cells continue to be added to the growing retina and tectum throughout life; these added cells must acquire their own positional labels, and the axons of the newly formed ganglion cells have to grow within an established optic nerve and become integrated into an already established, functional retinotectal projection. In these lower vertebrates it may be inappropriate to consider the system as ever reaching a "mature" state. The second point is that the chronological sequence actually runs counter to the historical sequence in which many of the key observations were made, and by dealing with these points chronologically there is a danger of obscuring the fact that many experiments bear on several issues. Thus, some of the most critical studies that were done to analyze target selection have provided the strongest evidence for the existence of positional markers. Similarly, many observations on the refinement of the retinotectal projection have both strengthened the evidence for target selection and set limits upon its role in the initial assembly of the projection. Finally, it should be frankly admitted that the molecular mechanisms underlying these various developmental events are, for the most part, unknown. A brief consideration of some of the avenues that are now being explored forms the last part of this chapter.

RETINAL AND TECTAL POLARITY

The notion that the retina is "polarized" derives from Sperry's early observations of rotated visuomotor behavior following 180° rotations of the eye in adult amphibians and of one-axis reversal of vision following left-to-right eye transplants. Following Harrison's (1918, 1921, 1936) studies on the polarization of amphibian limb and ear primordia, Sperry (1943b) predicted that rotating eyes at late embryonic stages, after the proposed polarization events but before the axons of the ganglion cells had formed connections with the brain, would result in a permanent reversal of vision. This prediction was confirmed in a general way in salamander embryos by Stone (1944, 1960), in two species of newt by Szekely (1954, 1966), and in *Xenopus* frog embryos by Jacobson (1968). In all four species, 180° rotation of the eye rudiment at late optic cup stages (when the retina contains at the most only a few hundred ganglion cells) gave rise to permanently reversed visuomotor reflexes. Electrophysiological studies also established that the topography and metrics of the adult retinotectal projection were normal although the map was completely rotated (Jacobson, 1968; see also Hunt and Jacobson, 1972a), strongly suggesting that a general developmental program governing the polarity of the entire retinotectal projection is established in the eye rudiment long before most of the ganglion cells are generated.

The embryonic rotation studies were initially interpreted by some as evidence that the eye rudiment is not polarized or has no positional instructions before late optic cup stages, and that whatever polarity cues are involved enter the eye cup only at about stages 30–31 in *Xenopus*. This inference was mistaken and was out of line with Harrison's original observations, which indicated that early limb and ear anlagen possessed all the necessary instructions for their future development from the earliest stages at which they could be identified and surgically manipulated, even though their polarities could for a time be modified by rotation or other surgical rearrangements. Evidence that this is true also of the developing retina was presented in 1973, when it was shown that *Xenopus* eye rudiments explanted *in vitro* at the earliest stage of surgical accessibility (at stage 22, when the optic vesicle first appears), and then reimplanted some days later in a rotated orientation into a host embryo, form orderly retinotectal projections whose polarity corresponds to the original orientation of the eye rudiment at the time it was explanted (Hunt and Jacobson, 1973). Subsequent studies have shown that rotated retinotectal projections can be found after a variety of early experimental procedures

(Hunt and Jacobson, 1974b), including simple *in situ* rotation at stages 22–26 (Hunt, 1975; Sharma and Hollyfield, 1980), and have also served to clarify the growth of the optic nerve and the formation of the first retinotectal connections. Optic fibers first leave the eye at stages 28–31 when it is still identifiable as an optic cup (Fisher and Jacobson, 1970; Grillo and Rosenbluth, 1972; Cima and Grant, 1980), and within a day or so (at stages 38–39) have given rise to terminal arbors that have a coarse retinotopic order within the tectum (Holt and Harris, 1983; Sakaguchi, 1984).

It would thus seem that the eye primordium carries within itself a developmental program which is sufficiently detailed to establish the topographic order of the entire retinotectal projection, and that this program is acquired very early in eye development—certainly well before the first optic fibers reach the brain, and often (as in *Xenopus*) some hours before they even leave the eye. Just how early the polarity-defining cues are acquired remains to be established, but there is suggestive evidence from the work of Jacobson (1959, 1964) that they may already be present at the neural plate stage. Certainly by late optic vesicle stages local cell groups within the eye rudiment have come under the influence of the cuing mechanism. This has been demonstrated experimentally in several ways (see Hunt and Berman, 1975; Hunt and Ide, 1977; Gaze et al., 1979a). Perhaps the best evidence comes from experiments in which wedges of embryonic eye tissue are transposed from one pole of a donor eye rudiment to the opposite pole of a host eye at stages 26–36 in *Xenopus*. Such transplants often give rise to sectors of retinal tissue that project autonomously to the tectum with the original polarities of the grafted tissue (Hunt and Ide, 1977; Conway et al., 1980; Willshaw et al., 1983).

One of the primary difficulties in pinpointing the origins of retinal polarity lies in the fact that the eye, like many other organ rudiments, seems to be able to adjust or change the initial polarity instructions it receives in response to a variety of surgical manipulations. For example, embryonic eye rudiments in amphibians can be shown to have adapted their retinotectal projections to match an altered orientation within the orbit or even on the body wall (Stone, 1944, 1960; Szekely, 1954, 1966; Jacobson, 1968; Jacobson and Hunt, 1973; Hunt and Piatt, 1978). Similarly, in *Xenopus* a single half-eye rudiment can reconstitute a whole larval eyeball and form a complete retinotectal projection (or a characteristic twinned projection) that covers the entire tectal surface (Gaze, 1970; Berman and Hunt, 1975; Feldman and Gaze, 1975; Hunt and Berman, 1975; Ling et al., 1979). And often, when small pieces of eye tissue are transposed or when two noncomplementary half-eyes are fused to form a "compound eye," the dissonant polarities of the graft and/or host eye tissue can be overridden so as to yield a single harmonious retinotectal projection from the chimeric retina (Hunt, 1975; Hunt and Frank, 1975; Ide et al., 1979; Conway et al., 1980; Gaze and Straznicky, 1980; Willshaw et al., 1983; Sullivan et al., 1984). This process, in which grafted embryonic tissues adopt new polarity instructions when placed in a new context, is well known in experimental embryology (Harrison, 1933, 1945) and has been variously termed "repolarization" or simply "pattern regulation." Some of the "rules" governing the repolarization of tissues have also been established; thus, for example, in *Xenopus* eye fragments, anterior portions of the eye tend to be dominant and to redirect the polarization of tissue from other regions (Hunt,

1975; Ide et al., 1979; Gaze and Straznicky, 1980). However, it is clear that the phenomenon of repolarization is a complex one and seems to depend not only on the specific combination of retinal tissues but also on a variety of less easily controlled experimental conditions including, among other things, the salinity of the solutions in which the tissues are maintained (Jacobson, 1976b; Gaze and Straznicky, 1980). Yet it is remarkable that a single embryonic eye fragment in *Xenopus* can apparently "remember" its original polarity when transplanted in one context and, in another context, respond to new polarity instructions (see Hunt et al., 1982a, for a review).

A good deal less attention has been paid to the question of the polarization of the developing optic tectum. On a priori grounds, we might expect that, like the retina, the tectum would become topographically polarized at some stage, and that the position-determining features which polarization imposes on the tectum would be necessary antecedents to the establishment of a set of connections with the ingrowing optic nerve fibers. As we shall see, a variety of studies concerned with axonal pathfinding and target recognition in the visual system have established beyond reasonable doubt that polarity cues (and, we believe, refined positional labels associated with them) must exist within the tectal lobe. Yet, it must be admitted that it has been very difficult to produce direct evidence for polarization in the developing tectum and to determine when it first appears. In large part this is attributable to technical factors. The embryonic tectum is exceedingly fragile, it is an integral part of the midbrain, and it is not easy to manipulate surgically; for these and other reasons it has not been possible to carry out the full range of grafting and rotation experiments used to characterize the development of polarity in the retina and other peripheral organs. Nevertheless, Levine and Jacobson (1974) did succeed in rotating the tectum in larval *Xenopus* tadpoles; when the operation was carried out at successively later stages, the frequency of rotated visuotectal projections progressively increased (see also Jacobson, 1978). Rotations of only the central portion of the adult tectum are, of course, easier to carry out and have now been done in several different species, with the common (but not invariant) result that the visual projection to the rotated segment was rotated to an appropriate degree (Yoon, 1973; Levine and Jacobson, 1974; Hope et al., 1976; Hunt, 1976).

A more critical test for the existence of polarity in the developing tectum was introduced in 1952 by Crelin and taken up again and explored in more detail by Chung and Cooke (1975, 1978). These workers rotated the whole optic tectum (with or without portions of the neighboring diencephalon or hindbrain) in *Xenopus* embryos shortly before the arrival of the first optic axons. Depending on the amount of tissue rotated, a variety of different retinotopic projections were established and a number of different anatomical patterns were seen. Among the latter were many that showed a reversal of the normal pattern of tectal histogenesis (with new cells being added at the rostral rather than the caudal pole of the tectum) and some in which the optic axons entered the tectum from behind rather than from in front. But most surprisingly, the polarity of the retinotectal projection in these animals did not seem to conform in a straightforward manner to the new orientation of the tectal lobes. Instead, it seemed to be most closely correlated with the position of the diencephalon and its relation to the inverted optic tectum. When the diencephalon was undisturbed, the projection of the retina onto the reversed tectum appeared to be essentially normal. By contrast, in those cases in which a portion of the diencephalon was attached to the reversed tectum (and hence was located between the tectum and the hindbrain), the fibers of the optic tract frequently entered the tectum from behind and the retinotectal map was usually inverted. These experiments are interesting in that they provide a powerful demonstration that under some circumstances the anatomical orientation of the tectum can be dissociated from whatever polarity cues and positional markers determine the ingrowth of optic fibers and direct them to their appropriate positions within the tectum. However, they leave undecided the questions of how and when the tectum normally acquires its polarity and the orderly set of positional markers that we think are involved in the establishment of the retinotectal map. The reason for this is that the observed results are amenable to a number of different interpretations. One interpretation (which was favored by Chung and Cooke) is that the diencephalon serves as a dominant polarizing focus that can override whatever polarity-determining mechanisms exist in the tectum. An equally plausible interpretation is that the tectum becomes polarized only at a relatively late stage and that the tectal markers that derive from this polarization become stable only at some stage after that at which Chung and Cooke's operations were performed (see Willshaw and von der Malsburg, 1976).

The establishment of polarity in the tectum remains as one of the major unresolved issues in our understanding of the development of the retinotectal system, and because of the technical difficulties involved it is not likely to be resolved in the near future. An extension of Chung and Cooke's experiments, preferably including cell markers and a wider range of embryonic stages, may help to clarify some of the interpretative difficulties that their work presents. It would also be desirable to see similar tectal rotation experiments done in other vertebrates. Again, this is likely to pose serious technical problems, at least in amniotes. One of us (W. M. C.), in collaboration with E. Wenger, attempted some years ago to completely rotate the midbrain in chick embryos at stages 9–10 of the Hamburger and Hamilton (1951) series. Unfortunately, none of the animals survived long enough to permit an analysis of the arrangement of the retinal projections onto the rotated tectal lobes, but it is significant that in nearly every case that was examined the normal pattern of histogenesis of the tectum appeared to have been reversed (see Cowan, 1971).

Many of these phenomena are strikingly similar to those described by Harrison (1918, 1921, 1936, 1945) and his students (Stultz, 1936; Swett, 1937; Yntema, 1955) in their studies of pattern determination in the amphibian forelimb, hindlimb, and ear. In several amphibian species, the maturing eye rudiment has been shown to lose its ability to respond to extraocular polarizing cues in two axial steps (Szekely, 1954, 1966; Jacobson, 1968; Hunt and Jacobson, 1972b; Hunt, 1975), with the anteroposterior axis leading the dorsoventral in much the same way as in limb and ear anlagen. More recent workers have stressed the capacity of eye rudiments to form mosaic retinotectal patterns, that is, their ability to retain their original polarity instructions despite an early rotation (Gaze et al., 1979b; Sharma and Hollyfield, 1980) or to derotate physically back to their normal anatomical orientation (Hunt and Jacobson, 1973; Hunt, 1975; Sharma and Hollyfield, 1980). These capacities were also described in detail by Harrison (1921, 1933), who pointed out that the failure of a grafted organ rudiment to adapt to a new orientation or position leads not to an absence of differentiation or a mixture of pattern elements, but rather to the expression of instructions that were originally present in the grafted tissue. Harrison and his students have documented the many factors which favor such mosaic expression. These include leaving a bit of the rudiment in the host wound or including nonorgan-forming "surround" tissue in the graft, which may help to stabilize the original instructions in the grafted rudiments.

If we know little about the general polarization of the retina, we know even less about the process by which individual ganglion cells acquire their addresses. Wolpert's (1971) concept of "positional information" is helpful in this regard, as it distinguishes the polarity cues that organize the pattern of address assignments from (1) positional information, which is a scalar property reflecting the address values assigned to individual cells, and from (2) the interpretation of these address values in the differentiation of the cell.

It seems clear that, even at embryonic stages, the instructions present in local groups of eye cells in *Xenopus* are highly refined and specify not only polarity but also a set of scalar values appropriate for the region of retina that is to be populated by the progeny of any given local cell group. The "autonomous" mapping onto the tectum of small wedges of eye tissue that were transposed from one part of the optic cup to another (Hunt and Ide, 1977; Conway et al., 1980; Willshaw et al., 1983) corresponds in a general way to the sectorlike regions of the retina that are populated by cells from the grafts. These regions of the retina project to the correct part of the tectum, that is, to the region of the tectum to which the ganglion cells would have projected if they had been allowed to develop in their normal position. The precise topography of these projections must mean that the transplanted tissue contained the necessary instructions for the orderly assignment of positional information during retinal growth, so that each new generation of ganglion cells added to the graft-derived region of the eye acquires the correct positional labels (Conway et al., 1980).

Further evidence of this kind derives from growth-suppressed early larval eyes which were allowed to coinnervate a single tectum with a fully grown eye (Hunt, 1976, 1977) and from early larval eyes transplanted into older larval hosts (Jacobson, 1978). In both instances, the resulting tiny eye was found to project only to the small zone appropriate to, and coinnervated by, the central region of the retina in the older eye. This suggests that the early larval retina contains only a restricted range of positional markers appropriate to their final positions in the central retina of the adult eye, and hence the range of positional values must expand in a manner commensurate with the addition of new cells at the retinal margins during growth. Hunt and Jacobson (1974a) differentiated two formal mechanisms by which such an expansion could be achieved. The first of these is a cell lineage model in which the stem cells at the retinal margin pass on the appropriate positional values to their daughters, the positional values being adjusted appropriately with each round of cell division. The second postulates a cell interaction mechanism in which each newborn ganglion cell interacts with the older neurons with which it is in contact and in this way acquires information about its address. Although there is no strong evidence to exclude either model, the second class has been explored in some detail (Bodenstein and Hunt, 1979; Hunt et al., 1982a).

A number of alternatives have been proposed for the way in which positional information may be coded across the retinal sheet. All investigators agree that some form of two-dimensional coding system is required to specify the position of cells in a two-dimensional sheet. Sperry (1943b, 1951) favored a system using two orthogonal reference axes in which each ganglion cell acquires separate information about its anteroposterior position and about its dorsoventral level. Several experimental findings are most easily explained by such a "Cartesian" code (reviewed in Hunt et al., 1982a), but other investigators have favored a polar coordinate system in which positional information is thought to mark the angular position of cells and also their radial distance from a center point (French et al., 1976; McDonald, 1977). This model has the advantage of fitting rather well the radial growth of the eye although, as Wolpert (1971) has recognized, embryos may well use a single positional coding scheme to control the spatial differentiation of cells in a variety of organ rudiments, many of which do not grow radially (see also Hunt et al., 1982a).

In short, the problem of polarity in the retina and tectum does not end with the acquisition of the first polarity cues. And clearly one of the major challenges to molecular studies is to characterize the processes by which positional information is conveyed to individual cells, the way it codes for cellular address in two dimensions, and the events by which it gives rise to stable positional markers on the neuronal surface.

OPTIC AXON GROWTH AND AXONAL PATHFINDING

Given that at some stage in their development the ganglion cells of the retina acquire an address (in the special sense that they become distinguishable from each other on the basis of their topographic positions in the retina), the next question that arises is how this information is expressed in the orderly outgrowth of their axons—first within the retina, then in the optic nerve, optic chiasm, and optic tract. The issues involved are obviously not unique to the visual system or even to the axons of neurons that form topographic projections, but historically the retinotectal projection has provided much of the information on which our current notions of axon guidance are based. A complete account of what is known about the growth of optic axons would have to consider each set of events involved, from the initial outgrowth of the axons from the somata of the ganglion cells, their morphology and the form of their growth cones, their traverse across the optic fiber layer of the retina to the optic disk, their fasciculation with other optic nerve fibers, their interactions with the various substrata encountered along the way, and the mechanisms underlying axon branching, to mention only a few. Here we can indicate, only in broadest outline, some of the directions that work on the growth of optic axons has taken in recent years and point out certain of the more important unanswered questions.

At the outset it should be said that little is known about the initial emergence of ganglion cell axons or the constraints that normally direct their outgrowth toward the vitreal surface of the retina. To date this issue has been examined only morphologically and has not, to our knowledge, been subject to experimental analysis. However, it is now known from several studies that the axons of neighboring ganglion cells tend to form distinct fascicles and converge upon the optic-nerve head in an orderly manner. The topographic arrangement of the ganglion cells is to a large extent maintained within the optic nerve, and this is especially clear in certain teleost fish (Scholes, 1979). Indeed, the high degree of topographic order that is evident in the optic nerves of cichlid fish has led some workers to suggest that this "morphogenetic factor" could, by itself, account for the orderly mapping of the retina upon the optic tectum (Horder and Martin, 1978). However, there is a good deal of evidence against this view, and certainly in the optic tract the axons of neighboring ganglion cells often follow markedly divergent courses en route to the diencephalon or midbrain. Even in cichlid fish, the orderly arrangement of the optic axons is completely lost as the fibers resort themselves prior to entering the tectum (Scholes, 1981). Moreover, as Sperry frequently emphasized, regenerating optic fibers in teleosts and amphibians are often completely scrambled at the site of the initial nerve section, yet they reestablish normal, or near normal, patterns of connections upon reaching the tectum.

It is important to make clear, however, that while "morphogenetics" (to use Horder and Martin's term) cannot by itself account for the orderly patterning of connections in the visual system—or, for that matter in any but the simplest systems—fiber–fiber interactions within the retina, optic nerve, and optic tract are probably of considerable importance and may well be mediated by the same type of molecular mechanism that serves to constrain the selection of postsynaptic targets. But the fact that the axons of neighboring ganglion cells can, and frequently do, diverge from one another as they grow from the retina to the tectum makes it clear that other factors must be involved and, in the last analysis, must be the critical elements in the establishment of the retinotectal map.

In this context it is also worth drawing attention to recent findings which indicate that there may be a considerable number of naturally occurring errors in axonal growth. These often manifest themselves in quite strikingly aberrant axon projections, for example, the growth of retinal axons from one eye across the optic chiasm into the optic nerve and retina of the contralateral eye (McLoon and Lund, 1982; O'Leary et al., 1983). Because most of these grossly aberrant projections seem to be eliminated at later stages in development (the retinoretinal fibers seen in early chick embryos are all eliminated by about the middle of the incubation period), they have usually been overlooked in discussions of axonal growth.

It seems reasonable to refer to these aberrant projections as errors of pathfinding, because it is difficult to believe that they could serve even a transient purpose in the development of the visual system. The transient projection of retinal fibers to the inferior colliculus and to the ventrobasal nucleus of the thalamus (the primary somatosensory relay nucleus) belongs in the same class (Frost, 1981; Insausti et al., 1984b), and even the finding that some of these fibers are topographically ordered and can persist if other afferents to the relevant (erroneous) targets are removed should probably not alter our judgment of them. However, we should probably be cautious in regarding all transient projections of this type as errors in pathfinding. At present we cannot exclude that at least some, which occur consistently in all brains at certain stages in development (like the initial widespread ipsilateral retinocollicular projection in rodents), may play an important role during those stages and are subsequently eliminated when they have served their as yet unknown purpose. We return to this issue when we consider refinements in the retinotectal projection.

A different set of issues concerns the clustering together within the optic nerve and tract of fibers of a particular functional class. The fibers that constitute the basal optic root are a case in point. In the avian brain these are known to arise from a special group of displaced retinal ganglion cells, and for most of their course to their target region in the basal midbrain they remain quite separate from the rest of the optic fibers (Karten et al., 1977). Presumably, surface features on these axons distinguish them from other ganglion cell axons and preferentially promote their selective fasciculation.

Lastly, we should point out that virtually nothing is known about the factors that are responsible for the rather consistent branching patterns of retinal axons at certain points along their course. For example, in mammals many retinal fibers give off branches to the lateral geniculate nucleus and to components of the pretectal complex while continuing on to the superior colliculus. What induces the axons to branch at these points? How do the branches sense the direction in which they should grow? Are there cues within the substratum along which the fibers grow that signal to the axons that they should branch? Or do the fibers branch repeatedly throughout their growth? Are most incipient branches retracted while a few that find a hospitable terrain or an appropriate attachment persist? It remains for later work to address these concerns, and until a significant number of individual retinal axons have been examined at various stages in development, further speculation about possible mechanisms seems fatuous.

For convenience, we have formally distinguished axon pathfinding from target recognition, although to some extent the two processes are interrelated. Moreover, many individual experiments that have involved the intentional misrouting of retinal fibers not only bear upon the issue of pathfinding but also address the question as to how axons identify their appropriate terminal zones within tectum. We return to these experiments in our discussion of target recognition, but in the present context they may be viewed as falling into two categories: those in which retinal axons were fortuitously or systematically induced to approach the tectum along some abnormal route; and those in which the normal schedule of fiber outgrowth from the retina *and/or* the schedule for their invasion of the tectum were disrupted.

There are now several reports in the literature of optic fibers that entered the brain along some grossly aberrant route (e.g., by way of the oculomotor nerve) yet seemed capable of forming a suitably ordered retinotopic map. Rather more informative have been the experiments in which the brachia of the optic tract were interrupted and the fibers within one brachium intentionally misrouted into another. The general conclusion drawn from all these experiments is that even when the optic fibers approach the tectum through an abnormal pathway they can still home in on their appropriate tectal targets. For example, eyes grafted to the dorsal midline approach the tectum from above (Sharma, 1972), whereas posteriorly placed eyes approach the tectum along its caudal border

(Harris, 1980b); yet, in both instances the retinotectal projection was established with appropriate internal order and normal polarity (see also Gaze, 1960; Beazley and Lamb, 1979; Hibbard, 1984). Similarly, when fibers from the lateral brachium of the optic tract were made to regrow through the medial brachium, they bypassed the noninnervated medial tectum and continued to grow until they reached their appropriate lateral half of the tectum (Arora and Sperry, 1962; Sperry, 1965).

Studies of this type provide compelling evidence that the normal visual pathways, and any cues they may provide for the topographic sorting of optic fibers, are not essential for the establishment of topographically ordered patterns of retinotectal connections. At the same time they provide strong, although indirect, evidence for intrinsic polarity markers within the tectum. Essentially the same conclusions derive from recent studies of internally rearranged retinae in which a small wedge of embryonic eye tissue is transposed into an abnormal position within a host-eye rudiment. In many such cases, the transposed portion of the retina forms an autonomously projecting sector within the retinotectal map (Hunt and Ide, 1977; Conway et al., 1980; Willshaw et al., 1983), which implies that the altered origin of the relevant fibers within the retina and their inverted mode of entry into the nerve was insufficient to disrupt the mechanisms that ultimately direct the axons to the appropriate tectal destination.

In several studies, the normal timing of retinotectal innervation has been intentionally or incidentally disrupted (Feldman et al., 1971; Hunt, 1976, 1977; Holt, 1984). For example, when the ventral half of a *Xenopus* optic cup was fused with a dorsal half-eye from earlier optic-vesicle stages, the normal schedule of initial retinotectal innervation was inverted. Thus ventral fibers, which normally arrive relatively late, were induced to enter the tectum before fibers from the dorsal retina. Despite this abnormal schedule, a normally oriented retinotectal map was formed (Holt, 1984). Similarly in chick embryos, when that part of the retina which normally innervates the rostral and lateral parts of the tectum (where optic nerve fibers first enter the tectum) was ablated, the ingrowing fibers from the surviving regions of the retina did not innervate the first region of the tectum that they encountered; instead, they grew over the surface and left the rostrolateral tectum uninnervated while going on to form synapses in their appropriate caudomedial quadrant (Crossland et al., 1974). Although these findings make it clear that the initial establishment of the retinotectal map is not critically dependent on the time of arrival of fibers from different regions of the retina, they do not exclude a role for timing in the establishment of connections in the visual pathway. Indeed, there is now good evidence in chick embryos that some fibers are selectively eliminated at later stages in development simply because their late arrival at the tectum seems to place them at a competitive disadvantage compared to those that arrive earlier. Thus, the modest ipsilateral retinotectal projection, which appears transiently in normal embryos and reaches the tectum about two days after its innervation by the contralateral eye, is completely eliminated during a phase of large-scale ganglion cell death (O'Leary et al., 1983). By contrast, if the region of the presumptive optic chiasm is disrupted, a significant portion of the fibers that would normally cross at the chiasm is now deflected into the ipsilateral optic tract; under these circumstances, the resulting projection to the ipsilateral tectum persists for the life of the animal. The difference between the normal and experimental situation seems to lie in the time of arrival of the fibers at the ipsilateral tectum. Whereas, as mentioned above, the ipsilateral fibers normally arrive about two days after the crossed fibers, the fibers from the two eyes reach the tectum at about the same time after chiasmal disruption. This by itself seems sufficient to account for the survival of the experimentally induced ipsilateral projection (Fawcett and Cowan, 1984).

In amphibians and fish, where new ganglion cells are continuously added at the retinal margin (and new optic axons continue to be added to an established retinotectal projection), it also seems likely that timing provides an important "helper effect." For example, in goldfish, optic fibers reach the tectum in orderly birthdate bundles with a degree of retinotopic order (see Easter et al., 1981, and Easter, this volume). This spatiotemporal ordering of cell birthdates and fiber arrivals minimizes the need for individual fibers to sample vast areas of the tectum before encountering their proper termination zone. It remains to be seen whether the varying degrees of retinotopy found in the optic nerve and tract of different vertebrate species represent a passive accident of ganglion cell positions and birthdates, as in the morphogenetic model of Horder and Martin, or if it results from an active process of discrimination and sorting in which the optic fibers use their position-dependent chemical labels to associate selectively with their neighbors and to select particular positions in the nerve and tract.

As yet comparatively little work has been done on the molecular mechanisms involved in the routing of optic nerve fibers. In fact, only recently has an experimental model for the study of this problem been developed. It consists essentially of growing strips of retinal tissue *in vitro* (chick retinae have been used, but the method is in principle applicable to any retina) and subsequently presenting the outgrowing fibers, which have been shown to be the axons of retinal ganglion cells, with a choice of two substrates derived from one or another region of the tectum. When confronted with a substratum derived from either a corresponding or a foreign region of the tectum, the retinal fibers show a decided preference for the former (Bonhoeffer and Huf, 1980, 1982). Of course, the interpretation of *in vitro* experiments of this type is complicated; it is always difficult to know to what extent the behavior of the cultured tissue reflects its inherent or natural tendencies and to what extent it has been modified by the inevitable experimental perturbations and the *in vitro* conditions.

TARGET RECOGNITION

Simply stated, the problems involved in target recognition are threefold. First, there is the question of how optic axons, having reached the tectum, identify the correct region (not merely the correct quadrant, but the precise locus appropriate for their origins within different parts of the retina). Second, having located the correct topographic locus, the fibers need to identify the appropriate functional class or subclass of tectal neurons with which to synapse. And third, they have to synapse upon the appropriate region of their target cells (e.g., the soma, or more proximal dendrites).

Sperry (1943b, 1951) addressed these problems with a single parsimonious solution: a system of position-dependent labels on the surfaces of growing ganglion cell axons which served to identify a set of complementary or matching labels on the surfaces of the cells in the tectum. Although his so-called chemoaffinity hypothesis remains the most widely accepted explanation for the formation of retinotectal connections, over the years a number of alternative hypotheses have been advanced, and for longer or shorter periods have found strong proponents. These include the following.

1. The functionalist theory, according to which the initial patterns of connections between the retina and the tectum, as in other parts of the nervous system, are rather diffuse, if not totally random, and out of this nonselective system of connections those that are behaviorally adaptive persist while the remainder are eliminated (see Weiss, 1924, 1941).
2. A vectorial growth hypothesis, according to which optic axons are programmed to follow a series of instructions of the form "grow a distance L in the direction D, then grow a distance L' in direction D'," and so forth. When correctly executed, such vectorial instructions could deliver the tip of the growing axon to its appropriate place within the tectum (Gaze, 1970).
3. The morphogenetic hypothesis, which holds that the axons of retinal ganglion cells fasciculate passively with their neighbors and thereafter maintain the same topographic relationship to their neighbors throughout their course through the optic nerve, optic chiasm, and optic tract until they reach the correct point within the tectum and there terminate (Horder and Martin, 1978; Bodick and Levinthal, 1980; Grant and Rubin, 1980).
4. The closely related temporal hypothesis, which proposes that the entire retinotectal map can be accounted for on the basis of the orderly sequence of generation of the ganglion cells and the temporal sequence in which their axons invade the tectum (Gaze, 1960). According to this, the first fibers that enter the tectum synapse with the first tectal neurons encountered, and fibers arriving later continue to grow until they reach an as yet uninnervated region.
5. Lastly, there are a number of variations on the chemoaffinity theme, some suggested by Sperry himself, others added from time to time as new observations have been made or in an attempt to combine the more attractive features of two or more earlier models. For example, one variant accepts the existence of position-dependent labels on optic axons but postulates that they serve principally to mediate self-sorting interactions among optic fibers (Prestige and Willshaw, 1975; Hope et al., 1976; Willshaw and von der Malsburg, 1976, 1979). In a second variant (Sperry, 1951, 1963; Gaze et al., 1963), the chemical matching between retinal axons and tectal cells is conceived of as being relative and spatially graded, so as to give each fiber a hierarchy of choices (second, third, and n^{th}) in the selection of synaptic targets within the tectum. By sampling potential choices at different timepoints or at different sites sampled by its rather broad growth cone, the fiber, according to this view, could direct its growth toward ever more preferred target sites (see Hunt and Jacobson, 1974a, for a review). A third large class of variants incorporates all the features of Sperry's original hypothesis but takes into account a number of additional variables and forces. Thus the postulated positional labels on optic axons are thought to mediate both fiber–fiber sorting and tectal-target recognition. By adding position-independent attractions and repulsions and accepting that functional interactions may serve to validate or

refine the initial retinotopic order, these variants seek to accommodate as many findings as possible (Prestige and Willshaw, 1975; Fraser, 1980; Fraser and Hunt, 1980a; Gierer, 1981, 1983; Meyer, 1982a; Whitelaw and Cowan, 1982).

The continuing attempt to formulate new models and to refine existing hypotheses has helped not only to clarify the issues involved in target recognition in the retinotectal system, but also to point up several sources of confusion in the literature. For example, the fact that the matching of retinal and tectal elements may show some degree of plasticity has often been cited, incorrectly, as evidence that tectal cells do not have positional labels, or that some factor other than recognition and affinity between retinal axons and tectal cells is the driving force behind the formation of a topographic pattern of connections. Similarly, it is now clear that some of the problems related to pathfinding have been mistaken for those related to target selection within the tectum; thus, the finding of a degree of retinotopic order within the optic nerve and tract has been incorrectly cited as evidence that there is no need to postulate a selective association between retinal and tectal elements. And finally, it may be noted that a number of the novel and insightful variants on Sperry's original hypothesis have been somewhat tarnished because they were miscast, not as variants, but as alternatives to the chemoaffinity hypothesis, or because they were derived from experimental findings that were mistakenly conceived of as reflecting Sperry's views (see Hunt and Jacobson, 1974a; Meyer, 1982a, 1984; Hunt and Cowan, 1984, for reviews). Since, in retrospect, it is clear that one important source of confusion has been the failure to recognize that experiments that were undertaken to answer one question or set of questions often bear significantly upon another, quite different, issue, in what follows we attempt to summarize briefly the various experimental paradigms that have been employed and indicate the issues that we think they really address.

One of the first and most informative experimental paradigms involves the rotation of whole eyes, either in adult animals (Sperry, 1943b, 1944, 1945) or in embryos (Stone, 1944, 1960, 1966; Szekely, 1954; Jacobson, 1968; Jacobson and Hunt, 1973; Hunt and Piatt, 1978), or of the embryonic optic tectum (Crelin, 1952; Chung and Cooke, 1975, 1978). As we have seen, such experiments speak first to the issue of retinal and tectal polarity, but they also establish that the cells in the rotated tissue can somehow identify and select their normal synaptic partners. A second experimental paradigm, designed originally to address the question of how positional information is distributed to successive generations of retinal ganglion cells, involves transposing small wedges of embryonic retinal tissue from one pole of the eye to another. Such experiments have shown that the transposed sector can project autonomously upon the tectum, which means that ectopic ganglion cells can somehow express their own local instructions and that, in spite of their unusual positions or the routes taken by their axons, they are able to track down their appropriate target cells on the opposite side of the brain (Hunt and Ide, 1977; Conway et al., 1980; Willshaw et al., 1983). A third paradigm, which was considered in our discussion of pathfinding, involves misrouting optic fibers into the tectum from abnormal directions (Gaze, 1960; Sharma, 1972; Harris, 1980b; see Hibbard, 1984, for a review). The fact that under these circumstances an essentially normal retinotectal projection can be formed suggests that there must be some type of target selection process at work. However, none of these paradigms provides compelling evidence for a strict recognition–matching process between individual optic axons and tectal cells as the sole mechanism. Optic fibers could, for example, use their positional labels primarily to sort themselves into retinotopic arrays and use only the coarsest of polarity cues in the tectum to orient them and to produce the connectional patterns reported in all three experimental settings (see Fraser and Hunt, 1980a; Meyer, 1982a). For this and other reasons, considerable weight has been placed on the fourth type of experimental paradigm, which has as its primary aim the analysis of target selection. This is often referred to as the size-disparity paradigm and involves either making retinal deletions and allowing the surviving partial retina to innervate a whole optic tectum or, alternatively, making partial tectal lesions and allowing a whole retina to innervate the surviving part of the tectum.

The retinal-deletion paradigm was introduced by Attardi and Sperry in 1963, when they made a number of different types of retinal lesions in adult goldfish and mapped the distribution of the regenerating optic fibers in protargol-stained preparations some time after crushing the optic nerve. They reported that in nearly every case the fibers from the remaining part of the retina had grown back to the appropriate region within the tectum and described how the fibers often bypassed uninnervated regions of the tectum in doing so. Later studies have confirmed, through electrophysiological mapping techniques, this place-specific pattern of reinnervation by partial goldfish retinae and have also shown that there is considerable retinotopic order within the restricted projection (Horder, 1971; Schmidt, 1978; see also Meyer, 1982a). However, an unexpected finding in these studies has been the discovery that, with time, the restricted projection gradually begins to expand and may come to occupy the entire tectal surface while maintaining its internal retinotopic order (Schmidt, 1978; Sharma and Tung, 1979; Meyer, 1982a). The plasticity evident in these expanded projections from partial retinae is similar to that seen when various compound eyes (which may be thought of as "selective retinal deletions") are constructed in *Xenopus* embryos. Thus, when two anterior half-eyes (or, for that matter, two posterior or two ventral half-eyes) are fused together at embryonic stage 32, each half of the resulting compound eye usually projects in an orderly manner to the entire tectum (Gaze et al., 1963; Straznicky et al., 1974; Berman and Hunt, 1975; Ide et al., 1979). Similarly, when the growth of early larval eyes is chemically suppressed so that in effect only the ganglion cells in the central retina are left to make connections, they often form an expanded projection over most of the tectal surface (Hunt, 1976).

Expanded projections of this type do not occur in all vertebrates. For example, in comparable partial retinal lesion experiments in chick embryos it has been found that the uninnervated region of the tectum undergoes a marked trans-neuronal atrophy, and the projection from the surviving retinal tissue never expands into this zone (Crossland et al., 1974). Nevertheless, such expanded projections are extremely informative and pose a number of questions about the nature of plasticity in the retinotectal projection. One plausible suggestion for the expansion of the retinal projection after partial eye lesions is that optic axons normally compete with each other for tectal space, and that by removing part of the retina the competitive pressure is reduced. Evidence in support of this view has come from a variety of experiments designed to increase the postulated competitive pressure. These include causing a whole retina to compete with the projection of a partial retina to the same tectal lobe (Hunt, 1976; Schmidt, 1978) or causing two retinae with complementary lesions to innervate the same lobe (Sharma and Tung, 1979). In both situations fibers that had expanded beyond their normal limits could be shown to be "driven out" from the inappropriate regions by the fibers that rightly belonged there.

A rather similar result has been obtained after partial tectal ablations and optic-nerve crushes in adult amphibians and fish. When this is done in goldfish, the surviving portion of the optic lobe is initially reinnervated only by regenerating fibers from the appropriate region of the retina (Jacobson and Gaze, 1965; Cook and Horder, 1977). In time, however, the entire retina comes to innervate the partial tectum by compressing its projection while still preserving appropriate retinotopic order (Gaze and Sharma, 1970; Cook and Horder, 1977). Interestingly, this gradual compression can also be observed even if the optic nerve is not crushed at the time of the partial tectal ablation (Gaze and Sharma, 1970). Furthermore, the phenomenon can be repeated more than once; thus, if the optic nerve is recrushed after the initial phase of compression is complete, the entire sequence of early place-specific reinnervation, followed by a secondary compression, may be recapitulated (Cook and Horder, 1977). More or less the same sequence of place-specific regeneration, followed by gradual compression, has been reported in ranid frogs following half-tectal ablations, although here the timecourse is considerably slower than in goldfish (Udin, 1977).

Perhaps more than any other experimental paradigm, these size-disparity experiments have generated considerable disagreement and misunderstanding. The acute, short-term results in amphibians and fish, like the long-term results in chicks, are relatively straightforward and provide compelling evidence that ingrowing optic fibers are capable of active discrimination among the many potential targets in the optic lobe. The only possible interpretative complication is that during regeneration optic fibers may be directed to their former sites of innervation by the axonal debris that persists after the initial nerve section; conceivably such debris could leave an enduring "scent," which marks the correct site for reinnervation (Murray, 1976; Schmidt, 1978; Romeskie and Sharma, 1980). Schmidt (1978) has, in fact, presented evidence that persistent debris may be significant; but against the view that it is always critical we may cite an appreciable body of evidence from experiments in which the debris would seem to be "malpositioned" and should lead the regenerating axons to form inappropriate patterns of connections. These include the optic-nerve crush experiments after an earlier phase of compression referred to above (Cook and Horder, 1974, 1977), experiments in which portions of the projection were transposed (Meyer, 1979), and others in which the retinotectal projection was disordered (Hunt, 1976).

The long-term results of the compression experiments are more difficult to interpret. They clearly demonstrate a measure of ordered plasticity in the matching of retinal axons and tectal targets. But contrary to the opinion of some, they do not necessarily imply that there are no positional labels in the tectum or that retinotectal affinities play no role in the selection of synaptic sites by ingrowing optic fibers (Gaze and Keating, 1972; Horder and Martin, 1978, 1982). An equally plausible explanation of the findings is that the distribution of chemical labels within the retina and tectum is "context sensitive" (Hunt and Jacobson, 1974a). By this we mean that a half-retina or a half-tectum may be able to expand its range of labels in much the same fashion as embryonic half-rudiments undergo pattern regulation to form a complete miniature organ (Harrison, 1918, 1945), or alternatively, that the preferences of individual optic fibers for tectal sites may be hierarchical and hence influenced by the presence or absence of other fibers that may be competing for the same tectal sites (see Hunt and Jacobson, 1974a; Meyer, 1982a, for reviews).

Yet another experimental paradigm has involved the rotation and/or transposition of portions of the adult tectum in goldfish (Gaze et al., 1966; Sharma and Gaze, 1971; Yoon, 1973, 1980), in *Xenopus* (Levine and Jacobson, 1974; Hunt, 1976; Jacobson, 1978; Rho and Hunt, 1980; Horder and Martin, 1982), and in the frog *Rana catesbiana* (Jacobson and Levine, 1975a,b). Because similar results have been obtained in different species they can be dealt with together in summary fashion. When a large, rectangular portion of the dorsal tectum is excised, rotated either 180 or 90°, and reimplanted, the regenerated projection (after optic-nerve crush) has been found to be of two types. In a majority of cases, the peripheral retina appears to project to the unmanipulated part of the tectum with normal orientation, but the central retinal projection to the region of the implant is rotated. In a minority of cases, the entire retinal projection is grossly normal and of uniform orientation. Again, these results require careful interpretation. The majority result is clearly compatible with the hypothesis that tectal neurons carry their own labels and that these are retained after the rotation of a sizable central region. But the minority result can be explained in several different ways, including: (1) the disappearance of the graft and its replacement by normally oriented tectal tissue; (2) physical derotation of the graft back to its normal orientation; (3) a repolarization of the rotated implant (through interactions with the surrounding unoperated region), resulting in a redeployment of tectal labels in a manner compatible with normal orientation; or (4) the reestablishment of retinotopic order by some mechanism other than retinal and tectal affinities [e.g., by self-sorting among the optic fibers without regard to the underlying tectal substrate (Hunt, 1976; Rho and Hunt, 1980)]. Recent studies by Rho (1978, 1982) using genetically labeled tectal grafts have effectively ruled out the first and second of these possibilities, but the others are less amenable to experimental test.

The transposition of two fragments of tectal tissue is generally a more sensitive assay for the ability of optic fibers to recognize positional labels on tectal cells. If the two fragments are implanted in their normal orientation, the retinal fibers must discriminate true positional differences between them, and the question of physical derotation of the graft can also be ruled out (Hope et al., 1976; Hunt, 1976). Rostral-to-caudal transpositions within the adult tectum of *Xenopus* (Hunt, 1976) and goldfish (Hope et al., 1976) have again given two different types of results. In some cases the regenerated map appears to be normal and the regenerated fibers apparently ignore the earlier tectal transposition. In others, two regions of the retina had clearly transposed their projections in keeping with the transposed positions of their target tissue. Similarly, lateral-to-medial transpositions in frogs and fish have produced both normal projections and projections showing the expected transposition (Rho and Hunt, 1980; Yoon, 1980). In still other cases, fibers appropriate for the transposed graft and others appropriate for the implantation site appeared to converge upon the region of the implant and to be competing there for synaptic sites (see also Jacobson and Levine, 1975a,b; Rho and Hunt, 1980; Horder and Martin, 1982). Again, the interpretation of these results is complicated. Some of the findings are obviously compatible with the view that adult tectal tissue bears a refined set of positional markers and that regenerating optic fibers can track down the appropriately labeled tissue at its new position within the optic lobe. The more complex cases are perhaps easiest to explain by saying that in some instances (and for unknown reasons) the regenerating axons use some mechanism other than tectal labels (e.g., fiber sorting among themselves) to reestablish topographically ordered connections with the transposed tectum.

As we pointed out earlier, the interpretation of regeneration experiments in adult animals is complicated on the one hand by the persistence of axonal debris (Murray, 1976) and on the other hand by the possibility of some form of "memory" in the tectal cells of their past synaptic associations with retinal fibers. Moreover, it has been suggested that if specific tectal labels are present in the adult tectum, they have been induced in the tectal cells by their association with retinal fibers (Willshaw and von der Malsburg, 1979). Attempts to address these problems have used two general strategies. The first has involved producing a tectum with an abnormal pattern of retinotectal connections (e.g., by surgical manipulations that cause an expanded or compressed retinotectal map), and then later challenging a normal cohort of optic fibers to regenerate to their appropriate sites in the presence of either abnormal patterns of debris or abnormal patterns of "induced" tectal labels. We have already alluded to the experiments of Cook and Horder (1974, 1977) involving the reinnervation of half-tecta in which there was a previously compressed retinal projection. Other experiments described in the literature make the same point in a rather different way, for example, by directing fascicles of fibers from the opposite eye into the ipsilateral tectum in goldfish (Meyer, 1978, 1979, 1984) or, in *Xenopus*, by making multiple embryonic eye transplants to produce a disordered retinotectal projection (Hunt, 1976). In none of these did regenerating fibers from a normal eye form erroneous connections that matched the earlier deranged projection. Instead, the regenerating fibers behaved as though they were unaffected by the immediate past history of the tectum. Together these studies seem to rule out the possibility that regenerating fibers are guided to their targets within the tectum by debris from an earlier set of axon terminals, and they argue strongly against the view that the labels on tectal cells are induced by their exposure to retinal afferents.

An alternative strategy for assessing the effects of prior innervation has involved attempts to prepare so-called virgin tectal lobes, that is, tecta which have not previously been innervated by retinal fibers. The first experiments of this kind involved removing one eye in *Xenopus* embryos and later deflecting the fibers from the surviving eye into the supposedly virgin tectum (Feldman et al., 1971). However, it is now known that in *Xenopus* (Fraser and Hunt, 1980b; Pardo and Rho, 1980), as in other vertebrates, early unilateral enucleation usually results in the remaining eye innervating both tectal lobes. More recent virgin tectum experiments have used genetic mutations in urodeles and bilateral eye enucleations in frogs and chicks. Eye transplants into genetically eyeless urodeles result in the formation of apparently normal retintotectal projections (Harris, 1982); in bilaterally enucleated frogs and chicks the remaining afferent and efferent connections of the truly virgin tecta show normal topographic order (Constantine-Paton and Ferrari-Eastman, 1981; O'Leary and Cowan, 1983). Although these studies provide only indirect evidence for the existence of tectal labels, they do make it clear that the tectum undergoes some form of polarized fieldlike differentiation completely independent of its innervation by the retina and that, if there are tectal labels, they are not induced by retinal fibers.

Over the years a number of attempts have been made to assay more directly for the presence of position-dependent cell surface markers on the tectum. These studies have ranged from demonstrating maturational gradients in the tectum (Boell et al., 1955; Cowan et al., 1968; Straznicky and Gaze, 1972) and gradients in glycolipid distributions (Balsamo et al., 1976; Marchase, 1977), through studies aimed at demonstrating the differential adhesion of cells from different parts of the retina to specific tectal quadrants (Barbera et al., 1973; Barbera, 1975; Gottlieb et al., 1976), to the electrophysiological demonstration that topographically ordered connections can develop *in vitro* between paired explants of mammalian retina and colliculus (Smalheiser and Crain, 1978). But perhaps the most direct assault on this problem is that of Bonhoeffer and Huf (1980, 1982). In their experiments strips of retina from different regions of the embryonic chick eye were confronted with monolayers of tectal cells from the related region or an unrelated region in such a way that the outgrowing retinal ganglion cell axons were forced to make a choice. In the majority of cases, the fibers appeared to show a distinct preference for the tectal cells from the region that they would normally innervate. Even allowing for the vagaries of *in vitro* experiments, these studies would seem to provide strong evidence for the notion that tectal cells have surface determinants that are regionally selective and recognizable in some way by the tips of growing retinal axons.

REFINEMENTS OF THE RETINOTECTAL PROJECTION

Although at one time it was thought that the initial invasion of the tectum by optic fibers resulted, from the beginning, in the formation of a precise and accurate rerepresentation of the retina upon the tectal surface, it is now clear that this is not the case. There is, of course, a considerable degree of order in the retinotectal map as it is first formed (Holt and Harris, 1983; Sakaguchi, 1984), but as it matures the projection becomes progressively refined and sharpened. These refinements result in neighboring ganglion cells projecting, with increasing precision, on neighboring tectal neurons and also in the elimination of targeting errors in the initial projection.

The earliest suggestion that there may be some reorganization in the retinotectal projection came from a consideration of the different patterns of neurogenesis in the retina and tectum of *Xenopus* (Gaze et al., 1972). As we have seen, the retina develops in an essentially concentric manner with successive generations of ganglion cells being added in annuli around the margins of the retina (Straznicky and Gaze, 1971). Retinal growth continues throughout the life of the animal, although it is now appreciated that the rate of cell accretion around the *Xenopus* eye becomes sharply biased toward the ventral pole from metamorphic climax onward (Jacobson, 1976a; Beach and Jacobson, 1979). The tectum, on the other hand, develops in a curvilinear fashion from its rostrolateral pole, with later-formed cells being added along its caudomedial edge. And apparently, the addition of new cells to the tectum abates to near zero after metamorphosis (Straznicky and Gaze, 1972). Thus, the growth patterns are spatially mismatched in the tadpole, and only the retina may continue to add significant numbers of new cells in the juvenile and adult. It was inferred from this that the retinotectal projection must be subject to continuous restructuring if it is to serve the functional needs of the organism and at the same time accommodate an ever-increasing population of retinal fibers (Gaze and Keating, 1972). To account for this, Gaze and his colleagues have suggested that there is a system of "sliding connections" between the retina and tectum, and they have adduced electrophysiological evidence that during the first several weeks of larval life the map of the retina upon the tectum changes quite markedly (Gaze et al., 1972). Anatomical studies aimed at confirming this hypothesis have relied on the demonstration of degenerating axon terminals in the rostral tectum of normal *Xenopus* larvae (Scott, 1973; Longley, 1978), but it is not clear that the synapses in question are derived from the retina or that their degeneration is involved in the shifting pattern of connections. A more direct test of the sliding connections hypothesis was made by Gaze et al. (1979a), who used [^{3}H]proline labeling and autoradiography to show that the axons of newly formed ganglion cells in the temporal retina displace existing axons from the older retinal regions. More recently, Fraser (1983a) used a double-lesioning strategy on individual tadpoles to show that the location of the central retinal projection shifts over a distance of 150 μm during metamorphosis. In *Rana pipiens*, where retinal growth remains radially symmetrical throughout life, the double-lesion paradigm (Fraser, 1983b) and HRP-labeling studies (Reh and Constantine-Paton, 1984) have revealed an even more dramatic rate of shift. Very recent studies suggest a similar pattern of shifting connections in the retinotectal projection of goldfish (Easter and Stuermer, 1984; Stuermer and Easter, 1984; see Easter, this volume). Nevertheless, a number of important questions remain unanswered. These include: (1) In what way does the arrival of new retinal fibers cause preexisting connections to shift? (2) Is the shifting of connections due to the expansion of existing terminal arbors, with subsequent pruning, or is it brought about by the generation of new axon collaterals? (3) Is it initiated by the detachment of presynaptic terminals followed by a transient phase of active growth of the fibers to their new locations, or are new synaptic contacts formed before the earlier ones are detached? Unfortunately, the techniques currently available to us are not well suited to answering detailed questions of this type.

It is worth mentioning that it may not be necessary to postulate the existence of sliding connections in other vertebrates. Thus in the chick, where the general patterns of retinal and tectal histogenesis are similar to those in fish and amphibians but the whole process is compressed in time to 10–12 days, it seems likely that the retinal fibers distribute themselves *ab initio* in the pattern found in the mature state; there is evidence from autoradiographic experiments that the pattern of retinotectal synapse formation, as opposed to the sequence in which the optic fibers enter the tectum, may parallel the roughly concentric pattern of ganglion cell proliferation. That is to say, the first synapses may be formed near the center of the tectal lobe; later synapses, formed by later-generated ganglion cells, appear to be laid down more or less concentrically from this central focus (Crossland et al., 1975; Rager and von Oeyenhausen, 1979; see Rager, 1980). Although these observations have been called into question by later experiments using axonal degeneration (McGraw and McLaughlin, 1980) and anterograde labeling techniques (McLoon, 1982), these methods are not without their own disadvantages. The whole issue merits reexamination with a marker that reveals the actual form of the retinal fibers as they penetrate the outer synaptic layers of the tectum and that does not readily escape from the growing fibers (see Gerfen et al., 1982).

As we discussed in the section on target recognition, some of the most compelling evidence that substantial reorganizations can occur in the retinotectal projection has come from size-disparity experiments. The fact that a half-retina can in time come to innervate the entire tectum in frogs and fish, or conversely, the finding that the projection of the entire retina can be compressed onto a half-tectum, makes it clear that this system is extremely plastic. Several control experiments that have been done to analyze what changes may be occurring during these reorganizations have provided strong confirmatory evidence that, in teleosts and amphibians, connections once formed can be changed repeatedly and that, at least in some circumstances, appropriate retinal fibers can displace inappropriate connections (Yoon, 1972; Cook and Horder, 1974, 1977; Schmidt, 1978; see Edds et al., 1979, for a review). In analogous experiments that have involved directing fibers from both retinae into a single tectal lobe, an important related phenomenon has been observed (Levine and Jacobson, 1975; Meyer, 1975; Constantine-Paton and Law, 1978, 1982; Fawcett and Willshaw, 1982; Fawcett and Cowan, 1984). This is the tendency for the projections from the two eyes to separate progressively to form distinctive stripes or patches, reminiscent in a way of the eye-dominance columns seen in layer IV of the visual cortex of monkeys and cats (Hubel et al., 1977; LeVay et al., 1978). As in the development of cortical eye dominance columns, the two retinotectal projections usually overlap extensively at first, and segregation into stripes occurs relatively slowly. The actual form and dimensions of the stripes vary from species to species (and even within a given species, depending on the way in which the experiments were carried out). In three-eyed frogs, they are relatively uniform and run rostrocaudally across the entire length of the tectum (Constantine-Paton and Law, 1978); in frogs with nasal–nasal or temporal–temporal compound eyes, they are more irregular (Fawcett and Willshaw, 1982; see also Ide et al., 1983); in chicks in which dual innervation of the tectum is produced by surgically disrupting the region of the optic chiasm before the outgrowth of optic axons, they can differ enormously in width and rostrocaudal extent (Fawcett and Cowan, 1984).

Together these observations suggest that the retinotectal projection is forged by one class of mechanisms and refined by another. As we have seen, the first seems to be dependent upon the existence of cytochemical labels on the retinal fibers and a matching system of labels on the surface of the tectal cells. It serves to generate the initial retinotectal map and to establish that fibers from different regions of the retina terminate in an orderly way within different parts of the tectum. The refinement mechanism ensures that the axons of neighboring ganglion cells terminate upon neighboring tectal neurons. In those situations where two retinae (or two identical partial retinae) innervate a single tectum, the fact that from the beginning each forms an orderly retinotopic map is attributable to the first mechanism; the fact that they subsequently form eye-specific stripes or patches must be attributed to a quite different mechanism.

This second mechanism, unlike the first, appears to be activity-dependent in the sense that the segregation of the retinal afferents into eye-specific groups does not occur if ganglion cell activity is blocked with the sodium channel-blocking drug tetrodotoxin (TTX) (Meyer, 1982b,c; Schmidt and Edwards, 1983; see Schmidt, this volume). That electrical activity probably plays little or no role in the early establishment of the retinotectal map, as has been postulated from time to time (Whitelaw and Cowan, 1982), has been convincingly demonstrated in an ingenious study that involved transplanting eyes from a TTX-sensitive species of salamanders into a newt species that is insensitive to TTX and actually has quite high levels of the drug in its circulation. Despite the fact that all activity was blocked in the transplanted eyes, they were able to form grossly normal retinotectal maps (Harris, 1980a) and, in a recent variant of the experiment, were able to do so via abnormal pathways as well (Harris, 1984).

Just how activity brings about this type of refinement in the retinotectal projection is not known. In fish and frogs, in which retinal axons seem to be able to move quite freely across the tectal surface, we may suppose that the axons of neighboring ganglion cells are able to relocate themselves actively within the tectum so as to innervate neighboring tectal neurons. In chicks, for which there is no evidence that fibers that have made connections can move across the tectal surface, it seems unlikely that the eye-dominance stripes can be formed in this way, especially since some stripes are as much as 600 μm or more in width. Here it seems more probable that ganglion cells whose axons terminate (erroneously) at a significant distance from those of their neighbors simply die during the phase of large-scale ganglion cell death (Cowan, 1973), which occurs during the period when eye-specific stripes are being formed (Rager and Rager, 1978; Fawcett and Cowan, 1984). If this is indeed so, it would suggest that, in amniotes at least, much of the secondary refinement of the retinotectal projection is brought about by the selective death of ganglion cells whose axons have grown to the wrong region of the tectum.

To date, the best evidence for this conclusion comes from studies of the developing retinotectal projection in rodents. In mature rats and hamsters, the retina projects principally to the superior colliculus of the opposite side, but there is a persistent small projection to the rostromedial part of the ipsilateral colliculus. In the immediate postnatal period, the ipsilateral retinocollicular projection is considerably larger and extends across the surface of the entire colliculus (Chalupa and Rhoades, 1979; Frost and Schneider, 1979; Insausti et al., 1984a, b). Long-term labeling experiments in which retrogradely transported tracers were injected into the colliculus have established that the postnatal restriction of the ipsilateral projection is due to the death of ganglion cells, rather than to the retraction of collateral branches (Fawcett et al., 1984; Insausti et al., 1984a). It seems likely that the cells die because of the inability of their axons to compete successfully against the much larger number of fibers from the contralateral eye, presumably for some trophic factor that is available in only limited amounts within the tectum (Cowan et al., 1984). The principal evidence for this is that if the contralateral eye is removed at birth the ipsilateral projection retains its early widespread distribution (see Insausti et al., 1984b, for references), and retrograde labeling experiments indicate that essentially all of the cells that usually die during the restriction of the projection can be "rescued" (Fawcett et al., 1984). This refinement of the retinocollicular projection appears to use some activity-dependent mechanism, because blocking impulses in the competing population of ganglion cell axons (by repeatedly injecting TTX into the contralateral eye) both limits the restriction of the ipsilateral projection and again rescues a significant proportion of the ganglion cells that would normally be expected to die (Fawcett et al., 1984; Holt and Thompson, 1984).

These findings on the ipsilateral retinocollicular projection raised the interesting question of whether, during normal development, some ganglion cells might send their axons to inappropriate positions within the contralateral tectum. It is, of course, widely accepted that occasional axons go astray; in the mammalian visual system there are several reports of aberrant retinocollicular fibers that grow beyond the superior colliculus to enter the inferior colliculus or even the ventrobasal thalamus (Frost, 1981; Insausti et al., 1984b). But the problem of targeting errors within the superior colliculus itself has not been addressed until quite recently. Using the long-term retrograde labeling procedures mentioned above, O'Leary et al. (1984) have shown that, in fact, a significant proportion (perhaps as many as 20%) of the ganglion cells in the retinae of newborn rats project to inappropriate regions of the tectum. Most of these targeting errors seem to be eliminated during the second postnatal week, again as a result of the selective death of the parent ganglion cells. The death of these cells, like that of the cells that give rise to the early widespread retinocollicular projection, is the result of some activity-dependent process and can be prevented by repeated injections of TTX into the eye (O'Leary et al., 1984).

Cowan et al. (1984) have recently proposed a mechanism that could account for the selective loss of ganglion cells whose axons project to the wrong region of the tectum and also for the persistence of neighboring ganglion cells whose axons contact neighboring tectal cells. The proposed mechanism is based on a hypothesis put forward some years ago by Hebb (1949) to account for some aspects of learning at the cellular level and which was later adopted by Stent (1973) to explain the effects of sensory deprivation in the mammalian visual system. It assumes: (1) that all neurons are dependent for their survival on the availability of some trophic factors within their projection field; (2) that the supply of the relevant trophic factors is normally quite limited; (3) that the factors are released (at least in part) when the target cells are depolarized beyond some threshold; and (4) that the factors are preferentially taken up by axon terminals shortly after their neurotransmitter is released (during the period of membrane recycling). It would seem probable that neighboring ganglion cells are more likely to fire synchronously than are cells that are widely separated, and that as a result of their synchronous firing, neighboring cells are more likely to affect significantly the postsynaptic target cells. This would confer on them a significant advantage in the competition for the trophic factors over distant cells whose axons had aberrantly grown into the target field. Unfortunately, it will be impossible to test this hypothesis directly until the postulated trophic agent has been identified.

Until then we shall have to rely on indirect evidence of the type discussed above to account for the considerable refinement that retinotectal projections must undergo during the later phases of their development (see also Schmidt, this volume).

MODELS, MOLECULAR MECHANISMS, AND CONCLUSIONS

The objective of this chapter was to review some of the major issues involved in: (1) the development of polarity in the retina and tectum; (2) the growth of optic fibers from the retina to the brain; (3) the selection by optic fibers of the appropriate terminal zones and target cells within the optic lobe; and (4) the progressive refinement of the initial pattern of retinotectal connections. In this section, while frankly acknowledging that the molecular bases of these various development processes remain largely unstudied, we point out some issues that now appear ripe for biochemical study and consider some of the newer molecular strategies that appear promising. In addition, we refer briefly to certain of the models that have been proposed to account for the observed experimental findings. For convenience, we do not follow the same roughly chronological sequence of the previous sections, but begin with the possible mechanisms involved in target selection and the chemical labels that are thought to mediate it. The first question we consider is: Can chemical matching between retinal and tectal cells by itself account for all the experimental data on the formation of an ordered retinotectal projection?

After the overthrow of the functionalist position, it was widely assumed that the chemoaffinity hypothesis by itself could adequately explain the formation of topographically ordered systems of connections. And it was generally forgotten that Sperry (1943b, 1951, 1965) himself always retained a role for function in both the establishment and the validation of neural circuitry. Recent studies have confirmed the soundness of his judgment. In the previous section we considered some of the evidence derived from TTX-blocking studies which have shown unequivocally that, following the initial establishment of the retinotectal projection, some form of activity-dependent mechanism is involved in refining or fine-tuning the connections. Similar studies in anurans using the cholinergic blocking agent α-bungarotoxin have further strengthened this idea, for when synaptic transmission in the tectum was blocked, regenerating optic fibers appeared to be unable to form enduring connections with their appropriate postsynaptic targets (Freeman, 1977; Oswald et al., 1980). It remains for further work to clarify the ways in which electrical activity and synaptic transmissions act to modify the formation of connections and to disclose what molecular mechanisms are implicated.

One unanticipated conclusion that can be drawn from the expansion and compression studies discussed earlier is that, at least in some forms, a great many more optic fibers can be accommodated in a tectum, or partial tectum, than are present in the normal retinotectal projection. By itself this should serve to shift the emphasis away from the view that axons compete for a limited number of postsynaptic sites toward some form of competition for trophic factors produced within the optic lobe. The nature of these postulated trophic agents (which, as we have seen, appear to regulate the total number of retinal ganglion cells and control the elimination of individual ganglion cells; see Fawcett et al., 1984) is quite unknown, and their identification remains an important challenge for future molecular studies.

In a previous section we considered a number of alternatives to chemical labeling and pointed out that patterns of functional activity, programs for vectorial growth, and the various morphogenetic and timing hypotheses that have been advanced over the years are insufficient to explain all, or even a significant proportion of, the experimental findings. A more intriguing class of what might be termed minimum information models recognizes the need for position-dependent chemical labels on optic nerve fibers but attempts to use these labels to achieve a retinotopic pattern without postulating the existence of positional labels on the tectum or the direct recognition of tectal cells by optic axons (cf. Prestige and Willshaw, 1975; Hope et al., 1976; Hunt, 1976; Willshaw and von der Malsburg, 1976, 1979). In a pioneering series of computer simulation studies, Prestige and Willshaw (1975) examined the connectional patterns formed between two neuronal arrays (such as the retina and tectum), where one or both arrays were marked with a single monotonic gradient of labels (or synaptic determinants). The simulations suggested that single gradients were insufficient to account for the restricted connections made by a half-retinal array within the correct portions of an intact tectal array. Hunt (1976) and Hope et al. (1976) considered an alternative cuing system within the optic lobe, according to which all tectal cells carry an intrinsic polarity marker but none is endowed with unique positional markers. These "arrows," as they were called by Hope et al. (1976), are sufficient to account for the development of a normal projection and even for a rotated projection to a rotated tectal implant, if it is assumed that the optic fibers (using positional markers or some other device) are able to sort themselves into a retinotopic pattern before reaching the tectum. However, local arrows are not sufficient to explain how optic fibers can identify their original synaptic sites when two pieces of the tectum have been transposed without inverting their orientation (Hope et al., 1976; Hunt, 1976; see also Jacobson and Levine, 1975b; Rho and Hunt, 1980; Yoon, 1980). Finally, we may mention that Willshaw and von der Malsburg (1976, 1979) have considered a variety of self-ordering mechanisms by which optic fibers might achieve retinotopic order during development, orient themselves with respect to crude polarity cues on the tectum, and thereafter induce in the postsynaptic cells a more permanent positional marker. Although, to date, their models have not been directly refuted, they are difficult to reconcile with the results of experiments in which abnormal patterns of retinal innervation failed to induce persistent changes in tectal markers (see also Hunt, 1976; Meyer, 1984). In the absence of discrete tectal labels it is difficult on a priori grounds to conceive how optic fibers could track down their proper synaptic partners in a translocated tectal graft (Jacobson and Levine, 1975b; Hope et al., 1976; Hunt, 1976), or how optic fibers from *any* half of a chick embryo retina can terminate selectively with its matching half of an intact optic lobe (Crossland et al., 1974).

In retrospect, these models have been useful in focusing attention on the possibility that the position-dependent labels on optic fibers might also promote an active sorting among the fibers themselves. In the past few years the experimental evidence for direct fiber–fiber ordering has become quite strong: Optic fibers can achieve a retinotopic order on the tectal surface that ignores the polarity of an underlying tectal implant (Rho, 1978; Horder and Martin, 1982) or even, in the case of transplanted fiber bundles, the polarity of an intact ipsilateral tectum (Meyer, 1979). Not surprisingly then, a number of more recent multivariate models have suggested a role for fiber–fiber sorting to complement chemical recognition by optic fibers of their appropriate postsynaptic neurons in the tectum (Fraser, 1980; Fraser and Hunt, 1980a; Gierer, 1981, 1983; Whitelaw and Cowan, 1982). The individual multivariate models that have been proposed differ from one another in the parameters used to achieve fiber sorting and in a number of other details. Gierer's (1981, 1983) model pits long-distance repulsions against shorter-range attractions to achieve both fiber sorting and expansion or compression of fiber sets when a size disparity has been created between the retina and the optic lobe. Whitelaw and Cowan (1982) use patterns of impulse activity as a means of ordering fibers whose retinal positions (and firing rates) are similar, while Fraser (1980, this volume) has used the same set of positional labels on optic fibers to mediate not only selective association among optic fibers (fibers from similar positions bearing similar positional labels), but also the association between optic fibers and their corresponding tectal neurons.

A common feature of these models is that, when they are used in computer simulations of certain kinds of biological experiments, they often pit fiber–fiber ordering *against* fiber–tectal associations. For example, in computer simulations involving rotated tectal grafts, the "optic fibers" cannot simultaneously achieve a uniform retinotopic order (normal pattern) and a perfect match with the underlying tectal tissue (in a projection whose central representation is rotated); the two forces in the model tend to drive the fiber "array" toward two distinct but relatively stable patterns, with one force or the other able to gain an edge in individual animals on the basis of such relatively serendipitous factors as healing scars (see Fraser and Hunt, 1980a, for a review; see Fraser, this volume). As we have seen, when these same experiments have been done on real frogs and fish they frequently give rise to mixed results within individual experimental series, including some that show a normally regenerated retinotectal projection and others that show the anticipated rotation in the projection from the central retina to the region of the tectal implant (Jacobson and Levine, 1975a,b; Hope et al., 1976; Hunt, 1976; Rho and Hunt, 1980).

The newer multivariate models have also pointed out the importance of position-independent forces when attempting to simulate the experimental findings. For example, Fraser (1980) has found that without a very strong position-independent attraction between all retinal fibers and all positions on the tectum, chemical labels mediating both fiber sorting and tectal recognition are still unable to explain how a whole retina can compress its projection onto a half-tectum. A generalized repulsive force among all optic terminals is similarly needed to explain how a half-retina would eventually expand its projection from one part of the tectum to cover the whole tectal surface (Fraser, 1980; Fraser and Hunt, 1980a; see also Prestige and Willshaw, 1975; Whitelaw and Cowan, 1982; Gierer, 1983). The implications of these findings are that there may be a range of ligands and surface moieties, on both optic fibers and tectal cells, which play a role in the formation of the retinotectal map. Unfortunately, while this may enrich our understanding, it is also likely to complicate the molecular studies needed to characterize the relevant determinant elements.

The nature of the position-dependent labels, of course, remains as the single most challenging feature of the retinotectal system for molecular studies. Sperry (1965) proposed a rather precise molecular model that he thought met the needs of his chemoaffinity hypothesis. According to this model, individual ganglion cells bear on their surfaces long-chain molecules, each of which is made up of varying ratios of two building blocks or subunits. One class of such molecules [A_xP_y] might mark cell position in the anteroposterior axis of the retina and be spatially graded in its ratio from one extreme of "all A" at the anterior pole to the other extreme of "all P" at the posterior pole, with cells along the vertical meridian bearing molecules made up of equal parts of the A and P subunits. A second class of molecules [D_aV_b] would specify in similar fashion the dorsoventral level of the cells. Comparable long-chain molecules, with stable ratios of two smaller subunits, were suggested as a means of making cellular addresses within the optic lobe and also within the various structures encountered by optic fibers in the course of their growth to the tectum (Sperry, 1965). The refined chemoaffinity hypothesis (Sperry, 1963, 1965) thus envisioned the growing optic axon tip as in some way sensing these local address cues and always extending toward the best possible chemical match until finally reaching its correct position within the tectum.

In the ensuing years a number of investigators have proposed schemes of chemical complementarity, not unlike that between an antibody and its antigen, for the labels on retinal axons and tectal cells (cf. Marchase et al., 1975). The difficulty with this alternative is that optic fibers from the same neighborhood in the retina cannot use such molecules to associate selectively with each other; indeed, in its most extreme form, complementarity of the type proposed by Marchase et al. (1975) might cause optic fibers specifically to avoid other fibers from the same region that bear similar chemical labels. Fraser's (1980, this volume) model is more in keeping with Sperry's original concept, and by using homophilic attractions the same molecules could mediate positional attraction between retinal and tectal cells and also mediate the predicted sorting operations among the optic fibers themselves.

Direct attempts to identify and isolate the molecules involved in positional marking in the retina and tectum have proved to be extremely difficult. The earliest attempts used tissue culture preparations, and it could be shown that intact retinal neurons are able to adhere fairly selectively to one quadrant of the tectum rather than others in a manner that was in keeping with the topographic associations seen *in vivo* (Barbera et al., 1973; Barbera, 1975; Marchase et al., 1975; Gottlieb et al., 1976). Although such whole-cell preparations can serve as useful assays to identify cell ligands present in graded distributions across the retina or tectum (cf. Balsamo et al., 1976; Marchase, 1977) the assays per se should not be viewed as providing direct evidence for chemoaffinity between optic fibers and tectal cells *in vivo*. Nor do they necessarily involve the same molecular ligands that are present on the growth cones of optic fibers or those that mediate selective cellular associations between optic fibers and tectal cells *in situ*. In most instances the culture and aggregation conditions have to be carefully adjusted in order to mimic the *in vivo* adhesivities; the use of whole tectal quadrants is complicated by the presence of meningeal and connective tissue coverings which may themselves serve as adhesive substrates but which are never encountered by growing optic fibers *in vivo*; and, as Gottlieb has found, cells from other brain structures, such as the telencephalon, often adhere more avidly to retinal cells than does the tectum or dissociated tectal cells (see Gottlieb and Glaser, 1981, for a review).

A more realistic approach has been to view such assays as simply a starting point for the isolation of a broad spectrum of interesting cell surface ligands, which may then be purified and individually characterized. Beginning with an assay for selective adhesion among retinal cells *in vitro*, Edelman and his colleagues (see Edelman, 1983, 1984) have succeeded in purifying a number of cell adhesion molecules (or CAMs) and in raising antibodies to these molecules that can be used to chart their spatial distribution, their ontogenic history, and their abilities to perturb a variety of developmental processes. As discussed in detail elsewhere (see Edelman, this volume), one of these cell adhesion molecules, designated N-CAM, undergoes a characteristic chemical modification during development. This has led to the interesting suggestion that maturational changes in such a single molecule (or perhaps in a small number of such molecules) could account for a rich variety of selective cellular aggregations in the developing brain and even for the establishment of specific fiber pathways. This represents an important new organizing principle; namely, that relatively few chemical determinants, by virtue of different topographic distributions on the cell surface and by their temporal modulation, can confer a wide range of specificities upon embryonic cells. Interestingly, there is already evidence from *in vivo* experiments that antibodies to at least one of the CAMs (purified from frog tissues) can induce disorder in the retinotectal projection of *Xenopus* when applied in implants to the tectal surface (Fraser et al., 1984).

A complementary strategy has been adopted by Trisler et al. (1981). This has involved screening large numbers of monoclonal antibodies directed against surface determinants on chick embryonic retinal cells. One such antibody has been found that recognizes an antigenic determinant with a graded distribution in the retina; the relative concentration of the antigen is about 30-fold higher at the dorsal pole of the retina than at its ventral margin. This antigen appears at the appropriate developmental stages and seems to be present on optic axons as well as on cell somata in the retina. Although its function is unknown, it has attracted considerable attention because it is the first well-documented example of a molecule with the smoothly graded distribution presupposed by Sperry's chemoaffinity hypothesis. Studies on similar surface determinants with graded distributions in the anteroposterior axis of the retina and in the tectum are an active focus of contemporary research.

Apart from the promising *in vitro* studies of Bonhoeffer and Huf (1980, 1982) alluded to earlier and the companion studies of Halfter et al. (1981), molecular studies on optic fiber pathfinding remain to be undertaken. This promises to be a difficult task because assay systems have to be developed to examine not only the selective association of optic axons with each other (and especially with other axons from the same region of the retina), but also the chemical cues present within the optic stalk, the optic chiasm, and the various structures encountered by retinal fibers in their growth to the tectum. Many of these tissues are present only in minuscule amounts, some exist for only a short time in development, and each is likely to contain many more molecules that have nothing to do with the growth and guidance of optic fibers than are implicated in this process.

As yet it seems that no molecular studies on the acquisition of polarity and positional markers in the retina and tectum have been reported, although in the case of the retina there are a number of interesting leads worth pursuing. One is that gap junctions are present in the central retina of *Xenopus* only up to optic-cup stages, but they persist (perhaps indefinitely) among the stem cells at the retinal margin (Dixon and Cronly-Dillon, 1972, 1974; Jacobson and Hunt, 1973; Fujisawa, 1982; Sakaguchi et al., 1984). Conceivably these junctions could provide channels for the passage of morphogens involved in the establishment of polarity and also in repolarization when this occurs. Another lead derives from recent studies in which two half-eyes were fused together across nucleopore filters. Recombinations that induce polarity reversal in one of the two half-eye fragments were observed to do so across filters whose pore diameters (0.015 μm) virtually rule out direct contact between cell membranes (Sullivan et al., 1984). This suggests that diffusible molecules may play a role at some stage in the maintenance and propagation of polarity cues in the retina. In this regard, computer simulation studies have shown that simple source–sink diffusion schemes (Crick, 1970) present difficulties when one attempts to simulate retinotectal nerve patterns, but reaction–diffusion models (Kauffman et al., 1978; Meinhardt, 1982; after Turing, 1952) based on two diffusible morphogens that influence each other's synthesis and degradation do rather well (Shoaf et al., 1984). The patterns they generate are stable over the twofold variation in embryonic eye size seen among individual *Xenopus* embryos, and they persist in the face of simulated cell deaths (Glucksmann, 1940), premature differentiations, and partial ablations followed by healing; in addition, the patterns that can be simulated from half-eye recombinations are strikingly similar to those observed in the regulated visual projections of real animals (Shoaf et al., 1984).

Finally, the application of cell markers to the embryonic eye in *Xenopus* has begun to reveal a wide variety of growth properties and cellular fates that appear to be under positional control in the growing eye (Conway et al., 1980; Hunt et al., 1982a,b; see also Stone, 1966). For example, the growth rates and patterns and the final sector domains populated by marked cell progeny are strikingly different when marked dorsal germinal cells are compared to germinal cells from the ventral part of the eye. The former contribute little or nothing to the iris of the eye and show a marked reduction in their rate of growth at metamorphosis in both neural retina and pigment retina; ventral germinal cells flanking the choroidal fissure behave quite differently, generating a vast number of iris cells and expanding their domains dorsally during larval growth of the retinal epithelia. When wedges of ventral optic cup are transposed at certain stages to a dorsal implantation site in a host eye, they often show characteristic ventral growth patterns and form an autonomous sector of retinotectal mapping. Other cases that show a regulated (i.e., normal) retinotectal projection usually also adapt their growth properties to match that of the dorsal part of the eye (Conway et al., 1980; Hunt et al., 1982a,b). Since the early signs of these differences between dorsal and ventral growth patterns can be recognized in the living eye within a few days of grafting, it may now be possible to use marked transplants in chemical perturbation experiments aimed at polarity and positional information without having to rely for a primary assay on the formation, some weeks or months later, of a retinotectal projection.

In summary, then, the retinotectal system poses a broad range of issues and questions that may now be suitable for molecular study. And if the present chapter appears to have raised a great many questions and provided few, if any, molecular answers, it can only be pleaded that this is the current state of the field. What is clear, however, is that the range of problems is now much better characterized in this system than in any other and much better characterized than even a few years ago. It is our hope and expectation that the newer molecular methods will be fruitfully applied to these problems during the latter half of this decade and that the characterization of chemoaffinity mechanisms at the level of specific molecules is within our reach.

ACKNOWLEDGMENTS

A special thanks is extended to Pat Thomas, Shelley Gambino, Crystal Burr, and Dr. Barbara Boss for their help in the preparation of the manuscript. The authors' research is supported by grants from the National Eye Institute (EY-03653) and the National Science Foundation (PCM-83-11082). W. M. C. is a Clayton Foundation Senior Investigator.

REFERENCES

Arora, H. L., and R. W. Sperry (1962) Optic nerve regeneration after surgical cross-union of medial and lateral optic tracts. *Am. Zool.* **2**:389.

Arora, H. L., and R. W. Sperry (1963) Color discrimination after optic nerve regeneration in the fish *Astronotus ocellatus*. *Dev. Biol.* **17**:234–243.

Attardi, D. G., and R. W. Sperry (1963) Preferential selection of central pathways by regenerating optic fibers. *Exp. Neurol.* **7**:46–64.

Balsamo, J., J. McDonough, and J. Lilien (1976) Retinal–tectal connections in the embryonic chick: Evidence for regionally specific cell surface components which mimic the pattern of innervation. *Dev. Biol.* **49**:338–346.

Barbera, A. J. (1975) Adhesive recognition between developing retinal cells and the optic tecta of the chick embryo. *Dev. Biol.* **46**:167–191.

Barbera, A. J., R. B. Marchase, and S. Roth (1973) Adhesive recognition and retinotectal specificity. *Proc. Natl. Acad. Sci. USA* **70**:2482–2486.

Beach, D. H., and M. Jacobson (1979) Patterns of cell proliferation in the developing retina of the clawed frog in relation to blood supply and position of the choroidal fissure. *J. Comp. Neurol.* **183**:625–632.

Beazley, L. D., and A. H. Lamb (1979) Rerouted optic axons in *Xenopus* tadpoles form normal visuotectal projections. *Brain Res.* **179**:373–378.

Berman, N., and R. K. Hunt (1975) Visual projections to the optic tecta in *Xenopus* after partial extirpation of the embryonic eye. *J. Comp. Neurol.* **162**:23–42.

Bodenstein, L., and R. K. Hunt (1979) Patterning in growing cell sheets: A model for neural retina. *Biophys. J.* **25**:84a.

Bodick, N., and C. Levinthal (1980) Growing optic nerve fibers follow neighbors during embryogenesis. *Proc. Natl. Acad. Sci. USA* **77**:4374–4378.

Boell, E. J., P. Greenfield, and S. C. Shen (1955) Development of cholinesterase in the optic lobes of the frog (*Rana pipiens*). *J. Exp. Zool.* **129**:415–452.

Bonhoeffer, F., and J. Huf (1980) Recognition of cell types by axonal growth cones *in vitro*. *Nature* **288**:162–164.

Bonhoeffer, F., and J. Huf (1982) *In vitro* experiments on axon guidance demonstrating an anterior–posterior gradient on the tectum. *EMBO J.* 1:427–431.

Chalupa, L. M., and R. W. Rhoades (1979) An autoradiographic study of the retinotectal projection in the golden hamster. *J. Comp. Neurol.* 186:561–569.

Chung, S.-H., and J. Cooke (1975) Polarity of structure and of ordered nerve connections in the developing amphibian brain. *Nature (Lond.)* 258:126–132.

Chung, S.-H., and J. Cooke (1978) Observations on the formation of the brain and of nerve connections following embryonic manipulation of the amphibian neural tube. *Proc. R. Soc. Lond. (Biol.)* 201:335–373.

Chung, S.-H., R. M. Gaze, and R. V. Stirling (1973) Abnormal visual function in *Xenopus* following stroboscopic illumination. *Nature (New Biol.)* 246:186–189.

Cima, C., and P. Grant (1980) Ontogeny of the retina and optic nerve of *Xenopus laevis*. IV. Ultrastructural evidence of early ganglion cell differentiation. *Dev. Biol.* 76:229–237.

Constantine-Paton, M., and P. Ferrari-Eastman (1981) Topographic and morphometric effects of bilateral embryonic eye removal on the optic tectum and nucleus isthmus of the leopard frog. *J. Comp. Neurol.* 196:645–661.

Constantine-Paton, M., and M. I. Law (1978) Eye-specific termination bands in tecta of three-eyed frogs. *Science* 202:639–641.

Constantine-Paton, M., and M. I. Law (1982) The development of maps and stripes in the brain. *Science* 247:62–70.

Conway, K., K. Feiock, and R. K. Hunt (1980) Polyclones and patterns in developing *Xenopus* eye. *Curr. Top. Dev. Biol.* 15:216–317.

Cook, J. E., and T. J. Horder (1974) Interaction between optic fibres in their regeneration to specific sites in the goldfish tectum. *J. Physiol. (Lond.)* 241:89–90.

Cook, J. E., and T. J. Horder (1977) The multiple factors determining retinotopic order in the growth of optic fibres into the optic tectum. *Philos. Trans. R. Soc. Lond. (Biol.)* 278:261–276.

Cowan, W. M. (1971) Studies on the development of the avian visual system. In *Cellular Aspects of Neural Growth and Differentiation*, D. Pease, ed., pp. 177–222, University of California Press, Berkeley.

Cowan, W. M. (1973) Neuronal death as a regulative mechanism in the control of cell number in the nervous system. In *Development and Aging in the Nervous System*, M. Rockstein and M. L. Sussman, eds., pp. 19–41, Academic, New York.

Cowan, W. M., A. H. Martin, and E. Wenger (1968) Mitotic patterns in the optic tectum of the chick during development and after early removal of the optic vesicle. *J. Exp. Zool.* 169:71–92.

Cowan, W. M., J. W. Fawcett, D. D. M. O'Leary, and B. B. Stanfield (1984) Regressive phenomena in the development of the vertebrate nervous system. *Science* 225:1258–1265.

Crelin, E. S. (1952) Excision and rotation of the developing *Amblystoma* optic tectum and subsequent visual recovery. *J. Exp. Zool.* 120:547–578.

Crick, F. H. C. (1970) Diffusion in embryogenesis. *Nature* 225:420–422.

Crossland, W. J., W. M. Cowan, L. A. Rogers, and J. P. Kelly (1974) The specification of the retinotectal projection in the chick. *J. Comp. Neurol.* 155:127–164.

Crossland, W. J., W. M. Cowan, and L. A. Rogers (1975) Studies on the development of the chick optic tectum. IV. An autoradiographic study of the development of retinotectal connections. *Brain Res.* 91:1–23.

Dixon, J. S., and J. R. Cronly-Dillon (1972) The fine structure of the developing retina in *Xenopus laevis*. *J. Embryol. Exp. Morphol.* 28:659–666.

Dixon, J. S., and J. R. Cronly-Dillon (1974) Intercellular gap junctions in pigment epithelium cells during retinal specification in *Xenopus laevis*. *Nature* 251:505.

Easter, S. S., Jr., and C. A. O. Stuermer (1984) An evaluation of the hypothesis of shifting terminals in goldfish optic tectum. *J. Neurosci.* 4:1052–1063.

Easter, S. S., Jr., A. C. Rusoff, and P. E. Kish (1981) The growth and organization of the optic nerve and tract in juvenile and adult goldfish. *J. Neurosci.* 1:793–811.

Edds, M. V., R. M. Gaze, G. E. Schneider, and L. N. Irwin (1979) Specificity and plasticity of retinotectal connections. *Neurosci. Res. Program Bull.* **17**:243–375.

Edelman, G. M. (1983) Cell adhesion molecules. *Science* **219**:450–457.

Edelman, G. M. (1984) Modulation of cell adhesion during induction, histogenesis, and perinatal development of the nervous system. *Annu. Rev. Neurosci.* **7**:337–339.

Fawcett, J. W., and W. M. Cowan (1984) On the formation of eye dominance stripes and patches in the doubly-innervated optic tectum of the chick. *Dev. Brain Res.* (in press).

Fawcett, J. W., and D. J. Willshaw (1982) Compound eyes project stripes on the optic tectum in *Xenopus*. *Nature* **296**:350–352.

Fawcett, J. W., D. D. M. O'Leary, and W. M. Cowan (1984) Activity and the control of ganglion cell death in the rat retina. *Proc. Natl. Acad. Sci. USA* **81**:5589–5593.

Feldman, J. D., and R. M. Gaze (1975) The development of half-eyes in *Xenopus* tadpoles. *J. Comp. Neurol.* **162**:13–22.

Feldman, J. D., R. M. Gaze, and M. J. Keating (1971) Delayed innervation of the optic tectum during development. *Exp. Brain Res.* **14**:16–23.

Fisher, S., and M. Jacobson (1970) Ultrastructural changes during early development of retinal ganglion cells in *Xenopus*. *Z. Zellforsch.* **104**:165–177.

Fraser, S. E. (1980) A differential adhesion approach to the patterning of nerve connections. *Dev. Biol.* **79**:453–464.

Fraser, S. E. (1983a) Fiber optic mapping of the *Xenopus* visual system: Shift in the retinotectal projection during development. *Dev. Biol.* **95**:505–511.

Fraser, S. E. (1983b) Plasticity in the retinotectal projection during normal development: Sliding connections in *Rana*. *Soc. Neurosci. Abstr.* **9**:760.

Fraser, S. E., and R. K. Hunt (1980a) Retinotectal specificity: Models and experiments in search of a mapping function. *Annu. Rev. Neurosci.* **3**:319–352.

Fraser, S. E., and R. K. Hunt (1980b) Retinotectal plasticity in *Xenopus*: Anomalous ipsilateral projection following late larval eye removal. *Dev. Biol.* **79**:444–452.

Fraser, S. E., B. A. Murray, C.-M. Chuong, and G. M. Edelman (1984) Alteration of the retinotectal map in *Xenopus* by antibodies to neural cell adhesion molecules. *Proc. Natl. Acad. Sci. USA* **81**:4222–4226.

Freeman, J. A. (1977) Possible regulatory function of acetylcholine receptor in maintenance of retinotectal synapses. *Nature* **269**:218–222.

French, V., P. J. Bryant, and S. V. Bryant (1976) Pattern regulation in epimorphic fields. *Science* **193**:969–981.

Frost, D. O. (1981) Orderly anomalous retinal projections to the medial geniculate, ventrobasal and lateral posterior nuclei of the hamster. *J. Comp. Neurol.* **203**:227–256.

Frost, D. O., and G. E. Schneider (1979) Postnatal development of retinal projections in Syrian hamsters: A study using autoradiographic and anterograde degeneration techniques. *Neuroscience* **4**:1649–1677.

Fujisawa, H. (1982) Formation of gap junctions by stem cells in the developing retina of the clawed frog (*Xenopus laevis*). *Anat. Embryol.* **165**:141–149.

Gaze, R. M. (1960) Regeneration of the optic nerve in amphibia. *Int. Rev. Neurobiol.* **2**:1–40.

Gaze, R. M. (1970) *The Formation of Nerve Connections*, Academic, New York.

Gaze, R. M., and M. J. Keating (1972) The visual system and "neuronal specificity." *Nature* **237**:375–378.

Gaze, R. M., and S. C. Sharma (1970) Axial differences in the reinnervation of the optic tectum by regenerating optic nerve fibres. *Exp. Brain Res.* **10**:171–181.

Gaze, R. M., and C. Straznicky (1980) Stable programming for map orientation in disarranged embryonic eyes in *Xenopus*. *J. Embryol. Exp. Morphol.* **55**:143–165.

Gaze, R. M., M. Jacobson, and G. Szekely (1963) The retinotectal projection in *Xenopus* with compound eyes. *J. Physiol. (Lond.)* **165**:484–499.

Gaze, R. M., M. Jacobson, and S. C. Sharma (1966) Visual responses from the goldfish brain following excision and reimplantation of the optic tectum. *J. Physiol. (Lond.)* **183**:38–39P.

Gaze, R. M., S.-H. Chung, and M. J. Keating (1972) The development of the retinotectal projection in *Xenopus*. *Nature (New Biol.)* **236**:133–135.

Gaze, R. M., M. J. Keating, A. Ostberg, and S.-H. Chung (1979a) The relationship between retinal and tectal growth in larval *Xenopus*: Implications for the development of the retinotectal projection. *J. Embryol. Exp. Morphol.* **53**:103–143.

Gaze, R. M., J. D. Feldman, J. Cook, and S.-H. Chung (1979b) The orientation of the visuotectal map in *Xenopus*: Developmental aspects. *J. Embryol. Exp. Morphol.* **53**:39–66.

Gerfen, C. R., D. D. M. O'Leary, and W. M. Cowan (1982) A note on the transneuronal transport of wheat germ agglutinin-conjugated horseradish peroxidase in the avian and rodent visual systems. *Exp. Brain Res.* **48**:443–448.

Gierer, A. (1981) Development of projections between areas of the nervous system. *Biol. Cybern.* **42**:69–78.

Gierer, A. (1983) Model for the retino-tectal projection. *Proc. R. Soc. Lond. (Biol.)* **218**:77–93.

Glucksmann, A. (1940) Development and differentiation of the tadpole eye. *Br. J. Ophthalmol.* **24**:153–178.

Gottlieb, D. I., and L. Glaser (1981) Cellular recognition during neural development. In *Studies in Developmental Neurobiology*, W. M. Cowan, ed., pp. 243–260, Oxford Univ. Press, New York.

Gottlieb, D. I., K. Rock, and L. Glaser (1976) A gradient of adhesive specificity in developing avian retina. *Proc. Natl. Acad. Sci. USA* **73**:410–414.

Grant, P., and E. Rubin (1980) Ontogeny of the retina and optic nerve in *Xenopus laevis*. II. Ontogeny of the optic fiber pattern in the retina. *J. Comp. Neurol.* **189**:671–698.

Grillo, M. A., and J. Rosenbluth (1972) Ultrastructure of the developing *Xenopus* retina before and after ganglion cell specification. *J. Comp. Neurol.* **145**:131–140.

Halfter, W., M. Claviez, and U. Schartz (1981) Preferential adhesion of tectal membranes to anterior embryonic chick retinal neurites. *Nature* **292**:67–70.

Hamburger, V., and H. L. Hamilton (1951) A series of normal stages in the development of the chick embryo. *J. Morphol.* **88**:49–92.

Harris, W. A. (1980a) The effects of eliminating impulse activity on the development of the retinotectal projection in salamanders. *J. Comp. Neurol.* **194**:303–317.

Harris, W. A. (1980b) Regions of the brain influencing the projection of developing optic tracts in the salamander. *J. Comp. Neurol.* **194**:319–333.

Harris, W. A. (1982) The transplantation of eyes to genetically eyeless salamanders: Visual projections and somatosensory interactions. *J. Neurosci.* **2**:339–353.

Harris, W. A. (1984) Axonal pathfinding in the absence of normal pathways and impulse activity. *J. Neurosci.* **4**:1153–1162.

Harrison, R. G. (1918) Experiments on the development of the forelimb of amblystoma, a self-differentiating equipotential system. *J. Exp. Zool.* **25**:413–461.

Harrison, R. G. (1921) On relations of symmetry in transplanted limbs. *J. Exp. Zool.* **32**:1–136.

Harrison, R. G. (1933) Some difficulties of the determination problem. *Am. Nat.* **67**:306–321.

Harrison, R. G. (1936) Relations of symmetry in the developing ear of *Amblystoma punctatum*. *Proc. Natl. Acad. Sci. USA* **22**:238–247.

Harrison, R. G. (1945) Relations of symmetry in the developing embryo. *Trans. Conn. Acad. Arts Sci.* **36**:277–330.

Hebb, D. O. (1949) *The Organization of Behavior*, Wiley, New York.

Hibbard, E. (1984) Retinotectal connections made through ectopic optic nerves. In *Brain Circuits and Functions of the Mind*, C. Trevarthen, ed., Cambridge Univ. Press, Cambridge (in press).

Holt, C. E. (1984) Does timing of axon outgrowth influence initial retinotectal topography in *Xenopus*? *J. Neurosci.* **4**:1130–1152.

Holt, C. E., and W. A. Harris (1983) Order in the initial retinotectal map in *Xenopus*: A new technique for labelling growing nerve fibres. *Nature* **301**:150–152.

Hope, R. A., B. J. Hammond, and R. M. Gaze (1976) The arrow model: Retinotectal specificity and map formation in the goldfish visual system. *Proc. R. Soc. Lond. (Biol.)* **194**:447–466.

Horder, T. J. (1971) Retention, by fish optic fibers regenerating to new terminal sites in the tectum, of "chemospecific" affinity for their original sites. *J. Physiol. (Lond.)* **216**:53–55P.

Horder, T. J., and K. A. C. Martin (1978) Morphogenetics as an alternative to chemospecificity in the formation of nerve connections. *Symp. Soc. Exp. Biol.* **32**:275–358.

Horder, T. J., and K. A. C. Martin (1982) Some determinants of optic terminal localization and retinotopic polarity within fibre populations in the tectum of goldfish. *J. Physiol. (Lond.)* **333**:481–509.

Hubel, D. H., T. N. Wiesel, and S. LeVay (1977) Plasticity of ocular dominance columns in monkey striate cortex. *Philos. Trans. R. Soc. Lond. (Biol.)* **278**:377–409.

Hunt, R. K. (1975) Developmental programming for retinotectal patterns. *Ciba Found. Symp.* **29**:131–159.

Hunt, R. K. (1976) Position-dependent differentiation of neurons. In *Developmental Biology*, D. McMahon and C. F. Fox, eds., pp. 227–256, W. A. Benjamin, Menlo Park, Cal.

Hunt, R. K. (1977) Competitive retinotectal mapping in *Xenopus*. *Biophys. J.* **17**:128a.

Hunt, R. K., and N. Berman (1975) Patterning of neuronal locus specificities in retinal ganglion cells after partial extirpation of the embryonic eye. *J. Comp. Neurol.* **162**:43–70.

Hunt, R. K., and W. M. Cowan (1984) The chemoaffinity hypothesis: An appreciation of Roger Sperry's contributions to developmental biology. In *Brain Circuits and Functions of the Mind*, C. Trevarthen, ed., Cambridge Univ. Press, Cambridge (in press).

Hunt, R. K., and E. Frank (1975) Neuronal locus specificity: Transrepolarization of *Xenopus* embryonic retina after the time of axial specification. *Science* **189**:563–565.

Hunt, R. K., and C. F. Ide (1977) Radial propagation of positional signals for retinotectal patterns in *Xenopus*. *Biol. Bull. (Woods Hole)* **153**:430–431.

Hunt, R. K., and M. Jacobson (1972a) Development and stability of positional information in *Xenopus* retinal ganglion cells. *Proc. Natl. Acad. Sci. USA* **69**:780–783.

Hunt, R. K., and M. Jacobson (1972b) Specification of positional information in retinal ganglion cells of *Xenopus*: Stability of the specified state. *Proc. Natl. Acad. Sci. USA* **69**:2860–2864.

Hunt, R. K., and M. Jacobson (1973) Specification of positional information in retinal ganglion cells of *Xenopus*: Assays for analysis of the unspecified state. *Proc. Natl. Acad. Sci. USA* **70**:507–511.

Hunt, R. K., and M. Jacobson (1974a) Neuronal specificity revisited. *Curr. Top. Dev. Biol.* **8**:203–258.

Hunt, R. K., and M. Jacobson (1974b) Specification of positional information in retinal ganglion cells of *Xenopus laevis*: Intraocular control of the time of specification. *Proc. Natl. Acad. Sci. USA* **71**:3616–3620.

Hunt, R. K., and J. Piatt (1978) Cross species axial signaling with reversal of retinal axes in embryonic *Xenopus* eyes. *Dev. Biol.* **62**:44–51.

Hunt, R. K., L. A. Faulkner, B. Szaro, K. Conway, and R. Tompkins (1982a) Mosaic and regulated patterns in retinal growth and retinotectal innervation following heterotopic grafting of genetically marked eye-bud fragments in *Xenopus*. *Neurosci. Abstr.* **8**:514.

Hunt, R. K., R. Tompkins, D. Reinschmidt, L. Bodenstein, and R. K. Murphey (1982b) Starting points for a developmental genetics of nerve patterns. *Am. Zool.* **22**:185–204.

Ide, C. F., B. Kosofsky, and R. K. Hunt (1979) Control of pattern duplication in the retinotectal system of *Xenopus*: Suppression of duplication by eye-fragment interactions. *Dev. Biol.* **69**:337–360.

Ide, C., S. Fraser, and R. L. Meyer (1983) Eye dominance columns from an isogenic double nasal frog eye. *Science* **221**:293–295.

Insausti, R., C. Blakemore, and W. M. Cowan (1984a) Ganglion cell death during the development of the ipsilateral retino-collicular projection in the golden hamster. *Nature* **308**:362–365.

Insausti, R., C. Blakemore, and W. M. Cowan (1984b) Postnatal development of the ipsilateral retinocollicular projection and the effects of unilateral enucleation in the golden hamster. *J. Comp. Neurol.* (in press).

Jacobson, C.O. (1959) The localization of the presumptive cerebral regions in the neural plate of the axolotl larva. *J. Embryol. Exp. Morphol.* **1**:1–21.

Jacobson, C. O. (1964) Motor nuclei, cranial nerve roots, and fibre pattern in the medulla oblongata after reversal experiments in the neural plate of axolotl larvae. *Zool. Bidr. Upps.* **36**:73–160.

Jacobson, M. (1968) Development of neuronal specificity in retinal ganglion cells of Xenopus. *Dev. Biol.* **17**:202–218.

Jacobson, M. (1976a) Histogenesis of retina in the clawed frog with implications for the pattern of development of retinotectal connections. *Brain Res.* **103**:541–545.

Jacobson, M. (1976b) Premature specification of the retina in embryonic Xenopus eyes treated with ionophore X537A. *Science* **191**:288–290.

Jacobson, M. (1978) *Developmental Neurobiology*, 2nd ed., Plenum, New York.

Jacobson, M., and R. M. Gaze (1965) Selection of appropriate terminations by regenerating optic fibres in the adult goldfish. *Exp. Neurol.* **13**:418–430.

Jacobson, M., and R. K. Hunt (1973) The origins of nerve–cell specificity. *Sci. Am.* **228**:26–32.

Jacobson, M., and R. L. Levine (1975a) Plasticity in the adult frog brain: Filling the visual scotoma after excision or translocation of parts of the optic tectum. *Brain Res.* **88**:339–345.

Jacobson, M., and R. L. Levine (1975b) Stability of implanted duplicate tectal positional markers serving as targets for optic axons in adult frogs. *Brain Res.* **92**:468–471.

Karten, H. J., K. V. Fite, and N. Brecha (1977) Specific projections of displaced retinal ganglion cells upon the accessory optic system in the pigeon (*Columba livia*). *Proc. Natl. Acad. Sci. USA* **74**:1753–1756.

Kauffman, S., R. Shymko, and K. Trabert (1978) Control of sequential compartment formation in Drosophila. *Science* **199**:259–270.

LeVay, S., M. P. Stryker, and C. J. Shatz (1978) Ocular dominance columns and their development in layer IV of the cat's visual cortex: A quantitative study. *J. Comp. Neurol.* **179**:223–244.

Levine, R., and M. Jacobson (1974) Deployment of optic nerve fibers is determined by positional markers in the frog's tectum. *Exp. Neurol.* **43**:527–538.

Levine, R., and M. Jacobson (1975) Discontinuous mapping of retina onto tectum by both eyes. *Brain Res.* **98**:172–176.

Ling, R. T., C. F. Ide, and R. K. Hunt (1979) Control of pattern duplication in the retinotectal system of Xenopus: Induction of duplication in eye fragments by secondary cuts. *Dev. Biol.* **69**:361–374.

Longley, A. (1978) Anatomical mapping of retino-tectal connections in developing and metamorphosed Xenopus: Evidence for changing connections. *J. Embryol. Exp. Morphol.* **45**:249–270.

Marchase, R. B. (1977) Biochemical studies of retino-tectal specificity. *J. Cell Biol.* **75**:237–257.

Marchase, R. B., A. J. Barbera, and S. Roth (1975) A molecular approach to retinotectal specificity. *Ciba Found. Symp.* **29**:315–341.

Matthey, R. (1926) Recuperation de la vue après graffe de l'oeil chez le triton adulte. *C. R. Soc. Biol.* **94**:4–5.

McDonald, N. (1977) A polar co-ordinate system for positional information in the vertebrate retina. *J. Theor. Biol.* **69**:153–165.

McGraw, C. F., and B. J. McLaughlin (1980) Fine structural studies of synaptogenesis in the superficial layers of the chick optic tectum. *J. Neurocytol.* **9**:79–93.

McLoon, S. C. (1982) Alterations in the precision of the crossed retinotectal projection during tectal development. *Science* **215**:1418–1420.

McLoon, S. C., and R. D. Lund (1982) Transient retinofugal pathways in the developing chick. *Exp. Brain Res.* **45**:277–284.

Meinhardt, H. (1982) *Models of Biological Pattern Formation*, Academic, London.

Meyer, R. L. (1975) Tests for field regulation in the retinotectal system of goldfish. In *Developmental Biology*, D. McMahon and C. F. Fox, eds., pp. 257–274, W. A. Benjamin, Menlo Park, Cal.

Meyer, R. L. (1978) Deflection of selected optic fibers into a denervated tectum in goldfish. *Brain Res.* **155**:213–227.

Meyer, R. L. (1979) Retinotectal projection in goldfish to an inappropriate region with a reversal in polarity. *Science* **205**:819–821.

Meyer, R. L. (1982a) Ordering of retinotectal connections: A multivariate operational analysis. *Curr. Top. Dev. Biol.* **17**:101–145.

Meyer, R. L. (1982b) Tetrodotoxin inhibits the formation of refined retinotopography in goldfish. *Dev. Brain Res.* **6**:293–298.

Meyer, R. L. (1982c) Tetrodotoxin blocks the formation of ocular dominance columns in goldfish. *Science* **218**:589–591.

Meyer, R. L. (1984) The case for chemoaffinity in the retinotectal system: Recent studies. In *Brain Circuits and Functions of the Mind*, C. Trevarthen, ed., Cambridge Univ. Press, Cambridge (in press).

Murray, M. (1976) Regeneration of retinal axons into the goldfish optic tectum. *J. Comp. Neurol.* **168**:175–196.

O'Leary, D. D. M., and W. M. Cowan (1983) Topographic organization of certain tectal afferent and efferent connections can develop normally in the absence of retinal input. *Proc. Natl. Acad. Sci. USA* **80**:6131–6135.

O'Leary, D. D. M., C. R. Gerfen, and W. M. Cowan (1983) The development and restriction of the ipsilateral retinofugal projection in the chick. *Dev. Brain Res.* **10**:93–109.

O'Leary, D. D. M., J. W. Fawcett, and W. M. Cowan (1984) Elimination of topographical targeting errors in the retinocollicular projection by ganglion cell death. *Soc. Neurosci. Abstr.* **10**:464.

Oswald, R. E., D. E. Schmidt, J. J. Norden, and J. A. Freeman (1980) Localization of alpha-bungarotoxin binding sites to the goldfish retinotectal projection. *Brain Res.* **187**:113–127.

Pardo, F. S., and J. H. Rho (1980) Visual projections after early eye removal in *Xenopus*. *Fed. Proc.* **39**:1613.

Prestige, M. C., and D. J. Willshaw (1975) On a role for competition in the formation of patterned neural connections. *Proc. R. Soc. Lond. (Biol.)* **190**:77–98.

Rager, G. H. (1980) Development of the retinotectal projection in the chicken. *Adv. Anat. Embryol. Cell Biol.* **63**:1–92.

Rager, G. H. and V. Rager (1978) Systems matching by degeneration. I. A quantitative electron microscopic study of the generation and degeneration of retinal ganglion cells in the chicken. *Exp. Brain Res.* **33**:65–78.

Rager, G. H., and B. von Oeyenhausen (1979) Ingrowth and ramification of retinal fibers in the developing optic tectum of the chick embryo. *Exp. Brain Res.* **35**:213–227.

Reh, T. A., and M. Constantine-Paton (1984) Retinal ganglion cell terminals change their projection sites during development of *Rana pipiens*. *J. Neurosci.* **4**:442–457.

Rho, J. H. (1978) Cell interactions in retinotectal mapping. *Biophys. J.* **21**:137a.

Rho, J. H. (1982) Restoration of the frog's retinotectal map after various surgical manipulations of the frog's optic tectum. Unpublished doctoral dissertation, The Johns Hopkins University, Baltimore.

Rho, J. H., and R. K. Hunt (1980) Visual projections to a "double lateral" tectum in *Xenopus*. *Dev. Biol.* **80**:436–453.

Romeskie, M., and S. C. Sharma (1980) Retinal projection to a 180 degree-rotated tectal reimplant following long-term tectal denervation in adult goldfish. *Brain Res.* **201**:202–205.

Sakaguchi, D. S. (1984) The retinotectal system of *Xenopus*: The development of ganglion cells and their projections to the brain. Unpublished doctoral dissertation, State University of New York, Albany.

Sakaguchi, D. S., R. K. Murphey, R. K. Hunt, and R. Tompkins (1984) The development of retinal ganglion cells in a tetraploid strain of *Xenopus laevis*. A morphological study utilizing intracellular dye injection. *J. Comp. Neurol.* **224**:231–251.

Schmidt, J. T. (1978) Retinal fibers alter tectal positional markers during the expansion of the half retinal projection in goldfish. *J. Comp. Neurol.* **177**:279–300.

Schmidt, J. T., and D. L. Edwards (1983) Activity sharpens the map during the regeneration of the retinotectal projection in goldfish. *Brain Res.* **269**:29–39.

Scholes, J. (1979) Nerve fibre topography in the retinal projection to the tectum. *Nature* **278**:620–624.

Scholes, J. (1981) Ribbon optic nerves and axonal growth patterns in the retinal projection to the tectum: In *Development in the Nervous System*, D. R. Garrod and J. D. Feldman, eds., pp. 181–214, Cambridge Univ. Press, Cambridge.

Scott, T. M. (1973) Degeneration of optic nerve terminals in the frog tectum. *J. Anat. (Lond.)* **114**:261–269.

Sharma, S. C. (1972) Retinotectal connexions of a heterotopic eye. *Nature New Biol.* **238**:286–287.

Sharma, S. C., and R. M. Gaze (1971) The retinotopic organization of visual responses from tectal reimplants in adult goldfish. *Arch. Ital. Biol.* **190**:357–366.

Sharma, S. C., and J. Hollyfield (1980) The determination of functional specificity in retinotectal connections in *Xenopus*. *J. Embryol. Exp. Morphol.* **55**:77–92.

Sharma, S. C., and Y. L. Tung (1979) Competition between nasal and temporal hemiretinal fibers in adult goldfish tectum. *Neuroscience* **4**:113–119.

Shoaf, S., K. Conway, and R. K. Hunt (1984) Application of reaction–diffusion models to cell patterning of *Xenopus* retina: Initiation of patterns and their biological stability. *J. Theor. Biol.* **109**:299–329.

Smalheiser, N. R., and S. M. Crain (1978) Formation of functional retinotectal connections in cocultures of fetal mouse explants. *Brain Res.* **148**:484–492.

Sperry, R. W. (1943a) Effect of 180 degree rotation of the retinal field on visuomotor coordination. *J. Exp. Zool.* **92**:263–279.

Sperry, R. W. (1943b) Visuomotor coordination in the newt (*Triturus viridescens*) after regeneration of the optic nerve. *J. Comp. Neurol.* **79**:33–55.

Sperry, R. W. (1944) Optic nerve regeneration with return of vision in anurans. *J. Neurophysiol.* **7**:57–69.

Sperry, R. W. (1945) Restoration of vision after crossing of optic nerves and after contralateral transplantation of eye. *J. Neurophysiol.* **8**:15–28.

Sperry, R. W. (1948) Orderly patterning of synaptic associations in regeneration of intracentral fiber tracts mediating visuomotor coordination. *Anat. Rec.* **102**:63–76.

Sperry, R. W. (1951) Regulative factors in the orderly growth of neural circuits. *Growth* **10**:63–87.

Sperry, R. W. (1963) Chemoaffinity in the orderly growth of nerve fiber patterns and connections. *Proc. Natl. Acad. Sci. USA* **50**:703–710.

Sperry, R. W. (1965) Embryogenesis of behavioral nerve nets. In *Organogenesis*, R. L. DeHaan and H. Ursprung, eds., pp. 161–186, Saunders, Philadelphia.

Stent, G. S. (1973) A physiological mechanism for Hebb's postulate of learning. *Proc. Natl. Acad. Sci. USA* **70**:997–1001.

Stone, L. S. (1944) Functional polarization in retinal development and its reestablishment in regenerating retinae of rotated grafted eyes. *Proc. Soc. Exp. Biol. N.Y.* **57**:13–14.

Stone, L. S. (1960) Polarization of the retina and development of vision. *J. Exp. Zool.* **154**:85–93.

Stone, L. S. (1966) Development, polarization and regeneration of the ventral iris clift (remnant of choroidal fissure) and protractor lentis muscle in Urodele eyes. *J. Exp. Zool.* **161**:95–108.

Straznicky, K., and R. M. Gaze (1971) The growth of the retina in *Xenopus laevis*: An autoradiographic study. *J. Embryol. Exp. Morphol.* **26**:67–79.

Straznicky, K., and R. M. Gaze (1972) The development of the optic tectum in *Xenopus laevis*: An autoradiographic study. *J. Embryol. Exp. Morphol.* **28**:87–115.

Straznicky, K., R. M. Gaze, and M. J. Keating (1974) The retinotectal projections from a double-ventral compound eye in *Xenopus laevis*. *J. Embryol. Exp. Morphol.* **31**:123–137.

Stuermer, C. A. O., and S. S. Easter, Jr. (1984) Rules of order in the retinotectal fascicles of goldfish. *J. Neurosci.* **4**:1045–1051.

Stultz, W. A. (1936) Relations of symmetry in the hindlimb of *Amblystoma punctatum*. *J. Exp. Zool.* **72**:317–367.

Sullivan, K., K. Conway, and R. K. Hunt (1984) Demonstration of a polarizing signal that reverses future retinotectal patterns across nucleopore filters, in *Xenopus* embryonic eyes. *Cell Differ.* **14**:33–45.

Swett, F. H. (1937) Determination of limb-axes. *Q. Rev. Biol.* **12**:322–339.

Szekely, G. (1954) Zur Ausbildung der lokalen funktionellen specifitat der retina. *Acta Biol. Acad. Sci. Hung.* **5**:157–167.

Szekely, G. (1966) Embryonic determination of neural connections. *Adv. Morphog.* **5**:181–219.

Turing, A. M. (1952) The chemical basis of morphogenesis. *Philos. Trans. R. Soc. Lond. (Biol.)* **237**:37–72.

Trisler, G. D., M. D. Schneider, and M. Nirenberg (1981) A topographic gradient of molecules in retina can be used to identify neuron position. *Proc. Natl. Acad. Sci. USA* **78**:2145–2149.

Udin, S. B. (1977) Rearrangements of the retinotectal projection in *Rana pipiens* after caudal half-tectum ablation. *J. Comp. Neurol.* **173**:561–681.

Weiss, P. (1924) Die Funktion transplantier fer Amphibien extremitaten. Aufstellung einer Resonanz-theorie der motorischen Nerventatigkeit auf Grund abgestimmter Endorgane. *Roux's Archiv.* **102**:635–672.

Weiss, P. (1941) Self-differentiation of the basic patterns of coordination. *Comp. Psychol. Monogr.* **17**:1–96.

Whitelaw, V. A., and J. D. Cowan (1982) Specificity and plasticity of retinal connections: A computational model. *J. Neurosci.* **1**:1369–1387.

Willshaw, D. J., J. W. Fawcett, and R. M. Gaze (1983) The visuotectal projections made by *Xenopus* "pie slice" compound eyes. *J. Embryol. Exp. Morphol.* **74**:29–45.

Willshaw, D. J., and C. von der Malsburg (1976) How patterned neural connections can be set up by self-organization. *Proc. R. Soc. Lond. (Biol.)* **194**:431–445.

Willshaw, D. J., and C. von der Malsburg (1979) A marker induction mechanism for the establishment of ordered neural mappings: Its application to the retinotectal problem. *Philos. Trans. R. Soc. Lond. (Biol.)* **287**:223–243.

Wolpert, L. (1971) Positional information and pattern formation. *Curr. Top. Dev. Biol.* **6**:183–224.

Yntema, C. L. (1955) Ear and nose. In *Analysis of Development*, B. H. Willier, P. Weiss, and V. Hamburger, eds., pp. 415–428, W. B. Saunders, Philadelphia.

Yoon, M. (1972) Reversibility of the reorganization of retinotectal projection in goldfish. *Exp. Neurol.* **35**:565–577.

Yoon, M. (1973) Retention of the original topographic polarity by the 180 degree rotated tectal reimplant in young adult goldfish. *J. Physiol. (Lond.)* **233**:575–588.

Yoon, M. (1980) Retention of topographic addresses by reciprocally translocated tectal reimplants in adult goldfish. *J. Physiol. (Lond.)* **308**:197–215.

Chapter 18

The Continuous Formation of the Retinotectal Map in Goldfish, with Special Attention to the Role of the Axonal Pathway

STEPHEN S. EASTER, JR.

ABSTRACT

The retinotectal projection of goldfish has been studied anatomically at many levels between retinal ganglion cell body and tectal terminal arbor. The ganglion cells are produced continually at the margin of the retina, over many years, so that the ages of the optic axons always span a wide range. The axons grow in along the most superficial level of the fiber layer and penetrate the center of the optic nerve head. They remain associated with axons of similar age throughout the nerve and tract and then course along the outermost edge of the tectum, ventral retinal axons dorsally, dorsal retinal axons ventrally. They exit systematically from their peripheral fascicle, temporal axons first, nasal ones last. Subsequent production of new tectal cells in a caudally apposed crescent creates a new tectal boundary along which a subsequent generation of axons will grow. This continuous formation of an ordered projection requires that the terminal arbors shift positions. Evidence for this shift is presented, along with speculative models of the process.

The study of the formation of the retinotectal map was initiated in the 1940s by Roger Sperry, who used experimental surgical methods to perturb, and behavioral methods to assess, the retinotectal projection (see Gaze, 1970, for a review). The experiments were remarkably ingenious, as they led to conclusions about microscopic structures (the terminals of retinal axons) from observations of the behavior of whole animals. In the late 1950s, Gaze (in Scotland) and Maturana and his collaborators (in the United States) simultaneously brought electrophysiological methods to bear on the problem. Low-resistance extracellular microelectrodes were used to record visually driven electrical activity in the tectum of anurans, and Sperry's inference from behavior was confirmed: when an optic nerve was cut, the axons grew back to terminate in the same region of tectum that they had previously occupied (Gaze, 1958, 1959; Maturana et al., 1959). The microelectrode had succeeded in bringing the experimental observations down to the microscopic level, but it was insensitive to axons of passage. Still unanswered was the important question: Did the regenerating axon grow directly toward its termination site, or did it wander and settle at the correct site only when it finally happened upon it?

Attardi and Sperry (1963) attempted to answer this question in the goldfish tectum by using an anatomical method that stained both the terminal and the axon. If the regenerated axons and their terminals could be visualized following crush of the optic nerve, it would be possible to infer whether growth was directed, provided that two additional pieces of information were available. First, it would be necessary to know where in the retina an axon had originated. This was established by making retinal lesions at the same time the nerve was crushed. If, for instance, peripheral retina was ablated, any regenerated axons in the tectum must have originated in central retina. Second, the normal trajectory and termination site in the tectum would have to be known. This information was provided in the "Materials and Procedures" section of Attardi and Sperry's paper and can be summarized as follows. Axons from posterior, dorsal, anterior, and ventral retina terminated in anterior, ventral, posterior, and dorsal tectum, respectively. The axons entered along the ventral or dorsal rim of the tectum and turned centrally in the "parallel layer," an array of fascicles that pointed away from the rims like the teeth on a comb and toward their termination sites. These pathways are best illustrated by the four pairs of sketches in Figure 1 that summarize the main observations of the paper, and although they represent regenerated axons, they illustrate what Attardi and Sperry believed to be the normal case—that the regenerated axons had reoccupied their original routes. This was taken as evidence that axons were guided toward their termination sites and did not wander en route. The second major conclusion was that they terminated at the same sites as before even though the retinal lesions had provided alternative denervated zones. For instance, in Figure 1B, ablation of posterior (temporal) hemiretina resulted in a total absence of input to anterior tectum, but the anterior (nasal) retinal axons grew right past the vacant area to terminate "correctly" in posterior tectum. Likewise in Figure 1C, central retinal axons grew across denervated peripheral tectum before terminating centrally. This selective reinnervation of particular regions, even when others were vacant, confirmed and considerably extended the electrophysiological results.

Attardi and Sperry's paper has been widely cited and these particular illustrations have often been reproduced over the intervening 20 years. The conclusions have been checked by a number of workers and modified considerably. The conclusion that the regenerating axons terminate in their old locations has been confirmed both anatomically and electrophysiologically, but with two interesting modifications. One is that the very first deployment of processes into the synaptic neuropil is not very well ordered, but it sharpens up over a few weeks (Meyer, 1980; Schmidt and Edwards, 1983). The other is that after several months the "correct" projection changes so that the partial retina projects to fill the entire tectum (Schmidt et al., 1978). [The converse occurs as well; when half the tectum is removed, the projection from the entire retina is compressed onto the reduced remnant (Gaze and Sharma, 1970).] The conclusion that the axons grow directly back to their termination sites along their original paths has been shown to be incorrect. Fujisawa et al. (1982) visualized both normal and regenerated individual axons in the newt and found that the regenerates took very tortuous paths, unlike the normals. Udin (1978) and Stuermer and Easter (1984a) used different methods but concluded similarly in frog and goldfish, respectively.

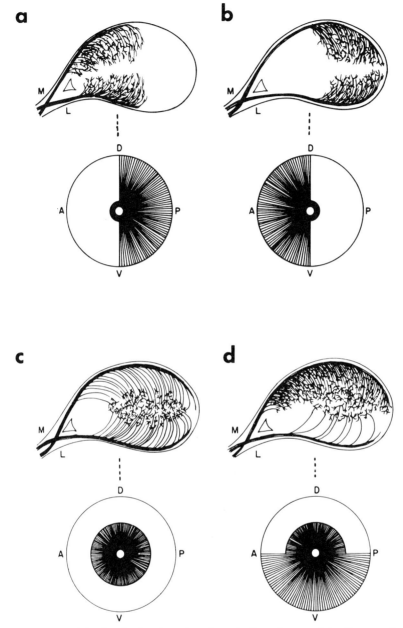

Figure 1. *Four pairs of sketches showing the paths and terminations of regenerated optic axons after partial ablations.* In all four pairs, the retina is shown below, the tectum above. M, medial; L, lateral; A, anterior; P, posterior; D, dorsal; V, ventral. (From Attardi and Sperry, 1963.)

In this chapter I propose a revision of a third aspect of Attardi and Sperry (1963), namely their description of the normal pathway, which had been given without any methodological details. C. A. O. Stuermer and I have checked it, concluded that it is incorrect, and produced an alternative description with interesting ramifications for development. I deal with this alternative in the next few pages; in the remainder of the chapter I attempt to incorporate our studies in the tectum with those of others in the retina, optic nerve, and optic tract in order to present a unified description of the retinotectal projection in goldfish.

RETINOTECTAL FASCICLES

Before developing an alternative to the description by Attardi and Sperry, it is important to establish that there is no dispute about the existence of the fascicles in stratum opticum. They are visible in the dissecting microscope as glistening white stripes. When the contralateral eye has been enucleated, they disappear within a few weeks (S. S. Easter, Jr., unpublished observations). They label anterogradely with horseradish peroxidase (HRP) applied to the contralateral optic nerve or retina (see below). In Figure 2, they have been visualized by briefly soaking a formalin-fixed brain in dilute OsO_4. Because the fascicles almost exclusively include myelinated fibers, which are heavily stained by the osmium ions, they appear as dark stripes. Notice in Figure 2A that they radiate from the rostral tectal pole like the rays of a curved fan that bend toward the tectal equator, the boundary between the dorsal and ventral halves. A symmetrically positioned array is found on the ventral hemitectum (Figure 2B), and both sets of fascicles can be visualized in a tectal whole-mount flattened after being hemisected along the equator (Figure 2C). Within either fan, the rays virtually never cross one another and therefore can be classified by their centroperipheral position. All originate near the rostral pole and curve to point normal to the equator. The rostral ones are short, run relatively close to the equator, and meet it rostrally. The more peripheral rays meet the equator more caudally. A fascicle loses axons along its course and disappears near the equator. The only substantial difference between the fascicles sketched in Figure 1 and those photographed in Figure 2 is their relation to the most peripheral fascicle. In Figure 1, the tectal fascicles look like branches from the peripheral trunk. In Figure 2, they look more like a set of independent trunks that are joined rostrally but diverge caudally.

Recall that the salient features of the description by Attardi and Sperry (1963) are, first, that an axon courses along the tectal rim and into a fascicle in the parallel layer (more commonly called the stratum opticum; Leghissa, 1955); and second, that the fascicle chosen is the one which leads to the axon's termination site. This requirement, that an axon terminate near its fascicle, is very difficult to reconcile with the facts of retinotectal growth. The difficulty stems from three observations. First, the retina grows appositionally by adding new cells around its edge in an annulus (Müller, 1952; Johns, 1977). Once a cell has made its terminal division, it stays in place and differentiates, and new cells are subsequently added peripherally. Second, the tectum also grows appositionally but along a contour topologically dissimilar to the retinal annulus. New cells are added in a C-shaped zone along the dorsal, ventral, and caudal rims, but not rostrally (Meyer, 1978; Raymond and Easter, 1983). Third, the projection of retina onto tectum is topographic; that is, ganglion cells from neighboring retinal regions project their terminals to neighboring tectal sites according to a standard set of rules that apply in both young and old animals. In particular, the peripheral retina maps to the peripheral tectum, and central retina to central tectum, as Figure 1 illustrates. These three observations—annular addition of new cells to

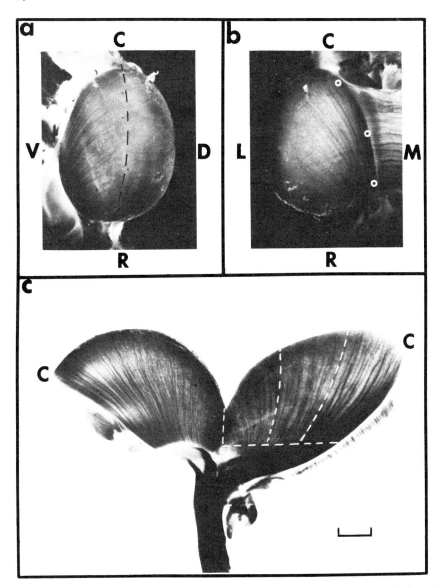

Figure 2. *Three photographs of the right tectal lobe of a goldfish after a brief soak in dilute OsO₄. a:* Ventrolateral aspect; the dashed line shows the equator. *b:* Dorsal aspect; the three white circles show the three sites of HRP application for Figure 4d. *c:* A flattened whole-mount, cut along the equator. The dashed white lines divide the dorsal tectal surface into the three zones of HRP application for Figure 4c. R, rostral; C, caudal; other abbreviations as in Figure 1. Calibration bar = 1 mm. (From Stuermer and Easter, 1984b.)

the retina, C-shaped addition of new cells to the tectum, and constancy of retinotectal topography—cannot easily be reconciled with the idea that retinal axons terminate near their tectal fascicle, as the following argument should make clear.

Consider a ganglion cell on the temporal margin of a young retina. By virtue of its position, it projects to the rostral edge of the tectum. But some years later, after many new annuli of retinal ganglion cells have been added more peripherally, it lies relatively closer to central retina and therefore projects to central tectum. Assume that it originally entered the tectum in a rostral fascicle and terminated nearby. Later in life, if it is to terminate near its fascicle, it must have shifted from a rostral to a more central fascicle. This possibility cannot be ruled out on logical grounds, but it is certainly unlikely, particularly since the retina grows continually, which would seem to require axons to withdraw frequently from one fascicle and to regenerate into a new one. An alternative would be to leave the axon in place and extend the terminal arbor to its appropriate retinotopic location.

This argument is derived from Gaze et al. (1972), who were the first to recognize that the topological dissimilarity of retinal and tectal growth (in *Xenopus*) probably required shifting terminals. I subscribe to that view, but I emphasize that the evidence in support of the hypothesis in *Xenopus* has never included a consideration of the pathways taken by the axons. In this respect, the model for shifting terminals in goldfish is different, for a detailed knowledge of the axonal pathways is essential, as will become evident below.

RETINOTECTAL FASCICLES REINTERPRETED

In my laboratory, we first began to develop an alternative interpretation of the tectal fascicles during the course of a study of the optic nerve and tract (Easter et al., 1981). We found that the newly added retinal fibers could be distinguished from the older ones by their lack of myelin, and we noticed that the most peripheral fascicles of stratum opticum were made up exclusively of nonmyelinated fibers. This made sense because the newest retinal ganglion cells—from peripheral retina—were known to terminate in peripheral tectum, so the most efficient way to arrive there would be to grow along the edge. If, in addition, one assumed that the rule followed by the current crop of new axons was the same rule followed by earlier generations, then the fascicular fan became interpretable in the context of growth. Specifically, if each new generation of axons grew in along the edge of tectum in two fascicles, axons from ventral hemiretina along the dorsal edge and those from dorsal hemiretina along the ventral edge, and if the fascicles were then enveloped by new tectal tissue added dorsally, caudally, and ventrally, the fascicular pattern of Figure 2 would result. Each bundle of fibers would mark the outer limit of the tectum at the time it grew in. The crescentic region between adjacent fascicles would be the portion of the tectum added between the times when the fascicles were formed. Later work (Raymond et al., 1983) showed that the rate of cellular proliferation in the tectal germinal zone depended on the arrival of new optic fibers.

In any case, our interpretation provided a clear and testable alternative to the scheme proposed by Attardi and Sperry. If individual fascicles could be labeled with HRP and the contralateral retinal ganglion cells filled retrogradely, then their distribution ought to permit the rejection of one or the other model. If the interpretation of Attardi and Sperry were correct, then the labeled ganglion cells ought to lie on a contour tectotopically related to the labeled fascicle (Figure 3). In contrast, our model predicts that the labeled cells should lie in a half-annulus

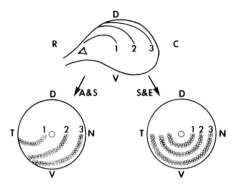

Figure 3. *The different predictions generated by the two competing models of fascicular organization.* When the fascicles on the dorsal tectum (top center) indicated by 1, 2, and 3 are labeled by HRP, the description by Attardi and Sperry predicts the patterns of labeled retinal ganglion cells shown on the lower left, and the description by Stuermer and Easter predicts the one on the lower right.

not necessarily tectotopically related to the labeled fascicle, and that the more rostral fascicles should originate from central retinal cells and the more peripheral ones from more peripheral retina (Figure 3). Notice also that according to our model, rostral fascicles should include axons from both nasal and temporal retina, whereas the arrangement of Attardi and Sperry would have fascicles that are restricted to rostral tectum made up of axons only from temporal retina. Both models predict that dorsal fascicles should originate from ventral retina.

We labeled individual fascicles, and the experimental results are summarized in Figure 4 (Stuermer and Easter, 1984b). They were in line with our model, and inconsistent with that of Attardi and Sperry. The labeled ganglion cells lay in a partial half-annulus in ventral retina (Figure 4B) and always extended into nasal retina, even when a very rostral fascicle was labeled. The position of the labeled fascicle also correlated with the position of the retinal half-annulus; the more rostral the fascicle (that is, the more rostral the fascicle's intersection with the tectal equator), the more central was the labeled contour of retinal cells (Figures 2C, 4C). We knew from an earlier study (Easter et al., 1981) that cells from all of ventral retina contribute to the dorsal brachium, and therefore we anticipated that we would see complete (180°) half-annuli of labeled cells. In fact, they were always much shorter than this, and the missing portion of the half-annulus always lay in temporal retina. This suggested that the axons from the most temporal cells had exited from the fascicle at locations rostral to the site at which the HRP was applied. Such an exit would be consistent with the retinotopic map because temporal retinal cells terminate in rostral tectum. We investigated this question further by labeling the outermost fascicle at one of the three sites indicated by the white circles in Figure 2C. The results, summarized in Figure 4D, show that the more rostral applications of HRP systematically produced longer retinal contours. On average, the rostral label produced contours that extended about 120° from the nasal boundary; the intermediate label, about 90°; and the caudal label, about 45°. We infer that the somata along the ventral retinal margin between 120 and 90° sent axons that exited from the fascicle between the rostral and intermediate application sites and presumably terminated

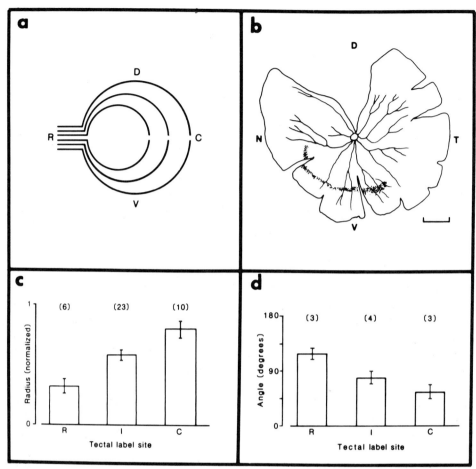

Figure 4. *Experimental assessment of fascicular organization.* a: Schematic representation of the caliperlike arrangement of the optic fascicles in the brachia and tectum, lateral view. Compare to Figures 2 and 3. b: Camera lucida drawing of a retinal whole-mount showing the partial annulus of labeled ganglion cells (dots). Compare to Figure 3, lower right. c: The radii of the partial annuli (normalized against the maximum) plotted versus the tectal label sites shown in Figure 2c. Means ± SEM. Numbers in brackets show the number of retinas. Mann-Whitney U test: $R < I$ ($p < 0.005$), $I < C$ ($p < 0.025$), $R < C$ ($p < 0.001$). d: The angular subtense of the partial annuli plotted versus the label sites shown by the white circles in Figure 2d. Conventions as in Figure 4c. Mann-Whitney U test: $C < R$ ($p < 0.025$), $I < R$ ($p < 0.05$), $C < I$ ($p < 0.1$). T, temporal; N, nasal; I, intermediate; other abbreviations as in Figure 2. Calibration bar = 1 mm. (From Stuermer and Easter, 1984b.)

nearby. Likewise, the axons from cells between 90 and 45° exited between the intermediate and caudal sites.

We offered the following interpretation of these results. All new axons follow the same two rules, which are (1) grow together along the edge of the tectum, and (2) exit to terminate nearby, temporal axons first, nasal axons last. These two steps constitute a sequential solution of a two-dimensional problem, one dimension per rule. Any tectal address can be represented as a pair of coordinates: r, the radial distance from the center of the tectum, and θ, the tectal sector. By growing along the edge, the axons have chosen the correct radial distance. By exiting sequentially, in temporonasal order, they have chosen the proper sector.

We are impressed by the efficiency of the axonal pathfinding. By following one group-specific rule (all new axons grow along the edge of the tectum), the individual axons are led onto a track that runs adjacent to a restricted tectal region which includes the appropriate termination sites. Once on this track, the axons must follow a more individualized rule (exit at the proper sector). We have no idea how individualized the process of sector selection might be. Neither do we know what signals the fibers use in deciding to exit, but the topographic arrangement in the brachia may play a role. Bunt (1982) has shown that temporal retinal fibers are at the center of the tract and nasal fibers at its dorsal and ventral peripheries at the point where the tract bifurcates into the brachia. If this stratification were maintained in the brachia, temporal axons would be closest to and nasal axons farthest from the tectal rim. If the order of exit were dictated by the proximity of the tectum, then temporal axons would precede nasal, as observed.

SHIFTING TERMINALS

If the retinotectal map is formed continually in peripheral tectum, what is happening simultaneously in central tectum? It will be recalled that the retina and tectum grow dissimilarly, yet the retinotectal topography is maintained constant. Therefore, whenever a new generation of axons arrives and terminates in peripheral tectum there are other terminations, already in place, that must be displaced. This is particularly evident in rostral tectum, where no new tectal cells are added but into which an ever-increasing number of temporal retinal terminals must be insinuated.

A corollary of our model is that axons which originally entered in rostral tectal fascicles must have been gradually displaced caudally from their initial termination near the fascicle of entry through the neuropil to their current site. This process is schematized in the sketch of Figure 5. There are two concrete predictions. The first is that retinal axons will be expected to have a tripartite morphology in the tectum. One part is in the fascicle through which they entered the tectum. Another is an extrafascicular segment directed caudally and passing through all those sites at which the axon once terminated. The third is the current terminal arbor.

We have visualized such axons by anterograde transport of HRP applied to the nerve or retina, and examples are shown in Figure 6. Cook et al. (1983) have described similar structures and interpreted them similarly. They provide solid evidence that some axons terminate far from their fascicle of entry. But how could the terminals have shifted? If each of the terminal branchlets initially produced were to extend caudally, the terminal arbor would be linked to the fascicular axon by many extrafascicular segments, one for each terminal branchlet, rather than by the single link shown in Figure 6. I believe that a more plausible mechanism would entail the independent loss (by either retraction or degeneration) of branchlets on one side and the outgrowth of new sprouts on the other. Figure 7 illustrates this hypothetical process. In each panel three fascicular axons are illustrated, one of which has exited the fascicle to terminate nearby. In Figure 7A, an early terminal arbor, the center of gravity of the four terminals (1, 2, 3,

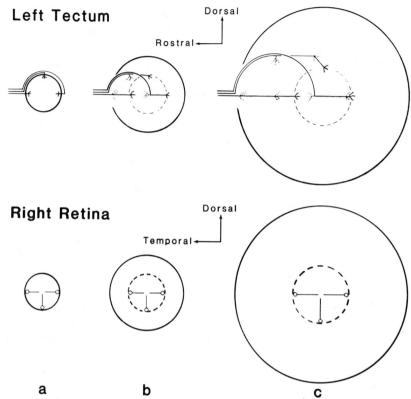

Figure 5. *An illustration of the events in shifting terminals.* a, b, and c show three successively older stages in the growth of the retina and tectum. In the lower series of diagrams, the outer boundary of the retina is shown by the dark circle; the dashed line gives the boundary of the earliest retina (the one in a); the three small circles are ganglion cell bodies; the three lines are their axons, converging on the optic-nerve head. In the upper series of diagrams, the contralateral tectum is seen from the perspective of Figure 4a. The terminals of the three ganglion cells project to retinotopic positions on the central dashed circle, but they enter through the same fascicle as originally. Successive positions in the tectum are shown by the solid forklike terminals. Previously occupied positions are shown by dotted terminals. The tectum is indicated as having stretched between b and c, consistent with Raymond and Easter (1983). (From Stuermer and Easter, 1984b.)

and 4), is marked by the x; the distance from the fascicle to the center of gravity is given by d_a. At a later time (Figure 7B) terminal 1 has disappeared, and terminal 5 has appeared. The new center of gravity is farther from the fascicle than before (d_b is greater than d_a). In Figure 7C, terminal 2 has disappeared, terminal 6 has appeared, and the new center of gravity has moved still farther. According to this idea, the individual terminal contacts have not shifted. They have either remained in place (3 and 4), appeared *de novo* (5 and 6), or disappeared (1 and 2). The center of gravity of the arbor has shifted.

The images of Figure 6 also enable us to estimate the overlap of terminal arbors on the tectal surface. A more complete description of these structures is given in Stuermer (1984), but let us take the examples of Figure 6 to be general and assume that the average terminal arbor extends over a circle about 150 μm in diameter, an area of 2.25×10^4 μm^2. The area of the total tectal surface is

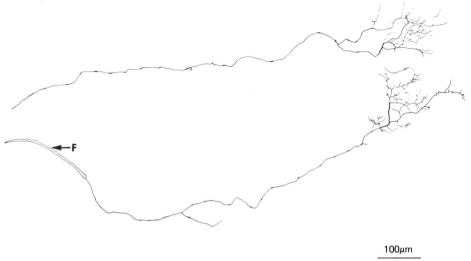

Figure 6. *A camera lucida drawing of two HRP-labeled extrafascicular segments of retinal axons in the tectum. Rostral is to the left, dorsal upward. F, fascicle of origin. (From Easter and Stuermer, 1984.)*

about 600×10^4 μm², and there are about 15×10^4 ganglion cells in these retinas. Therefore a given retinal axon—only 1/150,000 of all the retinal axons—covers about 1/300 of the tectal surface. If we assume that the terminal arbors ramify in one of five tectal layers (Sharma, 1972), and do so in equal numbers, then each layer must have 3×10^4 terminal arbors of aggregate area given by $3 \times 10^4 \times 2.25 \times 10^4 = 675 \times 10^6$ μm². This is roughly 100 times the tectal

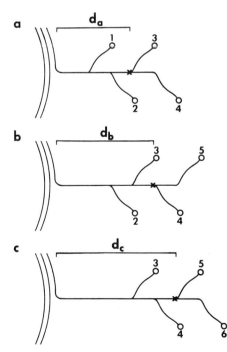

Figure 7. *Hypothetical shift of a terminal. a, b, and c illustrate my conception of how an arbor might shift its center of gravity (x) by losing terminals on one side (e.g., 1 and 2) while adding new ones on the other (e.g., 5 and 6). See text for details. d_a, d_b, and d_c are distances between the fascicle and the centers of gravity.*

surface; therefore there must be a lot of overlap, consistent with the impression from tectal whole-mounts (Stuermer and Easter, 1984b).

If these rather complicated axons have resulted from the processes that we have described it ought to be possible to predict, for any region of tectum, which retinal ganglion cells will have axons there. In principle any tectal site (that is, a thin column of tectum perpendicular to the surface) ought to have some fascicular axons in stratum opticum, and some extrafascicular axons and terminals in the synaptic layers. If the column is labeled by inserting HRP into it, we would expect to see: (1) a labeled patch of retinal ganglion cells in the tectotopic position (labeled through their terminals); (2) labeled cells in a portion of a half-annulus extending a variable distance away from the nasal boundary between dorsal and ventral hemiretinas (labeled through their fascicular axons); and (3) a row of labeled cells connecting the temporal end of the partial half-annulus to the terminally labeled patch (labeled through their extrafascicular axons). This last group comprises the cells that once terminated in the site that received the HRP but now terminate elsewhere. Figure 8 shows such a pattern obtained after inserting HRP into rostromedial tectum. The sickle-shaped pattern of labeled cells fulfills the prediction in all respects and is inconsistent with the other model. We have probed the predictive power further by inserting HRP at a variety of other tectal locations, and the retinal patterns of label varied as predicted (Easter and Stuermer, 1984).

To summarize, at any given time, the apparently static projection is actually changing; as new axons are added around the edge, their new arbors are insinuated nearby, and some of the preexisting arbors are forced to move to new regions.

How is it that the new fibers, particularly those that terminate in rostral tectum, can push out preexisting ones? And how do those erstwhile occupants of rostral tectum manage, in their turn, to displace others? Before speculating, it is useful to point out that there is good experimental evidence for displacement of some retinotectal terminals by others. Gaze and Sharma (1970) first showed that when the caudal half of the tectum was removed surgically, the entire retina succeeded in compressing its projection into the remaining half-tectum. Given enough time the projection became reasonably homogeneous, in the sense that the rostrocaudal tectal magnification factor (the number of micrometers along the rostrocaudal axis of the tectum per degree in the visual field) was roughly the same everywhere on the tectal remnant. Therefore, those retinal axons that at the time the tectum was halved terminated in the middle of the entire tectum (the caudal edge of the half-tectum) now projected to the middle of the half-tectum. They had been displaced rostrally, as had the other axons that had previously terminated even more rostrally. Notice that the array of terminals changed although the optic nerve was never sectioned. The terminals in rostral tectum had not been disturbed by the surgery, but they succumbed to the competitive pressures of those axons that had terminated caudally and all squeezed, roughly evenly spaced, into a reduced field. Behavioral work (Scott, 1977) showed that the compressed terminals formed functional synapses as well.

A second experimental demonstration of the displacement of terminals has come from hyperinnervation experiments in which both eyes were forced to innervate a single tectal lobe. This occurs after one tectal lobe is ablated; the optic fibers that had innervated it cross the midline and innervate the other.

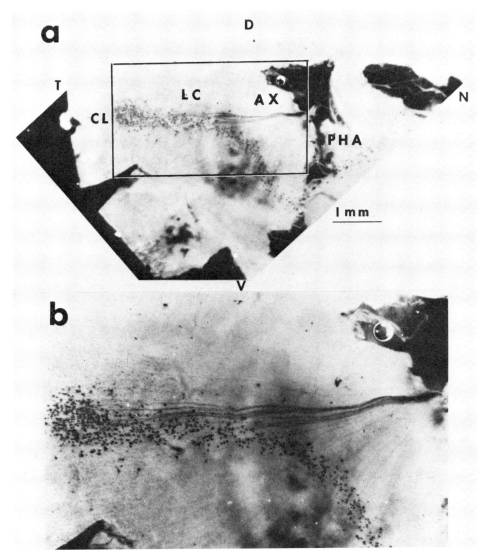

Figure 8. *Photomicrographs of retrogradely labeled right retinal whole-mount.* a: The partial half-annulus (PHA) is centered on the optic-nerve head and linked through the linear connector (LC) to the cluster (CL) of cells labeled through their terminals. Labeled axons (AX) radiate from the optic-nerve head. b: A detail of the field enclosed by the rectangle in a. (From Easter and Stuermer, 1984.)

The native innervation on that lobe is untouched by the surgeon, but within a month or so after the foreign nerve fibers have arrived, patches of the tectum begin to lose terminals from the native eye, to be replaced by foreign terminals (Levine and Jacobson, 1975; Springer and Cohen, 1981). Similar "ocular-dominance bands" have been heavily studied in anurans (Constantine-Paton and Law, 1978).

Given that intact arbors may be displaced, what sorts of rules could yield this result? Some years ago, we proposed a competitive model that could account for the experimentally altered projections (Schmidt et al., 1978), and it can account for the growth-related alterations, too. It contains three assumptions: (1) All axons tend to enlarge their arbors; (2) this enlargement is opposed by other arbors, competitively; (3) axons with larger arbors are poorer competitors. None of these assumptions has been evaluated independently, but there is indirect evidence to support all. The first seems reasonable in view of the expansion of the half-retinal projection. The second is supported by the evidence, cited in the preceding two paragraphs, of displacement of some terminals by others. The third assumption has been drawn from the neuromuscular literature, where it is used to explain the outcome of some experiments in which two nerves compete for synaptic space in the same muscle. When one of the two already innervates an additional muscle, it generally loses out to the competitor with no other terminal field.

When these three assumptions are applied to the conditions of normal growth, the following events would be expected to occur. The incoming axons exit from the fascicle and enter the neuropil. As they have no arbors, they are the best competitors and displace other terminals nearby. Insofar as they remain smaller than their competitors, they will continue to enlarge. Meanwhile, the axons that have lost the rostral portions of their arbors to the interlopers are made better competitors than their caudal neighbors, and their arbors enlarge caudally at the expense of these neighbors. They, in their turn, become better competitors against *their* caudal neighbors, displace more caudal terminals, and so on. Therefore the spacing of retinal terminals is in a continual state of flux, and the center of gravity of an individual arbor changes as it shrinks on one side in response to the more effective competitors there, and enlarges on the other at the expense of the less effective competitors in that position. Such local competitive interactions should maintain the orderly retinotectal map, given the initial conditions: (1) that the array of retinal arbors is already retinotopic; and (2) that the new fibers enter at retinotopically correct locations. Maintenance of this order would require only that the new arbors keep their positions relative to those that were there when they arrived. Although the example above was from temporal retina and rostral tectum, a similar dynamic would be predicted for everywhere on the tectal surface; only the directions of the displacements would vary. Even in caudal tectum, where the most new cells are added, the same process would be expected to yield a stable result. The most caudal arbors already in tectum would enlarge in the direction of the new, uninnervated cells, and they would compete with the newly arrived axons for this territory. In this model, no arbor is isolated. The size and location of each is influenced directly by its neighbors and indirectly by its neighbors' neighbors.

One problem with this scheme is that the day-to-day formation of the retinotectal map depends on the prior existence of a map. When the nerve regenerates after being cut or crushed, the terminals reappear at their old locations without the benefit of either an existing array of terminals to join or the orderly path to the correct tectal annulus. One must conclude from this that there is information on the tectum that the regenerating retinal axons can use in the deployment of their arbors, and it would be foolish to suggest that the naturally ingrowing ones could not do the same. It seems probable that the regenerating axons are sensing the same signs of positional information as the axons that normally grow in around the edge. The advantage of growing in at the edge is that the search is minimized by leading the axon to the tectal periphery where it is sure to pass close to its correct termination site. In contrast, the regenerating axons wander widely before reaching home.

These speculations aside, the data shown in Figure 4 indicate strongly that the axons arrive at the tectum in the company of others of similar age. Where did this age-relatedness originate? In order to answer that, I now review some work at the level of the optic nerve and the retina.

OPTIC NERVE AND TRACT

Three independent lines of evidence indicate that the retinal axons are ordered by age in the optic nerve. First, electron microscopy of normal nerves always reveals a few clustered fascicles of nonmyelinated axons in a sea of myelinated ones. The former arise from the peripheral retinal ganglion cells; therefore, even in the nerve, the new axons are clustered (Easter et al., 1981). Second, when a pin coated with HRP is inserted into the nerve it labels a bundle of fibers, and the retrogradely labeled cell bodies in the retina lie in an annulus centered on the optic disk (Rusoff and Easter, 1980). Figure 9 shows one such annulus. Note that it is a middle-aged generation of ganglion cells, well separated from the retinal margin. This result indicates that the fascicular clustering by age, established early in axonogenesis, is maintained throughout life. Third, this same conclusion follows from an experiment in which the nerve was partially sectioned and the eye was injected with radioactive proline. Because the retinotectal projection is topographic, unlabeled regions in the tectal neuropil ought to correspond retinotopically to the locations of the ganglion cells whose axons were severed. Figure 10 shows the results of such an experiment, in which both eyes were injected. The neuropil of the control side (C) is labeled throughout. On the experimental side (E), there is a clear annular void (curved arrows), thus confirming that axons of the same generation run together in the nerve.

A minor controversy has developed over whether the nerve is organized according to age or retinal position. Scholes (1979) was the first to show age-related order in the ribbon-shaped optic nerves of cichlid fish. Dawnay (1979) called this a "chronotopic" organization. Later, Bunt (1982) made small retinal lesions in the goldfish and traced the degenerating axonal debris in the nerve. The retinal lesions axotomized the ganglion cells in a wedge-shaped region that extended from the lesion out to the retinal margin. This produced a contiguous strip of debris in the cross section of the nerve, which Bunt interpreted to mean that the retinal positions of ganglion cells were conserved topographically by their axons in the cross section of the nerve. This was consistent with both Scholes (1979) and Easter et al. (1981), who concurred in showing that the radial direction (r) is laid out in one direction and the sector addresses (θ) in the other, so that most axons ran in the immediate vicinity of axons from neighboring ganglion cells. Nonetheless, Bunt (1982) asserted that retinotopy was inconsistent with chronotopy and claimed that his results argued against the latter. There seems to be little dispute now about the facts, as all workers agree, but the issue of whether r and θ are equally important is an interesting one. Clearly, in a geometrical context both must be specified to establish a cell's position, and

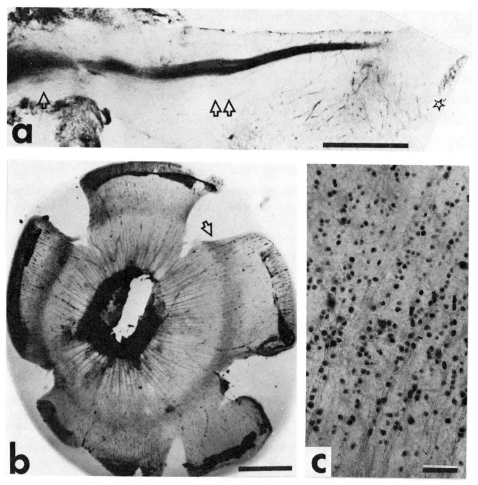

Figure 9. *Results of labeling a bundle of optic fibers with HRP.* a: Longitudinal section through the optic nerve and tract. HRP was applied at the left, near the retina, and labeled a group of axons in the nerve (single arrow) and tract (double arrow), which extended into the tectum (star). b: A retinal whole-mount showing the annulus (arrow) of labeled ganglion cells. c: Detail of labeled annulus showing individual cells. Calibration bars = 500 μm, 1 mm, and 25 μm in a, b, and c, respectively. (From Rusoff and Easter, 1980, by permission of the AAAS.)

both are equally important. But what about the biological context? Here, several observations suggest that r is treated as a more important variable than θ. First, when HRP is applied to the nerve it never labels wedges of retinal ganglion cells, only annuli or partial annuli. This cannot be interpreted unambiguously, but if both dimensions were equally important, wedge labeling would seem to be as likely to occur as annulus labeling. The fact that wedge labeling has not been reported suggests that the label is restricted by diffusion barriers around compartments of age-related axons, in such a way that whenever the label enters one such compartment it becomes available to all its occupants but inaccessible to others. Second, partial section of the nerve followed by intraocular tritiated proline always produced unlabeled regions in the tectum with a strong annular component (Figure 10). The cuts were made with scissors; perhaps the scissor points could not penetrate the fascicles but could only push them between or outside the blades to be cut or spared, respectively. If this interpretation is correct, the axons were evidently bundled by similarity of age rather than sector. Third, the age-related order of the fascicles in the tectum suggests that the sector address is of secondary importance. The axons do not run in wedge-specific fascicles and exit sequentially according to age; the reverse is true.

RETINA

Most recently, we have attempted to understand how the age-related order in the nerve could arise in the retina (Easter et al., 1984). The retinal axons seem to be arranged in a way that would exclude age-relatedness, as most axons course rather directly from soma toward the optic disk, like the rays on an umbrella. The most peripheral ganglion cells must, therefore, send their axons in association with the axons of more central somata. Thus retinal fascicles contain axons of all ages that originate from the same sector: How could the age-related order arise from this arrangement?

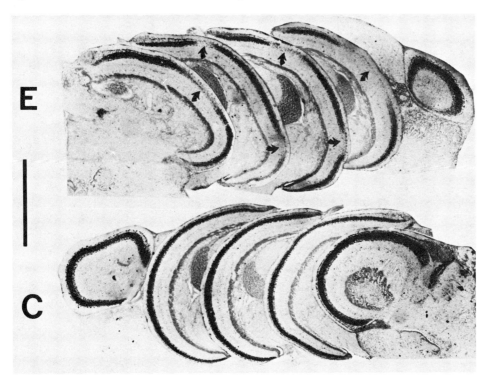

Figure 10. *Radioautograms showing five sections from the tectal lobes of a goldfish.* The right optic nerve had been partially cut five days before sacrifice, and both eyes were injected with tritiated proline one day before sacrifice. The experimental side (E) has unlabeled regions in the optic afferent layer (curved arrows) in the form of an annulus. The control side (C) shows no such voids. Dorsal, upward; medial, left in E, right in C; the more rostral sections are to the left in E, to the right in C. Calibration bar = 1 mm. (From Easter et al., 1981.)

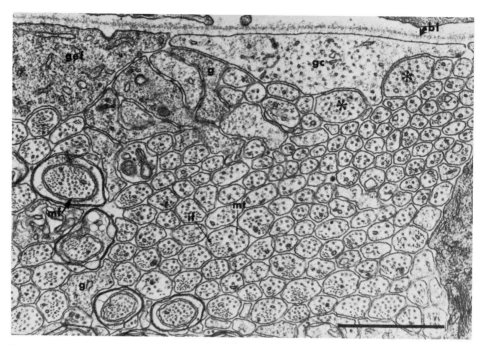

Figure 11. *Electron micrograph of a transverse section through a growth cone (gc) in the retina.* Immediately below it are two large nonmyelinated fibers marked by asterisks. mf, myelinated fibers; mt, microtubules; if, intermediate filaments; gef, glial end foot; g, glial processes; bl, basal lamina. Calibration bar = 10 μm. (From Easter et al., 1984.)

It turns out that the axons are stratified within each fascicle according to age. The most peripheral somata send growth cones into the most superficial lamina of the optic fiber layer, just beneath the basal lamina of the inner limiting membrane. Figure 11 shows a radial section of one such process, cut perpendicular to its long axis. It is in direct contact with the basal lamina superficially, with other optic axons deeply, and very slightly with the glial end feet laterally. A reconstruction of one growth cone is shown in Figure 12. It has a narrow filopodium on its leading edge and a rather thick neurite connecting it to its soma near the retinal margin. Both the leading and trailing edges lie deep to glial end feet, separated from the basal lamina, but most of the flattened intervening portion is in direct contact with the basal lamina. This static image suggests that the broad intermediate zone has pushed the glial end feet aside to adhere to the basal lamina directly, perhaps because the two are mutually adhesive. Growth cones have been seen in direct contact with the basal lamina elsewhere in the developing nervous system, too (e.g., Nordlander and Singer, 1982), so it is tempting to suppose that the basal lamina might be a favorable surface for neuritic elongation. But in other places, including the retina (Krayanek and Goldberg, 1981) and the optic nerve (Rakic and Riley, 1983), growth cones have been seen separated by end feet from the basal lamina, so if it is important for axonogenesis in the goldfish retina, it probably is not so generally.

If the basal lamina does not guide axonal outgrowth, what does? The glial end feet seem unlikely candidates because the growth cone appears to push them aside. The remaining possibility is the array of nonmyelinated fibers, which, as Figure 13 illustrates, are relatively rare. Despite this, the growth cones always coursed on them, suggesting an active selection. The nonmyelinated fibers originate from ganglion cells near the margin and therefore probably include the cohort of axons that just preceded the growth cones. If they provided the most favorable substrate for axonal elongation an age-related order would result in that the most recently produced axons in any retinal sector would course together, all in the most superficial lamina of the optic-fiber layer. We have traced these axons to the optic disk and have confirmed that they remain superficial and dive into the center of the nerve adjacent to the central ophthalmic artery, as summarized in Figure 14. The consequence of this very orderly and stereotyped outgrowth is that the most recently produced axons from all around the retina course toward the very center of the disk, where they occupy a common annulus around the artery, segregated from older axons more peripherally located in the nerve. As the nerve leaves the eye, the artery leaves the nerve, and the central annulus of new axons is opened and becomes a small cluster, as illustrated in Figure 15.

In this way successive generations of axons from the retinal periphery cluster from the beginning with other axons of the same generation and remain together in the retina, the optic nerve and tract, and the stratum opticum of the tectum. Once in the tectum, they systematically exit and arborize and begin the trek from their initial site toward a succession of later ones, each appropriate for a short time but gradually made inappropriate and then changed as a result of the steady competition of new arrivals.

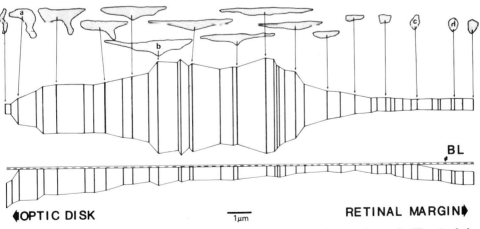

Figure 12. *A reconstruction of a retinal growth cone from 45 quasiserial electron micrographs.* The stippled figures above show the shapes of the process in cross section. The middle figure shows the process as viewed from the vitreous in which each vertical line is scaled to show the maximum width of the process. The lower figure shows this same growth cone, viewed from the side in relation to the basal lamina (BL). Each vertical line is scaled to show the maximum depth of the process. (From Easter et al., 1984.)

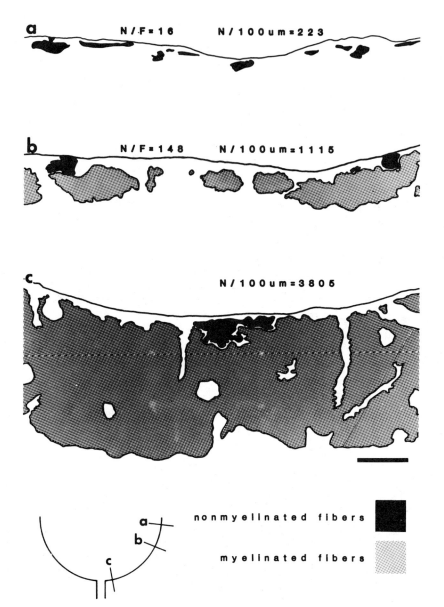

Figure 13. *Tracings of electron-microscope photomosaics.* Radial sections through the retinal fiber layer at three levels (see lower left) were traced and both nonmyelinated and myelinated fibers were outlined and coded (see lower right). N/F, mean number of fibers per fascicle; N/100 μm, number of fibers per 100 μm of inner limiting membrane. Calibration bar = 10 μm. (From Easter et al., 1984.)

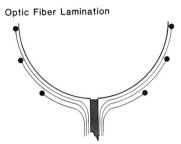

Figure 14. *Sketch of the lamination in the retinal fiber layer and the adjacent optic nerve.* The most peripheral cells project their axons most superficially on the retina and most centrally in the nerve. Successively more central somata send axons more deeply in the retina and more peripherally in the nerve. The stippled region is the ophthalmic artery (see Figure 15). (From Easter et al., 1984.)

CONCLUSION

What has any of this to do with the molecular bases of neural development? On the face of it, nothing, but the description suggests that three regions might be profitably examined to learn more about the molecular bases of selective innervation. They are (1) the superficial region of the retina and nerve, where new fibers systematically grow; (2) the vicinity of the bifurcation of the tract, where fibers must choose one brachium or the other, and do so quite selectively; and (3) the rim of the tectum, where fibers exit systematically to form the incremental retinotectal projection. At these three locations, the new fibers are demonstrably selective. In the tectum proper, in contrast, we have very little insight as to how the terminals are rearranged, and for that reason molecular questions about selective innervation within the tectum cannot be framed with any clarity.

From a different perspective, one could ask what attributes are striking or exaggerated or unique in the goldfish retinotectal system and therefore might best be studied there. I find the three most impressive features to be: (1) the continuous proliferation of neurons; (2) the high degree of spatial order in the pathway; and (3) the continuous relocation of terminal arbors, accompanied by gradual elongation of the axons. The first of these is very striking indeed, particularly to those individuals who work on mammals, but my intuition tells me that control of the cell cycle is better studied elsewhere. When it has been clarified in, say, yeast cells, then it might be ripe for investigation in the nervous system. The spatially ordered pathway has attracted a lot of attention, but spatial order per se does not seem to be essential to the formation of a spatially ordered set of terminals. My intuition tells me that the spatial order is pretty well accounted for by the spatiotemporal order of the neurogenesis and subsequent association of axons of similar ages. I think the spatial order of the pathway may be a useful attribute for experimental purposes, but it probably is not sufficiently general to warrant further investigation. The continuous relocation of terminal arbors strikes me as the most interesting phenomenon of the three, and the one with the greatest leverage for neurobiology in general. Most theories of development, learning, and memory involve alterations in connectivity, and the goldfish tectum

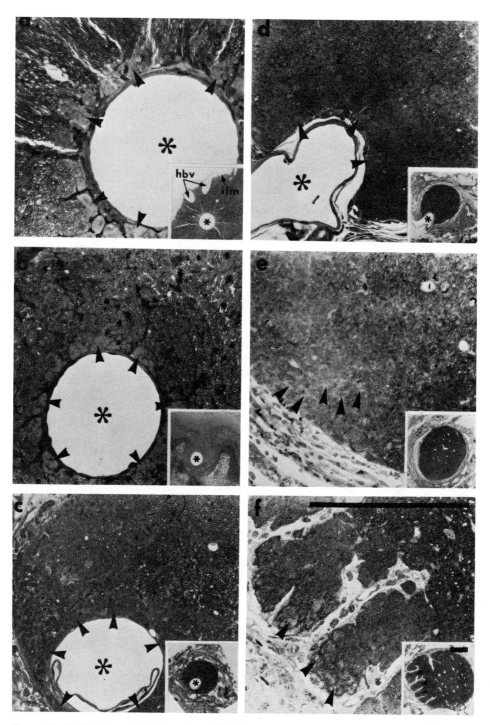

Figure 15. *Light micrographs of successive sections through the optic-nerve head and optic nerve.* The larger micrograph in *a–f* is a detail of its inset. Ventral is toward 8 o'clock. The ophthalmic artery is indicated by an asterisk; arrowheads show the nonmyelinated fibers. Panel a is the most distal section, through the retinal fiber layer, and the remainder of the panels are successively deeper, ending outside the eye in panel f. hbv, hyaloid blood vessel; ilm, inner limiting membrane. Calibration bars = 100 μm. (From Easter et al., 1984.)

seems to be the most alterable tissue yet described. Therefore it should prove useful for studies of plasticity; a good place to start would be to characterize new synapses ultrastructurally or biochemically or both. But such characterizations are the stuff of future volumes.

ACKNOWLEDGMENTS

I have been privileged to have had many excellent collaborators and students, and much of the work described in this chapter was done by them. I thank Drs. P. A. Raymond, A. C. Rusoff, J. T. Schmidt, and C. A. O. Stuermer, and Messrs. S. S. Scherer, P. E. Kish, and B. Bratton. The research was supported by grant EY-00168 from the National Eye Institute. The manuscript was written while in residence, on sabbatical leave, at The Salk Institute, where Dr. W. M. Cowan and his associates were most gracious hosts, for which I thank them.

REFERENCES

Attardi, D. G., and R. W. Sperry (1963) Preferential selection of central pathways by regenerating optic fibers. *Exp. Neurol.* **7**:46–64.

Bunt, S. M. (1982) Retinotopic and temporal organization of the optic nerve and tracts in the adult goldfish. *J. Comp. Neurol.* **206**:209–226.

Constantine-Paton, M., and M. I. Law (1978) Eye-specific termination bands in tecta of three-eyed frogs. *Science* **202**:639–641.

Cook, J. E., E. C. C. Rankin, and H. P. Stevens (1983) A pattern of optic axons in the normal goldfish tectum consistent with the caudal migration of optic terminals during development. *Exp. Brain Res.* **52**:147–151.

Dawnay, N. A. H. (1979) "Chronotopic" organization of goldfish optic pathway. *J. Physiol. (Lond.)* **296**:13–14P.

Easter, S. S., Jr., and C. A. O. Stuermer (1984) An evaluation of the hypothesis of shifting terminals in goldfish optic tectum. *J. Neurosci.* **4**:1052–1063.

Easter, S. S., Jr., A. C. Rusoff, and P. E. Kish (1981) The growth and organization of the optic nerve and tract in juvenile and adult goldfish. *J. Neurosci.* **1**:793–811.

Easter, S. S., Jr., B. Bratton, and S. S. Scherer (1984) Growth-related order of the retinal fiber layer in goldfish. *J. Neurosci.* **4**:2173–2190.

Fujisawa, H., N. Tani, K. Watanabe, and Y. Ibata (1982) Branching of regenerating retinal axons and preferential selection of appropriate branches for specific neuronal connection in the newt. *Dev. Biol.* **90**:43–57.

Gaze, R. M. (1958) The representation of the retina on the optic lobe of the frog. *Q. J. Exp. Physiol.* **43**:209–214.

Gaze, R. M. (1959) Regeneration of the optic nerve in *Xenopus laevis*. *Q. J. Exp. Physiol.* **44**:290–308.

Gaze, R. M. (1970) *The Formation of Nerve Connections*, Academic, London.

Gaze, R. M., and S. C. Sharma (1970) Axial differences in the reinnervation of the optic tectum by regenerating optic nerve fibres. *Exp. Brain Res.* **10**:171–181.

Gaze, R. M., S.-H. Chung, and M. J. Keating (1972) The development of the retinotectal projection in *Xenopus*. *Nature* **236**:133–135.

Johns, P. R. (1977) Growth of the adult goldfish eye. III. Source of the new retinal cells. *J. Comp. Neurol.* **176**:343–357.

Krayanek, S., and S. Goldberg (1981) Oriented extracellular channels and axonal guidance in the embryonic chick retina. *Dev. Biol.* **84**:41–50.

Leghissa, S. (1955) La struttura microscopica e la citoarchitettonica del tetto ottico dei pesci teleostei. *Z. Anat. Entwicklungsgesch.* **118**:427–463.

Levine, R. L., and M. Jacobson (1975) Discontinuous mapping of retina onto tectum innervated by both eyes. *Brain Res.* **98**:172–176.

Maturana, H. R., J. Y. Lettvin, W. S. McCulloch, and W. H. Pitts (1959) Evidence that cut optic nerve fibers in a frog regenerate to their proper places in the tectum. *Science* **130**:1709–1710.

Meyer, R. L. (1978) Evidence from thymidine labeling for continuing growth of retina and tectum in juvenile goldfish. *Exp. Neurol.* **59**:99–111.

Meyer, R. L. (1980) Mapping the normal and regenerating retinotectal projection of goldfish with autoradiographic methods. *J. Comp. Neurol.* **189**:273–289.

Müller, H. (1952) Bau and Wachstum der Netzhaut des Guppy (*Lebistes reticulatus*). *Zool. Jahrb. Abt. Allg. Zool. Physiol. Tiere* **63**:275–324.

Nordlander, R. H., and M. Singer (1982) Morphology and position of growth cones in the developing *Xenopus* spinal cord. *Dev. Brain Res.* **4**:181–193.

Rakic, P., and K. P. Riley (1983) Overproduction and elimination of retinal axons in the fetal rhesus monkey. *Science* **219**:1441–1444.

Raymond, P. A., and S. S. Easter, Jr. (1983) Postembryonic growth of the optic tectum in goldfish. I. Location of germinal cells and numbers of neurons produced. *J. Neurosci.* **3**:1077–1091.

Raymond, P. A., S. S. Easter, Jr., J. A. Burnham, and M. K. Powers (1983) Postembryonic growth of the optic tectum in goldfish. II. Modulation of cell proliferation by retinal fiber input. *J. Neurosci.* **3**:1092–1099.

Rusoff, A. C., and S. S. Easter, Jr. (1980) Order in the optic nerve of goldfish. *Science* **208**:311–312.

Schmidt, J. T., C. M. Cicerone, and S. S. Easter, Jr. (1978) Expansion of the half retinal projection to the tectum in goldfish: An electrophysiological and anatomical study. *J. Comp. Neurol.* **177**:257–278.

Schmidt, J. T., and D. L. Edwards (1983) Activity sharpens the map during the regeneration of the retinotectal projection in goldfish. *Brain Res.* **269**:29–39.

Scholes, J. (1979) Nerve fibre topography in the retinal projection to the tectum. *Nature* **278**:620–624.

Scott, M. Y. (1977) Behavioral tests of compression following partial tectal ablation in goldfish. *Exp. Neurol.* **54**:579–590.

Sharma, S.C. (1972) The retinal projection in the goldfish: An experimental study. *Brain Res.* **39**:213–223.

Springer, A. D., and S. M. Cohen (1981) Optic fiber segregation in goldfish with two eyes innervating one tectal lobe. *Brain Res.* **225**:23–36.

Stuermer, C. A. O. (1984) Rules for retinotectal terminal arborizations in the goldfish optic tectum. A whole-mount study. *J. Comp. Neurol.* **229**:214–232.

Stuermer, C. A. O., and S. S. Easter, Jr. (1984a) A comparison of the normal and regenerated retinotectal pathways of goldfish. *J. Comp. Neurol.* **223**:57–76.

Stuermer, C. A. O., and S. S. Easter, Jr. (1984b) Rules of order in the retinotectal fascicles of goldfish. *J. Neurosci.* **4**:1045–1051.

Udin, S. B. (1978) Permanent disorganization of the regenerating optic tract in the frog. *Exp. Neurol.* **58**:455–470.

Chapter 19

Factors Involved in Retinotopic Map Formation: Complementary Roles for Membrane Recognition and Activity-Dependent Synaptic Stabilization

JOHN T. SCHMIDT

ABSTRACT

Sperry's ideas of chemospecific target adhesion have been modified to include other interactions such as selective fiber–fiber adhesion, selective fiber–pathway interactions, and interfiber competition at the target. These seem sufficient to orient and organize a crude map, but they are not sufficient to produce the high degree of order in the mature projection. The final sharpening process appears to depend on the correlated activity of optic fibers from neighboring ganglion cells. In areas where their arbors overlap, the resultant summation of postsynaptic responses seems to cause a selective stabilization of those synapses. These two classes of mechanisms—the differential affinity of membranes, and the activity-dependent stabilization of synapses—complement each other in producing the final map.

Topographic maps are an extremely important part of neural organization, since much of the brain contains maps of the body surfaces and of auditory and visual space. It is not surprising, then, that the mechanisms of formation of such maps have received great attention. In particular, many studies have focused on the development or regeneration of the retinotectal projection of fish and amphibians as a simple system for analysis. Several features contribute to the simplicity of the retinotectal projection. First, the retina and tectum are essentially hemispheric sheets of cells directly connected by a bundle of fibers called the optic nerve. Particularly in goldfish, there is a roughly uniform density of ganglion cells over the retina (Johns and Easter, 1977) and thus no overrepresentation of a fovea or area centralis in the retinotopic map on the tectum. Second, most of these animals have a field of view of about 180°, which can readily be approximated for mapping purposes by an external hemisphere centered on the eye. The lens images each point on the hemisphere to a corresponding point on the retinal surface, and electrophysiological recording can demonstrate the mapping of the hemisphere's surface onto the tectal surface. Third, the ease of access to both retina and tectum completes the set of features that have made the system so convenient for both electrophysiological and anatomical studies of neuronal connectivity.

Perhaps the seeming simplicity of the projection led Sperry (1963) to propose a single mechanism of map formation. He postulated that each retinal and tectal neuron derives a unique set of membrane markers from two orthogonal gradients during development. The matching of these markers would then determine the synaptic interactions. For many years, experiments have tested this proposal with mixed results. Nevertheless, many neurobiologists have persisted in interpreting the projection formed after experimental manipulations almost exclusively in terms of membrane marker interactions. Recently, however, it has become clear that several layers of mechanisms may interact to produce the final projection. Although position-dependent differences between cells still seem to be a factor at the most basic level, the fiber-to-target matching model has had to be modified to allow for several additional mechanisms: (1) a competition between fibers for tectal space (probably synaptic sites); (2) selective fiber–fiber adhesion between neighbors from the retina; and (3) selective fiber–pathway interactions. These modifications to chemoaffinity are considered in the first section of this chapter. In addition, a mechanism not at all dependent upon surface interactions, that of activity-dependent stabilization of synapses, now appears to play a great role in sharpening the final projection. This evidence is discussed in the second section. Other examples of activity-dependent stabilization are considered in the third section, to lend some perspective to the activity-dependent sharpening of the map.

A BRIEF SUMMARY OF CRUCIAL EXPERIMENTS ON CHEMOAFFINITY

Sperry's Experiments

In the early 1940s, Sperry used rotation of the newt eye combined with regeneration of the optic nerve to establish that the orientation of the regenerated projection was fixed. His newts consistently made reversed behavioral responses upon regeneration; the visual experience of the newt with the upside-down eye did not compensate to reorient the map. Later, Attardi and Sperry's (1963) anatomical study of regenerating goldfish optic fibers appeared to suggest a very selective mode of regrowth and prompted Sperry to claim that each cell must have unique cytochemical markers that must be matched in order to allow the formation of connections. Much of the succeeding 20 years was spent testing just how selective the regenerating fibers are under various conditions. Some of these experimental tests are outlined below.

Evidence for Competitive Interactions between Fibers

Electrophysiological studies have confirmed that selective reinnervation occurs after partial retinal removal (Schmidt et al., 1978). However, this reinnervation is followed some months later by an expansion of the projection over the tectal

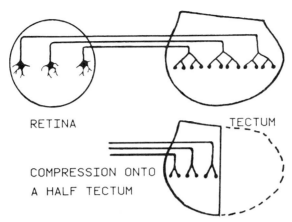

Figure 1. *Schematic diagram showing the compression of the retinal projection onto a half-tectum. Three retinal ganglion cells are shown in the retina with their arbors in the tectum both before and after the compression. The conservation of synaptic density in the remaining rostral half combined with the full set of fibers means that each fiber has roughly half its usual number of synaptic boutons.*

surface. After half-tectal removal, regeneration produces a similar sequence. Initially there is a selective innervation of the remaining half-tectum by the corresponding half-retina, but this is followed by a compression of the tectal map to accommodate the whole projection (Gaze and Sharma, 1970; see Figure 1). The expanded or compressed maps in the two cases mentioned above generally maintain good retinotopic order, even though each fiber then terminates over a different tectal site. The formation of such projections is inconsistent with the matching of unique position-dependent retinal and tectal markers, but not with the use of general gradients to orient the map. However, even if the optic fibers have only a relative preference for one spot on the gradient, one must postulate competitive interactions to force the spreading of the half-retinal projection and the compression of the full projection onto a remaining half-tectum. In the half-retinal case, the fibers move away from the preferred spot merely to gain more territory. Schmidt et al. (1978) postulated that each optic fiber has an intrinsic tendency to expand its arbor, that this tendency is opposed by that of nearby arbors, and that the tendency becomes weaker as the arbor grows larger. Recent evidence indicates that this competition may be for a fixed number of postsynaptic sites. In an electron-microscope study of synaptic density, Murray et al. (1982) concluded that each fiber in the compressed projection on a half-tectum must occupy approximately half the usual number of synaptic sites (Figure 1).

Selective Fiber–Fiber Adhesion

The experiments just described also prompted the suggestion that optic fibers might interact with each other to maintain retinotopic order independent of any markers on the tectal surface (Willshaw and von der Malsburg, 1976). This idea was driven home forcefully when Meyer (1979) showed that deflection of a few fiber bundles across the midline sometimes resulted in a partial map of opposite polarity within the frame of a normally oriented map. The polarity was opposite that to be expected from any tectal polarity cues, so the orderly but reversed partial map could only have formed by fiber–fiber interactions. This result and similar ones (Sharma, 1975; Horder and Martin, 1983) implied more than mere compression or expansion along the same polar axis of the tectum. They suggest that selective interfiber affinity based on the closeness of the two cells in the retina may be another mechanism in organizing a retinotopic map. This type of interaction has recently been included in models of the retinotectal projection (see Fraser, this volume).

Of course, fiber-to-fiber interactions need not be confined to the tectum but may occur along the pathway to the tectum as well. The organization of the optic nerve has now been studied extensively, and a great deal of selective fasciculation of fibers from the same area of the retina has been found (see Easter, this volume). However, it is not entirely clear that order along the pathway is always necessary for the formation of an orderly map. After regeneration of the optic nerve in goldfish, order in the pathway is poor (Stuermer and Easter, 1982) and yet the retinotopic map is restored (Schmidt, 1978). Much of the final order in the map derives from the activity-dependent sharpening.

Changeable Tectal Markers

But how does one account for the selective regeneration that occurs reproducibly after partial removal of the retina? Several experiments suggest that it is due to the influences or remnants of the previous projection. For example, after the half-retina has expanded its projection over the tectum, a second regeneration is no longer selective for the corresponding half of the tectum but instead immediately reinstates the expanded projection (Schmidt, 1978). Likewise, regeneration after compression onto a half-tectum immediately restores the compressed projection (Schmidt, 1983).

Subsequent experiments have shown that it is the positional markers on the tectum, and not those on the retina, that are changed during reorganizations such as the expansion of the half-retinal projection. This was demonstrated by allowing the fibers from a whole eye to innervate a tectum that contained the expanded half-retinal projection. The corresponding half of the intact eye also formed an expanded half-retinal projection, and the other half was not represented. Thus the positional markers on the tectum changed during the expansion. The positional markers on the half-retina, however, did not change during the expansion, as it still selectively innervated the appropriate half of a normal tectum (for further discussion, see Schmidt, 1982). Similar findings have been reported in regenerating projections from surgically created compound eyes (Gaze and Straznicky, 1980). The main conclusion to be drawn is that tectal positional markers differ from retinal markers in being changeable. Evidently, tectal positional markers are induced by the retinal fibers. This does not mean that in the absence of retinal innervation the tectum is a complete tabula rasa. The tectum or its boundaries with surrounding structures provide sufficient cues to establish the polarity of the map but not to direct each retinal fiber to a specific location. Thus after long-term denervation to allow for removal of the traces of the previous projection, regeneration still results in the formation of an orderly retinotopic map. However, if half of the retina is removed, the regenerated map is immediately expanded (Schmidt, 1978). Likewise, if half of the tectum is removed, the projection is immediately compressed (Sharma and Romeskie, 1977). Both the expanded and compressed maps are normally oriented. Thus the tectum has enough information to orient a map independent of the previous projection, but it does not have exact positional information explicitly directing the termination of each fiber. This same conclusion can be reached by examining the results of experiments on tectal grafts.

Tectal Transplantation Experiments as Tests for Tectal Markers

These experiments fall into two general classes: translocations and rotations. In translocation experiments, grafts are removed from two sites within the same tectum, interchanged, and reimplanted with the same orientation. Such experiments are designed to test for absolute positional markers on the tectal surface. In the rotation experiments, on the other hand, a single piece of tectum is excised, rotated, and reimplanted at the same site. These experiments test for polarity information in the tectal graft. If the tectum contained only polarity cues, each fiber would not be seeking a single site, but a group of fibers could organize themselves in correct order, regardless of what tectal site any individual fibers would then occupy.

In adult fish and frogs, regenerating optic fibers often seek out their "correct" tectal sites even though they have been translocated or rotated (Yoon, 1973). These findings in the adult are consistent with positional markers being left behind by the previous projection (see Schmidt, 1982).

In contrast to these experiments, which clearly demonstrate both polarity and tectal positional markers in the adult tectum, graft experiments in embryonic or larval *Xenopus* have not produced any evidence for tectal markers. Before the arrival of the optic fibers in early development, the tectal area of the neural tube is too small for the translocation of tectal grafts. However, the entire tectum (dorsal mesencephalon of the neural tube) can be rotated, and if the incoming fibers ignore the rotation, the tectum has neither absolute positional information nor stable polarity information. This is what Chung and Cooke (1978) concluded after a long series of rotations of most of the tectal surface at stages 22–24 or at stage 37, which is just before the arrival of the optic fibers. Only when the diencephalon was included in the rotated portion was the map formed by the optic fibers reversed. In fact, Chung and Cooke generated several other configurations of misplaced diencephalon, and the polarity of the retinotectal map was always found to correspond to the position of the diencephalon (Figure 2). Jacobson (1978) also found no evidence for tectal markers at stages well after the arrival of optic fibers at the tectum. Only gradually, after the projection had matured, were results such as those in the adults obtained in a fraction of cases.

Indeed, in the context of the disparate modes of growth of the retina and tectum and the sliding of the connections between them, it is difficult to see how exact tectal positional markers could possibly play a prominent role in the formation of the early projections (Gaze et al., 1972). The central retina forms first and annular rings are added around it. The rostrolateral corner of the tectum, however, is the first to form and initially receives the projection from central retina. In the final projection the fibers from the central retina eventually will move their arbors caudally to occupy central tectum. However, the initial in-

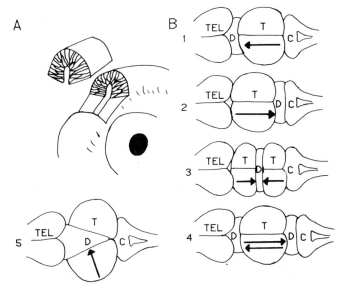

Figure 2. *Diagram of the results of* Xenopus *rotation experiments of Chung and Cooke (1978)*. A: The surgical rotation of presumptive tectal region (dorsal mesencephalon) of the neural tube at stages 22–24 or at stage 37. B: Five classes of results obtained from such rotations. 1: Tectal area alone rotated without any effect on the polarity of the retinal map (shown by the arrow). 2: When the diencephalic region was also included in the rotation so that it developed behind the tectum, the polarity of the retinal map was reversed. 3: Rotations of the anterior half-tectum plus diencephalon resulted in a diencephalon between two smaller tecta on each side. The polarity of the map on each was consistently oriented toward the tectodiencephalic boundary. 4: In a few cases, two diencephalons formed, one rostral and one caudal to the tectum, which then contained two maps of opposite polarity. 5: When the diencephalon developed medially a corresponding rotation of map polarity of about 80° was shown. Tel, telencephalon; D, diencephalon; T, tectum; C, cerebellum.

nervation occurs several weeks before those cells in central tectum are even formed. It is difficult to conceive, therefore, that the initial retinal fibers search for specifically labeled sites in the tectum. Instead they seem to react to polarity cues to form a rudimentary map.

Retinal Grafts

It is interesting to compare the results of tectal grafts with those of retinal grafts, because in larval *Xenopus* the retinal grafts produce results nearly opposite to those of tectal grafts. Surgically created compound eyes, produced before the outgrowth of optic fibers by fusing two nasal halves, for example, nearly always retain the identities of the halves and produce overlapping double maps of the half-eyes in the tectum. In contrast, compound tecta, produced even in adults, receive normal maps (Sharma, 1975). This means that the retinal projection ignores the reversed polarity of the grafted half-tectum. More recently, small grafts have been inserted in embryonic eyes (Conway et al., 1980; Willshaw et al., 1983), and these also tend to retain their identity. Thus there is a fundamental difference between grafts of retinal and tectal tissue early in development. Whereas retinal tissue retains the identity of its origin from very early stages, tectal grafts do not do so reliably until long after retinal innervation. Retinotopic maps therefore appear to be formed not by the matching of position-specific markers between retinal fibers and tectal cells but primarily through fiber–fiber interactions, using retinal markers to maintain order, and through more general orienting cues in the brain. These orienting cues might consist of weak relative gradients on the tectum that do not survive the grafting procedure of Chung and Cooke (1978). Alternatively the map might be oriented by a single organizing region such as the tectodiencephalic border.

Pathway Interactions in the Diencephalon

Several lines of evidence point to the importance of the diencephalon in organizing the first approximation to the retinotopic map on the tectum. First, the graft rotation experiments of Chung and Cooke (1978) point to the diencephalic border as the determinant of the polarity of the tectal map. Rotated maps were seen only when the diencephalon was displaced relative to the tectum (see Figure 2). Four classes of results were generated in which the diencephalon appeared medially, caudally, both rostrally and caudally, and in the middle of the tectum. There was always a good correspondence between the tectodiencephalic boundary and the polarity of the retinotectal map. The importance of this same border is also signaled by the orientation of other retinal maps in the adult frog (Scalia and Fite, 1974). The posterior thalamic nucleus, the nucleus belonci, and the corpus geniculatum all contain retinotopic maps, yet in all three of these thalamic nuclei the nasotemporal axis of the map is inverted relative to the tectal map. In the tectal map, temporal retina (nasal field) projects most rostrally, nearest the tectodiencephalic boundary. Yet in these other three maps temporal retina projects most *caudally* or nearest to the tectodiencephalic boundary. Thus this boundary seems to produce the nasotemporal orientation of the retinotopic maps. Second, the diencephalon, through which the optic fibers pass on the way to the tectum, is the region where the optic tract is reorganized. Fibers from ventral retina pass dorsomedially and fibers from dorsal retina pass more ventrolaterally for entry into the appropriate parts of the tectum (Scholes, 1979); this has recently been demonstrated autoradiographically in larval *Xenopus* (Holt and Harris, 1983). These pathways cannot be due merely to mechanical constraints, because nearly all of the fibers from double-ventral compound eyes pass dorsomedially, and few if any take the ventrolateral route. Fibers from the ventral half placed in the dorsal position follow a path that reflects their origin rather than their new position in the eye. Thus this sorting according to dorsoventral origin could orient the corresponding dimension of the tectal map.

By use of the two orienting cues outlined above (tectodiencephalic boundary for rostrocaudal orientation and optic-tract bifurcation for mediolateral orientation), the two axes of the tectal map might be aligned. If orienting cues are demonstrably present in the diencephalon, should they not also be present to some extent in the tectum itself? This would seem to be a reasonable possibility, yet the graft experiments seem to indicate that if orienting cues are present in the tectum before the arrival of optic fibers, either they are too unstable at early stages to survive grafting or they are so weak that they cannot override the orienting influence of the diencephalon.

At later stages of development the polarity of the tectum becomes stronger, even without retinal innervation. Straznicky (1978) created "virgin" tecta by removing one eye before innervation. After metamorphosis he tested the polarity of the tectum by rotating grafts and inducing a delayed innervation by the remaining eye. The rotated graft retained its polarity. This result stands in contrast to the negative results of Chung and Cooke (1978), where rotations were made before innervation in larval *Xenopus*.

Other Topographic Tectal Maps

The topographic projections between the nucleus isthmi and the two tecta have recently received considerable attention. The nucleus isthmi on each side projects to and receives a projection from the ipsilateral tectum. Experiments tracing the projections of the nucleus isthmi in eyeless frogs have demonstrated normal orientation of the projections in the absence of retinal input (Constantine-Paton and Ferrari-Eastman, 1981). This has been interpreted as evidence for the original Sperry (1963) concept of absolute positional markers and against the idea that such positional markers are imposed on the tectum by retinal fibers. However, experimental studies (Sharma and Romeskie, 1977; Schmidt, 1978) and theoretical models (Hope et al., 1976; Willshaw and von der Malsburg, 1976) have shown that absolute markers are not needed to form a normal map, as polarity cues are sufficient. Absolute positional markers must be demonstrated by interchanging grafts as outlined above. Polarity cues sufficient to orient a developing retinal map should also be available for orienting the other maps such as those from the nucleus isthmi or the somatosensory system. In addition, Dunn-Meynell and Sharma (1983) have recently shown in goldfish that the isthmic map is changeable and undergoes a compression paralleling that of the retinal map when the caudal half-tectum is ablated. The isthmic fibers probably do not search for absolute tectal positions but align themselves in relative order in their map. The evidence favors the presence of orienting cues, but not absolute positional markers, in the tectum.

Other cells of the frog's nucleus isthmi receive the same topographic projection from ipsilateral tectum but project to the contralateral tectum. They subserve an indirect ipsilateral retinotectal pathway that has backward polarity relative to the direct contralateral pathway: A point in rostral tectum relays its information through nucleus isthmi to a point in the caudal portion of the contralateral tectum; a second point in caudal tectum relays to the rostral portion. This reversed polarity is necessary to bring the images of the outside world seen through the two eyes into register on one tectum. The formation of such maps means that the fibers from nucleus isthmi, when projecting contralaterally, follow an ordering opposite to that when projecting ipsilaterally. The contralateral map therefore cannot be made by matching the same positional markers as for receiving the map from the ipsilateral tectum. The contralateral isthmic maps in *Xenopus* have been examined in great detail and found to be modifiable according to the position of the two eyes (Keating, 1975). The modification depends on visual experience, and it may therefore depend on the effects of correlated activity in stabilizing synaptic connections.

These same theoretical considerations (backward rostrocaudal polarity) also hold for direct ipsilateral retinotectal projections found in mammals. Finlay et al. (1979) found that ipsilateral retinotectal projections in the hamster which develop in the absence of the contralateral projection tended to follow the inappropriate contralateral ordering. This contralateral ordering is always found in surgically created, direct ipsilateral projections in goldfish and leads to visually reversed behavior (Easter and Schmidt, 1977). In the mammal, it is clear that normally there is some sort of interaction between the ipsilateral fibers and the contralateral projection to reverse the usual ordering and bring into register the two images seen through the two eyes.

ACTIVITY-DEPENDENT SHARPENING OF THE MAP

The Diffuse Organization of Early Projections

Recent experiments have used anatomical techniques to focus on the degree of order in the earliest projections formed either during regeneration or during development. In both cases studies have shown that the initial degree of order is much less than that of the final map. Regeneration studies using more modern techniques directly contradict the results of Attardi and Sperry (1963), obtained by using the Bodian silver stain, and they remove the basis for Sperry's (1963) claim of unique cytochemical markers.

Goldfish. In goldfish, regenerating optic fibers reach the rostral tectum within 15 days after nerve crush, but the projection cannot be mapped electrophysiologically until approximately 34 days (Schmidt and Edwards, 1983). As a result anatomical techniques are better for studying the early period of regeneration. Meyer (1980) investigated the early order by making small retinal lesions and labeling the rest of the retinal projection autoradiographically. Normally a small retinal lesion would produce a correspondingly small unlabeled zone on the tectal surface. However, in the early stages of regeneration more than half of the retina had to be removed before any tectal areas were left unlabeled. This implies that during the early stages of regeneration each area of the retina projects widely over the tectum. Stuermer and Easter (1982) demonstrated the same thing with retrograde horseradish peroxidase (HRP) transport. After a punctate tectal injection, ganglion cells were labeled over half or more of the retina instead of over a small, discrete area. Electron-microscope studies indicate that each regenerated axon produces abundant collaterals, at least four in the optic nerve and perhaps as many as 20 in the tectum (Murray and Edwards, 1982). The early diffuse projection might be the result of scattered branches of single fibers as well as errors in targeting of single fibers.

Newt. The exact morphology of these regenerated axons is best seen in tectal whole-mounts. Fujisawa et al. (1982) were the first to use HRP to label individual regenerating optic axons in newt and trace them in tectal whole-mounts. They found that shortly after regeneration the fibers often took aberrant or circuitous routes and made many branches. At later times after regeneration they found smaller arbors in appropriate regions of the tectum, suggesting that most of the branches were eventually retracted.

Frog. In the frog the sharpening of the map could be followed with the electrophysiological mapping technique (Adamson and Grobstein, 1982; Humphrey and Beazley, 1982). Recording at each tectal point early in regeneration yielded many units responding to stimulation of a wide area of the retina or visual field, indicating only a very crude level of retinotopic organization on the tectum. Over the next 20 to 30 days these large responsive areas shrank to the normal size for multiunit receptive fields and a normal map emerged. The Australian tree frog (Humphrey and Beazley, 1982) showed a pronounced rostral-to-caudal progression of the sharpening. In addition to the direct retinotectal projection, frogs have an intertectal relay through the nucleus isthmi that is easily recorded. This projection allowed Adamson and Grobstein to assess, in the relay projection, the postsynaptic effects of the crude topography of the direct retinotectal projection. They reported that early in regeneration the receptive fields of the relay fibers in the normal tectum were grossly enlarged, reflecting the crude map on the opposite tectum. Thus the crude maps were still present for some time after synaptic transmission was established. We shall see later, in the studies of goldfish regeneration, that synaptic transmission appears to play a role in the sharpening process.

Several groups are also studying single optic axons in early larval *Xenopus*. Piper et al. (1980) found the arbors to be approximately 250 μm across, very large with respect to the tectal dimensions at early larval stages. D. S. Sakaguchi and R. K. Murphey (unpublished observations; Sakaguchi, 1984) also found that early arbors (embryonic stages 45–50) were large, often covering half to two-thirds of the rostrocaudal length of the tectal neuropil. The arbors tended to be slightly more compact in the mediolateral dimension. Thus early retinal projections appear to be very diffuse in both development and regeneration, and a later process of sharpening must take place to produce the final orderly result. In development, there may be an ongoing need for sharpening the map, because the disparate modes of growth of the retina and the tectum cause a sliding of connections over many hundreds of micrometers of tectal surface (Gaze et al., 1972).

Chick. In chick embryos McLoon (1982) used quadrantal retinal lesions combined with HRP filling of the remaining projection to demonstrate a diffuse projection at 10 days of incubation. The tectal quadrant that corresponded topographically to the retinal lesion was not denervated and still received retinal innervation at 10 days. By 14 days the map had sharpened and the corresponding tectal quadrant was denervated. Thus a gradual sharpening of the projection begins several days after the fibers arrive at the tectum.

Activity-Dependent Sharpening of the Retinotectal Map in Goldfish

The sharpening of the initially diffuse early projection regenerated in adult goldfish has been found to be activity-dependent. Anatomical (Meyer, 1983) and electrophysiological (Schmidt and Edwards, 1983) studies of this process appeared virtually simultaneously.

Anatomical Studies. Meyer's study used the same autoradiographic and lesion technique he had previously employed to show that initial regeneration is diffuse. At various times after nerve crush he determined how large an area of nasal retina had to be removed in order to create an unlabeled zone in the caudal tectum. At 40 days and before, half of the retina had to be removed to denervate any portion of the corresponding caudal tectum, but by 90 days removal of a small sector would produce an unlabeled zone. In this qualitative way he defined the progression back to normal topography. Blocking activity with injections of tetrodotoxin (TTX), either from 32 to 80 days or from 42 to 80 days, prevented the return to a normal topography at 80 days as assessed by the lack of an unlabeled area after small lesions. Even after the sharpening process had been blocked out to 80 days Meyer found sharpening if the fish were allowed 24 more days of activity after TTX was discontinued. Thus, in small goldfish, activity could still sharpen the map several months after the fibers reached the tectum.

Electrophysiological Studies of Sharpening. Schmidt and Edwards (1983) used TTX blocking during regeneration and assessed the effects with electrophysiological recordings of retinotectal maps in large goldfish. Blocking activity from zero to 28 days after nerve crush produced dramatic effects on the maps recorded at 35

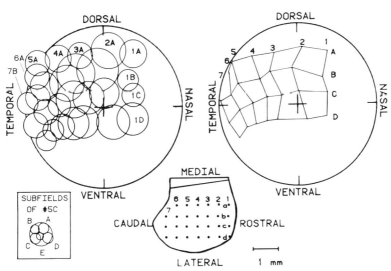

Figure 3. *Retinotectal map recorded 63 days after optic-nerve crush in a fish blocked intraocularly with TTX from zero to 27 days.* The two large circles are representations of the visual field that contain the outlines of the receptive fields and the positions of their centers. Centered between the large circles is a drawing of the tectal surface as viewed from above. Each point in the tectal array is an electrode penetration. The receptive fields recorded at these points fall into an orderly array in the visual field similar to the array on the tectal surface, and are numbered accordingly. The inset at the lower left diagrams the relationship between the multiunit receptive field at position 5C in the array and several of its single-unit components, isolated with a spike-height discriminator and mapped separately. (From Schmidt and Edwards, 1983.)

days and thereafter. Typically, recordings made at 35 days were normal in organization and in the size of the tectally recorded receptive fields, which averaged 11° at each tectal point. In the fish that had been blocked for the first 28 days, however, the multiunit receptive fields recorded at each tectal point were grossly enlarged, averaging around 30° (Figure 3). The centers of the these enlarged fields were approximately where they should normally be, indicating that the gross organization of the map was correct. The enlarged, multiunit receptive fields are a direct reflection of the convergence onto one tectal point of the arbors from retinal ganglion cells distributed over a wide area of retina (Figure 4). On the other hand, the high degree of overlap of the enlarged receptive fields recorded at neighboring tectal points shows the divergence of fibers from one retinal point to a wide area of tectum. In other words, the size of the multiunit receptive fields is a direct quantitative measure of the inherent error in the targeting of the regenerating optic arbors. As the map becomes more orderly there is less convergence and divergence, and the projection approaches the theoretical point-to-point map. It is never really point-to-point, however, because of the size of the receptive fields of ganglion cells (11°) and the size of their axonal arbors in the tectum (100–250 μm).

This interpretation of the enlarged multiunit receptive field was supported by recordings from single ganglion cells in the retina which demonstrated that their receptive fields were not enlarged (Schmidt and Edwards, 1983). In addition, the presynaptic origin of the tectal recordings was upheld by showing that the units persisted during a postsynaptic block with α-bungarotoxin (α-BTX). The tectal recordings yielded many units at each site. When individual units recorded in the tectum could be isolated by amplitude, their receptive fields were of normal

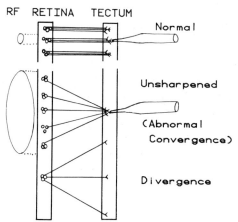

Figure 4. *Schematic diagram showing the interpretation of the enlarged multiunit receptive fields.* The two vertical bars represent the retina and the tectum and show projections between them. At the top is the normal orderly projection. The electrode inserted into the tectum records several optic arbors, but because of the orderly nature of the projection, these come from retinal ganglion cells in a tight cluster so that their multiunit receptive field (RF) is small (11°). After regeneration without sharpening, the electrode records several optic arbors coming from retinal ganglion cells scattered over a wide area of the retina. Consequently, the composite multiunit receptive field is very large. This spatial convergence is due to an abnormal divergence of optic arbors from each cluster of retinal ganglion cells, a result of the uncorrected errors in targeting of the regenerated fibers.

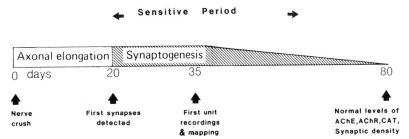

Figure 5. *Timetable of the regeneration of the retinotectal projection showing the sensitive period (lined).* After crush of the optic nerve, the axons must first grow back to the tectum, and blocking activity during this phase does not affect the map. The first synapses are detected around day 20. Synaptogenesis proceeds until 80–100 days by several criteria including synaptic density in electron micrographs and levels of acetylcholinesterase (AChE), choline acetyltransferase (CAT), and acetylcholine receptors (AChR) as assessed by α-BTX binding. The sensitive period corresponds to the time of synaptogenesis, particularly early synaptogenesis.

size (Figure 3). The only remaining interpretation is that the enlarged multiunit receptive fields reflect an inherent error in the targeting of the many regenerated arbors converging on that point. These errors are normally corrected in an activity-dependent manner.

Correlation of Sharpening with Synaptogenesis

The error in targeting was attributable to the lack of activity during the period of synaptogenesis (Figure 5). Schmidt et al. (1983) showed that the first synapses, as assessed by recording field potentials, were detectable on day 20 (Figure 6).

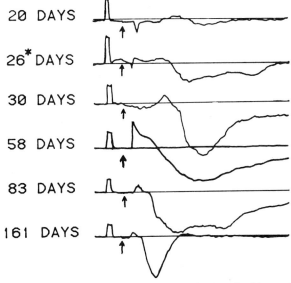

Figure 6. *Representative traces recorded from superficial tectum (150 μm depth) at various times after optic-nerve crush. Asterisk indicates a fish kept at 25°C, although the time has been corrected to the equivalent at 20°C. Calibration pulses: 2 msec × 0.5 mV for the upper four traces; 2 msec × 1 mV for the lower two traces. Trace length is 80 msec. (From Schmidt et al., 1983.)*

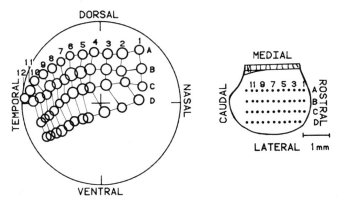

Figure 7. *Retinotectal map recorded 38 days after nerve crush in a fish blocked intraocularly with TTX from zero to 14 days. The array of receptive fields is regular and the map is normal. Conventions as in Figure 3. (From Schmidt and Edwards, 1983.)*

Blocking before this time did not prevent sharpening, and normal maps were present at 35 days (Figure 7). Blocking from 14 to 34 days, however, was extremely effective, producing fields enlarged an average of 40°. The period of susceptibility therefore correlates well with the period of synaptogenesis.

Synaptogenesis proceeds at a reduced rate beyond 35 days (when the map can first be recorded), as shown in the increasing amplitude of the field potentials (Figure 6) or by rebounds in the levels of choline acetyltransferase, acetylcholinesterase, and α-BTX binding (Francis and Schechter, 1979; Schechter et al., 1979) or in levels of total synaptic density (Murray and Edwards, 1982). J. T. Schmidt and L. E. Eisele (unpublished observations) have now assessed the effects of blocking activity from 35 to 50 days postcrush and found that the initially small receptive fields become enlarged. The effect of the 35–50 day block is, however, somewhat less than that created by blocking during early synaptogenesis (20–35 days). In the mature projection, where synaptogenesis should be minimal, blocks of up to 28 days do not cause any enlargement of the tectally recorded receptive fields. So far the results suggest a parallel between the rate of synaptogenesis and the degree of susceptibility to TTX blocking.

TTX blocking does not seem to cause a decreased rate of synaptogenesis, as judged by the field potentials elicited by optic-nerve shock. A similar lack of effect of TTX was noted in neuromuscular synapse formation *in vitro* (Obata, 1977). In the fish tectum the amplitude of the field potentials from eyes blocked during regeneration was not decreased when they were recorded just after the block wore off (Schmidt et al., 1983). For the low-amplitude field potentials in early regeneration, amplitude is likely to be a reasonably sensitive measure of the number of synapses formed. The synaptic currents generating the field potentials depend on the number of channels opened at the synapses and on the transmembrane potential. With only a few synapses the membrane potential does not change radically, so that the field potentials reflect primarily the number of open channels. These results suggest that the blocking of activity does not affect the quantity of synapses made, but rather that it interferes with the deployment of the synapses in the retinotopically correct order.

A Model Based on Correlated Activity

Schmidt and Edwards (1983) proposed a model in which the cue for sharpening the map depended on the correlated firing of neighboring ganglion cells and the resultant summation of their excitatory postsynaptic potentials (EPSPs) in the postsynaptic tectal cells. It appears likely (Willshaw and von der Malsburg, 1976) that neighboring cells which view the same part of the visual world fire with a high degree of correlation if they are of the same type (ON or OFF, etc.). Arnett (1978) has in fact demonstrated that this occurs even in absolute darkness when only spontaneous activity is present. The model (shown in Figure 8) also assumes that the arbors in the initial diffuse projection are large. This allows a high degree of overlap of the arbors. A summation of the postsynaptic EPSPs would appear in neighboring ganglion cells that have some overlap of their arbors and fire in synchrony. Finally, as postulated by others (Hebb, 1949; Changeux and Danchin, 1976) if the most effective synapses are differentially stabilized and retained, the correlated activity, resulting in larger EPSPs, would serve as a cue for convergence of the arbors in the region of overlap. Of course arbors from distant ganglion cells would also overlap, but the probability of their firing in synchrony is very low, so there would be little summation of their EPSPs and therefore no cue for their convergence.

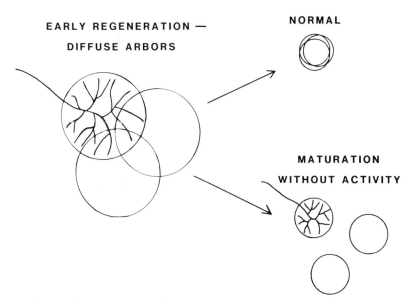

Figure 8. *Schematic diagram of the model for activity-dependent sharpening of the retinotectal map.* On the left is the situation early in regeneration, showing three neighboring retinal ganglion cells with diffuse and inaccurately targeted arbors. Within the area of overlap, synchronous firing of the three cells would lead to summation of their EPSPs and to stabilization of their synapses. This in turn would lead to the retraction of their other branches and the concentration of the arbors within the smaller area (see normal at upper right). The final arbor is pictured as smaller than that initially regenerated. Without activity a similar contraction would take place, but the arbors would merely retract toward their centers, leading to a loss of overlap and therefore to a loss of the capability for activity-dependent convergence later on. (From Schmidt and Edwards, 1983.)

Testing the Model

Recent experiments have yielded considerable support for the major features of the model. So far two predictions have been tested. First, the model proposes that the correlated firing of neighboring ganglion cells is the cue; sharpening should be disrupted even if ganglion cells are allowed activity as long as this special correlation is disrupted. Two methods were used—stroboscopic illumination and dark rearing. The simplest way to remove the relationship between correlated firing and neighboring ganglion cells was to use stroboscopic illumination to synchronize the firing of all ganglion cells. Tests showed that OFF cells as well as ON cells were effectively driven by the strobe light. Therefore all cells, not just near neighbors, fired with a high degree of correlation, and correlated firing was no longer a cue for finding neighbors from the retina. When the projection regenerated under stroboscopic illumination, the same enlarged multiunit receptive fields were recorded in tectum as those that occurred after TTX blocking; they averaged 33.2° in diameter. Again, the projection was sensitive only during regeneration. These experiments rule out any effect due to the gross deficit of activity and show that sharpening depends on the pattern of activity, presumably the aforementioned correlation in firing between neighbors in the retina.

Another way to remove the normal visually driven pattern of activity without using TTX was to place the fish in complete darkness so that only a very low level of spontaneous activity persisted. A group of fish that regenerated under these conditions also had enlarged multiunit receptive fields averaging 28.7° in diameter. This is somewhat smaller than the average of 33.2° under strobe-light illumination or 40° with the equivalent TTX blocking. This may reflect a slight degree of sharpening due to correlation in the low spontaneous activity of neighboring ganglion cells that is found even in absolute darkness (Arnett, 1978). This low level of correlated activity is not sufficient for complete sharpening of the map.

The second prediction from the model is that the correlated activity must be transmitted to a postsynaptic tectal cell to allow summation of the EPSPs in the postsynaptic cell. To test this prediction, activity was blocked during the period of early synaptogenesis (20–35 days). An osmotic minipump was attached to the fish's head to produce a continuous infusion of α-BTX. Previous experiments (Freeman et al., 1980; Schmidt and Freeman, 1980) have shown that the toxin blocks synaptic transmission by binding postsynaptically. The fish were kept in a normal visual environment so that the regenerating optic fibers would have normal levels and patterns of activity. However, the maps were the same as those seen in TTX-blocked fish, with enlarged fields averaging 28.5° in diameter. This experiment demonstrated that fibers with correlated activity do not interact directly but must interact through the transmission of their signals to the postsynaptic tectal cells, probably through the summation of EPSPs. The slightly smaller average receptive field size relative to the strobe-raised and TTX-blocked fish may reflect an incomplete block of synaptic transmission. At the present time we cannot determine exactly what form of activity in the postsynaptic cell (EPSPs or spike activity) may be important in the stabilization of synaptic contacts. However, some inferences about the stabilization process can be drawn from the local α-BTX experiments discussed below.

Sharpening after Release from TTX Block

One difference between the results of Schmidt and Edwards (1983) and those of Meyer (1983) is in the ability of the projection to sharpen after the TTX block was discontinued. In Meyer's small (5–7 cm) fish, sharpening apparently occurred readily when the block was discontinued at 80 days. In Schmidt and Edwards' large (10–13 cm) fish, however, the effects of a block discontinued at 28 days persisted until 145 days with only a few signs of improvement. Recently I recorded fish after a year or more and found that many multiunit receptive fields eventually became small and sharp. Others were still large but had small subfields whose spikes were much larger and dominated the response (J. T. Schmidt, unpublished observations). This indicates that the sensitive period is not completely confined to the short period after the regeneration in which the map is first set up. But there is still one basic difference with Meyer's result concerning the time course of the sharpening during regeneration. At 40 days postcrush, a normal orderly map has long been established, yet his autoradiographs indicate only the grossest topography at this point. One problem with autoradiography is that the silver grains do not distinguish between synaptic terminals on the one hand and axons of passage and bare collaterals on the other. For reasons of geometry and current generation the electrodes probably do not record the latter. Resolution of this difference may require the filling of single arbors with the anterograde HRP technique.

A second reason to fill single arbors is the need to unravel the exact cause for the enlarged multiunit receptive fields seen in the experiments with TTX, strobe-rearing, dark-rearing and α-BTX. On the one hand, the enlarged fields may reflect enlarged optic arbors, each centered on the appropriate tectal area. Calculations based on the magnification factor of the retinotectal map show that such arbors would have to be approximately 700–800 μm in diameter to produce overlap sufficient for the electrode at each point to pick up fields scattered over 30–40°. On the other hand, the arbors might be of normal size but deployed with a certain error distribution around the appropriate tectal site. Arguing against the first possibility is the unlikelihood that each ganglion cell would be capable of supporting such a grossly enlarged arbor over long periods of time. In addition, if each fiber did maintain large arbors it would still have a high degree of overlap with its neighbors, and the current model would predict a continued ability for activity-dependent sharpening. The model of Schmidt and Edwards (1983) therefore assumes that the initially enlarged arbors gradually revert to a normal size so that there is a decreasing capability for activity-dependent sharpening late in regeneration. Nevertheless, one can easily see that several features of the sharpening process can be addressed only when we have good anatomical data on the sizes and shapes of the regenerated arbors to correlate with the sensitivity to activity. This is no small problem because there appear to be three separate classes of optic fibers and arbors, based on conduction velocity (Schmidt, 1979) or arbor size (J. T. Schmidt, unpublished observations). For the class with the largest arbors, our fills have so far failed to turn up any enlarged arbors of fibers regenerating without activity (J. T. Schmidt, unpublished observations), but many more will have to be examined before a firm conclusion can be drawn.

ACTIVITY AND SYNAPSE ELIMINATION ELSEWHERE

The production of excess synaptic connections and the subsequent elimination of many of these connections is a familiar pattern in neural development. A review of all such cases is beyond the scope of this discussion (see Purves and Lichtman, 1980). In most cases, the role of activity in synapse elimination is not easily studied. In some cases, circumstantial evidence suggests a possible role, but in a few cases the evidence is more direct. These suggestive examples include visual callosal connections and the autonomic ganglia. Visual callosal fibers initially arise from a large set of cortical neurons, but after synapse elimination only a small subset of these neurons continues to send fibers across the callosum. The selection of these remaining neurons may depend on patterned visual activity, as a different subset retains its callosal connections when the animal is raised in the dark (Innocenti, 1981). In autonomic ganglia (Purves and Hume, 1981) the complexity of dendritic structures is highly correlated with the number of inputs retained after the period of synapse elimination. Here again the evidence is only mildly suggestive and other interpretations are certainly plausible.

Some evidence seems more direct. In this section I discuss several of the cases that suggest mechanisms similar to those of our model for sharpening the map. These lines of evidence include (1) the activity-dependent segregation into eye-specific stripes; (2) the destabilization of ineffective synapses after local α-BTX treatment in the tectum; (3) the activity-dependent segregation of receptive field types in the lateral geniculate nucleus of the cat; (4) the alignment of auditory and visual maps in the tectum of the owl; and (5) the elimination of polyneuronal innervation at the neuromuscular junction. In concluding I draw some inferences about the features of activity-dependent synaptic stabilization and its possible widespread applicability.

Activity-Dependent Segregation of Ocular-Dominance Patches

The segregation of visual afferents into eye-specific patches or stripes is well known in cat and monkey visual cortex (LeVay et al., 1978). The segregation occurs within the first few months of visual experience after eye opening. Although the retinotectal projection of fish and frogs is normally totally crossed so that fibers from the two eyes do not directly interact in the tectum, various experimental manipulations can cause dual innervation of the tectum. This leads to the spontaneous formation of ocular-dominance stripes very similar to those in mammalian visual cortex. Such manipulations include the grafting of an extra eyecup in the embryonic frog (Constantine-Paton and Law, 1978), the formation of compound eyes in embryonic *Xenopus* (Fawcett and Willshaw, 1982), the deflection of a bundle of optic fibers in goldfish (Meyer, 1982), and the surgical removal of one tectum in the adult fish to force regeneration and sharing of the remaining tectum by the two eyes (Boss and Schmidt, 1982, 1984). As in the visual cortex, the two populations of fibers first mingle and cover the entire area before they withdraw into eye-specific stripes. Because the projections are direct instead of multisynaptic, they are easily traced with anterograde transport of HRP or [^{3}H]proline. Consequently tectal stripes have become a convenient model for studying the formation of cortical stripes. The activity-dependence of this segregation has now been established in several studies that employed intraocular TTX (Boss and Schmidt, 1982, 1984; Meyer, 1982; Constantine-Paton and Reh, 1983), and the mechanism may be similar to that which drives the sharpening of the retinotectal map.

Such studies have demonstrated that using TTX to block the competing eyes prevents the segregation into ocular-dominance stripes. Stryker (1981) has reported the same result in cat visual cortex, so the activity-dependence of such segregation appears to be a general phenomenon. Boss and Schmidt (1982, 1984) also reported a slowing of the segregation when only one eye was blocked. Meyer (1982) has shown that a period of activity after discontinuing the block allowed delayed segregation to take place. In addition, Constantine-Paton and Reh (1983) showed that the segregation is reversible in tadpoles. Blocking after the stripes had formed eventually caused the two sets of fibers to mix again.

In the goldfish it is interesting to compare the time course of the sharpening of the map with that of the segregation into ocular-dominance patches. The sharpening begins early (20–35 days at 20°C), whereas there is a long delay before segregation occurs. In our fish it began around day 50 at 30°C, which is roughly equivalent to 100 days at 20°C (Q_{10} of 2.0; Springer and Agranoff, 1977). Sharpening therefore precedes segregation into eye-specific patches. This would be predicted by the model, as each eye must first deploy its synapses in a highly retinotopic manner to increase their effectiveness before it can begin excluding those of the other eye. Fibers from cells far apart in the same eye should not have any better correlation in firing than do fibers from different eyes. In addition the density of synapses may have to reach normal levels before exclusion can take place. In regeneration this takes 80–100 days by the criteria listed above. The exclusion is evident in autoradiographs and HRP-filled material, but it is not absolute. Fibers can be seen running between the dense terminal bands (Springer and Cohen, 1981), and units from both eyes can still be recorded from most tectal sites (Schmidt, 1978). These fibers and presynaptic recordings need not reflect the actual density of synapses, however. To assess synaptic distribution we have examined the amplitude of field potentials in striped tecta and found that they vary periodically. The placement of electrolytic lesions at recording sites showed a positive correlation with the anatomical stripes (Boss and Schmidt, 1982). Thus the anatomical stripes reflect the relative segregation of functional synapses.

A model that fits both the sharpening of the map and the segregation into patches is shown in Figure 9. Initially the terminals from the two areas are mixed, so the stippled and unstippled retinal ganglion cells could represent cells from distant sites in the same retina or from opposite eyes. On some tectal cells there is a larger input from one cluster of retinal cells than from the others. This might happen either from the bias of the crude map established or merely statistically by chance in the case of two eyes. The correlated activity in these neighboring retinal cells could result in the summation of their postsynaptic responses and lead to the stabilization of their synapses. The synapses from other retinal cells that do not have correlated firing would not be stabilized. If each retinal cell periodically removes its least stable synapses and moves them elsewhere, the eventual result should be tight clustering of synapses from the same retinal site or, if the projections are from two eyes, segregation into ocular-dominance stripes. Again, the model depends on correlated activity between

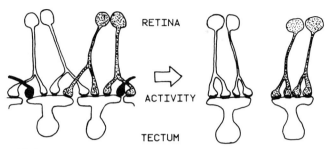

Figure 9. *A model of activity-dependent synaptic stabilization.* The two pairs of retinal cells (stippled and unstippled) represent neighbors from two sites in a retina or neighbors in opposite retinas. In the initial diffuse tectal projection their synapses are mixed together. If a tectal cell receives two inputs from the same cluster, their correlated activity results in the summation of responses and stabilization of those synapses. Others firing alone are not stabilized. If the unstabilized synapses are periodically moved at random, they will eventually be stabilized within separate clusters according to their origin, as this arrangement provides the greatest stabilization.

neighbors from the same retina and on effective synaptic transmission to allow the postsynaptic summation of inputs.

Several of the experiments that were effective in the blocking of sharpening have not yet been carried out on the segregation into stripes. For instance, stimulating both eyes at the same time should, by analogy with the sharpening results, block the segregation. Our first attempts with stroboscopic illumination did not do so (J. T. Schmidt, unpublished observations). However, we soon realized that the latencies of the arrival of signals at the tectum were quite different, because the ipsilateral nerve was recently regenerated and conducted quite slowly and because its path was substantially longer (Easter et al., 1978). Therefore simultaneous activation of the eyes did not result in simultaneous activation at the tectum, as shown in Figure 10. When both eyes regenerate we find some evidence for decreased segregation. The larger issue remains, however: How close in time must the two inputs be in order for them to stabilize each other's synapses? Judged by field potential recordings, the EPSP is very brief, ending in less than 10 msec (Figure 10). The strobe experiment suggests that we may be dealing with such a brief time frame. If so, this is a very different time frame from that of associative learning, where the optimal interval can range from hundreds of milliseconds to seconds.

Figure 10. *Superimposition of tectally recorded field potentials elicited by stimulation of the left and right (regenerated) optic nerves.* The regenerated nerve's volley arrives later because of its longer path and slower conduction. Stimulus onset is marked by the arrow. Depth of recording: 150 μm.

Retraction of Ineffective Synapses

The selective stabilization of the most effective synapses implies that ineffective ones are destabilized; that is, they are more likely to be removed. Freeman (1977) demonstrated this effect by using α-BTX to block postsynaptic responses in a small, localized region of toad tectum. One week after such a block, the optic terminals that normally terminate within the blocked area were recorded just outside, leaving the blocked area silent. By two to three weeks the arbors moved back into the treated area, probably as new acetylcholine receptors replaced the blocked ones (Oswald and Freeman, 1980). I replicated these results in the goldfish and extended the findings, as sketched in Figure 11. In the additional experiments I showed that the retraction is activity-dependent, as no retraction from the toxin-blocked area occurred if activity in the optic fibers was blocked simultaneously with intraocular TTX. This rules out any nonspecific effect of α-BTX and shows that the optic fibers must have activity to respond to the local postsynaptic block. If there is no activity in any of the optic fibers there is no synaptic transmission (other than spontaneous release) anywhere on the tectum, either within or outside the blocked area, and no movement occurs. Thus fibers with activity appear to be able to respond to differences in the responsiveness of postsynaptic cells. A third set of experiments confirmed the idea of a relative comparison of responsiveness. In these experiments, α-BTX was applied over all the tectal surface. After a week of uniform tectal block, completely normal retinal maps were recorded. The fibers therefore do not automatically retract from blocked synapses, but do so only if they have the opportunity to make

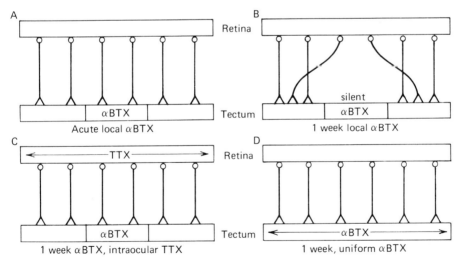

Figure 11. *Schematic diagrams showing the acute and chronic effects of α-BTX applied to the goldfish tectum.* In each case, the upper bar represents the retina with the retinal ganglion cells projecting to the tectum (lower bar). A: There is no acute effect on the map even though the synapses in one area are blocked. B: One week after local application, the fibers originally projecting to the blocked area now project just outside. C: One week after local application of intraocular TTX to prevent activity in the optic fibers for the first six days. The retinal fibers do not move out of the blocked area. D: One week after uniform application of toxin over the tectal surface. There is no movement of terminals, which can still be recorded all over the tectum.

effective synapses nearby. The results support the idea that optic arbors can move over short distances to maximize their synaptic transmission. The distances involved are somewhat larger than the total width of the arbors, so that such movements might take place either by random sampling or by a sprouting response to a short-range extracellular factor.

Segregation of Receptive Field Types in Lateral Geniculate

In the cat visual system Archer et al. (1982) have demonstrated that activity is necessary for the normal development of inputs to geniculate cells. Normally, all geniculate cells, like the retinal fibers, respond either to the onset of a stimulus or to its cessation, but not to both ON and OFF. Without activity the majority end up with mixed ON and OFF inputs. Likewise, the lack of activity causes abnormal mixing of X and Y inputs and a lack of retinotopic sharpening. If, as seems likely, the initial innervation is not selective by receptive field type, the TTX block may be interfering with synapse elimination, preventing the capture of cells of the lateral geniculate nucleus by ON or OFF fibers exclusively. This segregation of parallel streams of incoming information may represent another important role of activity-dependent synaptic stabilization in the development of highly ordered neural connections.

The Aligning of Auditory and Visual Maps

A fourth area of activity-dependent synaptic changes, and one in which the timing of activity also seems to be the important variable, is that of the alignment of the visual and auditory maps in the tectum of the barn owl (Knudsen, 1983). The visual map is a direct retinotopic projection, but the auditory spatial map is formed centrally through an analysis of binaural difference cues. The horizontal dimension is based primarily on interaural latency differences, and the vertical dimension is based primarily on interaural intensity differences. If a plug is inserted into one ear of very young owls these cues (longer latency and lower intensity) are altered systematically as the owls grow to adulthood. Recordings of bimodal tectal cells in these adult owls showed that, with the plugs in, the auditory and visual maps were in precise register. The connections somewhere along the auditory pathway had adjusted for the distortion caused by the plugs. However, removal of the plugs produced a predictable shift in the auditory map such that the two maps were about 10° out of alignment. The misaligned maps persisted in adult owls after the plugs had been removed, indicating that the mechanism for adjustment is either lost or vastly reduced in the adult nervous system, similar to much of the visual system plasticity outlined above. In owls, the exact site at which the altered connectivity takes place is not known, but it is tempting to speculate that it may be at the level of the tectal cells, where the convergence between the two modalities occurs. If so, those auditory fibers that fire simultaneously with the visual fibers reporting the image of the object might, via an enhanced postsynaptic response, selectively stabilize their connections during the early life of the owl. Such a mechanism would be an intermodality stabilization analogous to the intramodality stabilization seen in the sharpening of the retinotopic map in goldfish.

Synapse Elimination at the Neuromuscular Junction

The final example is the neuromuscular junction, which is by far the most thoroughly studied synapse. The pre- and postsynaptic elements are extremely accessible and manipulable, which has permitted a thorough analysis of the activity-dependence of synapse elimination. Early in development the end plate has several terminals, and in most vertebrate muscles it loses all but one of these inputs (see Van Essen, 1982, for a review). Although some of this retraction may occur simply because each axon is unable to support all of its initially large arbor (Betz et al., 1980), much is tied to the activity transmitted at the end plate. First, TTX blocks of motor nerves result in continued polyneuronal innervation (Thompson et al., 1979), paralleling the retention of the diffuse map in the retinotectal projection. Second, blocking postsynaptically also results in retention of polyneuronal innervation (Sohal et al., 1979; Magchielse and Meeter, 1982). This also parallels the results on sharpening the map. Third, electrical stimulation of the motor nerve increased the rate of synapse elimination (O'Brien et al., 1978; Srihari and Vrbova, 1978). This is actually the opposite of the effect of synchronous activation (stroboscopic illumination) on the sharpening of the map in the retinotectal system and so constitutes a fundamental difference. However, it is not surprising that synchronous firing does not promote convergence at the neuromuscular junction, which is destined to be singly innervated anyway. Different activity-dependent mechanisms may furnish the needs of different systems, as we can see in this direct comparison.

CONCLUSION

Activity-dependent mechanisms for the selective stabilization or elimination of synaptic connections are likely to have widespread applicability. Experiments suggest their occurrence at many levels, from neuromuscular junction to tectum and cortex. It may be a relatively efficient mechanism in development, particularly if the excess initial synapses are not degenerated but are resorbed, as has been suggested, at the neuromuscular junction. This compares favorably with the rather wasteful mechanism of cell death.

This class of mechanisms bears a complementary relationship to differential affinity mechanisms commonly referred to as chemoaffinity. It may be that the exact specification of higher order or detailed connections would be too costly in the number of genes and gene products necessary to code each fiber uniquely. These numbers multiply rapidly as we consider the various cell types in the visual system: ON versus OFF; X, Y, and W; color coding, and so forth. However, the molecular cost might not be as relevant as the fact that such a developmentally rigid process would not be flexible enough to succeed. Small mistakes at one site would not be able to be corrected at later points in the pathway. Such factors as the variable geometry of the head, the separation between the two ears, the two eyes, and so forth, might demand flexibility in the process of aligning auditory and visual maps or of aligning visual maps from the two eyes. Indeed, if things were rigidly specified in the way that Sperry envisaged, functionally ordered ipsilateral retinal projections might be impossible, because they would not have the necessary reverse ordering discussed above.

Activity-dependent mechanisms do have a definite limitation in creating, on their own, the reproducibly ordered structures we find in the nervous system. This type of process only forms a reproducible result when the correct sets of fibers have been brought reasonably close together so that they can interact effectively. Thus the reproducible polarity of the map still depends on the ability of membrane affinity mechanisms to bring a statistically greater number of temporal versus nasal retinal fibers to the rostral area of the tectum. A totally random mix of retinal fibers would result in randomly oriented maps.

The complementary roles of chemoaffinity and activity-dependent mechanisms is another way of saying that activity-dependent stabilization of synapses must be controlled by the initial starting conditions. A second method of control lies in the window of sensitivity during development, after which the ability to change is drastically damped. In the adult, too much ability to change might interfere with the reliable functioning of the nervous system.

The unsolved problems, some of which I have already pointed out, are many, and we have only a very rough outline. We do not know, for instance, what form of postsynaptic activity is important (EPSP or spike), and this may have geometrical implications for cell morphology (Purves and Hume, 1981). We do not know the exact level of synchrony of inputs necessary for the interaction postsynaptically. The feedback signal is not yet known, which points out that the area has not yet joined with the growing molecular research on synapse formation. This calls to mind other known mechanisms for controlling synaptic efficacy—presynaptic modulation of release via inhibition, habituation, and facilitation—which have been only partially explored at the molecular level. It is interesting to note, in this regard, the recent demonstration in *Aplysia* that such modifications of synaptic strength are followed by parallel morphological changes in the presynaptic terminal (Bailey and Chen, 1983).

ACKNOWLEDGMENTS

I thank David Tieman and Rod Murphey for their useful comments on the manuscript. This work was supported by NIH grant EY-03736 and a Sloan Foundation Fellowship.

REFERENCES

Adamson, J. R., and P. Grobstein (1982) Reestablishment of the ipsilateral oculotectal projection after optic nerve crush in the frog: Evidence for synaptic remodeling during regeneration. *Neurosci. Abstr.* **8**:514.

Archer, S. M., M. Dubin, and L. A. Stark (1982) Abnormal development of kitten retinogeniculate connectivity in the absence of action potentials. *Science* **217**:743–745.

Arnett, D. W. (1978) Statistical dependence between neighboring retinal ganglion cells in goldfish. *Exp. Brain Res.* **32**:49–53.

Attardi, D. G., and R. W. Sperry (1963) Preferential selection of central pathways by regenerating optic fibers. *Exp. Neurol.* **7**:46–64.

Bailey, C. H., and M. Chen (1983) Morphological basis of long-term habituation and sensitization in *Aplysia*. *Science* **220**:91–93.

Betz, W. J., J. H. Caldwell, and R. R. Ribchester (1980) The effects of partial denervation at birth on the development of muscle fibres and motor units in rat lumbrical muscle. *J. Physiol. (Lond.)* **303**:265–279.

Boss, V., and J. T. Schmidt (1982) Tests for a role of activity in the formation of ocular dominance patches. *Neurosci. Abstr.* **8**:668.

Boss, V., and J. T. Schmidt (1984) Activity and the segregation of ocular dominance patches in goldfish tectum. *J. Neurosci.* (in press).

Changeux, J.-P., and A. Danchin (1976) Selection stabilisation of developing synapses as a mechanism for the specification of neuronal networks. *Nature* **264**:705–712.

Chung, S.-H., and J. Cooke (1978) Observations on the formation of the brain and of nerve connections following embryonic manipulation of the amphibian neural tube. *Proc. R. Soc. Lond. (Biol.)* **201**:335–373.

Constantine-Paton, M., and P. Ferrari-Eastman (1981) Topographic and morphometric effects of bilateral embryonic eye removal on the optic tectum and nucleus isthmi of the leopard frog. *J. Comp. Neurol.* **196**:645–661.

Constantine-Paton, M., and M. I. Law (1978) Eye-specific bands in the tecta of three-eyed frogs. *Science* **202**:639–641.

Constantine-Paton, M., and T. A. Reh (1983) Eye-specific stripes in the tectal lobes of three-eyed frogs are dependent on neural activity. *Neurosci. Abstr.* **9**:760.

Conway, K., K. Feiock, and R. K. Hunt (1980) Polyclones and patterns in growing *Xenopus* eye. *Curr. Top. Dev. Biol.* **15**:216–317.

Dunn-Meynell, A. A., and S. C. Sharma (1983) Changes in the projection of the nucleus isthmus to the tectum with retinotectal compression in goldfish. *Neurosci. Abstr.* **9**:57.

Easter, S. S., Jr., and J. T. Schmidt (1977) Reversed visuomotor behavior mediated by induced ipsilateral retinal projections in goldfish. *J. Neurophysiol.* **40**:1245–1254.

Easter, S. S., Jr., J. T. Schmidt, and S. M. Leber (1978) The paths and destinations of the induced ipsilateral retinal projection in goldfish. *J. Embryol. Exp. Morphol.* **45**:145–159.

Fawcett, J. W., and D. J. Willshaw (1982) Compound eyes project stripes on the optic tectum in *Xenopus*. *Nature* **296**:350–352.

Finlay, B. L., K. G. Wilson, and G. E. Schneider (1979) Anomalous ipsilateral retinotectal projections in Syrian hamsters with early lesions: Topography and functional capacity. *J. Comp. Neurol.* **183**:721–740.

Francis, A., and N. Schechter (1979) Activity of choline acetyltransferase and acetylcholinesterase in the goldfish tectum after disconnection. *Neurochem. Res.* **4**:547–556.

Freeman, J. A. (1977) Possible regulatory function of acetylcholine receptor in maintenance of retinotectal synapses. *Nature* **269**:218–222.

Freeman, J. A., J. T. Schmidt, and R. E. Oswald (1980) Effect of α-bungarotoxin on retinotectal synaptic transmission in the goldfish and the toad. *Neuroscience* **5**:929–942.

Fujisawa, H., N. Tani, K. Watanabe, and Y. Ibata (1982) Branching of regenerating retinal axons and preferential selection of appropriate branches for specific neuronal connection in the newt. *Dev. Biol.* **90**:43–57.

Gaze, R. M., and S. C. Sharma (1970) Axial differences in the reinnervation of the goldfish optic tectum by regenerating optic nerve fibers. *Exp. Brain Res.* **10**:171–181.

Gaze, R. M., and C. Straznicky (1980) Regeneration of optic nerve fibers from a compound eye to both tecta in *Xenopus*: Evidence relating to the state of specification of the eye and the tectum. *J. Embryol. Exp. Morphol.* **60**:125–140.

Gaze, R. M., M. Jacobson, and G. Szekely (1963) The retinotectal projection in *Xenopus* with compound eyes. *J. Physiol. (Lond.)* **165**:484–499.

Gaze, R. M., S.-H. Chung, and M. J. Keating (1972) The development of the retinotectal projection in *Xenopus*. *Nature (New Biol.)* **236**:133–135.

Hebb, D. O. (1949) *Organization of Behavior*, Wiley, New York.

Holt, C., and W. A. Harris (1983) Order in the initial retinotectal map in *Xenopus*: A new technique for labelling growing nerve fibres. *Nature* **301**:150–152.

Hope, R. A., B. J. Hammond, and R. M. Gaze (1976) The arrow model: Retinotectal specificity and map formation in the goldfish visual system. *Proc. R. Soc. Lond. (Biol.)* **194**:447–466.

Horder, T. J., and K. A. C. Martin (1983) Some determinants of optic terminal localization and retinotopic polarity within fibre populations in the tectum of goldfish. *J. Physiol. (Lond.)* **333**:481–509.

Humphrey, M. F., and L. D. Beazley (1982) An electrophysiological study of early retinotectal projection patterns during optic nerve regeneration in *Hyla moorei*. *Brain Res.* **239**:595–602.

Innocenti, G. M. (1981) Growth and reshaping of axons in the establishment of visual callosal connections. *Science* **212**:824–826.

Jacobson, M. (1978) *Developmental Neurobiology*, 2nd ed., Plenum, New York.

Johns, P. R., and S. S. Easter, Jr. (1977) Growth of the adult goldfish eye. II. Increase in retinal cell number. *J. Comp. Neurol.* **176**:331–342.

Keating, M. J. (1975) The time course of experience-dependent synaptic switching of visual connections in *Xenopus laevis*. *Proc. R. Soc. Edinb. (Sect. B)* **189**:603–610.

Knudsen, E. I. (1983) Early auditory experience aligns the auditory map of space in the optic tectum of the barn owl. *Science* **222**:939–941.

LeVay, S., M. P. Stryker, and C. J. Shatz (1978) Ocular dominance columns and their development in layer IV of the cat's visual cortex: A quantitative study. *J. Comp. Neurol.* **179**:223–244.

Magchielse, T., and E. Meeter (1982) Reduction of polyneuronal innervation of muscle cells in tissue culture after long-term indirect stimulation. *Dev. Brain Res.* **3**:130–133.

McLoon, S. (1982) Alterations in precision of the crossed retinotectal projection during chick development. *Science* **218**:1418–1420.

Meyer, R. L. (1979) Retinotectal projection in goldfish to an inappropriate region with a reversal in polarity. *Science* **205**:819–821.

Meyer, R. L. (1980) Mapping the normal and regenerating retinotectal projection of goldfish with autoradiographic methods. *J. Comp. Neurol.* **189**:273–289.

Meyer, R. L. (1982) Tetrodotoxin blocks the formation of ocular dominance columns in goldfish. *Science* **218**:589–591.

Meyer, R. L. (1983) Tetrodotoxin inhibits the formation of refined retinotopography in goldfish. *Dev. Brain Res.* **6**:293–298.

Murray, M., and M. A. Edwards (1982) A quantitative study of the reinnervation of the goldfish optic tectum following optic nerve crush. *J. Comp. Neurol.* **209**:363–373.

Murray, M., S. C. Sharma, and M. P. Edwards (1982) Target regulation of synaptic density in the compressed retinotectal projection of goldfish. *J. Comp. Neurol.* **209**:374–385.

Obata, K. (1977) Development of neuromuscular transmission in culture with a variety of neurons and in the presence of cholinergic substances and tetrodotoxin. *Brain Res.* **119**:141–153.

O'Brien, R. A., A. J. C. Ostberg, and G. Vrbova (1978) Observations on the elimination of polyneuronal innervation in developing mammalian skeletal muscle. *J. Physiol. (Lond.)* **282**:571–582.

Oswald, R. E., and J. A. Freeman (1980) Degradation rate of goldfish brain nicotinic acetylcholine receptor. *Brain Res.* **187**:499–503.

Oswald, R. E., J. T. Schmidt, J. J. Norden, and J. A. Freeman (1980) Localization of α-bungarotoxin binding sites to the goldfish retinotectal projection. *Brain Res.* **187**:113–128.

Piper, E. A., J. G. Steedman, and R. V. Stirling (1980) Three-dimensional computer reconstruction of cobalt-stained optic fibers in whole brains of *Xenopus* tadpoles. *J. Physiol. (Lond.)* **300**:13–14P.

Purves, D., and R. I. Hume (1981) The relation of postsynaptic geometry to the number of presynaptic axons that innervate autonomic ganglion cells. *J. Neurosci.* **1**:441–452.

Purves, D., and J. W. Lichtman (1980) Elimination of synapses in the developing nervous system. *Science* **210**:153–157.

Sakaguchi, D. S. (1984) The retinotectal system of *Xenopus*. The development of ganglion cells and their projections to the brain. Unpublished doctoral dissertation, State University of New York, Albany.

Scalia, F., and K. Fite (1974) A retinotopic analysis of the central connections of the optic nerve in the frog. *J. Comp. Neurol.* **158**:455–478.

Schechter, N., A. Francis, D. G. Deutsch, and M. S. Gazzaniga (1979) Recovery of tectal nicotinic–cholinergic receptor sites during optic nerve regeneration in goldfish. *Brain Res.* **166**:57–64.

Schmidt, J. T. (1978) Retinal fibers alter tectal positional markers during the expansion of the half-retinal projection in goldfish. *J. Comp. Neurol.* **177**:279–300.

Schmidt, J. T. (1979) The laminar organization of optic nerve fibers in the tectum of goldfish. *Proc. R. Soc. Lond. (Biol.)* **205**:287–306.

Schmidt, J. T. (1982) The formation of retinotectal projections. *Trends Neurosci.* **5**:111–116.

Schmidt, J. T. (1983) Regeneration of the retinotectal projection following compression onto a half-tectum in goldfish. *J. Embryol. Exp. Morphol.* **77**:39–51.

Schmidt, J. T., and D. L. Edwards (1983) Activity sharpens the map during the regeneration of the retinotectal projection in goldfish. *Brain Res.* **269**:29–39.

Schmidt, J. T., and J. A. Freeman (1980) Electrophysiologic evidence that the retinotectal projection in the goldfish is nicotinic–cholinergic. *Brain Res.* **187**:129–142.

Schmidt, J. T., C. M. Cicerone, and S. S. Easter, Jr. (1978) Expansion of the half retinal projection to the tectum in goldfish: An electrophysiological and anatomical study. *J. Comp. Neurol.* **177**:257–278.

Schmidt, J. T., D. L. Edwards, and C. A. O. Stuermer (1983) The reestablishment of synaptic transmission by regenerating optic axons in goldfish: Time course and effects of blocking activity by intraocular injection of tetrodotoxin. *Brain Res.* **269**:15–27.

Scholes, J. H. (1979) Nerve fiber topography in the retinal projection to the tectum. *Nature* **278**:620–624.

Sharma, S. C. (1975) Visual projection in surgically created "compound" tectum in adult goldfish. *Brain Res.* **93**:497–501.

Sharma, S. C., and M. Romeskie (1977) Immediate "compression" of the goldfish retinal projection to a tectum devoid of degenerating debris. *Brain Res.* **133**:367–370.

Sohal, G. S., T. L. Creazzo, T. G. Oblak (1979) Effects of chronic paralysis with α-bungarotoxin on development of innervation. *Exp. Neurol.* **66**:619–628.

Sperry, R. W. (1943) Effect of 180 degree rotation of the retinal field on visuomotor coordination. *J. Exp. Zool.* **92**:263–279.

Sperry, R. W. (1963) Chemoaffinity in the orderly growth of nerve fiber patterns and connections. *Proc. Natl. Acad. Sci. USA* **50**:703–709.

Springer, A. D., and B. W. Agranoff (1977) Effect of temperature on rate of goldfish nerve regeneration: A radioautographic and behavioral study. *Brain Res.* **128**:405–415.

Springer, A. D., and S. M. Cohen (1981) Optic fiber segregation in goldfish with two eyes innervating one tectal lobe. *Brain Res.* **255**:23–36.

Srihari, T., and G. Vrbova (1978) The role of muscle activity in the differentiation of neuromuscular junctions in slow and fast chick muscles. *J. Neurocytol.* **7**:529–540.

Straznicky, C. (1978) The acquisition of tectal positional specification in *Xenopus*. *Neurosci. Lett.* **9**:177–184.

Straznicky, C., R. M. Gaze, and T. J. Horder (1979) Selection of appropriate medial branch of the optic tract by fibers of ventral retinal origin during development and in regeneration: An autoradiographic study in *Xenopus*. *J. Embryol. Exp. Morphol.* **50**:253–267.

Stryker, M. P. (1981) Late segregation of geniculate afferents to the cat's visual cortex after recovery from binocular impulse blockade. *Neurosci. Abstr.* **7**:842.

Stuermer, C., and S. S. Easter, Jr. (1982) Regenerating optic fibers of goldfish do not follow their old pathways in tectum. *Invest. Ophthalmol. Visual Sci. (Suppl.)* **22**:45.

Thompson, W., D. P. Kuffler, and J. K. S. Jansen (1979) The effect of prolonged, reversible block of nerve impulses on the elimination of polyneuronal innervation of newborn rat skeletal muscle fibers. *Neuroscience* **4**:271–281.

Van Essen, D. C. (1982) Neuromuscular synapse elimination. In *Neuronal Development*, N. C. Spitzer, ed. pp. 333–376, Plenum, New York.

Willshaw, D. J., and C. von der Malsburg (1976) How patterned neural connections can be set up by self-organization. *Proc. Roy. Soc. Lond. (Biol.)* **194**:431–445.

Willshaw, D. J., J. W. Fawcett, and R. M. Gaze (1983) The visuotectal projection made by *Xenopus* "pie-slice" compound eyes. *J. Embryol. Exp. Morphol.* **74**:29–45.

Yoon, M. G. (1973) Retention of the original topographic polarity by the 180° rotated tectal reimplant in young adult goldfish. *J. Physiol. (Lond.)* **233**:575–588.

Chapter 20

Cell Interactions Involved in Neuronal Patterning: An Experimental and Theoretical Approach

SCOTT E. FRASER

ABSTRACT

Experimental and theoretical investigations of the nerve connections in the retinotectal system are discussed. A model for the patterning of nerve connections, based largely on adhesive interaction between cells, is described and tested against several classes of experiments on the lower vertebrate visual system. The analysis shows that a wide range of data can be simulated with this simple model; with the addition of an activity-dependent competition between nerve terminals, the model can fit more of the literature. A simple, biophysically based mechanism for such synaptic competition is also described and analyzed. The combination of these two models generates a hybrid, with all the capabilities of the original nerve patterning model, but with the added ability to fit the experiments on ocular-dominance columns in lower vertebrates.

The patterning of cells, tissues, and organs during development remains one of the most intriguing mysteries of biology. Yet our knowledge of the mechanisms that bring about this patterning is incomplete at best. A large data base that describes the appearance of order in the embryo and the responses of the patterns to experimental intervention has accumulated over the past decades. More recently, biochemical and immunological techniques have been focused on developing differential markers for embryonic cells. Thus we find ourselves in a situation in which a great deal of descriptive information is available, ranging from classical observation to biochemical characterization, but nevertheless surprisingly little is known about the underlying mechanisms that might guide the development of cell patterns.

This lack of data on mechanism becomes particularly limiting when considering the nervous system. The developing nervous system shows all of the patterned development characteristic of embryogenesis, but as an added complication, these groups of patterned cells interconnect with one another via synaptic contacts in precise and well-defined ways. Thus another level of patterning is nested with the patterned differentiation seen in other tissues. The challenge, therefore, becomes clear—to develop insights into the mechanisms that might pattern the nervous system and its interconnections. In some laboratories, an experimental approach has been employed to help define those interactions important in development. For example, the role of activity-dependent refinement of the retinotectal projection is being assessed by the use of tetrodotoxin (TTX) to silence neuronal activity (Harris, 1980; Meyer, 1983; Schmidt and Edwards, 1983). Another approach, which I discuss here, involves the use of computer simulations to test the capabilities of different hypotheses to explain patterning. The advantages of this theoretical approach are that it can quickly test the feasibility of a mechanism, direct attention to the possible shortcomings of experiments, and suggest further experiments.

Our efforts have concentrated on the patterning of connections in the lower vertebrate visual system, particularly in the retinotectal projection. The retinotectal system was selected because it provides a well-ordered and experimentally approachable preparation that offers the challenge of demonstrating both excellent neuronal specificity together with profound neuronal plasticity. We have developed a model for the patterning of the retinotectal projection based largely on adhesive interactions between cells; this model is capable of fitting the majority of the large body of experimental data on the retinotectal system (Fraser, 1980b). In order to explain the model and the testing it has received, I first discuss the retinotectal system, the assays used to study it, and the data on the experimental behaviors of the system.

BACKGROUND

The Retinotectal Projection

The major projection of the visual system in lower vertebrates is the link between the eye and the optic tectum (the retinotectal projection). The optic nerve crosses the midline and projects to the contralateral tectum with little or no projection to the ipsilateral tectum. The projection is ordered over the surface of the tectum such that a "map" of the visual field of the eye is conveyed to the tectum. The order and the orientation of the projection are very regular and preserve the near-neighbor relationships of the cells in the retina. That is, cells that are neighbors in the retina have neighboring termination sites in the tectum. The dorsal retina projects to the ventral (or lateral) tectum and the anterior (or nasal) retina projects to the posterior tectum.

Physiological Mapping of the Projection. Extracellular electrophysiology is typically used to assay the order of the retinotectal projection. A metal microelectrode is lowered into the superficial neuropil of the tectum, and a visual perimeter is used to provide stimuli to the eye. The electrodes used for extracellular recording appear to record preferentially from the terminal arbors of the optic nerve fibers, so the technique measures the distribution of optic nerve terminals, not functional connections. For each of about 25 electrode positions, the region of visual field that elicits activity at the electrode tip is determined. The set of electrode positions and the corresponding responsive areas are usually presented as shown in Figure 1A,B. This map is often referred to as the retinotectal map, although the term

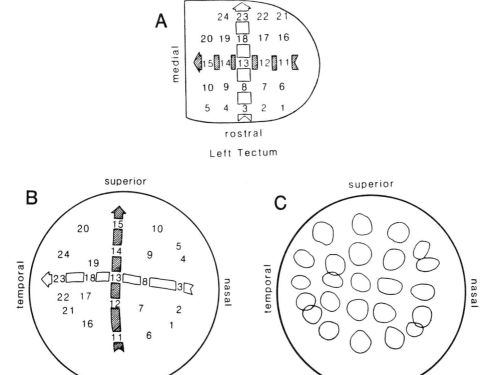

Figure 1. *A typical retinotectal map for* Xenopus. The electrode was inserted into the tectal neuropil at each of the numbered electrode positions (A), and the region of visual field that elicits activity at each electrode position was determined (B). The center of each region is typically denoted by the electrode position number. The entire visual field region (shown in C) that elicits activity at the electrode tip, which is the sum of signals from many optic nerve fiber terminals, is termed a multiunit receptive field.

"visuotectal map" more accurately reflects the method of stimulation. While there are limitations to this technique, it has the advantages of being quick, reliable, and reproducible.

As normally presented, the retinotectal map displays only the center of the area responsive to light for each electrode position. This unfortunately sacrifices some additional information that could be derived from the size of the multiunit receptive field (the total responsive area). The size of the multiunit receptive field (MURF) is typically 20° or larger (see Figure 1C). In contrast, the single-unit receptive fields that make up the MURF are about 5° or smaller. The larger size of the MURF is probably due in part to the electrode-sensing activity that is not immediately at its tip and in part to the large size of the terminal arbors of the optic nerve fibers. However, the major contribution is probably imprecision in the order of the projection. As a result of this slight disorder cells that are not true near neighbors in the retina terminate together and thereby increase the size of the responsive area. Thus MURF size can be used successfully as an assay of the short-range disorder in the projection pattern (see Schmidt, this volume).

Limitations of Physiological Mapping. There are some limitations to the extracellular electrophysiological methods described above. The recorded signals are thought to originate from the terminal arbors of the optic nerve fibers, so the technique assays the ordering of the highly branched terminals but gives no information as to the pattern of functional connections with the tectum. In normal, mature frogs it appears that the projection pattern and connection pattern coincide, but in experimentally altered animals the possibility exists that not all the terminal arbors are functionally connected. Thus the connection pattern must be independently assayed by isolating postsynaptic units in the tectum. This is typically accomplished with techniques very similar to those described above, but the usual electrode is replaced with a smaller-tipped, higher-resistance electrode. Some researchers assay the connectivity of the projection by measuring the laminar field potentials produced by a large shock to the optic nerve or a flash of light to the eye. While this provides an assay for the presence of functional synapses, it sacrifices any information about the pattern of connections because of the global nature of the stimulation.

The manner in which the eye is stimulated in a physiological experiment presents another limitation of the technique. For some experimental questions the exact regions of the retina that are stimulated should be known. Unfortunately, the stimulus is presented in the visual field of the eye, and the optics of the eye are therefore interposed between the stimulus and the retina. Because these optics change during amphibian development, as does the geometry of the eye, any developmental changes observed in the map could be produced or modified by the optics and their changes. To circumvent these difficulties, we have developed two techniques to stimulate the retina directly, thereby allowing a true retinotectal projection to be determined. These two techniques use fiber optics to stimulate the retina or to measure and compensate for the distortions from the optics of the eye, and hence they are called light-pipe techniques (Fraser, 1983a). Though difficult and time-consuming, the techniques allow a direct assay of the retinotectal map for experiments that require a high degree of precision.

Anatomical Assays of the Projection. In addition to the physiological assays for the order of the retinotectal projection, several anatomical techniques can be employed. These include horseradish peroxidase (HRP) and cobalt backfilling of the fibers and autoradiographic techniques.

Autoradiography has long been used to assay the extent and density of tectal innervation. This is accomplished by injecting radiolabeled amino acid into the vitreous of the eye, allowing time for the isotope to be incorporated into proteins and transported down the optic nerve, and then fixing and processing the tissue. In addition to this standard approach, two relatively new anatomical approaches have been developed to assay the order of the retinotectal projection. In the first technique, Meyer (1983) made small lesions in the retina before injecting the isotope. Because the damaged regions cannot incorporate or transport the isotope, the regions of the tectum to which the lesioned retina projected lack autoradiographic grains. The technique also yields a measure of the degree of disorder in the projection, since the blank region will be less pronounced if the projection is less ordered. The second technique, from Holt and Harris (1983), uses embryonic grafting to label only a subportion of the retina. A fragment of the eyebud is excised, soaked in labeled proline, and then reimplanted. The projection site of the labeled cells can then be determined by autoradiography.

Single-fiber tracing techniques using HRP or cobalt have been successfully employed by several laboratories (see Easter, Schmidt, this volume). Local application of the tracer to the cut end of the nerve, to the retina, or to the optic tract is used to label a subpopulation of the optic nerve fibers. The position in the tectum of this small group of labeled fibers can thus be directly assayed.

Limitations of Anatomical Techniques. A limitation of these techniques is that they allow the projection site of only one group of fibers to be assayed for each animal. In addition the techniques require that the animal be sacrificed. Thus time-course experiments can be performed only by collecting data from different animals. This is unlike the physiological techniques that can be repeated on the same animal at different times to build up a time course (see Fraser, 1983a). For the anatomical techniques, as for the typical physiological techniques, the functional connection pattern can only be inferred from the data. Electron microscopy can be used to confirm that the labeled neurons make synapses. However, using electron microscopy to document that each part of the labeled arbor seen at the light microscope level is functionally connected would represent a Herculean effort and is therefore not a feasible assay.

Specificity in the Retinotectal Projection

Several lines of evidence point toward the existence of some form of positional differences in the retina that help to guide the optic nerve fibers to the correct region of the tectum. For example, a wedge of retina can be grafted to an ectopic site on the eyebud. This fragment can grow with the host eyebud and form a wedge of the adult eye tissues. In some cases this wedge of tissue projects to the tectum in a manner appropriate to its position of origin, not to its final position in the eye (see Fraser and Hunt, 1980, for a review). In addition, the work of Holt, Harris, and their coworkers (cf. Holt and Harris, 1983) demonstrated that the eye forms a projection ordered from the earliest stages along at least the dorsoventral dimension of the tectum. Attempts to disrupt this order by grafting fragments of the eyebud to ectopic sites or to animals of an inappropriate age all failed. Since the fibers can still find their correct dorsoventral position in the absence of spatiotemporal growth cues or activity patterns, positional differences in the eyebud appear to be inherent and to help guide the fibers to the correct position on the tectum.

Additional evidence suggests the presence of some form of information on the surface of the tectum that helps to guide the fibers (see Fraser and Hunt, 1980, for a review). Pieces of tectum can be interchanged, or a single fragment can be rotated and reinserted. After such a surgical translocation of a piece of tectum, the optic nerve fibers are able to find the ectopic tectal fragment and to terminate in proper order on it. Additional evidence for positional cues on the tectum comes from the work of Holt, Harris and their coworkers (1983); in the grafts mentioned above, the fibers found their correct dorsoventral position on the tectum even with a disrupted order of arrival at the tectum. Similar experiments have used eyeless axolotls to demonstrate that the optic tract is not necessary for the formation of a normal map. In these cases the fibers were still able to locate the appropriate target area in the tectum despite having arrived by an abnormal route (Harris, 1982). Thus the data seem to indicate that both in regeneration and in development some form of positional difference that helps guide the optic nerve fibers is present on the surface of the tectum.

Plasticity in the Retinotectal Projection

In light of the specificity observed in the retinotectal system, the great plasticity of the projection may seem somewhat surprising. Following removal of a portion of the retina or the tectum, the remaining fragments project in an expanded or compressed fashion to compensate for the surgical manipulation. If half of the retina is removed, the optic nerve fibers from the remaining half will expand their representation to occupy the whole tectum. This expansion will take place even in cases in which controls show there is no regeneration of the missing fragment or regulation of the remaining tissue (see Fraser and Hunt, 1980; Meyer, 1982a, for reviews). Similarly, compression of the projection from a whole eye can be induced by the removal of half of the tectum. Here again regeneration can be ruled out. There remains some question as to whether the "labels" on the surface of the tectum might later regulate when innervated by these unusual expanded or compressed projections (cf. Schmidt, 1978; see Meyer, 1982a, for a review).

The importance of plasticity in normal development has now been shown for several lower vertebrates. The eyes and the tectum of frogs and fish continue to grow as the animals mature. Gaze and his coworkers realized that the pattern of retinal growth did not match the pattern of tectal growth, yet the retinotectal projection remained reasonably ordered throughout development and growth (see Chung, 1974). Thus they proposed that the optic nerve fibers must continually shift their connection sites to compensate for the growth mismatch. Evidence for such a slide comes from anatomical tracing of optic nerve fibers and physiological mapping experiments. Anatomical tracing of the fibers using autoradiography in *Xenopus* of different ages (Gaze et al., 1979) or using HRP in *Rana* of different ages (Reh and Constantine-Paton, 1984) indicates a shift in the projection over the tectum. Fiber-tracing experiments in goldfish indicate neuron trajectories consistent with such a shift (see Easter, this volume). Physiological measurements using light-pipe techniques documented a shift in the projection pattern in *Xenopus* (Fraser, 1980a, 1983a) and a shift in the projection and connection patterns in *Rana* (Fraser, 1983b). Thus the plasticity first observed after experimental manipulations of the retina or tectum has now proven to be an important feature of the normal development of the animal.

Regeneration and Refinement of the Retinotectal Map

The lower vertebrate visual system first attracted attention as an experimental system because of its regenerative ability. Reexamination of the details of this regeneration is now yielding new data on the nature of the ordering of the retinotectal map. After crushing or severing the optic nerve, it regenerates within a few weeks to form a well-ordered projection over the tectum. The projection initially demonstrates a good bit of disorder, as judged by both physiological and anatomical techniques (cf. Meyer, 1983; Schmidt and Edwards, 1983). Over the next couple of weeks the map becomes more and more refined in its regularity and accuracy until it is nearly as precise as the original map. Interestingly, this refinement of the projection appears to be activity-dependent as intraocular application of tetrodotoxin (TTX) blocks the refinement. TTX blocks the sodium channels of nerves and thus silences the propagated nerve activity from the retina down the optic nerve. The favored hypothesis is that the correlated neural activity of neighboring cells in the retina may help refine the projection as abolishing local differences by strobe illumination of the animals appears to block the refinement (see Schmidt, this volume). These results are in contrast to several experiments showing that the map can form, at least crudely, in the absence of activity (Harris, 1980). Thus the fibers may be able to find their approximate synaptic site without activity but may then require patterned activity to refine themselves into a more exact order.

Ocular-Dominance Columns

Recent work from several laboratories has documented that the innervation of a single tectum by two eyes leads to the segmentation of the projection into ocular-dominance columns (Levine and Jacobson, 1975; Constantine-Paton and Law, 1978; Meyer, 1979a). In frogs these ocular-dominance columns are organized as stripes preferentially oriented along the anteroposterior extent of the tectum. During regeneration, the fibers from the two eyes initially appear to make two overlapping projections which then segregate into columns. Several mechanisms—ranging from competitive interactions to inherent recurrent circuits in the tectum (see Constantine-Paton, 1982, for a review)—have been proposed for column formation. Some of these possibilities have now been ruled out. For example, one suggested cause was a global label that marked the fibers from the two eyes on the basis of their handedness or animal of origin. This was a viable suggestion because the right and left eyes of the same animal, or the eyes of two different animals, were used to doubly innervate the tectum in all experiments. To address this possibility an embryonic operation that caused a single eye rudiment to duplicate was performed. Since these two projections arise from the same side of the same animal, there are no right/left or histocompatibility differences between the two projections; the formation of stripes by these duplicate projections shows that global labels are not necessary (Fraser et al., 1982; Ide et al., 1983). Recent work from other laboratories indicates that the segregation of the fibers into stripes may be dependent on patterned neuronal activity, because intraocular injections of TTX abolish stripe formation (Meyer, 1982b; Schmidt, this volume).

Summary of Key Issues

The data on the retinotectal system of frogs and fish, briefly reviewed above, indicates the rich set of behaviors demonstrated by the system and the many means available for assaying the projection. The order of the projection can be assayed by extracellular electrophysiology, and the details of this map (i.e., receptive field sizes) yield information about the precision of the projection. Anatomical methods can document the rough topography of the projection, as well as measure its refinement, by means of local lesions in the retina. Experiments employing these techniques have indicated the following:

1. The normal projection can regenerate to near-normal precision following trauma to the nerve.
2. Optic nerve fibers possess some inherent information about their position in the retina that helps to guide their terminals.
3. Optic nerve fibers can locate and synapse on a transposed piece of tectum, indicating that the tectum has some form of positional information.
4. The system can be very plastic after surgical removal of a portion of the tectum or the retina (compression or expansion).
5. The system demonstrates considerable plasticity during its own normal development, probably to compensate for the continued and mismatched growth of the retina and the tectum.
6. Optic nerve fibers can find their appropriate region of the tectum in the absence of activity, but refinement of the map may require patterned neuronal activity.
7. When two eyes innervate the same tectum the optic nerve fibers from the two eyes segregate into ocular-dominance columns, the formation of which may require neuronal activity.

The data summarized here present a challenge to develop a consistent view that can explain the full range of the data. Our approach has been to develop and test models by using computer simulations, with the goal of defining a simple set of interactions that yield behavior complex enough to fit the observations. In the next section I outline our model and some of the testing it has undergone.

AN ADHESIVE MODEL FOR NEURONAL PATTERNING

The model we have developed and tested owes much to the concepts of differential adhesion refined by Steinberg (1970) and his coworkers. It proposes that the fibers are guided to the tectum and on the tectum by adhesive cues but that there is also a strong competition or repulsion between optic nerve terminals (Fraser, 1980b). The model has the ability to fit both the specificity and the plasticity data on the retinotectal system and also predicts the mixed results obtained from some experiments. It is not incompatible with some earlier models (Sperry, 1963; Hope et al., 1976), as it collapses in some experimental settings to become indistinguishable from these earlier models. It therefore represents an improvement over the previous efforts, drawing on their strengths and circumventing their difficulties. Before discussing computer tests of the model, I shall describe its rules and some thought experiments that should help to clarify its behavior.

Rules of the Model

The model consists of three adhesive interactions and a competitive interaction, listed in order of decreasing strength below.

> C represents *a strong, position-independent adhesion* between the optic nerve fibers and the tectum. An optic nerve fiber would adhere to any region of the tectum equally well by means of this mechanism, and hence C carries no positional information except to define the boundaries of the tectum and to help the fibers locate the tectum. The simulations of the model work best if the adhesion is mediated by a homophilic molecule (i.e., one that adheres to an identical molecule instead of to a distinct receptor or ligand). While the proposal of a homophilic molecule may depart from the views of many (e.g., Marchase et al., 1975) concerning adhesive molecules, experiments on N-CAM (see Edelman, this volume) demonstrate that examples of such a chemistry can be found in nature.
>
> R represents *a strong repulsion or competition* between optic nerve fiber terminals. This interaction is independent of the position on the tectum, but would drive the terminals toward a more uniform density of tectal innervation. The simulations work best if this interaction is initiated at the onset of synaptogenesis in the tectum.
>
> DV represents *a weak position-dependent adhesion* between optic nerve fiber terminals and the tectum, oriented along the dorsoventral axis of the eye and the tectum. This adhesion should also be mediated by homophilic adhesive molecules, because the same position-dependent adhesion that causes the fibers to prefer a certain site on the tectum would then also cause the fibers to adhere preferentially to their near neighbors (with respect to the retina). Thus the same position-dependent adhesion can generate both fiber–tectum guidance and fiber–fiber guidance.
>
> AP represents *a second, slightly weaker, position-dependent adhesion*, oriented along the anteroposterior axis of the eye and the tectum. Again, if the molecules that subserve this adhesion stick to one another in a homophilic manner, then fiber–fiber guidance would also be present.

The model thus consists of two strong position-independent interactions and a set of weak position-dependent adhesions. To best fit the data on lower vertebrates, these interactions are ranked in strength as: $C >> R >> DV > AP$. The model states that these interactions guide the optic nerve fibers in their search for their "best-fit site" on the tectum. In terms of the model a best-fit site is the location at which the fiber has maximized its adhesive interactions and minimized its competitive interactions. A fiber can search only its near vicinity for the best-fit site and cannot climb large energy barriers to find a distant site that is energetically more favorable. Thus each fiber searches for its own local minimum-energy site, which for some settings may not be the global energy minimum.

The nature of *AP* and *DV* deserves further explanation. The position-dependent adhesion could be produced in many ways; one of the simplest schemes is shown

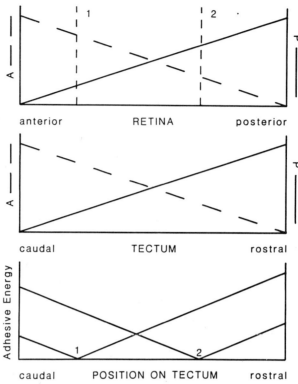

Figure 2. *A simple set of interactions to produce* AP *and* DV. The retina and tectum could be labeled with a pair of homophilic adhesive molecules (A and P), as shown in the top and middle panels. Thus each position along the anterior–posterior extent of the tectum is identified by a unique ratio, A/P. If two positions in the retina are considered (#1 and #2 at top), the pattern of adhesive interactions with the tectum can be determined, producing the two triangular energy wells shown in the bottom panel. Each position in the retina has a corresponding best-fit site in the tectum, defined as the apex of the triangular energy wells.

in Figure 2. In this scheme *AP* and *DV* are each produced by a pair of adhesive gradients across each of the tissues. For example, *AP* could be a pair of opposing gradients of the substances *A* and *P*, which are homophilic adhesive molecules (molecules that stick to each other, not to a receptor). The number of matched adhesive sites determines the adhesive strength between any two cells. We can select one cell from the set of retinal cells and determine the strength of its adhesion to different regions on the tectum by counting the number of matched *A* and *P* sites at each location. This has been done for cells from two different positions in Figure 2 (bottom panel), and in each case the result is a triangular energy well with the apex (minimum) at an appropriate best-fit site. In this scenario the positional information is contained in the *A/P* ratio, which determines the position of the apex of the triangular energy well. Increasing the total amount of both *A* and *P* while keeping the ratio constant would alter the depth and steepness of the energy well but would leave the best-fit site unaltered.

Thought Experiments

Simple thought experiments can be used to explain the characteristics of the model and its behavior. The C interaction is by far the strongest driving force the fibers experience. The fibers can maximize this interaction by being in contact with the tectum; because the interaction is position-independent, the contact can be anywhere on the tectal surface. The C interaction would clearly be the dominant force following any maneuver that deprived the fibers of contact with the tectum. After removal of a portion of the tectum, for example, a group of optic nerve fibers would be "orphaned" and could only regain their C interaction by invading regions of the tectum that they normally would not innervate. Both the R and DV–AP interactions would oppose this encroachment, but the C interaction is so much more powerful that the encroachment would still occur. Once the fibers are on the tectum, their C interaction is satisfied; now the R and DV–AP interactions become important in ordering the compressed projection. The R interaction would drive the system toward a uniform distribution of optic nerve fibers. This would mean displacing some fibers from their best-fit site on the tectum, which would be opposed by the DV–AP interactions. However, the R interaction is considerably stronger and would prevail. The R interaction is also position-independent, so the final ordering of the compressed fibers is the result of AP and DV. The fiber–fiber interactions mediated by DV–AP are sufficient to maintain near-neighbor order in the compressed projection; the fiber–tectum interactions play a role in this ordering as well, but more importantly they orient the projection properly. Thus this thought experiment, which considers the interactions, their magnitudes, and the cell movements required to maximize them, indicates that an ordered, compressed projection should result from the removal of part of the tectum.

Similar thought experiments can be used to consider the removal of a portion of the retina. In this situation no fibers are deprived of their C interaction and thus no movement of the fibers is necessary to satisfy this major force. The R interaction, however, would act as a driving force to spread the fibers, since those fibers at the border of the denervated region of the tectum would experience competition from one side only. The DV–AP interactions would oppose this expansion, but the larger size of the R interaction assures that it would dominate. The R interaction is limited in that it cannot force any of the fibers off the edge of the tectum, as this is strongly opposed by the possible loss of C interactions by those fibers. Through this interplay between R and C, the model avoids the dilemma (recognized by Prestige and Willshaw, 1975) of having to adjust the strength of the competition between nerve terminals in a context-dependent manner. Since the R interaction is position-independent its only concern is the density of nerve terminals. The DV and AP interactions act to order this expanded projection. As before, the fiber–fiber interactions help maintain the internal order of the projection, while the fiber–tectum interactions help in both the ordering and the orienting of the projection. It is possible to make the prediction, based on these thought experiments, that the degree of order in an expanded projection should decrease as the degree of expansion increases. This would be expected, because a greater expansion would lead to less overlap between terminal arbors and hence less fiber–fiber guidance. This prediction is borne out in computer simulations (S. E. Fraser, unpublished observations) and experiments (Meyer, 1978). Thus by considering the ranking of the interactions of the model, it is possible to predict the outcomes of some experiments. For more exact predictions based on the model we must turn to digital computer simulations.

Computer Tests of the Model

Each of the interactions of the model can be easily expressed in equation form as its effect on the free energy of each fiber. This makes computer tests of the model very straightforward. The C interaction is a large, square energy well whose depth is the strength of the interaction and whose width and length are the size of the tectum. The flat bottom of this energy well reflects our assumption that the binding energy is position-independent. The R interaction is a much smaller energy hill that surrounds every nerve terminal and raises the energy of neighboring arbors as a function of the amount of overlap between the two terminal arbors. The energy hills of all fibers are the same size and are independent of the fibers' positions on the tectum. The AP and DV interactions each appear as a triangular energy well, with the position of the apex of the energy well dependent on the position of origin of the nerve terminal. When AP and DV are summed, they form a shallow, cone-shaped energy well whose apex defines the best-fit site for that fiber or group of fibers.

Computer simulations of the model are performed by defining a large computer array to represent the tectal cells and allowing another large array, representing the retinal cells, to project to it. Based on the anatomy of the projection, the retinal elements possess a terminal arbor that contacts many of the tectal elements at any one time. The "cells" that make up the retinal array are individually polled to determine whether, using the energy profiles given above, one of their neighboring sites on the tectum is more energetically favorable. If so, the fiber is moved to that site. The simulation defines a neighboring site as one with which the terminal arbor of the optic nerve fiber is in contact. A fiber is moved by allowing it to sprout in the favorable direction and pruning away branches from the opposite side. This simulation iterates until all of the fibers have found sites on the tectum at which they are stable; the sites represent their local minimum energy. The fibers can determine only their local minimum, because they can explore only their immediate environment and hence would be unaware of a better site a great distance away.

The progress of the simulation is followed by asking the computer program to perform the equivalent of a physiological mapping experiment on the retinal and tectal arrays. An "electrode" is lowered into a site in the tectal array and the "optic nerve fibers" terminating in the near vicinity are determined. Based on the group of optic nerve terminals terminating in the region, a "multiunit receptive field" is drawn in the program's "visual field." The procedure is repeated, just as in a real mapping experiment, building up a retinotectal map from the electrode positions and multiunit receptive fields. The results are then displayed in the familiar format of a retinotectal map.

These computer simulations show that a normal map is the final result of the projection between a normal eye and a normal tectum. If the "optic nerve" is then "crushed," the initial, crudely ordered projection is later refined as the

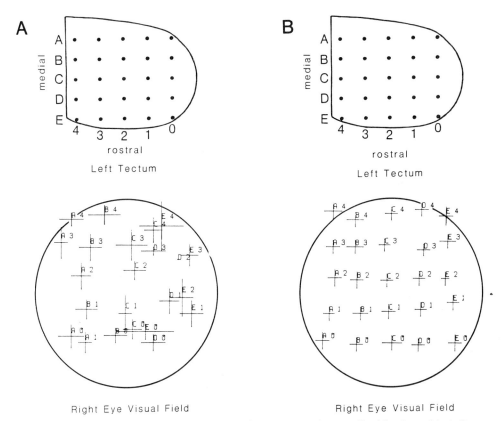

Figure 3. *The refinement of the retinotectal map after a nerve crush, as predicted by the model.* A: Soon after the fibers have started to order themselves on the tectum. B: After the fibers have refined their order to near normal. The refinement can be followed by both the increased order of the projection and the smaller multiunit receptive fields. The computer program generated these retinotectal maps by sampling the "optic nerve fibers" near the "electrode positions" shown on the tectal outline. The extent of each multiunit receptive field is given by the crossed "error bars."

arbors mature. This refinement results partly from the growth of the terminal arbors, which thereby increases the number of possible interactions. Because both fiber–fiber and fiber–tectum interactions guide the fibers, as the fibers become more ordered the effective energy well resulting from DV and AP becomes steeper and deeper. A simulation showing this progression from a crudely ordered to a more refined projection is shown in Figure 3.

Computer tests show that the model fits both the specificity and plasticity data. The ability to produce the specificity data can be seen in Figure 4, which compares a real and a simulated tectal graft experiment. One-dimensional simulations can be seen in Figure 5A,B. As predicted in the thought experiments, the model can also produce the plasticity data. Figure 5C shows the compression of the projection following removal of half of the tectum.

Mixed Results

Several experiments on the retinotectal system yield more than one result, all of which appear to be stable final states of the optic nerve fibers. Computer tests

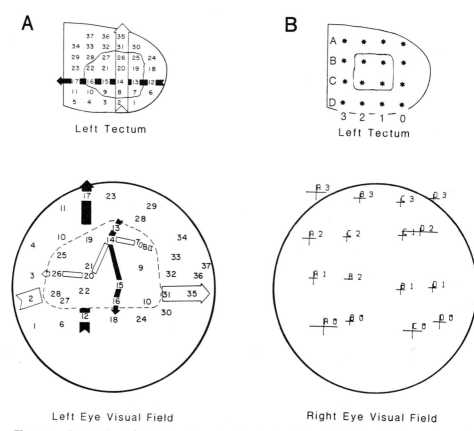

Figure 4. *A comparison of a tectal graft experiment and a simulation of a tectal graft.* An experiment in which a portion of the tectum (A) has been excised, rotated 180°, and reinserted. The fibers follow the grafted tissue just as they do in (B), a simulation of a 90° tectal graft rotation. The retinotectal simulation can be performed for a large set of electrode positions, and indicates that the transition between the fibers outside the graft and those that follow the graft is quite sharp. The coarse map is shown here for clarity. (A: Data from an experiment by R. K. Hunt; based on a figure in Fraser and Hunt, 1980.)

of the model show its ability to produce mixed results in the same settings in which experiments produce mixed results. For example, a projection of reversed polarity (a backward map) may be obtained in the goldfish retinotectal system by surgical procedures that confront ingrowing nerve fibers with a partially innervated tectum (Bunt et al., 1979; Meyer, 1979b). Since under very similar conditions the system will produce a properly oriented projection, this backward projection is a form of mixed result. In the model simulations, these same procedures also yield a backward projection. Such mixed results are somewhat easier to understand in one-dimensional computer simulations. Figure 5 shows some one-dimensional simulations of the specificity and plasticity experiments as well as some of the mixed results that can be obtained. Figure 5D shows the pattern from a simulation of the reversed projection pattern. In this case each fiber is in a local minimum, although far from its global minimum. We can compare the energies involved by comparing the free energies calculated by the computer simulation (expressed in arbitrary units). The difference in free energy for the two organizations (normal versus backward map; Figure 5A,D) is surprisingly small, only 200. In contrast, the energy peak separating the two states is greater than 1000. The size of the energy peak is difficult to calculate exactly because it is very path-dependent.

Tectal grafting experiments have also generated mixed results (cf. Hope et al., 1976) that are observed in model simulations. The experimental results fall into two categories. In the majority result, the fibers follow the grafted tissue; in the minority result, the fibers ignore the graft. In simulations the energy barrier separating the two states is quite large, with only a rather small difference in the energy between the majority and minority configurations (barrier = 500 arbitrary units; difference = 20 units). The two results are easy to understand in terms of the model since *AP* and *DV* interactions mediate both fiber–fiber and fiber–tectum interactions. The majority result represents the fibers maximizing their fiber–tectum interactions whereas the minority result represents the fibers maximizing their fiber–fiber interactions. Thus a competition between two organizing principles—the fiber–fiber and fiber–tectum affinities—generates the two different organizations. Not surprisingly the favored distribution varies with the size of the tectal graft. For small grafts the ratio of the perimeter to the surface

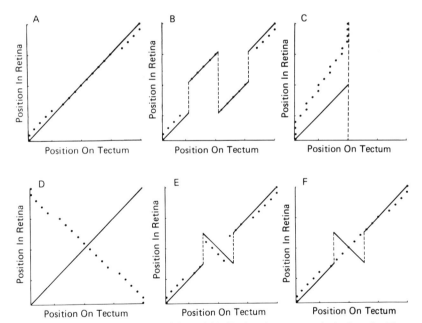

Figure 5. *One-dimensional simulations of the model indicating the presence of mixed results.* The two axes represent the position of origin of the optic nerve fibers and the position in which they terminate in the tectum. The straight line indicates the expected pattern for a perfect "specificity-type" model of patterning. The dots represent the centers of the optic nerve fiber terminal arbors, which extend almost two positions to each side. *A*: The normal map is a stable final state of the model; the slight deviations from the straight line result from edge effects present when small numbers of cells are used. *B*: The exchange of two pieces of tectum shows that the fibers can follow the transposed tectal fragments. *C*: Removal of half of the tectum (cut at the dashed line) leads to the compression of the optic nerve fibers onto the remaining half-tectum. *D*: A backward projection is also a stable result of the model in some conditions. The tectal graft rotation produces both the majority result (*E*), in which the fibers follow the graft, and the minority result (*F*), in which the fibers ignore the graft.

area is quite large; thus the fiber–fiber interactions, which occur at the perimeter of the graft, dominate. Larger grafts decrease this ratio and thus favor the fiber-tectum interactions that occur on the surface of the graft. As a result, the larger the graft, the more likely it is that the fibers will follow the grafted tissue. Therefore, it may well be that experiments in which progressively smaller grafts were used to determine the "minimum unit of tectal information" measured instead the consequences of varying this ratio of fiber–fiber to fiber–tectum guidance (Jacobson and Levine, 1975).

The differences in free energy between states plays a role in determining the majority result. However, not all states are equally accessible, so several other factors can come into play. For example, experimental conditions could make the highest-energy metastable state the most accessible and therefore the most likely experimental outcome. As a result, factors that are not central to the ordering of the fibers can play a large role in determining the final pattern of the projection. These factors may include fibers already innervating the tectum, the amount of debris remaining from previous innervation, and the path of fiber ingrowth. Interestingly, this means that several interactions, typically viewed as the antithesis of this sort of model, may have a large effect on the experimental result merely by determining which of the states is most accessible.

Ocular-Dominance Columns

Ocular-dominance columns have been the object of a great deal of experimental analysis and are a natural product of the model. The columns, like the mixed results reviewed above, are produced by a competition between the fiber–fiber and fiber–tectum adhesions of the model. The driving force to segregate into columns comes from the adhesion of the fibers to one another as well as to the tectum; therefore, in a doubly innervated tectum, the fibers can maximize their adhesive state by maximizing either their fiber–tectum or fiber–fiber adhesions. Normally, if the fibers were to clump together to maximize their fiber–fiber interactions gaps in the innervation pattern would be generated. This would be disfavored by the R interaction. However, when two eyes coinnervate a tectum fibers from the second eye could fill in these gaps by clumping together themselves. The fibers could thus maintain an even overall density on the tectum while aggregating into eye-specific regions, thereby avoiding the opposing effects of R.

In order to form columns the model requires that each fiber be able to identify the fibers that originated from the same retina as it did. In the simulations either a global label (which identifies the eye of origin of the fibers) or a local label (which identifies only those fibers from neighboring sites in the retina) is sufficient. As an example, the fibers could be labeled so that fibers from the same eye share a minor adhesive affinity (a global label); alternatively, the fibers could compete less strongly with their near neighbors, perhaps due to more coincident neural activity (a local label). Recent experiments indicate that global labels are very unlikely candidates for causing column formation (Ide et al., 1983). Thus attention is now being turned toward a local label that would identify nerve fibers that originate from near-neighboring ganglion cells in the retina. Since toxin treatments that abolish activity also abolish column formation (see "BACKGROUND" and

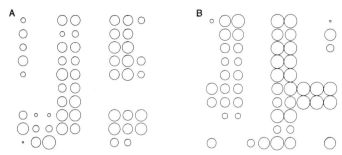

Figure 6. *The nerve patterning model can produce ocular-dominance columns if the nerve terminals can recognize their near neighbors when terminating in the tectum.* In this simulation, local correlation of nerve activity was allowed to decrease the R interaction by 25%, which produces stable and well-defined columns. A and B: The density of innervation of a subregion of tectum by the left and right eyes, respectively, is shown. The density of innervation at a particular position in the region from a given eye is shown by the size of the circle.

also Schmidt, this volume), the correlated activity of neighboring ganglion cells has become a prime candidate for this local label.

A computer simulation of the ocular-dominance columns in *Xenopus* appears in Figure 6. The figure shows a subregion of the tectum; the density of innervation of the tectum is shown by the size of the circle at each location. Such computer simulations have yielded several interesting insights into the mechanism that may produce columns. The size and the orientation of the columns are independent of the mechanism used by the optic nerve fibers to recognize their eye of origin. Instead the presence of columns is dictated by the strengths of the fiber–fiber and fiber–tectum adhesions, as well as by the strength of the eye of origin or neighbor marker. The spacing of the columns is largely determined by the anatomy of the terminal arbors of the optic nerve fibers. The preferential orientation of the columns is due to the greater strength of DV as compared to AP.

Dependence on Surface Area of Arbor

The formation of ocular-dominance columns and the refinement of the projection are both strongly dependent on fiber–fiber interactions, and hence on the total surface area of the optic-nerve fiber arbor. In the case of the refinement of the map it is easy to see that smaller arbors would offer less overlapping surface area to interact with neighboring arbors. Since fiber–fiber guidance plays a large role in the refinement process, this reduced interaction leads to less guidance and a more disordered projection. In the case of ocular-dominance columns it is the interplay between fiber–fiber and fiber–tectum adhesions that is important. Larger fiber–fiber interactions favor the formation of columns. As the arbors are reduced in size the interaction between arbors is reduced more severely because the amount of fiber–fiber interaction follows a square law. For example, reducing the size of the arbor by 50% would reduce fiber–tectum interactions by 50% but would reduce fiber–fiber interactions by 75%. In some simulations as little as a 50% decrease in surface area can change the pattern from one containing well-organized columns to one in which there are no columns. Thus the computer simulations clearly show that if experimental treatments affect arbor size, their effects on columns and topography may well be mediated by this rather nonspecific effect rather than by the more specific effects now being attributed to them. In light of these simulation results, it now appears prudent to examine the total arbor area after experimental treatments that reduce the topography of the map or that prevent the formation of ocular-dominance columns. This is an example of the way in which theoretical work can help to refine experimental design by calling attention to unforeseen complications or shortcomings.

A MODEL FOR ACTIVITY-DEPENDENT COMPETITION

The work described above indicates that our adhesive model for nerve patterning can simulate a wide range of experimental findings. Ocular-dominance columns present a challenge to any model of nerve patterning; computer simulations show that with the addition of some marker for neighborliness or eye of origin, our adhesive model can generate these columns. Experimental evidence now points to an activity-dependent marker for near neighbors in the retina as the permissive factor in column formation, perhaps due to the correlated activity of neighboring ganglion cells (cf. Schmidt, this volume). A possible mechanism through which this could be realized is to reduce the competition between nerve terminals slightly if there is correlated neural activity. If neighboring retinal ganglion cells tended to fire action potentials at the same time, they would compete with one another less strongly in the tectum and thereby preferentially group together.

The challenge now becomes one of developing a model that accounts for the activity-dependent or activity-modulated competition between synapses. Although many ideas have been put forward, most are not expressed in terms that can be incorporated in our nerve patterning model and tested. Together with Poo and his coworkers, we have been developing a model for synaptic competition, the "electromigration hypothesis," that draws upon the biophysics of the embryonic muscle cell (Fraser and Poo, 1982). While it is inappropriate to give a detailed description of this model here, the summary below outlines its key features and gives a broad view of its merits.

My discussion of synaptic competition centers on the neuromuscular junction. This choice is appropriate because much of what is known about the function and the development of synapses has come from experiments on this model system. The neuromuscular junction demonstrates a well-characterized form of synaptic competition during development. Each muscle fiber is initially innervated by several neurons; this polyinnervation is later reduced to the adult pattern in which each muscle fiber receives input from only one motor neuron (see Dennis, 1981 and Van Essen, 1982, for reviews). While this choice of system leads us away from our discussion of the central nervous system, the wealth of data it provides for our consideration justifies the detour.

The Biophysics of the Embryonic Muscle Cell

The electromigration model builds upon the biophysical measurements performed on embryonic muscle cells by several workers, most notably Poo and his collaborators (see Poo, 1981, for a review). These measurements show that the acetylcholine receptor (AChR) is quite mobile in the membrane for both cultured cells and cells *in vivo* (Poo, 1982; Young and Poo, 1982). The AChR was found to diffuse through the membrane at a rate only slightly slower than the rate measured for rhodopsin in the membrane of photoreceptor disks (Poo and Cone, 1974). This range of diffusion rates is consistent with relatively unhampered diffusion of the protein through the membrane lipids.

In addition to this rapid diffusion, the AChR responds to weak electric fields in the culture medium surrounding the cells. The receptors will accumulate to the cathode-facing pole of the cell for fields on the order of 0.1 V/cm or larger within hours. This electric field is rather small on cellular dimensions, representing about a 0.2 mV voltage drop across the diameter of the typical cultured muscle cell. Intermittent electric fields are also effective in relocalizing the AChR (S. E. Fraser and M.-m. Poo, unpublished observations). Interestingly, the AChR accumulates on the cathodal pole of the cell; on the basis of its charge it would be expected to accumulate on the anodal pole of the cell. This reverse migration may well indicate an overwhelming contribution of electroosmotic (solvent drag) effects to the movement of the AChR (McLaughlin and Poo, 1981).

In light of the rapid diffusion of the AChR and the speed with which electric fields can redistribute the receptors, it has been surprising to find that the AChR will not diffuse back into a uniform distribution after the removal of the electric field (see Poo, 1981, for a review). Instead the receptor aggregate remains stable for several hours. However, the aggregate is not completely immobile; a second electric field applied at 90° to the first field will move the aggregate en masse toward the new cathodal pole of the cell. It remains unclear whether this aggregating capacity is an inherent ability of the receptors or whether some aggregating factor colocalizes with the receptors and then stabilizes the AChR cluster.

The Electromigration Hypothesis

The electromigration hypothesis combines the behaviors of the AChR described above into a model for use-dependent synaptic competition. The basic premise is that the electric fields produced by synaptic potentials cause the redistribution of the AChR (Fraser and Poo, 1982). This redistribution can be rather long-lived because of the inherent stability of electric field–induced AChR aggregates. The most active synapse would have the largest time-averaged electric field, and hence would be the most effective at localizing the receptors. Since the total pool of receptors is limited, a stronger localization of receptors by one synapse would occur at the expense of the other synapses. Thus the model provides a mechanism for an activity-dependent competition between synapses. The electric field experienced by each AChR is the sum of the fields from all active synapses. Therefore the electric fields from two synapses that fire synchronously would partially cancel each other's ability to localize the receptors; a third synapse firing asynchronously would be at a disadvantage compared to the paired field produced by the synchronous synapses. The model therefore predicts that cells with correlated activity would compete less strongly with one another. This would provide a possible means for cells that were neighbors in the retina to "recognize" one another when synapsing in the tectum. Furthermore, multiple synaptic terminals from the same nerve would compete least strongly among themselves. While the hypothesis is couched in terms of a competition for transmitter receptors, the arguments would remain unchanged if the synapses were competing for a trophic or synapse-stabilizing substance that shared some of the mobility properties of the AChR.

Feasibility of the Model. Synaptic potentials, by necessity, produce an electric field along the membrane; this component is responsible for the spread of the synaptic potential. While it is difficult to calculate exactly those electric fields outside a cell that are produced by an active synapse, one can easily calculate the lower limit of the electric fields expected. For such a calculation we consider a cell in isolation in the extracellular medium with a synaptic current flowing through one synapse. This current must flow through the resistance of the medium surrounding the cell and thereby produce an electric field. Estimates indicate that electric fields during an end-plate potential (EPP) in skeletal muscle are greater than 0.1 V/cm for 50 μm around the synapse. Because in reality the extracellular space is filled to some extent by other cells, electrical current would be restricted to flowing through a smaller cross section of extracellular fluid; therefore these figures may greatly underestimate the electric field strengths. Yet even this lower limit of the electric field strength indicates that such fields are capable of rearranging the distribution of the AChR.

The intracellular electric fields produced by synaptic currents potentially could have an effect on the distribution of membrane proteins as well; however, the effects of intracellular fields on the distribution of the AChR or any other plasma membrane protein remain unexplored. Both theoretical and experimental evidence indicate that the intracellular electric fields should match or exceed the extracellular field in both magnitude and extent. Though it is tempting to consider these intracellular fields, our current ignorance of the effects of intracellular electric fields prevents further consideration at this time.

A Variant Based on Spontaneous Transmitter Release

One of the dangers of the electromigration model as described above is that it centers on the evoked release of transmitter (i.e., the EPP), which is directly dependent upon nerve activity. Thus a few very active neurons eventually could outcompete the bulk of the neurons in a large assemblage. For example, a muscle could be dominated by an active motoneuron that continues to extend its motor unit size, which seems at odds with experimental results. As a potential solution to this dilemma, I have recently concentrated on a variant of the electromigration model. In this modification of the model, competition is based largely on the spontaneous, quantal release of transmitter from the nerve terminal. It is this spontaneous, quantal release of transmitter that is termed the miniature end-plate potential (MEPP) in muscle. While MEPPs are considerably smaller than EPPs, they can generate a significant time-averaged electric field because of their constant presence. Calculations based on the same assumptions as the EPP treatment above indicate that the electric field generated by a MEPP will be greater than 0.1 V/cm within 5 μm of the transmitter release site. Thus the electric field in the near vicinity of a MEPP is sufficient to redistribute the AChR. Since the multiple neuronal inputs on developing muscle fibers are at the same end plate (Brown et al., 1976), the short-range effects of the MEPP do not present a difficulty in considering the reduction of polyinnervation. Therefore the MEPP may provide a means to avoid the possible overdependence of the electromigration hypothesis on nerve activity.

The advantage of coupling synaptic competition to spontaneous transmitter release is that the MEPPs are more constant and less dependent on nerve activity. If the frequency or size of MEPPs varies with the strength of a synapse, MEPPs may provide a direct and stable marker of "synaptic strength." It should be noted that the electric fields produced by EPPs are still a significant fraction of the total electric field at the synapse; therefore the argument made earlier about lessened competition between synapses that fire in a correlated manner is still valid. The features of the model are therefore as follows: (1) the synapse with the largest time-averaged electric field competes most strongly; (2) the time-averaged field in the near vicinity of the synapse is due to both MEPPs and EPPs; (3) MEPPs are somewhat more important than EPPs in determining the time-averaged field in the near vicinity of the synapse; (4) EPPs provide the dominant electric field farther away from the synapse as the MEPP contribution decays to subthreshold values with distance.

Several factors, including the amount of synaptic contact and the biosynthetic capacity of the neuron, would most likely sum to determine the frequency of MEPPs at any one synapse. Since MEPPs reflect the release of transmitter from the neuron, the biosynthetic capacity of the neuron must set an upper limit on the MEPP frequency (summed over all the terminals of the neuron). Thus the model automatically limits the number of synapses that any one neuron can support. This allows the model to escape from the previous dilemma in which a few cells dominate the competition, because a single neuron has a finite biosynthetic capacity, and thus can only support a limited amount of total "synaptic strength." Interestingly, this brings the biophysically based electromigration model into agreement with the purely theoretical effort of Willshaw (1981). His work better defined the general characteristics of a model that can account for the reduction in polyinnervation in developing muscle. These characteristics were: (1) that the neurons competed with one another for a limited substance on (or in) the muscle fibers; and (2) that the total synaptic strength of each neuron was limited. By way of comparison, in the MEPP-based electromigration hypothesis the AChR (or some other membrane protein) is competed for on the basis of MEPP frequency, which is limited for each neuron.

Implications of the Spontaneous Release Model. This model has implications for the reduction in polyinnervation seen in skeletal muscle. The model predicts that the reduction would be synaptic activity–dependent but not completely neuronal activity–dependent. Competition between nearby synapses is dominated by the spontaneous release of transmitter, whereas the competition between more distant synapses is dominated by evoked transmitter release. Thus the model predicts different characteristic results from blocking either all synaptic function or only evoked transmitter release.

Simple thought experiments can be used to predict the effects of blocking all synaptic function. The application of some toxins (curare, α-bungarotoxin, botulinus toxin) can block all synaptic currents and hence all electric fields generated by the currents; therefore these treatments would be expected to block the reduction of polyinnervation. In experiments using botulinus toxin treatments the reduction in polyinnervation is slowed (Hopkins et al., 1981), but this finding may be complicated by the increased sprouting of the nerve terminals. However, even this increased sprouting of the nerves may be consistent with the electromigration model. For example, the membrane protein for which the synapses are competing could also be involved in stabilizing synapses and hence slowing nerve terminal sprouting. Since the poisoned synapses are no longer able to localize the stabilizing factor, its influence would be lost and increased sprouting would result.

Similar thought experiments can be used to consider the effects of abolishing nerve activity. Blockade of nerve activity appears to slow the reduction of polyinnervation (cf. Thompson et al., 1979). However, the use of TTX on a subset of the fibers innervating a muscle did not appear to put those fibers at a disadvantage during the reduction of polyinnervation (Gordon, 1983). The multiple nerve terminals in a polyinnervated muscle are all within a short distance of one another and are therefore within the effective range of MEPP-produced electric fields. Neither TTX nor the other treatments can abolish the spontaneous transmitter release; therefore, the dominant electric field would persist and the competition between the terminals could proceed largely unaltered. The elimination of evoked transmitter release would be expected to reduce the effective range of the synaptic electric fields substantially, however, and may thereby slow the process of synaptic elimination or stabilization. The model predicts a different effect of TTX if the competing synapses were separated by a large distance. This is because the MEPP-based competition decays with distance such that EPP-based competition becomes dominant. TTX blockage of both nerves that innervate a muscle would therefore be expected to slow considerably the elimination of distant supernumerary terminals, whereas TTX application to only one nerve would be expected to put that nerve at a severe disadvantage.

Applications to the Central Nervous System

This variant of the electromigration model can be applied to the central nervous system and in particular to the formation of ocular-dominance columns discussed previously. The model predicts that optic nerve terminals would compete with one another over short distances on the basis of both their spontaneous and evoked transmitter release. The component of the competition due to evoked transmitter release would cause lesser competition between nerve terminals that are concurrently active. Thus the modified electromigration model still meets the criteria (see "Ocular-Dominance Columns") for a marker of near neighborliness. Optic nerve fibers from nearby ganglion cells would tend to fire coincidently and thus would compete among themselves less avidly. Blocking the propagation of action potentials with TTX would be expected to block the formation of columns because this would eliminate the marker of neighborliness. However, the TTX would not eliminate all competition, as it does in some models (Whitelaw and Cowan, 1981), because the spontaneous release would continue. Interestingly, the application of TTX to one optic nerve would not be expected to abolish column formation, which is in agreement with recent findings (Meyer, 1983). The unblocked nerve would still compete less strongly with itself and more strongly with the blocked nerve, which is sufficient for the segregation of optic nerve terminals.

By modifying the relative importance of spontaneous and evoked transmitter release, the properties of the model can be dramatically changed. For example, if the spontaneous release of transmitter is made insignificant or if synapses were very distant, the model might lose the ability to fit some of the results discussed above. Instead, the model produces a strong use-dependent modulation of synaptic strength. That is, an active synapse will become stronger than a less active synapse. If the receptors self-aggregate in a stable manner, then this use-dependent modification of synaptic efficacy could be very long-lived. The different abilities of this simple mechanism under different conditions make the electromigration model both powerful and adaptable to settings in the nervous system.

CONCLUSION

The goal of this chapter was to present some theoretical efforts in nerve patterning centering on the retinotectal system. Experimental findings and the techniques used to obtain some of these findings were reviewed to help give the reader a better understanding of the range of behaviors of the system, as well as an appreciation for the limitations of these experiments. While any review of the experimental literature of this sort is destined to be incomplete, the aim of this effort was not to give a comprehensive review of the field, but rather to provide a foundation for a discussion of a theoretical approach to these issues. Those readers interested in a more detailed discussion of the field should turn to some recent review articles on the system (Fraser and Hunt, 1980; Constantine-Paton, 1982; Meyer, 1982; Cowan and Hunt, this volume).

The adhesive model for nerve patterning presented above has the advantage of fitting both the plasticity and specificity literature on the retinotectal system of lower vertebrates. In addition, for several experimental settings, the model predicts the observed mixed results. Thus it becomes important that researchers in the field remain open-minded concerning differences in experimental results from the "same" experiment performed in two different laboratories. The modeling shows that subtle differences in experimental methods can make different states of the fibers more favorable. For example, very similar experiments in which a portion of both the retina and the tectum are removed can generate either a shifted but correctly ordered projection or a backward projection of the remaining eye-fragment on the remaining tectum. A slight difference in the position of the tectal lesion determines which of these states is favored in the model, and it appears that such a difference is the cause of the different results in the two experiments as well (Bunt et al, 1979; Meyer, 1979b). Therefore neither result can be taken as evidence for the presence or absence of tectal labels. Instead, our analysis suggests that a single model that uses tectal labels is compatible with both results.

The modeling also offers some predictions about the types of cell interactions that might underlie the ordering of the retinotectal projection. The position-dependent interactions in the adhesive nerve patterning model are rather small; the dominant interactions, C and R, are position-independent. The model will still function well even if the size of the position-dependent interactions, AP and DV, are made dramatically small in comparison to C and R. Thus subtle positional differences may be sufficient to produce all of the patterning seen in the system. This prediction is in agreement with other theoretical efforts (Whitelaw and Cowan, 1981) and contrasts with the view that the order of the projection must result from large positional differences that guide the fibers. Because such subtle differences in the positional markers might be sufficient, a direct approach to isolating these positional labels on the tectum or the retina may be extremely difficult and perhaps impossible with present technology. The fact that many laboratories have been unable to show any functional positional differences in either the retina or the tectum might be taken as evidence in favor of this view.

Computer simulations of the adhesive model show that it is able to generate the ocular-dominance columns seen in a doubly innervated tectum if optic nerve fibers can recognize their neighbors from the retina. A label of neighborliness could be provided by a synaptic competition in which cells with correlated activity compete less strongly. A simple mechanism based on the mobility of the acetylcholine receptor and other membrane proteins was outlined. The advantage of the electromigration model (Fraser and Poo, 1982) is that it builds upon the known behaviors of membrane proteins. The variant of the electromigration model based on spontaneous transmitter release can fit a variety of data on the reduction of polyinnervation in developing muscle. It offers predictions about the treatments that should and should not block the reduction of polyinnervation in developing muscle, many of which are experimentally testable. In addition, by combining the electromigration model with the nerve patterning model, a hybrid that can successfully address the formation of ocular-dominance columns in the retinotectal system is produced. For example, the hybrid model predicts that binocular silencing of neuronal activity would block the formation of columns. However, monocular silencing would still allow column formation. Since these predictions are in agreement with experiment observations, this hybrid model will be further refined and tested.

Each of the models considered represents a different type of theoretical effort. The nerve patterning model considers a set of experimental findings and attempts to predict the types of cell interactions that might underlie the findings. In contrast, the synaptic competition model considers some very basic experimental observations and assembles these known biophysical features into a model that addresses synaptic competition. The nerve patterning model therefore represents an effort at top-to-bottom modeling, in which a set of previously unobserved interactions are assembled into a workable model. In contrast, the synaptic competition model represents bottom-to-top modeling, in which known facets of the system are melded into a working hypothesis. Each approach has its own advantages: Top-to-bottom modeling can dramatically alter one's insight into the experimental system by drawing attention to new hypothetical mechanisms, whereas bottom-to-top modeling enjoys a firm experimental basis and can demonstrate the abilities of the assemblage of a set of known behaviors. Each approach also has some disadvantages: Top-to-bottom modeling relies heavily on imagination, and bottom-to-top modeling may make predictions that are quite hard to test. Thus we are left with two very different approaches, each with its own characteristic set of advantages and disadvantages. A synthesis of these two approaches, as has been attempted above, may offer a powerful tool with which to attack numerous questions in developmental neurobiology.

ACKNOWLEDGMENTS

I wish to thank the many people who helped, through a great many discussions, to refine the ideas presented in this chapter. Among these are R. A. Cone, M. Steinberg, D. Petersen, M.-m. Poo, J. Cowan, G. M. Edelman, H. Gordon, D. Van Essen, D. Willshaw, and C. von der Malsburg. I thank J. Edelman for assistance with some of the computer programs, V. Bayer for drawing the figures, and M. Bronner-Fraser, N. O'Rourke, and J. Coulombe for helpful comments on the manuscript. Some of the modeling was refined during my stay as a Fellow at The Neurosciences Institute of the Neurosciences Research Program, and I gratefully acknowledge the opportunities provided there. Portions of this work were supported by grants from the National Science Foundation and the National Institutes of Health.

REFERENCES

Brown, M. C., J. K. S. Jansen, and D. C. Van Essen (1976) Polyneuronal innervation of skeletal muscle in newborn rats and its elimination during maturation. *J. Physiol. (Lond.)* **261**:387–422.

Bunt, S. M., T. J. Horder, and K. A. C. Martin (1979) The nature of the nerve fiber guidance mechanism responsible for the formation of an orderly central visual projection. In *Developmental Neurobiology of Vision*, R. D. Freeman, ed., pp. 331–344, Plenum, New York.

Chung, S.-H. (1974) In search of the rules for nerve connections. *Cell* **3**:201–205.

Constantine-Paton, M. (1982) The retinotectal hookup: The process of neural mapping. In *Developmental Order: Its Origin and Regulation*, S. Subtelny and P. B. Green, eds., pp. 317–349, Alan R. Liss, New York.

Constantine-Paton, M., and M. I. Law (1978) Eye-specific termination bands in tecta of three-eyed frogs. *Science* **202**:639–641.

Dennis, M. J. (1981) Development of the neuromuscular junction: Inductive interactions between cells. *Annu. Rev. Neurosci.* **4**:43–68.

Fraser, S. E. (1980a) Light-pipe mapping of the *Xenopus* retinotectal projection. *J. Physiol. (Lond.)* **305**:113P.

Fraser, S. E. (1980b) A differential adhesion approach to the patterning of nerve connections. *Dev. Biol.* **79**:453–464.

Fraser, S. E. (1983a) Fiber-optic mapping of the *Xenopus* visual system: Shift in the retinotectal projection during development. *Dev. Biol.* **95**:505–511.

Fraser, S. E. (1983b) Plasticity in the retinotectal projections during normal development: Sliding connections in *Rana*. *Soc. Neurosci. Abstr.* **9**:760.

Fraser, S. E., and R. K. Hunt (1980) Retinotectal specificity: Models and experiments in search of a mapping function. *Annu. Rev. Neurosci.* **3**:319–352.

Fraser, S. E., and M.-m. Poo (1982) Development, maintenance, and modulation of patterned membrane topography: Models based on the acetylcholine receptor. *Curr. Top. Dev. Biol.* **17**:77–100.

Fraser, S. E., C. F. Ide, and R. L. Meyer (1982) Eye dominance columns formed by an isogenic double nasal frog eye. *Soc. Neurosci. Abstr.* **8**:450.

Gaze, R. M., M. J. Keating, A. Ostberg, and S.-H. Chung (1979) The relationship between retinal and tectal growth in larval *Xenopus*. Implications for the development of the retinotectal projection. *J. Embryol. Exp. Morphol.* **53**:103–143.

Gordon, H. (1983) Postnatal development of motor units in rabbit and rat soleus muscles. Unpublished doctoral dissertation, California Institute of Technology, Pasadena.

Harris, W. A. (1980) The effects of eliminating impulse activity on the development of the retinotectal projection in salamanders. *J. Comp. Neurol.* **194**:303–317.

Harris, W. A. (1982) The transplantation of eyes to genetically eyeless salamanders: Visual projections and somatosensory interactions. *J. Neurosci.* **2**:339–353.

Holt, C., and W. A. Harris (1983) Order in the initial retinotectal map in *Xenopus*: A new technique for labeling growing nerve fibers. *Nature* **301**:150–152.

Hope, R. A., B. J. Hammond, and R. M. Gaze (1976) The arrow model: Retinotectal specificity and map formation in the goldfish visual system. *Proc. R. Soc. Lond. (Biol.)* **194**:447–466.

Hopkins, W. G., M. C. Brown, and R. J. Keynes (1981) Persistence of multiple innervation in mouse muscles paralysed neonatally with botulinum toxin. *Soc. Neurosci. Abstr.* **7**:70.

Ide, C. F., S. E. Fraser, and R. L. Meyer (1983) Eye dominance columns formed by an isogenic double-nasal frog eye. *Science* **221**:293–295.

Jacobson, M., and R. Levine (1975) Stability of implanted duplicate tectal positional markers serving as targets for optic axons in adult frogs. *Brain Res.* **92**:468–471.

Levine, R., and M. Jacobson (1975) Discontinuous mapping of retina into tectum innervated by both eyes. *Brain Res.* **98**:172–176.

Marchase, R. B., A. J. Barbera, and S. Roth (1975) A molecular approach to retinotectal specificity. *Ciba Found. Symp.* **29**:315–341.

McLaughlin, S., and M.-m. Poo (1981) The role of electro-osmosis in the electric field induced movement of charged macromolecules on the cell surface. *Biophys. J.* **34**:85–93.

Meyer, R. L. (1978) Deflection of selected optic fibers into a denervated tectum in goldfish. *Brain Res.* **155**:213–227.

Meyer, R. L. (1979a) Extra optic fibers exclude normal fibers from tectal regions in goldfish. *J. Comp. Neurol.* **183**:883–901.

Meyer, R. L. (1979b) Retinotectal projection in goldfish with a reversal in polarity. *Science* **205**:819–821.

Meyer, R. L. (1982a) Ordering of retinotectal connections: A multivariate operational analysis. *Curr. Top. Dev. Biol.* **17**:101–145.

Meyer, R. L. (1982b) Tetrodotoxin blocks the formation of ocular dominance columns in goldfish. *Science* **218**:589–591.

Meyer, R. L. (1983) Tetrodotoxin inhibits the formation of refined retinotopography in goldfish. *Dev. Brain Res.* **6**:293–298.

Poo, M.-m. (1981) *In situ* electrophoresis of membrane components. *Annu. Rev. Biophys. Bioeng.* **10**:245–276.

Poo, M.-m. (1982) Rapid lateral diffusion of acetylcholine receptors in the embryonic muscle cell membrane. *Nature* **295**:332–334.

Poo, M.-m., and R. A. Cone (1974) Lateral diffusion of rhodopsin in the photoreceptor membrane. *Nature* **247**:438–441.

Prestige, M. C., and D. J. Willshaw (1975) On a role for competition on the formation of patterned neural connections. *Proc. R. Soc. Lond. (Biol.)* **190**:77–98.

Reh, T., and M. Constantine-Paton (1984) Retinal ganglion cell terminals change their projection sites during larval development of *Rana pipiens*. *J. Neurosci.* **4**:442–457.

Schmidt, J. T. (1978) Retinal fibers alter tectal positional markers during the expansion of the half retinal projection in goldfish. *J. Comp. Neurol.* **177**:279–300.

Schmidt, J. T., and D. L. Edwards (1983) Activity sharpens the map during the regeneration of the retinotectal projection in goldfish. *Brain Res.* **269**:29–39.

Sperry, R. (1963) Chemoaffinity in the orderly growth of nerve fiber patterns and connections. *Proc. Natl. Acad. Sci. USA* **50**:703–710.

Steinberg, M. S. (1970) Does differential adhesion govern self-assembly processes in histogenesis? Equilibrium configurations and the emergence of hierarchy among populations of embryonic cells. *J. Exp. Zool.* **73**:395–434.

Thompson, W., D. P. Kuffler, and J. K. S. Jansen (1979) The effect of prolonged, reversible block of nerve impulses on the elimination of polyneuronal innervation of new-born rat skeletal muscle fibers. *Neuroscience* **4**:271–281.

Van Essen, D. C. (1982) Neuromuscular synapse elimination. In *Neuronal Development*, N. Spitzer, ed., pp. 333–376, Plenum, New York.

Whitelaw, V. A., and J. D. Cowan (1981) Specificity and plasticity of retinotectal connections: A computational model. *J. Neurosci.* **1**: 1369–1387.

Willshaw, D. J. (1981) The establishment and the subsequent elimination of polyneuronal innervation of developing muscle. *Proc. R. Soc. Lond. (Biol.)* **212**:233–252.

Young, S., and M.-m. Poo (1982) Rapid lateral diffusion of the extrajunctional acetylcholine receptors in the developing muscle membrane of *Xenopus* tadpole. *J. Neurosci.* **3**:225–231.

Section 6

Gene Expression, Primary Processes, and Behavior

In this final section, we return to fundamental issues connected with the regulation of primary processes of development and finally touch upon the connection of such basic events with molar processes of behavior. Behavior itself can be considered to depend essentially upon developmental events occurring in populations of synapses, as well as upon hormonal expressions and other regulatory phenomena.

One of the key issues at the molecular level is to develop key analytic examples of the gene expression of neuromodulators and neuropeptides. Scheller uses the nervous system of the gastropod mollusk *Aplysia californica* as his model for understanding the order, function, and expression of neuropeptide genes. Using recombinant DNA technology, he and his colleagues have isolated relevant clones differentially expressed in bag cells and in identified neurons. The adopted strategy is then to analyze the nucleotide sequences of the clones in order to define the structure of precursor proteins. Based on these sequences, synthetic peptides are made and tested in the organism and are used as antigens to generate antibodies to probe for the entire protein. DNA clones have been made that encode a protein corresponding to the prevalent 12-kD protein of R3-R14 neurons, abdominal neurons that modulate cardiac output. In addition, a set of clones corresponding to egg-laying hormone (ELH) has been isolated. These molecular genetic analyses permit the synthesis of appropriate peptides which are being used to study the complex of behaviors associated with egg laying. Such studies indicate that the active peptides are components of larger precursors, often containing more than one active molecule. Expression of the genes for the precursors can be used as markers for the expression of a particular phenotype and, besides helping to understand behavior, are useful in lineage and differentiation studies.

One of the major primary processes of development, cell death, is often ignored in considering the development of the nervous system, despite the fact that it is seen prominently in both vertebrate and invertebrate species. Truman considers the subject in his chapter, emphasizing the usefulness of hormonal studies of programmed death. In the emergence of adult *Manduca sexta* moths, there is a programmed degeneration of abdominal muscles and half of the abdominal interneurons and motoneurons. Neuron death is on a characteristic program, and death of both kinds of cells is triggered by steroid hormones (ecdysteroids). In other related silk moth species, only muscle death occurs and is dependent upon steroids and a particular neuropeptide. By timed application of ecdysteroids, Truman dissociated the deaths of neurons and muscle cells and showed that the temporal program of death is intrinsic to the central nervous system: Cells that die have a maturational commitment point beyond which steroid replacement is no longer preservative. The use of protein synthesis inhibitors suggested that programmed degeneration depends upon synthesis of new messenger RNA. Truman thus emphasizes the need to distinguish the molecular basis of programmed death from the triggering signals that initiate the program. In the cases he has analyzed, the degeneration response resides with the affected cell's reaction to the hormonal signal and not to complex cell–cell interactions. Comparison of the similarities and differences in these programmed deaths with cell death in vertebrates, some of which may be nonprogrammed, is an important task for the future.

Genetic approaches to the analysis of neuropeptides in insects such as *Drosophila* are potentially of great power. White and Valles point out that genetic variants resulting in the elimination of synthesis of a specific molecule can be used to understand the role of that molecule in development and function. To that end, these workers have begun to map neurotransmitters and neuropeptides using immunohistochemical methods. Antisera to serotonin, substance P, and FMRFamide were used to stain whole-mounts of *Drosophila* larval nervous systems. Each specific serum revealed a characteristic neuropil and peripheral staining pattern involving discrete groups of cells. By using a temperature-sensitive mutant in the gene *Ddc* (encoding the enzyme dopa decarboxylase), a decrease in serotoninlike immunoreactivity was detected. Such studies are likely to provide benchmarks for analyzing the development of identifiable neurons, for studies of neural reorganization in metamorphosis, and for analysis of mutations affecting development of the central nervous system.

While we are a long way from understanding genotype-phenotype interactions in complex vertebrate species, the *Drosophila* model provides key guidelines for the genetics of behavior.

In the final chapter, Hall takes up studies of behavioral and neurochemical mutations that may affect several forms of conditioning in *Drosophila*, with emphasis on reproductive behavior. Hall first reviews the effects of learning mutants upon courtship behavior and considers how genetic mosaic analysis of these mutationally induced learning defects can be used to localize behavioral bases in particular regions of the animal's phenotype. In this fashion, the question of whether different learning mutations affect the same or different parts of the brain or of the life cycle can be approached. Similar genetic methodology can be used to analyze biological rhythm mutants that show altered courtship patterns. Hall describes a beginning analysis of the clock gene *per*, which is involved in courtship song rhythms.

It is perhaps fitting that this final section ends with an open question—whether it will be possible to use molecular approaches directly to understand the fundamentals of the development of a complex behavior. While such approaches are certainly essential to dissecting out the necessary components, it may be that the sufficiency will be supplied only by understanding the complex phenotypic interactions in a given species and environment. The least we can expect, however, is that molecular developmental approaches will provide us with means to modulate behavior if not to understand it fully.

Chapter 21

Gene Expression in *Aplysia* Peptidergic Neurons

RICHARD H. SCHELLER

ABSTRACT

The nervous system of the gastropod mollusk Aplysia californica *is a useful model system for study of the organization, expression, and function of neuropeptide genes. Differential screening of cDNA and genomic libraries allows the isolation of recombinant clones differentially expressed in bag cells as well as in neurons R3–R14, L11, and R15. Analysis of the nucleotide sequences of these clones allows the definition of the structure of the precursor proteins. This information is employed to generate synthetic peptides that are useful for studying the processing, distribution, and physiological activities of the peptides. Some neuropeptide genes are expressed specifically in the brain and therefore provide an opportunity to study neuronal development and the control of gene expression at the molecular level.*

The brain receives input from the external and internal environment, interprets this information, and responds so as to ensure survival of the species. This process relies on an extensive communication network between nerve cells and target tissues. The activity of one cell can influence the state of another cell by either of two general mechanisms. First, a cell can send an axon to a target and release a chemical messenger at a specialized membrane structure, the synapse. The messenger crosses the synapse and binds to receptors on the postsynaptic cell. The evoked response can be affected by a conformational change in an ion channel, which results in altered ion conductances, or alternatively the response may be mediated by second messengers such as calcium or cyclic nucleotides. Synaptic communications are usually short-lived (on the millisecond time scale), and the chemical messenger is often specifically degraded or resorbed into the presynaptic cell.

Chemical messengers can also be released into diffuse regions of the brain or the circulatory system, finding their targets some distance from the site of release. In this way intercellular communication can be achieved without the requirement of an axon extending to the target. This form of communication is therefore less constrained spatially. In addition the evoked responses are frequently long-lived, sometimes modulating the excitability of the target cell for minutes to hours.

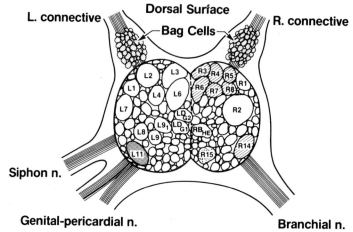

Figure 1. *Schematic representation of the dorsal surface of the abdominal ganglion.* Cells are labeled L or R (designating left or right hemiganglion) and carry identifying numbers (Frazier et al., 1967). Peptidergic neurons are designated by various shadings. Bag cell neurons are situated on the rostral side of the abdominal ganglion. Each cluster contains about 400 cells. Cells R3–R8 and R14 each have a single large axon that exits the ganglion via the branchial nerve and terminates on the efferent vein of the gill at the base of the auricle. R14 is anatomically distinct in that it has additional axons which terminate on the vasculature close to the ganglion. R15 is a cell thought to be involved in controlling the salt and water balance of the animal. This neuron sends out numerous processes and receives input from the osphradium. L11 is a cholinergic cell that also uses one or more neuropeptides as an extracellular messenger.

The longer duration of the response is thought to be due, at least in part, to the stability of this type of messenger.

Nerve cells have evolved a vast array of chemical structures for use as extracellular messengers. These include, for instance, serotonin, gamma-aminobutyric acid, and acetylcholine. Sometimes neurons also utilize molecules with other cellular functions such as glycine or adenosine triphosphate as chemical messengers. By far the most diverse extracellular messengers are the peptides. Chains of amino acids containing as few as two and as many as a few hundred residues have been shown to mediate intercellular communication in both the nervous and the endocrine systems.

Invertebrates such as insects and mollusks are particularly advantageous for studying cellular and molecular aspects of the nervous system (Kandel, 1976). Numerical simplifications, along with the convenient organization of neurons, makes these preparations experimentally manipulable. The gastropod mollusk *Aplysia* has a central nervous system consisting of about 20,000 neurons. These cells are organized into four pairs of symmetric ganglia in the head and a single asymmetric abdominal ganglion. The neurons can reach giant proportions, some being up to a millimeter in diameter. In addition the neurons occur in reproducibly identifiable positions, and many of their electrical activities have been correlated with specific physiological and behavioral events (Frazier et al., 1967).

The abdominal ganglion of *Aplysia* consists of about 2000 nerve cells and governs a number of reflex and fixed action patterns. One well-defined ganglion circuit that controls the withdrawal of the gill serves as a model system for studying simple forms of learning, including habituation, sensitization, and more recently classical conditioning (Kandel and Schwartz, 1982). The ganglion also governs several visceral functions, including excretion, respiration, and cardiac output. Since the excitability of many of the neurons that control these functions is modulated by neuropeptides, the abdominal ganglion serves as an excellent model system for studying these molecules.

We use a multifaceted approach to elucidate the function of neuropeptides, relying on both molecular and cell biological techniques. Using recombinant DNA techniques we have isolated a variety of genes that encode the precursors for peptide extracellular messengers utilized by neurons in the abdominal ganglion. These peptides are thought to mediate a number of specific physiological and behavioral events. Figure 1 is a diagram of the abdominal ganglion highlighting the neurons we are studying.

To isolate genes encoding neuropeptides we took advantage of the fact that some neurons devote a substantial fraction of their biosynthetic capacity to the synthesis of these products (Loh and Gainer, 1975; Berry, 1976a,b; Loh et al., 1977; Aswad, 1978). Individual or clusters of peptidergic neurons were dissected and stored frozen until used. From as few as 50 cells, 1–2 μg of poly(A)$^+$ RNA can routinely be obtained (Nambu and Scheller, 1983). The cells shown in Figure 1 were used as sources of mRNA. Copies of these poly(A)$^+$ RNAs were generated by using labeled nucleotide triphosphates and reverse transcriptases. These cDNAs were hybridized to duplicate replica filters of an abdominal ganglion cDNA library, and clones were selected that are expressed differentially in particular cell types. Three such clones are shown in Figure 2. The hybridizations were done under conditions such that only prevalent mRNAs were detected. Nucleotide sequencing, *in vitro* translations, and immunocytochemistry have demonstrated

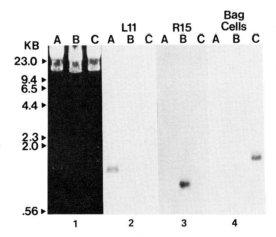

Figure 2. *Restriction enzyme and blot hybridization analysis of three neuron-specific genes.* Three recombinant phages from an abdominal ganglion cDNA library were selected for their ability to hybridize to the cDNA from identified abdominal ganglion neurons. Lane 1: 2 μg of phage DNA were digested with the restriction enzyme Eco RI, electrophoresed on a 0.7% agarose gel, and stained with ethidium bromide. Lane 2: The set of clones was transferred to nitrocellulose and hybridized to ^{32}P-labeled cDNA from cell L11. Lane 3: Hybridization with ^{32}P-labeled cDNA from cell R15. Lane 4: Hybridization with ^{32}P-labeled cDNA from the bag cells.

that all of the clones obtained are likely to encode the precursors for secreted peptides.

NEURONS R3–R14

The abdominal ganglion "white cells" R3–R14 are thought to modulate cardiac output (Price and McAdoo, 1979; Rozsa et al., 1980). Neurons R3–R14 each send a large axon to the efferent vein of the gill at the base of the auricle. R14 is anatomically distinct in that it sends additional processes to the arteries close to the ganglion. These neurons take up the amino acid glycine from the hemolymph 20–40 times more efficiently than do other abdominal ganglion neurons. This phenomenon is specific for glycine and for no other amino acid. Furthermore, the tritiated glycine absorbed by the cells becomes localized in vesicles and is transported to axon terminals as free glycine (Price et al., 1979; Price and McAdoo, 1981). When glycine is applied to the heart it modulates cardiac output, causing multiple contractions of enhanced force in response to an excitatory input (Sawada et al., 1981). These observations suggest that neurons R3–R14 may use glycine as a chemical messenger.

The R3–R14 neurons also synthesize a prevalent 12-kilodalton (kD) protein that is cleaved to smaller products with time. Since these cells have a white color characteristic of neurosecretory cells and the resultant cleavage products are localized to a vesicle fraction, the peptides are thought to be extracellular messengers. We isolated cDNA and genomic clones on the basis of their specific expression in the R3–R14 neurons. These clones encode a protein that has the physical characteristics of the prevalent product in R3–R14.

The structure of the single R3–R14 peptide gene is shown in Figure 3 (Nambu et al., 1983). Exon one encodes the 5' untranslated region of the mRNA with an intron positioned two nucleotides before the initiator methionine. The second exon encodes the first 43 amino acids of the protein that contain the "signal sequence" and the bulk of the negative charge in the precursor. A final exon encodes the remaining 65 amino acids and the 3' untranslated region (Taussig

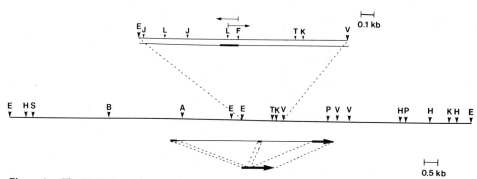

Figure 3. *The R3-R14 peptide gene.* A genomic clone encoding the R3-R14 gene is diagramed. The primary transcript and mRNA are indicated by arrows that point in the direction of transcription. The expanded region above encodes the middle exon and the regions sequenced are represented by arrows above this area. Restriction enzyme sites: E, Eco RI; B, Bam HI; H, Hind III; K, Kpn I; P, Pst I; A, Sal I; S, Sma I; V, Pvu II; T, Sst I; J, Ava II; L, Hae II; F, Hinf I.

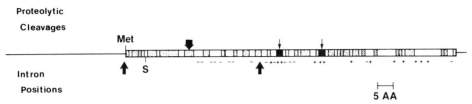

Figure 4. *Schematic diagram of the R3–R14 peptide precursor.* The coding region is derived from the only in-phase amino acid reading frame that matches the molecular weight of the *in vitro* translation product. The first AUG in the cDNA clone is the initiator methionine residue. Positions of hydrophobic residues (stippled boxes), histidine residues (diagonally lined boxes), a cysteine (S), dibasic (Lys-Arg and Arg-Arg) residues (black boxes), and charged amino acids (+, −) are noted. The thick arrow indicates the position of a putative cleavage site of the signal sequence; thin arrows indicate potential internal proteolytic cleavages. Arrows below the sequence indicate the positions of the intervening sequences.

et al., 1984). Both introns are quite large, at least 2.5 kilobases (kb), and generate a primary transcript from this gene of at least 6 kb. The fact that we find only a single copy of this gene in the *Aplysia* haploid genome suggests that this sequence is expressed in anatomically distinct neurons.

The cDNA clone we isolated is 1199 nucleotides long, excluding the poly(A) track. The first methionine, located at nucleotide position 174, is followed by a histidine residue at amino acid position five. Of the next 15 amino acids, 13 are hydrophobic and none is charged. This region is therefore likely to constitute the signal sequence necessary for insertion of the nascent protein chain into the lumen of the endoplasmic reticulum (ER) (Blobel and Dobberstein, 1975). ER cleavages are usually made at amino acid residues with small side chains such as alanine, glycine, or serine (Vlasuk et al., 1983). It is therefore likely that this cleavage occurs at the alanine found at position 20.

Previous studies have shown that the R3–R14 proprecursor is processed into at least two products of molecular weights 5 and 3.8 kD. These cleavages usually occur at basic residues and frequently are made at dibasic sequences. The primary sequence of the R3–R14 peptide precursor revealed two such dibasic residues. If cleaved they would generate three products, a 5-kD basic protein, a 3.4-kD acidic peptide, and a 12-amino-acid acidic peptide. The sizes of the two large products match the *in vivo* sizes. The 12-amino-acid peptide would not, however, have been resolved in the electrophoretic analysis (Figure 4).

To facilitate further analysis of this precursor we have synthesized peptides from its various regions. Three peptides have been generated. They consist of the first 16 amino acids that follow the signal sequence cleavage, the 12 amino acids between the pair of dibasics, and the carboxy-terminal nine-amino-acid residues. Solid-state synthesis is performed in such a way that 50% of the material is produced as native peptide and the other 50% contains an amino-terminal cysteine residue. This material can then easily be coupled to a protein carrier and used as an antigen to generate antibodies (Kreiner et al., 1984).

These synthetic peptides are being used to pursue three lines of study. First, it is important for further work to know exactly how the precursor is cleaved. To this end we are labeling the cells with tritiated amino acids and studying the resultant protein patterns. The products are being fractionated on acrylamide gels and by high-performance liquid chromatography (HPLC). Preliminary results demonstrate that the R3–R14 neurons generate the 12-amino-acid peptide found between the dibasic residues *in vivo* (R. R. Kaldany and R. H. Scheller, unpublished observations). Second, we are using the synthetic peptides to study the physiological activities of the R3–R14 chemical messengers. Again, preliminary evidence suggests that the 12-amino-acid peptide has the ability to alter the firing patterns of abdominal ganglion neurons. Third, rabbit and guinea-pig antibodies are being raised to the synthetic peptides. These antisera are being used to study the distribution of cells and processes that express the R3–R14 peptide gene and are helping to determine the identity of HPLC-fractionated pulse-chase cleavage products.

In summary, the precise function of the R13–R14 neurons remains obscure. It appears that peptides may be released centrally to affect the activity of neurons and peripherally to modulate heart rate and vasoconstriction. It is tempting to speculate that the multiple chemical messengers have effects on distinct yet functionally related targets. One might also wonder if the various inputs to the heart and vasculature might be affected by independent substances released by R3–R14. We hope to understand eventually how glycine and the peptides are packaged into vesicles and released. We would also like to know how the actions of these extracellular messengers which arise via independent biosynthetic pathways generate a coordinated response.

THE ELH GENE FAMILY AND THE EGG-LAYING FIXED ACTION PATTERN

Aplysia culminate their annual life cycle with the laying of large egg masses containing up to a million fertilized oocytes. A stereotyped pattern of behavior accompanies each egg-laying episode: The animal ceases locomotion and feeding and experiences an increase in cardiac output. As the egg string is extruded from the gonopore, the *Aplysia* grabs the string in its mouth and coils it into an irregular pile on the ocean floor. Presumably this arrangement of the egg mass increases the potential for fruitful hatching of the embryo, thus ensuring survival of the species. It is thought that egg laying is brought about by the activation of a specific neural network. This activation is likely to be the result of the actions of a number of neuropeptides on central neurons and peripheral tissues (Branton et al., 1978a,b; Dudek and Tobe, 1979).

Egg laying is preceded or accompanied by discharge of the bag cell neurons, two clusters of 800 electrically coupled neurons situated on the rostral side of the abdominal ganglion (Figure 1; Kupfermann and Kandel, 1970; Kupfermann, 1972). These cells release a battery of neuropeptides, the best characterized of which is the egg-laying hormone (ELH). ELH is a 36-amino-acid basic peptide that affects a number of neuronal and peripheral tissues (Chiu et al., 1979).

A second, less-understood tissue that synthesizes peptides related to ELH is the atrial gland. An exocrine gland, it is situated at the distal portion of the hermaphroditic duct and forms a continuous lumen with this tissue (Arch et al., 1978). The atrial gland synthesizes large amounts of two peptides called A and B. These 34-amino-acid peptides differ from each other by four amino acid substitutions and have no homology to ELH (Heller et al., 1980). When applied to bag cell bodies *in vitro*, either of these peptides initiates a bag cell discharge.

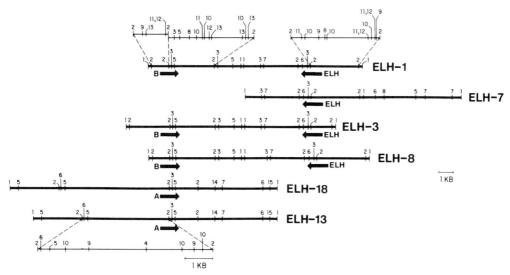

Figure 5. *Restriction enzyme maps of ELH recombinant clones.* Restriction enzyme maps were determined by a combination of single, partial, and double digests of intact clones and isolated fragments. The arrows indicate the position of mRNA homologous sequences and point in the direction of transcription as determined from R-loop analysis and DNA sequencing. Restriction enzyme sites: 1, Eco RI; 2, Pst I; 3, Xho I; 4, Stu I; 5, Pvu II; 6, Hind III; 7, Bgl I; 8, Xba I; 9, Ava II; 10, Hinc II; 11, Hae II; 12, Hha I; 13, Hpa II; 14, Bam HI; 15, Sal I.

Using differential hybridization techniques similar to those described above, we isolated a small family of genes encoding ELH and the A and B peptides (Scheller et al., 1982). Figure 5 shows the arrangement of some of the sequences in this gene family. The B peptide genes are linked to the ELH genes in every case observed. Furthermore, the DNA surrounding the genes is homologous for several kilobases on either side of the coding region. We have no evidence that the A peptide gene is located in the vicinity of the ELH or B genes, so this issue remains open.

Recent investigations have demonstrated that A, B, and ELH mRNAs contain exons in the 5' untranslated region that are not encoded within the clones shown in Figure 5 (A. C. Mahon and R. H. Scheller, unpublished observations). This implies that the initiation of transcription occurs at least several thousand nucleotides distal to the main body of the coding region. The identification of these chromosomal regions, and the elucidation of the mechanisms whereby ELH genes are expressed in the nervous system and A and B peptide genes are expressed in the atrial gland, await further investigation.

Two techniques have been used to uncover a network of interneurons in the *Aplysia* nervous system that express members of the ELH gene family (Figures 6, 7; McAllister et al., 1983). Indirect immunofluorescence with antibodies to ELH and the A peptide reveals the locations of immunoreactive molecules. While this technique is very sensitive and useful, one cannot be sure that a cell which contains the peptide is a site of primary synthesis of the product. For this information we rely on the technique of *in situ* hybridization. Cloned DNA fragments encoding the peptides are labeled radioactively by nick translation with [^{125}I]dCTP. These fragments are then hybridized to 5-μm sections of fixed tissue. After washing off the unhybridized radioactivity, the sections are coated

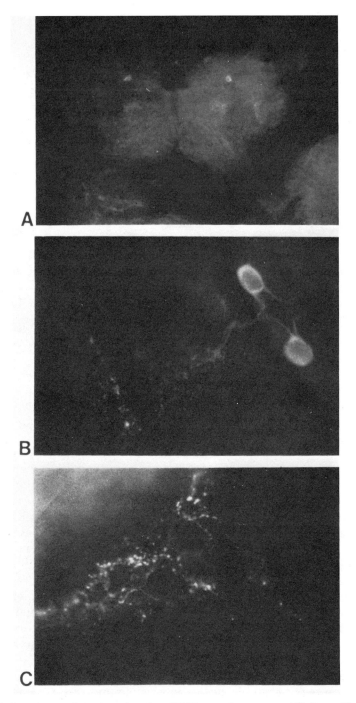

Figure 6. *Indirect immunofluorescence detection of ELH-expressing neurons.* Affinity-purified anti-ELH antibodies were reacted with whole nervous systems of 10–50-day postmetamorphic animals. Rhodamine-coupled goat anti-rabbit antibodies were used to visualize the reactive sites. *A*: The cerebral ganglion. *B*: The abdominal ganglion. *C*: The arborization of the pedal ganglion.

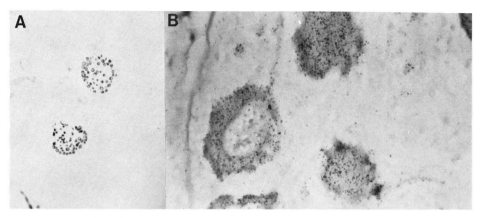

Figure 7. In situ *hybridization to the bag cells*. Radiolabeled DNA was hybridized to fixed abdominal ganglion tissue sections. A: An autoradiogram of a section through a bag cell cluster ($\times 15$). B: A single bag cell ($\times 450$).

with liquid emulsion and exposed. The sections are then developed, stained with a histological reagent, and visualized. Grains indicate the positions of sequences homologous to the probe. Nonspecific DNAs, as well as treatment with ribonuclease, are used to control the specificity of the reaction. A large number of grains can be seen in the cytoplasm of several neurons, indicating the presence of mRNA transcribed from the ELH gene family (Figure 7).

The approximate positions of the cells and processes found using the techniques described above are indicated in Figure 8. All ganglia, with the exception of the pleural and pedal, have immunoreactive cells in stereotyped positions. We never observed any cell bodies in the pedal ganglion, and the cells of the pleural ganglia were highly variable in both number and position. Whereas many of the processes seemed to enter either the sheath or the neuropil, others seemed to surround cell bodies, giving rise to gaps in the arborization patterns shown in Figure 6. Since varicosities are observed in all of these positions, it is likely that peptides are being released at the neuropil, onto cell bodies of other neurons, and into the vascularized sheath.

Although receptors for various chemical messengers may exist on neuron cell bodies, it should be noted that very few bona fide axosomatic contacts are found in the *Aplysia* central nervous system. The distribution of release sites again implies that the extracellular messengers utilized by these cells have central and peripheral targets. Bag cells are strongly implicated in the egg-laying process; however, no similar evidence exists for the other ELH-expressing cells. Many components of egg-laying behavior are governed by neural circuits in the head, and it is likely that ELH-producing neurons in the head ganglia are a source of the peptides that modulate these activities.

Reuptake or degradation of classical transmitters released at synapses ensures only short-range activities. As discussed earlier, neuropeptides such as ELH can modulate the activities of neurons at distances far from their sites of release. This suggests that the complete network of ELH-producing neurons should be active only during egg laying. If ELH were released from these cells at other times, its contact with the ovotestis would result in the inexpedient release of

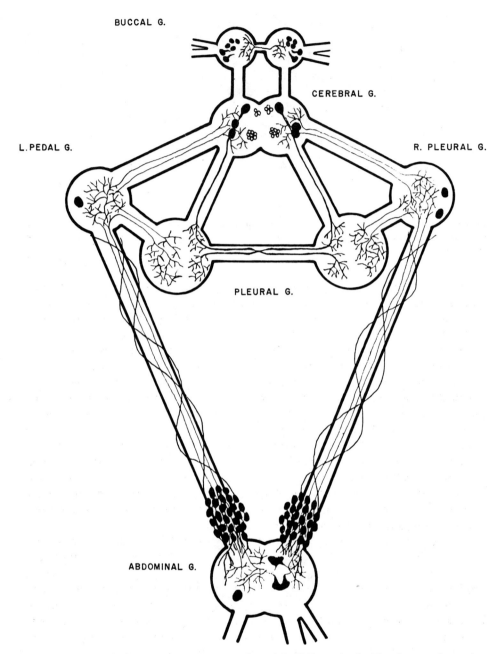

Figure 8. *A network of interneurons expresses members of the ELH gene family.* This diagram shows the approximate positions of the interneurons that express members of the ELH gene family. The positions of these cells were determined by the immunofluorescence and *in situ* hybridization techniques described in Figures 6 and 7.

eggs. The resolution of these issues awaits further experiments. Unfortunately, studies of this kind are not likely to be straightforward because of the difficulty in readily identifying these cells in the absence of immunocytochemical procedures.

We have used the immunocytochemical and *in situ* hybridization techniques described above to study the developmental origin of ELH gene-expressing neurons and the atrial gland (McAllister et al., 1983). *Aplysia* develop in several stages. Early development occurs in the egg capsule. After about 12 days a free-swimming and feeding larva, the veliger, hatches. This animal already has a complex nervous system consisting of cells organized into all the major ganglia. The veliger metamorphoses after 35 days or so and undergoes dramatic changes into the juvenile stage. Over a subsequent period of 30–60 days the juvenile matures into the adult form. As early as we have observed, which is stage three, or just after hatching, cells can be seen to be expressing ELH. The precise localization of these cells has been difficult because of the small size of the organism at premetamorphic stages.

Bag cells and the atrial gland arise after metamorphosis and therefore are convenient structures to study developmentally. Bag cells are the last neurons to develop and in a sense define a mature adult (Kreigstein, 1977). We took advantage of this and fixed whole 10–40-day postmetamorphic animals in Bouin's fluid, embedded them in paraffin, and generated 5-μm tissue sections. These sections were then reacted with DNA or antibody reagents. Cells that expressed members of the ELH gene family—as many as 10–25 cells—were observed along the inner lining of the body wall in the caudal region of the animal. Moving from this area rostrally we observed a number of cells along the body wall. Around the region of the abdominal ganglion, cells were observed on fibrous strands connecting the body wall with the ganglion or the pleural abdominal connective. Current methodology does not allow us to follow the migration of single neurons, but we propose that bag cells originate in proliferative zones in the caudal ectoderm and migrate to their eventual positions in the central nervous system. No mitotic structures have ever been observed in the ganglion or in migrating neurons, which suggests that all cell division occurs at or near the proliferative zone.

The situation within the atrial gland was quite different. Sections through the distal region of the hermaphroditic duct did not reveal any expressing cells until about 40–60 days postmetamorphosis. At this time both a highly infolded region of the lumen and patchy hybridization to cells in this area were observed. This distribution suggests that the atrial gland arises from a thickened epithelium of the duct itself. No expressing cells have ever been seen migrating to the duct in surrounding tissue.

To summarize the developmental expression of these genes as we understand it, two distinct modes of proliferation give rise to bag cell neurons and the atrial gland. Neurites arise from the ectoderm in the caudal region of the animal and then migrate into the central nervous system. The fact that these cells are expressing ELH genes at this early stage, while still along the body wall, suggests that they are destined to become bag cells before their migration. Cells do not migrate into the nervous system, sense their position, and undergo an inductive event that causes them to differentiate into particular neurons expressing unique characteristics. The function of the gene product in the migrating cells is unknown.

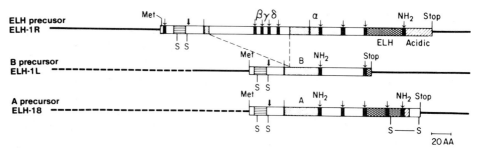

Figure 9. *Comparison of the precursors encoding ELH, A, and B peptides.* Coding regions are derived by comparison to the *in vitro* translation molecular weights and analysis of cDNA clones. Each of the three proteins is initiated by a methionine followed by a hydrophobic region (horizontally lined boxes). An S below the coding region indicates the location of a cysteine residue. Thick arrows represent the putative site of cleavage of the signal sequence. Vertical lines extending above the sequences represent potential cleavages at single arginine residues; thin arrows represent potential or known cleavages at dibasic, tribasic, or tetrabasic residues. If carboxy-terminal amidation is believed to occur NH_2 is written above the arrow. The A and B peptide homology is represented by stippled boxes, the ELH homology by crosshatched boxes, and the acidic peptide homology by diagonally lined boxes. Solid horizontal lines represent sequenced noncoding regions and interrupted horizontal lines those regions not yet sequenced.

Atrial gland tissue forms a continuous lumen with the cells of the white hermaphroditic duct by arising as a specialization of the duct epithelium. How the proper members of the ELH gene family are expressed in the different tissues and how this relates to the developmental origin of the cells remains a mystery.

The rest of this chapter concerns the organization of peptides in the A, B, and ELH precursors and the activities of these extracellular messengers (Scheller et al., 1983a,b). Figure 9 is a schematic of the three precursors as determined from the nucleotide sequence of the genes. All of the precursors begin with a methionine residue, which is followed by a stretch of largely hydrophobic residues. These regions are the signal sequences used by all membrane-bound or secreted proteins for entrance of the nascent protein chain into the ER. Following the hydrophobic region in the atrial gland genes, we observe the 102 nucleotides that encode either the A or B peptides. The active peptides are flanked on the amino-terminal side by a single arginine residue and on the carboxy-terminal end by the sequence Gly-Lys-Arg. The dibasic residue, Lys-Arg, is the signal for proteolytic cleavage. The glycine is a signal for the transamidation of the carboxy-terminal amino acid of the A and B peptides. Amidation protects the peptide from degradation by carboxypeptidases, thus stabilizing the product and allowing action at a distance.

A nucleotide sequence encoding peptides very similar to ELH is found 141 nucleotides from the end of the A and B peptides' coding regions. In the B peptide precursor a single base deletion alters the reading frame, putting a stop codon after the sixth amino acid of ELH. However, this deletion is not found in the A peptide gene. As a consequence of translating beyond the sixth amino acid, several additional changes become significant. A single base change substitutes a cysteine residue at position 25 of the ELH peptide. Also, substitution of arginine at position 22 results in an additional potential cleavage site in the A peptide gene. Moving 3' farther in the A peptide gene, a cysteine residue one amino acid prior to the stop codon is noted. It is likely that a disulfide bridge will be formed between the two cysteines in this region of the molecule.

The ELH gene expressed in the nervous system is more complex than either of the atrial genes. After the sixth amino acid of the A and B peptides' coding regions, a sequence of 240 nucleotides is present that does not exist in the atrial genes. The sequence TCATCA is found on either side of this 80-amino-acid region, which is most likely a deletion in the A peptide gene with respect to the ELH gene (we cannot formally rule out the possibility that this region is an insertion). This stretch of DNA encodes four pairs of dibasic residues, potential proteolytic processing sites. These dibasics flank the related pentapeptides Arg-Leu-Arg-Phe-His and Arg-Leu-Arg-Phe-Asp and an unrelated heptapeptide. The presence of this element in one gene but not another ensures that the production of these peptides will follow the tissue specificity of expression of the genes.

Moving toward the 3' end, five amino acid substitutions in a row are observed. One of them creates an arginine residue that serves as the cleavage site for another active peptide, nine amino acids in length. The carboxy-terminal cleavage of this peptide is the same as that of the A and B peptides, with one important difference: An arginine residue has been substituted for the glycine residue that signals amidation. This implies that the peptide is not amidated and therefore that it is less stable than either the A or B peptides. We have named the four peptides that potentially arise from this region the α, β, γ, and δ bag cell peptides (α, β, γ, δ-BCP).

Farther 3' in the precursor, the 108 nucleotides encoding ELH can be seen. As with the A and B peptides, the carboxyl terminus is flanked by glycine preceding the dibasic residue. Directly after the dibasic residue is the sequence of another peptide known to be released from bag cells. This 27-amino-acid "acidic peptide" is named to reflect its pK value. The nucleotide sequence of this area of the precursor precisely matches that of the peptide isolated from bag cells (Scheller et al., 1983a).

The many peptides cleaved from the three precursors can be grouped into classes according to their primary sequences. Three main divisions can be formed consisting of the acidic peptide homology, the ELH peptide homology, and the A and B peptide homology. About 70% of the amino acids within each of these groups are in identical positions. Less striking but greater than random homologies are observed between the three groups, implying that the precursor polyprotein arose from internal duplications within the transcription unit of a single primordial gene. Following the internal duplications large regions of the chromosome must have duplicated to give rise to a family of genes, each of which encodes multiple peptides. Divergence following the duplications then generated the unique sets of peptides we see today, and less well understood changes gave rise to the tissue specificity of expression.

Intracellular recordings have shown that alterations in the firing patterns of a variety of neurons accompany bag cell discharge (Figure 10; Mayeri et al., 1979a,b; Brownell and Schaefer, 1982; Mayeri and Rothman, 1982; K. Sigvardt and colleagues, unpublished observations). These activities presumably arise from the actions of one or more of the peptides encoded in the bag cell ELH precursor. Recent studies have been concerned with the actions of specific peptides that have either been purified from the ganglion or synthesized chemically. These peptides are perfused into the abdominal ganglion through the vasculature leading to the ganglion while intracellular recordings from various neurons are

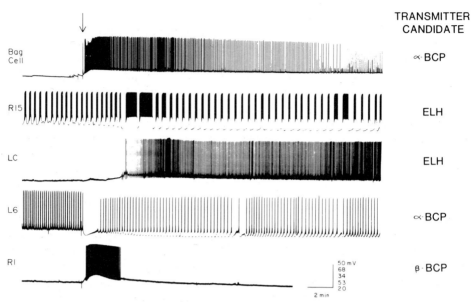

Figure 10. *Electrical activities of bag cells and the four types of firing pattern changes generated during afterdischarge.* Intracellular recordings from bag cells and four identified target neurons in the abdominal ganglion. The bag cells fired for a period of about 20 min, after which they entered a refractory period. Each target neuron showed one of four types of response: burst augmentation (R15); prolonged excitation (LC); slow inhibition (L6); and transient excitation (R1). The column at the right indicates the peptide transmitter candidates for the corresponding response.

made simultaneously. It appears as though several of the individual bag cell peptides have transmitter activities which together roughly mimic the activities seen during a bag cell burst. ELH is quite stable, but the action of other peptides such as α-BCP can be seen only if a number of protease inhibitors are included in the perfusion cocktail. The candidate peptide transmitters responsible for some of the activities seen during the discharge are listed next to their respective intracellular recordings in Figure 10.

Our model is based on the assumption that these peptide transmitters modulate the activity of cells in the central nervous system to generate the behavior accompanying egg laying. ELH is presumed to diffuse into the hemolymph, from which it is dispersed throughout the animal. Upon reaching the ovotestis this peptide induces contraction of smooth muscle and release of the egg string (Rothman et al., 1982). In this way the combined actions of the peptides results in a coordinated physiological response. The role of the atrial gland is not clear at this time.

In summary, this discussion has focused on the genes, distribution, actions, and developmental expression of neuropeptides in the *Aplysia* central nervous system. Several prominent themes emerge from these studies. The active peptides are synthesized as components of larger precursors that often contain more than one active molecule. The combined actions of the multiple substances used as transmitters and hormones by these peptidergic neurons result in coordinated physiological and behavioral responses. Genes encoding multiple peptide messengers can evolve by exon shuffling or internal duplications, followed by divergence. The genes for peptide precursors are often expressed in small networks of neurons that may or may not act as functional groups. Finally, the expression of these genes can be used as a marker for the commitment of cells to a particular adult phenotype. Experiments of this type suggest that bag cell neurons arise from the ectodermal lining of the body wall and migrate to their final position in the central nervous system. Since the cells are expressing members of the ELH gene family while still on the body wall, it is likely that the neurites are committed to the bag cell lineage before their migration and that interactions after migration are not required to determine this lineage.

ACKNOWLEDGMENTS

The author would like to thank Marina Picciotto and especially Mark Schaefer for the critical reading of this manuscript. The work discussed in this chapter is supported by grants to R. H. S. from the National Institutes of Health, the McKnight Foundation, and The March of Dimes Foundation.

REFERENCES

Arch, S., T. Smock, R. Gurvis, and C. McCarthy (1978) Atrial gland induction of egg-laying response in *Aplysia californica*. *J. Comp. Physiol.* **128**:67–70.

Aswad, D. W. (1978) Biosynthesis and processing of presumed neurosecretory proteins in single identified neurons of *Aplysia californica*. *J. Neurobiol.* **9**:267–284.

Berry, R. W. (1976a) A comparison of the 12,000 dalton proteins synthesized by *Aplysia* neurons L11 and R15. *Brain Res.* **115**:457–466.

Berry, R. W. (1976b) Processing of low molecular weight proteins by identified neurons of *Aplysia*. *J. Neurochem.* **26**:229–231.

Blobel, G., and B. Dobberstein (1975) Transfer of proteins across membranes. I. Presence of proteolytically processed and unprocessed nascent immunoglobulin light chains on membrane-bound ribosomes of murine myeloma. *J. Cell Biol.* **67**:835–851.

Branton, W. D., S. Arch, T. Smock, and E. Mayeri (1978a) Evidence for mediation of a neuronal interaction by a behaviorally active peptide. *Proc. Natl. Acad. Sci. USA* **75**:5732–5736.

Branton, W. D., E. Mayeri, P. Brownell, and S. Simon (1978b) Evidence for local hormonal communication between neurons in *Aplysia*. *Nature* **274**:70–72.

Brownell, P. H., and M. E. Schaefer (1982) Activation of a long lasting motor program by the bag cell neurons in *Aplysia*. *Soc. Neurosci. Abstr.* **8**:736.

Chiu, A. Y., M. W. Hunkapiller, E. Heller, D. K. Stuart, L. E. Hood, and F. Strumwasser (1979) Purification and primary structure of the neuropeptide egg-laying hormone of *Aplysia californica*. *Proc. Natl. Acad. Sci. USA* **76**:6656–6660.

Dudek, F. E., and S. S. Tobe (1979) Bag cell peptides act directly on ovotestis of *Aplysia californica*: Basis for an *in vivo* bioassay. *Gen. Comp. Endocrinol.* **36**:618–627.

Frazier, W. T., E. R. Kandel, I. Kupfermann, R. Waziri, and R. E. Coggeshall (1967) Morphological and functional properties of identified neurons in the abdominal ganglion of *Aplysia californica*. *J. Neurophysiol.* **30**:1288–1351.

Heller, E., L. K. Kaczmarek, M. W. Hunkapiller, L. E. Hood, and F. Strumwasser (1980) Purification and primary structure of two neuroactive peptides that cause bag cell afterdischarge and egg-laying in *Aplysia*. *Proc. Natl. Acad. Sci. USA* **77**:2328–2332.

Kandel, E. R. (1976) *Cellular Basis of Behavior*. Freeman, San Francisco.

Kandel, E. R., and J. H. Schwartz (1982) Molecular biology of learning: Modulation of transmitter release. *Science* **218**:433–443.

Kreigstein, A. R. (1977) Development of the central nervous system of *Aplysia californica*. *Proc. Natl. Acad. Sci. USA* **74**:375.

Kreiner, P., J. B. Rothbard, G. K. Schoolnik, and R. H. Scheller (1984) Antibodies to synthetic peptides defined by cDNA cloning reveal a network of peptidergic neurons in *Aplysia*. *J. Neurosci.* **10**:2581–2584.

Kupfermann, I. (1972) Studies on the neurosecretory control of egg laying in *Aplysia*. *Am. Zool.* **12**:513–519.

Kupfermann, I., and E. R. Kandel (1970) Electrophysiological properties and functional interconnections of two symmetrical neurosecretory clusters (bag cells) in abdominal ganglion of *Aplysia*. *J. Neurophysiol.* **33**:865–876.

Loh, L. P., and H. Gainer (1975) Low molecular weight specific proteins in identified molluscan neurons. I. Synthesis and storage. *Brain Res.* **92**:181–192.

Loh, L. P., Y. Sarne, M. P. Daniels, and H. Gainer (1977) Subcellular fractionation studies related to the processing of neurosecretory proteins in *Aplysia* neurons. *J. Neurochem.* **29**:135–139.

McAllister, L. B., R. H. Scheller, E. R. Kandel, and R. Axel (1983) Gene expression in the nervous system: *In situ* hybridization as a marker for the origins and fates of identified neurons. *Science* **222**:800–808.

Mayeri, E., and B. Rothman (1982) Nonsynaptic peptidergic neurotransmission in the abdominal ganglion of *Aplysia*. In *Neurosecretion—Molecules, Cells and Systems*, D. S. Farner and K. Lederis, eds., pp. 307–318, Plenum, New York.

Mayeri, E., P. Brownell, and W. D. Branton (1979a) Multiple, prolonged actions of neuroendocrine bag cells on neurons in *Aplysia*. II. Effects on beating pacemaker and silent neurons. *J. Neurophysiol.* **42**:1185–1197.

Mayeri, E., P. Brownell, W. D. Branton, and S. B. Simon (1979b) Multiple prolonged actions of the neuroendocrine bag cells on neurons in *Aplysia*. *J. Neurophysiol.* **42**:1165–1184.

Nambu, J. R., and R. H. Scheller (1983) Molecular cloning and characterization of neuropeptide genes from identified *Aplysia* neurons. In *Molecular Approaches to the Nervous System*, R. D. McKay, ed., pp. 110–121, Society for Neuroscience Press, Bethesda, Md.

Nambu, J. R., R. Taussig, A. C. Mahon, and R. H. Scheller (1983) Gene isolation with cDNA probes from identified *Aplysia* neurons: Neuropeptide modulators of cardiovascular physiology. *Cell* **35**:47–56.

Price, C. H., and D. J. McAdoo (1979) Anatomy and ultrastructure of the axons and terminals of neurons R3-14 in *Aplysia*. *J. Comp. Neurol.* **188**:647–678.

Price, C. H., and D. J. McAdoo (1981) Localization of axonally transported 3H-glycine in vesicles of identified neurons. *Brain Res.* **219**:307–315.

Price, C. H., D. J. McAdoo, W. Farr, and R. Okuda (1979) Bidirectional axonal transport of free glycine in identified neurons R3-14 of *Aplysia*. *J. Neurobiol.* **10**:551–571.

Rothman, B. S., G. Weir, and F. E. Dudek (1982) Direct action of egg-laying hormone on ovotestis of *Aplysia*. *Science* **197**:490–493.

Rozsa, S., K. J. Salanki, M. Vero, and D. Konjevic (1980) Neural network regulating heart activity in *Aplysia depilans* and its comparison with other gastropod species. *Comp. Biochem. Physiol. A. Comp. Physiol. (Oxford)* **65A**:61–68.

Sawada, M., D. J. McAdoo, J. E. Blankenship, and C. H. Price (1981) Modulation of arterial muscle contraction in *Aplysia* by glycine and neuron R14. *Brain Res.* **207**:486–490.

Scheller, R. H., J. F. Jackson, L. B. McAllister, J. H. Schwartz, E. R. Kandel, and R. Axel (1982) A family of genes that codes for ELH, a neuropeptide eliciting a stereotyped pattern of behavior in *Aplysia*. *Cell* **28**:707–719.

Scheller, R. H., J. F. Jackson, L. B. McAllister, B. S. Rothman, E. Mayeri, and R. Axel (1983a) A single gene encodes multiple neuropeptides mediating a stereotyped behavior. *Cell* **32**:7–22.

Scheller, R. H., B. S. Rothman, and E. Mayeri (1983b) A single gene encodes multiple peptide-transmitter candidates involved in a stereotyped behavior. *Trends Neurosci.* **6**:340–345.

Taussig, R., R. R. Kaldany, and R. H. Scheller (1984) A cDNA clone encoding neuropeptides isolated from *Aplysia* neuron L11. *Proc. Natl. Acad. Sci. USA* **81**:4988–4992.

Vlasuk, G. P., S. Inouye, H. Ito, K. Itakura, and M. Inouye (1983) Effects of the complete removal of basic amino acid residues from the signal sequence on secretion of lipoprotein in *Escherichia coli*. *J. Biol. Chem.* **258**:7141–7148.

Chapter 22

Hormonal Approaches to the Study of Cell Death in a Developing Nervous System

JAMES W. TRUMAN

ABSTRACT

The emergence of the adult Manduca sexta *moth is followed by the programmed degeneration of its abdominal muscles and about 50% of its abdominal interneurons and motoneurons. The neurons die according to a precise temporal program, with individual cells each having a characteristic time of death. Both neuron and muscle death are triggered by the disappearance of the steroid hormones, ecdysteroids, at the end of metamorphosis. In a related group of silkmoths only muscle death occurs after metamorphosis, and this degeneration requires the absence of the steroids as well as the presence of a neuropeptide. In both groups of moths, cell–cell interactions appear capable of retarding the endocrine-triggered death to a minor extent.*

The deaths of motoneurons, interneurons, and muscles in Manduca *can be dissociated from one another by carefully timed application of ecdysteroids, thereby indicating that the hormone is acting separately on all three groups of cells. In vitro studies involving cultured ganglia show that the temporal program of cell death is intrinsic to the central nervous system and can be blocked by addition of steroids to the medium. The cells that die exhibit a "commitment point" after which death can no longer be blocked by steroid replacement. Overt signs of degeneration appear about 10 hours after the commitment point. The processes that go on during this interval are unknown, but they likely involve the synthesis of new RNA and protein essential for the implementation of the degeneration program.*

Programmed cell death is an important force in the developing nervous system. It occurs in a number of developmental contexts, the most familiar of which is the adjustment of cell numbers between various regions of the central nervous system (CNS) or between the CNS and the periphery. Indeed, in many regions of the CNS of both vertebrates (Hamburger and Oppenheim, 1982) and invertebrates (Nordlander and Edwards, 1968), neurons are initially overproduced and then interactions with pre- or postsynaptic cells determine which neurons will survive and which will die. A second context for neuronal death is as a byproduct of the production of cells of a specific phenotype. This phenomenon is illustrated most clearly by studies on the nematode *Caenorhabditis elegans*. Neurons in the ventral CNS of this worm are produced according to a strict lineage, and in order to produce neurons with specific characteristics it is necessary to produce the earlier cells in the lineage, even though the latter might not be needed in that particular region of the animal (White et al., 1976; Horvitz et al., 1982). A third context for cell death is the elimination of neurons that have only a transient function during the life of the animal. One such example is the Rohon-Beard cells found in the spinal cords of embryonic fish and amphibians. These first primary sensory neurons die as their function is taken over by the dorsal root ganglion cells (Hughes, 1957; Lamborghini, 1981). Cells exhibiting transient functions during CNS development are also seen during the embryogenesis of invertebrates (e.g., in locusts; Bate et al., 1981).

Over the past few years studies on a number of invertebrate preparations have provided insights into the phenomenon of neuronal death (see Truman, 1984, for a review). Features of these systems such as the presence of unique, identifiable neurons that can be located and studied in successive individuals, coupled with the detailed knowledge of nervous system development in a few species, has allowed neuronal death to be studied in a cellular context that is not yet possible in vertebrate systems. One system that has been actively studied over the past four years is the postmetamorphic cell death that occurs in a number of moths. Degeneration in these insects is under endocrine regulation, a feature that has provided a tool for probing some of the features that regulate cell death.

THE PROGRAM OF CELL DEATH IN MOTHS

The process of metamorphosis in insects terminates with the ecdysis (or eclosion) of the newly formed adult from the old skin of the pupal stage. In the tobacco hornworm moth, *Manduca sexta*, this event signals a period of extensive cell death. Numerous abdominal muscles that were used for this behavior die over the succeeding 24–36 hours. The best-studied groups of muscles are the intersegmental muscles that are longitudinally arranged in segments 4–6 of the abdomen. These muscles are embryonically derived and are used during the larval stages and through metamorphosis. They represent the major muscle groups in the abdomen and generate many of the movements seen during ecdysis. By 16 hours after ecdysis these muscles are noncontractile, and by 30 hours they are essentially gone (Schwartz and Truman, 1983).

This programmed muscle degeneration is accompanied by the death of about 40% of the neurons in the segmental abdominal ganglia (Taylor and Truman, 1974; Truman, 1983). Interneurons begin to degenerate about two hours before ecdysis. Peak numbers of dying cells are evident at 12 hours after ecdysis and by 30 hours degeneration of these cells is essentially finished. Motoneurons also die beginning at about eight hours after ecdysis. By about 36 hours the last motoneurons that are fated to die have begun the process. Remains of these degenerating cells can be found up through about 90 hours after ecdysis.

The motoneurons that die are large cells that have characteristic locations in the ganglion. Since many of them can be reproducibly identified in sectioned material, it is possible to follow the fate of specific neurons through the degeneration process (Truman, 1983). Two aspects of the degeneration of the motoneurons are of interest. First, death does not occur randomly in the motoneuron population; specific cells invariably die, whereas other cells always live through the remainder

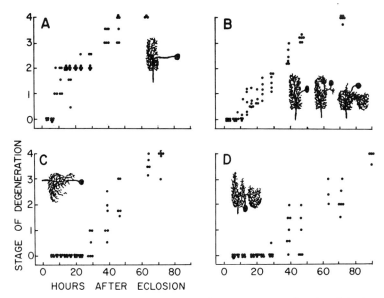

Figure 1. *Progression of cell death in various identified motoneurons as a function of the time after adult ecdysis.* A: MN-11; B: D-IV motoneurons; C: MN-2; D: MN-12. Each point shows the average stage of degeneration of the respective cells in a single abdominal ganglion. Insets show the central morphology of the neurons. Degeneration stages as described in Figure 2. (Modified from Truman, 1983.)

of adult life. Second, as illustrated in Figure 1, the motoneurons that die do not do so at the same time; rather, there is a characteristic sequence of degeneration with each cell dying at its own particular time in this sequence. The first motoneuron to die is MN-11, which commences degeneration at eight hours postecdysis. The D-IV cells, three pairs of ventral intersegmental muscle motoneurons situated in the anterior dorsal medial region of the ganglion, begin to die at 12 hours. MN-2 initiates degeneration at about 24 hours and MN-12 at about 36 hours postecdysis. Thus, the neurons die according to a precise temporal program that is reproducible from individual to individual.

It is not clear whether the interneurons likewise die according to such a temporal program, since it is not possible to follow the time course of death of specific identified individuals. However, the ganglia reproducibly show regional differences as to where cells are dying at various times. These characteristic patterns suggest that the interneurons also exhibit cell-specific times of degeneration.

The ability to identify the same cell in successive preparations allows a precise description of the timing of events during cell death (Figure 2; Stocker et al., 1978). We concentrated on the D-IV cells because their cell bodies are located in a region of the ganglion that is shared by only four other neurons. Up through 12 hours after ecdysis the ultrastructural appearance of these cells is identical to that of neurons that live through the life of the adult. At about 12 hours, the endoplasmic reticulum releases its ribosomes and breaks up into small channels. A few hours later cellular organelles such as mitochondria begin to swell and attain a rounder shape. By about 25 hours the first ultrastructural changes are

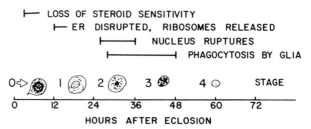

Figure 2. *Time course of the events that occur during the degeneration of D-IV motoneurons.* Drawings show the appearance of the cell bodies at various degeneration stages: 0, a healthy cell with round nucleus and dark perinuclear cytoplasm; 1, nuclear shape distorted, the cytoplasm turned pale; 2, nucleus collapsed and condensed in center of cell; 3, entire cell collapsed and darkly staining; 4, cell considerably shrunken and lightly staining. ER, endoplasmic reticulum.

noted in the nucleus with the abrupt rupture of the nuclear membrane. The surrounding glial cells then begin to engulf pieces of the degenerating neurons. The cell shrinks and by about 60–70 hours is reduced to a ball of tightly wrapped membranes. Thus the death of these neurons appears to be a very orderly process during which the cell is systematically dismantled. This type of orderly programmed cell death has been termed apoptosis (Wyllie, 1981) to distinguish it from the more disorderly types of degeneration seen when such external factors as toxins or starvation cause cellular necrosis.

The process of postecdysial muscle atrophy or degeneration occurs commonly in the insects, being reported in true bugs (Wigglesworth, 1956), a number of lepidoptera (Finlayson, 1956; Lockshin and Williams, 1965a), and flies (Cottrell, 1962). Whether this muscle death is also accompanied by degeneration of central neurons has not been examined for these species. Although neuronal death occurs in *Manduca*, in a related family of moths, the giant silkmoths, the abdominal motoneurons persist through the life of the adult even though the muscles they innervate die after ecdysis. In these moths there is a limited amount of interneuron degeneration but not on the scale that occurs in *Manduca* (J. W. Truman, unpublished observations).

THE TRIGGER FOR CELL DEGENERATION

Endocrine Factors

Neuronal survival can depend on a number of factors, including the formation of appropriate contacts with pre- or postsynaptic cells, the availability of certain growth factors, and the presence or absence of hormones. In the case of muscle and neuron degeneration in *Manduca*, the changing steroid levels that occur around the time of adult ecdysis play a major role in the regulation of cell survival (Schwartz and Truman, 1983; Truman and Schwartz, 1984). The first evidence that the time of cell death could be manipulated experimentally came from experiments in which the abdomens of developing insects were separated from the head and thorax by means of a hemostat clamped at the thorax–abdomen juncture. When abdomens were isolated in this manner two to three days before the normal time of ecdysis, the abdominal neurons and muscles showed a precocious onset of degeneration. The transection of the nerve cord at the abdomen–thorax junction at this time did not accelerate death, suggesting that disruption of humoral communication with the head and thorax, rather than interruption of neural connections, was responsible for degeneration. An important humoral factor in the blood during this time is the molting hormone 20-hydroxyecdysone (20-HE). Levels of this ecdysteroid are high early in adult development but decline progressively as development proceeds toward completion (Schwartz and Truman, 1983). The hormone is made by glands in the thorax, and isolation of the abdomen results in a precocious drop in steroid levels in this region of the body. This observation suggested that the ecdysteroid decline that occurs at the end of development might signal the death of neurons and muscle.

The hypothesis of the steorid regulation of neuron and muscle death was tested by injecting various dosages of 20-HE into isolated abdomens or intact insects late in development (Schwartz and Truman, 1983; Truman and Schwartz, 1984). The steroid treatment evoked delay in the time of both neuron and muscle death. Significant delays were obtained after injection of physiological levels of steroid, and the length of the delay increased with increasing dosages of 20-HE. Maintenance of moderate levels of circulating steroids by continuous infusion of 20-HE resulted in the preservation of the muscles and neurons for the duration of the treatment (up to one week). Thus the steroid decline that occurs at the end of development appears to be the trigger for postecdysial death in this insect.

All of the motoneurons that die are ones that supply muscles that also die. In most cases the muscle degeneration commences prior to the degeneration of the motoneurons that supply them. Consequently it was important to determine whether the steroid effect on neuronal death represented a direct effect on the CNS or an indirect effect through the death of peripheral targets. One argument against the latter hypothesis is the finding that neurons in implanted ganglia undergo degeneration in concert with those of their host (Truman and Schwartz, 1984). However, at least some of these cells hyperinnervate muscles in the region of the implant, and these improper targets may provide some information to the motoneurons. Other evidence suggesting a direct action of ecdysteroids on the CNS comes from experiments in which ecdysteroids are given to the insects at various times late in development. For each system that is affected by ecdysteroids one can identify a critical time before which ecdysteroids can block or delay the effect but after which the treatment is without effect. We have selected the time when 50% of the cells are committed to die and hence cannot be saved by steroid treatment. We call this the commitment point. Interestingly, the commitment point for the ventral intersegmental muscles is about 16 hours before adult ecdysis, whereas that for the D-IV motoneurons that innervate these muscles is three hours after ecdysis (Figure 3). When given between these times, ecdysteroid treatment results in the normal degeneration of the muscles but the preservation of their motoneurons. These results argue that the motoneurons do not degenerate as a direct result of the death of their target muscles.

A conclusive demonstration that ecdysteroids act directly on the CNS to regulate cell death has come from recent *in vitro* studies (K. Bennet and J. W. Truman, unpublished observations). Single abdominal ganglia that were removed from the insects about 10 hours before ecdysis and placed in culture for 24–48

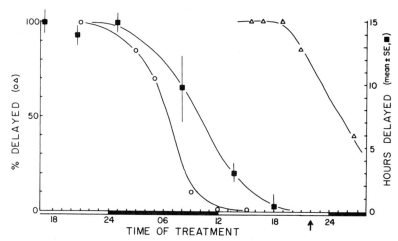

Figure 3. *Effects of the time of ecdysteroid administration on the ability of the treatment to delay ecdysis behavior (circles), intersegmental muscle degeneration (squares), and D-IV motoneuron degeneration (triangles). Insects were treated with 20-HE (5 μg/gm) at the times indicated. Black and white bars refer to the nights and days, respectively, of the last 32 hours of adult development and the first six hours of adult life. Arrow shows normal time of adult ecdysis. (From Truman and Schwartz, 1984.)*

hours subsequently showed neuronal death. The program of degeneration was slowed *in vitro* as compared to that *in vivo*. The distribution of dying interneurons in the cultured ganglia was similar to that seen during normal degeneration in the insects, suggesting that the appropriate neurons were dying. In a few ganglia the degeneration program had progressed to the motoneurons. Importantly, motoneuron death was confined only to the cells that normally die *in vivo*. Although the number of ganglia that progressed to the stage of motoneuron death were few ($N = 3$), it is of interest that the stages of degeneration of the various identified motoneurons suggest that they were dying according to the expected sequence. These results indicate that the specificity of cell degeneration is retained in isolated ganglia maintained in culture.

Equally significant was the fact that the degeneration in the cultured ganglia could be modified by steroid treatment. Control ganglia cultured for 24 hours showed an average of 16 ± 3 ($\pm$ SEM; $N = 8$) dying cells, whereas those cultured for an equivalent time with a physiological dosage of 0.1 μg 20-HE/ml showed only 3 ± 1 ($N = 8$). After 48 hours in culture the difference was much more striking, being 51 ± 15 ($N = 4$) and 2 ± 1 ($N = 2$) respectively. Thus ecdysteroids can act directly on the CNS to regulate the death of central neurons.

The endocrine trigger for postecdysial cell death used by *Manduca* is not used universally by all insects. For example, in giant silkmoths degeneration of the intersegmental muscles requires the decline in ecdysteroids as well as an additional factor. In silkmoths such as *Antheraea polyphemus* the simple manipulation of abdomen isolation does not advance degeneration but rather prevents it (Lockshin, 1969). This prevention disappears around the time of adult ecdysis, after which abdomen isolation is followed by normal degeneration. Adult ecdysis is triggered by a peptide, eclosion hormone, that is released from the moth CNS (Truman, 1980). Studies on isolated abdomens showed that this peptide was the trigger for the degeneration of the muscles, and it did so as a result of the direct action of the peptide on the muscles (Schwartz and Truman, 1982). Thus in these species there is an endocrine signal that actively promotes the death of the cells. In order for the muscles to respond to the peptide, however, the muscles must also have experienced the prior decline in the ecdysteroid titer.

Other Factors Influencing Cell Death

Although the signal for postecdysial cell death appears to be primarily endocrine in nature, cellular interactions can apparently prolong the life of cells that have been signaled to die. One example of this is seen in the death of the intersegmental muscles in silkmoths. As stated above, the degeneration of these muscles is caused by the direct action of eclosion hormone (Schwartz and Truman, 1982). This conclusion was based on the finding that muscles chronically denervated by the removal of the abdominal CNS soon after pupal ecdysis nevertheless persisted through metamorphosis and developed hormonal sensitivity such that they could then be caused to degenerate by exposure to the peptide. This finding was of interest, because Lockshin and Williams (1965b,c) had earlier demonstrated that chronic stimulation of single motor roots resulted in the local preservation of the stimulated muscle even though its companions in other segments died. Also, treatment of insects with drugs having anticholinesterase activity induced hyperactivity that resulted in muscle preservation, but only if the CNS was present. These experiments were originally thought to suggest that cessation of motor activity was the trigger for muscle death. More likely, however, the enhanced motor activity to the muscles somehow overrides the signal provided by the hormone.

Another example in which cellular interactions can apparently maintain cells is seen in the case of neuronal death in *Manduca* (Truman, 1983). If newly emerged moths are not allowed to expand their wings immediately, but instead are forced to perform sustained digging movements (a normal happening for these insects that normally pupate underground), degeneration of some of their motoneurons and interneurons is delayed. Interestingly, the cell preservation is selective. For example, MN-11 dies at its normal time, whereas the death of the D-IV group is delayed for 6–8 hours (Figure 4). Since D-IV neurons are

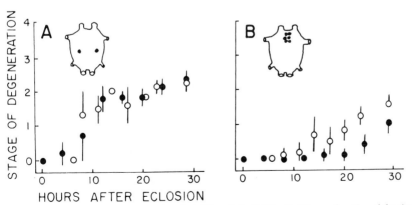

Figure 4. *Average stage of degeneration of MN-11 (A) and the D-IV cells (B) as a function of the time after eclosion.* Open circles, ganglia from moths that inflated their wings immediately after eclosion; filled circles, ganglia from moths that were forced to dig. Mean ($\pm$SD) for approximately five animals per point. Degeneration stages as in Figure 2. (From Truman, 1983.)

involved in the movements shown during digging, it is reasonable to suggest that the delay might be due to the sustained activity of these cells. Thus the fates of the motoneurons are somewhat plastic and may be modified by conditions that the insect experiences after it emerges from the pupal cuticle.

THE SPATIAL SELECTIVITY OF THE DEGENERATION RESPONSE

An intriguing aspect of neuronal cell death in *Manduca* is the selectivity of the response. Even though the motoneurons that die after adult ecdysis are those whose target muscles also die, the data described above show clearly that this relationship is correlative rather than causative. Indeed, culture experiments show that steroid disappearance is the only signal from the periphery required for the death to occur. The determination of which cells die may result from cellular interactions within the CNS, but we have no evidence to date that is consistent with this hypothesis. Ultrastructural studies of the death of the D-IV neurons indicate that the glia do not become involved in this process until late in degeneration (Stocker et al., 1978). Hence, the glia do not appear to be "killer" cells. The same conclusion has been reached for programmed cell death in the CNS of the nematode *C. elegans* (Horvitz et al., 1982). Another possible type of intraganglionic interaction that could be involved in determining which cells die is an interaction in which the deaths of some neurons lead to the deaths of other cells, and so forth. Indeed, the previous section presented evidence which suggested that enhanced activity in particular cells resulted in the delay of their death. This domino relationship is unlikely in the case of possible interactions of motoneurons with other motoneurons. When ecdysteroids are applied at various times during the last day of development and the fate of particular cells followed, it is clear that the motoneurons differ in the times of their commitment point (Truman and Schwartz, 1984). For example, the commitment point of MN-11 occurs at five hours before ecdysis, whereas that for the D-IV motoneurons is eight hours later. Those for MN-2 and for MN-12 are still later. The result of this relationship is that ecdysteroid treatment at times between the commitment points of various cells is followed by the death of those motoneurons whose times have passed, while the later cells are delayed in their degeneration. Indeed, the sequence of motoneuron death can be arrested at essentially any point in the sequence. It should be stressed that in these cases the cells that have passed their commitment point at the time of the treatment continue to degenerate in a normal fashion irrespective of the presence of the steroid. These results strongly argue that the death of a particular motoneuron is not directly triggered by the loss of the motoneurons that died earlier.

The relationship between the death of interneurons and motoneurons is more difficult to study because we have not been able to identify specific individual interneurons in the process of dying. Presumably many of the interneurons that degenerate are presynaptic to the motoneurons. It is of interest that the commitment points for about 80% of the interneurons occur prior to ecdysis, whereas those for most of the motoneurons occur after ecdysis (Truman and Schwartz, 1984). Consequently application of ecdysteroids at ecdysis results in the death of most of the interneurons that normally die, but blocks the death of most of the motoneurons. The deaths of most of the interneurons and motoneurons can be dissociated in time, suggesting that one is not dependent on the other. However, since we cannot account for individual interneurons, it is possible that a small number of key cells may have late commitment points at times similar to those of the motoneurons. Another observation pertaining to the relationship between the interneurons and motoneurons is that numerous synaptic potentials can still be recorded from D-IV motoneurons that have already begun degeneration (Truman and Levine, 1980). Although this does not exclude the possibility that these cells lose a critical input essential for their survival, it does show that there is not a massive withdrawal of synaptic inputs prior to the start of cell death.

In light of this evidence it seems likely that the fate of a particular neuron is intrinsically controlled and is not imposed by interactions with other cells either inside or outside the CNS at the time that the endocrine signal is given. We do not know the nature of the differences between various cells that result in one cell's dying whereas its neighbors continue to live. One factor that probably does not play a role in determining which will die is the neuroblast of origin of a particular cell. Although the lineages of particular neurons have not been established in *Manduca* as they have been in the locust (Goodman and Bate, 1981), it is likely that groups of adjacent motoneurons sharing similar neurite projections in the CNS and similar peripheral targets are likely to be progeny from the same neuroblast. There are several such groups of motoneurons in the CNS of *Manduca*, and these groups are very heterogeneous in terms of the fates of the cells that comprise them.

Another point of interest is that although particular neurons may die in one segment, apparently homologous neurons in other segments may not die. This is especially true in contrasting the fates of cells in thoracic versus abdominal ganglia. This segment specificity has also been elegantly demonstrated for cells of defined lineages during embryonic development in the locust (Bate et al., 1981).

The difference between cells that live and those that die also cannot be explained in terms of one group of cells being exposed to the signal and the other not. Ecdysteroids have access to all cells in the CNS. The difference between the two groups of cells may be in how they are "programmed" to respond to the hormonal signal or it may be that at the time in question one group of cells does not have sufficient numbers of receptors to respond. This latter hypothesis is open to experimental testing.

The specificity of which cells die also has relevance when considering other species of moths. As mentioned above, the motoneurons of the giant silkmoths also lose their target muscles after adult ecdysis, yet the cells continue to live through the life of the adult. At this time it is not clear in what way the silkmoth cells differ so as to allow them to live after ecdysis whereas the homologous neurons in *Manduca* die. It may be important that the two groups of moths differ not only in the fate of their respective neurons but also in the endocrine cues that serve as the trigger for postecdysial cell death.

TEMPORAL PATTERNS OF CELL DEATH

Neuronal death in *Manduca* shows specificity not only with regard to which cells die but also when they die. Two alternate hypotheses could explain the temporal sequence of degeneration that is seen after ecdysis. One is that the sequence is set up by cell–cell interactions. The other is that is arises from intrinsic differences in the response properties of the cells involved.

Neither hypothesis can be ruled out at present, but in its simplest form the cell–cell interaction hypothesis does not seem likely. The fact that the degeneration program can be interrupted at essentially any point by application of ecdysteroid (Truman and Schwartz, 1984) indicates that the death of a particular cell is not directly triggered by the loss of the cells that degenerated earlier. However, since to date we have not been able to alter the order of cell death by any of our manipulations, we cannot exclude the possibility that the death of the early cells is necessary, although not sufficient, for the death of the later cells.

The second hypothesis is that the sequence is due to intrinsic differences in how the various cells respond to ecdysteroids. Consequently, one must explain how the motoneurons can show a difference of more than 24 hours in the time that they begin to degenerate although they are exposed to the same ecdysteroid levels. Since the ecdysteroid titer is slowly declining during this period, one option would be that the cells have different thresholds and thus are triggered to die at different stages of the ecdysteroid decline. A second option is that all of the cells respond to essentially the same threshold titer of hormone but differ in their respective response latencies.

An interesting finding of the *in vivo* experiments has been the commitment points described above. The time of the commitment point has been determined most accurately for the D-IV cells and for MN-11. In both cases it occurs approximately 10 hours before the first obvious signs of degeneration (Truman and Schwartz, 1984). It is harder to determine a precise commitment point for the various interneurons because they cannot be followed as individuals, but analysis of the population data suggests a value of about 10 hours for these cells as well (Truman and Schwartz, 1984; J. W. Truman, unpublished observations). The sequence of the commitment points matches the sequence of death of the respective cells; thus the times of the commitment points may indicate when each has passed its respective threshold as the ecdysteroid titer slowly declines. Unfortunately, another explanation is possible in that an irreversible event leading to degeneration may occur about 10 hours before the cells first show symptoms of dying. This event could occur at different latencies after the cells are signaled to die, but until it occurs the cell may retain its steroid sensitivity. Because alternative explanations are possible, the *in vivo* experiments performed to date cannot discriminate between these two options.

The *in vitro* situation should give more insight into the programming of the degeneration sequence because the ecdysteroid concentration can be controlled precisely. For example, in the experiments that have been done to date the ganglia that have been explanted have experienced an abrupt step-down in the ecdysteroid titer rather than the gradual decay that is invariably found *in vivo* even in the case of isolated abdomens. The few ganglia that have shown motoneuron death *in vitro* have been interesting with respect to the fates of the D-IV cells and MN-12 (K. Bennet and J. W. Truman, unpublished observations). The ganglia were explanted about eight hours before ecdysis, well before the critical points for these two groups of cells. In the case of the two ganglia that had progressed far enough in their program to show the degeneration of the

D-IV motoneurons, the MN-12 pairs had not yet begun to die. Thus the two sets of neurons maintained their appropriate relationship after exposure to an abrupt step-down in the steroid concentration. This preliminary observation suggests that latency differences play an important role in ordering the onset of death of the various cells.

CRITICAL PERIODS FOR CELL DEATH

In developing systems cells often exhibit critical periods during which decisions are made as to which developmental pathway to follow. For example, in neurons that require specific pre- or postsynaptic contacts in order to survive, the contacts usually must occur when the neuron is immature, around the time that processes are being sent out (e.g., Macagno, 1979). Once the cell is mature and stable contacts have formed, it becomes resistant to the effects of the removal of these contacts. Thus even in the case of cell death, the cells must be in a receptive state to respond to the signals that lead to their destruction. These critical periods are found not only in immature neurons but also in fully differentiated cells that choose the programmed death pathway. What are the factors that render a cell responsive to a signal that will trigger the death program?

Postecdysial death in *Manduca* occurs when the ecdysteroids disappear but this signal in itself does not explain the response. Under certain conditions at the preceding pupal stage, the insect enters a prolonged state of developmental arrest called diapause that is maintained by the chronic absence of ecdysteroids. Even though the steroids are absent the intersegmental muscles and their motoneurons survive for months (J. W. Truman, unpublished observations). The absence of ecdysteroids becomes lethal for the cells only after the animal has subsequently undergone adult development, a process that requires 18 days and is caused by the appearance of ecdysteroids. Consequently the action of these steroids at the pupal stage renders the neurons dependent on these hormones for their continued survival. This alteration occurs through a direct action of the steroids on the CNS, because implanted ganglia deprived of their normal targets also develop the steroid requirement.

Another aspect of critical periods relates to muscle degeneration in silkmoths (Schwartz and Truman, 1982). As described above, the death of the intersegmental muscles in silkmoths is caused by eclosion hormone. These muscles experience pulses of the hormone at earlier stages in the life of the insect, but it is only after adult differentiation that the insect responds to the signal with muscle degeneration. In this case also, the action of the ecdysteroids during adult differentiation renders the tissue sensitive to the hormone.

Interestingly, many systems, including the eclosion-hormone release system (Truman et al., 1983), are inhibited by ecdysteroids late in development, and each has its characteristic commitment point. Treatment with ecdysteroids at certain times allows the release of eclosion hormone and the behavioral response of the CNS to it but blocks the ability of the muscle to respond to the signal. In these cases the muscles are refractory when the hormone is released, and hence they escape their programmed degeneration and persist into the adult (Schwartz and Truman, 1982).

BIOCHEMICAL ASPECTS OF CELL DEATH

The biochemical events of primary interest are the early events leading to cell degeneration. The timing of the commitment points provides important information as to when the cells have made irrevocable commitments to die. For MN-11 and the D-IV cells this decision is made about 10 hours before the cells show the first overt signs of degeneration. Unfortunately, at this time we know nothing of the nature of the biochemical events that underlie this decision. Obviously the fact that steroids are involved and the relatively long latency until their effects are manifested would suggest that alterations in genomic activity are involved.

Thus far most of the information on the biochemical aspects of postecdysial cell death in moths has dealt with the degeneration of the intersegmental muscles in silkmoths. As one would expect from the peptide activation of the degeneration program in these moths, cyclic nucleotides appear to mediate the response, specifically cyclic GMP (Truman and Schwartz, 1982). The action of the peptide in triggering cell death can be mimicked by treatment with exogenous cyclic GMP as well as by such drugs as the methylxanthines and sodium nitroprusside, which allow a buildup in endogenous cyclic GMP levels. Other evidence for the involvement of cyclic GMP in the response to eclosion hormone is that treatment with this peptide results in a rapid, 20- to 30-fold increase in cyclic GMP levels in the muscles, whereas cyclic AMP levels remain essentially unchanged.

Besides cyclic GMP, early studies by Lockshin (1969) also implicated the synthesis of new protein and RNA in the degeneration response of intersegmental muscles. The use of the inhibitors actinomycin D and cycloheximide to block RNA and protein synthesis, respectively, showed that the former was effective in blocking death if given as late as two hours after ecdysis, whereas the latter was effective up to 5–6 hours. This finding is consistent with the notion that the programmed death of the cells is a distinct developmental program which requires the mobilization of new genetic information for its expression. In this respect, it is of interest that in *C. elegans* a single gene has been implicated in controlling the activation of the cell death program (Horvitz et al., 1982). One would expect that similar loci are present to regulate cell death in moths. In silkmoth muscle they apparently are activated through the action of eclosion hormone and the accompanying increase in cyclic GMP. In the neurons and muscles in *Manduca* the steroid drop alone apparently activates this program.

CONCLUSIONS

The studies on *Manduca*, as well as on *C. elegans*, indicate that neuronal death is a carefully controlled process during which the cell systematically dismantles itself. It should be thought of as a program of differentiation, just as when a cell extends an axon or produces a particular transmitter. The molecular mechanism(s) through which the cell brings about its own destruction are not understood, but it is likely that new genes must be expressed in order for this to take place.

It should be stressed that the molecular basis of the programmed degeneration of cells should be considered separately from the signals that trigger the implementation of this program. These triggering signals may be the presence or absence of hormones or other trophic molecules as well as other factors related to interactions with pre- or postsynaptic cells. One would expect that these triggering signals would vary according to the context of the cell death but that the molecular events involved in the playing out of the death program might be much more conservative.

In the insect systems that have been considered here some of the factors that trigger the degeneration are well understood. The neuron and muscle death that occurs in *Manduca* is caused by the disappearance of steroids, whereas muscle death in silkmoths requires both the steroid decline and the presence of a neuropeptide. The endocrine approach to this problem of cell death has also been of use in gaining insight into the basis of the timing of cell death. The temporal relationships of interneuron, motoneuron, and muscle death are consistent with the idea that cell–cell interactions are responsible for the temporal sequence of cell death. However, by the careful timing of endocrine applications, the degeneration of the respective cell groups can be partially or completely dissociated from one another. Indeed, even within the cell groups, endocrine treatment can effectively fractionate the response. Thus the key to the degeneration response appears to reside primarily with the responses of the affected cells to the endocrine signal rather than to the more complex cell–cell interactions. However, it should be noted that there is some evidence that direct cell–cell interactions may play a minor role in delaying the degeneration program, as in the digging response of *Manduca* (Figure 4).

At this time we are only beginning to attempt to confront the more difficult problems of the molecular events initiated by these signals. The fact that the ecdysteroids act through genomic mechanisms in most systems that have been studied (Riddiford, 1980) would suggest that their action in regulating neuronal death involves the activation of new genes. The 10-hour latency between the commitment point and the overt onset of death is consistent with such a mechanism. Obvious questions to be resolved relate to the exact nature of the genes that are turned on in order to cause death. Equally important are questions relating to the specificity of the response—namely, why do some cells but not others respond to the signal, and why do the cells that do respond do so only late in the insect's life, even though they are exposed to the same endocrine cues at earlier times?

ACKNOWLEDGMENTS

The unpublished work reported here was supported by grants from the McKnight Foundation, the National Institutes of Health, and the National Science Foundation.

REFERENCES

Bate, M., C. S. Goodman, and N. C. Spitzer (1981) Embryonic development of identified neurons: Segment-specific differences in the H cell homologues. *J. Neurosci.* 1:103–106.

Cottrell, C. B. (1962) The imaginal ecdysis of blowflies. Observations on the hydrostatic mechanisms involved in digging and expansion. *J. Exp. Biol.* **39**:431–448.

Finlayson, L. H. (1956) Normal and induced degeneration of abdominal muscles during metamorphosis in the lepidoptera. *Q. J. Microsc. Sci.* **97**:215–234.

Goodman, C. S., and M. Bate (1981) Neuronal development in the grasshopper. *Trends Neurosci.* **4**:163–169.

Hamburger, V., and R. W. Oppenheim (1982) Naturally occurring neuronal death in vertebrates. *Neurosci. Comment.* **1**:39–55.

Horvitz, H. R., H. M. Ellis, and P. W. Sternberg (1982) Programmed cell death in nematode development. *Neurosci. Comment.* **1**:56–65.

Hughes, A. F. (1957) The development of the primary sensory system in Xenopus laevis (Davdin). *J. Anat.* **91**:323–338.

Lamborghini, J. E. (1981) Kinetics of Rohon-Beard neuron disappearance in Xenopus laevis. *Soc. Neurosci. Abstr.* **7**:291.

Lockshin, R. A. (1969) Programmed cell death. Activation of lysis by a mechanism involving the synthesis of protein. *J. Insect Physiol.* **15**:1505–1516.

Lockshin, R. A., and C. M. Williams (1965a) Programmed cell death. I. Cytology of degeneration in the intersegmental muscles of the Pernyi silkmoth. *J. Insect Physiol.* **11**:123–133.

Lockshin, R. A., and C. M. Williams (1965b) Programmed cell death. III. Neural control of the breakdown of the intersegmental muscles of silkmoths. *J. Insect Physiol.* **11**:601–610.

Lockshin, R. A., and C. M. Williams (1965c) Programmed cell death. IV. The influence of drugs on the breakdown of the intersegmental muscles of silkmoths. *J. Insect Physiol.* **11**:803–809.

Macagno, E. R. (1979) Cellular interactions and pattern formation in the development of the visual system of Daphnia magna (Crustacea, Branchiopoda). *Dev. Biol.* **73**:206–238.

Nordlander, R. H., and J. S. Edwards (1968) Morphological cell death in the postembryonic development of the insect optic lobes. *Nature* **218**:780–781.

Riddiford, L. M. (1980) Insect endocrinology: Action of hormones at the cellular level. *Annu. Rev. Physiol.* **42**:511–528.

Schwartz, L. M., and J. W. Truman (1982) Peptide and steroid regulation of muscle degeneration in an insect. *Science* **215**:1420–1421.

Schwartz, L. M., and J. W. Truman (1983) Hormonal control of rates of metamorphic development in the tobacco hornworm Manduca sexta. *Dev. Biol.* **99**:103–114.

Stocker, R. F., J. S. Edwards, and J. W. Truman (1978) Fine structure of degenerating moth abdominal motor neurons after eclosion. *Cell Tissue Res.* **191**:317–331.

Taylor, H. M., and J. W. Truman (1974) Metamorphosis of the abdominal ganglia of the tobacco hornworm, Manduca sexta: Changes in populations of identified motor neurons. *J. Comp. Physiol.* **90**:367–388.

Truman, J. W. (1980) Eclosion hormone: Its role in coordinating ecdysial events in insects. In *Insect Biology in the Future*, M. Locke and D. S. Smith, eds., pp. 385–401, Academic, New York.

Truman, J. W. (1983) Programmed cell death in the nervous system of an adult insect. *J. Comp. Neurol.* **216**:445–452.

Truman, J. W. (1984) Cell death in invertebrate nervous systems. *Annu. Rev. Neurosci.* **7**:171–188.

Truman, J. W., and R. B. Levine (1980) Programmed cell death in the nervous system of an insect: Histological and physiological aspects. *Soc. Neurosci. Abstr.* **6**:668.

Truman, J. W., and L. M. Schwartz (1982) Insect systems for the study of programmed neuronal death. *Neurosci. Comment.* **1**:66–72.

Truman, J. W., and L. M. Schwartz (1984) Steroid regulation of neuronal death in the moth nervous system. *J. Neurosci.* **4**:274–280.

Truman, J. W., D. B. Rountree, S. E. Reiss, and L. M. Schwartz (1983) Ecdysteroids regulate the release and action of eclosion hormone in the tobacco hornworm, Manduca sexta. *J. Insect Physiol.* **29**:895–900.

White, J., E. Southgate, N. Thomson, and S. Brenner (1976) The structure of the ventral nerve cord of Caenorhabditis elegans. *Philos. Trans. R. Soc. Lond. (Biol.)* **275**:327–348.

Wigglesworth, V. B. (1956) Formation and involution of striated muscle fibres during the growth and moulting cycles of *Rhodnius prolixus* (Hemiptera). *Q. J. Microsc. Sci.* **97**:465–480.

Wyllie, A. H. (1981) Cell death: A new classification separating apoptosis from necrosis. In *Cell Death in Biology and Pathology*, I. D. Bowen and R. A. Lockshin, eds., pp. 9–34, Chapman and Hall, London.

Chapter 23

Immunohistochemical and Genetic Studies of Serotonin and Neuropeptides in *Drosophila*

KALPANA WHITE
ANA M. VALLES

ABSTRACT

During the last few years immunological methods have localized several neurotransmitters and neuropeptides in the insect nervous system. Powerful genetic techniques available in Drosophila melanogaster *potentially permit manipulation of the levels of a transmitter or a peptide. Genetic variants eliminating synthesis of a specific molecule can be used to probe the impact of that molecule on the development or function of the nervous system. Therefore we have initiated immunohistochemical mapping of identified neurotransmitters and neuropeptides in the fruit fly.*

Sera raised against serotonin, substance P, and FMRFamide conjugates show immunoreactivity in the Drosophila *central nervous system. In whole-mounts of the larval central nervous system, each serum reproducibly revealed a discrete group of cells and a specific neuropilar and peripheral pattern. Functional implications of the observed immunoreactivities as well as the potential of the identified immunochemical cell markers in addressing questions about neural development in Drosophila are discussed.*

A decrease in serotoninlike immunoreactivity was demonstrated through the use of a temperature-sensitive mutant in the gene Ddc, *the structural gene encoding the enzyme dopa decarboxylase. Strategies to achieve manipulation of peptide synthesis through genetic variants are outlined.*

Immunological tools have become increasingly important in the study of the nervous system. Antisera raised against well-defined molecules of neurobiological interest have been and will continue to be extremely useful in understanding the biological role of these molecules. The arrival of hybridoma technology has further sharpened the immunological approach (Kohler and Milstein, 1975). Monoclonal antibodies make possible the study of specific domains of a molecule (Tzartos and Lindstrom, 1980) and allow the study of a neural cell component without prior knowledge of any of its properties (Zipser and McKay, 1981; Fujita et al., 1982; White et al., 1983). By allowing visualization of the distribution of the molecule, antisera and monoclonal antibodies have also aided in the understanding of the cellular and subcellular organization of the nervous system.

To investigate whether a particular molecule is causally involved in neural development will involve further strategies. A strategy uniquely possible in the fruit fly *Drosophila melanogaster* is to identify genes that specify antigen(s) of interest. Then, through the use of mutations, the amount of antigen can be controlled in a precise manner during development. It should be possible to utilize defined genetic variants to correlate an antibody specificity with an abnormality *in vivo* and thus assign a developmental function to the antigen. This approach will greatly augment conventional uses of antibodies in understanding cellular function and differentiation.

Neurotransmitters and neuropeptides are some of the more interesting molecules in the nervous system. In addition to their more conventional role as chemical messengers important in the control of neural activity, some neurotransmitters and neuropeptides may also function decisively in neural development. In this context it is interesting to note that the head-activator hormone, a classical morphogen isolated from *Hydra attenuata*, is a peptide with an amino acid sequence identical to that of a peptide found in human hypothalamus (Bodenmuller and Schaller, 1981).

In invertebrate systems discrete immunoreactive patterns have been observed by using antisera raised against neurotransmitters and neuropeptides (Bishop and O'Shea, 1982; Beltz and Kravitz, 1983). Such characterization of neurotransmitter and neuropeptide systems facilitates the localization of immunoreactivity in central nervous system (CNS) cells, neuropil, nerve roots, and the periphery. Immunoreactive cellular labels allow the study of specific subsets of cells throughout development. A comparative study of immunoreactivity patterns of several transmitters and peptide analogues in the same system provides insights into the global organization of the CNS.

We have initiated the characterization of immunoreactive patterns of some of the known neurotransmitters and neuropeptides in *Drosophila* because of the potential of genetically manipulating transmitter or peptide levels. The ability to perturb levels of a neurotransmitter or a neuropeptide in order to study its role in nervous system development and function is valuable. Through use of mutations in the gene(s) involved in the synthesis of a given molecule such perturbations can be achieved. If a gene encoding a neuropeptide can be mutated to eliminate that peptide, the effect of its absence on neural development or function can then be directly assessed. An example of this neurogenetic approach is found in the study of Greenspan et al. (1980). In *Drosophila*, neuropil chronically deprived of the enzyme acetylcholinesterase shows a morphological derangement, implying a role for acetylcholine metabolism in development and/or maintenance of the nervous system (Greenspan et al., 1980). The effects of deprivation of acetylcholinesterase activity were demonstrable in genetic mosaics for *Ace*, the structural gene for acetylcholinesterase.

We have found immunoreactive cells in the CNS of *Drosophila* by using antisera raised against the monoamine neurotransmitter serotonin, the invertebrate neuropeptide FMRFamide, and the vertebrate neuropeptide substance P. In each case we find a stereotypic pattern of immunoreactivity. In this chapter we describe and compare the three patterns of immunoreactivity. Further, we outline possible genetic strategies to control levels of these antigens and discuss how selective labeling of subsets of neural cells will be advantageous in addressing a number of questions about development which are currently under investigation.

CNS ORGANIZATION AND DEVELOPMENT IN *DROSOPHILA*

The CNS of *Drosophila melanogaster* consists of segmental ganglia. A fused cephalic and a compound thoracic ganglion comprise the adult CNS. As in other insects, the CNS is organized such that the cell bodies are peripheral and the neuropil is central, with the neuropil consisting of regions that can be roughly delineated as being primarily synaptic or primarily fiber tracts. The neuropil contains few cell bodies.

The development of the nervous system in *Drosophila* is similar to that in other holometabolous insects. Neurogenesis in *Drosophila* occurs during the embryonic period and is resumed during the postembryonic larval and pupal stages. The larval CNS consists of functioning larval neurons in close association with neuroblasts, mother cells, preganglion cells, and differentiating ganglion cells. The adult CNS consists of daughters of neuroblasts whose cell divisions occurred during larval stages of development and at least some larval neurons (White and Kankel, 1978).

For the studies presented in this chapter, we used whole-mount immunocytochemistry on the dissected larval CNS to localize immunoreactive cells. The CNS was dissected at the late third instar stage. The major ganglia present during larval stages are the paired brain hemispheres and the segmental ventral ganglion. The brain hemispheres are fused ventrally, and developing optic lobes flank them laterally. In the medial area transverse commissures connecting the two lobes are observed. The transverse commissures encircle the aorta that passes through the brain lobes dorsal to the ventral ganglion.

In the fused ventral ganglion at least 12 major ganglia can be distinguished—the subesophageal ganglion, the three thoracic ganglia, and the eight abdominal ganglia. The individual ganglia in the ventral nervous system are connected by longitudinal fibers that run along its length in an anteroposterior direction. Each hemiganglion is connected to its symmetrical pair by transverse commissures that cross the midline.

Fourteen major pairs of nerves are associated with the larval CNS. The optic nerve and the antennal nerve are associated with the brain lobes; the maxillary nerve is associated with the subesophageal ganglia; and the remaining 11 nerves are associated with the thoracic and abdominal segments (Hertweck, 1931). In addition to the major nerves, Hertweck (1931) describes an undetermined number of very fragile nerves originating at the dorsal midline of the ventral ganglion. The peripheral target(s) of these dorsal nerves are not known.

The larval-to-adult transition occurs during the pupal period. The CNS undergoes dramatic morphological changes, which are readily evidenced in the overall differences in morphology of the larval and the adult CNS (see Kankel et al., 1980, for a review). Within the neuropil there is extensive synapse formation, and the preganglion cells differentiate for the first time. At least some of the functioning larval neurons in *Drosophila* undergo restructuring during metamorphosis (Technau and Heisenberg, 1982).

MAPPING OF SEROTONINLIKE IMMUNOREACTIVITY IN THE LARVAL CNS

Cell Bodies

Whole-mount immunohistochemistry with antibody raised against serotonin reproducibly stained an extensive system of cell bodies in the brain hemispheres and in the ventral ganglion (Figure 1A,B). Preincubation of antisera with 10^{-4} M serotonin drastically reduced the intensity of staining to almost background level.

Figure 1 shows the distribution of serotoninlike immunoreactive cells. Approximately 20 immunoreactive cell bodies arranged in a bilaterally symmetrical fashion were observed in the brain lobes. Each of these cells seems to send a fine process medially toward the central commissures. Four regions within which the immunoreactive cells were reliably found could be delineated in each brain lobe—anteromedial, medial, posteromedial, and mediolateral. In the anteromedial region a single immunoreactive cell body was observed. Posterior to the single

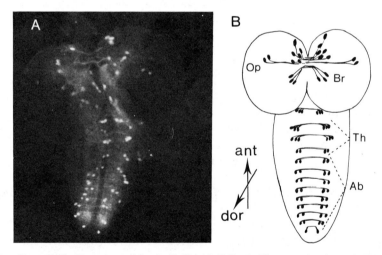

Figure 1. *Serotoninlike immunoreactivity in the larval CNS.* A: Photograph after whole-mount immunohistochemical processing. The whole-mount procedure was adapted from Beltz and Kravitz (1983). Third instar larval CNS was dissected in cold *Drosophila* Ringers solution and fixed for four hours at 4°C in 4% paraformaldehyde in 0.1 M phosphate buffer (pH 7.3). For all washes, 0.1 M phosphate buffer (pH 7.2) plus 0.3% Triton X-100 was used. Samples were thoroughly washed and then incubated at 4°C in 1:200 dilution of anti-serotonin antibody (Immuno Nuclear Corporation) for four hours. Before adding the secondary antibody, the tissue was thoroughly rinsed. The secondary antibody was goat anti-rabbit IgG labeled with FITC. Tissues were incubated in 1:20 dilution of secondary antibody at 4°C for about two hours. The samples were subjected to multiple washes and then mounted as in Beltz and Kravitz (1983). Axes as shown in Figure 1B. ×80. B: Schematic diagram of serotoninlike immunoreactive cell bodies and major fibers. Immunoreactivity in the neuropil and fibers leaving the CNS are not drawn. Ab, abdominal ganglia; ant, anterior; Br, brain lobe; dor, dorsal; Op, optic lobe; Th, thoracic ganglia.

immunoreactive cell, in the medial region close to the midline, five stained cells were usually seen. Posterior to the central commissures a group of three cells was present. The mediolateral group is composed of two cells that are just proximal to the developing optic lobes that flank the midbrain region. Some variability in the number of stained cells was found in different animals, but stained cells were always observed in all four regions.

In the ventral ganglion approximately 46 immunoreactive cell bodies were observed. As in the brain they are arranged in perfect bilateral symmetry (Figure 1). In the region of the subesophageal and thoracic ganglia, three cell groups each containing three cells were observed per side. Immediately posterior to these three groups, nine consecutive pairs of cells were observed. At the very posterior tip, only one cell was seen. The arrangement of cells appears to be segmental, with a cell group present in each of the thoracic and abdominal segments. Each of the abdominal segments, with the exception of the last, contains a pair of immunoreactive cells. Studies to establish boundaries between the subesophageal, the thoracic, and the abdominal segments are in progress. An immunoreactively stained transverse fiber tract corresponding to each symmetrical pair of cells was observed. Cells from both halves of the bilateral pair contribute fibers to the transverse tract.

Neuropil

A meshwork of fine immunoreactive processes is observed in all ganglia of the larval CNS. The only exceptions are the developing optic lobes. However, in preliminary studies we have observed serotonergic innervations in the adult optic lobe. In the neuropil the intensity of immunoreactive processes and their branching shows a characteristic pattern.

Peripheral Projections

Immunoreactive fibers were observed leaving the brain lobes. These fibers were traced to two main targets. Fibers filled with immunoreactive varicosities were seen to project dorsocephally into the ring gland, the larval endocrine organ (Poulson, 1945). These fibers originated medially in the brain. The corpus allatum–corpus cardiacum complex of the adult ring gland is innervated by a pair of nerves originating in the brain (King et al., 1966).

The maxillary nerve also contains immunoreactive fibers. These fibers, which emerge from the subesophageal region, were traced into the larval cephalopharyngeal apparatus. This apparatus contains the set of muscles that moves the pharynx and the mouth hooks. Near the pharynx the immunoreactive fibers running laterally within each maxillary nerve merge together with fibers from a third central nerve. The fibers in the central nerve are highly immunoreactive and a densely stained immunoreactive plexus is observed along the nerve.

GENETIC PERTURBATION OF SEROTONINLIKE IMMUNOREACTIVITY

Mutations in structural genes encoding enzymes in the synthetic pathway of a neurotransmitter can be used to reduce the synthesis of the neurotransmitter

(Greenspan, 1980; Livingstone and Tempel, 1983). Livingstone and Tempel (1983) have demonstrated a reduction in the synthesis of serotonin and dopamine in adult flies expressing a temperature-sensitive mutation in the structural gene that encodes the enzyme dopa decarboxylase (*Ddc*). In order to test if the serotoninlike immunoreactivity in CNS cells is also affected by mutations in the *Ddc* gene, we used the temperature-sensitive allele Ddc^{ts2}. The Ddc^{ts2} allele encodes a thermolabile enzyme (Wright et al., 1982). Homozygous Ddc^{ts2} larvae were raised at the permissive temperature (18°C) and shifted to the nonpermissive temperature (29°C) during midsecond instar stage. After three days at 29°C, at the late third instar larval stage, the CNSs were dissected and processed for serotoninlike immunohistochemistry. For controls, we processed the CNSs of wild-type larvae raised at 29°C and the CNSs of homozygous Ddc^{ts2} larvae raised at 18°C to comparable developmental stages.

Figure 2A and B shows the ventral ganglion of the Ddc^{ts2} whole-mount and the comparable whole-mount of the normal CNS. When the experimental whole-mounts were compared to the control whole-mounts, serotoninlike immunoreactivity was depressed in the former but was still detectable. For example, at the posterior tip of the abdominal ganglia, four cells in the mutant continued to be stained at a much higher intensity than did other abdominal cells in the

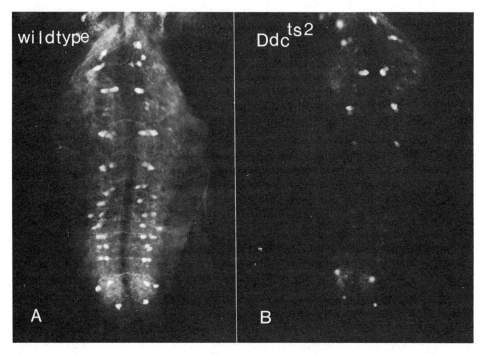

Figure 2. *Serotoninlike immunoreactivity in the ventral ganglion of wild-type larva and Ddc^{ts2} mutant larva.* Homozygous Ddc^{ts2} and wild-type larvae were raised at permissive temperature (18°C) and then shifted to nonpermissive temperature (29°C) during the midsecond instar stage. After three days at 29°C the CNS was dissected and processed for serotonin immunoreactivity as in Figure 1. A: Wild-type ventral nervous system. B: Mutant ventral nervous system. Note the differential depression of immunoreactivity. Cells in the anterior abdominal segments show decreased immunoreactivity compared to wild-type cells, as well as the four posterior and the thoracic cells in the mutant which continue to be stained. Axes as in Figure 1B. ×150.

same animal. Also, the immunoreactivity in subesophageal and thoracic cells was more resistant to the 29°C treatment than brain lobe cells and a majority of the abdominal cells. This staining pattern was reliably seen in all the mutant whole-mounts.

The results from the Ddc^{ts2} whole-mounts imply that the immunoreactivity observed is in fact due to serotonin, and that the levels of serotonin in the CNS cells do decrease at nonpermissive temperature. We do not know if the serotonin that remains after heat treating the mutant larvae is due to stored serotonin from before the temperature shift, is a result of residual dopa decarboxylase activity (i.e., not turned off by the heat treatment), or is caused by the synthesis of serotonin by an aromatic acid decarboxylase activity other than that coded by the Ddc gene.

MAPPING OF FMRFamidelike IMMUNOREACTIVITY IN THE LARVAL CNS

Cell Bodies

When the larval CNS was processed with antibody raised against the molluscan peptide FMRFamide, immunoreactive cells were discerned in the ventral ganglion as well as in the brain lobes. In the brain lobes approximately a dozen cells were immunoreactive; they varied in intensity in a stereotypic fashion. The details of their pattern are currently under investigation. Preincubation of the antisera with 10^{-4} M FMRFamide resulted in a complete block of the staining.

The immunoreactive cells in the ventral ganglion fall into two classes—those that were intensely stained (eight cell bodies) and those that were weakly stained (approximately twelve). As can be seen in Figure 3A and B, the most anterior of the highly immunoreactive cells is a pair located close to the midline in the subesophageal ganglion. Each of the three thoracic hemiganglia also contains a single cell that is highly immunoreactive and is located laterally (Figure 3). Each of the lateral thoracic immunoreactive cells sends a single process toward the midline; the primary process does not appear to ramify until it reaches the midline. The subesophageal cells also send primary processes toward the midline.

Four weakly immunoreactive cells are in the subesophageal ganglion: two are close to the midline, slightly anterior to the highly immunoreactive cell pair; the other two occupy more lateral positions. A cluster of weakly immunoreactive cells (two or three cell bodies) was found between the primary unbranched processes of the highly immunoreactive lateral cell bodies of the meso- and metathoracic segments. A pair of weakly immunoreactive cell bodies was discerned between the central longitudinal fibers in a location posterior to the point where the fiber from the metathoracic immunoreactive cell reaches the longitudinal fiber. A pair of weakly immunoreactive cell bodies also was observed in the last abdominal segment.

Neuropil

Immunoreactive neuropil was observed in the medial region of the brain lobes. In the ventral nervous system a grid of longitudinal parallel fibers was seen.

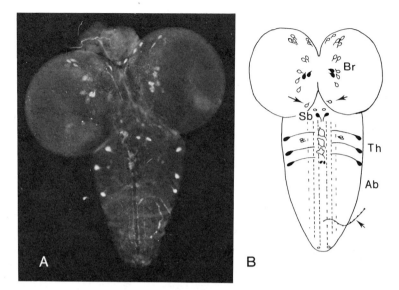

Figure 3. *FMRFamidelike immunoreactivity in the larval CNS.* A: Photograph after whole-mount immunohistochemistry. Procedure as in Figure 1, except that the primary antibody was 1:200 dilution of anti-FMRFamide antibody (Cambridge Research Biochemicals). ×76. B: Schematic diagram of FMRFamidelike immunoreactive cell bodies and major fibers giving a dorsal view with brain lobes positioned above the subesophageal ganglion. The neuropil and fibers in the brain lobes are not shown. The two types of cell bodies represent the two classes of cells: highly immunoreactive (filled) and weakly immunoreactive (open). Arrows in the brain lobes point to two open cell bodies in the subesophageal ganglion. Arrow outside the CNS points to a single fiber that exits from the dorsal midline in the abdominal segments. Note the three thoracic immunoreactive plexuses along the thoracic dorsal midline. Axes as in Figure 1B. Ab, abdominal ganglia; Br, brain lobe; Sb, subesophageal ganglion; Th, thoracic ganglia.

The processes from each of the lateral cells were traced to the pair of longitudinal fibers closest to the midline. The other fiber pairs flank it laterally. Short transverse fibers were observed between the grid of longitudinal fibers. All fibers show a punctated pattern of immunoreactivity.

Peripheral Nerves

A single fiber containing FMRFamidelike immunoreactive material leaving the ventral nervous system was observed. This fiber originates at the dorsal midline in the abdominal ganglia. In addition, in the thoracic segments and in close proximity to the dorsal midline of the ventral nervous system, three immunoreactive plexuses dense with varicosities were always seen (Figure 3). An immunoreactive process connecting each of these clublike structures to the longitudinal connective was observed. We have not yet determined whether nerves associated with the brain lobes contain FMRFamidelike immunoreactivity.

MAPPING OF SUBSTANCE P-LIKE IMMUNOREACTIVITY IN THE LARVAL CNS

Cell Bodies

Whole-mount immunohistochemistry of the larval CNS using antisera raised against the vertebrate peptide substance P revealed eight immunoreactive cell bodies (Figure 4A,B). The distribution of the eight intensely stained cell bodies is interesting in that it is identical to that observed for the highly stained FMRFamidelike immunoreactive ventral ganglion cells described earlier (cf. Figures 3, 4). As can be seen in Figure 4, the substance P-like immunoreactive subesophageal pair is also close to the midline and the three thoracic pairs are similarly located laterally. Each of these highly immunoreactive cell bodies sends a single fiber toward the midline, as in the case of the FMRFamidelike immunoreactive cells. Preincubation of the anti-substance P antisera with 10^{-4} M substance P blocked the staining. Because of the similarity in staining to the FMRFamidelike cells, we also preincubated the anti-substance P antisera with 10^{-4} M FMRFamide. FMRFamide did not block the substance P-like immunoreactive staining.

Immunoreactive cells equal in intensity of staining to the ventral ganglion cells were not observed in the brain hemispheres. In some preparations a small cluster of weakly stained cells in the anteromedial position was seen. The staining in this anteromedial cell cluster was variable and in all cases much lower than that seen in the cells of the ventral ganglion. We do not know if the differences

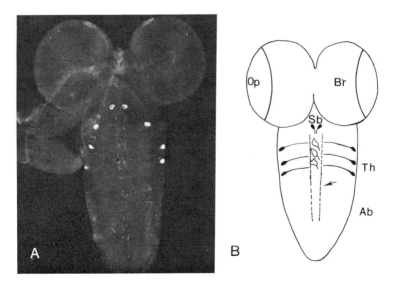

Figure 4. *Substance P-like immunoreactivity in the larval CNS.* A: Photograph after whole-mount immunohistochemistry. Procedure as in Figure 1, except that the primary antibody was 1:100 dilution of anti-substance P antibody (Immuno Nuclear Corporation). ×110. B: Schematic diagram of substance P-like immunoreactivity. Arrow points to the longitudinal punctated fibers. Axes as in Figure 1B. Compare with Figure 3B. Ab, abdominal ganglia; Br, brain lobe; Op, optic lobe; Sb, subesophageal ganglion; Th, thoracic ganglia.

in staining between the cells of the ventral ganglion and those of the brain is due to different concentrations of the same substance P-like antigen or different substance P-like antigens.

Neuropil

Two immunoreactive parallel longitudinal fibers were observed in the ventral ganglion in a location similar to that for the most medial of the FMRFamidelike immunoreactive fibers (as judged from the distance between them). The primary process of each of the ventral ganglion immunoreactive cell bodies can be traced to the immunoreactive longitudinal fiber on the same side as the cell. The longitudinal fibers show a punctated pattern of immunoreactivity along their entire length.

Peripheral Nerves

We did not observe any immunoreactivity along the major 14 nerves. However, along the dorsal midline in the thoracic region are three intensely staining plexuses in positions identical to that observed for FMRFamidelike immunoreactivity.

Substance P Immunoreactive Cells through Postembryonic Development

We have processed, with anti-substance P antibody, the thoracic ganglion of one- and three-day-old pupae and newly emerged male adults. In all cases we have observed six immunoreactive cells, one cell pair per thoracic segment. These preliminary studies suggest that the thoracic substance P-like immunoreactive cells persist throughout metamorphosis and are immunoreactive throughout the pupal period.

Putative Colocalization of FMRFamidelike and Substance P-like Immunoreactivity

Because of the striking similarity of patterns between the eight intensely stained FMRFamidelike cells and the substance P immunoreactive cells, we processed larval CNS with a mixture of anti-FMRFamide and anti-substance P antibodies. The immunoreactive cells revealed by the mixed antibody incubation were the same as those stained by anti-FMRFamide antibody alone. Since only eight highly immunoreactive cells were observed, we conclude that in these cells FMRFamidelike and substance P-like antigens are likely to be colocalized. In the combined antisera incubations only three intensely immunoreactive plexuses along the thoracic dorsal midline were observed.

IMPLICATIONS OF AND CONCLUSIONS FROM IMMUNOREACTIVITY PATTERNS

The Whole-Mount Immunohistochemical Technique

The limitations of the immunohistochemical method are well acknowledged (Bishop and O'Shea, 1982; Hökfelt et al., 1980). When we describe a class of immunoreactive cells, for instance, FMRFamide immunoreactive cells, we do not claim that the cells contain FMRFamide per se, that all the cells contain the same cross-reacting antigen, or that we have identified all the cells that may contain FMRFamide, cross-reacting antigens, or both. Our descriptions of immunoreactivity patterns are minimal estimates of parts of the *Drosophila* central and peripheral nervous systems that contain substances potentially detectable by these antibodies.

Serotoninlike Immunoreactivity

Mapping of putative serotonergic neurons, their neuropil arborizations, and their peripheral projections is helpful in identifying possible sites of serotonin action. For instance, in the stomatogastric ganglion of the lobster *Homarus americanus*, serotonin, serotoninlike staining, and the physiological effects of serotonin are all demonstrable (Beltz et al., 1984).

We find serotonin immunoreactive cells in all the ganglia of the larval CNS with the exception of precursors of the adult optic lobes. The dendritic arborizations of the serotonergic cells consist of fine processes that are found in all the neuropil regions. The widespread distribution of serotonergic neurons in the CNS is similar to what has been observed in other invertebrates (Beltz and Kravitz, 1983; Bishop and O'Shea, 1983).

In the periphery we find putative serotonergic innervation of the ring gland and the pharyngeal apparatus. The ring gland innervation suggests a role for serotonin in the regulation of endocrine activity. The pharyngeal innervation points to a putative role in the feeding behavior of the larvae. In the leech *Hirudo medicinalis* Lent et al. (1983) have demonstrated that serotonin affects feeding behavior.

FMRFamidelike Immunoreactivity

FMRFamide was first isolated as a cardioexcitatory peptide from molluscan tissue (Price and Greenberg, 1977). Although FMRFamidelike immunoreactivity has been observed in both vertebrates and invertebrates, at least some of this immunoreactivity is likely to be due to other related peptides (see Greenberg and Price, 1983, for a review). In *Drosophila melanogaster* anti-Met-enkephalin and anti-Leu-enkephalin sera also reveal a set of immunoreactive cells (A. Ferrus, personal communication). We do not yet know if different cells or the same cells are revealed with each antibody. A detailed comparative analysis will be necessary to define the specificities of these sets of immunoreactive cells. To find the peptide responsible for FMRFamidelike immunoreactivity it will be essential to determine, biochemically, the native *Drosophila* peptide(s).

Immunoreactive localization to the CNS implies a role for an FMRFamidelike peptide in the CNS. Additionally, a single peptide-containing fiber leaves the ventral ganglion at the dorsal midline. We have not yet traced this fiber to its target, the presumptive site of FMRFamidelike peptide action. In insects, neurosecretory products from the ventral nerve cord are delivered to neurohemal storage and release sites—the perisympathetic organs (see Raabe, 1982, for a review). The presence of the peptide in the structures in close apposition to the dorsal midline of the thoracic ganglia suggests that these metameric structures may be the thoracic perisympathetic organs of *Drosophila*. Of course, the cells and fibers we have mapped may be due to more than one FMRFamidelike peptide, and each FMRFamide analogue may serve a distinct physiological function.

Substance P-like Immunoreactivity

Substance P-like immunoreactivity is observed in a wide variety of organisms (Greenberg and Price, 1983). At present the relation between substance P, an undecapeptide present in the vertebrate CNS, and the antigen responsible for the immunoreactive pattern observed in the *Drosophila* CNS is not known. A logical peptide candidate is the invertebrate peptide eledoisin (Erspamer, 1981).

In the whole-mounts substance P-like immunoreactivity is confined to the CNS except for the putative thoracic neurohemal release sites. Jan and Jan (1982) have localized substance P-like immunoreactivity in the male reproductive system. Whether the CNS and the male reproductive system immunoreactivities are due to the same or cross-reacting antigens is not known. Localization of FMRFamidelike and substance P-like immunoreactivity to the same cells indicates that these cells are specialized neuroendocrine cells. Many instances of colocalization within the same neuron of a classical neurotransmitter and a neuropeptide or of co-localization of two different neuropeptides have been discovered in vertebrate systems (Hökfelt et al., 1980). Recently, colocalization instances have also been reported in invertebrates (Adams and Phelps, 1983; Siwicki and Kravitz, 1983).

GENETIC CONTROL OF ANTIGENIC LEVELS

Serotonin

Tryptophan hydroxylase and aromatic acid decarboxylase are the two enzymes in the synthetic biochemical pathway of serotonin. Mutations eliminating enzyme activity in either of the structural genes will affect serotonin levels. In the fruit fly a structural gene for tryptophan hydroxylase has not been identified as yet. However, *Ddc*, the structural gene encoding the enzyme dopa decarboxylase, has been extensively studied both genetically and molecularly (Wright et al., 1976; Scholnick et al., 1983). Dopa decarboxylase activity is found in the hypoderm and in the nervous system. Null and temperature-sensitive mutations that eliminate dopa decarboxylase activity already exist. Flies homozygous for a null mutation in the gene *Ddc* die as unhatched larvae (Wright et al., 1976). Flies carrying a temperature-sensitive mutation survive when raised under permissive conditions, and can be shifted to high temperature to study the effects of elimination of dopa decarboxylase activity.

To study the effect of *Ddc* mutations on monoamine neurotransmitter synthesis, Livingstone and Tempel (1983) utilized the temperature-sensitive allele Ddc^{ts1}. They measured dopamine, serotonin, and octopamine synthesis, and L-dopa, 5-hydroxytryptophan, and tyrosine decarboxylase activities in the heads of wild-type and mutant flies. Both sets of flies had been reared at a permissive temperature (20°C) and then shifted to a nonpermissive temperature (29°C) for three days.

Livingstone and Tempel suggest that two aromatic acid decarboxylase activities are responsible for monoamine synthesis in *Drosophila*. First, dopa decarboxylase activity decarboxylates 5-hydroxytryptophan and L-dopa to synthesize serotonin and dopamine. Second, tyrosine decarboxylase activity decarboxylates tyrosine to tyramine in the pathway to octopamine. Our results showing a decrease in *in situ* levels of immunologically detectable serotonin in the CNS of temperature-sensitive mutants of the *Ddc* gene are consistent with those of Livingstone and Tempel (1983).

Under our experimental conditions a comparison of serotoninlike immunoreactivity in control and Ddc^{ts2} CNS whole-mounts leads to two striking conclusions. One, serotoninlike staining is decreased but not blocked; two, the staining in certain cells is more resistant. The residual staining may be due to stored serotonin (from before the temperature shift) or to residual dopa decarboxylase activity at the nonpermissive temperature. It is also likely that within the serotonergic cells of the mutant 5-hydroxytryptophan may rise to abnormally high levels. Increased concentrations of 5-hydroxytryptophan may allow its decarboxylation by another decarboxylase with a lower affinity for 5-hydroxytryptophan. The nonuniformity in staining reduction observed in the Ddc^{ts2} CNS whole-mounts is intriguing. It may be due to the differential amount of stored serotonin, to the differential rate of release or uptake of serotonin, or to the presence in the resistant cells of the hypothetical decarboxylase that would not be encoded by *Ddc*.

The expression of *Ddc* is subject to both temporal and tissue-specific regulation; dopa decarboxylase activity is found in the hypoderm and in the nervous system. The *Ddc* gene has been molecularly cloned (Hirsch and Davidson, 1981); by use of DNA-mediated transformation, a 7-kb fragment of DNA within which the gene resides has been shown to contain all the information necessary for its regulation (Scholnick et al., 1983). Experiments to further define DNA sequences that are responsible for the regulation of gene expression are in progress (J. Hirsch, personal communication). Therefore it is possible that within a couple of years, genetic variants constructed from *in vitro* engineered pieces of DNA that selectively abolish dopa decarboxylase activity in the nervous system, but not in the hypoderm, will be available. The immunohistochemical labeling of the serotonergic neurons will aid in the analysis of the expression of the gene in the serotonergic cells of the transformed flies.

The cellular distribution of dopa decarboxylase in the nervous system is not yet known. It is reasonable to assume that in the CNS the *Ddc* gene must be expressed at least in the serotonergic and dopaminergic neurons. The localization of *Ddc* gene expression is now possible through the technique of *in situ* hybridization, which allows detection of RNA transcripts by using DNA probes on sectioned tissue (Hafen et al., 1983).

Peptides

In order to manipulate genetically the concentration of a peptide, it is essential to identify the structural gene encoding the peptide of interest. In both vertebrate and invertebrate systems in which nucleotide sequences encoding small neuropeptides have been characterized, one generalization seems to have emerged. The peptide is initially synthesized as a large precursor that contains a family of related peptides (e.g., Roberts and Herbert, 1977; Scheller et al., 1982). In fact, several previously unidentified peptides have been discovered by sequencing the nucleotide precursor (e.g., Nakanishi et al., 1979). The discovery of peptide families raises many questions related to the function of individual peptide members and to the organization of peptide precursor genes. If the peptide genes in *Drosophila* are organized similarly, the identification of a peptide gene family will provide a unique opportunity to study the function and synthesis of its individual members.

Several molecular and genetic approaches to identifying chromosomal locations of peptide genes are possible in *Drosophila*. Where nucleotide sequences have already been identified in other organisms, putative DNA clones from the gene may be isolated based on possible homologies (Fyrberg et al., 1980; Royden et al., 1982). The success of this strategy will depend on the degree of homology. An alternative strategy is to make a synthetic nucleotide probe incorporating prevalent *Drosophila* codons based on the sequence of the native *Drosophila* peptide. Such a probe may be used to select possible homologous sequences for cloning. DNA probes from the cloned sequences can then be used to determine a gene's cytogenetic location on the salivary chromosome (Fyrberg et al., 1980).

A genetic technique to identify a chromosomal segment in which a gene resides is available in *Drosophila* for genes whose products can be assayed quantitatively and exhibit gene dosage effect (see O'Brien and MacIntyre, 1978, for a review). In practice, screening for dosage-sensitive chromosomal regions involves the generation of aneuploids (flies carrying one or three doses of various chromosome segments) and assaying the gene product. Determination of the dosage-sensitive region preceded the mutational identification of the *Ddc* gene encoding the enzyme dopa decarboxylase (Hodgetts, 1974; Wright et al., 1976) and the genes *Ace* and *Cha* encoding enzymes acetylcholinesterase and choline acetyltransferase, respectively (Hall et al., 1979).

Once the chromosomal location of a gene or the segment in which it resides is determined, mutations mapping within the gene can be isolated using the strategies reviewed in Hall et al. (1979). Mutational variants in the gene allow the generation of precise genetic lesions to probe the consequences of eliminating the gene product.

Identification of a gene encoding a peptide family, and more importantly the isolation of genetic variants within that gene, will permit specific questions about the consequences of elimination of all or one or subsets of peptides.

USE OF IMMUNOREACTIVE CELL LABELS IN ADDRESSING SPECIFIC NEURODEVELOPMENTAL QUESTIONS IN *DROSOPHILA*

Development of Identifiable Neurons

Developmental study of individual neurons has been greatly impeded in *Drosophila* due to the small size of its nervous system and the neurons. The immunoreactive cell labels described in this chapter are a first step toward providing reliable labels for specific cells. It is reasonable to assume that in time a large library of specific immunological cell labels will be available. The antigenic labels will make it feasible to define the developmental time at which the antigen is present at an immunologically detectable level. It will also be feasible to follow the cell through subsequent development. A comparative developmental study of cells with different immunochemical labels will be instructive in defining functional developmental sequences within the CNS.

Colocalization of a neurotransmitter and a neuropeptide or two neuropeptides within a cell raises the following question: Is the acquisition of two functional specificities during development independent or is it causally related? If synthesis of a peptide or a neurotransmitter can be genetically eliminated (using genetic variants), an analysis of causal interrelationships of these specificities may be possible.

Neural Reorganization during Metamorphosis

In *Drosophila*, major reorganization occurs in the CNS during metamorphosis. Remodeling of the CNS in holometabolous insects is not surprising, as the adults differ from the larvae in morphology and in behavior. Unlike many other organ systems the CNS is not completely histolyzed. Although the dynamic changes taking place during metamorphosis have long been recognized (Edwards, 1969; Truman and Reiss, 1976), the underlying mechanisms are not understood. Below, we briefly describe two studies that use different methodologies to reveal events during metamorphosis, and then discuss a potential use in this context for the immunoreactive cell markers.

The developmentally best-studied substructure in *Drosophila* central brain is the mushroom body (Technau and Heisenberg, 1982). Ultrastructural analysis of the cellular and neuropilar region of the mushroom bodies during early pupal stages has revealed that extensive degeneration occurs in the neuropil and that many of the fibers are broken down and made anew (Technau and Heisenberg, 1982). Interestingly, the Kenyon cells, which are the intrinsic cells of the mushroom bodies, remain alive during this period, and the overall morphology of the mushroom bodies remains unaltered.

We have investigated the behavior during metamorphosis of SYN1 and UNA1, two neuronal antigens that we have defined by using monoclonal antibodies (White et al., 1983). SYN1 is concentrated in the synaptic areas of the CNS of larvae and adults; UNA1 is more ubiquitously distributed, being present in both cortical and synaptic regions. The behavior of these two neuronal antigens, one ubiquitous and the other more restrictive, differs markedly during metamorphosis. The concentration of SYN1, as judged from the immunoreactive staining of sectioned material, decreases substantially during the first day of pupation. It is at its lowest between 20 and 30 hours after puparium formation and then increases to attain its adult levels. In contrast UNA1 concentration does not show dramatic changes (White et al., 1983).

The behavior of SYN1 concentration parallels the degeneration observed in the ultrastructural studies of Technau and Heisenberg (1982). It is therefore likely that the observed modulation of SYN1 during metamorphosis is due to degradation concomitant with breakdown of larval synaptic neuropil, followed by resynthesis as the adult synaptic neuropil is formed.

The ultrastructural analysis of Technau and Heisenberg (1982), as well as our immunological study (White et al., 1983), provide a general view of the CNS changes. They do not, however, bear on the behavior of individual neurons. The availability of immunological cell markers will make possible the study of individual cells during metamorphosis. Indeed, it is likely that different neuronal cell classes will show distinct developmental patterns.

Effect of Mutations Affecting Pattern on CNS

Genetic analysis of development in *Drosophila* has far surpassed that of any other organism. Mutations that alter development of the pattern have revealed a segmental specification program (Nusslein-Volhard and Wieschaus, 1980). Also of great developmental interest are the homeotic mutations that transform one segment into another (Lewis, 1978). The segment-specific epidermal pattern, rich in a variety of easily recognizable landmarks, has been essential to the initial recognition and subsequent analysis of these pattern mutations. Investigating the effects of pattern mutations on CNS development will be greatly aided by the availability of segment-specific cellular landmarks. The segment-specific pattern of putative serotonin cells within the adult thoracic ganglia has already been used to demonstrate the transformation of this pattern by expression of mutations within the homeotic bithorax complex (Martel, 1983).

Analysis of Mutations Affecting CNS Development

Drosophila melanogaster has been the organism of choice for investigators interested in using genetic tools for analyzing neural development and behavior. The cumulative efforts of neurogeneticists have resulted in the identification of a large number of genes that affect neural development or function (see Hall, 1982). Because of the small size of the neurons in the fruit-fly brain, developmental studies with individual neurons or neuron classes have not been possible. The availability of cellular labels will now allow investigators to analyze the effects of mutations on defined classes of neural cells.

ACKNOWLEDGMENTS

We thank J. Hall for many useful comments on the manuscript, T. Hurteau for technical assistance, and E. Marder for stimulating discussions and for being a good yenta. This work was supported by NIH grant #GM-31503.

REFERENCES

Adams, M. E., and M. N. Phelps (1983) Co-localization of Bursicon bioactivity and proctolin in identified neurons. *Soc. Neurosci. Abstr.* **9**:313.

Beltz, B. S., and E. A. Kravitz (1983) Mapping of serotonin-like immunoreactivity in the lobster nervous system. *J. Neurosci.* **3**:585–602.

Beltz, B. S., J. S. Eisen, R. Flamm, R. M. Harris-Warrick, S. L. Hooper, and E. Marder (1984) Serotonergic innervation and modulation of the stomatogastric ganglion of three decapod crustaceans (*Panulirus interruptus, Homarus americanus* and *Cancer irroratus*). *J. Exp. Biol.* **109**:35–54.

Bishop, C. A., and M. O'Shea (1982) Neuropeptide proctolin (H-Arg-Tyr-Leu-Pro-Thr-OH): Immunocytochemical mapping of neurons in the central nervous system of the cockroach. *J. Comp. Neurol.* **207**:223-238.

Bishop, C. A., and M. O'Shea (1983) Serotonin immunoreactive neurons in the central nervous system of an insect (*Periplaneta americana*). *J. Neurobiol.* **14**:251-269.

Bodenmuller, H., and H. C. Schaller (1981) Conserved amino acid sequence of a neuropeptide, the head activator, from coelenterates to humans. *Nature* **293**:579-580.

Edwards, J. S. (1969) Postembryonic development and regeneration of the insect nervous system. *Adv. Insect Physiol.* **6**:97-137.

Erspamer, V. (1981) The tachykinin peptide family. *Trends Neurosci.* **4**:187-204.

Fujita, S. C., S. L. Zipursky, S. Benzer, A. L. Ferrus, and S. L. Shotwell (1982) Monoclonal antibodies against *Drosophila* nervous system. *Proc. Natl. Acad. Sci. USA* **79**:7929-7933.

Fyrberg, E. A., K. L. Kindle, N. Davidson, and A. Sodja (1980) The actin genes of *Drosophila*: A dispersed multigene family. *Cell* **19**:365-378.

Greenberg, M. J., and D. A. Price (1983) Invertebrate neuropeptides: Native and naturalized. *Annu. Rev. Physiol.* **45**:271-288.

Greenspan, R. J. (1980) Mutations of choline acetyltransferase and associated neural defects in *Drosophila melanogaster*. *J. Comp. Physiol.* **137**:83-92.

Greenspan, R. J., A. Finn, Jr., and J. C. Hall (1980) Acetylcholinesterase mutants in *Drosophila* and their effects on the structure and function of the central nervous system. *J. Comp. Neurol.* **189**:741-744.

Hafen, E., M. Levine, R. L. Garber, and W. J. Gehring (1983) An improved *in situ* method for detection of cellular RNAs in *Drosophila* tissue sections and its application for localizing transcripts of homeotic *Antennapedia* gene complex. *EMBO J.* **2**:617-623.

Hall, J. C. (1982) Genetics of the nervous system in *Drosophila*. *Q. Rev. Biophys.* **15**:223-479.

Hall, J. C., R. J. Greenspan, and D. R. Kankel (1979) Neural defects induced by genetic manipulation of acetylcholine metabolism in *Drosophila*. *Soc. Neurosci. Symp.* **4**:1-42.

Hertweck, H. (1931) Anatomie und Variabilität des Nerven-system und der Sinnesorganne von *Drosophila melanogaster* (Meigen). *Z. Wiss. Zool.* **139**:559-663.

Hirsch, J., and N. Davidson (1981) Isolation and characterization of the dopa decarboxylase gene of *Drosophila melanogaster*. *Mol. Cell. Biol.* **1**:475-485.

Hodgetts, R. B. (1974) The response of dopa decarboxylase activity to variations in gene dosage in *Drosophila*: A possible location of the structural gene. *Genetics* **79**:45-54.

Hökfelt, T., O. Johansson, A. Ljungdahl, J. M. Lundberg, and M. Schultzberg (1980) Peptidergic neurones. *Nature* **284**:515-521.

Jan, L. Y., and Y. N. Jan (1982) Genetic and immunological studies of the nervous system of *Drosophila melanogaster*. *CIBA Found. Symp.* **88**:221-239.

Kankel, D. R., A. Ferrus, S. H. Garen, P. J. Harte, and P. E. Lewis (1980) The structure and development of the nervous system. In *The Genetics and Biology of Drosophila*, Vol. 2, M. Ashburner and T. R. F. Wright, eds., pp. 295-368, Academic, New York.

King, R. C., S. K. Aggarwal, and D. Bodenstein (1966) The comparative submicroscopic cytology of the corpus allatam-corpus cardiacum complex of *Drosophila melanogaster*. *J. Exp. Zool.* **161**:151-176.

Kohler, G., and C. Milstein (1975) Continuous cultures of fused cells secreting antibody of predefined specificity. *Nature* **256**:495-497.

Lent, C. M., M. H. Dickinson, and C. G. Marshall (1983) Serotonin controls feeding behavior in the medicinal leech. *Soc. Neurosci. Abstr.* **9**:913.

Lewis, E. B. (1978) A gene complex controlling segmentation in *Drosophila*. *Nature* **276**:565-570.

Livingstone, M. S., and B. L. Tempel (1983) Genetic dissection of monoamine neurotransmitter synthesis in *Drosophila*. *Nature* **303**:67-70.

Martel, E. (1983) Bithorax complex mutants transform the segment-specific pattern of 5-HT containing neurons in *Drosophila*. *Soc. Neurosci. Abstr.* **9**:832.

Nakanishi, S., A. Inoue, T. Kita, M. Nakamura, A. C. Y. Chang, S. N . Cohen, and S. Numa (1979) Nucleotide sequence of cloned cDNA for bovine corticotropin-β-lipotropin precursor. *Nature* **278**:423–427.

Nusslein-Volhard, C., and E. Wieschaus (1980) Mutations affecting segment number and polarity in *Drosophila*. *Nature* **287**:795–801.

O'Brien, S. J., and R. J. MacIntyre (1978) Genetics and biochemistry of enzymes and specific proteins of *Drosophila*. In *The Genetics and Biology of Drosophila*, Vol. 2, M. Ashburner and T. R. F. Wright, eds., pp. 396–557, Academic, New York.

Poulson, D. F. (1945) On the origin and nature of the ring gland (Weismann's ring) of the higher diptera. *Trans. Conn. Acad. Arts Sci.* **36**:449–487.

Price, D. A., and M. J. Greenberg (1977) The structure of a molluscan cardioexcitatory neuropeptide. *Science* **197**:670–671.

Raabe, M. (1982) *Insect Neurohormones*, Plenum, New York.

Roberts, J. L., and E. Herbert (1977) Characterization of a common precursor to corticotropin and β-lipotropin: Cell-free synthesis of the precursor and identification of corticotropin peptides in the molecule. *Proc. Natl. Acad. Sci. USA* **74**:4826–4830.

Royden, C. S., P. S. O'Farrell, E. Herbert, M. Uhler, Y. N. Jan, and L. Y. Jan (1982) Identification of an adrenocorticotropic hormone-like substance in *Drosophila melanogaster* by DNA hybridization and immunocytochemistry. *Soc. Neurosci. Abstr.* **8**:703.

Scheller, R. H., J. F. Jackson, L. B. McAllister, J. H. Schwartz, E. R. Kandel, and R. Axel (1982) A family of genes that codes ELH, a neuropeptide eliciting a stereotyped pattern of behavior in *Aplysia*. *Cell* **28**:707–719.

Scholnick, S. B., B. A. Morgan, and J. Hirsh (1983) The cloned dopa decarboxylase gene is developmentally regulated when reintegrated into the *Drosophila* genome. *Cell* **34**:37–45.

Siwicki, K. K., and E. A. Kravitz (1983) Proctolin in lobsters: General distribution and co-localization with serotonin. *Soc. Neurosci. Abstr.* **9**:313.

Technau, G., and M. Heisenberg (1982) Neural reorganization during metamorphosis of the corpora pendunculata in *Drosophila melanogaster*. *Nature* **295**:405–407.

Truman, J. W., and S. E. Reiss (1976) Dendritic reorganization of an identified motorneurone during metamorphosis of the tobacco hornworm moth. *Science* **192**:477–479.

Tzartos, S. J., and J. M. Lindstrom (1980) Monoclonal antibodies used to probe acetylcholine receptor structure: Localization of the main immunogenic region and detection of similarities between subunits. *Proc. Natl. Acad. Sci. USA* **77**:755–759.

White, K., and D. R. Kankel (1978) Patterns of cell division and cell movement in the formation of the imaginal nervous system in *Drosophila melanogaster*. *Dev. Biol.* **65**:296–321.

White, K., A. Pereira, and L. E. Cannon (1983) Modulation of a neural antigen during metamorphosis in *Drosophila melanogaster*. *Dev. Biol.* **98**:239–244.

Wright, T. R. F., G. C. Bewley, and A. F. Sherald (1976) The genetics of dopa decarboxylase in *Drosophila melanogaster*. II. Isolation and characterization of dopa-decarboxylase-deficient mutants and their relationship to the α-methyl-dopa-decarboxylase-hypersensitive mutants. *Genetics* **84**:287–310.

Wright, T. R. F., B. C. Black, C. P. Bishop, J. L. Marsh, E. S. Pentz, R. Steward, and E. Y. Wright (1982) The genetics of dopa decarboxylase in *Drosophila melanogaster*. V. *Ddc* and 1(21)*amd* alleles: Isolation, characterization and intragenic complementation. *Mol. Gen. Genet.* **188**:18–26.

Zipser, B., and R. McKay (1981) Monoclonal antibodies distinguish identifiable neurones in the leech. *Nature* **289**:549–554.

Chapter 24

Neural and Developmental Implications of the Genetic and Molecular Analysis of Behavior

JEFFREY C. HALL

ABSTRACT

Behavioral and neurochemical mutations affect several forms of conditioning in Drosophila. *Some of the relevant experiments involve experience-dependent features of reproductive behavior. Certain elements of conditioned male courtship may bear relationships to the brain development that takes place after the fly becomes an adult. Methods for demonstrating conditioning in courting* Drosophila *have also permitted a genetic mosaic approach to determining parts of the nervous system that are influenced by different learning and memory genes. These results can in turn be compared to information that is emerging on the biochemical cell differentiation of portions of the central nervous system in which the products of these genes are detectable. A similar strategy is at hand with regard to the neural tissues in which a clock gene is expressed in its control of circadian rhythms and a short-term rhythm in courtship. Results from mosaic focusing and from an assessment of the tissue distribution of the primary transcript coded for by this genetic locus can be critically compared. The latter determination is facilitated by using molecular clones. We are analyzing cloned DNA segments from the X chromosome of* Drosophila melanogaster *and attempting to identify which of them contains the* period *locus, one of the most salient clock genes in this species.*

Neural control of an animal's behavior has developmental implications. This truism becomes meaningful when we address behavioral questions from a genetic point of view. How do specific genes participate in the development of a nervous system's pattern, and how do separate portions of that system differentiate their own array of gene actions? Behavioral experiments can help one approach such questions, and as answers begin to emerge they will almost certainly allow us to know more about the structural/functional relationships which underlie an animal's actions.

This chapter summarizes some of the work that my colleagues and I have been doing in the area of reproductive behavior in *Drosophila*. Our approach is genetic and has recently begun to take on molecular overtones as well. The latter is essential if we are ever to know just how the genes we are investigating become active at defined stages of development, possibly within particular cells in certain neural ganglia, and how the products of these genes control the

LEARNING AND MEMORY MUTANTS IN COURTSHIP

Behavior and Biochemistry of Mutations Affecting Conditioning

Fruit flies learn and remember certain of their courtship experiences (see Hall, 1984; Siegel et al., 1984, for reviews). Several of the relevant experiments have used general conditioning mutants (whose origin and properties are reviewed in Aceves-Piña et al., 1983; Quinn and Greenspan, 1984). For example, a male exposed to and directing courtship at a previously fertilized female exhibits weak courtship of a subsequent female (another mated one, or even a virgin) for at least two hours afterward (Siegel and Hall, 1979). Three of the learning variants that disturb this aftereffect are *dunce (dnc)*, *rutabaga (rut)*, and *turnip (tur)*. Males expressing any of these mutations will vigorously court a female in spite of their previous "training" experience with mated females (Gailey et al., 1984). That this information storage and retrieval involves the male's associations between a mated female's antiaphrodisiac pheromones, the aphrodisiac odors she also emanates, and the female as an object of courtship (i.e., the odors alone are insufficient training stimuli) need not be discussed in detail here (see Richmond et al., 1984 and Tompkins, 1984, for reviews). Instead, the focus of this chapter is on the neurochemical correlates of these three types of conditioning mutants. The *dnc* gene codes for a specific form of cyclic AMP phosphodiesterase (cAMP PDE) (see Davis and Kauvar, 1984, for a review); *rut* appears to code for a portion of the multisubunit adenylate cyclase enzyme (Livingstone et al., 1984); the *tur* mutation affects the affinity of material that binds to serotonin in the fly (R. F. Smith, unpublished observations; cited in Aceves-Piña et al., 1983).

Another class of mutation—discovered initially on biochemical rather than behavioral criteria—has turned out to affect learning. These are temperature-sensitive dopa decarboxylase (Ddc^{ts}) mutants (e.g., Wright et al., 1982). When Ddc^{ts} flies are raised to adulthood at low temperature, hence bypassing the developmental-lethal stages they would encounter if reared at high temperature, and if this "permissive" rearing is followed by a few days of heat treatment, the learning ability of these adults is turned off in parallel with a drop in the enzyme activity (Tempel et al., 1984). As this dopa decarboxylase activity drops, so do rates of synthesis of dopamine and serotonin (Livingstone and Tempel, 1983). Most of the behavioral effects of the Ddc^{ts} mutations are assessed by applying avoidance learning tests involving electric shocks and odors (the procedures used to isolate the aforementioned conditioning mutants such as *dnc*) or reward learning tests involving sugars and odors (cf. Tempel et al., 1983). In addition, heat-treated Ddc^{ts} males are deficient in the learning that occurs when individual flies are trained by exposure to and courtship of fertilized females (Tempel et al., 1984).

One reason for bringing out these genetically induced learning defects in the context of the associated biochemical abnormalities is that all of the functions just mentioned seem to have a relatively wide tissue distribution in *Drosophila*.

The degradative and synthesizing activities controlling cAMP levels are found in essentially all body segments (Shotwell, 1983; Livingstone et al., 1984). The putative serotonin receptors that may be defined by *tur* are likely to be widely distributed as well, as inferred from the fact that serotonin-containing cells are in all or most ganglia in developing and adult *Drosophila* (see Martel, 1983; White, this volume). Finally, dopa decarboxylase has been known for some time to be present from head to tail throughout development and into adulthood (Wright, 1977).

Genetic Mosaic Analysis of Mutationally Induced Learning Defects

With respect to their participation in the control of conditioned behavior, the molecules just discussed may exert a far more *local* influence on neural development and function in normal flies than might be inferred from their tissue distribution. These substances may affect the brain only, or a very restricted portion of it, which might be some sort of learning center. The mere description of the broad distribution of these neurochemical functions in wild-type tissues is thus misleading or at least inadequate. To look into this matter, a mosaic analysis of the mutated genes is essential. A series of flies can be created such that each animal is part mutant and part normal with respect to the expression of a *dnc*, *rut*, or *tur* mutation. Several previous studies have described and made use of the technology for generating mosaics of this type—that is, for X-chromosomal mutations such as these three—and for marking the genotype of external and internal tissues (see Hall, 1978, 1982, for reviews). Once produced by these genetic procedures, the mosaics need to be tested behaviorally one by one because each mosaic is unique. For example, it can be mutant anteriorally and normal posteriorally; mutant on the left side and normal on the right side; only part of the head mutant, and so forth (see Hall, 1978). Tests of groups of mosaic flies, analogous to the originally and most commonly applied conditioning tests in *Drosophila* (Quinn and Greenspan, 1984), would be problematic. Our current experiments on individual males conditioned by prior courtship of mated females can allow for all-or-none determination of normal versus mutant learning, at least in the case of the *dnc* mutations. That is, every dnc^+ male trained with a mated female and then tested with another such female has shown poor courtship afterward; yet every male expressing dnc^1 or an independently isolated allele, dnc^2, exhibited relatively vigorous courtship of the second (tester) mated female (S. Kulkarni and J. C. Hall, unpublished observations).

We have already applied the genetic materials, principles, and techniques just discussed to a different kind of conditioned male courtship. Naive, mature males of *D. melanogaster* court immature males very vigorously (e.g., Jallon and Hotta, 1979). After this experience the older male avoids courtship of a subsequent young male (Gailey et al., 1982). This kind of aftereffect is impaired by the conditioning mutations; for *dnc* mutants, it can be categorized in the same bimodal mutant versus normal fashion as in the conditioning effects elicited by mated females (Gailey et al., 1982). Against this background of genetic and behavioral results, we have constructed and tested approximately 20 $dnc//dnc^+$ mosaics; they have been analyzed behaviorally and in terms of the histochemical marking of mutant and normal cells in the central nervous system (D. A. Gailey and J.

C. Hall, unpublished observations). We have found that the genotype of the brain determines whether a mosaic will exhibit weak courtship of the second young male or the anomalous vigorous courtship of the tester fly—that is, strong versus weak learning, respectively.

Neural and Molecular Implications of Learning Defects in Mosaics

The preliminary results from the *dnc* mosaics are perhaps as would be expected: The fly's brain must be involved somehow in its ability to learn and to remember. There are, in addition, several other useful conclusions, questions, and new pieces of information related to this tentative "brain focusing" of the *dnc*-induced defect in conditioned courtship; eight of these are considered in the following paragraphs.

1. The product of the *dnc* gene is widely distributed, but the significance of it—at least for this mode of learning in a courtship context—is much more localized.
2. The particular cAMP PDE coded for by *dnc* is, in normal flies, relatively concentrated in the head (Shotwell, 1983), where the focus is. Therefore this particular enzyme could be brain-specific in its significance for the control of conditioned behavior, even though it is present in all major subdivisions of the animal and, when aberrant or absent, leads to female sterility. The latter phenotype is almost certainly related to a direct influence of the mutations on the female's reproductive system (Byers, 1979).

These quasipleiotropic effects of *dnc* variants encourage one to mention certain of the nonlearning phenotypes associated with some of the other genes currently under discussion. The *tur* mutant seems to be generally debilitated in its behavior (Booker and Quinn, 1981). Mutant *rut* males cannot discriminate between mated and normal females (Gailey et al., 1984) as opposed to normal males, and those expressing other of the conditioning mutations in the presence of the first mated female they encounter court her significantly less vigorously than they court virgin females. Finally, *Ddc* mutants are extremely and generally abnormal in phenotype (including being dead if they develop in the absence of this enzyme) because of the well-known role of dopa decarboxylase in cuticle hardening during several developmental stages and after eclosion (Wright, 1977).
3. The possibly surprising fact that a cAMP PDE-minus mutant can survive at all, let alone have less than overt or global behavioral problems, is rationalized by supposing that a variety of basic features of cyclic nucleotide metabolism is taken care of by other forms of this enzyme that are not affected by *dnc* mutations (see Davis and Kauvar, 1984). A further speculation is that the *dnc* gene product, but not these other cyclic nucleotide PDEs, is present primarily in a ganglion of the brain, or even in a part of such a tissue that is central to the control of learning (cf. Shotwell, 1983).
4. This hypothetical learning center could be approached experimentally not only by completing the mosaic studies, but also by a direct and relatively high-resolution localization of the *dnc*-specific cAMP PDE in tissue sections. This could be done by immunohistochemical detection of the protein coded for by *dnc* through experiments that would commence with the still-to-be-accomplished

purification of the enzyme, or by applying a method even closer to reality—detection of transcripts from the *dnc* gene by using DNA sequences from this locus which have been isolated by Davis and Davidson (1984). Techniques of *in situ* hybridization of radioactive DNA probes to messenger RNAs transcribed by *Drosophila* genes (e.g., Akam, 1983; Hafen et al., 1983, 1984) are becoming advanced and refined (see Scheller, this volume, for examples from *Aplysia*). Note that a failure to detect such transcripts from the *dnc* locus—for example, in the brain of adult flies—would not necessarily be strange. The reason is that *dnc* and its product might play a purely developmental role in the establishment of neural centers involved in learning. Thus RNAs complementary to *dnc* would eventually be looked for in embryonic through pupal stages, and in nonneural tissues as well as in the developing brain and thoracic nervous system.

5. Do the different learning mutations affect the same part of the nervous system and the same stage of the life cycle? Separate portions of the fly's brain could mediate information storage and retrieval depending on the conditioned behavior in question, or even in terms of the neural control of one kind of learning task. The different learning mutations affect molecules whose effects are related to one another, as inferred from work on other organisms. Therefore each gene's product could function in the same set of neurons as it acts to influence learning and interacts with other molecules. Another alternative is that two given genes might be expressed in nonoverlapping parts of the fly's brain, where "expressed" relates to the presence of the function in a relatively restricted subset of the animal—that which is involved in conditioned behavior—realizing that the gene product could be present, as well, in many other tissues. The two functions specified by this pair of genes would not, if their respective mutant foci do not overlap, be intimately related in their intracellular control of learning. A further possibility is that one gene's focus of action is narrowly local, whereas the other gene does not have a defined focus, in that a large proportion of the brain's volume must be mutant (or normal) for mutant (or normal) behavior to occur; or in that several discrete and relatively large brain regions, if genetically normal, are sufficient to allow for normal conditioning. Therefore it will be essential for us to map, in mosaics, the separate foci for the different learning mutations in order to determine whether a given focus is restricted or diffuse, and whether this focus overlaps another, including the possibility that the former is included within the latter.

We must also keep in mind that some of these genes are expressed transiently, during only certain stages of development, and that others are present throughout the life cycle. (Note: Each gene that has been discussed does code for a product present in adults, although it is not known if any of them except for dopa decarboxylase is relevant to learning at this stage.) Of course, we hope that the particular brain region or regions which are mapped in this manner are themselves of some heuristic value as to what the neural structures to be identified are actually doing to participate in the control of conditioned behavior.

The molecular techniques currently available for manipulating *Drosophila* genes will allow us to map, using mosaics, the learning focus of autosomal mutations such as those defining the *Ddc* locus. Essentially all the genetic techniques for producing mosaics depend on induced somatic loss of an X chromosome early in development (see Hall et al., 1976, for a review), which is accompanied by simultaneous loss of the normal alleles of both the "behavioral gene" and marker genes, the latter being necessary to score the genotypes of external and internal tissues. Since *Ddc* is situated on chromosome 2, not the X chromosome, these kinds of mosaics are ostensibly impossible to produce, in part because somatic loss of an autosome early in development is invariably lethal to the zygote. Yet a normal allele of this gene has been moved to the X chromosome by DNA-mediated transformation of cloned *Ddc* sequences (Scholnick et al., 1983). Ddc^+ in this anomalous location proved to be fully functional, not only in terms of the gross level of activity, but also in regard to the normal stage-specific rises and falls in its levels, as well as in the standard tissue distribution (Scholnick et al., 1983). Transformation of cloned genes is providing an extraordinarily powerful method for studying genes and their functions in *Drosophila* (see Flavell, 1983, for a review), and its utility goes all the way down to the technical level, as mentioned here.

6. Are different modes of learning controlled from the same neural centers? The combined mosaic/molecular strategies for studying the role that conditioning genes play in learning are important in considering whether there could be core mechanisms and structures that could control all features of experience-dependent behavior, or whether the neural underpinnings for various modes of conditioning could be learning-task-specific. Our approach to these questions is to compare learning foci for a given gene across a series of learning modalities in addition to making comparisons for a given task across the array of relevant genes. Two of the pertinent learning systems have been described, that is, aftereffects on a *Drosophila* male's courtship that occur after prior courtship of either a mated female or an immature male. These two types of aftereffects are not different merely because they involve overtly different behavioral situations.

In addition, the learning and memory mutations—each one known affects both phenomena—perturb "mated female learning" in a different way than they disrupt "young male learning." For example, the *cabbage (cab)* mutant is aberrant in its learning per se with respect to mated females (Gailey et al., 1984), whereas it is defective in the memory of its experience with young males (Gailey et al., 1982). These cases are related to the general fact that several of the conditioning mutants in *Drosophila* appear not to have learned, although they probably did. This means that the effects of training cannot be demonstrated, no matter how soon afterward the flies are tested (i.e., in the original sorts of shock-odor learning experiments; see Quinn and Greenspan, 1984). Yet a manipulation of the training/testing methods or of the mutant genotypes suggests that mutants such as *dnc, rut, tur*, and *cab* usually can have no learning demonstrated because they forget so rapidly that training effects are difficult to reveal (Dudai, 1979, 1983; Tempel et al., 1983).

In conditioned courtship then, it is significant that *cab* males tested with females just after mated-female-mediated training show an aberrant aftereffect (i.e., they court vigorously). The same mutant exposed to this first immature male shows robust learning (very weak courtship) when tested with the second such male immediately after training; this aftereffect decays back to strong courtship within approximately 30 min, whereas normal males are depressed in their courtship for at least four hours afterward (Gailey et al., 1982). Another example involves the memory mutant *amnesiac (amn)*. This was isolated, as usual, on the basis of defective conditioning in shock-odor experiments; groups of mutant flies forgot that a given odor was coupled with shock within 30 min of the training session, instead of remembering for at least six hours as did wild-type flies (Quinn et al., 1979). In conditioned courtship involving mated females, *amn* males are mutant in memory, forgetting to "avoid" the second female within 15–30 min (Siegel and Hall, 1979). But in the experiments with young males, *amn* males either are aberrant in learning per se or are superfast forgetters. They show no aftereffect when tested immediately after training (Gailey et al., 1982).

Mated-female and young-male learning phenomena do not cross-react. Males trained by exposure to mated females will then court young males vigorously; males trained by exposure to immature males then court females vigorously (summarized by Siegel et al., 1984). This implies that the fly's brain can effectively store and retrieve both kinds of information with no apparent mutual interference. A corollary is that a male trained sequentially in the presence of, first, a mated female and, second, a young male will, in subsequent sequential tests with another female and another male, show the normal weak courtship of both kinds of courtship objects (Siegel et al., 1984).

Mated-female learning seems to involve associative conditioning, whereas young-male learning resembles habituation. This has been concluded from experiments on reproductive pheromones in *Drosophila* alluded to above, and in the present context is related to two approaches. First, training a male to avoid courtship of a female requires that he has previously courted a fly in the presence of pheromones from mated females (Tompkins et al., 1983), which are known to contain antiaphrodisiac compounds (Jallon et al., 1981; Mane et al., 1983). Attempts to train such males only with the extracted antiaphrodisiacs lead to no aftereffect (Tompkins et al., 1983). Second, a male can be made to exhibit depressed courtship of a young male by prior exposure solely to odor traces from other young males (Gailey et al., 1982). In general, such materials include aphrodisiacs (Tompkins et al., 1980), and these seem to be sufficient—they do not need to be coupled with a courtship object to decrease male responsiveness to an object that would elicit a strong response from a naive male (Gailey et al., 1982).

Against these several pieces of background information then, it will be informative to ask, using our mosaic methods, if wholly separate neural substrates mediate these different types of learning and memory—at least insofar as the action of a given gene is concerned—or if associative conditioning and the possibly simpler case of habituation are physically connected in a mechanistic fashion by virtue of their being controlled by one neural center.

Another kind of apparent associative conditioning in *Drosophila* that involves tests of individuals, and hence is amenable to mosaic analysis, is affected by learning mutations (Booker and Quinn, 1981). The experiences involve electric shock-elicited conditioning of leg position. Booker and Quinn (summarized in Aceves-Piña et al., 1983) have concluded from studies of mosaics marked externally as to the distribution of *dnc*-mutant versus normal tissues that the head contains the focus for the mutationally induced behavioral defect. It will be informative to deepen this analysis and to compare the apparent focus in the brain to those foci that will be mapped with respect to the neural foci that influence conditioned courtship.

7. It appears that the mechanisms underlying higher and simple learning are related. This assumption is made not only because a given mutation in *Drosophila* can affect associatively conditioned courtship and, in a wholly separate situation, the decreased responsiveness which appears to be a case of nonassociative habituation. Some more general cases of simple learning in this insect are perturbed by the same mutations, that is, habituation and sensitization as demonstrated in experiments using sugar stimuli (Duerr and Quinn, 1982). Also, there is a feature of conditioned reproductive behavior (Schilcher, 1976) that involves the enhanced responsiveness of female *Drosophila* previously exposed to acoustical courtship stimuli (described later in the section of this chapter on behavioral rhythms). On the face of it, these aftereffects seem to be another case of sensitization. Strikingly, the learning and memory mutations expressed in females impair the aftereffects in a manner that parallels the effects of these genetic variants on other forms of associative and nonassociative conditioning (Kyriacou and Hall, 1984). It is well known that certain types of simple, experience-dependent behavior in mollusks—which are intensively investigated cases of cellular learning—are intimately associated with intercellular and intracellular communication as mediated by monoamines and cAMP (see Kandel and Schwartz, 1982, for a review). These are of course just the molecules whose levels are aberrant in several of the conditioning mutants in *Drosophila*. Therefore we now have a genetic connection between these pieces of neurochemistry and not only simple learning, but higher, associative conditioning as well. Those who study the relevant neurochemistry, physiology, and behavior in *Aplysia* are currently in the process of establishing similar connections without using mutants (Hawkins et al., 1983; Ocorr et al., 1983; Walters and Byrne, 1983). In *Drosophila*, the possibility of showing that a portion of a given ganglion is influenced by more than one of the relevant behavioral/neurochemical mutations, such that both types of learning are disrupted, should allow one to approach an understanding of the putatively intimate relationship, at the cellular and molecular levels, between the mechanisms controlling these ostensibly different kinds of conditioning. By inference, the components of the underlying control of these behaviors might be highly conserved, and one starts to wonder if there could be a "universal" learning mechanism, several parts of which are defined by the products of these *Drosophila* genes. Of course, the entirety of the learning machinery probably will not turn out to be the enzymes or receptors related to the levels or actions of the cyclic nucleotides or monoamines under discussion here. In this respect, it is important to discover the nature of the products of genes such as *cab* and *amn*, which as yet have no known biochemical correlate.

8. There are some hints as to the anatomical correlates of mutationally disrupted learning in *Drosophila* and as to the implications of these brain/behavioral relationships. Fruit flies have "mushroom bodies" in their dorsal brain, as do many insects and other invertebrates (Heisenberg, 1980). In these bodies, the calyces and lobes—and nerve cell bodies plus the intrinsic and extrinsic fibers which they comprise—have long been implicated as "association centers" at which a variety of behavioral stimuli converge (see Howse, 1975, for a review). Mushroom bodies of bees, for example, are strongly implicated as participants in the control of certain types of learning (e.g., Menzel and Erber, 1978). In *Drosophila* then, the role of the mushroom bodies in at least some aspects of learning comes from

the behavioral analysis of several brain mutants induced by Heisenberg and his colleagues. Remarkably, the mutants were isolated after chemical mutagenesis and brute-force anatomical screens (Heisenberg, 1980). One mutant is *mushroom-bodies-deranged (mbd)*, with its aberrantly large calyces, depleted fiber number in the peduncles that normally project from the calyces, and absence of mushroom body lobes. These anatomical abnormalities occur because of a late developmental failure: Peduncular fibers normally disappear after the larval stage and are then rebuilt during metamorphosis; the brains of *mbd* mutants develop and appear normal through the larval stage, but the regrowth of the mushroom body peduncles is much less than normal (Technau and Heisenberg, 1982). Groups of adult flies expressing the *mbd* mutation exhibit no detectable learning in tests using coupled shock-odor training stimuli (Heisenberg et al., 1984). Another relevant case is the separate mutant, *mushroom-bodies-miniature (mbm)*. Anatomically, *mbm* is inherently interesting in that it preferentially or exclusively affects female mushroom bodies (reducing them in size, by definition), with little or no apparent effect on males (Heisenberg et al., 1984). In parallel, *mbm* females, but not males, are aberrant in shock-odor learning (Heisenberg et al., 1984).

Mushroom bodies, in *Drosophila* and many other insects, are constantly invoked as being important centers for the "higher processing" of olfactory information (e.g. Howse, 1975). The learning which is defective in both the *mbd* and *mbm* mutants involves olfactory stimuli. Perhaps then, the mushroom bodies influence conditioned behavior in *Drosophila*, but in a paradigm-dependent manner. That is, mated female learning is normal in tests of *mbd* males (Heisenberg, 1980), although this type of conditioning does in part involve olfactory stimuli (Tompkins et al., 1983). Yet the memory of a male's exposure to a mated female has not yet been tested for a possible *mbd*-induced defect.

Behavioral Implications of Brain Development in Adult *Drosophila*

Before leaving the discussion of learning and mutants, some additional developmental issues related to some of the matters just raised will be discussed. Also relevant is a question left over from the discussion of conditioned courtship of immature males: Why does a young male emanate aphrodisiac pheromones that stimulate older males to court him so vigorously? This phenomenon seems maladaptive, although the fact that a mature male learns not to keep courting additional young males does appear to be adaptively significant. What if a young male stores some meaningful input that he receives because of the high-level courtship that he elicits? One of the most conspicuous of such stimuli comes from a song that courting *Drosophila* males routinely produce as they extend and vibrate their wings (see Ewing, 1983, for a review). This song was alluded to previously and will form an important part of the later discussion of a rhythmic component of the relevant acoustical output. Is it conceivable that a young male is aided in his later singing behavior because he heard so much singing when he was immature? (Young males stimulate courtship for only one to two days posteclosion, and it is during this stage that they begin to be "able or willing" to court other flies; e.g., Jallon and Hotta, 1979; Tompkins et al., 1980.) This outlandish suggestion implies that flies are in some ways like birds, which in certain cases must, as youngsters, be exposed to mature songs in order to store acoustical information and eventually produce a complete song with its proper dialect on their own (e.g., Konishi and Gurney, 1982). To shorten part of the current discussion: *Drosophila* males raised and stored in isolation do sing to females when the formers' isolation is finally ended. Therefore the males are "hard-wired" for their ability to perform this behavior in the most basic manner. But whether all elements of the singing are normal, including the song rhythm, is not yet known.

If isolated rearing tests of this question prove encouraging, certain neurodevelopmental phenomena must be considered in the context of postdevelopmental maturation of courtship behavioral actions. The *Drosophila* mushroom bodies continue to develop after the adult stage is reached (Technau, 1984). The fiber number in the peduncles increases by approximately 15%, compared with the number present after pupation. It is also striking that this increase depends on environmental inputs: Stimulus-deprived adults did not exhibit the usual increase in peduncular fiber numbers (Technau, 1984). The nature of the relevant stimuli is at present not fully understood, although visual stimuli seem to be irrelevant (Technau, 1984). Therefore, the fly's brain—possibly in regions beyond the mushroom bodies as well as within them—is not hard-wired during embryonic, larval, and pupal stages. Posteclosion brain development, analogous to that which occurs in *Drosophila*, is not unprecedented in insects (e.g., the results of Coss and Brandon, 1982, from bees). It is not known if an increase in complexity of functionally meaningful pattern formation accompanies these quantitative changes in the development of the adult brain of the fruit fly.

Modulation of this neuroanatomy relatively late in *Drosophila*'s life cycle could be influenced in part by the acoustical input received by a young male fly and elicited by his ostensibly anomalous aphrodisiac pheromones. Such modulations, in turn, could be part of a stimulus-induced programming of the functional anatomy which underlies the animal's ability to perform part of his repertoire of higher behaviors.

Behavioral Implications of Sexual Development and Its Genetic Control

The development of sex-specific characteristics, including the behavioral ones that are central to the current discussion, is a broadly studied subject in *Drosophila* genetics and biology. One aspect of these investigations is the manipulation of specific genes involved in sex determination (see Baker and Belote, 1983, for a review). Some of the relevant mutations transform genotypically female flies into phenotypic males. We have used one of these variants—the *transformer* (*tra*) mutation of Sturtevant (1945)—to turn our behaviorally analyzed mosaics completely into males (e.g., Hall et al., 1980). If we had not done so, a fly which was mosaic for expression of a *dnc* mutation, for example, would have been a diplo-X//haplo-X gynandromorph whose courtship behavior frequently involves no male actions (e.g., Hall, 1979) or defective male-specific behaviors (Schilcher and Hall, 1979). This situation, which is of intrinsic interest because a haplo-X genotype in certain portions of the central nervous system is correlated with male-specific actions, would of course ruin one's ability to correlate mutant versus normal conditioned male courtship with *dnc* versus *dnc$^+$* neural tissues.

Another sex-transforming mutant is of more substantive importance for considering the brain mechanisms which control courtship behavior. This is a temperature-sensitive mutant allele of the *tra-2* gene, which is not linked to the original *tra* mutation. When raised at low temperature, diplo-X *tra-2^{ts1}* (temperature-sensitive sex-transformed) mutant individuals develop as essentially normal phenotypic females; animals of the same genotype raised at high temperature are phenotypic males (Belote and Baker, 1982). Baker and Belote (1983) have demonstrated the striking fact that a *tra-2^{ts1}* mutant female that has been raised to adulthood at low temperature can be turned into a male-behaving fly simply by exposing it to high temperature for approximately six days. Hence the female-appearing mutants court other females, including extending and vibrating their wings and attempting to copulate with these (normal) females. Perhaps the easiest interpretation of this result would be to suppose that some sort of mutated, thermolabile substance is changed in its level of effectiveness of action by the heat treatment. Such material would furthermore be imagined somehow to mediate or to allow the release of male-specific actions once the substance reaches its normal level—that which is present in orthodox haplo-X males. The substance might also be thought of as acting via the control of neural function only—against a background of brain structure which one would assume to be unaffected by manipulating the posteclosion action of the *tra-2* gene.

Yet we now must recall that the fly's brain continues to develop during preadult stages, and so a very different type of interpretation of the behavioral results from this mutant can be entertained. It is possible that the posteclosion increase in mushroom body fibers (Technau, 1984) is related to sex-specific behavior in the general sense, including the hypothetical maturation of courtship singing behavior discussed earlier. This would mean, in the first place, that these anatomical changes occur in a male-specific manner in haplo-X adults, in nonconditionally mutant *tra* or *tra-2* flies which are diplo-X, or—most interestingly—in diplo-X *tra-2^{ts1}* adults that are being exposed to the male-determining conditions after their development is ostensibly completed. In the second place, it is useful to keep in mind that the only portion of the central nervous system so far known to keep developing in the adult fly is sexually dimorphic in morphology: Females have more fibers in their mushroom body peduncles (Technau, 1984). This kind of difference is found in other species of insect as well (e.g., Mobbs, 1982). Such sexual dimorphisms of anatomy tempt one to think further about the possible role of these neural association centers in the control of reproductive behavior.

BIOLOGICAL RHYTHM MUTANTS IN COURTSHIP

Basic Elements of the Variants and Their Effects on Courtship Song Rhythms

The courtship song of *D. melanogaster* consists most conspicuously of a series of tone pulses; there is also a low-amplitude hum which blends into or out of a pulse train (see Ewing, 1983, for a review). Our neurogenetic investigations of the male's singing behavior commenced with the discovery that the interpulse intervals (IPIs) of approximately 35 msec in the song of this species oscillate in a regular manner (Kyriacou and Hall, 1980). If one records several minutes of courtship song and then computes average IPIs for a consecutive series of, say, 10-sec time frames, one sees that these means fluctuate in a sinusoidal fashion.

The range between high and low IPIs for a given *D. melanogaster* male's song is about 4 msec, and the period of the oscillation is approximately 55 sec.

The song parameters are species-specific, in that *D. simulans* males produce pulse songs which oscillate around approximately 45–50-msec IPIs and have approximately 7–8-msec amplitudes with approximately 35-sec periods (Kyriacou and Hall, 1980). The species differences suggest that the pulse songs and their accompanying rhythms might generally stimulate females to accept males' mating attempts and help such females to distinguish between conspecific and foreign males. This was shown to be so by using electronically simulated songs played to wingless males who were courting normal females. To effect the strongest enhancement it was necessary to program the simulator with IPI values and rhythm periods characteristic of the species of the flies to which the sounds were being played (Kyriacou and Hall, 1982).

These same kinds of artificial songs were used in the aforementioned prestimulation experiments that involved normal or learning-mutant females (Kyriacou and Hall, 1984). Effective prestimulation of mating success was achieved when normal *D. melanogaster* females had the courtship hums played to them (cf. Schilcher, 1976) or were exposed to the oscillating pulse songs (55-sec period). In the latter case, pulse songs with invariant IPIs were ineffective (cf. Schilcher, 1976), as were songs with a *D. simulans*-like period of oscillation. Parenthetically it is important to mention that the blocked or attenuated aftereffects induced by the conditioning variants do *not* occur because these mutations affect the song-rhythm's oscillator. Consider that a *dnc* or *rut* mutation, for instance, could lead to an altered song rhythm. Recall that these mutants have aberrant cAMP levels, and note that this small molecule has been implicated in the control of biological rhythms in general (e.g., Eskin et al., 1982; Eskin and Takahashi, 1983; Feldman and Dunlap, 1983). The hypothetically aberrant rhythm in the song of *dnc* or *rut* males might carry over to the female's processing of acoustic input, in the sense that there could be "genetic coupling" between oscillators involved in song production and reception (cf. Pollack and Hoy, 1979). Therefore, a *dnc* or *rut* female would be defective in the "storage" of rhythmic song prestimulations for reasons not really related to the defective learning and memory induced by these mutations. This straw man collapsed when we analyzed the songs of males expressing the conditioning mutations and found them to be normal in rhythm parameters as well as in the basic IPI values (Kyriacou and Hall, 1984). Also, *dnc* variants have been tested for circadian rhythmicity and have been found to exhibit normal periodicity (R. J. Konopka, F. R. Jackson, and C. P. Kyriacou, unpublished observations).

Genetic Connections between Short-Term and Long-Term Rhythms

Some mutations do affect the courtship song rhythms in a dramatic manner. These are not arbitrarily chosen behavioral or neural mutants but instead are variants originally isolated by virtue of their aberrant circadian rhythms. Konopka and colleagues have induced several X-chromosomal mutations that lead to circadian rhythms with altered periods or to a collapse of such rhythms. Most of the mutations are in one gene, called *period* (*per*); two other variants, called *clock* (*clk*) and *andante* (*and*), involve changes at other genetic loci on the X chromosome (see Konopka, 1984, for a review). Jackson (1983) has isolated three clock mutants which seem to be in two separate autosomal genes, *phase-angle-2* (*psi-2*) and *phase-angle-3* (*psi-3*). These newly isolated circadian mutations are in chromosomes 2 and 3 respectively and lead to long-period rhythm phenotypes as well as aberrantly early (i.e., phase-shifted) eclosion. All of these mutations, except *and*, lead to parallel changes in the song rhythm's period (Kyriacou and Hall, 1980; C. P. Kyriacou and F. R. Jackson, unpublished observations; W. A. Zehring and J. C. Hall, unpublished observations). For example, long-period circadian mutants (e.g., per^L, *psi-2*, and *psi-3*) have approximately 62–82-sec pulse-song periods (depending on the mutation); short-period circadian mutants (e.g., per^s and *clk*) have about 35–42-sec periods in the behavioral rhythm; arrhythmic circadian mutants such as per^o, and flies expressing the *gate* (*gat*) mutation (which appears to be an allele of *psi-2*), sing courtship songs that have no regular fluctuations of IPIs.

Therefore the mechanisms underlying what seem to be very different kinds of biological rhythms share certain components: the presumed products coded for by four of the genes just noted. The control of the two rhythmic characters is not thoroughly congruent however, because the fifth gene, *and*, and the long-period circadian phenotype that this mutation causes, so far cannot be related to the control of the song period.

Neural and Molecular Analysis of the Clock Genes

We wish to determine if the connection between the long-term and the short-term rhythms extends to the cellular level. Is there one neural center that, if mutant in its expression of a gene like *per*, would simultaneously lead to a defective circadian rhythm and an aberrant song rhythm? This question can be refined somewhat, in that Konopka et al. (1983) have shown, by using mosaics marked externally as to their distribution of per^s versus normal tissue, that the circadian phenotype is linked to the genotype of the head. This suggests that the brain, or a portion of it, must be per^+ for the normal circadian period to be obtained (in this case, measured with respect to locomotor activity of individual flies).

We have extended this kind of experiment in two ways (J. C. Hall, R. J. Konopka, and C. P. Kyriacou, unpublished observations). First, we set up a new series of $per^s//per^+$ mosaics to express an internal neural marker as well as an external one (cf. the previous section on learning). Second, we tested each mosaic for its period of song rhythm in addition to its circadian period (meaning that each mosaic here was, as in the case of those tested for conditioned male courtship, a diplo-X//haplo-X fly transformed into a male by the expression of *tra*). A complete analysis of the behavioral and histochemical marking data from a portion of the approximately 150 mosaics that were generated and analyzed in this manner has led to the conclusion that the brain's genotype indeed matters as to the circadian phenotype. In fact, it appears as if the dorsal brain needs to be normal for normal behavior and mutant for the short-period phenotype to occur. Yet a mosaic in the former category often has a mutant song rhythm, and the reciprocal class of mixed-behaving mosaics was found as well. The song phenotype (short versus normal rhythm period) has so far been correlated one-to-one with the genotype of the thoracic nervous system. More specifically, the existence of mutant neurons on only one side (left or bilaterally symmetrical right) of the ventral nervous system is sufficient to cause both left and right wings to sing with the short-period oscillation.

The analysis of these mosaics must be completed to see if the thoracic neural focus for the control of the song rhythm will hold up, and also to determine if only a relatively small portion of the ventral nervous system (say, the mesothoracic ganglion) will contain the entirety of the song's pacemaker, as defined by the effects of the *per* gene. It is already clear, however, that there is no apparent "master clock" which exerts control over both types of rhythm. Another way to state a conclusion from this experiment is to note that *per* has direct influence on the brain with regard to the circadian oscillator, and it also directly affects the development or function of the thoracic nervous system in terms of the song clock. A priori, the thoracic ganglia must be influenced by *per*, since the final neural output which runs the wing vibrations are from these ganglia. Without the mosaics one could not know whether this effect of the gene is indirect or due to the fact that the *per* must be expressed properly, within the thoracic nervous system, if the song rhythm is to be normal.

It is essential to investigate the expression of the *per* locus in another manner, and for this reason we have initiated a molecular approach to the study of biological clocks. At the outset, we considered a question which suggested itself from the mosaic experiment just described: Is *per* expressed not only in a specific part of the brain plus a ganglion of the thoracic nervous system (only in one of its ganglia?), but also in a variety of other tissues as well? For example, the gene could be active in cells of the heart, since the normally rhythmic heartbeat of the larva is at least quasiarrhythmic in per^o animals (Livingstone, 1981). This begs the question as to what sort of expanding array of clocklike functions will turn out to be influenced by this gene. The array has already begun to embrace phenotypes that are not overtly in the realm of rhythmic ones: Certain mutant *per* alleles—which lead to longer-than-normal periods or, metaphorically, slowly ticking clocks—induce a relatively poor ability of males to be conditioned by mated females or by young males (Jackson et al., 1983). It is as if these mutant flies are storing or retrieving information too slowly. Hence, one of the foci for *per*'s influence on neural development or function could overlap with the sites of action of the "learning genes."

Where, then, is the *per* gene active? Unlike what is possible when one asks the analogous question with regard to a gene like *dnc*, it cannot be asked here by planning to assess the tissue distribution of *per*'s protein product (or products), since the nature of that material is of course a total mystery. Therefore we have isolated DNA sequences from—and probably including all of—the *per* locus (Reddy et al., 1984). Similar findings have also been reported by Bargiello and Young (1984).

We first obtained three recombinant bacteriophages containing a series of overlapping segments from the vicinity of *per*. These clones had been isolated by Pirrotta et al. (1983) by the mind-boggling techniques that depend on microexcision of defined parts of the salivary gland chromosomes of *D. melanogaster* larvae. This is but one of the approaches in *Drosophila* which allow the isolation of a gene without any prior knowledge about its product (see Artavanis-Tsakonas et al., 1983, and Scott, 1984, for other examples of neurogenetic interest).

Our clones near the *per* locus were initially placed on the X-chromosomal map by relating portions of three overlapping recombinant phage to "breakpoints" of deletions from, and a translocation involving, this chromosome (Pirrotta et al., 1983; Reddy et al., 1984). These aberrations had previously been determined to be quite near *per* (Young and Judd, 1978; Smith and Konopka, 1981). Therefore we were able to define an approximately 25 kb subset of our cloned material within which *per* is likely to lurk.

We are attempting to identify completely the parts of this segment of the X chromosome that contain the *per* locus. One way of approaching this problem uses DNA-mediated transformation in *Drosophila*, which was alluded to earlier in the discussion of the *Ddc* gene and the neurochemical mutations at this locus. For identifying *per*, we have generated a series of transformation vectors (cf. Rubin and Spradling, 1982), containing various "unknown" (i.e., per^+ or not-per^+, as the case may be) segments of DNA, subcloned from the material now known to be in and around this clock gene. The vectors also carry a selectable marker, a normal allele of the alcohol dehydrogenase (*Adh*) gene (see Goldberg et al., 1983). Such plasmids also, and crucially, contain sequences derived from the transposable "P elements" (see O'Hare and Rubin, 1983). These sequences allow the vectors to integrate into *D. melanogaster*'s genome in a variety of locations, as long as a "transposase" function is supplied from a complete element (Spradling and Rubin, 1982; Rubin and Spradling, 1982). Practically, this means that a series of individual embryos are injected with the vectors carrying DNA sequences to be tested, plus Adh^+; simultaneously, one injects cloned vectors carrying most or all of the P-element sequences. Our host embryos were homozygous for an *Adh*-null (Adh^n) allele (cf. Goldberg et al., 1983). Among the progeny of the injectees that survived to fertile adulthood were phenotypically Adh^+ flies, selected by survival of an ethanol treatment which kills Adh^n animals (cf. Goldberg et al., 1983). The several transformed lines we established in this manner included examples from each of six unknown subclones that had been set up to be tested for containing per^+. We introduced an arrhythmic per^o allele into these lines and have tested individuals from them for two types of rhythmicity, that is, by monitoring their general locomotor activity over the course of several 24-hour cycles and also by recording the courtship songs of transformed males. So far, two overlapping DNA fragments partially "rescue" the mutant phenotypes, in that approximately 20–60% of the individuals from a given strain, carrying either of these inserts, are rhythmic in their circadian or singing behavior (Zehring et al., 1984).

The various transformed lines have turned out to be cases of DNA inserted somewhere on the X chromosome or on one of the fly's autosomes. Most of these locations are anomalous, since *per* is X chromosomal, but still allow at least quasi-normal per^+ functioning. This is correlated with earlier data showing that unnatural genomic locations for other cloned loci in *Drosophila* allow normal or near-normal levels of gene products, plus normal temporal control and tissue distribution of the relevant enzyme-encoding genes (e.g., Goldberg et al., 1983; Scholnick et al., 1983; Spradling and Rubin, 1983). Thus, transformed segments of DNA are typically (but not always) regulated normally when they find themselves in unusual chromosomal positions.

Given our preliminary success at a molecular plus behavioral identification of *per*, we have begun an analysis of the regulation and expression of this clock gene. These studies commenced with an examination of putative *per* transcripts, by labeling *per*-region DNA and using it to probe RNAs extracted from animals at various stages of the life cycle and from various portions of the circadian cycle. We have identified four transcripts which are complementary to separate subsegments of the original 25 kb of X-chromosomal DNA which almost certainly includes the *per* locus. One of these RNAs is at dramatically subnormal levels in adult flies expressing either of two independently isolated per^0 mutations (Reddy et al., 1984). In normal adults, this same RNA species fluctuates strikingly in its levels during a given circadian cycle, being much more abundant during the middle of the day than the middle of the night; such cycling behavior of the transcription or stability of this putative messenger RNA continues when the light/dark cycles are terminated, that is, when the flies are subsequently kept in constant darkness (Reddy et al., 1984), conditions in which circadian rhythms of locomotor activity persist (e.g., Konopka and Benzer, 1971; Smith and Konopka, 1981).

Once our further experiments have identified the DNA sequences and transcripts homologous to them that seem to comprise all of this clock gene and its primary molecular functions, then we hope to assess the complete tissue distribution of *per*'s expression. Such studies will involve *in situ* hybridization of labeled clones to RNAs in sectioned material (a technique mentioned earlier in the discussion of the *dnc* gene; see Scott, 1984, for a review of the method and its application). Attention must be paid, at least initially, to the *in situ* localization of messages transcribed from portions of the *per* region beyond that which gives rise to the RNA whose abundance fluctuates. The reason is that transcript might not be the entirety of this clock gene's expression. Such a caveat stems from the fact that the transcript in question is not detectable until the late pupal stage of development (Reddy et al., 1984).

This finding introduces several elements of *per*'s temporal expression. A priori, this locus must be active during some part of the fly's development, since the original mutant phenotypes involved aberrant eclosion rhythms or their absence (Konopka and Benzer, 1971). Yet this does not address the question as to whether certain components of the locus might function during most or all of the developmental sequence, beginning even in the embryo. In fact we predict that important elements of *per* are transcribed and translated at many stages of the life cycle. Recall that a larval rhythm is abnormal in a per^0 mutant tested during this relatively early developmental stage (Livingstone, 1981). Another larval phenotype affected (although not markedly) by per^0 is a rhythmic fluctuation in membrane potential (indirectly measured) of salivary gland cells (Weitzel and Rensing, 1981). Embryonic cells might also be influenced by this gene, in the sense that *in vitro* cultures established from such cells divide in a rhythmic manner when subjected to specific temperature fluctuations; this rhythm is aberrant in cells derived from per^s embryos (R. J. Konopka and S. Wells, unpublished observations). Returning to the nervous system: Neurosecretory cells in the adult brain seem to have developed abnormally in per^0 animals, because clusters of such cells have an elevated frequency of ectopic locations compared to what is observed in the normal brain (Konopka and Wells, 1980). Two RNA species transcribed from within (or near) the *per* locus are present throughout the life cycle (Reddy et al., 1984), and at least one of them could be involved in the developmental phenotypes exhibited by these *per* mutants.

As for adult functioning of the gene, it is significant that a temperature-sensitive *per* allele, called *per*^{L2}, has a long-period circadian phenotype when raised to adulthood and stored at low temperatures; but if one then turns up the temperature, the flies gradually become arrhythmic in their locomotor activity (Orr, 1982). It has not been asked whether *per*^{L2} is also sensitive to heat pulses during larval, pupal, or even embryonic stages with regard to the eventual behavior of the adult. But the fact that one can "turn off" this *per* allele's control of rhythmic (albeit abnormal) behavior by treating adults implies that the gene maintains its expression during this stage of the life cycle. Indeed, all the RNAs transcribed from the *per* region are readily detectable in adults (Bargiello and Young, 1984; Reddy et al., 1984), and it is difficult at this point in the analysis to rule out the relevance of any of them.

Can we go beyond investigating *per*'s expression at the transcriptional level to delve into the nature of the protein or proteins coded for by the locus? We should be able to coax our cloned material into manufacturing at least some of the amino acid sequence information, putatively encoded within it, into polypeptides (see techniques of Gray et al., 1982, and the relevant extensions of such methods as discussed by these authors). Will the information from the analysis of such molecules provide clues as to what a clock function could be at the molecular level? This is problematic in molecular biology in general. For instance, almost none of the genes in *Drosophila* that have been cloned solely by application of cytogenetic and molecular/genomic materials and methods are understood as to what the protein products (if any) are doing in their contributions to cellular structure and physiology. It cannot be guaranteed that one or more of the polypeptides putatively encoded within the *per* locus will yield to attempts at understanding their exact functions. Nevertheless we can make some educated guesses as to what sort of roles the product might play to control pacemaking.

A protein produced by *per* could be a membrane-associated molecule. This suggestion stems from the extensive background of facts and thoughts revolving around membrane models of biological rhythms (reviewed by Engelmann and Schrempf, 1980). Konopka and Orr (1980) discuss how the *per* gene product(s) could be thought of as controlling ionic fluxes across the neural membranes of pacemaker cells. Another possibility is that *per* codes for a neurohumoral factor, given that transplants of a *per*^{s} brain into the abdomen of a *per*^{o} adult cause the latter to take on an appropriate short-period rhythmic character (Handler and Konopka, 1979). This means, first of all, that arrhythmicity is not irreversibly set by a lack of *per*'s action during development. The results could also mean that the *per* product is a readily diffusible peptide which, at least in the case of this transplant experiment, can modulate the function of brain cells after moving some distance to find them. Such a molecule could even be a very small one, in spite of the fact that transcripts from the DNA which includes the *per* locus range from approximately 1 to 4.5 kb in length (Reddy et al., 1984). This is not necessarily informative with respect to the size of the final active product. For instance, consider that mRNAs from which neuropeptides themselves are translated can be much longer than the coding capacity necessary to synthesize the protein molecule in question (see Scheller, this volume).

per might encode an orthodox catalytic function. This possibility can have a specific hypothesis attached to it, given the results of Livingstone and Tempel (1983) on octopamine production in *per^o* flies: The rate of such synthesis is depressed; a pertinent enzyme, tyrosine decarboxylase, is decreased in concentration. Could the effects of a clock gene be understood by imagining that the product of the locus is such a nonnovel component of monoamine metabolism? Livingstone and Tempel (1983) discuss genetic evidence which suggests that the *per^o* mutation affects tyrosine decarboxylase by some kind of indirect route. In any event, cloned sequences from *per* will allow us to test whether the gene does in fact code for this "mundane" enzyme: Antibodies raised against polypeptides artificially synthesized from all or part of *per* (see discussion in Gray et al., 1982) can be used to see whether they will inhibit tyrosine decarboxylase activity.

We should not forget the other clock genes. They, too, must be dealt with in a manner which eventually includes the kind of neural and molecular analysis that has been outlined in this section. Where, for instance, does an X-chromosomal gene like *clk* affect the nervous system such that both the circadian and the song clocks go awry when the locus is mutated? Does a gene such as *and* influence only the brain, and a portion of it related to *per*'s focus, in its mutationally induced change of the circadian period—whereby, as far as the effects of the one extant mutant allele are concerned, the song rhythm is normal? Since both *clk* and *and* are X-linked genes, they are amenable to a determination via internally marked mosaics of their presumptive neural foci. Clock genes such as these should also be subjected to the same sort of molecular isolation and characterization as that which has been achieved and is ongoing with *per*. Ultimately it will be possible to study the temporal and spatial features of *clk*'s and *and*'s expression—plus the intrinsic nature of the products of these genes—and compare them with the corresponding information obtained from our emerging and expanding work on the molecular clones from the *per* gene.

CONCLUSIONS

We have found it both useful and essential to use genetic variants and genetic techniques in our analysis of complex behavior in *Drosophila*. Otherwise we could not have effected disruptions in conditioned behavior and endogenous rhythms and studied these defects at will in intact animals—those which, moreover, were not burdened by nonspecific debilitations that can occur when other kinds of treatments, such as drugs or surgery, are used in attempts to disrupt the fly's nervous system. The mutants and other genetic variants have circumvented these potential difficulties for us, and in addition have allowed us frequently to induce deep and localized neural lesions in behaviorally studied mosaics.

For these and many other reasons, it has been imperative to study the basic elements of higher behavior from a genetic standpoint. We have been surprised by several of the genetic connections which have now been established among mechanisms that underlie ostensibly very different kinds of complex functions. The interrelationships among courtship, conditioning, and rhythmic behaviors, as established by the effects of a given mutation on more than one or even all of these types of behavioral activities, are so far formalistic. Yet, as we proceed in our investigation of the neural and molecular natures of these behavioral defects, we believe that the manner in which the phenomena are related will be increasingly understood in literal terms. In this way, studying the molecular biology of neurally and behaviorally interesting genes will become a more and more important element of genetic analysis of the development and functioning of the nervous system.

ACKNOWLEDGMENTS

I am extremely grateful to the following colleagues for contributing so mightily to the ideas behind these experiments, and to the interpretation of the data stemming from them: Don Gailey, Dick Siegel, Laurie Tompkins, Shankar Kulkarni, Bambos Kyriacou, Pranhitha Reddy, Will Zehring, David Wheeler, Ron Konopka, and Michael Rosbash. I thank others for providing crucial materials and information: Chip Quinn, Rob Jackson, Vincent Pirrotta, Christopher Hadfield, Jim Posakony, and Jay Hirsh. The work in my laboratory has been supported by grants GM-21473 and GM-33205 from the National Institutes of Health.

REFERENCES

Aceves-Piña, E. O., R. Booker, J. S. Duerr, M. S. Livingstone, W. G. Quinn, R. F. Smith, P. P. Sziber, B. L. Tempel, and T. P. Tully (1983) Learning and memory in *Drosophila*, studied with mutants. *Cold Spring Harbor Symp. Quant. Biol.* **48**:831–840.

Akam, M. (1983) The location of *Ultrabithorax* transcripts in *Drosophila* tissue sections. *EMBO J.* **2**:2075–2084.

Artavanis-Tsakonas, S., M. A. T. Muskavitch, and B. Yedvobnick (1983) Molecular cloning of *Notch*, a locus affecting neurogenesis in *Drosophila melanogaster*. *Proc. Natl. Acad. Sci. USA* **80**:1977–1981.

Baker, B. S., and J. M. Belote (1983) Sex determination and dosage compensation in *Drosophila melanogaster*. *Annu. Rev. Genet.* **17**:345–393.

Bargiello, T. A., and M. W. Young (1984) Molecular genetics of a biological clock in *Drosophila*. *Proc. Natl. Acad. Sci. USA* **81**:2142–2146.

Belote, J. M., and B. S. Baker (1982) Sex determination in *Drosophila melanogaster*: Analysis of transformer-2, a sex-transforming locus. *Proc. Natl. Acad. Sci. USA* **79**:1568–1572.

Booker, R., and W. G. Quinn (1981) Conditioning of leg position in normal and mutant *Drosophila*. *Proc. Natl. Acad. Sci. USA* **78**:3940–3944.

Byers, D. (1979) Studies on learning and cyclic AMP phosphodiesterase of the *dunce* mutant of *Drosophila melanogaster*. Unpublished doctoral dissertation, California Institute of Technology, Pasadena.

Coss, R. G., and J. G. Brandon (1982) Rapid changes in dendritic spine morphology during honey bee's first orientation flight. In *The Biology of Social Insects*, M. D. Breed, C. D. Michener, and H. E. Evans, eds., pp. 338–342, West-View Press, Boulder, Col.

Davis, R. L., and N. Davidson (1984) Isolation of the *Drosophila melanogaster* dunce chromosomal region and recombinational mapping of dunce sequences with restriction site polymorphisms as genetic markers. *Mol. Cell Biol.* **4**:358–367.

Davis, R. L., and L. M. Kauvar (1984) *Drosophila* cyclic nucleotide phosphodiesterases. *Adv. Cyclic Nucleotide Res.* **16**:393–402.

Dudai, Y. (1979) Behavioral plasticity in a *Drosophila* mutant, dunce^{DB276}. *J. Comp. Physiol.* **130**:271–275.

Dudai, Y. (1983) Mutations affect storage and usage of memory differentially in *Drosophila*. *Proc. Natl. Acad. Sci. USA* **80**:5445–5448.

Duerr, J., and W. G. Quinn (1982) Three *Drosophila* mutations that block associative learning also affect habituation and sensitization. *Proc. Natl. Acad. Sci. USA* **79**:3646–3650.

Engelmann, W., and M. Schrempf (1980) Membrane models of circadian rhythms. *Photochem. Photobiol.* **5**:49–86.

Eskin, A., and J. S. Takahashi (1983) Adenylate cyclase activation shifts the phase of a circadian pacemaker. *Science* **220**:82–84.

Eskin, A., G. Corrent, C.-Y. Lin, and D. J. McAdoo (1982) Mechanism for shifting the phase of a circadian rhythm by serotonin: Involvement of cyclic AMP. *Proc. Natl. Acad. Sci. USA* **79**:660–664.

Ewing, A. W. (1983) Functional aspects of *Drosophila* courtship. *Biol. Rev.* **58**:275–292.

Feldman, J. F., and J. C. Dunlap (1983) *Neurospora crassa*: A unique system for studying circadian rhythms. *Photochem. Photobiol. Rev.* **7**:319–368.

Flavell, A. (1983) *Drosophila* takes off. *Nature* **305**:96–97.

Gailey, D. A., F. R. Jackson, and R. W. Siegel (1982) Male courtship in *Drosophila*: The conditioned response to immature males and its genetic control. *Genetics* **102**:771–782.

Gailey, D. A., F. R. Jackson, and R. W. Siegel (1984) Conditioning mutations in *Drosophila melanogaster* affect an experience-dependent behavioral modification in courting males. *Genetics* **106**:613–623.

Goldberg, D. A., J. W. Posakony, and T. Maniatis (1983) Correct developmental expression of a cloned alcohol dehydrogenase gene transduced into the *Drosophila* germ line. *Cell* **34**:59–73.

Gray, M. R., H. V. Colot, L. Guarente, and M. Rosbash (1982) Open reading frame cloning: Identification, cloning, and expression of open reading frame DNA. *Proc. Natl. Acad. Sci. USA* **79**:6598–6602.

Hafen, E., M. Levine, R. L. Garber, and W. J. Gehring (1983) An improved *in situ* hybridization method for the detection of cellular RNAs in *Drosophila* tissue sections and its application for localizing transcripts of the homeotic *Antennapedia* gene complex. *EMBO J.* **2**:617–623.

Hafen, E., M. Levine, and W. J. Gehring (1984) Regulation of *Antennapedia* transcript distribution by the bithorax complex in *Drosophila*. *Nature* **307**:287–289.

Hall, J. C. (1978) Behavioral analysis in *Drosophila* mosaics. In *Genetic Mosaics and Cell Differentiation*, W. J. Gehring, ed., pp. 259–305, Springer-Verlag, Berlin.

Hall, J. C. (1979) Control of male reproductive behavior by the central nervous system of *Drosophila*: Dissection of a courtship pathway by genetic mosaics. *Genetics* **92**:437–457.

Hall, J. C. (1982) Genetics of the nervous system in *Drosophila*. *Q. Rev. Biophys.* **15**:223.

Hall, J. C. (1984) Mutants of biological rhythms and conditioned behavior in *Drosophila* courtship. In *Evolutionary Genetics of Invertebrate Behavior*, M. D. Huettel, ed., Plenum, New York (in press).

Hall, J. C., W. M. Gelbart, and D. R. Kankel (1976) Mosaic systems. In *The Genetics and Biology of Drosophila*, M. Ashburner and E. Novitski, eds., pp. 265–314, Academic, London.

Hall, J. C., L. Tompkins, C. P. Kyriacou, R. W. Siegel, F. v. Schilcher, and R. J. Greenspan (1980) Higher behavior in *Drosophila* analyzed with mutants that disrupt the structure and function of the nervous system. In *Development and Neurobiology of Drosophila*, O. Siddiqi, P. Babu, L. M. Hall, and J. C. Hall, eds., pp. 425–455, Plenum, New York.

Handler, A. M., and R. J. Konopka (1979) Transplantation of a circadian pacemaker in *Drosophila*. *Nature* **279**:236–238.

Hawkins, R. D., T. W. Abrams, T. J. Carew, and E. R. Kandel (1983) A cellular mechanism of classical conditioning in *Aplysia*: Activity-dependent amplification of presynaptic facilitation. *Science* **219**:400–405.

Heisenberg, M. (1980) Mutants of brain structure and function: What is the significance of the mushroom bodies for behavior? In *Development and Neurobiology of Drosophila*, O. Siddiqi, P. Babu, L. M. Hall, and J. C. Hall, eds., pp. 373–390, Plenum, New York.

Heisenberg, M., A. Borst, S. Wagner, and D. Byers (1984) *Drosophila* mushroom body mutants are deficient in olfactory learning. *J. Neurogenet.* (in press).

Howse, P. E. (1975) Brain structure and behavior in insects. *Annu. Rev. Entomol.* **20**:359–379.

Jackson, F. R. (1983) The isolation of biological rhythm mutations on the autosomes of *Drosophila melanogaster. J. Neurogenet.* **1**:3–15.

Jackson, F. R., D. A. Gailey, and R. W. Siegel (1983) Biological rhythm mutations affect an experience-dependent modification of male courtship behavior in *Drosophila melanogaster. J. Comp. Physiol.* **151**:545–552.

Jallon, J.-M., and Y. Hotta (1979) Genetic and behavioral studies of female sex appeal in *Drosophila. Behav. Genet.* **9**:257–275.

Jallon, J.-M., C. Anthony, and O. Benamar (1981) Une antiaphrodisiaque produit par les mâles *Drosophila melanogaster* et transféré aux femelles lor des la copulation. *C. R. Séances Acad. Sci. (III)* **292**:1147–1149.

Kandel, E. R., and J. H. Schwartz (1982) Molecular biology of learning: Modulation of transmitter release. *Science* **218**:433–443.

Konishi, M., and M. E. Gurney (1982) Sexual differentiation of brain and behavior. *Trends Neurosci.* **5**:20–23.

Konopka, R. J. (1984) Neurogenetics of *Drosophila* circadian rhythms. In *Evolutionary Genetics of Invertebrate Behavior*, M. D. Huettel, ed., Plenum, New York (in press).

Konopka, R.J., and S. Benzer (1972) Clock mutants of *Drosophila melanogaster. Proc. Natl. Acad. Sci. USA* **68**:2112–2116.

Konopka, R. J., and D. Orr (1980) Effects of a clock mutation on the subjective day—implications for a membrane model of the *Drosophila* circadian clock. In *Development and Neurobiology of Drosophila*, O. Siddiqi, P. Babu, L. M. Hall, and J. C. Hall, eds., pp. 409–416, Plenum, New York.

Konopka, R. J., and S. Wells (1980) *Drosophila* clock mutations affect the morphology of a brain neurosecretory cell group. *J. Neurobiol.* **11**:411–415.

Konopka, R.J., S. Wells, and T. Lee (1983) Mosaic analysis of a *Drosophila* clock mutant. *Mol. Gen. Genet.* **190**:284–288.

Kyriacou, C. P., and J. C. Hall (1980) Circadian rhythm mutations in *Drosophila melanogaster* affect short-term fluctuations in the male's courtship song. *Proc. Natl. Acad. Sci. USA* **77**:6729–6733.

Kyriacou, C. P., and J. C. Hall (1982) The function of courtship song rhythms in *Drosophila. Anim. Behav.* **30**:794–801.

Kyriacou, C. P., and J. C. Hall (1984) Learning and memory mutations impair acoustic priming of mating behaviour in *Drosophila. Nature* **308**:62–65.

Livingstone, M. S. (1981) Two mutations in *Drosophila* affect the synthesis of octopamine, dopamine and serotonin by altering the activities of two different amino-acid decarboxylases. *Neurosci. Abstr.* **7**:351.

Livingstone, M. S., and B. L. Tempel (1983) Genetic dissection of monoamine synthesis in *Drosophila. Nature* **303**:67–70.

Livingstone, M. S., P. P. Sziber, and W. G. Quinn (1984) Loss of calcium/calmodulin responsiveness in the adenylate cyclase of *rutabaga*, a *Drosophila* learning mutant. *Cell* **37**:205–215.

Mane, S. D., L. Tompkins, and R. C. Richmond (1983) Male esterase 6 catalyzes the synthesis of a sex pheromone in *Drosophila melanogaster* females. *Science* **222**:419–421.

Martel, E. (1983) Bithorax complex mutants transform the segment-specific pattern of 5-HT containing neurons in *Drosophila. Neurosci. Abstr.* **9**:832.

Menzel, R., and J. Erber (1978) Learning memory in bees. *Sci. Am.* **239** *(No. 1)*:103–110.

Mobbs, P. G. (1982) The brain of the honeybee *Apis mellifera*. I. The connections and spatial organization of the mushroom bodies. *Philos. Trans. R. Soc. Lond. (Biol.)* **298**:309–354.

Ocorr, K. A., E. T. Walters, and J. H. Byrne (1983) Associative conditioning analog in *Aplysia* tail sensory neurons selectively increases cAMP content. *Neurosci. Abstr.* **9**:169.

O'Hare, K., and G. M. Rubin (1983) Structures of P transposable elements and their sites of insertion and excision in the *Drosophila melanogaster* genome. *Cell* **34**:35–45.

Orr, D. P.-Y. (1982) Behavioral neurogenetic studies of a circadian clock in *Drosophila melanogaster*. Unpublished doctoral dissertation, California Institute of Technology, Pasadena.

Pirrotta, V., C. Hadfield, and G. H. J. Pretorius (1983) Microdissection and cloning of the white locus and the 3B1–3C2 region of the *Drosophila* X chromosome. *EMBO J* **2**:927–934.

Pollack, G. S., and R. R. Hoy (1979) Temporal pattern as a cue for species-specific calling song recognition. *Science* **204**:429–432.

Quinn, W. G., and R. J. Greenspan (1984) Learning and courtship in *Drosophila*: Two stories with mutants. *Annu. Rev. Neurosci.* **7**:67–93.

Quinn, W. G., P. P. Sziber, and R. Booker (1979) The *Drosophila* memory mutant *amnesiac*. *Nature* **277**:212–214.

Reddy, P., W. A. Zehring, D. A. Wheeler, V. Pirrotta, C. Hadfield, J. C. Hall, and M. Rosbash (1984) Molecular analysis of the *period* locus in *Drosophila melanogaster* and identification of a transcript involved in biological rhythms. *Cell* **38**:701–710.

Richmond, R. C., S. D. Mane, and L. Tompkins (1984) The behavioral effects of a carboxylesterase in *Drosophila*. In *Evolutionary Genetics of Invertebrate Behavior*, M. D. Huettel, ed., Plenum, New York (in press).

Rubin, G. M., and A. C. Spradling (1982) Genetic transformation of *Drosophila* with transposable element vectors. *Science* **218**:348–353.

Schilcher, F. v. (1976) The function of pulse song and sine song in the courtship of *Drosophila melanogaster*. *Anim. Behav.* **24**:622–625.

Schilcher, F. v., and J. C. Hall (1979) Neural topography of courtship song in sex mosaics of *Drosophila melanogaster*. *J. Comp. Physiol.* **129**:85–95.

Scholnick, S. B., B. A. Morgan, and J. Hirsh (1983) The cloned dopa decarboxylase gene is developmentally regulated when reintegrated into the *Drosophila* genome. *Cell* **34**:37–45.

Scott, M. P. (1984) Homoeotic gene transcripts in the neural tissue of insects. *Trends Neurosci.* **7**:221–223.

Shotwell, S. L. (1983) Cyclic adenosine 3':5'-monophosphate phosphodiesterase and its role in learning in *Drosophila*. *J. Neurosci.* **3**:739–747.

Siegel, R. W., and J. C. Hall (1979) Conditioned responses in courtship of normal and mutant *Drosophila*. *Proc. Natl. Acad. Sci. USA* **76**:3430–3434.

Siegel, R. W., J. C. Hall, D. A. Gailey, and C. P. Kyriacou (1984) Genetic elements of courtship in *Drosophila*: Mosaics and learning mutants. *Behav. Genet.* **14**:425–452.

Smith, R. F., and R. J. Konopka (1981) Circadian clock phenotypes of chromosome aberrations with a breakpoint at the *per* locus. *Mol. Gen. Genet.* **183**:243–251.

Spradling, A. C., and G. M. Rubin (1982) Transposition of cloned P elements into *Drosophila* germ line chromosomes. *Science* **218**:341–347.

Spradling, A. C., and G. M. Rubin (1983) The effect of chromosomal position on the expression of the *Drosophila* xanthine dehydrogenase gene. *Cell* **34**:47–57.

Sturtevant, A. H. (1945) A gene in *Drosophila melanogaster* that transforms females into males. *Genetics* **30**:297–299.

Technau, G. (1984) Fiber number in the mushroom bodies of adult *Drosophila melanogaster* depends on age, sex, and experience. *J. Neurogenet.* **1**:113–126.

Technau, G., and M. Heisenberg (1982) Neural reorganization during metamorphosis of the corpora pedunculata in *Drosophila melanogaster*. *Nature* **295**:405–407.

Tempel, B. T., N. Bonini, D. R. Dawson, and W. G. Quinn (1983) Reward learning in normal and mutant *Drosophila*. *Proc. Natl. Acad. Sci. USA* **80**:1482–1486.

Tempel, B. T., M. S. Livingstone, and W. G. Quinn (1984) Mutations in the dopa decarboxylase affect learning in *Drosophila*. *Proc. Natl. Acad. Sci. USA* **81**:3577–3581.

Tompkins, L. (1984) Genetic analysis of sex appeal in *Drosophila melanogaster*. *Behav. Genet.* **14**:453–482.

Tompkins, L., J. C. Hall, and L. M. Hall (1980) Courtship-stimulating volatile compounds from normal and mutant *Drosophila*. *J. Insect Physiol.* **26**:689–697.

Tompkins, L., R. W. Siegel, D. A. Gailey, and J. C. Hall (1983) Conditioned courtship in *Drosophila* and its mediation by association of chemical cues. *Behav. Genet.* **13**:565–578.

Walters, E. T., and J. H. Byrne (1983) Associative conditioning of single sensory neurons suggests a cellular mechanism for learning. *Science* **219**:405–408.

Weitzel, G., and L. Rensing (1981) Evidence for cellular circadian rhythms in isolated fluorescent dye-labelled salivary glands of wild type and an arrhythmic mutant of *Drosophila melanogaster*. *J. Comp. Physiol.* **143**:229–235.

Wright, T. R. F. (1977) The genetics of dopa decarboxylase and α-methyl dopa sensitivity in *Drosophila melanogaster*. *Am. Zool.* **17**:707–721.

Wright, T. R. F., B. C. Black, C. P. Bishop, L. Marsh, E. S. Pentz, R. Steward, and E. Y. Wright (1982) The genetics of dopa decarboxylase in *Drosophila melanogaster*. V. *Ddc* and *l(2)amd* alleles: Isolation, characterization and intragenic complementation. *Mol. Gen. Genet.* **188**:18–26.

Young, M. W., and B. H. Judd (1978) Nonessential sequences, genes, and the polytene chromosome bands of *Drosophila melanogaster*. *Genetics* **88**:723–742.

Zehring, W. A., D. A. Wheeler, P. Reddy, R. J. Konopka, C. P. Kyriacou, M. Rosbash, and J. C. Hall (1984) P-element transformation with *period* locus DNA restores rhythmicity to arrhythmic *Drosophila melanogaster*. *Cell* **39**:369–376.

Contributors and Participants

Hirohiko Aoyama
Centre National de la Recherche Scientifique
Institut D'Embryologie
49bis, av. de la Belle Gabrielle
94130 Nogent-Sur-Marne
France

Michael J. Bastiani
Department of Biological Sciences
Stanford University
Stanford, California 94305

Darwin K. Berg
Department of Biology
University of California, San Diego
La Jolla, California 92093

Floyd E. Bloom
Division of Preclinical Neuroscience and Endocrinology
Research Institute of Scripps Clinic
10666 North Torrey Pines Road
La Jolla, California 92037

Jeremy P. Brockes
MRC Cell Biophysics Unit
26 Drury Lane
London WC2B 5RL
England

Frank Collins
Department of Anatomy
The University of Utah College of Medicine
Salt Lake City, Utah 84132

W. Maxwell Cowan
The Salk Institute
P.O. Box 85800
San Diego, California 92138

Katheryn Cusick
The Neurosciences Institute/NRP
1230 York Avenue
New York, New York 10021

Stephen S. Easter, Jr.
Division of Biological Sciences
The University of Michigan
830 North University
Ann Arbor, Michigan 48109

Gerald M. Edelman
The Rockefeller University
1230 York Avenue
New York, New York 10021

John S. Edwards
Department of Zoology
University of Washington
Seattle, Washington 98195

Justin R. Fallon
Department of Neurobiology
Stanford University School of Medicine
Stanford, California 94305

Sergey Fedoroff
Department of Anatomy
University of Saskatchewan College of Medicine
Saskatoon, Saskatchewan
Canada S7N 0W0

Leif H. Finkel
The Rockefeller University
1230 York Avenue
New York, New York 10021

Scott E. Fraser
Department of Physiology and Biophysics
California College of Medicine
University of California, Irvine
Irvine, California 92717

W. Einar Gall
The Rockefeller University
1230 York Avenue
New York, New York 10021

Michael D. Gershon
Department of Anatomy and Cell
 Biology
College of Physicians and Surgeons
Columbia University
630 West 168th Street
New York, New York 10032

Marcie Glicksman
Department of Anatomy and
 Neurobiology
Washington University School of
 Medicine
600 South Euclid Avenue
St. Louis, Missouri 63110

Corey S. Goodman
Department of Biological Sciences
Stanford University
Stanford, California 94305

Jeffrey C. Hall
Department of Biology
Brandeis University
415 South Street
Waltham, Massachusetts 02254

William A. Harris
Department of Biology
University of California, San Diego
La Jolla, California 92092

Susan Hassler
The Neurosciences Institute/NRP
1230 York Avenue
New York, New York 10021

Margaret Hollyday
Department of Pharmacological and
 Physiological Sciences
University of Chicago
947 East 58th Street
Chicago, Illinois 60637

John D. Houle
Department of Anatomy
University of Saskatchewan College of
 Medicine
Saskatoon, Saskatchewan
Canada S7N 0W0

R. Kevin Hunt
The Salk Institute
P.O. Box 85800
San Diego, California 92138

Michele H. Jacob
Department of Biology
University of California, San Diego
La Jolla, California 92093

Roger J. Keynes
Department of Anatomy
University of Cambridge
Downing Street
Cambridge CB2 3DY
England

Chris R. Kintner
MRC Cell Biophysics Unit
26 Drury Lane
London WC2B 5RL
England

Nicole M. Le Douarin
Centre National de la Recherche
 Scientifique
Institut D'Embryologie
49bis, av. de la Belle Gabrielle
94130 Nogent-Sur-Marne
France

Greg Erwin Lemke
Institute of Cancer Research
College of Physicians and Surgeons
Columbia University
701 West 168th Street
New York, New York 10032

Paul C. Letourneau
Department of Anatomy
University of Minnesota
Minneapolis, Minnesota 55455

Jon M. Lindstrom
The Salk Institute
P.O. Box 85800
San Diego, California 92138

Joseph F. Margiotta
Department of Biology
University of California, San Diego
La Jolla, California 92093

Contributors and Participants

U. Jackson McMahan
Department of Neurobiology
Stanford University School of Medicine
Stanford, California 94305

Karina Meiri
Department of Anatomy and
 Neurobiology
Washington University School of
 Medicine
600 South Euclid Avenue
St. Louis, Missouri 63110

Mark R. Meyer
Department of Zoology
University of Washington
Seattle, Washington 98195

Rae Nishi
Department of Biology
University of California, San Diego
La Jolla, California 92093

Ralph M. Nitkin
Department of Neurobiology
Stanford University School of Medicine
Stanford, California 94305

Gail Proos
The Neurosciences Institute/NRP
1230 York Avenue
New York, New York 10021

Pasko Rakic
Section of Neuroanatomy
Yale University School of Medicine
333 Cedar Street
New Haven, Connecticut 06510

Jonathan A. Raper
Department of Biological Sciences
Stanford University
Stanford, California 94305

Taube P. Rothman
Department of Anatomy and Cell
 Biology
College of Physicians and Surgeons
Columbia University
630 West 168th Street
New York, New York 10032

Donald S. Sakaguchi
Department of Biological Sciences
State University of New York at Albany
Albany, New York 12222

Richard H. Scheller
Department of Biological Sciences
Stanford University
Stanford, California 94305

John T. Schmidt
Department of Biological Sciences
State University of New York at Albany
Albany, New York 12222

Eric M. Shooter
Department of Neurobiology
Stanford University School of Medicine
Stanford, California 94305

Martin A. Smith
Department of Biology
University of California, San Diego
La Jolla, California 92093

Nicholas C. Spitzer
Department of Biology
University of California, San Diego
La Jolla, California 92093

Jes Stollberg
Department of Biology
University of California, San Diego
La Jolla, California 92093

Jean-Paul Thiery
Centre National de la Recherche
 Scientifique
Institut D'Embryologie
49bis, av. de la Belle Gabrielle
94130 Nogent-Sur-Marne
France

John B. Thomas
Department of Biological Sciences
Stanford University
Stanford, California 94305

James W. Truman
Department of Zoology
University of Washington
Seattle, Washington 98195

Gordon Tucker
Centre National de la Recherche
 Scientifique
Institut D'Embryologie
49bis, av. de la Belle Gabrielle
94130 Nogent-Sur-Marne
France

Ana M. Valles
Department of Biology
Brandeis University
415 South Street
Waltham, Massachusetts 02254

Bruce G. Wallace
Department of Neurobiology
Stanford University School of Medicine
Stanford, California 94305

Anne E. Warner
Department of Anatomy and
 Embryology
University College London
Gower Street
London WC1E 6BT
England

Kalpana White
Department of Biology
Brandeis University
415 South Street
Waltham, Massachusetts 02254

Mark B. Willard
Department of Anatomy and
 Neurobiology
Washington University School of
 Medicine
600 South Euclid Avenue
St. Louis, Missouri 63110

Robert W. Williams
Section of Neuroanatomy
Yale University School of Medicine
333 Cedar Street
New Haven, Connecticut 06510

Index

Abdominal ganglion:
  dorsal surface of, 514
  "white cells," 516
Acetylcholine receptors:
  aggregating molecules, 62–65
  in cricket synapses, 330
  in formation of synaptic contacts, 80–81
  in muscle fiber basal lamina, 61–65
  at regenerating neuromuscular junctions, 61–65
Actinomycin:
  and neurite motility, 272–274
  in RNA synthesis, 83
Adrenomedullary cells, levels of origin, 164
Aganglionosis in bowel, 231–234
Amphibian limbs:
  axonal growth in, 248–249
  glial growth factor, 119–135
  motor innervation in, 248–249
  neuronal control of cell division, 119–135
  regeneration in, 119–135
Amphibians:
  embryonic eye rudiments, 394
  glial growth factor in, 128–130
  limb regeneration, 119–135
  neuronal dependence of limb regeneration, 126–128
  neuronal differentiation in, 20–22
  neuronal excitability of, 67–69
  neurotransmitter sensitivity, 78–79
  retinal-deletion experiments, 406
  retinal polarity in, 393–394
  retinotectal development, 392–393
  retinotectal patterns in, 396–397
Animal behavior, genetic and molecular analysis, 565–583
Antigenic levels, genetic control of, 558–560
Aplysia:
  bag cell development, 523
  egg-laying hormone in, 518–527
  gene expression in peptidergic neurons, 513–527
Arbors, surface area and neuronal patterning, 497–498
Associative conditioning, in genetic mosaics, 571
Astroblasts, composition of, 98
Astrocytes:
  early development, 101
  fibrous, 98
  gold sublimate staining, 93
  lineages, 91–93, 96–98
  morphology of, 95
  plasma membrane in, 100
  response to injury, 100
  transitional, 107
  see also Astroglial cells
Astroglial cells:
  lineages, 95–101
  radial glia relation to lineage, 101
  terminology for lineage, 97
Ataxia, in mutant animals, 48–49
Atrial gland:
  development of, 523
  in neuropeptide synthesis, 518
Auditory tectal maps, aligning of, 474
Autonomic ganglia:
  differentiation of, 169–172
  levels of origin, 164
  neural crest and placodal anlagen for, 164–165
Autonomic nervous system, divisions of, 213–214
Autonomic-sensory cell lines, segregation hypothesis, 174
Avian embryo:
  differentiation of peripheral nervous system, 163–178
  ectoderm structure, 182
  gangliogenesis in, 181–207
Axolotl:
  glial growth factor activity, 129
  neurulation in, 27
Axonal growth:
  in aberrant nerve pathways, 259–260
  in amphibian limbs, 248–249
  attractant chemical cues, 259
  and chemotrophic agents, 253–254
  in different motor pools, 245–246
  distribution of specific cues, 255–256
  and guidance, 269–289
  guidance of, 447
  interactions and projection patterns, 260–261
  in limb bud, 255–256
  in nerve pathways, 256–257
  neurofilament expression, 356–358
  neuronal pathways, 351–353
  nonspecific cues, 250–253
  optic, 398–402
  proximal *vs.* distal pathways, 258–259
  retinotectal projection models, 441
  shifting terminals in, 437–442

Axonal growth (*Continued*)
    specific cues, 253–260
    in vertebrate limbs, 245, 250–260
Axonal pathfinding:
    errors in, 399–400
    and optic system, 398–402
    vs. target recognition, 400
Axon elongation:
    growth-associated proteins and, 349–351
    in growth state, 348
    in neuronal development, 346
    in pregrowth state, 353–354
    regulation of, 341–358
    in stationary state, 346–347

Bacteriophage λ:
    gene expression in, 342–344
    structure of, 342
Bag cells, firing pattern changes, 526
Behavior:
    brain development implications, 573–574
    conditioned reproductive, 572
    genetic and molecular analysis, 565–583
    genetic mosaics, 570
    molecular processes of, 509–511
    of mutations affecting conditioning, 566–567
    neural and developmental implications, 565–583
    sexual development and, 574–575
Bergmann glial fibers, granule cell interactions, 150
Biological clock:
    gene types, 576, 582
    in gliogenesis, 99
    neural and molecular analysis, 577–581
Biological rhythm mutants, in courtship, 575–582
Birds:
    gangliogenesis in, 181–207
    nervous system in, 12
    see also Avian embryo
Blastema:
    cells after amputation, 131–133
    glial growth factor activity, 128–129
    mitogenic growth factor, 127
    monoclonal antibodies to, 130–134
    "reciprocal" expression, 131, 134
Bloodhound model, of axonal pathfinding, 259
Bowel:
    aganglionic colonization by neural crest-derived cells, 235
    colonization by enteric neuronal precursors, 225–227
    neuronal determination in, 216
Brain:
    behavioral implications of development, 573–574
    cell adhesion molecules by region, 48

"focusing" in conditioned courtship, 568
immunoreactive fibers in lobes, 551
organogenesis, 47–51
α-Bungarotoxin:
    binding sites, 373–377
    in cholinergic synapse formation, 371
    extrasynaptic distribution, 374–377
    independent regulation, 373–374
    in nicotinic ligand binding experiments, 332–334

Calcium:
    blocks and synaptic contacts, 80–81
    dependent impulses and neuronal differentiation, 78–80
    intracellular concentrations, 24–25
    sodium pump effects, 27
Cancer, cell adhesion molecules and, 53–54
Carbohydrate, in cell–cell binding, 42–43
Cardiotonic steroids, and sodium pumping, 19
Cat, visual system, 474
Catecholaminergic cells:
    in developing gut, 227–229
    immunocytochemistry of, 228–229
    neuroblast development by, 227–229
Cell:
    learning hypothesis, 413
    migration and neural form, 90
    recognition during neuronal development, 295–314
Cell adhesion:
    assays, 37–40
    detection and perturbation of, 37–40
Cell adhesion molecule (CAM):
    in adult animals, 51–54
    assays, 37–40
    binding and local cell surface modulation, 42–43
    electrophoretic patterns, 43
    expression of, 206
    fate maps, 43–47
    modulation in neural morphogenesis, 36–37
    regulator hypothesis, 55–57
    in retinotectal development, 418
    sequences in neural patterning, 54–57
    structure and specificity, 40–42
    types of, 41–42, 54
Cell death:
    biochemical aspects of, 542
    cellular interactions in, 537
    contexts for, 531–532
    critical periods for, 541
    degeneration triggers, 534–537
    in developing nervous system, 531–543
    factors influencing, 537
    hormonal studies, 510, 531–543
    interneurons *vs.* motoneurons, 538–539
    motoneuron expression during, 247–249

muscle degeneration and, 532
postecdysial, 537
program in moths, 532–534
"programmed," 539
progression in motoneurons, 533
temporal patterns of, 539–540
timing of events, 533–534
see also Neuronal death
Cell degeneration:
    endocrine factors in, 534–536
    hormonal triggers, 534–537
    spatial selectivity in, 538–539
Cell interactions:
    hypothesis for neuronal death, 539–540
    in neuronal patterning, 481–505
    in retinotectal projection, 504
Cell junctional molecule (CJM), in adult animals, 53
Cell surface:
    local modulation, 42–43
    modulation as developmental mechanism, 36–37
    transient antigen expression, 310–312
Central nervous system (CNS):
    analysis of mutations affecting development, 562
    cell death during development, 531–543
    central cholinergic receptors in, 330–335
    of crickets, 324
    deafferentation effects, 330–335
    electromigration applications to, 502–503
    FMRFamidelike immunoreactivity mapping in, 553–554
    of grasshopper, 296–297
    of insects, 317
    larval, 550–551
    neural reorganization during metamorphosis, 561
    neuropeptides in, 548
    neurotransmitters in, 548
    organization and development, 549
    peripheral ganglia, 174–175
    positional relationships, 174–175
    serotoninlike immunoreactivity mapping in, 550–551
    substance P-like immunoreactivity mapping in, 555–556
Cephalic levels, neural crest cell migration at, 184–187
Cerebellar cortex:
    basic cytological organization, 140–141
    cellular layers of, 141
    embryonic and postnatal ages, 142
    growth cones in molecular layer, 147
    major cellular events, 141–143
    migrating granule cells in, 144–145
    morphogenetic transformation of granule cells, 143–150

neurite types, 150
neuronal migration mechanisms, 139–158
neuron sources, 141
Chemoaffinity experiments, summary of, 454–461
Chemoaffinity hypothesis, 5–6
    alternatives to, 403
    molecular model, 417
    and motor pool differentiation, 245
    in retinotectal development, 390
    in vertebrate limb regeneration, 244
Chemotaxis, and neurite guidance, 287–289
Chemotropic agents, as axonal growth cues, 253–254
Chick embryo:
    aggregation of neural crest cells, 204
    blastodisc fate maps, 45–46
    cholinergic synapses, 363–380
    ciliary ganglion neurons, 363–380
    diffuse organization of early retinal projections, 462
    motoneuron differences in wing, 247–248
Choline acetyltransferase (CAT):
    levels in ciliary ganglion neurons, 366–367
    stimulating activity, 368–369
Cholinergic binding sites:
    comparative pharmacology of, 331
    density following deafferentation, 332
    in terminal ganglion homogenates, 331
Cholinergic receptors:
    for ciliary ganglion neurons, 371–373
    deafferentation effects, 330–335
    in electromigration hypothesis, 499
    extrasynaptic distribution, 374–377
    independent regulation, 373–374
    in synapse formation, 371–373
Cholinergic synapses, in ciliary ganglion neurons, 363–380
Chronotopic organization, in retinotectal maps, 443
Ciliary ganglion neurons, 364–365
    α-bungarotoxin binding sites, 373–377
    cholinergic development and identification, 363–380
    cholinergic receptors, 371–373
    evoked synaptic transmission, 370
    fractionation by gel filtration, 367
    long-term growth and development, 366–369
    neurite growth in, 288
    survival in cell culture, 365
    synapse formation, 369–370
    synaptic components, 363–380
Circadian rhythms:
    genetic connections between, 576–577
    short-term *vs.* long-term, 576–577
Cloning:
    of cell adhesion molecule genes, 57
    of neuropeptide genes, 515

Cockroach, motor neurons of, 321
Collagen, in neurite growth, 286
Computer simulations:
  of neuronal patterning, 482
  of ocular-dominance columns, 497
  in retinotectal experiments, 416
Computer tests, of adhesive model of neuronal patterning, 492–493
Conditioning, behavior and biochemistry of mutations, 566–567
Courtship:
  biological rhythm mutants in, 575–582
  learning and memory mutants in, 566–575
  song rhythms, 575–576
Cranial sensory ganglia, neural crest and placodal anlagen for, 164–165
Crayfish, learning capacity of, 317
Crickets:
  developmental interactions of neurons, 317–336
  giant interneurons in, 322–323
  metabolic interactions in cercal sensory–giant interneuron pathway, 324–335
  motor neurons of, 321
  neuropharmacology of, 330–334
  sensory neurons of, 318
Cycloheximide, in protein synthesis, 83–84

Deafferentation:
  biochemical correlates of, 326–329
  effects on central cholinergic receptors, 330–335
  giant interneuron effects, 322–323
  with labeled muscarinic ligands, 330–332
  with labeled nicotinic ligands, 332–335
  morphological correlates of, 324–326
Diencephalon, pathway interactions in, 459–460
Differential adhesivity, and neurite growth, 285–286
Diseases, cell adhesion molecules and, 53–54
Distal neurite, membrane expansion at, 277–278
Deoxyribonucleic acid (DNA):
  in neuronal development, 342–343
  recombinant technology, 509–510, 515
  sequences in biological clock genes, 579–580
Dopamine:
  mutations and serotonin activity, 558–559
  synthesis of, 552
Dorsal root ganglia:
  neurofilament protein immunoreactivity, 175
  origin of, 166
Drosophila:
  behavioral implications of brain development, 573–574
  behavioral implications of sexual development, 574–575
  cell recognition in, 295–314
  CNS organization and development, 549
  genetic mosaic analysis of mutationally induced learning defects, 567–568
  genetic studies, 547–562
  immunohistochemical studies, 547–562
  immunoreactive cell labels in, 560–562
  learning and memory mutants in courtship, 566–575
  learning capacity of, 317
  neural reorganization during metamorphosis, 561
  neuronal development in, 295–314
  neurotransmitters in, 548
  selective fasciculation in, 298, 313–314
  sensory neurons of, 320
  serotonin and neuropeptides in, 547–562
  specific neurodevelopmental questions, 560–562
Duodenum, serotonin uptake in, 22

Ecdysteroids, in cell degeneration, 535–536
Ectoderm cells:
  electrical coupling of, 16, 17
  electron microscopy of, 14
  mesoderm cell interaction with, 11–12
  neural induction in, 16–31
Egg-laying hormone (ELH):
  gene family and fixed action pattern, 518–527
  indirect immunofluorescence detection, 519–521
  interneuron network, 522
  recombinant clones, 519
  restriction enzyme maps, 519
Electrical fields, and neurite growth, 289
Electromigration hypothesis:
  defined, 499–500
  feasibility of model, 500
  for synaptic competition, 498–500
Embryonic muscle cell, biophysics of, 498–499
Endocrine factors, in cell degeneration, 534–536
Enteric nervous system (ENS):
  derivation of, 217–236
  ganglion cells of, 217
  glia precursors in gut, 224
  gut microenvironment, 217–219
  mature, 214–216
  microenvironment changes during ontogeny, 229–230
  neural components, 216
  neuronal determination in, 213–236
  neuronal phenotypic expression, 217–219, 224
  neuronal precursors in gut, 221–224
  phenotypic diversity of neurons, 216
  phenotypic expression by neurons in gut, 219–221
Enteric plexus, ontogeny of, 188–190

Epidermal growth factor (EGF), 120
Epithelial cells, conversion of, 16
Extracellular matrix:
  adhesion and migration components, 194–198
  as critical substrate for neural crest cell migration, 192–198
  molecular organization, 192–193
  and neural crest migration, 236
  proteins, 236
Eye transplants, and retinotectal projection, 409

Fasciculation:
  in *Drosophila*, 298
  selective in grasshopper embryo, 296–299
Fate maps:
  of cell adhesion molecules, 43–47
  of chick blastodisc, 46
  of embryonic peripheral nervous system structures, 163–166
  topological *vs.* topographic, 45
Fibronectin, 193–194
  and cell adhesion, 206
  distribution before individualization of neural crest cells, 194
  migration and directionality, 197
  neural crest cell interaction, 193–194
  structure and function, 193
  in vitro migration of neural crest cells on, 196
Fibrous astrocytes, formation of, 98
Filopodia:
  adhesion and motile behavior, 275
  defined, 296
  differential adhesivity of, 285–286
  and growth cones, 300–304
  and neurite growth, 282–283
  pulse-chase experiments, 273
  selective adhesion of, 300–302
  selective insertion, 302–304
  tension on neurites, 274
Fish:
  optic nerve crush, 463
  retinal-deletion experiments, 405
  retinotectal development, 392–393
  retinotectal map formation, 429–451
  retinotectal system of, 412, 484–487
  tectum rotation experiments, 407
FMRFamidelike immunoreactivity:
  implications and conclusions, 557
  mapping in larval CNS, 553–554
  putative colocalization of, 556
Frog:
  diffuse organization of early retinal projections, 462
  ocular-dominance columns in, 487
  retinotectal system of, 412, 484–487
  topographic tectal maps, 460

Fruit fly, see *Drosophila*
Functionalist theory, of retinotectal projection, 403

Ganglia:
  cell differentiation, 169
  maturation of, 200–205
Gangliogenesis, 198–205
  in avian embryo, 181–207
  junctions in, 199
  and neural crest cell aggregation, 198–200
Gap junctions:
  calcium concentration in, 24–25
  in neural plate, 15
  permeability of, 24–25
  in retinotectal development, 419
Gene expression, 509–511
  in bag cells, 523–524
  and egg-laying hormone, 518–527
  in peptidergic neurons, 513–527
Genetic control, of sexual development, 574–575
Genetic mosaics:
  analysis of learning defects, 567–568
  molecular techniques, 569–570
  of mutationally induced learning defects, 567–568
  neural and molecular implications, 568–573
Genetic perturbation, of serotoninlike immunoreactivity, 551–553
Giant interneurons:
  deafferentation effects, 322–324
  metabolic interactions in crickets, 324–335
  in orthopteroid insects, 322
  protein metabolism in, 327, 330
Glial cells:
  in amphibian limb regeneration, 119–135
  development of, 89–90
  earliest detection of glial fibrillary acidic protein (GFAP), 101
  immunological markers, 93
  neuronal control of cell division, 119–135
  and neuronal guidance, 89–90
  proliferation studies, 119–120
  transitional, 107–108
Glial growth factor (GGF):
  in amphibian limb regeneration, 119–135
  blastema activity, 128–129
  bovine pituitary *vs.* caudate nucleus, 125
  cell-type specificity, 125
  large-scale purification of bovine pituitary, 124
  and neuronal control of cell division, 119–135
  presence of activity in amphibians, 128–130
  regional distribution, 125
Glioblasts:
  composition of, 96–97
  lineage in culture, 96–97

Gliogenesis:
  in postnatal animals, 95–96
  studies of, 95
Goldfish:
  activity-dependent sharpening of retinotectal map, 462–465
  anatomical studies, 463
  diffuse organization of early retinal projections, 461
  electrophysiological studies of retinotectal sharpening, 463–465
  ocular-dominance patches in, 471–472
  retinotectal map formation, 429–451
Granule cells:
  axons in molecular layer, 146
  binding molecules and, 156–157
  identification of, 144
  layer of cerebellar cortex, 141–142
  migrating, 144–145, 148–149
  migration and axonal growth, 143–144
  morphogenetic transformation of, 143–150
  in neonatal monkey, 148–149
  neuron–glia interaction, 152–153
  in newborn monkey, 144–145
  postmitotic, 150
Grasshopper:
  cell recognition in, 295–314
  learning capacity of, 317
  neuronal development in, 295–314
  selective fasciculation in, 298
Grasshopper embryo:
  cell recognition in, 296
  motile activity at neurite trip, 270–275
  motor neurons of, 320
  selective fasciculation in, 296–299
Growth-associated protein (GAP):
  in axon elongation, 348–349
  expression in axon growth, 351–353
  function of, 349–351
  and growth cones, 350
  in mammals, 349
Growth cones:
  abnormal behavior of, 308
  defined, 296
  of fasciculating grasshopper neurons, 296–299
  motility and neurite growth, 280–284
  in retina, 446–447
  structure during motility, 270–274
  ultrastructure and filopodia, 299
Growth-promoting activity (GPA), of ciliary ganglion neurons, 367–368
Guinea pig, myenteric plexus of, 215–216
Gut:
  enteric glia precursors in, 224
  enteric neuronal precursors, 221–224
  microenvironment and phenotypic expression, 217–219
  norepinephrine localization, 221
  phenotypic expression by enteric neurons, 219–221

Helper effect, in retinotectal development, 402
Hemopoietic cells, lineages, 92
Honeybee, learning capacity of, 317
Human bowel, aganglionosis in, 231–233

Immunofluorescence assays, of gene expression, 354
Immunoreactive cell labels, in neuronal development, 560–562
Inductive signal:
  nature of, 13
  transmission of, 13–15
Insect embryos:
  motor neurons of, 318–319
  neural growth patterns, 318–319
Insects:
  developmental interactions of neurons, 317–336
  identified interneurons in, 321–323
  metamorphosis in, 532
  motor neurons of, 320–321
  nervous system of, 514–515
  neuronal development in, 295–296
  neuropeptides in, 510
  neuropharmacology of, 330–334
  neurotransmission in, 330
  postembryonic neuronal growth patterns, 319
  sensory neurons of, 318–320
Interneurons, growth and form of, 321–323
Intrinsic response hypothesis, for neuronal death, 540
Invertebrates, cell death in, 531–532

Labeled pathways hypothesis:
  experimental test of, 305–308
  monoclonal antibody correlate of, 309–310
  of neurite formation, 266
  of neuronal development, 304
Lateral geniculate, segregation of receptive field types, 474
Learning:
  anatomical correlates, 572–573
  higher vs. simple, 572
  mutationally disrupted, 572–573
  neural centers of, 570
Learning defects:
  development stages and, 569
  genetic mosaic analysis of, 567–573
  mutationally induced, 567–568
  neural and molecular implications, 568–573
  quasipleiotropic effects, 568
Learning mutations:
  in conditioned courtship, 570–571
  in courtship, 566–575

Limb bud:
  axonal growth cues, 255–256
  axonal pathfinding in, 258–259
Lithium ions, in neuronal experiments, 14
Locust, sensory neurons of, 318
Lysogenic pathway, in neuronal development, 342–343

Macroglial cells:
  lineages, 91–110
  origin of, 93–95
  transitional, 108
Magnesium, sodium pump effects, 26
Male–female learning, patterns of, 571
Mammals, axon elongation in, 349
Manduca sexta:
  cell death in, 531–543
  cell degeneration in, 534–536
Manganese, sodium pump effects, 26
Map formation, in retinotectal system, 385–387
Marker genes, in genetic mosaics, 570
Membrane recognition, and activity-dependent synaptic stabilization, 453–476
Memory mutants, in courtship, 566–575
Mesoderm cells:
  in amphibian embryo, 12
  ectoderm cell interaction with, 11–12
Metamorphosis, neural reorganization during, 561
Mice:
  bowel aganglionosis, 233–234
  neuronal precursor cells in terminal bowel, 234
Microglial cells:
  "resting," 108
  silver carbonate staining, 93
  transitional, 108
Microtubule-associated proteins, in neurite microtubules, 278
Microtubules:
  dynamic cytoskeletal framework of, 278–279
  in elongating neurites, 278
  and neurite growth, 277, 284
Miniature end-plate potential in spontaneous transmitter release, 500–501
Mitogenic growth factors, importance of, 120
Modulation hypothesis, and brain organogenesis, 47
Molecular layer, of cerebellar cortex, 141
Molecular regulation, of neural morphogenesis, 35–57
Mollusks:
  egg-laying hormone in, 518–527
  gene expression in peptidergic neurons, 513–527
  nervous system of, 514–515
Monoclonal antibodies:
  antigen binding to astrocytes, 100
  antigen binding to blastema, 130–134
  antigen binding to ciliary ganglion neurons, 378–379
  in immunohistochemical studies, 547–548
  and labeled pathways hypothesis, 309–310
  properties of, 131
  staining techniques, 132–133
Morphogenesis:
  cell adhesion molecules in, 54–57
  molecular basis of, 53
Morphogenetic hypothesis, for retinotectal projection, 403
Morphogenetics, and retinotectal development, 399
Moths:
  cell death in, 531–543
  sensory neurons of, 320
Motile activity:
  and growth-cone structure, 270–274
  at neurite tip, 270–275
Motoneurons:
  during axonal outgrowth, 247–249
  during cell death, 247–249, 532–533
  differences among, 247–249
  differentiation in motor pools, 245–247
  monoclonal antibody staining, 310
  muscle positions after limb rotations, 252
Motor innervation:
  in amphibian limbs, 248–249
  nonspecific growth cues, 250–253
  sequential pattern of, 259
  in vertebrate limbs, 243–261
Motor pools:
  motoneuron differentiation, 245–247
  positions and muscle innervation in vertebrate limbs, 246
Motor systems, of insect embryos, 318–321
Multiunit receptive field, in retinotectal maps, 483
Murine bowel, aganglionosis in, 231–233
Murine small intestine, serotonin immunoreactivity, 220
Muscarinic ligands, deafferentation studies with, 330–332
Muscle death, steroid regulation of, 535–536
Muscle fiber:
  acetylcholine receptor aggregation in, 61–65
  basal lamina components, 61–65
Mushroom bodies:
  as "associative centers," 572–573
  in *Drosophila* metamorphosis, 561
Mutants, biological rhythm, 575–582
Mutations:
  analysis of CNS development, 562
  behavior and biochemistry affecting conditioning, 566–567
  behavioral and neurochemical, 511

Mutations (*Continued*)
  effect on gene pattern, 562
  sex-transforming, 574–575
Myelin, in retinotectal fascicles, 434
Myenteric plexus:
  enteric glial cells in, 215
  of enteric nervous system, 214–216
  noradrenergic and serotonergic elements of, 231
Myofibers, in acetylcholine receptor aggregation, 51

Nerve growth factor (NGF):
  discovery of, 6
  and intrinsic enteric neurons, 230–231
Nerve–muscle synapses, in ciliary ganglion neurons, 369
Nerve–nerve synapses, in ciliary ganglion neurons, 369–370
Nerve pathways:
  aberrant, 259–260
  axonal growth points, 256–257
  and local specific growth cues, 259–260
  proximal *vs.* distal choices, 258–259
Nervous system:
  in amphibian early embryo, 12
  anatomical organization of, 31
  factors controlling early development, 11–32
  fiber tracts and maps, 36
  gap junctions in, 53
  glial fibrillary acidic protein in, 101
  see also Central nervous system (CNS)
Neural cell adhesion molecule (N-CAM), 156–157
  embryonic to adult conversion in ganglia, 205
  expression in neural crest cells, 202–203
Neural connections:
  during morphogenesis, 47–51
  in mutant animals, 48–49
Neural crest:
  back-transplantation of cells, 168–169
  cell ontogeny in peripheral nervous system, 167–174
  craniocaudal regionalization, 166
  development of periphery, 161–162
  differentiation of sensory neurons, 174–175
  heterogeneity of cell population, 175–177
  homogeneous distribution of developmental capabilities, 166–167
Neural crest cells:
  adhesion and migration of, 194–198
  adhesive properties of, 206
  aggregation in gangliogenesis, 198–200
  aggregation in quail tissue, 204
  cephalic level migration, 184–187
  early phase of migration, 183–188
  enteric plexus and, 188–190
  expression of neuronal phenotypes by, 176–177
  extracellular matrix components, 194–198
  extracellular matrix as critical migration substrate, 192–198
  fibronectin interaction, 193–194
  final location of, 189, 192
  individualization of, 183
  maturation of ganglia, 200–205
  migration:
    and adhesion properties, 181–207
    along extracellular matrix, 236
    and intermodulation, 53
    pathways, 183–192
  sensory ganglia and sympathetic chains, 190–191
  specificity of localization, 191–192
  trunk level migration, 187–188
  vagal level migration, 187–188
  in vitro adhesion studies, 194–197
  in vitro migration studies, 196
  in vivo adhesion studies, 197–198
Neural development, molecular basis of, 449–451
Neural groove, appearance of, 16
Neural growth, embryonic patterns, 318–319
Neural induction, 11–12
  and cell adhesion molecules, 43–47
  consequences of, 16–31
  in ectoderm cells, 16–31
  mechanism of, 12–13
  signal types, 15
  sodium pumping and, 17–30
Neural mapping, individual markers, 50–51
Neural morphogenesis, molecular regulation of, 35–57
Neural patterning:
  cell adhesion molecules in, 54–57
  and local cell surface modulation, 50
Neural plate:
  border definition of, 15–16
  cell adhesion molecules in, 45
  cells and inductive signal, 14
  electrical coupling in, 24–25
  extracellular calcium in, 29
  intracellular sodium in, 28
  membrane potential changes in, 18–20
  sodium pumping in, 17–30
  structure of, 71
Neural reorganization, during *Drosophila* metamorphosis, 561
Neural retina cells, aggregation in chick tissue, 204
Neural tube:
  formation of, 93–94
  generation of, 16–17
  in mouse embryos, 94–95

Neurite growth:
  adhesion and motile behavior, 275
  "contact guidance," 286
  dynamic cone structure during motility, 270–274
  extrinsic factors, 285–289
  vs. growth cone motility, 280–284
  and guidance, 269–289
  localized, 281–284
  as localized accumulation of materials, 276–280
  and microtubules, 278–279
  motile activity and, 270–275
  restrictive effects on, 283
  see also Axonal growth
Neurites:
  adhesive affinities of, 286–287
  chemotactic mechanisms of, 287
  cytoskeleton of, 270
  differential adhesivity and, 285–286
  electrical fields and, 289
  elongation of, 276–277
  formation of, 265–267
  growth cone in, 268–270
  motile activity at tip, 270–275
  neurofilament stability in, 279–280
  oriented growth toward cathode, 289
  soluble gradient guidance, 287–289
Neuroepithelium, primitive, 92
Neurofilaments:
  cross-linking of, 355–356
  developmental expression, 356–358
  from enteric neuronal precursors, 225–227
  gene expression, 355–358
  structure and function, 355–356
Neurogenesis:
  and cell adhesion molecule fate maps, 43–47
  and regulator hypothesis, 57
Neuromodulators, gene expression of, 509
Neuromuscular junction(s):
  acetylcholine receptors at, 61–65
  synapse elimination at, 475
Neuronal death:
  cellular interactions in, 537
  critical periods for, 541
  degeneration triggers, 534–537
  in developing nervous system, 531–543
  hormonal studies, 510
  spatial selectivity of, 538–539
  steroid regulation of, 535–536
  temporal patterns of, 539–540
  see also Cell death
Neuronal determination, in enteric nervous system, 213–236
Neuronal development:
  in absence of calcium-dependent impulses, 78–80
  cell–cell interactions, 363–364
  cell recognition during, 295–314
  center and periphery interdependence during, 319–321
  changes of state during, 341–358
  in chick ciliary ganglia, 363–380
  cholinergic synapses in, 363–380
  control of excitability, 67–85
  early differentiation, 78–80
  gene switching during, 346
  growth state, 346–349
  implications of genetic and molecular analysis of behavior, 565–583
  interactions in insects, 317–336
  in vitro and *in vivo*, 69–77
  ionic dependence of action potential, 73, 74–75
  neurofilament expression, 356–358
  neurofilament polypeptides in, 355–358
  postembryonic patterns, 319
  pregrowth state, 346–349, 353–354
  primary processes, 509–511
  states of, 345–346
  stationary state, 346–349
Neuronal differentiation:
  early, 78–80
  measurement of, 20–23
  perturbation experiments, 78
Neuronal excitability:
  control of development, 67–85
  developmental stages, 72
  developmental time tables, 67–68
  perturbation experiments, 73
Neuronal migration:
  adhesion molecules in, 154
  cell penetration during, 157–158
  defined, 139
  downward displacement mechanism, 156
  and gangliogenesis, 181–207
  along glial fibers, 154–155
  mechanisms in developing cerebellar cortex, 139–158
  possible molecular mechanisms, 154–158
Neuronal patterning:
  activity-dependent competition model, 498–503
  adhesive model for, 488–498
  arbor surface area dependence, 497–498
  cell interactions in, 481–505
  computer tests of adhesive model, 492–493
  thought experiments using adhesive model, 491–492
Neuronal phenotypes, expression by cultured crest cells, 176–177
Neuron–glia interaction(s):
  during granule cell migration, 152–153
  during morphogenesis, 47–51
  in mutant animals, 48–50
  significance and incidence, 150–154

Neurons:
  in insects, 317–336
  peptidergic, 516–518
  "trophic" growth of, 326
Neuropeptide genes, 516–518
Neuropeptides:
  function of, 515
  gene expression of, 509, 513–527
  gene isolation, 515
  genetic control of antigenic levels, 559–560
  genetic studies, 547–562
  immunohistochemical studies, 547–562
  organization in egg-laying hormone precursors, 524–527
  and serotonin, 547–562
Neuropharmacology, of insects, 330–334
Neuropil:
  in FMRFamidelike immunoreactivity mapping, 553–554
  in larval CNS, 551, 553–554, 556
  in substance P-like immunoreactivity mapping, 556
Neurotransmission, pharmacology and physiology, 330
Neurotransmitters:
  impulse-evoked release, 80–82
  in neuronal development, 548
  neuronal sensitivity to, 72–73
  reversal potential and ionic dependence, 76–77
  synthesis of, 551–552
Neurulation:
  frequency histograms during, 23
  morphogenetic movements, 11–12
Newt:
  blastema, 131
  diffuse organization of early retinal projections, 461–462
  limb regeneration, 127
  neural induction in, 14–15
Nicotinic ligands, deafferentation studies with, 332–335
Nodose ganglion, of chimeric chick embryo, 170–173
Norepinephrine, localization in fetal rat gut, 221
Notochord cells, electrical coupling with, 15
Nucleopore filters, in neural induction experiments, 14–15
Nucleotides, in cell death, 542

Ocular-dominance columns:
  in adhesive model of neuronal patterning, 496–497
  formation of, 487
  nerve patterning model, 497
  in retinotectal projection, 487

Ocular-dominance patches, activity-dependent segregation of, 470–472
Oligodendroblasts:
  differentiation of, 105
  maturation stages, 103
Oligodendrocytes:
  genesis stages, 107
  lineages, 91–93
  neuronal dependence, 105
  silver carbonate staining, 93
  transitional, 107
Oligodendroglial cells:
  immunological markers, 104
  lineage, 102–107
  in culture, 103–106
    vs. in vivo, 106–107
  in vivo, 102–103
  postnatal development, 102–103
  types of, 104
Optic axons:
  growth and pathfinding, 398–402
  molecular pathfinding studies, 418
  position-dependent labels on, 403
  regeneration of, 430–431
  target recognition in, 402–403
Optic nerve:
  age *vs.* retinal position, 443
  composition of, 94
  crush experiments, 406–407, 430
  fiber arbors, 497–498
  glial cells in, 94
  gliogenesis in, 95
  regeneration, 389–390
  retinal fiber lamination, 449
  retinotectal map formation, 443–445
Orthopteroid insects, giant interneurons of, 322

Peptide hormones, 122
Peptidergic neurons:
  egg-laying hormone and, 518–521
  gene expression in, 513–527
Perikaryon, neurite microtubule-organizing centers in, 278–279
Period (*per*) mutation:
  regulation and expression of, 580–581
  and song rhythms, 576–779
Peripheral nerves:
  FMRFamidelike immunoreactivity mapping in, 554
  in larval CNS, 554, 556
  nonspecific growth cues, 253
  substance P-like immunoreactivity mapping in, 556
Peripheral nervous system (PNS):
  development of, 161–162
  differentiation in avian embryo, 163–178
  embryonic fate maps of structures, 163–166
  enteric region, 214

ganglia and central nervous system, 174–175
ganglia types, 169
gangliogenesis in avian embryo, 181–207
homogeneous distribution of developmental capabilities in neural crest, 166–167
neural crest and, 161–162
ontogenesis from vagal and trunk levels, 188–192
ontogeny of ganglia, 164–166
ontogeny and neural crest cells, 167–174
positional relationships, 174–175
sensory ganglia and sympathetic chains, 190–191
Peripheral projections, in larval CNS, 551
Phenotypes, in neuronal development, 78–79
Pheromones, in conditioned courtship, 566–573
Physiological mapping:
limitations of, 484
of retinotectal system, 482–483
Placodes, 165
fate map of, 166
Plasmalemma, and neurite growth, 277, 282
Plasminogen, migrating granule cells and, 157–158
Platelet-derived growth factor (PDGF), 120
Plexus formation, nonspecific cues for, 250–252
Potassium:
concentration and neurulation, 23–24
sodium pump effects, 27–30
Primary induction, 11. *See also* Neural induction
Proastroblasts, cytoskeleton of, 100
Protein growth factors, and cell division, 120
Protein hormones, 122
Protein metabolism, in giant interneurons, 327
Proteins, in biological clock genes, 581–582
Protein synthesis:
and cell death, 542
and neuronal excitability, 82–85
Purkinje cell layer, of cerebellar cortex, 141

Quail, aggregation of neural crest cells, 204
Quail embryos, back-transplantation of neural crest cells, 168–169
Quail ganglia, cell distribution patterns, 169–170

Radial glia, and astroglial cell lineage, 101
Rat, control of proliferation of cultured Schwann cells in, 121–123
Reeler mutant, 48–49
Regenerating muscles, acetylcholine receptors in, 61–65
Regeneration, of retinotectal map, 483–487
Repolarization, in retinotectal development, 394–395

Retina:
axonal pathfinding in, 399–400
axon growth in, 398–399
"context sensitive" labeling, 407
fascicular organization, 445–446
fiber photomosaics, 448
growth cone in, 446–447
histogenesis of, 391–392
lamination in fiber layer, 449
molecular polarity studies, 419
and optic tectum, 389
polarity of, 393–398
positional marking in, 417
retinotectal map formation, 445–449
selective cell adhesion, 418
selective regeneration, 456
shifting terminals in, 438
Retina cells, aggregation of, 204
Retinal axons:
arrangement of, 445
branching patterns of, 400
extrafascicular segments, 439
topographic features, 401
Retinal-deletion paradigm, 405
Retinal ganglion cells:
labeling experiments, 434–437
"positional information" concept, 397
and tectal neurons, 392
Retinal grafts, *vs.* tectal grafts, 458
Retinotectal fascicles, 432–437
axon growth and, 432
labeling experiments, 434–437
organization models, 435–436
reinterpretations, 434–437
Retinotectal maps:
activity-dependent sharpening of, 462–465
anatomical studies, 463
axonal pathway and, 429–451
"chronotopic" organization, 443
correlated activity model, 467
early experiments, 429–430
electrophysiological studies of sharpening, 463–465
formation in goldfish, 429–451
multiunit receptive fields, 464–465
after optic nerve crush, 463, 466
optic nerve and tract, 443–445
refinement of, 391
regeneration and refinement of, 486–487
shifting terminals in, 437–442
tetrodotoxin blocking, 463–465
typical, 483
Retinotectal projection, 482–485
aberrant mapping, 400–401
activity-dependent mechanism, 412
alternative theories, 403
anatomical assays of, 484–485
anatomical experiments, 488
axonal elongation and synaptogenesis, 465

Retinotectal projection (*Continued*)
  axonal pathfinding and, 390
  cell markers in, 419
  chemical complementarity, 417, 418
  chemical labeling, 415
  coding systems, 398
  compression studies, 414–415
  cytochemical labels, 411–412
  defined, 482
  development of, 389–420
  in goldfish, 429–451
  key issues, 390–391
  models of, 414–419
  molecular mechanisms, 391, 414–419
  physiological mapping of, 482–483
  plasticity in, 486
  positional forces, 416–417
  refinements of, 409–413
  reorganization in, 409–410
  rotation experiments, 393–394, 404–405
  schematic diagram, 455
  size-disparity experiments, 411
  specificity in, 485–486
  target recognition and, 402–409
  target selection in, 391
  topographic features, 392
Retinotectal system:
  axon growth and pathfinding, 398–402
  chemoaffinity hypothesis, 390–391
  electrophysiological mapping, 405
  in goldfish, 449
  innervation patterns, 401
  key issues, 487–488
  map formation in, 385–387
  neuronal patterning in, 481–505
  "pattern regulation" in, 394–395
  size-disparity experiments, 405–406
  target recognition in, 402–409
Retinotopic maps:
  activity-dependent sharpening of, 461–469
  auditory *vs.* visual, 474
  changeable tectal markers, 456
  competitive fiber interactions, 454–455
  diffuse organization of early projections, 461–462
  expanded *vs.* compressed, 455
  fiber-to-target matching model, 454
  formation of, 453–476
  pathway interactions in diencephalon, 459–460
  retinal grafts and, 458–459
  selective fiber–fiber adhesion, 455–456
  see also Retinotectal maps
Retroregulation:
  defined, 345
  gene expression and, 344–345
Ribonucleic acid (RNA):
  and cell death, 542
  in neuronal development, 342, 344–345
  and neuronal excitability, 82–85
  transcripts in biological clock genes, 580–581
Rodents, retinotectal projection in, 412–413
Rohon-Beard neurons, 12
  development of, 67–85
  membrane properties, 67–69
  neurotransmitter sensitivity of, 72–73
  size and position, 70
  temporal sequence, 68

Schwann cells:
  agents tested for proliferation effect, 122
  control of proliferation, 121–123
Selective fasciculation:
  in *Drosophila* embryo, 313–314
  in grasshopper embryo, 296–299
  model, 312–313
  pattern of, 299
Sensory ganglia:
  center and periphery independence, 319–320
  differentiation of, 169–172
  of insect embryos, 318–320
  and sympathetic chains, 190–191
Serotonergic neurons, 220
Serotonin:
  dopamine mutations and, 558–559
  enzyme activity in biochemical pathway, 558
  genetic control of antigenic levels, 558–559
  genetic studies, 547–562
  immunohistochemical studies, 547–562
  immunoreactivity in murine small intestine, 220
  neuron labeling by, 223
  and neuropeptides, 547–562
  synthesis of, 552
  uptake in duodenum, 222
  see also Neurotransmitters
Serotoninlike immunoreactivity:
  genetic perturbation of, 551–553
  mapping in larval CNS, 550–551
Sex determination, genetic control of, 574–575
Sex-transforming mutants, 574–575
Sexual behavior, genetic control of, 574–575
Sexual development, behavioral implications of, 574–575
Shifting terminals:
  hypothesis, 439
  in retinotectal maps, 437–442
Sialic acid, in cell–cell binding, 42–43
Silkmoths:
  critical periods for cell death, 541
  prolonging cell death in, 537
Sliding connections hypothesis, for retinotectal development, 410–411
Sodium pump:
  electrogenic contribution, 17–19
  inhibition effects, 24–25

Index

and neural induction, 17–30
stoichiometry of, 19
Soluble gradients, in neurite guidance, 287–289
Song rhythms:
and biological clock genes, 577–581
courtship, 575–576
effects of mutations, 576–577
effects of variants on, 575–576
neural and molecular analysis, 577–581
Songs, courtship, 573–574
Sperry's chemoaffinity hypothesis, 390
Spinal neurons:
excitability of, 67–85
neurotransmitter sensitivity, 78–79
sodium-dependent action potential in, 84
Spontaneous transmitter release:
electromigration model based on, 500–502
model implications, 501–502
Staggerer mutant, 48–49
Stationary axons, morphological features, 347
Strict neural addressing, chemoaffinity and, 50
Strontium, sodium pump effects, 26
Strophanthidin:
calcium effect on, 25
inhibitory effect of, 25
and sodium pumping, 19
Subependymal cells, in postnatal animals, 100
Submucosal plexus, of enteric nervous system, 214
Substance P-like immunoreactivity:
mapping in larval CNS, 555–556
postembryonic cell development, 556
putative colocalization of, 556
Substrate adhesion molecule (SAM), 53
Substrates, cell surface, 9–10
Synapse elimination, 470–475
activity-dependent segregation, 470–472
at neuromuscular junction, 475
of ocular-dominance patterns, 470–472
Synapses:
cell–cell interactions and, 364
formation of, 265–267, 364–365, 369–370
retraction of ineffective, 473–474
Synaptic antigens:
in ciliary ganglion neurons, 377–380
immunological cross-reaction, 378
ultrastructural distribution, 378–380
Synaptic competition, electromigration model, 498–503
Synaptic components:
cholinergic development and identification, 363–380
of ciliary ganglion neurons, 363–380
Synaptic contacts:
in absence of impulse-evoked transmitter release, 80–82
formation of, 80–82

Synaptic stabilization:
activity-dependent, 453–476
membrane recognition, 453–476
Synaptogenesis, 80–82
correlation of retinotectal sharpening with, 465–466
tetrodotoxin blocking in, 466

Tadpoles:
neural development in, 20–23
strophanthidin treatment, 20
Target recognition:
eye rotation experiments, 404–405
and retinotectal projection, 402–409
Tectal fibers, competitive interactions, 454–455
Tectal graft:
computer simulation of, 494
experiments, 494–495
Tectal lobes, "virgin," 408–409
Tectum:
"autonomous" mapping, 397
axonal pathfinding and, 395
axon growth in, 430
changeable markers, 456
"context sensitive" labeling, 407
curvilinear development, 410
diencephalon position, 395–396
graft experiments, 457
histogenesis of, 391–392
intrinsic polarity markers, 401
molecular polarity studies, 419
polarity of, 393–398
positional marking in, 417
position-dependent markers, 409
retinotectal map formation, 430, 437–438
rotation experiments, 395–396, 407, 457–458
topographic maps, 460–461
translocation experiments, 457
transplantation experiments, 457–458
Temporal hypothesis, for retinotectal projection, 403
Tetrodotoxin blocking:
in anatomical studies, 463
in electrophysiological studies, 463–465
retinotectal map sharpening after, 469
[^{3}H]Thymidine:
in cell labeling, 92
oligodendrocyte incorporation of, 105
tissue specificity, major determinants of, 53
Toads, axon elongation in, 348
Topographic maps:
in neural development, 453–454
tectal, 460–461
Torpedo electric organ, acetylcholine receptor aggregation in, 62–65
Transitional cells, 107–108
astrocyte/oligodendrocyte, 107
macroglia/microglia, 108

Transplantation experiments, as tectal marker tests, 457–458
Trunk level:
　intersomitic migration pathway at, 191
　neural crest cell migration at, 186
　ontogenesis of peripheral nervous system from, 188–192
Turning response, of sensory neurite tip, 288

Uvomorulin, and morphogenesis, 53

Vagal level:
　neural crest cell migration at, 184–187
　ontogenesis of peripheral nervous system from, 188–192
Vectoral growth hypothesis, for retinotectal projection, 403
Ventral ectoderm, membrane potential changes in, 18
Ventral pathway, axonal growth in, 258
Ventricular cells:
　lineages, 92
　of optic stalk, 94
Vertebrate limbs:
　growth cues for peripheral nerves, 253
　innervation patterns, 253–254
　motoneuron differences in, 247–249
　motor innervation in, 243–261
　motor pool position, 254
　motor pools, 244–245
　muscle innervation in, 254
　nerve pathways in, 256–257
　neuronal connections in, 244
　nonspecific growth cues, 250–253
　plexus formation in, 250–252
Vertebrates:
　retinal-deletion experiments, 405–406
　retinotectal projection in, 391, 482
Visual map, aligning of, 474
Visuotectal map, 483

Weaver mice, granule cell migration in, 152
Weaver mutant, 48–49
Whole-mount immunohistochemical technique, 549–550

Xenopus embryos:
　neuronal differentiation of, 20–32
　and sodium pump inhibition, 20–32